T0229032

Second Edition

Computability, Complexity, and Languages

Fundamentals of Theoretical Computer Science

This is a volume in
COMPUTER SCIENCE AND SCIENTIFIC COMPUTING
A Series of Monographs and Textbooks
Editor: Werner Rheinboldt
A complete list of titles in this series is available from the publisher upon request.

Second Edition

Computability, Complexity, and Languages

Fundamentals of Theoretical Computer Science

Martin D. Davis

Department of Computer Science
Courant Institute of Mathematical Sciences
New York University
New York, New York

Ron Sigal

Departments of Mathematics and Computer Science
Yale University
New Haven, Connecticut

Elaine J. Weyuker

Department of Computer Science
Courant Institute of Mathematical Sciences
New York University
New York, New York

Morgan Kaufmann is an imprint of Academic Press

A Harcourt Science and Technology Company

San Diego San Francisco New York Boston
London Sydney Tokyo

This book is printed on acid free paper

Morgan Kaufmann Publishers
340 Pine Street, Sixth Floor, San Francisco, California 94104-3205
http://www.mkp.com

Transferred to Digital Printing 2003
Academic Press
An Elsevier Science Imprint
525 B Street, Suite 1900, San Diego, California 92101-4495, U.S.A.
http://www.academicpress.com

Academic Press
84 Theobalds Road, London WC1X 8RR, UK
http://www.academicpress.com

Library of Congress Cataloging-in-Publication Data
Davis, Martin 1928
Computability, complexity, and languages: fundamentals of Theoretical computer science / Martin D. Davis, Ron Sigal, Elaine J. Weyuker. –2nd ed.
p. cm. –(Computer science and applied mathematics)
Includes bibliographical references and index.
ISBN 0-12-206382-1
1. Machine theory. 2. Computational complexity. 3. Formal Languages. I. Sigal, Ron. II. Weyuker, Elaine J. III. Title. IV. Series.
QA267.D38 1994
511.3–dc20 93-26807
CIP

To the memory of Helen and Harry Davis
and to
Hannah and Herman Sigal
Sylvia and Marx Weyuker

Virginia Davis, Dana Latch, Thomas Ostrand
and to
Rachel Weyuker Ostrand

Contents

Preface

Theoretical computer science is the mathematical study of models of computation. As such, it originated in the 1930s, well before the existence of modern computers, in the work of the logicians Church, Gödel, Kleene, Post, and Turing. This early work has had a profound influence on the practical and theoretical development of computer science. Not only has the Turing machine model proved basic for theory, but the work of these pioneers presaged many aspects of computational practice that are now commonplace and whose intellectual antecedents are typically unknown to users. Included among these are the existence in principle of all-purpose (or universal) digital computers, the concept of a program as a list of instructions in a formal language, the possibility of interpretive programs, the duality between software and hardware, and the representation of languages by formal structures, based on productions. While the spotlight in computer science has tended to fall on the truly breathtaking technological advances that have been taking place, important work in the foundations of the subject has continued as well. It is our purpose in writing this book to provide an introduction to the various aspects of theoretical computer science for undergraduate and graduate students that is sufficiently comprehensive that the professional literature of treatises and research papers will become accessible to our readers.

We are dealing with a very young field that is still finding itself. Computer scientists have by no means been unanimous in judging which parts of the subject will turn out to have enduring significance. In this situation, fraught with peril for authors, we have attempted to select topics that have already achieved a polished classic form, and that we believe will play an important role in future research.

In this second edition, we have included new material on the subject of programming language semantics, which we believe to be established as an important topic in theoretical computer science. Some of the material on computability theory that had been scattered in the first edition has been brought together, and a few topics that were deemed to be of only peripheral interest to our intended audience have been eliminated. Numerous exercises have also been added. We were particularly pleased to be able to include the answer to a question that had to be listed as open in the first edition. Namely, we present Neil Immerman's surprisingly straightforward proof of the fact that the class of languages accepted by linear bounded automata is closed under complementation.

We have assumed that many of our readers will have had little experience with mathematical proof, but that almost all of them have had substantial programming experience. Thus the first chapter contains an introduction to the use of proofs in mathematics in addition to the usual explanation of terminology and notation. We then proceed to take advantage of the reader's background by developing computability theory in the context of an extremely simple abstract programming language. By systematic use of a macro expansion technique, the surprising power of the language is demonstrated. This culminates in a universal program, which is written in all detail on a single page. By a series of simulations, we then obtain the equivalence of various different formulations of computability, including Turing's. Our point of view with respect to these simulations is that it should not be the reader's responsibility, at this stage, to fill in the details of vaguely sketched arguments, but rather that it is our responsibility as authors to arrange matters so that the simulations can be exhibited simply, clearly, and completely.

This material, in various preliminary forms, has been used with undergraduate and graduate students at New York University, Brooklyn College, The Scuola Matematica Interuniversitaria–Perugia, The University of California–Berkeley, The University of California–Santa Barbara, Worcester Polytechnic Institute, and Yale University.

Although it has been our practice to cover the material from the second part of the book on formal languages after the first part, the chapters on regular and on context-free languages can be read immediately after Chapter 1. The Chomsky–Schützenberger representation theorem for context-free languages in used to develop their relation to pushdown automata in a way that we believe is clarifying. Part 3 is an exposition of the aspects of logic that we think are important for computer science and can also be read immediately following Chapter 1. Each of the chapters of Part 4 introduces an important theory of computational complexity, concluding with the theory of **NP**-completeness. Part 5, which is new to the second edition, uses recursion equations to expand upon the notion of computability developed in Part 1, with an emphasis on the techniques of formal semantics, both denotational and operational. Rooted in the early work of Gödel, Herbrand, Kleene, and others, Part 5 introduces ideas from the modern fields of functional programming languages, denotational semantics, and term rewriting systems.

Because many of the chapters are independent of one another, this book can be used in various ways. There is more than enough material for a full-year course at the graduate level on *theory of computation*. We have used the unstarred sections of Chapters 1–6 and Chapter 9 in a successful one-semester junior-level course, Introduction to Theory of Computation, at New York University. A course on *finite automata and formal languages* could be based on Chapters 1, 9, and 10. A semester or quarter course on *logic for computer scientists* could be based on selections from Parts 1 and 3. Part 5 could be used for a third semester on the theory of computation or an introduction to *programming language semantics*. Many other arrangements and courses are possible, as should be apparent from the dependency graph, which follows the Acknowledgments. It is our hope, however, that this book will help readers to see theoretical computer science not as a fragmented list of discrete topics, but rather as a unified subject drawing on powerful mathematical methods and on intuitions derived from experience with computing technology to give valuable insights into a vital new area of human knowledge.

Note to the Reader

Many readers will wish to begin with Chapter 2, using the material of Chapter 1 for reference as required. Readers who enjoy skipping around will find the *dependency graph* useful.

Sections marked with an asterisk (*) may be skipped without loss of continuity. The relationship of these sections to later material is given in the dependency graph.

Exercises marked with an asterisk either introduce new material, refer to earlier material in ways not indicated in the dependency graph, or simply are considered more difficult than unmarked exercises.

A reference to Theorem 8.1 is to Theorem 8.1 of the chapter in which the reference is made. When a reference is to a theorem in another chapter, the chapter is specified. The same system is used in referring to numbered formulas and to exercises.

Acknowledgments

It is a pleasure to acknowledge the help we have received. Charlene Herring, Debbie Herring, Barry Jacobs, and Joseph Miller made their student classroom notes available to us. James Cox, Keith Harrow, Steve Henkind, Karen Lemone, Colm O'Dunlaing, and James Robinett provided helpful comments and corrections. Stewart Weiss was kind enough to redraw one of the figures. Thomas Ostrand, Norman Shulman, Louis Salkind, Ron Sigal, Patricia Teller, and Elia Weixelbaum were particularly generous with their time, and devoted many hours to helping us. We are especially grateful to them.

Acknowledgments to Corrected Printing

We have taken this opportunity to correct a number of errors. We are grateful to the readers who have called our attention to errors and who have suggested corrections. The following have been particularly helpful: Alissa Bernholc, Domenico Cantone, John R. Cowles, Herbert Enderton, Phyllis Frankl, Fred Green, Warren Hirsch, J. D. Monk, Steve Rozen, and Stewart Weiss.

Acknowledgments to Second Edition

Yuri Gurevich, Paliath Narendran, Robert Paige, Carl Smith, and particularly Robert McNaughton made numerous suggestions for improving the first edition. Kung Chen, William Hurwood, Dana Latch, Sidd Puri, Benjamin Russell, Jason Smith, Jean Toal, and Niping Wu read a preliminary version of Part 5.

Acknowledgments to Reprint of Second Edition

We are grateful to the following people for their careful reading of the Second Edition: John Case, P. Klingsberg, Ken Klein, Eugenio Omodeo, David Schedler, John David Stone, and Lenore Zuck.

Dependency Graph

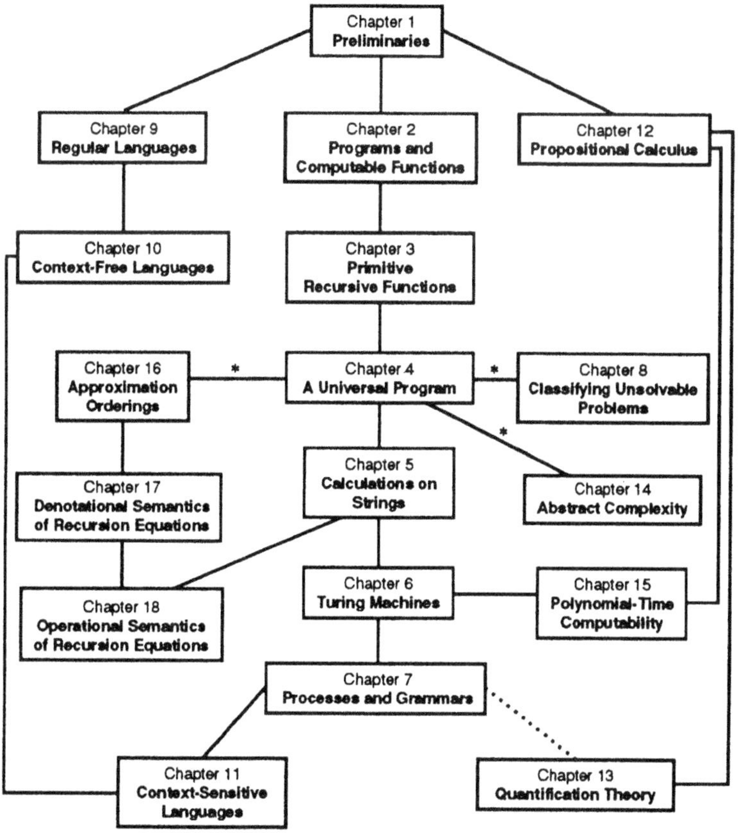

A solid line between two chapters indicates the dependence of the unstarred sections of the higher numbered chapter on the unstarred sections of the lower numbered chapter. An asterisk next to a solid line indicates that knowledge of the starred sections of the lower numbered chapter is also assumed. A dotted line shows that knowledge of the unstarred sections of the lower numbered chapter is assumed for the starred sections of the higher numbered chapter.

1

Preliminaries

1. Sets and *n*-tuples

We shall often be dealing with *sets* of objects of some definite kind. Thinking of a collection of entities as a *set* simply amounts to a decision to regard the whole collection as a single object. We shall use the word *class* as synonymous with *set*. In particular we write N for the set of *natural numbers* $0, 1, 2, 3, \ldots$. In this book the word *number* will always mean *natural number* except in contexts where the contrary is explicitly stated.

We write

$$a \in S$$

to mean that a belongs to S or, equivalently, is a member of the set S, and

$$a \notin S$$

to mean that a does not belong to S. It is useful to speak of the *empty set*, written $\varnothing$, which has no members. The equation $R = S$, where R and S are sets, means that R and S are *identical as sets*, that is, that they have exactly the same members. We write $R \subseteq S$ and speak of R as a *subset* of S to mean that every element of R is also an element of S. Thus, $R = S$ if and only if $R \subseteq S$ and $S \subseteq R$. Note also that for any set R, $\varnothing \subseteq R$ and $R \subseteq R$. We write $R \subset S$ to indicate that $R \subseteq S$ but $R \neq S$. In this case R

is called a *proper subset* of S. If R and S are sets, we write $R \cup S$ for the *union* of R and S, which is the collection of all objects which are members of either R or S or both. $R \cap S$, the *intersection* of R and S, is the set of all objects that belong to both R and S. $R - S$, the set of all objects that belong to R and do not belong to S, is the *difference* between R and S. S may contain objects not in R. Thus $R - S = R - (R \cap S)$. Often we will be working in contexts where all sets being considered are subsets of some fixed set D (sometimes called a *domain* or a *universe*). In such a case we write $\bar{S}$ for $D - S$, and call $\bar{S}$ the *complement* of S. Most frequently we shall be writing $\bar{S}$ for $N - S$. The De Morgan identities

$$\overline{R \cup S} = \bar{R} \cap \bar{S},$$

$$\overline{R \cap S} = \bar{R} \cup \bar{S}$$

are very useful; they are easy to check and any reader not already familiar with them should do so. We write

$$\{a_1, a_2, \ldots, a_n\}$$

for the set consisting of the n objects $a_1, a_2, \ldots, a_n$. Sets that can be written in this form as well as the empty set are called *finite*. Sets that are not finite, e.g., N, are called *infinite*. It should be carefully noted that a and $\{a\}$ are not the same thing. In particular, $a \in S$ is true if and only if $\{a\} \subseteq S$. Since two sets are equal if and only if they have the same members, it follows that, for example, $\{a, b, c\} = \{a, c, b\} = \{b, a, c\}$. That is, the order in which we may choose to write the members of a set is irrelevant. Where order is important, we speak instead of an n-tuple or a *list*. We write n-tuples using parentheses rather than curly braces:

$$(a_1, \ldots, a_n).$$

Naturally, the elements making up an n-tuple need not be distinct. Thus $(4, 1, 4, 2)$ is a 4-tuple. A 2-tuple is called an *ordered pair*, and a 3-tuple is called an *ordered triple*. Unlike the case for sets of one object, we *do not distinguish between the object a and the 1-tuple* (a). The crucial property of n-tuples is

$$(a_1, a_2, \ldots, a_n) = (b_1, b_2, \ldots, b_n)$$

if and only if

$$a_1 = b_1, \quad a_2 = b_2, \quad \ldots, \quad \text{and} \quad a_n = b_n.$$

If $S_1, S_2, \ldots, S_n$ are given sets, then we write $S_1 \times S_2 \times \cdots \times S_n$ for the set of all n-tuples $(a_1, a_2, \ldots, a_n)$ such that $a_1 \in S_1, a_2 \in S_2, \ldots, a_n \in S_n$.

$S_1 \times S_2 \times \cdots \times S_n$ is sometimes called the *Cartesian product* of $S_1, S_2, \ldots, S_n$. In case $S_1 = S_2 = \cdots = S_n = S$ we write S^n for the Cartesian product $S_1 \times S_2 \times \cdots \times S_n$.

2. Functions

Functions play an important role in virtually every branch of pure and applied mathematics. We may define a function simply as a set f, all of whose members are ordered pairs and that has the special property

$$(a, b) \in f \text{ and } (a, c) \in f \quad \text{implies} \quad b = c.$$

However, intuitively it is more helpful to think of the pairs listed as the rows of a table. For f a function, one writes $f(a) = b$ to mean that $(a, b) \in f$; the definition of function ensures that for each a there can be at most one such b. The set of all a such that $(a, b) \in f$ for some b is called the *domain* of f. The set of all $f(a)$ for a in the domain of f is called the *range* of f.

As an example, let f be the set of ordered pairs (n, n^2) for $n \in N$. Then, for each $n \in N$, $f(n) = n^2$. The domain of f is N. The range of f is the set of perfect squares.

Functions f are often specified by *algorithms* that provide procedures for obtaining $f(a)$ from a. This method of specifying functions is particularly important in computer science. However, as we shall see in Chapter 4, it is quite possible to possess an algorithm that specifies a function without being able to tell which elements belong to its domain. This makes the notion of a so-called *partial function* play a central role in computability theory. A *partial function on a set* S is simply a function whose domain is a subset of S. An example of a partial function on N is given by $g(n) = \sqrt{n}$, where the domain of g is the set of perfect squares. If f is a partial function on S and $a \in S$, then we write $f(a)\downarrow$ and say that $f(a)$ is *defined* to indicate that a is in the domain of f; if a is not in the domain of f, we write $f(a)\uparrow$ and say that $f(a)$ is *undefined.* If a partial function on S has the domain S, then it is called *total.* Finally, we should mention that the empty set $\varnothing$ is itself a function. Considered as a partial function on some set S, *it is nowhere defined.*

For a partial function f on a Cartesian product $S_1 \times S_2 \times \cdots \times S_n$, we write $f(a_1, \ldots, a_n)$ rather than $f((a_1, \ldots, a_n))$. A partial function f on a set S^n is called an *n-ary partial function on S*, or a function of n variables on S. We use *unary* and *binary* for 1-ary and 2-ary, respectively. For n-ary partial functions, we often write $f(x_1, \ldots, x_n)$ instead of f as a way of showing explicitly that f is n-ary.

Sometimes it is useful to work with particular kinds of functions. A function f is *one–one* if, for all x, y in the domain of f, $f(x) = f(y)$ implies $x = y$. Stated differently, if $x \neq y$ then $f(x) \neq f(y)$. If the range of f is the set S, then we say that f is an *onto* function with respect to S, or simply that f is *onto* S. For example, $f(n) = n^2$ is one–one, and f is onto the set of perfect squares, but it is not onto N.

We will sometimes refer to the idea of *closure*. If S is a set and f is a partial function on S, then S is *closed under* f if the range of f is a subset of S. For example, N is closed under $f(n) = n^2$, but it is not closed under $h(n) = \sqrt{n}$ (where h is a total function on N).

3. Alphabets and Strings

An *alphabet* is simply some finite nonempty set A of objects called *symbols*. An n-tuple of symbols of A is called a *word* or a *string* on A. Instead of writing a word as $(a_1, a_2, \ldots, a_n)$ we write simply $a_1 a_2 \cdots a_n$. If $u = a_1 a_2 \cdots a_n$, then we say that n is the length of u and write $|u| = n$. We allow a unique null word, written 0, of length 0. (The reason for using the same symbol for the number zero and the null word will become clear in Chapter 5.) The set of all words on the alphabet A is written A^*. Any subset of A^* is called a *language on A* or a *language with alphabet A*. We do *not* distinguish between a symbol $a \in A$ and the word of length 1 consisting of that symbol. If $u, v \in A^*$, then we write $\widehat{u\,v}$ for the word obtained by placing the string v after the string u. For example, if $A = \{a, b, c\}$, $u = bab$, and $v = caa$, then

$$\widehat{uv} = babcaa \qquad \text{and} \qquad \widehat{vu} = caabab.$$

Where no confusion can result, we write uv instead of $\widehat{u\,v}$. It is obvious that, for all u,

$$u0 = 0u = u,$$

and that, for all u, v, w,

$$u(vw) = (uv)w.$$

Also, if either $uv = uw$ or $vu = wu$, then $v = w$.

If u is a string, and $n \in N$, $n > 0$, we write

$$u^{[n]} = \underbrace{u\,u \cdots u}_{n}.$$

We also write $u^{[0]} = 0$. We use the square brackets to avoid confusion with numerical exponentiation.

If $u \in A^*$, we write u^R for u written backward; i.e., if $u = a_1 a_2 \cdots a_n$, for $a_1, \ldots, a_n \in A$, then $u^R = a_n \cdots a_2 a_1$. Clearly, $0^R = 0$ and $(uv)^R = v^R u^R$ for $u, v \in A^*$.

4. Predicates

By a *predicate* or a *Boolean-valued function* on a set S we mean a *total* function P on S such that for each $a \in S$, either

$$P(a) = \text{TRUE} \quad \text{or} \quad P(a) = \text{FALSE},$$

where TRUE and FALSE are a pair of distinct objects called *truth values*. We often say *$P(a)$ is true* for $P(a) = \text{TRUE}$, and *$P(a)$ is false* for $P(a) = \text{FALSE}$. For our purposes it is useful to identify the truth values with specific numbers, so we set

$$\text{TRUE} = 1 \quad \text{and} \quad \text{FALSE} = 0.$$

Thus, a predicate is a special kind of function with values in N. Predicates on a set S are usually specified by expressions which become statements, either true or false, when variables in the expression are replaced by symbols designating fixed elements of S. Thus the expression

$$x < 5$$

specifies a predicate on N, namely,

$$P(x) = \begin{cases} 1 & \text{if} \quad x = 0, 1, 2, 3, 4 \\ 0 & \text{otherwise.} \end{cases}$$

Three basic operations on truth values are defined by the tables in Table 4.1. Thus if P and Q are predicates on a set S, there are also the predicates $\sim P$, $P \,\&\, Q$, $P \vee Q$. $\sim P$ is true just when P is false; $P \,\&\, Q$ is true when both P and Q are true, otherwise it is false; $P \vee Q$ is true when either P or Q or both are true, otherwise it is false. Given a predicate P

Table 4.1

p	$\sim p$
0	1
1	0

p	q	$p \,\&\, q$	$p \vee q$
1	1	1	1
0	1	0	1
1	0	0	1
0	0	0	0

on a set S, there is a corresponding subset R of S, namely, the set of all elements $a \in S$ for which $P(a) = 1$. We write

$$R = \{a \in S | P(a)\}.$$

Conversely, given a subset R of a given set S, the expression

$$x \in R$$

defines a predicate on S, namely, the predicate defined by

$$P(x) = \begin{cases} 1 & \text{if } x \in R \\ 0 & \text{if } x \notin R. \end{cases}$$

Of course, in this case,

$$R = \{x \in S | P(x)\}.$$

The predicate P is called the *characteristic function* of the set R. The close connection between sets and predicates is such that one can readily translate back and forth between discourse involving one of these notions and discourse involving the other. Thus we have

$$\{x \in S \mid P(x) \,\&\, Q(x)\} = \{x \in S \mid P(x)\} \cap \{x \in S \mid Q(x)\},$$

$$\{x \in S \mid P(x) \vee Q(x)\} = \{x \in S \mid P(x)\} \cup \{x \in S \mid Q(x)\},$$

$$\{x \in S \mid \sim P(x)\} = S - \{x \in S \mid P(x)\}.$$

To indicate that two expressions containing variables define the same predicate we place the symbol $\Leftrightarrow$ between them. Thus,

$$x < 5 \Leftrightarrow x = 0 \vee x = 1 \vee x = 2 \vee x = 3 \vee x = 4.$$

The De Morgan identities from Section 1 can be expressed as follows in terms of predicates on a set S:

$$P(x) \,\&\, Q(x) \Leftrightarrow \sim(\sim P(x) \vee \sim Q(x)),$$

$$P(x) \vee Q(x) \Leftrightarrow \sim(\sim P(x) \,\&\, \sim Q(x)).$$

5. Quantifiers

In this section we will be concerned exclusively with predicates on N^m (or what is the same thing, m-ary predicates on N) for different values of m. Here and later we omit the phrase "on N" when the meaning is clear.

Thus, let $P(t, x_1, \ldots, x_n)$ be an $(n + 1)$-ary predicate. Consider the predicate $Q(y, x_1, \ldots, x_n)$ defined by

$$Q(y, x_1, \ldots, x_n) \Leftrightarrow P(0, x_1, \ldots, x_n) \vee P(1, x_1, \ldots, x_n) \vee \cdots \vee P(y, x_1, \ldots, x_n).$$

Thus the predicate $Q(y, x_1, \ldots, x_n)$ is true just in case there is a value of $t \leq y$ such that $P(t, x_1, \ldots, x_n)$ is true. We write this predicate Q as

$$(\exists t)_{\leq y} P(t, x_1, \ldots, x_n).$$

The expression "$(\exists t)_{\leq y}$" is called a *bounded existential quantifier*. Similarly, we write $(\forall t)_{\leq y} P(t, x_1, \ldots, x_n)$ for the predicate

$$P(0, x_1, \ldots, x_n) \,\&\, P(1, x_1, \ldots, x_n) \,\&\, \cdots \,\&\, P(y, x_1, \ldots, x_n).$$

This predicate is true just in case $P(t, x_1, \ldots, x_n)$ is true for *all* $t \leq y$. The expression "$(\forall t)_{\leq y}$" is called a *bounded universal quantifier*. We also write $(\exists t)_{< y} P(t, x_1, \ldots, x_n)$ for the predicate that is true just in case $P(t, x_1, \ldots, x_n)$ is true for at least one value of $t < y$ and $(\forall t)_{< y} P(t, x_1, \ldots, x_n)$ for the predicate that is true just in case $P(t, x_1, \ldots, x_n)$ is true for all values of $t < y$.

We write

$$Q(x_1, \ldots, x_n) \Leftrightarrow (\exists t) P(t, x_1, \ldots, x_n)$$

for the predicate which is true if there exists some $t \in N$ for which $P(t, x_1, \ldots, x_n)$ is true. Similarly, $(\forall t) P(t, x_1, \ldots, x_n)$ is true if $P(t, x_1, \ldots, x_n)$ is true for all $t \in N$.

The following generalized De Morgan identities are sometimes useful:

$$\sim (\exists t)_{\leq y} P(t, x_1, \ldots, x_n) \Leftrightarrow (\forall t)_{\leq y} \sim P(t, x_1, \ldots, x_n),$$

$$\sim (\exists t) P(t, x_1, \ldots, x_n) \Leftrightarrow (\forall t) \sim P(t, x_1, \ldots, x_n).$$

The reader may easily verify the following examples:

$$(\exists y)(x + y = 4) \Leftrightarrow x \leq 4,$$

$$(\exists y)(x + y = 4) \Leftrightarrow (\exists y)_{\leq 4}(x + y = 4),$$

$$(\forall y)(xy = 0) \Leftrightarrow x = 0,$$

$$(\exists y)_{\leq z}(x + y = 4) \Leftrightarrow (x + z \geq 4 \,\&\, x \leq 4).$$

6. Proof by Contradiction

In this book we will be calling many of the assertions we make *theorems* (or *corollaries* or *lemmas*) and providing *proofs* that they are correct. Why are proofs necessary? The following example should help in answering this question.

Recall that a number is called a *prime* if it has *exactly two distinct divisors*, itself and 1. Thus 2, 17, and 41 are primes, but 0, 1, 4, and 15 are not. Consider the following assertion:

$$n^2 - n + 41 \text{ is prime for all } n \in N.$$

This assertion is in fact *false*. Namely, for $n = 41$ the expression becomes

$$41^2 - 41 + 41 = 41^2,$$

which is certainly not a prime. However, the assertion is true (readers with access to a computer can easily check this!) for all $n \leq 40$. This example shows that inferring a result about all members of an infinite set (such as N) from even a large finite number of instances can be very dangerous. A proof is intended to overcome this obstacle.

A proof begins with some initial statements and uses logical reasoning to infer additional statements. (In Chapters 12 and 13 we shall see how the notion of logical reasoning can be made precise; but in fact, our *use* of logical reasoning will be in an informal intuitive style.) When the initial statements with which a proof begins are already accepted as correct, then any of the additional statements inferred can also be accepted as correct. But proofs often cannot be carried out in this simple-minded pattern. In this and the next section we will discuss more complex proof patterns.

In a *proof by contradiction*, one begins by supposing that the assertion we wish to prove is false. Then we can feel free to use the negation of what we are trying to prove as one of the initial statements in constructing a proof. In a proof by contradiction we look for a pair of statements developed in the course of the proof which *contradict* one another. Since both cannot be true, we have to conclude that our original supposition was wrong and therefore that our desired conclusion is correct.

We give two examples here of proof by contradiction. There will be many in the course of the book. Our first example is quite famous. We recall that every number is either even (i.e., $= 2n$ for some $n \in N$) or odd (i.e., $= 2n + 1$ for some $n \in N$). Moreover, if m is even, $m = 2n$, then $m^2 = 4n^2 = 2 \cdot 2n^2$ is even, while if m is odd, $m = 2n + 1$, then $m^2 = 4n^2 + 4n + 1 = 2(2n^2 + 2n) + 1$ is odd. We wish to prove that the equation

$$2 = (m/n)^2 \tag{6.1}$$

has no solution for $m, n \in N$ (that is, that $\sqrt{2}$ is not a "rational" number). We suppose that our equation has a solution and proceed to derive a contradiction. Given our supposition that (6.1) has a solution, it must have a solution in which m and n are not both even numbers. This is true because if m and n are both even, we can repeatedly "cancel" 2 from numerator and denominator until at least one of them is odd. On the other hand, we shall prove that for every solution of (6.1) m and n must both be even. The contradiction will show that our supposition was false, i.e., that (6.1) has no solution.

It remains to show that in every solution of (6.1), m and n are both even. We can rewrite (6.1) as

$$m^2 = 2n^2,$$

which shows that m^2 is even. As we saw above this implies that m is even, say $m = 2k$. Thus, $m^2 = 4k^2 = 2n^2$, or $n^2 = 2k^2$. Thus, n^2 is even and hence n is even. ■

Note the symbol ■, which means "the proof is now complete."

Our second example involves strings as discussed in Section 3.

Theorem 6.1. Let $x \in \{a, b\}^*$ such that $xa = ax$. Then $x = a^{[n]}$ for some $n \in N$.

Proof. Suppose that $xa = ax$ but x contains the letter b. Then we can write $x = a^{[n]}bu$, where we have explicitly shown the first (i.e., leftmost) occurrence of b in x. Then

$$a^{[n]}bua = aa^{[n]}bu = a^{[n+1]}bu.$$

Thus,

$$bua = abu.$$

But this is impossible, since the same string cannot have its first symbol be both b and a. This contradiction proves the theorem. ■

Exercises

1. Prove that the equation $(p/q)^2 = 3$ has no solution for $p, q \in N$.
2. Prove that if $x \in \{a, b\}^*$ and $abx = xab$, then $x = (ab)^{[n]}$ for some $n \in N$.

7. Mathematical Induction

Mathematical induction furnishes an important technique for proving statements of the form $(\forall n)P(n)$, where P is a predicate on N. One proceeds by proving a pair of auxiliary statements, namely,

$$P(0)$$

and

$$(\forall n)(\textit{If } P(n) \textit{ then } P(n+1)). \tag{7.1}$$

Once we have succeeded in proving these auxiliary statements we can regard $(\forall n)P(n)$ as also proved. The justification for this is as follows.

From the second auxiliary statement we can infer each of the infinite set of statements:

$$\textit{If } P(0) \textit{ then } P(1),$$
$$\textit{If } P(1) \textit{ then } P(2),$$
$$\textit{If } P(2) \textit{ then } P(3), \ldots .$$

Since we have proved $P(0)$, we can infer $P(1)$. Having now proved $P(1)$ we can get $P(2)$, etc. Thus, we see that $P(n)$ is true for all n and hence $(\forall n)P(n)$ is true.

Why is this helpful? Because sometimes it is much easier to prove (7.1) than to prove $(\forall n)P(n)$ in some other way. In proving this second auxiliary proposition one typically considers some fixed but arbitrary value k of n and shows that if we assume $P(k)$ we can prove $P(k+1)$. $P(k)$ is then called the *induction hypothesis*. This methodology enables us to use $P(k)$ as one of the initial statements in the proof we are constructing.

There are some paradoxical things about proofs by mathematical induction. One is that considered superficially, it seems like an example of circular reasoning. One seems to be assuming $P(k)$ for an arbitrary k, which is exactly what one is supposed to be engaged in proving. Of course, one is not really assuming $(\forall n)P(n)$. One is assuming $P(k)$ for some *particular* k in order to show that $P(k+1)$ follows.

It is also paradoxical that in using induction (we shall often omit the word *mathematical*), it is sometimes easier to prove statements by first making them "stronger." We can put this schematically as follows. We wish to prove $(\forall n)P(n)$. Instead we decide to prove the *stronger* assertion $(\forall n)(P(n) \,\&\, Q(n))$ (which of course implies the original statement). Proving the stronger statement by induction requires that we prove

$$P(0) \,\&\, Q(0)$$

and

$$(\forall n)[\textit{If } P(n) \,\&\, Q(n) \textit{ then } P(n+1) \,\&\, Q(n+1)].$$

In proving this second auxiliary statement, we may take $P(k) \,\&\, Q(k)$ as our induction hypothesis. Thus, although strengthening the statement to be proved gives us more to prove, it also gives us a stronger induction hypothesis and, therefore, more to work with. The technique of deliberately strengthening what is to be proved for the purpose of making proofs by induction easier is called *induction loading*.

It is time for an example of a proof by induction. The following is useful in doing one of the exercises in Chapter 6.

Theorem 7.1. For all $n \in N$ we have $\sum_{i=0}^{n}(2i + 1) = (n + 1)^2$.

Proof. For $n = 0$, our theorem states simply that $1 = 1^2$, which is true.

Suppose the result known for $n = k$. That is, our induction hypothesis is

$$\sum_{i=0}^{k} (2i + 1) = (k + 1)^2.$$

Then

$$\begin{aligned}\sum_{i=0}^{k+1} (2i + 1) &= \sum_{i=0}^{k} (2i + 1) + 2(k + 1) + 1 \\ &= (k + 1)^2 + 2(k + 1) + 1 \\ &= (k + 2)^2.\end{aligned}$$

But this is the desired result for $n = k + 1$. ∎

Another form of mathematical induction that is often very useful is called *course-of-values induction* or sometimes *complete induction*. In the case of course-of-values induction we prove the single auxiliary statement

$$(\forall n)[\textit{If } (\forall m)_{m<n} P(m) \textit{ then } P(n)], \tag{7.2}$$

and then conclude that $(\forall n)P(n)$ is true. A potentially confusing aspect of course-of-values induction is the apparent lack of an initial statement $P(0)$. But in fact there is no such lack. The case $n = 0$ of (7.2) is

$$\textit{If } (\forall m)_{m<0} P(m) \textit{ then } P(0).$$

But the "induction hypothesis" $(\forall m)_{m<0}P(m)$ is entirely vacuous because there is no $m \in N$ such that $m < 0$. So in proving (7.2) for $n = 0$ we really are just proving $P(0)$. In practice it is sometimes possible to give a single proof of (7.2) that works for all n including $n = 0$. But often the case $n = 0$ has to be handled separately.

To see why course-of-values induction works, consider that, in the light of what we have said about the $n = 0$ case, (7.2) leads to the following infinite set of statements:

$$P(0),$$
$$\textit{If } P(0) \textit{ then } P(1),$$
$$\textit{If } P(0) \,\&\, P(1) \textit{ then } P(2),$$
$$\textit{If } P(0) \,\&\, P(1) \,\&\, P(2) \textit{ then } P(3),$$
$$\vdots$$

Here is an example of a theorem proved by course-of-values induction.

Theorem 7.2. There is no string $x \in \{a, b\}^*$ such that $ax = xb$.

Proof. Consider the following predicate: *If* $x \in \{a, b\}^*$ and $|x| = n$, *then* $ax \neq xb$. We will show that this is true for all $n \in N$. So we assume it true for all $m < k$ for some given k and show that it follows for k. This proof will be by contradiction. Thus, suppose that $|x| = k$ and $ax = xb$. The equation implies that a is the first and b the last symbol in x. So, we can write $x = aub$. Then

$$aaub = aubb,$$

i.e.,

$$au = ub.$$

But $|u| < |x|$. Hence by the induction hypothesis $au \neq ub$. This contradiction proves the theorem. ■

Proofs by course-of-values induction can always be rewritten so as to involve reference to the principle that if some predicate is true for some element of N, then there must be a least element of N for which it is true. Here is the proof of Theorem 7.2 given in this style.

Proof. Suppose there is a string $x \in \{a, b\}^*$ such that $ax = xb$. Then there must be a string satisfying this equation of minimum length. Let x be such a string. Then $ax = xb$, but, if $|u| < |x|$, then $au \neq ub$. However, $ax = xb$ implies that $x = aub$, so that $au = ub$ and $|u| < |x|$. This contradiction proves the theorem. ■

Exercises

1. Prove by mathematical induction that $\sum_{i=1}^{n} i = n(n + 1)/2$.
2. Here is a "proof" by mathematical induction that if $x, y \in N$, then $x = y$. What is wrong?

Let

$$\max(x, y) = \begin{cases} x & \text{if } x \geq y \\ y & \text{otherwise} \end{cases}$$

for $x, y \in N$. Consider the predicate

$$(\forall x)(\forall y)[\textit{If } \max(x, y) = n, \textit{then } x = y].$$

For $n = 0$, this is clearly true. Assume the result for $n = k$, and let $\max(x, y) = k + 1$. Let $x_1 = x - 1$, $y_1 = y - 1$. Then $\max(x_1, y_1) = k$. By the induction hypothesis, $x_1 = y_1$ and therefore $x = x_1 + 1 = y_1 + 1 = y$.

3. Here is another incorrect proof that purports to use mathematical induction to prove that all flowers have the same color! What is wrong?

   Consider the following predicate: If S is a set of flowers containing exactly n elements, then all the flowers in S have the same color. The predicate is clearly true if $n = 1$. We suppose it true for $n = k$ and prove the result for $n = k + 1$. Thus, let S be a set of $k + 1$ flowers. If we remove one flower from S we get a set of k flowers. Therefore, by the induction hypothesis they all have the same color. Now return the flower removed from S and remove another. Again by our induction hypothesis the remaining flowers all have the same color. But now both of the flowers removed have been shown to have the same color as the rest. Thus, all the flowers in S have the same color.

4. Show that there are no strings $x, y \in \{a, b\}^*$ such that $xay = ybx$.

5. Give a "one-line" proof of Theorem 7.2 that does not use mathematical induction.

Part 1

Computability

2

Programs and Computable Functions

1. A Programming Language

Our development of computability theory will be based on a specific programming language $\mathscr{S}$. We will use certain letters as variables whose values are *numbers*. (In this book the word *number* will always mean nonnegative integer, unless the contrary is specifically stated.) In particular, the letters

$$X_1 \ X_2 \ X_3 \ \cdots$$

will be called the *input variables* of $\mathscr{S}$, the letter Y will be called the *output variable* of $\mathscr{S}$, and the letters

$$Z_1 \ Z_2 \ Z_3 \ \cdots$$

will be called the *local variables* of $\mathscr{S}$. The subscript 1 is often omitted; i.e., X stands for X_1 and Z for Z_1. Unlike the programming languages in actual use, there is no upper limit on the values these variables can assume. Thus from the outset, $\mathscr{S}$ must be regarded as a purely theoretical entity. Nevertheless, readers having programming experience will find working with $\mathscr{S}$ very easy.

In $\mathscr{S}$ we will be able to write "instructions" of various sorts; a "program" of $\mathscr{S}$ will then consist of a *list* (i.e., a finite sequence) of

Table 1.1

Instruction	Interpretation
$V \leftarrow V + 1$	Increase by 1 the value of the variable V.
$V \leftarrow V - 1$	If the value of V is 0, leave it unchanged; otherwise decrease by 1 the value of V.
IF $V \neq 0$ GOTO L	If the value of V is nonzero, perform the instruction with label L next; otherwise proceed to the next instruction in the list.

instructions. For example, for each variable V there will be an instruction:

$$V \leftarrow V + 1$$

A simple example of a program of $\mathscr{S}$ is

$$X \leftarrow X + 1$$
$$X \leftarrow X + 1$$

"Execution" of this program has the effect of increasing the value of X by 2. In addition to variables, we will need "labels." In $\mathscr{S}$ these are

$$A_1\ B_1\ C_1\ D_1\ E_1\ A_2\ B_2\ C_2\ D_2\ E_2\ A_3\ \cdots.$$

Once again the subscript 1 can be omitted. We give in Table 1.1 a complete list of our instructions. In this list V stands for any variable and L stands for any label.

These instructions will be called the *increment*, *decrement*, and *conditional branch* instructions, respectively.

We will use the special convention that *the output variable Y and the local variables Z_i initially have the value* 0. We will sometimes indicate the value of a variable by writing it in lowercase italics. Thus x_5 is the value of X_5.

Instructions may or may not have labels. When an instruction is labeled, the label is written to its left in square brackets. For example,

$$[B] \qquad Z \leftarrow Z - 1$$

In order to base computability theory on the language $\mathscr{S}$, we will require formal definitions. But before we supply these, it is instructive to work informally with programs of $\mathscr{S}$.

2. Some Examples of Programs

(a) Our first example is the program

$$
\begin{array}{ll}
[A] & X \leftarrow X - 1 \\
 & Y \leftarrow Y + 1 \\
 & \text{IF } X \neq 0 \text{ GOTO } A
\end{array}
$$

If the initial value x of X is not 0, the effect of this program is to copy x into Y and to decrement the value of X down to 0. (By our conventions the initial value of Y is 0.) If $x = 0$, then the program halts with Y having the value 1. We will say that this program *computes* the function

$$f(x) = \begin{cases} 1 & \text{if} \quad x = 0 \\ x & \text{otherwise.} \end{cases}$$

This program halts when it executes the third instruction of the program with X having the value 0. In this case the condition $X \neq 0$ is not fulfilled and therefore the branch is not taken. When an attempt is made to move on to the nonexistent fourth instruction, the program halts. A program will also halt if an instruction labeled L is to be executed, but there is no instruction in the program with that label. In this case, we usually will use the letter E (for "exit") as the label which labels no instruction.

(b) Although the preceding program is a perfectly well-defined program of our language $\mathscr{S}$, we may think of it as having arisen in an attempt to write a program that copies the value of X into Y, and therefore containing a "bug" because it does not handle 0 correctly. The following slightly more complicated example remedies this situation.

$$\begin{array}{ll} [A] & \text{IF } X \neq 0 \text{ GOTO } B \\ & Z \leftarrow Z + 1 \\ & \text{IF } Z \neq 0 \text{ GOTO } E \\ [B] & X \leftarrow X - 1 \\ & Y \leftarrow Y + 1 \\ & Z \leftarrow Z + 1 \\ & \text{IF } Z \neq 0 \text{ GOTO } A \end{array}$$

As we can easily convince ourselves, this program does copy the value of X into Y for all initial values of X. Thus, we say that it computes the function $f(x) = x$. At first glance Z's role in the computation may not be obvious. It is used simply to allow us to code an *unconditional branch*. That is, the program segment

$$\begin{array}{l} Z \leftarrow Z + 1 \\ \text{IF } Z \neq 0 \text{ GOTO } L \end{array} \tag{2.1}$$

has the effect (ignoring the effect on the value of Z) of an instruction

GOTO L

such as is available in most programming languages. To see that this is true we note that the first instruction of the segment guarantees that Z has a nonzero value. Thus the condition $Z \neq 0$ is always true and hence the next instruction performed will be the instruction labeled L. Now GOTO L is not an instruction in our language $\mathscr{S}$, but since we will frequently have use for such an instruction, we can use it as an abbreviation for the program segment (2.1). Such an abbreviating pseudoinstruction will be called a *macro* and the program or program segment which it abbreviates will be called its *macro expansion*.

The use of these terms is obviously motivated by similarities with the notion of a macro instruction occurring in many programming languages. At this point we will not discuss how to ensure that the variables local to the macro definition are distinct from the variables used in the main program. Instead, we will manually replace any such duplicate variable uses with unused variables. This will be illustrated in the "expanded" multiplication program in (e). In Section 5 this matter will be dealt with in a formal manner.

(c) Note that although the program of (b) does copy the value of X into Y, in the process the value of X is "destroyed" and the program terminates with X having the value 0. Of course, typically, programmers want to be able to copy the value of one variable into another without the original being "zeroed out." This is accomplished in the next program. (Note that we use our macro instruction GOTO L several times to shorten the program. Of course, if challenged, we could produce a legal program of $\mathscr{S}$ by replacing each GOTO L by a macro expansion. These macro expansions would have to use a local variable other than Z so as not to interfere with the value of Z in the main program.)

$[A]$	If $X \neq 0$ GOTO B
	GOTO C
$[B]$	$X \leftarrow X - 1$
	$Y \leftarrow Y + 1$
	$Z \leftarrow Z + 1$
	GOTO A
$[C]$	IF $Z \neq 0$ GOTO D
	GOTO E
$[D]$	$Z \leftarrow Z - 1$
	$X \leftarrow X + 1$
	GOTO C

In the first loop, this program copies the value of X into both Y and Z, while in the second loop, the value of X is restored. When the program terminates, both X and Y contain X's original value and $z = 0$.

We wish to use this program to justify the introduction of a macro which we will write

$$V \leftarrow V'$$

the execution of which will replace the contents of the variable V by the contents of the variable V' while leaving the contents of V' unaltered. Now, this program (c) functions correctly as a copying program only under our assumption that the variables Y and Z are initialized to the value 0. Thus, we can use the program as the basis of a macro expansion of $V \leftarrow V'$ only if we can arrange matters so as to be sure that the corresponding variables have the value 0 whenever the macro expansion is entered. To solve this problem we introduce the macro

$$V \leftarrow 0$$

which will have the effect of setting the contents of V equal to 0. The corresponding macro expansion is simply

```
[L]     V ← V − 1
        IF V ≠ 0 GOTO L
```

where, of course, the label L is to be chosen to be different from any of the labels in the main program. We can now write the macro expansion of $V \leftarrow V'$ by letting the macro $V \leftarrow 0$ precede the program which results when X is replaced by V' and Y is replaced by V in program (c). The result is as follows:

```
        V ← 0
[A]     IF V' ≠ 0 GOTO B
        GOTO C
[B]     V' ← V' − 1
        V ← V + 1
        Z ← Z + 1
        GOTO A
[C]     IF Z ≠ 0 GOTO D
        GOTO E
[D]     Z ← Z − 1
        V' ← V' + 1
        GOTO C
```

With respect to this macro expansion the following should be noted:

1. It is unnecessary (although of course it would be harmless) to include a $Z \leftarrow 0$ macro at the beginning of the expansion because, as has already been remarked, program (c) terminates with $z = 0$.
2. When inserting the expansion in an actual program, the variable Z will have to be replaced by a local variable which does not occur in the main program.

3. Likewise the labels A, B, C, D will have to be replaced by labels which do not occur in the main program.
4. Finally, the label E in the macro expansion must be replaced by a label L such that the instruction which follows the macro in the main program (if there is one) begins $[L]$.

(d) A program with two inputs that computes the function

$$f(x_1, x_2) = x_1 + x_2$$

is as follows:

$$
\begin{array}{ll}
 & Y \leftarrow X_1 \\
 & Z \leftarrow X_2 \\
[B] & \text{IF } Z \neq 0 \text{ GOTO } A \\
 & \text{GOTO } E \\
[A] & Z \leftarrow Z - 1 \\
 & Y \leftarrow Y + 1 \\
 & \text{GOTO } B
\end{array}
$$

Again, if challenged we would supply macro expansions for "$Y \leftarrow X_1$" and "$Z \leftarrow X_2$" as well as for the two unconditional branches. Note that Z is used to preserve the value of X_2.

(e) We now present a program that multiplies, i.e. that computes $f(x_1, x_2) = x_1 \cdot x_2$. Since multiplication can be regarded as repeated addition, we are led to the "program"

$$
\begin{array}{ll}
 & Z_2 \leftarrow X_2 \\
[B] & \text{IF } Z_2 \neq 0 \text{ GOTO } A \\
 & \text{GOTO } E \\
[A] & Z_2 \leftarrow Z_2 \dot{-} 1 \\
 & Z_1 \leftarrow X_1 + Y \\
 & Y \leftarrow Z_1 \\
 & \text{GOTO } B
\end{array}
$$

Of course, the "instruction" $Z_1 \leftarrow X_1 + Y$ is not permitted in the language $\mathscr{S}$. What we have in mind is that since we already have an addition program, we can replace the macro $Z_1 \leftarrow X_1 + Y$ by a program for computing it, which we will call its macro expansion. At first glance, one might wonder why the pair of instructions

$$
\begin{array}{l}
Z_1 \leftarrow X_1 + Y \\
Y \leftarrow Z_1
\end{array}
$$

was used in this program rather than the single instruction

$$Y \leftarrow X_1 + Y$$

since we simply want to replace the current value of Y by the sum of its value and x_1. The sum program in (d) computes $Y = X_1 + X_2$. If we were to use that as a template, we would have to replace X_2 in the program by Y. Now if we tried to use Y also as the variable being assigned, the macro expansion would be as follows:

$$\begin{array}{ll}
 & Y \leftarrow X_1 \\
 & Z \leftarrow Y \\
[B] & \text{IF } Z \neq 0 \text{ GOTO } A \\
 & \text{GOTO } E \\
[A] & Z \leftarrow Z - 1 \\
 & Y \leftarrow Y + 1 \\
 & \text{GOTO } B
\end{array}$$

What does this program actually compute? It should not be difficult to see that instead of computing $x_1 + y$ as desired, this program computes $2x_1$. Since X_1 is to be added over and over again, it is important that X_1 not be destroyed by the addition program. Here is the multiplication program, showing the macro expansion of $Z_1 \leftarrow X_1 + Y$:

$$\begin{array}{lll}
 & Z_2 \leftarrow X_2 & \\
[B] & \text{IF } Z_2 \neq 0 \text{ GOTO } A & \\
 & \text{GOTO } E & \\
[A] & Z_2 \leftarrow Z_2 - 1 & \\
 & Z_1 \leftarrow X_1 & \\
 & Z_3 \leftarrow Y & \\
[B_2] & \text{IF } Z_3 \neq 0 \text{ GOTO } A_2 & \text{Macro Expansion of} \\
 & \text{GOTO } E_2 & Z_1 \leftarrow X_1 + Y \\
[A_2] & Z_3 \leftarrow Z_3 - 1 & \\
 & Z_1 \leftarrow Z_1 + 1 & \\
 & \text{GOTO } B_2 & \\
[E_2] & Y \leftarrow Z_1 & \\
 & \text{GOTO } B &
\end{array}$$

Note the following:

1. The local variable Z_1 in the addition program in (d) must be replaced by another local variable (we have used Z_3) because Z_1 (the other name for Z) is also used as a local variable in the multiplication program.

2. The labels A, B, E are used in the multiplication program and hence cannot be used in the macro expansion. We have used A_2, B_2, E_2 instead.
3. The instruction GOTO E_2 terminates the addition. Hence, it is necessary that the instruction immediately following the macro expansion be labeled E_2.

In the future we will often omit such details in connection with macro expansions. All that is important is that our infinite supply of variables and labels guarantees that the needed changes can always be made.

(f) For our final example, we take the program

```
      Y ← X1
      Z ← X2
[C]   IF Z ≠ 0 GOTO A
      GOTO E
[A]   IF Y ≠ 0 GOTO B
      GOTO A
[B]   Y ← Y − 1
      Z ← Z − 1
      GOTO C
```

If we begin with $X_1 = 5$, $X_2 = 2$, the program first sets $Y = 5$ and $Z = 2$. Successively the program sets $Y = 4$, $Z = 1$ and $Y = 3$, $Z = 0$. Thus, the computation terminates with $Y = 3 = 5 - 2$. Clearly, if we begin with $X_1 = m$, $X_2 = n$, where $m \geq n$, the program will terminate with $Y = m - n$.

What happens if we begin with a value of X_1 less than the value of X_2, e.g., $X_1 = 2$, $X_2 = 5$? The program sets $Y = 2$ and $Z = 5$ and successively sets $Y = 1$, $Z = 4$ and $Y = 0$, $Z = 3$. At this point the computation enters the "loop":

```
[A]   IF Y ≠ 0 GOTO B
      GOTO A
```

Since $y = 0$, there is no way out of this loop and the computation will continue "forever." Thus, if we begin with $X_1 = m$, $X_2 = n$, where $m < n$, the computation will never terminate. In this case (and in similar cases) we will say that the program computes the *partial function*

$$g(x_1, x_2) = \begin{cases} x_1 - x_2 & \text{if} \quad x_1 \geq x_2 \\ \uparrow & \text{if} \quad x_1 < x_2. \end{cases}$$

(Partial functions are discussed in Chapter 1, Section 2.)

Exercises

1. Write a program in $\mathscr{S}$ (using macros freely) that computes the function $f(x) = 3x$.
2. Write a program in $\mathscr{S}$ that solves Exercise 1 using no macros.
3. Let $f(x) = 1$ if x is even; $f(x) = 0$ if x is odd. Write a program in $\mathscr{S}$ that computes f.
4. Let $f(x) = 1$ if x is even; $f(x)$ undefined if x is odd. Write a program in $\mathscr{S}$ that computes f.
5. Let $f(x_1, x_2) = 1$ if $x_1 = x_2$; $f(x_1, x_2) = 0$ if $x_1 \neq x_2$. Without using macros, write a program in $\mathscr{S}$ that computes f.
6. Let $f(x)$ be the greatest number n such that $n^2 \leq x$. Write a program in $\mathscr{S}$ that computes f.
7. Let $\gcd(x_1, x_2)$ be the greatest common divisor of x_1 and x_2. Write a program in $\mathscr{S}$ that computes gcd.

3. Syntax

We are now ready to be mercilessly precise about the language $\mathscr{S}$. Some of the description recapitulates the preceding discussion.

The symbols

$$X_1 \ X_2 \ X_3 \ \cdots$$

are called *input variables*,

$$Z_1 \ Z_2 \ Z_3 \ \cdots$$

are called *local variables*, and Y is called the *output variable* of $\mathscr{S}$. The symbols

$$A_1 \ B_1 \ C_1 \ D_1 \ E_1 \ A_2 \ B_2 \ \cdots$$

are called *labels* of $\mathscr{S}$. (As already indicated, in practice the subscript 1 is often omitted.) A *statement* is one of the following:

$$V \leftarrow V + 1$$
$$V \leftarrow V - 1$$
$$V \leftarrow V$$
$$\text{IF } V \neq 0 \text{ GOTO } L$$

where V may be any variable and L may be any label.

Note that we have included among the statements of $\mathscr{S}$ the "dummy" commands $V \leftarrow V$. Since execution of these commands leaves all values unchanged, they have no effect on what a program computes. They are included for reasons that will not be made clear until much later. But their inclusion is certainly quite harmless.

Next, an *instruction* is either a statement (in which case it is also called an *unlabeled* instruction) or $[L]$ followed by a statement (in which case the instruction is said to have L as its label or to be labeled L). A *program* is a list (i.e., a finite sequence) of instructions. The length of this list is called the *length* of the program. It is useful to include the *empty program* of length 0, which of course contains no instructions.

As we have seen informally, in the course of a computation, the variables of a program assume different numerical values. This suggests the following definition:

A *state of a program* $\mathscr{P}$ is a list of equations of the form $V = m$, where V is a variable and m is a number, including an equation for each variable that occurs in $\mathscr{P}$ and including no two equations with the same variable. As an example, let $\mathscr{P}$ be the program of (b) from Section 2, which contains the variables X Y Z. The list

$$X = 4, \qquad Y = 3, \qquad Z = 3$$

is thus a state of $\mathscr{P}$. (The definition of *state* does not require that the state can actually be "attained" from some initial state.) The list

$$X_1 = 4, \qquad X_2 = 5, \qquad Y = 4, \qquad Z = 4$$

is also a state of $\mathscr{P}$. (Recall that X is another name for X_1 and note that the definition permits inclusion of equations involving variables not actually occurring in $\mathscr{P}$.) The list

$$X = 3, \qquad Z = 3$$

is *not* a state of $\mathscr{P}$ since no equation in Y occurs. Likewise, the list

$$X = 3, \qquad X = 4, \qquad Y = 2, \qquad Z = 2$$

is *not* a state of $\mathscr{P}$: there are two equations in X.

Let σ be a state of $\mathscr{P}$ and let V be a variable that occurs in σ. The *value of V at σ* is then the (unique) number q such that the equation $V = q$ is one of the equations making up σ. For example, the value of X at the state

$$X = 4, \qquad Y = 3, \qquad Z = 3$$

is 4.

Suppose we have a program $\mathscr{P}$ and a state σ of $\mathscr{P}$. In order to say what happens "next," we also need to know which instruction of $\mathscr{P}$ is about to be executed. We therefore define a *snapshot* or *instantaneous description* of a program $\mathscr{P}$ of length n to be a pair (i, σ) where $1 \le i \le n + 1$, and σ is a state of $\mathscr{P}$. (Intuitively the number i indicates that it is the ith instruction which is about to be executed; $i = n + 1$ corresponds to a "stop" instruction.)

If $s = (i, \sigma)$ is a snapshot of $\mathscr{P}$ and V is a variable of $\mathscr{P}$, then the *value of V at s* just means the value of V at σ.

A snapshot (i, σ) of a program $\mathscr{P}$ of length n is called *terminal* if $i = n + 1$. If (i, σ) is a nonterminal snapshot of $\mathscr{P}$, we define the *successor* of (i, σ) to be the snapshot (j, τ) defined as follows:

Case 1. The ith instruction of $\mathscr{P}$ is $V \leftarrow V + 1$ and σ contains the equation $V = m$. Then $j = i + 1$ and τ is obtained from σ by replacing the equation $V = m$ by $V = m + 1$ (i.e., the value of V at τ is $m + 1$).

Case 2. The ith instruction of $\mathscr{P}$ is $V \leftarrow V - 1$ and σ contains the equation $V = m$. Then $j = i + 1$ and τ is obtained from σ by replacing the equation $V = m$ by $V = m - 1$ if $m \neq 0$; if $m = 0$, $\tau = \sigma$.

Case 3. The ith instruction of $\mathscr{P}$ is $V \leftarrow V$. Then $\tau = \sigma$ and $j = i + 1$.

Case 4. The ith instruction of $\mathscr{P}$ is IF $V \neq 0$ GOTO L. Then $\tau = \sigma$, and there are two subcases:

Case 4a. σ contains the equation $V = 0$. Then $j = i + 1$.

Case 4b. σ contains the equation $V = m$ where $m \neq 0$. Then, if there is an instruction of $\mathscr{P}$ labeled L, j is the *least number* such that the jth instruction of $\mathscr{P}$ is labeled L. Otherwise, $j = n + 1$.

For an example, we return to the program of (b), Section 2. Let σ be the state

$$X = 4, \qquad Y = 0, \qquad Z = 0$$

and let us compute the successor of the snapshots (i, σ) for various values of i.

For $i = 1$, the successor is $(4, \sigma)$ where σ is as above. For $i = 2$, the successor is $(3, \tau)$, where τ consists of the equations

$$X = 4, \qquad Y = 0, \qquad Z = 1.$$

For $i = 7$, the successor is $(8, \sigma)$. This is a terminal snapshot.

A *computation* of a program $\mathscr{P}$ is defined to be a sequence (i.e., a list) $s_1, s_2, \ldots, s_k$ of snapshots of $\mathscr{P}$ such that s_{i+1} is the successor of s_i for $i = 1, 2, \ldots, k - 1$ and s_k is terminal.

Note that we have not forbidden a program to contain more than one instruction having the same label. However, our definition of successor of a snapshot, in effect, interprets a branch instruction as always referring to the *first* statement in the program having the label in question. Thus, for example, the program

```
[A]    X ← X − 1
       IF X ≠ 0 GOTO A
[A]    X ← X + 1
```

is equivalent to the program

```
[A]    X ← X − 1
       IF X ≠ 0 GOTO A
       X ← X + 1
```

Exercises

1. Let $\mathscr{P}$ be the program of (b), Section 2. Write out a computation of $\mathscr{P}$ beginning with the snapshot $(1, \sigma)$, where σ consists of the equations $X = 2$, $Y = 0$, $Z = 0$.
2. Give a program $\mathscr{P}$ such that for every computation $s_1 = (1, \sigma), \ldots, s_k$ of $\mathscr{P}$, $k = 5$.
3. Give a program $\mathscr{P}$ such that for any $n > 0$ and every computation $s_1 = (1, \sigma), s_2, \ldots, s_k$ of $\mathscr{P}$ that has the equation $X = n$ in σ, $k = 2n + 1$.

4. Computable Functions

We have been speaking of the function computed by a program $\mathscr{P}$. It is now time to make this notion precise.

One would expect a program that computes a function of m variables to contain the input variables $X_1, X_2, \ldots, X_m$, and the output variable Y, and to have all other variables (if any) in the program be local. Although this has been and will continue to be our practice, it is convenient not to make it a formal requirement. According to the definitions we are going to present, any program $\mathscr{P}$ of the language $\mathscr{S}$ can be used to compute a function of one variable, a function of two variables, and, in general, for each $m \geq 1$, a function of m variables.

Thus, let $\mathscr{P}$ be any program in the language $\mathscr{S}$ and let $r_1, \ldots, r_m$ be m given numbers. We form the state σ of $\mathscr{P}$ which consists of the equations

$$X_1 = r_1, \qquad X_2 = r_2, \qquad \ldots, \qquad X_m = r_m, \qquad Y = 0$$

together with the equations $V = 0$ for each variable V in $\mathscr{P}$ other than $X_1, \ldots, X_m, Y$. We will call this the *initial state*, and the snapshot $(1, \sigma)$, the *initial snapshot*.

Case 1. *There is a computation* $s_1, s_2, \ldots, s_k$ of $\mathscr{P}$ *beginning with the initial snapshot*. Then we write $\psi_{\mathscr{P}}^{(m)}(r_1, r_2, \ldots, r_m)$ for the value of the variable Y at the (terminal) snapshot s_k.

Case 2. *There is no such computation;* i.e., there is an *infinite* sequence $s_1, s_2, s_3, \ldots$ beginning with the initial snapshot where each s_{i+1} is the successor of s_i. In this case $\psi_{\mathscr{P}}^{(m)}(r_1, \ldots, r_m)$ is undefined.

Let us reexamine the examples in Section 2 from the point of view of this definition. We begin with the program of (b). For this program $\mathscr{P}$, we have

$$\psi_{\mathscr{P}}^{(1)}(x) = x$$

for all x. For this one example, we give a detailed treatment. The following list of snapshots is a computation of $\mathscr{P}$:

$$\begin{aligned}
&(1, \{X = r, Y = 0, Z = 0\}),\\
&(4, \{X = r, Y = 0, Z = 0\}),\\
&(5, \{X = r - 1, Y = 0, Z = 0\}),\\
&(6, \{X = r - 1, Y = 1, Z = 0\}),\\
&(7, \{X = r - 1, Y = 1, Z = 1\}),\\
&(1, \{X = r - 1, Y = 1, Z = 1\}),\\
&\vdots\\
&(1, \{X = 0, Y = r, Z = r\}),\\
&(2, \{X = 0, Y = r, Z = r\}),\\
&(3, \{X = 0, Y = r, Z = r + 1\}),\\
&(8, \{X = 0, Y = r, Z = r + 1\}).
\end{aligned}$$

We have included a copy of $\mathscr{P}$ showing line numbers:

$[A]$	IF $X \neq 0$ GOTO B	(1)
	$Z \leftarrow Z + 1$	(2)
	IF $Z \neq 0$ GOTO E	(3)
$[B]$	$X \leftarrow X - 1$	(4)
	$Y \leftarrow Y + 1$	(5)
	$Z \leftarrow Z + 1$	(6)
	IF $Z \neq 0$ GOTO A	(7)

For other examples of Section 2 we have

$$\text{(a)} \quad \psi^{(1)}(r) = \begin{cases} 1 & \text{if } r = 0 \\ r & \text{otherwise,} \end{cases}$$

$$\text{(b), (c)} \quad \psi^{(1)}(r) = r,$$

$$\text{(d)} \quad \psi^{(2)}(r_1, r_2) = r_1 + r_2,$$

$$\text{(e)} \quad \psi^{(2)}(r_1, r_2) = r_1 \cdot r_2,$$

$$\text{(f)} \quad \psi^{(2)}(r_1, r_2) = \begin{cases} r_1 - r_2 & \text{if } r_1 \geq r_2 \\ \uparrow & \text{if } r_1 < r_2. \end{cases}$$

Of course in several cases the programs written in Section 2 are abbreviations, and we are assuming that the appropriate macro expansions have been provided.

As indicated, we are permitting each program to be used with any number of inputs. If the program has n input variables, but only $m < n$ are specified, then according to the definition, the remaining input variables are assigned the value 0 and the computation proceeds. If on the other hand, m values are specified where $m > n$ the extra input values are ignored. For example, referring again to the examples from Section 2, we have

$$\text{(c)} \quad \psi_{\mathscr{P}}^{(2)}(r_1, r_2) = r_1,$$

$$\text{(d)} \quad \psi_{\mathscr{P}}^{(1)}(r_1) = r_1 + 0 = r_1,$$

$$\psi_{\mathscr{P}}^{(3)}(r_1, r_2, r_3) = r_1 + r_2.$$

For any program $\mathscr{P}$ and any positive integer m, the function $\psi_{\mathscr{P}}^{(m)}(x_1, \ldots, x_m)$ is said to be *computed* by $\mathscr{P}$. A given partial function g (of one or more variables) is said to be *partially computable* if it is computed by some program. That is, g is partially computable if there is a program $\mathscr{P}$ such that

$$g(r_1, \ldots, r_m) = \psi_{\mathscr{P}}^{(m)}(r_1, \ldots, r_m)$$

for all $r_1, \ldots, r_m$. Here this equation must be understood to mean not only that both sides have the same value when they are defined, but also that when either side of the equation is undefined, the other is also.

As explained in Chapter 1, a given function g of m variables is called *total* if $g(r_1, \ldots, r_m)$ is defined for *all* $r_1, \ldots, r_m$. A function is said to be *computable* if it is both partially computable and total.

Partially computable functions are also called *partial recursive*, and computable functions, i.e., functions that are both total and partial recursive, are called *recursive*. The reason for this terminology is largely historical and will be discussed later.

Our examples from Section 2 give us a short list of partially computable functions, namely: $x, x + y, x \cdot y$, and $x - y$. Of these, all except the last one are total and hence computable.

Computability theory (also called recursion theory) studies the class of partially computable functions. In order to justify the name, we need some evidence that for every function which one can claim to be "computable" on intuitive grounds, there really is a program of the language $\mathscr{S}$ which computes it. Such evidence will be developed as we go along.

We close this section with one final example of a program of $\mathscr{S}$:

$$\begin{array}{ll} [A] & X \leftarrow X + 1 \\ & \text{IF } X \neq 0 \text{ GOTO } A \end{array}$$

For this program $\mathscr{P}$, $\psi_{\mathscr{P}}^{(1)}(x)$ is undefined for all x. So, the nowhere defined function (see Chapter 1, Section 2) must be included in the class of partially computable functions.

Exercises

1. Let $\mathscr{P}$ be the program

$$\begin{array}{ll} & \text{IF } X \neq 0 \text{ GOTO } A \\ [A] & X \leftarrow X + 1 \\ & \text{IF } X \neq 0 \text{ GOTO } A \\ [A] & Y \leftarrow Y + 1 \end{array}$$

What is $\psi_{\mathscr{P}}^{(1)}(x)$?

2. The same as Exercise 1 for the program

$$\begin{array}{ll} [B] & \text{IF } X \neq 0 \text{ GOTO } A \\ & Z \leftarrow Z + 1 \\ & \text{IF } Z \neq 0 \text{ GOTO } B \\ [A] & X \leftarrow X \end{array}$$

3. The same as Exercise 1 for the empty program.

4. Let $\mathscr{P}$ be the program

$$\begin{array}{ll} & Y \leftarrow X_1 \\ [A] & \text{IF } X_2 = 0 \text{ GOTO } E \\ & Y \leftarrow Y + 1 \\ & Y \leftarrow Y + 1 \\ & X_2 \leftarrow X_2 - 1 \\ & \text{GOTO } A \end{array}$$

What is $\psi_{\mathscr{P}}^{(1)}(r_1)$? $\psi_{\mathscr{P}}^{(2)}(r_1, r_2)$? $\psi_{\mathscr{P}}^{(3)}(r_1, r_2, r_3)$?

5. Show that for every partially computable function $f(x_1, \ldots, x_n)$, there is a number $m \geq 0$ such that f is computed by infinitely many programs of length m.

6. **(a)** For every number $k \geq 0$, let f_k be the constant function $f_k(x) = k$. Show that for every k, f_k is computable.

(b) Let us call an $\mathscr{S}$ program a *straightline program* if it contains no (labeled or unlabeled) instruction of the form IF $V \neq 0$ GOTO L. Show by induction on the length of programs that if the length of a straightline program $\mathscr{P}$ is k, then $\psi_{\mathscr{P}}^{(1)}(x) \leq k$ for all x.

(c) Show that, if $\mathscr{P}$ is a straightline program that computes f_k, then the length of $\mathscr{P}$ is at least k.

(d) Show that no straightline $\mathscr{S}$ program computes the function $f(x) = x + 1$. Conclude that the class of functions computable by straightline $\mathscr{S}$ programs is contained in but is not equal to the class of computable functions.

7. Let us call an $\mathscr{S}$ program $\mathscr{P}$ *forward-branching* if the following condition holds for each occurrence in $\mathscr{P}$ of a (labeled or unlabeled) instruction of the form IF $V \neq 0$ GOTO L. If IF $V \neq 0$ GOTO L is the ith instruction of $\mathscr{P}$, then either L does not appear as the label of an instruction in $\mathscr{P}$, or else, if j is the least number such that L is the label of the jth instruction in $\mathscr{P}$, then $i < j$. Show that the functions computed by straightline programs [see Exercise 6] are a proper subset of the functions computed by forward-branching programs.

8. Let us call a unary function $f(x)$ *partially n-computable* if it is computed by some $\mathscr{S}$ program $\mathscr{P}$ such that $\mathscr{P}$ has no more than n instructions, every variable in $\mathscr{P}$ is among $X, Y, Z_1, \ldots, Z_n$, and every label in $\mathscr{P}$ is among $A_1, \ldots, A_n, E$.

(a) Show that if a unary function is computed by a program with no more than n instructions, then it is partially n-computable.

(b) Show that for every $n \geq 0$, there are only finitely many distinct partially n-computable unary functions.

(c) Show that for every $n \geq 0$, there are only finitely many distinct unary functions computed by $\mathscr{S}$ programs of length no greater than n.

(d) Conclude that for every $n \geq 0$, there is a partially computable unary function which is not computed by any $\mathscr{S}$ program of length less than n.

5. More about Macros

In Section 2 we gave some examples of computable functions (i.e., $x + y$, $x \cdot y$) giving rise to corresponding macros. Now we consider this process in general.

Let $f(x_1, \ldots, x_n)$ be some partially computable function computed by the program $\mathscr{P}$. We shall assume that the variables that occur in $\mathscr{P}$ are all included in the list $Y, X_1, \ldots, X_n, Z_1, \ldots, Z_k$ and that the labels that occur in $\mathscr{P}$ are all included in the list $E, A_1, \ldots, A_l$. We also assume that for each instruction of $\mathscr{P}$ of the form

$$\text{IF } V \neq 0 \text{ GOTO } A_i$$

there is in $\mathscr{P}$ an instruction labeled A_i. (In other words, E is the only "exit" label.) It is obvious that, if $\mathscr{P}$ does not originally meet these conditions, it will after minor changes in notation. We write

$$\mathscr{P} = \mathscr{P}(Y, X_1, \ldots, X_n, Z_1, \ldots, Z_k; E, A_1, \ldots, A_l)$$

in order that we can represent programs obtained from $\mathscr{P}$ by replacing the variables and labels by others. In particular, we will write

$$\begin{aligned} \mathscr{Q}_m = \mathscr{P}(&Z_m, Z_{m+1}, \ldots, Z_{m+n}, Z_{m+n+1}, \ldots, Z_{m+n+k}; \\ &E_m, A_{m+1}, \ldots, A_{m+l}) \end{aligned}$$

for each given value of m. Now we want to be able to use macros like

$$W \leftarrow f(V_1, \ldots, V_n)$$

in our programs, where $V_1, \ldots, V_n, W$ can be any variables whatever. (In particular, W might be one of $V_1, \ldots, V_n$.) We will take such a macro to be an abbreviation of the following expansion:

$$\begin{array}{ll} & Z_m \leftarrow 0 \\ & Z_{m+1} \leftarrow V_1 \\ & Z_{m+2} \leftarrow V_2 \\ & \vdots \\ & Z_{m+n} \leftarrow V_n \\ & Z_{m+n+1} \leftarrow 0 \\ & Z_{m+n+2} \leftarrow 0 \\ & \vdots \\ & Z_{m+n+k} \leftarrow 0 \\ & \mathscr{Q}_m \\ [E_m] & W \leftarrow Z_m \end{array}$$

Here it is understood that the number m is chosen so large that none of the variables or labels used in $\mathscr{Q}_m$ occur in the main program of which the expansion is a part. Notice that the expansion sets the variables corresponding to the output and local variables of $\mathscr{P}$ equal to 0 and those corresponding to $X_1, \ldots, X_n$ equal to the values of $V_1, \ldots, V_n$, respectively. Setting the variables equal to 0 is necessary (even though they are all local variables automatically initialized to 0) because the expansion may be part of a loop in the main program; in this case, at the second and subsequent times through the loop the local variables will have whatever values they acquired the previous time around, and so will need to be reset. Note that when $\mathscr{Q}_m$ terminates, the value of Z_m is $f(V_1, \ldots, V_n)$, so that W finally does get the value $f(V_1, \ldots, V_n)$.

If $f(V_1, \ldots, V_n)$ is undefined, the program $\mathscr{Q}_m$ will never terminate. Thus if f is not total, and the macro

$$W \leftarrow f(V_1, \ldots, V_n)$$

is encountered in a program where $V_1, \ldots, V_n$ have values for which f is not defined, the main program will never terminate.

Here is an example:

$$Z \leftarrow X_1 - X_2$$
$$Y \leftarrow Z + X_3$$

This program computes the function $f(x_1, x_2, x_3)$, where

$$f(x_1, x_2, x_3) = \begin{cases} (x_1 - x_2) + x_3 & \text{if} \quad x_1 \geq x_2 \\ \uparrow & \text{if} \quad x_1 < x_2 . \end{cases}$$

In particular, $f(2,5,6)$ is undefined, although $(2 - 5) + 6 = 3$ is positive. The computation never gets past the attempt to compute $2 - 5$.

So far we have augmented our language $\mathscr{S}$ to permit the use of macros which allow assignment statements of the form

$$W \leftarrow f(V_1, \ldots, V_n),$$

where f is any partially computable function. Nonetheless there is available only one highly restrictive conditional branch statement, namely,

$$\text{IF } V \neq 0 \text{ GOTO } L$$

We will now see how to augment our language to include macros of the form

$$\text{IF } P(V_1, \ldots, V_n) \text{ GOTO } L$$

where $P(x_1, \ldots, x_n)$ is a computable predicate. Here we are making use of the convention, introduced in Chapter 1, that

$$\text{TRUE} = 1, \qquad \text{FALSE} = 0.$$

Hence predicates are just total functions whose values are always either 0 or 1. And therefore, it makes perfect sense to say that some given *predicate* is or is not computable.

Let $P(x_1, \ldots, x_n)$ be any computable predicate. Then the appropriate macro expansion of

$$\text{IF } P(V_1, \ldots, V_n) \text{ GOTO } L$$

is simply

$$Z \leftarrow P(V_1, \ldots, V_n)$$
$$\text{IF } Z \neq 0 \text{ GOTO } L$$

Note that P is a computable function and hence we have already shown how to expand the first instruction. The second instruction, being one of the basic instructions in the language $\mathscr{S}$, needs no further expansion.

A simple example of this general kind of conditional branch statement which we will use frequently is

$$\text{IF } V = 0 \text{ GOTO } L$$

To see that this is legitimate we need only check that the predicate $P(x)$, defined by $P(x) = \text{TRUE}$ if $x = 0$ and $P(x) = \text{FALSE}$ otherwise, is computable. Since TRUE = 1 and FALSE = 0, the following program does the job:

$$\text{IF } X \neq 0 \text{ GOTO } E$$
$$Y \leftarrow Y + 1$$

The use of macros has the effect of enabling us to write much shorter programs than would be possible restricting ourselves to instructions of the original language $\mathscr{S}$. The original "assignment" statements $V \leftarrow V + 1$, $V \leftarrow V - 1$ are now augmented by general assignment statements of the form $W \leftarrow f(V_1, \ldots, V_n)$ for any partially computable function f. Also, the original conditional branch statements IF $V \neq 0$ GOTO L are now augmented by general conditional branch statements of the form IF $P(V_1, \ldots, V_n)$ GOTO L for any computable predicate P. The fact that any function which can be computed using these general instructions could already have been computed by a program of our original language $\mathscr{S}$ (since the general instructions are merely abbreviations of programs of $\mathscr{S}$) is powerful evidence of the generality of our notion of computability.

Our next task will be to develop techniques that will make it easy to see that various particular functions are computable.

Exercises

1. (a) Use the process described in this section to expand the program in example (d) of Section 2.
   (b) What is the length of the $\mathscr{S}$ program expanded from example (e) by this process?
2. Replace the instructions

$$Z_1 \leftarrow X_1 + Y$$
$$Y \leftarrow Z_1$$

in example (e) of Section 2 with the instruction $Y \leftarrow X_1 + Y$, and expand the result by the process described in this section. If $\mathscr{P}$ is the resulting $\mathscr{S}$ program, what is $\psi_{\mathscr{P}}^{(2)}(r_1, r_2)$?
3. Let $f(x)$, $g(x)$ be computable functions and let $h(x) = f(g(x))$. Show that h is computable.
4. Show by constructing a program that the predicate $x_1 \leq x_2$ is computable.
5. Let $P(x)$ be a computable predicate. Show that the function f defined by

$$f(x_1, x_2) = \begin{cases} x_1 + x_2 & \text{if } P(x_1 + x_2) \\ \uparrow & \text{otherwise} \end{cases}$$

is partially computable.
6. Let $P(x)$ be a computable predicate. Show that

$$EX_P(r) = \begin{cases} 1 & \text{if there are at least } r \text{ numbers } n \text{ such that } P(n) = 1 \\ \uparrow & \text{otherwise} \end{cases}$$

is partially computable.
7. Let π be a computable permutation (i.e., one–one, onto function) of N, and let π^{-1} be the inverse of π, i.e.,

$$\pi^{-1}(y) = x \qquad \text{if and only if} \qquad \pi(x) = y.$$

Show that π^{-1} is computable.
8. Let $f(x)$ be a partially computable but not total function, let M be a finite set of numbers such that $f(m)\uparrow$ for all $m \in M$, and let $g(x)$ be

an arbitrary partially computable function. Show that

$$h(x) = \begin{cases} g(x) & \text{if } x \in M \\ f(x) & \text{otherwise} \end{cases}$$

is partially computable.

9. Let $\mathscr{S}^+$ be a programming language that extends $\mathscr{S}$ by permitting instructions of the form $V \leftarrow k$, for any $k \geq 0$. These instructions have the obvious effect of setting the value of V to k. Show that a function is partially computable by some $\mathscr{S}^+$ program if and only if it is partially computable.

10. Let $\mathscr{S}'$ be a programming language defined like $\mathscr{S}$ except that its (labeled and unlabeled) instructions are of the three types

$$V \leftarrow V'$$
$$V \leftarrow V + 1$$
$$\text{If } V \neq V' \text{ GOTO } L$$

These instructions are given the obvious meaning. Show that a function is partially computable in $\mathscr{S}'$ if and only if it is partially computable.

3

Primitive Recursive Functions

1. Composition

We want to combine computable functions in such a way that the output of one becomes an input to another. In the simplest case we combine functions f and g to obtain the function

$$h(x) = f(g(x)).$$

More generally, for functions of several variables:

Definition. Let f be a function of k variables and let $g_1, \ldots, g_k$ be functions of n variables. Let

$$h(x_1, \ldots, x_n) = f(g_1(x_1, \ldots, x_n), \ldots, g_k(x_1, \ldots, x_n)).$$

Then h is said to be obtained from f and $g_1, \ldots, g_k$ by *composition*.

Of course, the functions $f, g_1, \ldots, g_k$ need not be total. $h(x_1, \ldots, x_n)$ will be defined when all of $z_1 = g_1(x_1, \ldots, x_n), \ldots, z_k = g_k(x_1, \ldots, x_n)$ are defined and also $f(z_1, \ldots, z_k)$ is defined.

Using macros it is very easy to prove

Theorem 1.1. If h is obtained from the (partially) computable functions $f, g_1, \ldots, g_k$ by composition, then h is (partially) computable.

The word *partially* is placed in parentheses in order to assert the correctness of the statement with the word included or omitted in both places.

Proof. The following program obviously computes h:

$$
\begin{aligned}
&Z_1 \leftarrow g_1(X_1, \dots, X_n)\\
&\vdots\\
&Z_k \leftarrow g_k(X_1, \dots, X_n)\\
&Y \leftarrow f(Z_1, \dots, Z_k)
\end{aligned}
$$

If $f, g_1, \dots, g_k$ are not only partially computable but are also total, then so is h. ■

By Section 4 of Chapter 2, we know that $x, x + y, x \cdot y$, and $x - y$ are partially computable. So by Theorem 1.1 we see that $2x = x + x$ and $4x^2 = (2x) \cdot (2x)$ are computable. So are $4x^2 + 2x$ and $4x^2 - 2x$. Note that $4x^2 - 2x$ is total, although it is obtained from the nontotal function $x - y$ by composition with $4x^2$ and $2x$.

2. Recursion

Suppose k is some fixed number and

$$
\begin{aligned}
h(0) &= k,\\
h(t + 1) &= g(t, h(t)),
\end{aligned}
\tag{2.1}
$$

where g is some given *total* function of two variables. Then h is said to be obtained from g by *primitive recursion*, or simply *recursion*.[1]

Theorem 2.1. Let h be obtained from g as in (2.1), and let g be computable. Then h is also computable.

Proof. We first note that the constant function $f(x) = k$ is computable; in fact, it is computed by the program

$$
\left.
\begin{aligned}
&Y \leftarrow Y + 1\\
&Y \leftarrow Y + 1\\
&\vdots\\
&Y \leftarrow Y + 1
\end{aligned}
\right\} \quad k \text{ lines}
$$

[1] Primitive recursion, characterized by Equations (2.1) and (2.2), is just one specialized form of recursion, but it is the only one we will be concerned with in this chapter, so we will refer to it simply as *recursion*. We will consider more general forms of recursion in Part 5.

Hence we have available the macro $Y \leftarrow k$. The following is a program that computes $h(x)$:

$$
\begin{array}{ll}
 & Y \leftarrow k \\
[A] & \text{IF } X = 0 \text{ GOTO } E \\
 & Y \leftarrow g(Z, Y) \\
 & Z \leftarrow Z + 1 \\
 & X \leftarrow X - 1 \\
 & \text{GOTO } A
\end{array}
$$

To see that this program does what it is supposed to do, note that, if Y has the value $h(z)$ before executing the instruction labeled A, then it has the value $g(z, h(z)) = h(z+1)$ after executing the instruction $Y \leftarrow g(Z, Y)$. Since Y is initialized to $k = h(0)$, Y successively takes on the values $h(0), h(1), \ldots, h(x)$ and then terminates. ■

A slightly more complicated kind of recursion is involved when we have

$$
\begin{aligned}
h(x_1, \ldots, x_n, 0) &= f(x_1, \ldots, x_n), \\
h(x_1, \ldots, x_n, t+1) &= g(t, h(x_1, \ldots, x_n, t), x_1, \ldots, x_n).
\end{aligned}
\tag{2.2}
$$

Here the function h of $n + 1$ variables is said to be obtained by *primitive recursion*, or simply *recursion*, from the total functions f (of n variables) and g (of $n + 2$ variables). The recursion (2.2) is just like (2.1) except that parameters $x_1, \ldots, x_n$ are involved. Again we have

Theorem 2.2. Let h be obtained from f and g as in (2.2) and let f, g be computable. Then h is also computable.

Proof. The proof is almost the same as for Theorem 2.1. The following program computes $h(x_1, \ldots, x_n, x_{n+1})$:

$$
\begin{array}{ll}
 & Y \leftarrow f(X_1, \ldots, X_n) \\
[A] & \text{IF } X_{n+1} = 0 \text{ GOTO } E \\
 & Y \leftarrow g(Z, Y, X_1, \ldots, X_n) \\
 & Z \leftarrow Z + 1 \\
 & X_{n+1} \leftarrow X_{n+1} - 1 \\
 & \text{GOTO } A
\end{array}
$$

■

3. PRC Classes

So far we have considered the operations of composition and recursion. Now we need some functions on which to get started. These will be

$$s(x) = x + 1,$$

$$n(x) = 0,$$

and the *projection functions*

$$u_i^n(x_1, \ldots, x_n) = x_i, \qquad 1 \le i \le n.$$

[For example, $u_3^4(x_1, x_2, x_3, x_4) = x_3$.] The functions s, n, and u_i^n are called the *initial functions*.

Definition. A class of total functions $\mathscr{C}$ is called a *PRC*[2] *class* if

1. the initial functions belong to $\mathscr{C}$,
2. a function obtained from functions belonging to $\mathscr{C}$ by either composition or recursion also belongs to $\mathscr{C}$.

Then we have

Theorem 3.1. The class of computable functions is a PRC class.

Proof. By Theorems 1.1, 2.1, and 2.2, we need only verify that the initial functions are computable.

Now this is obvious; $s(x) = x + 1$ is computed by

$$Y \leftarrow X + 1$$

$n(x)$ is computed by the empty program, and $u_i^n(x_1, \ldots, x_n)$ is computed by the program

$$Y \leftarrow X_i$$

■

Definition. A function is called *primitive recursive* if it can be obtained from the initial functions by a finite number of applications of composition and recursion.

It is obvious from this definition that

[2] This is an abbreviation for "primitive recursively closed."

Corollary 3.2. The class of primitive recursive functions is a PRC class.

Actually we can say more:

Theorem 3.3. A function is primitive recursive if and only if it belongs to every PRC class.

Proof. If a function belongs to every PRC class, then, in particular, by Corollary 3.2, it belongs to the class of primitive recursive functions.

Conversely let a function f be a primitive recursive function and let $\mathscr{C}$ be some PRC class. We want to show that f belongs to $\mathscr{C}$. Since f is a primitive recursive function, there is a list $f_1, f_2, \ldots, f_n$ of functions such that $f_n = f$ and each f_i in the list is either an initial function or can be obtained from preceding functions in the list by composition or recursion. Now the initial functions certainly belong to the PRC class $\mathscr{C}$. Moreover the result of applying composition or recursion to functions in $\mathscr{C}$ is again a function belonging to $\mathscr{C}$. Hence each function in the list $f_1, \ldots, f_n$ belongs to $\mathscr{C}$. Since $f_n = f$, f belongs to $\mathscr{C}$. ■

Corollary 3.4. Every primitive recursive function is computable.

Proof. By the theorem just proved, every primitive recursive function belongs to the PRC class of computable functions. ■

In Chapter 4 we shall show how to obtain a computable function that is not primitive recursive. Hence it will follow that the set of primitive recursive functions is a proper subset of the set of computable functions.

Exercises

1. Let $\mathscr{C}$ be a PRC class, and let g_1, g_2, g_3, g_4 belong to $\mathscr{C}$. Show that if

$$h_1(x, y, z) = g_1(z, y, x),$$

$$h_2(x) = g_2(x, x, x), \text{ and}$$

$$h_3(w, x, y, z) = h_1(g_3(w, y), z, g_4(2, g_4(y, z))),$$

then h_1, h_2, h_3 also belong to $\mathscr{C}$.

2. Show that the class of all total functions is a PRC class.
3. Let $n > 0$ be some given number, and let $\mathscr{C}$ be a class of total functions of no more than n variables. Show that $\mathscr{C}$ is not a PRC class.

4. Let $\mathscr{C}$ be a PRC class, let h belong to $\mathscr{C}$, and let

$$f(x) = h(g(x)) \quad \text{and}$$
$$g(x) = h(f(x)).$$

Show that f belongs to $\mathscr{C}$ if and only if g belongs to $\mathscr{C}$.

5. Prove Corollary 3.4 directly from Theorems 1.1, 2.1, 2.2, and the proof of Theorem 3.1.

4. Some Primitive Recursive Functions

We proceed to make a short list of primitive recursive functions. Being primitive recursive, they are also computable.

1. $x + y$

To see that this is primitive recursive, we have to show how to obtain this function from the initial functions using only the operations of composition and recursion.

If we write $f(x, y) = x + y$, we have the recursion equations

$$f(x,0) = x,$$
$$f(x, y + 1) = f(x, y) + 1.$$

We can rewrite these equations as

$$f(x,0) = u_1^1(x),$$
$$f(x, y + 1) = g(y, f(x, y), x),$$

where $g(x_1, x_2, x_3) = s(u_2^3(x_1, x_2, x_3))$. The functions $u_1^1(x), u_2^3(x_1, x_2, x_3)$, and $s(x)$ are primitive recursive functions; in fact they are initial functions. Also, $g(x_1, x_2, x_3)$ is a primitive recursive function, since it is obtained by composition of primitive recursive functions. Thus, the preceding is a valid application of the operation of recursion to primitive recursive functions. Hence $f(x, y) = x + y$ is primitive recursive.

Of course we already knew that $x + y$ was a computable function. So we have only obtained the additional information that it is in fact primitive recursive.

2. $x \cdot y$

The recursion equations for $h(x, y) = x \cdot y$ are

$$h(x,0) = 0,$$
$$h(x, y + 1) = h(x, y) + x.$$

This can be rewritten

$$h(x,0) = n(x)$$
$$h(x,y+1) = g(y, h(x,y), x).$$

Here, $n(x)$ is the zero function,

$$g(x_1, x_2, x_3) = f(u_2^3(x_1, x_2, x_3), u_3^3(x_1, x_2, x_3)),$$

$f(x_1, x_2)$ is $x_1 + x_2$, and $u_2^3(x_1, x_2, x_3), u_3^3(x_1, x_2, x_3)$ are projection functions. Notice that the functions $n(x)$, $u_2^3(x_1, x_2, x_3)$, and $u_3^3(x_1, x_2, x_3)$ are all primitive recursive functions, since they are all initial functions. We have just shown that $f(x_1, x_2) = x_1 + x_2$ is primitive recursive, so $g(x_1, x_2, x_3)$ is a primitive recursive function since it is obtained from primitive recursive functions by composition. Finally, we conclude that

$$h(x,y) = x \cdot y$$

is primitive recursive.

3. $x!$

The recursion equations are

$$0! = 1,$$
$$(x+1)! = x! \cdot s(x).$$

More precisely, $x! = h(x)$, where

$$h(0) = 1,$$
$$h(t+1) = g(t, h(t)),$$

and

$$g(x_1, x_2) = s(x_1) \cdot x_2.$$

Finally, g is primitive recursive because

$$g(x_1, x_2) = s(u_1^2(x_1, x_2)) \cdot u_2^2(x_1, x_2)$$

and multiplication is already known to be primitive recursive.

In the examples that follow, we leave it to the reader to check that the recursion equations can be put in the precise form called for by the definition of the operation of recursion.

4. x^y

The recursion equations are

$$x^0 = 1,$$
$$x^{y+1} = x^y \cdot x.$$

Note that these equations assign the value 1 to the "indeterminate" 0^0.

5. $p(x)$

The *predecessor function* $p(x)$ is defined as follows:

$$p(x) = \begin{cases} x - 1 & \text{if } x \neq 0 \\ 0 & \text{if } x = 0. \end{cases}$$

It corresponds to the instruction in our programming language $X \leftarrow X - 1$.
The recursion equations for $p(x)$ are simply

$$p(0) = 0,$$
$$p(t + 1) = t.$$

Hence, $p(x)$ is primitive recursive.

6. $x \dot{-} y$

The function $x \dot{-} y$ is defined as follows:

$$x \dot{-} y = \begin{cases} x - y & \text{if } x \geq y \\ 0 & \text{if } x < y. \end{cases}$$

This function should not be confused with the function $x - y$, which is undefined if $x < y$. In particular, $x \dot{-} y$ is total, while $x - y$ is not.

We show that $x \dot{-} y$ is primitive recursive by displaying the recursion equations:

$$x \dot{-} 0 = x,$$
$$x \dot{-} (t + 1) = p(x \dot{-} t).$$

7. $|x - y|$

The function $|x - y|$ is defined as the absolute value of the difference between x and y. It can be expressed simply as

$$|x - y| = (x \dot{-} y) + (y \dot{-} x)$$

and thus is primitive recursive.

8. $\alpha(x)$

The function $\alpha(x)$ is defined as

$$\alpha(x) = \begin{cases} 1 & \text{if } x = 0 \\ 0 & \text{if } x \neq 0. \end{cases}$$

$\alpha(x)$ is primitive recursive since

$$\alpha(x) = 1 \dot{-} x.$$

Or we can simply write the recursion equations:

$$\alpha(0) = 1,$$
$$\alpha(t + 1) = 0.$$

Exercises

1. Give a detailed argument that x^y, $p(x)$, and $x \dot{-} y$ are primitive recursive.
2. Show that for each k, the function $f(x) = k$ is primitive recursive.
3. Prove that if $f(x)$ and $g(x)$ are primitive recursive functions, so is $f(x) + g(x)$.
4. Without using $x + y$ as a macro, apply the constructions in the proofs of Theorems 1.1, 2.2, and 3.1 to give an $\mathscr{S}$ program that computes $x \cdot y$.
5. For any unary function $f(x)$, the nth *iteration* of f, written f^n, is

$$f^n(x) = f(\cdots f(x) \cdots),$$

where f is composed with itself n times on the right side of the equation. (Note that $f^0(x) = x$.) Let $\iota_f(n, x) = f^n(x)$. Show that if f is primitive recursive, then ι_f is also primitive recursive.

6.* (a) Let $E(x) = 0$ if x is even, $E(x) = 1$ if x is odd. Show that $E(x)$ is primitive recursive.
(b) Let $H(x) = x/2$ if x is even, $(x - 1)/2$ if x is odd. Show that $H(x)$ is primitive recursive.

7.* Let $f(0) = 0$, $f(1) = 1$, $f(2) = 2^2$, $f(3) = 3^{3^3} = 3^{27}$, etc. In general, $f(n)$ is written as a stack n high, of n's as exponents. Show that f is primitive recursive.

8.* Let k be some fixed number, let f be a function such that $f(x+1) < x+1$ for all x, and let

$$h(0) = k$$
$$h(t+1) = g(h(f(t+1))).$$

Show that if f and g belong to some PRC class $\mathscr{C}$, then so does h. [*Hint:* Define $f'(x) = \min_{t \le x} f^t(x) = 0$. See Exercise 5 for the definition of $f^t(x)$.]

9.* Let $g(x)$ be a primitive recursive function and let $f(0, x) = g(x)$, $f(n+1, x) = f(n, f(n, x))$. Prove that $f(n, x)$ is primitive recursive.

10.* Let COMP be the class of functions obtained from the initial functions by a finite sequence of compositions.

(a) Show that for every function $f(x_1, \ldots, x_n)$ in COMP, either $f(x_1, \ldots, x_n) = k$ for some constant k, or $f(x_1, \ldots, x_n) = x_i + k$ for some $1 \le i \le n$ and some constant k.

(b) An n-ary function f is *monotone* if for all n-tuples $(x_1, \ldots, x_n)$, $(y_1, \ldots, y_n)$ such that $x_i \le y_i$, $1 \le i \le n$, $f(x_1, \ldots, x_n) \le f(y_1, \ldots, y_n)$. Show that every function in COMP is monotone.

(c) Show that COMP is a proper subset of the class of primitive recursive functions.

(d) Show that the class of functions computed by straightline $\mathscr{S}$ programs is a proper subset of COMP. [See Exercise 4.6 in Chapter 2 for the definition of straightline programs.]

11.* Let $\mathscr{P}_1$ be the class of all functions obtained from the initial functions by any finite number of compositions and no more than one recursion (in any order).

(a) Let $f(x_1, \ldots, x_n)$ belong to COMP. [See Exercise 10 for the definition of COMP.] Show that there is a $k > 0$ such that $f(x_1, \ldots, x_n) \le \max\{x_1, \ldots, x_n\} + k$.

(b) Let

$$h(0) = c$$
$$h(t+1) = g(t, h(t)),$$

where c is some given number and g belongs to COMP. Show that there is a $k > 0$ such that $h(t) \le tk + c$.

(c) Let

$$h(x_1, \ldots, x_n, 0) = f(x_1, \ldots, x_n)$$
$$h(x_1, \ldots, x_n, t+1) = g(t, h(x_1, \ldots, x_n, t), x_1, \ldots, x_n),$$

where f, g belong to COMP. Show that there are $k, l > 0$ such that $h(x_1, \ldots, x_n, t) \le tk + \max\{x_1, \ldots, x_n\} + l$.

(d) Let $f(x_1, \ldots, x_n)$ belong to $\mathscr{P}_1$. Show that there are $k, l > 0$ such that $f(x_1, \ldots, x_n) \le \max\{x_1, \ldots, x_n\} \cdot k + l$.

(e) Show that $\mathscr{P}_1$ is a proper subset of the class of primitive recursive functions.

5. Primitive Recursive Predicates

We recall from Chapter 1, Section 4, that predicates or Boolean-valued functions are simply total functions whose values are 0 or 1. (We have identified 1 with TRUE and 0 with FALSE.) Thus we can speak without further ado of primitive recursive predicates.

We continue our list of primitive recursive functions, including some that are predicates.

9. $x = y$

The predicate $x = y$ is defined as 1 if the values of x and y are the same and 0 otherwise. Thus we wish to show that the function

$$d(x, y) = \begin{cases} 1 & \text{if} \quad x = y \\ 0 & \text{if} \quad x \neq y \end{cases}$$

is primitive recursive. This follows immediately from the equation

$$d(x, y) = \alpha(|x - y|).$$

10. $x \le y$

This predicate is simply the primitive recursive function $\alpha(x \dot{-} y)$.

Theorem 5.1. Let $\mathscr{C}$ be a PRC class. If P, Q are predicates that belong to $\mathscr{C}$, then so are $\sim P$, $P \vee Q$, and $P \,\&\, Q$.[3]

Proof. Since $\sim P = \alpha(P)$, it follows that $\sim P$ belongs to $\mathscr{C}$. (α was defined in Section 4, item 8.)

---

[3] See Chapter 1, Section 4.

Also, we have

$$P \;\&\; Q = P \cdot Q,$$

so that $P \;\&\; Q$ belongs to $\mathscr{C}$.

Finally, the De Morgan law

$$P \vee Q \Leftrightarrow \sim(\sim P \;\&\; \sim Q)$$

shows, using what we have already done, that $P \vee Q$ belongs to $\mathscr{C}$. ■

A result like Theorem 5.1 which refers to PRC classes can be applied to the two classes we have shown to be PRC. That is, taking $\mathscr{C}$ to be the class of all primitive recursive functions, we have

Corollary 5.2. If P, Q are primitive recursive predicates, then so are $\sim P$, $P \vee Q$, and $P \;\&\; Q$.

Similarly taking $\mathscr{C}$ to be the class of all computable functions, we have

Corollary 5.3. If P, Q are computable predicates, then so are $\sim P$, $P \vee Q$, and $P \;\&\; Q$.

As a simple example we have

11. $x < y$

We can write

$$x < y \Leftrightarrow x \leq y \;\&\; \sim(x = y),$$

or more simply

$$x < y \Leftrightarrow \sim(y \leq x).$$

Theorem 5.4 (Definition by Cases). Let $\mathscr{C}$ be a PRC class. Let the functions g, h and the predicate P belong to $\mathscr{C}$. Let

$$f(x_1, \ldots, x_n) = \begin{cases} g(x_1, \ldots, x_n) & \text{if} \quad P(x_1, \ldots, x_n) \\ h(x_1, \ldots, x_n) & \text{otherwise.} \end{cases}$$

Then f belongs to $\mathscr{C}$.

This will be recognized as a version of the familiar "if ... then ..., else ..." statement.

Proof. The result is obvious because

$$f(x_1, \ldots, x_n) \\ = g(x_1, \ldots, x_n) \cdot P(x_1, \ldots, x_n) + h(x_1, \ldots, x_n) \cdot \alpha(P(x_1, \ldots, x_n)).$$

■

Corollary 5.5. Let $\mathscr{C}$ be a PRC class, let n-ary functions $g_1, \ldots, g_m, h$ and predicates $P_1, \ldots, P_m$ belong to $\mathscr{C}$, and let

$$P_i(x_1, \ldots, x_n) \,\&\, P_j(x_1, \ldots, x_n) = 0$$

for all $1 \leq i < j \leq m$ and all $x_1, \ldots, x_n$. If

$$f(x_1, \ldots, x_n) = \begin{cases} g_1(x_1, \ldots, x_n) & \text{if } P_1(x_1, \ldots, x_n) \\ \vdots & \vdots \\ g_m(x_1, \ldots, x_n) & \text{if } P_m(x_1, \ldots, x_n) \\ h(x_1, \ldots, x_n) & \text{otherwise,} \end{cases}$$

then f also belongs to $\mathscr{C}$.

Proof. We argue by induction on m. The case for $m = 1$ is given by Theorem 5.4, so let

$$f(x_1, \ldots, x_n) = \begin{cases} g_1(x_1, \ldots, x_n) & \text{if } P_1(x_1, \ldots, x_n) \\ \vdots & \vdots \\ g_{m+1}(x_1, \ldots, x_n) & \text{if } P_{m+1}(x_1, \ldots, x_n) \\ h(x_1, \ldots, x_n) & \text{otherwise,} \end{cases}$$

and let

$$h'(x_1, \ldots, x_n) = \begin{cases} g_{m+1}(x_1, \ldots, x_n) & \text{if } P_{m+1}(x_1, \ldots, x_n) \\ h(x_1, \ldots, x_n) & \text{otherwise.} \end{cases}$$

Then

$$f(x_1, \ldots, x_n) = \begin{cases} g_1(x_1, \ldots, x_n) & \text{if } P_1(x_1, \ldots, x_n) \\ \vdots & \vdots \\ g_m(x_1, \ldots, x_n) & \text{if } P_m(x_1, \ldots, x_n) \\ h'(x_1, \ldots, x_n) & \text{otherwise,} \end{cases}$$

and h' belongs to $\mathscr{C}$ by Theorem 5.4, so f belongs to $\mathscr{C}$ by the induction hypothesis. ■

Exercise

1. Let us call a predicate *trivial* if it is always TRUE or always FALSE. Show that no nontrivial predicates belong to COMP (see Exercise 4.10 for the definition of COMP.)

6. Iterated Operations and Bounded Quantifiers

Theorem 6.1. Let $\mathscr{C}$ be a PRC class. If $f(t, x_1, \ldots, x_n)$ belongs to $\mathscr{C}$, then so do the functions

$$g(y, x_1, \ldots, x_n) = \sum_{t=0}^{y} f(t, x_1, \ldots, x_n)$$

and

$$h(y, x_1, \ldots, x_n) = \prod_{t=0}^{y} f(t, x_1, \ldots, x_n).$$

A common error is to attempt to prove this by using mathematical induction on y. A little reflection reveals that such an argument by induction shows that

$$g(0, x_1, \ldots, x_n), g(1, x_1, \ldots, x_n), \ldots$$

all belong to $\mathscr{C}$, but not that the function $g(y, x_1, \ldots, x_n)$, one of whose arguments is y, belongs to $\mathscr{C}$.

We proceed with the correct proof.

Proof. We note the recursion equations

$$g(0, x_1, \ldots, x_n) = f(0, x_1, \ldots, x_n),$$
$$g(t+1, x_1, \ldots, x_n) = g(t, x_1, \ldots, x_n) + f(t+1, x_1, \ldots, x_n),$$

and recall that since $+$ is primitive recursive, it belongs to $\mathscr{C}$.

Similarly,

$$h(0, x_1, \ldots, x_n) = f(0, x_1, \ldots, x_n),$$
$$h(t+1, x_1, \ldots, x_n) = h(t, x_1, \ldots, x_n) \cdot f(t+1, x_1, \ldots, x_n). \quad \blacksquare$$

Sometimes we will want to begin the summation (or product) at 1 instead of 0. That is, we will want to consider

$$g(y, x_1, \ldots, x_n) = \sum_{t=1}^{y} f(t, x_1, \ldots, x_n)$$

or

$$h(y, x_1, \ldots, x_n) = \prod_{t=1}^{y} f(t, x_1, \ldots, x_n).$$

Then the initial recursion equations can be taken to be

$$g(0, x_1, \ldots, x_n) = 0,$$
$$h(0, x_1, \ldots, x_n) = 1,$$

with the equations for $g(t + 1, x_1, \ldots, x_n)$ and $h(t + 1, x_1, \ldots, x_n)$ as in the preceding proof. Note that we are implicitly defining a vacuous sum to be 0 and a vacuous product to be 1. With this understanding we have proved

Corollary 6.2. If $f(t, x_1, \ldots, x_n)$ belongs to the PRC class $\mathscr{C}$, then so do the functions

$$g(y, x_1, \ldots, x_n) = \sum_{t=1}^{y} f(t, x_1, \ldots, x_n)$$

and

$$h(y, x, \ldots, x_n) = \prod_{t=1}^{y} f(t, x_1, \ldots, x_n).$$

We have

Theorem 6.3. If the predicate $P(t, x_1, \ldots, x_n)$ belongs to some PRC class $\mathscr{C}$, then so do the predicates[4]

$$(\forall t)_{\leq y} P(t, x_1, \ldots, x_n) \qquad \text{and} \qquad (\exists t)_{\leq y} P(t, x_1, \ldots, x_n).$$

Proof. We need only observe that

$$(\forall t)_{\leq y} P(t, x_1, \ldots, x_n) \Leftrightarrow \left[\prod_{t=0}^{y} P(t, x_1, \ldots, x_n)\right] = 1$$

and

$$(\exists t)_{\leq y} P(t, x_1, \ldots, x_n) \Leftrightarrow \left[\sum_{t=0}^{y} P(t, x_1, \ldots, x_n)\right] \neq 0. \qquad \blacksquare$$

Actually for the universal quantifier it would even have been correct to write the equation

$$(\forall t)_{\leq y} P(t, x_1, \ldots, x_n) = \prod_{t=0}^{y} P(t, x_1, \ldots, x_n).$$

---

[4] See Chapter 1, Section 5.

Sometimes in applying Theorem 6.3 we want to use the quantifier

$$(\forall t)_{<y} \qquad \text{or} \qquad (\exists t)_{<y}.$$

That the theorem is still valid is clear from the relations

$$(\exists t)_{<y}P(t, x_1, \ldots, x_n) \Leftrightarrow (\exists t)_{\le y}[t \ne y \;\&\; P(t, x_1, \ldots, x_n)],$$

$$(\forall t)_{<y}P(t, x_1, \ldots, x_n) \Leftrightarrow (\forall t)_{\le y}[t = y \vee P(t, x_1, \ldots, x_n)].$$

We continue our list of examples.

12. $y \mid x$

This is the predicate "y is a divisor of x." For example,

$$3 \mid 12 \qquad \text{is true}$$

while

$$3 \mid 13 \qquad \text{is false}.$$

The predicate is primitive recursive since

$$y|x \Leftrightarrow (\exists t)_{\le x}(y \cdot t = x).$$

13. Prime(x)

The predicate "x is a prime" is primitive recursive since

$$\text{Prime}(x) \Leftrightarrow x > 1 \;\&\; (\forall t)_{\le x}\{t = 1 \vee t = x \vee \sim (t|x)\}.$$

(A number is a *prime* if it is greater than 1 and it has no divisors other than 1 and itself.)

Exercises

1. Let $f(x) = 2x$ if x is a perfect square; $f(x) = 2x + 1$ otherwise. Show that f is primitive recursive.

2. Let $\sigma(x)$ be the sum of the divisors of x if $x \ne 0$; $\sigma(0) = 0$ [e.g., $\sigma(6) = 1 + 2 + 3 + 6 = 12$]. Show that $\sigma(x)$ is primitive recursive.

3. Let $\pi(x)$ be the number of primes that are $\le x$. Show that $\pi(x)$ is primitive recursive.

4. Let SQSM(x) be true if x is the sum of two perfect squares; false otherwise. Show that SQSM(x) is primitive recursive.

5. Let $\mathscr{C}$ be a PRC class, let $P(t, x_1, \ldots, x_n)$ be a predicate in $\mathscr{C}$, and let

$$g(y, z, x_1, \ldots, x_n) = (\forall t)_{y \le t \le z} P(t, x_1, \ldots, x_n) \text{ and}$$
$$h(y, z, x_1, \ldots, x_n) = (\exists t)_{y \le t \le z} P(t, x_1, \ldots, x_n),$$

(where $(\forall t)_{y \le t \le z} P(t, x_1, \ldots, x_n)$ and $(\exists t)_{y \le t \le z} P(t, x_1, \ldots, x_n)$ mean that $P(t, x_1, \ldots, x_n)$ is true for all t (respectively, for some t) from y to z). Show that g, h also belong to $\mathscr{C}$.

6. Let $\text{RP}(x, y)$ be true if x and y are relatively prime (i.e., their greatest common divisor is 1). Show that $\text{RP}(x, y)$ is primitive recursive.

7. Give a sequence of compositions and recursions that shows explicitly that $\text{Prime}(x)$ is primitive recursive.

7. Minimalization

Let $P(t, x_1, \ldots, x_n)$ belong to some given PRC class $\mathscr{C}$. Then by Theorem 6.1, the function

$$g(y, x_1, \ldots, x_n) = \sum_{u=0}^{y} \prod_{t=0}^{u} \alpha(P(t, x_1, \ldots, x_n))$$

also belongs to $\mathscr{C}$. (Recall that the primitive recursive function α was defined in Section 4.) Let us analyze this function g. Suppose for definiteness that for some value of $t_0 \le y$,

$$P(t, x_1, \ldots, x_n) = 0 \qquad \text{for} \quad t < t_0,$$

but

$$P(t_0, x_1, \ldots, x_n) = 1,$$

i.e., that t_0 *is the least value of* $t \le y$ *for which* $P(t, x_1, \ldots, x_n)$ *is true*. Then

$$\prod_{t=0}^{u} \alpha(P(t, x_1, \ldots, x_n)) = \begin{cases} 1 & \text{if} \quad u < t_0 \\ 0 & \text{if} \quad u \ge t_0. \end{cases}$$

Hence,

$$g(y, x_1, \ldots, x_n) = \sum_{u < t_0} 1 = t_0,$$

so that $g(y, x_1, \ldots, x_n)$ is the least value of t for which $P(t, x, \ldots, x_n)$ is true. Now, we define

$$\min_{t \le y} P(t, x_1, \ldots, x_n) = \begin{cases} g(y, x_1, \ldots, x_n) & \text{if} \quad (\exists t)_{\le y} P(t, x_1, \ldots, x_n) \\ 0 & \text{otherwise}. \end{cases}$$

Thus, $\min_{t \le y} P(t, x_1, \ldots, x_n)$ *is the least value of* $t \le y$ *for which* $P(t, x_1, \ldots, x_n)$ *is true, if such exists; otherwise it assumes the* (default) *value* 0. Using Theorems 5.4 and 6.3, we have

Theorem 7.1. If $P(t, x_1, \ldots, x_n)$ belongs to some PRC class $\mathscr{C}$ and $f(y, x_1, \ldots, x_n) = \min_{t \le y} P(t, x_1, \ldots, x_n)$, then f also belongs to $\mathscr{C}$.

The operation "$\min_{t \le y}$" is called *bounded minimalization.*
Continuing our list:

14. $\lfloor x / y \rfloor$

$\lfloor x/y \rfloor$ is the "integer part" of the quotient x/y. For example, $\lfloor 7/2 \rfloor = 3$ and $\lfloor 2/3 \rfloor = 0$. The equation

$$\lfloor x/y \rfloor = \min_{t \le x} [(t+1) \cdot y > x]$$

shows that $\lfloor x/y \rfloor$ is primitive recursive. Note that according to this equation, we are taking $\lfloor x/0 \rfloor = 0$.

15. $R(x, y)$

$R(x, y)$ is the *remainder* when x is divided by y. Since

$$\frac{x}{y} = \lfloor x/y \rfloor + \frac{R(x, y)}{y},$$

we can write

$$R(x, y) = x \dot{-} (y \cdot \lfloor x/y \rfloor),$$

so that $R(x, y)$ is primitive recursive. [Note that $R(x, 0) = x$.]

16. p_n

Here, for $n > 0$, p_n is the nth prime number (in order of size). So that p_n be a total function, we set $p_0 = 0$. Thus, $p_0 = 0$, $p_1 = 2$, $p_2 = 3$, $p_3 = 5$, etc.

Consider the recursion equations

$$p_0 = 0,$$
$$p_{n+1} = \min_{t \le p_n! + 1} [\text{Prime}(t) \mathbin{\&} t > p_n].$$

To see that these equations are correct we must verify the inequality

$$p_{n+1} \le (p_n)! + 1. \tag{7.1}$$

To do so note that for $0 < i \le n$ we have

$$\frac{(p_n)!+1}{p_i} = K + \frac{1}{p_i},$$

where K is an integer. Hence $(p_n)!+1$ is not divisible by any of the primes $p_1, p_2, \ldots, p_n$. So, either $(p_n)!+1$ is itself a prime or it is divisible by a prime $> p_n$. In either case there is a prime q such that $p_n < q \le (p_n)!+1$, which gives the inequality (7.1). (This argument is just Euclid's proof that there are infinitely many primes.)

Before we can confidently assert that p_n is a primitive recursive function, we need to justify the interleaving of the recursion equations with bounded minimalization. To do so, we first define the primitive recursive function

$$h(y, z) = \min_{t \le z}[\text{Prime}(t) \mathbin{\&} t > y].$$

Then we set

$$k(x) = h(x, x!+1),$$

another primitive recursive function. Finally, our recursion equations reduce to

$$p_0 = 0,$$

$$p_{n+1} = k(p_n),$$

so that we can conclude finally that p_n is a primitive recursive function.

It is worth noting that by using our various theorems (and appropriate macro expansions) we could now obtain explicitly a program of $\mathscr{S}$ which actually computes p_n. Of course the program obtained in this way would be extremely inefficient.

Now we want to discuss minimalization when there is no bound. We write

$$\min_y P(x_1, \ldots, x_n, y)$$

for the least value of y for which the predicate P is true *if there is one. If there is no value of y for which $P(x_1, \ldots, x_n, y)$ is true, then $\min_y P(x_1, \ldots, x_n, y)$ is undefined.* (Note carefully the difference with bounded minimalization.) Thus unbounded minimalization of a predicate can easily produce a function which is not total. For example,

$$x - y = \min_z [y + z = x]$$

is undefined for $x < y$. Now, as we shall see later, there are primitive recursive predicates $P(x, y)$ such that $\min_y P(x, y)$ is a total function which is *not* primitive recursive. However, we can prove

Theorem 7.2. If $P(x_1, \ldots, x_n, y)$ is a computable predicate and if

$$g(x_1, \ldots, x_n) = \min_y P(x_1, \ldots, x_n, y),$$

then g is a partially computable function.

Proof. The following program obviously computes g:

$$\begin{array}{ll} [A] & \text{IF } P(X_1, \ldots, X_n, Y) \text{ GOTO } E \\ & Y \leftarrow Y + 1 \\ & \text{GOTO } A \end{array}$$

■

Exercises

1. Let $h(x)$ be the integer n such that $n \leq \sqrt{2}x < n + 1$. Show that $h(x)$ is primitive recursive.
2. Do the same when $h(x)$ is the integer n such that

$$n \leq (1 + \sqrt{2})x < n + 1.$$

3. p is called a *larger twin prime* if p and $p - 2$ are both primes. (5, 7, 13, 19 are larger twin primes.) Let $T(0) = 0$, $T(n) =$ the nth larger twin prime. It is widely believed, but has not been proved, that there are infinitely many larger twin primes. Assuming that this is true prove that $T(n)$ is computable.
4. Let $u(n)$ be the nth number in order of size which is the sum of two squares. Show that $u(n)$ is primitive recursive.
5. Let $R(x, t)$ be a primitive recursive predicate. Let

$$g(x, y) = \max_{t \leq y} R(x, t),$$

i.e., $g(x, y)$ is the largest value of $t \leq y$ for which $R(x, t)$ is true; if there is none, $g(x, y) = 0$. Prove that $g(x, y)$ is primitive recursive.

6. Let $\gcd(x, y)$ be the greatest common divisor of x and y. Show that $\gcd(x, y)$ is primitive recursive.
7. Let $\text{lcm}(x, y)$ be the least common multiple of x and y. Show that $\text{lcm}(x, y)$ is primitive recursive.

8. Give a computable predicate $P(x_1, \ldots, x_n, y)$ such that the function $\min_y P(x_1, \ldots, x_n, y)$ is not computable.

9.* A function is *elementary* if it can be obtained from the functions s, n, u_j^i, $+$, $\dot{-}$ by a finite sequence of applications of composition, bounded summation, and bounded product. (By application of bounded summation we mean obtaining the function $\sum_{t=0}^{y} f(t, x_1, \ldots, x_n)$ from $f(t, x_1, \ldots, x_n)$, and similarly for bounded product.)

(a) Show that every elementary function is primitive recursive.

(b) Show that $x \cdot y$, x^y, and $x!$ are elementary.

(c) Show that if $n + 1$-ary predicates P and Q are elementary, then so are $\sim P$, $P \vee Q$, $P \;\&\; Q$, $(\forall t)_{\leq y} P(t, x_1, \ldots, x_n)$, $(\exists t)_{\leq y} P(t, x_1, \ldots, x_n)$, and $\min_{t \leq y} P(t, x_1, \ldots, x_n)$.

(d) Show that Prime(x) is elementary.

(e) Let the binary function $\exp_y(x)$ be defined

$$\exp_0(x) = x$$

$$\exp_{y+1}(x) = 2^{\exp_y(x)}.$$

Show that for every elementary function $f(x_1, \ldots, x_n)$, there is a constant k such that $f(x_1, \ldots, x_n) \leq \exp_k(\max\{x_1, \ldots, x_n\})$. [*Hint:* Show that for every n there is an $m \geq n$ such that $x \cdot \exp_n(x) \leq \exp_m(x)$ for all x.]

(f) Show that $\exp_y(x)$ is not elementary. Conclude that the class of elementary functions is a proper subset of the class of primitive recursive functions.

8. Pairing Functions and Gödel Numbers

In this section we shall study two convenient coding devices which use primitive recursive functions. The first is for coding pairs of numbers by single numbers, and the second is for coding lists of numbers.

We define the primitive recursive function

$$\langle x, y \rangle = 2^x(2y + 1) \dot{-} 1.$$

Note that $2^x(2y + 1) \neq 0$ so

$$\langle x, y \rangle + 1 = 2^x(2y + 1).$$

If z is any given number, there is a *unique* solution x, y to the equation

$$\langle x, y \rangle = z, \tag{8.1}$$

namely, x is the largest number such that $2^x \mid (z+1)$, and y is then the solution of the equation

$$2y + 1 = (z+1)/2^x;$$

this last equation has a (unique) solution because $(z+1)/2^x$ must be odd. (The twos have been "divided out.") Equation (8.1) thus defines functions

$$x = l(z), \qquad y = r(z).$$

Since Eq. (8.1) implies that $x, y < z + 1$ we have

$$l(z) \le z, \qquad r(z) \le z.$$

Hence we can write

$$l(z) = \min_{x \le z} [(\exists y)_{\le z}(z = \langle x, y \rangle)],$$

$$r(z) = \min_{y \le z} [(\exists x)_{\le z}(z = \langle x, y \rangle)],$$

so that $l(z), r(z)$ are primitive recursive functions.

The definition of $l(z), r(z)$ can be expressed by the statement

$$\langle x, y \rangle = z \Leftrightarrow x = l(z) \;\&\; y = r(z).$$

We summarize the properties of the functions $\langle x, y \rangle$, $l(z)$, and $r(z)$ in

Theorem 8.1 (Pairing Function Theorem). The functions $\langle x, y \rangle$, $l(z)$, and $r(z)$ have the following properties:

1. they are primitive recursive;
2. $l(\langle x, y \rangle) = x$, $r(\langle x, y \rangle) = y$;
3. $\langle l(z), r(z) \rangle = z$;
4. $l(z), r(z) \le z$.

We next obtain primitive recursive functions that encode and decode arbitrary finite sequences of numbers. The method we use, first employed by Gödel, depends on the prime power decomposition of integers.

We define the *Gödel number* of the sequence $(a_1, \ldots, a_n)$ to be the number

$$[a_1, \ldots, a_n] = \prod_{i=1}^{n} p_i^{a_i}.$$

Thus, the Gödel number of the sequence (3, 1, 5, 4, 6) is

$$[3, 1, 5, 4, 6] = 2^3 \cdot 3^1 \cdot 5^5 \cdot 7^4 \cdot 11^6.$$

For each fixed n, the function $[a_1, \ldots, a_n]$ is clearly primitive recursive.

Gödel numbering satisfies the following uniqueness property:

Theorem 8.2. If $[a_1, \ldots, a_n] = [b_1, \ldots, b_n]$, then

$$a_i = b_i, \qquad i = 1, \ldots, n.$$

This result is an immediate consequence of the uniqueness of the factorization of integers into primes, sometimes referred to as the *unique factorization theorem* or the *fundamental theorem of arithmetic*. (For a proof, see any elementary number theory textbook.)

However, note that

$$[a_1, \ldots, a_n] = [a_1, \ldots, a_n, 0] \tag{8.2}$$

because $p_{n+1}^0 = 1$. This same result obviously holds for any finite number of zeros adjoined to the right end of a sequence. In particular, since

$$1 = 2^0 = 2^0 3^0 = 2^0 3^0 5^0 = \cdots,$$

it is natural to regard 1 as the Gödel number of the "empty" sequence of length 0, and it is useful to do so.

If one adjoins 0 to the left end of a sequence, the Gödel number of the new sequence will not be the same as the Gödel number of the original sequence. For example,

$$[2,3] = 2^2 \cdot 3^3 = 108,$$

and

$$[2,3,0] = 2^2 \cdot 3^3 \cdot 5^0 = 108,$$

but

$$[0,2,3] = 2^0 \cdot 3^2 \cdot 5^3 = 1125.$$

We will now define a primitive recursive function $(x)_i$ so that if

$$x = [a_1, \ldots, a_n],$$

then $(x)_i = a_i$. We set

$$(x)_i = \min_{t \le x} (\sim p_i^{t+1} \mid x).$$

Note that $(x)_0 = 0$, and $(0)_i = 0$ for all i.

We shall also use the primitive recursive function

$$\mathrm{Lt}(x) = \min_{i \le x} ((x)_i \neq 0 \,\&\, (\forall j)_{\le x} (j \le i \vee (x)_j = 0)).$$

(Lt stands for "length.") Thus, if $x = 20 = 2^2 \cdot 5^1 = [2, 0, 1]$, then $(x)_3 = 1$, but $(x)_4 = (x)_5 = \cdots = (x)_{20} = 0$. So, $\mathrm{Lt}(20) = 3$. Also, $\mathrm{Lt}(0) = \mathrm{Lt}(1) = 0$.

If $x > 1$, and $\mathrm{Lt}(x) = n$, then p_n divides x but no prime greater than p_n divides x. Note that $\mathrm{Lt}([a_1, \ldots, a_n]) = n$ if and only if $a_n \neq 0$.

We summarize the key properties of these primitive recursive functions.

Theorem 8.3 (Sequence Number Theorem).

a. $([a_1, \ldots, a_n])_i = \begin{cases} a_i & \text{if } 1 \le i \le n \\ 0 & \text{otherwise.} \end{cases}$

b. $[(x)_1, \ldots, (x)_n] = x \qquad \text{if } n \ge \mathrm{Lt}(x).$

Our main application of these coding techniques is given in the next chapter. The following exercises indicate that they can also be used to show that PRC classes are closed under various interesting and useful forms of recursion.

Exercises

1. Let $f(x_1, \ldots, x_n)$ be a function of n variables, and let $f'(x)$ be a unary function defined so that $f'([x_1, \ldots, x_n]) = f(x_1, \ldots, x_n)$ for all $x_1, \ldots, x_n$ and $f'(x)\uparrow$ if $Lt(x) > n$. Show that f' is partially computable if and only if f is partially computable.
2. Define $\mathrm{Sort}([x_1, \ldots, x_n] - 1) = [y_1, \ldots, y_n] - 1$, where $n = \mathrm{Lt}([x_1, \ldots, x_n])$ and $y_1, \ldots, y_n$ is a permutation of $x_1, \ldots, x_n$ such that $y_1 \le y_2 \le \cdots \le y_n$. Show that Sort($x$) is primitive recursive.
3. Let $F(0) = 0$, $F(1) = 1$, $F(n + 2) = F(n + 1) + F(n)$. [$F(n)$ is the nth so-called Fibonacci number.] Prove that $F(n)$ is primitive recursive.
4. (Simultaneous Recursion) Let

$$h_1(x,0) = f_1(x),$$
$$h_2(x,0) = f_2(x),$$
$$h_1(x,t+1) = g_1(x, h_1(x,t), h_2(x,t)),$$
$$h_2(x,t+1) = g_2(x, h_1(x,t), h_2(x,t)).$$

Prove that if f_1, f_2, g_1, g_2 all belong to some PRC class $\mathscr{C}$, then h_1, h_2 do also.

5.* (Course-of-Values Recursion)

(a) For $f(n)$ any function, we write

$$\tilde{f}(0) = 1, \tilde{f}(n) = [f(0), f(1), \ldots, f(n-1)] \text{ if } n \neq 0.$$

Let

$$f(n) = g(n, \tilde{f}(n))$$

for all n. Show that if g is primitive recursive so is f.

(b) Let

$$f(0) = 1, \qquad f(1) = 4, \qquad f(2) = 6,$$

$$f(x + 3) = f(x) + f(x + 1)^2 + f(x + 2)^3.$$

Show that $f(x)$ is primitive recursive.

(c) Let

$$h(0) = 3$$

$$h(x + 1) = \sum_{t=0}^{x} h(t).$$

Show that h is primitive recursive.

6.* (Unnested Double Recursion) Let

$$f(0, y) = g_1(y)$$

$$f(x + 1, 0) = g_2(x)$$

$$f(x + 1, y + 1) = h(x, y, f(x, y + 1), f(x + 1, y)).$$

Show that if g_1, g_2, and h all belong to some PRC class $\mathscr{C}$, then f also belongs to $\mathscr{C}$.

4

A Universal Program

1. Coding Programs by Numbers

We are going to associate with each program $\mathscr{P}$ of the language $\mathscr{S}$ a number, which we write $\#(\mathscr{P})$, in such a way that the program can be retrieved from its number. To begin with we arrange the variables in order as follows:

$$Y\ X_1\ Z_1\ X_2\ Z_2\ X_3\ Z_3 \ldots .$$

Next we do the same for the labels:

$$A_1\ B_1\ C_1\ D_1\ E_1\ A_2\ B_2\ C_2\ D_2\ E_2\ A_3 \ldots .$$

We write $\#(V)$, $\#(L)$ for the position of a given variable or label in the appropriate ordering. Thus $\#(X_2) = 4$, $\#(Z_1) = \#(Z) = 3$, $\#(E) = 5$, $\#(B_2) = 7$.

Now let I be an instruction (labeled or unlabeled) of the language $\mathscr{S}$. Then we write

$$\#(I) = \langle a, \langle b, c \rangle \rangle$$

where

1. if I is unlabeled, then $a = 0$; if I is labeled L, then $a = \#(L)$;
2. if the variable V is mentioned in I, then $c = \#(V) - 1$;

3. if the statement in I is

$$V \leftarrow V \qquad \text{or} \qquad V \leftarrow V + 1 \qquad \text{or} \qquad V \leftarrow V - 1,$$

then $b = 0$ or 1 or 2, respectively;
4. if the statement in I is

$$\text{IF } V \neq 0 \text{ GOTO } L'$$

then $b = \#(L') + 2$.

Some examples:
The number of the unlabeled instruction $X \leftarrow X + 1$ is

$$\langle 0, \langle 1, 1 \rangle \rangle = \langle 0, 5 \rangle = 10,$$

whereas the number of the instruction

$$[A] \quad X \leftarrow X + 1$$

is

$$\langle 1, \langle 1, 1 \rangle \rangle = \langle 1, 5 \rangle = 21.$$

Note that for any given number q there is a unique instruction I with $\#(I) = q$. We first calculate $l(q)$. If $l(q) = 0$, I is unlabeled; otherwise I has the $l(q)$th label in our list. To find the variable mentioned in I, we compute $i = r(r(q)) + 1$ and locate the ith variable V in our list. Then, the statement in I will be

$$\begin{array}{ll} V \leftarrow V & \text{if} \quad l(r(q)) = 0, \\ V \leftarrow V + 1 & \text{if} \quad l(r(q)) = 1, \\ V \leftarrow V - 1 & \text{if} \quad l(r(q)) = 2, \\ \text{IF } V \neq 0 \text{ GOTO } L & \text{if} \quad j = l(r(q)) - 2 > 0 \end{array}$$

and L is the jth label in our list.

Finally, let a program $\mathscr{P}$ consist of the instructions $I_1, I_2, \ldots, I_k$. Then we set

$$\#(\mathscr{P}) = [\#(I_1), \#(I_2), \ldots, \#(I_k)] - 1. \tag{1.1}$$

Since Gödel numbers tend to be very large, the number of even rather simple programs usually will be quite enormous. We content ourselves with a simple example:

$$\begin{array}{ll} [A] & X \leftarrow X + 1 \\ & \text{IF } X \neq 0 \text{ GOTO } A \end{array}$$

The reader will recognize this as the example given in Chapter 2 of a program that computes the nowhere defined function. Calling these instructions I_1 and I_2, respectively, we have seen that $\#(I_1) = 21$. Since I_2 is unlabeled,

$$\#(I_2) = \langle 0, \langle 3, 1 \rangle\rangle = \langle 0, 23 \rangle = 46.$$

Thus, finally, the number of this short program is

$$2^{21} \cdot 3^{46} - 1.$$

Note that the number of the unlabeled instruction $Y \leftarrow Y$ is

$$\langle 0, \langle 0, 0 \rangle\rangle = \langle 0, 0 \rangle = 0.$$

Thus, by the ambiguity in Gödel numbers [recall Eq. (8.2), Chapter 3], the number of a program will be unchanged if an unlabeled $Y \leftarrow Y$ is tacked onto its end. Of course this is a harmless ambiguity; the longer program computes exactly what the shorter one does. However, we remove even this ambiguity by adding to our official definition of program of $\mathcal{S}$ the harmless stipulation that *the final instruction in a program is not permitted to be the unlabeled statement $Y \leftarrow Y$.*

With this last stipulation each number determines a unique program. As an example, let us determine the program whose number is 199. We have

$$199 + 1 = 200 = 2^3 \cdot 3^0 \cdot 5^2 = [3, 0, 2].$$

Thus, if $\#(\mathcal{P}) = 199$, $\mathcal{P}$ consists of 3 instructions, the second of which is the unlabeled statement $Y \leftarrow Y$. We have

$$3 = \langle 2, 0 \rangle = \langle 2, \langle 0, 0 \rangle\rangle$$

and

$$2 = \langle 0, 1 \rangle = \langle 0, \langle 1, 0 \rangle\rangle.$$

Thus, the program is

$$\begin{aligned} [B]\#Y &\leftarrow Y \\ Y &\leftarrow Y \\ Y &\leftarrow Y + 1 \end{aligned}$$

a not very interesting program that computes the function $y = 1$.

Note also that the empty program has the number $1 - 1 = 0$.

Exercises

1. Compute $\#(\mathcal{P})$ for $\mathcal{P}$ the programs of Exercises 4.1, 4.2, Chapter 2.
2. Find $\mathcal{P}$ such that $\#(\mathcal{P}) = 575$.

2. The Halting Problem

In this section we want to discuss a predicate HALT(x, y), which we now define. For given y, let $\mathscr{P}$ be the program such that $\#(\mathscr{P}) = y$. Then HALT(x, y) *is true if* $\psi_{\mathscr{P}}^{(1)}(x)$ is defined and false if $\psi_{\mathscr{P}}^{(1)}(x)$ is undefined. To put it succinctly:

$$\text{HALT}(x, y) \Leftrightarrow \text{program number } y \text{ eventually halts on input } x.$$

We now prove the remarkable

Theorem 2.1. HALT(x, y) is not a computable predicate.

Proof. Suppose that HALT(x, y) were computable. Then we could construct the program $\mathscr{P}$:

$$[A] \quad \text{IF HALT}(X, X) \text{ GOTO } A$$

(Of course $\mathscr{P}$ is to be the macro expansion of this program.) It is quite clear that $\mathscr{P}$ has been constructed so that

$$\psi_{\mathscr{P}}^{(1)}(x) = \begin{cases} \text{undefined} & \text{if} \quad \text{HALT}(x, x) \\ 0 & \text{if} \quad \sim \text{HALT}(x, x). \end{cases}$$

Let $\#(\mathscr{P}) = y_0$. Then using the definition of the HALT predicate,

$$\text{HALT}(x, y_0) \Leftrightarrow \sim \text{HALT}(x, x).$$

Since this equivalence is true for all x, we can set $x = y_0$:

$$\text{HALT}(y_0, y_0) \Leftrightarrow \sim \text{HALT}(y_0, y_0).$$

But this is a contradiction. ■

To begin with, this theorem provides us with an example of a function that is not computable by any program in the language $\mathscr{S}$. But we would like to go further; we would like to conclude the following:

> *There is no algorithm that, given a program of $\mathscr{S}$ and an input to that program, can determine whether or not the given program will eventually halt on the given input.*

In this form the result is called the *unsolvability of the halting problem*. We reason as follows: if there were such an algorithm, we could use it to check the truth or falsity of HALT(x, y) for given x, y by first obtaining program $\mathscr{Q}$ with $\#(\mathscr{Q}) = y$ and then checking whether $\mathscr{Q}$ eventually halts on input x. But we have reason to believe that *any algorithm for computing on* *numbers can be carried out by a program of $\mathscr{S}$.* Hence this would contradict the fact that HALT(x, y) is not computable.

The last italicized assertion is a form of what has come to be called *Church's thesis*. We have already accumulated some evidence for it, and we will see more later. But, since the word *algorithm* has no general definition separated from a particular language, Church's thesis cannot be proved as a mathematical theorem.

In fact, we will use Church's thesis freely in asserting the nonexistence of algorithms whenever we have shown that some problem cannot be solved by a program of $\mathscr{S}$.

In the light of Church's thesis, Theorem 2.1 tells us that there really is no algorithm for testing a given program and input to determine whether it will ever halt. Anyone who finds it surprising that no algorithm exists for such a "simple" problem should be made to realize that it is easy to construct relatively short programs (of $\mathscr{S}$) such that nobody is in a position to tell whether they will ever halt. For example, consider the assertion from number theory that every even number ≥ 4 is the sum of two prime numbers. This assertion, known as *Goldbach's conjecture*, is clearly true for small even numbers: $4 = 2 + 2$, $6 = 3 + 3$, $8 = 3 + 5$, etc. It is easy to write a program $\mathscr{P}$ of $\mathscr{S}$ that will search for a counterexample to Goldbach's conjecture, that is, an even number $n \geq 4$ that is not the sum of two primes. Note that the test that a given even number n is a counterexample only requires checking the primitive recursive predicate

$$\sim (\exists x)_{\leq n} (\exists y)_{\leq n} [\text{Prime}(x) \mathbin{\&} \text{Prime}(y) \mathbin{\&} x + y = n].$$

The statement that $\mathscr{P}$ never halts is equivalent to Goldbach's conjecture. Since the conjecture is still open after 250 years, nobody knows whether this program $\mathscr{P}$ will eventually halt.

Exercises

1. Show that HALT(x, x) is not computable.
2. Let $\overline{\text{HALT}}(x, y)$ be defined

   $$\overline{\text{HALT}}(x, y) \Leftrightarrow \text{program number } y \text{ never halts on input } x.$$

   Show that $\overline{\text{HALT}}(x, y)$ is not computable.
3. Let $\text{HALT}^1(x)$ be defined $\text{HALT}^1(x) \Leftrightarrow \text{HALT}(l(x), r(x))$. Show that $\text{HALT}^1(x)$ is not computable.

4. Prove or disprove: If $f(x_1, \ldots, x_n)$ is a total function such that for some constant k, $f(x_1, \ldots, x_n) \le k$ for all $x_1, \ldots, x_n$, then f is computable.

5. Suppose we claim that $\mathscr{P}$ is a program that computes HALT(x, x). Give a counterexample that shows the claim to be false. That is, give an input x for which $\mathscr{P}$ gives the wrong answer.

6. Let

$$f(x) = \begin{cases} x & \text{if Goldbach's conjecture is true} \\ 0 & \text{otherwise.} \end{cases}$$

Show that $f(x)$ is primitive recursive.

3. Universality

The negative character of the results in the previous section might lead one to believe that it is not possible to compute in a useful way with numbers of programs. But, as we shall soon see, this belief is not justified.

For each $n > 0$, we define

$$\Phi^{(n)}(x_1, \ldots, x_n, y) = \psi_{\mathscr{P}}^{(n)}(x_1, \ldots, x_n), \qquad \textit{where} \quad \#(\mathscr{P}) = y.$$

One of the key tools in computability theory is

Theorem 3.1 (Universality Theorem). For each $n > 0$, the function $\Phi^{(n)}(x_1, \ldots, x_n, y)$ is partially computable.

We shall prove this theorem by showing how to construct, for each $n > 0$, a program $\mathscr{U}_n$ which computes $\Phi^{(n)}$. That is, we shall have for each $n > 0$,

$$\psi_{\mathscr{U}_n}^{(n+1)}(x_1, \ldots, x_n, x_{n+1}) = \Phi^{(n)}(x_1, \ldots, x_n, x_{n+1}).$$

The programs $\mathscr{U}_n$ are called *universal*. For example, $\mathscr{U}_1$ can be used to compute *any* partially computable function of one variable, namely, if $f(x)$ is computed by a program $\mathscr{P}$ and $y = \#(\mathscr{P})$, then $f(x) = \Phi^{(1)}(x, y) = \psi_{\mathscr{U}_1}^{(2)}(x, y)$. The program $\mathscr{U}_n$ will work very much like an interpreter. It must keep track of the current snapshot in a computation and by "decoding" the number of the program being interpreted, decide what to do next and then do it.

In writing the programs $\mathscr{U}_n$ we shall freely use macros corresponding to functions that we know to be primitive recursive using the methods of Chapter 3. We shall also freely ignore the rules concerning which letters may be used to represent variables or labels of $\mathscr{S}$.

In considering the state of a computation we can assume that all variables which are not given values have the value 0. With this understanding, we can code the state in which the ith variable in our list has the value a_i and all variables after the mth have the value 0, by the Gödel number $[a_1, \ldots, a_m]$. For example, the state

$$Y = 0, \qquad X_1 = 2, \qquad X_2 = 1$$

is coded by the number

$$[0, 2, 0, 1] = 3^2 \cdot 7 = 63.$$

Notice in particular that the input variables are those whose position in our list is an *even* number.

Now in the universal programs, we shall allocate storage as follows:

K will be the number such that the Kth instruction is about to be executed;

S will store the current state coded in the manner just explained.

We proceed to give the program $\mathscr{U}_n$ for computing

$$Y = \Phi^{(n)}(X_1, \ldots, X_n, X_{n+1}).$$

We begin by exhibiting $\mathscr{U}_n$ in sections, explaining what each part does. Finally, we shall put the pieces together. We begin:

$$Z \leftarrow X_{n+1} + 1$$

$$S \leftarrow \prod_{i=1}^{n} (p_{2i})^{X_i}$$

$$K \leftarrow 1$$

If $X_{n+1} = \#(\mathscr{P})$, where $\mathscr{P}$ consists of the instructions $I_1, \ldots, I_m$, then Z gets the value $[\#(I_1), \ldots, \#(I_m)]$ [see Eq. (1.1)]. S is initialized as $[0, X_1, 0, X_2, \ldots, 0, X_n]$, which gives the first n input variables their appropriate values and gives all other variables the value 0. K, the instruction counter, is given the initial value 1 (so that the computation can begin with the first instruction). Next,

$$[C] \quad \text{IF } K = \text{Lt}(Z) + 1 \vee K = 0 \text{ GOTO } F$$

If the computation has ended, GOTO F, where the proper value will be output. (The significance of $K = 0$ will be explained later.) Otherwise, the current instruction must be decoded and executed:

$$U \leftarrow r((Z)_K)$$

$$P \leftarrow p_{r(U)+1}$$

$(Z)_K = \langle a, \langle b, c \rangle\rangle$ is the number of the Kth instruction. Thus, $U = \langle b, c \rangle$ is the code for the *statement* about to be executed. The variable mentioned in the Kth instruction is the $(c + 1)$th, i.e., the $(r(U) + 1)$th, in our list. Thus, its current value is stored as the exponent to which P divides S:

$$\begin{aligned}&\text{IF } l(U) = 0 \text{ GOTO } N\\ &\text{IF } l(U) = 1 \text{ GOTO } A\\ &\text{IF } \sim(P \mid S) \text{ GOTO } N\\ &\text{IF } l(U) = 2 \text{ GOTO } M\end{aligned}$$

If $l(U) = 0$, the instruction is a dummy $V \leftarrow V$ and the computation need do nothing to S. If $l(U) = 1$, the instruction is of the form $V \leftarrow V + 1$, so that 1 has to be added to the exponent on P in the prime power factorization of S. The computation executes a GOTO A (for *A*dd). If $l(U) \neq 0, 1$, then the current instruction is either of the form $V \leftarrow V - 1$ or IF $V \neq 0$ GOTO L. In either case, if P is not a divisor of S, i.e., if the current value of V is 0, the computation need do *nothing* to S. If $P \mid S$ and $l(U) = 2$, then the computation executes a GOTO M (for *M*inus), so that 1 can be subtracted from the exponent to which P divides S. To continue,

$$\begin{aligned}&K \leftarrow \min_{i \leq \text{Lt}(Z)} [l((Z)_i) + 2 = l(U)]\\ &\text{GOTO } C\end{aligned}$$

If $l(U) > 2$ and $P \mid S$, the current instruction is of the form IF $V \neq 0$ GOTO L where V has a nonzero value and L is the label whose position in our list is $l(U) - 2$. Accordingly the next instruction should be the first with this label. That is, K should get as its value the least i for which $l((Z)_i) = l(U) - 2$. If there is no instruction with the appropriate label, K gets the value 0, which will lead to termination the next time through the main loop. In either case the GOTO C causes a "jump" to the beginning of the loop for the next instruction (if any) to be processed. Continuing,

$$\begin{aligned}[M] \quad & S \leftarrow \lfloor S/P \rfloor\\ & \text{GOTO } N\\ [A] \quad & S \leftarrow S \cdot P\\ [N] \quad & K \leftarrow K + 1\\ & \text{GOTO } C\end{aligned}$$

1 is subtracted or added to the value of the variable mentioned in the current instruction by dividing or multiplying S by P, respectively. The

```
        Z ← X_{n+1} + 1
        S ← ∏_{i=1}^{n} (p_{2i})^{X_i}
        K ← 1
[C]     IF K = Lt(Z) + 1 ∨ K = 0 GOTO F
        U ← r((Z)_K)
        P ← p_{r(U)+1}
        IF l(U) = 0 GOTO N
        IF l(U) = 1 GOTO A
        IF ~(P | S) GOTO N
        IF l(U) = 2 GOTO M
        K ← min_{i ≤ Lt(Z)} [l((Z)_i) + 2 = l(U)]
        GOTO C
[M]     S ← ⌊S/P⌋
        GOTO N
[A]     S ← S · P
[N]     K ← K + 1
        GOTO C
[F]     Y ← (S)_1
```

Figure 3.1. Program $\mathscr{U}_n$, which computes $Y = \Phi^{(n)}(X_1, \ldots, X_n, X_{n+1})$.

instruction counter is increased by 1 and the computation returns to process the next instruction. To conclude the program,

$$[F] \quad Y \leftarrow (S)_1$$

On termination, the value of Y for the program being simulated is stored as the exponent on $p_1(= 2)$ in S. We have now completed our description of $\mathscr{U}_n$ and we put the pieces together in Fig. 3.1.

For each $n > 0$, the sequence

$$\Phi^{(n)}(x_1, \ldots, x_n, 0), \Phi^{(n)}(x_1, \ldots, x_n, 1), \ldots$$

enumerates all partially computable functions of n variables. When we want to emphasize this aspect of the situation we write

$$\Phi_y^{(n)}(x_1, \ldots, x_n) = \Phi^{(n)}(x_1, \ldots, x_n, y).$$

It is often convenient to omit the superscript when $n = 1$, writing

$$\Phi_y(x) = \Phi(x, y) = \Phi^{(1)}(x, y).$$

A simple modification of the programs $\mathscr{U}_n$ would enable us to prove that the predicates

$$\begin{aligned} \mathrm{STP}^{(n)}(x_1, \ldots, x_n, y, t) &\Leftrightarrow \text{Program number } y \text{ halts after } t \text{ or fewer steps on inputs } x_1, \ldots, x_n \\ &\Leftrightarrow \text{There is a computation of program } y \text{ of length } \le t+1, \text{ beginning with inputs } x_1, \ldots, x_n \end{aligned}$$

are computable. We simply need to add a counter to determine when we have simulated t steps. However, we can prove a stronger result.

Theorem 3.2 (Step-Counter Theorem). For each $n > 0$, the predicate $\mathrm{STP}^{(n)}(x_1, \ldots, x_n, y, t)$ is primitive recursive.

Proof. The idea is to provide numeric versions of the notions of snapshot and successor snapshot and to show that the necessary functions are primitive recursive. We use the same representation of program states that we used in defining the universal programs, and if z represents state σ, then $\langle i, z \rangle$ represents the snapshot (i, σ).

We begin with some functions for extracting the components of the ith instruction of program number y:

$$\mathrm{LABEL}(i, y) = l((y+1)_i)$$

$$\mathrm{VAR}(i, y) = r(r((y+1)_i)) + 1$$

$$\mathrm{INSTR}(i, y) = l(r((y+1)_i))$$

$$\mathrm{LABEL}'(i, y) = l(r((y+1)_i)) \dot{-} 2$$

Next we define some predicates that indicate, for program y and the snapshot represented by x, which kind of action is to be performed next.

$$\begin{aligned} \mathrm{SKIP}(x, y) \Leftrightarrow {} & [\mathrm{INSTR}(l(x), y) = 0 \;\&\; l(x) \le \mathrm{Lt}(y+1)] \\ & \vee \left[\mathrm{INSTR}(l(x), y) \ge 2 \;\&\; \sim\left(p_{\mathrm{VAR}(l(x), y)} \mid r(x)\right)\right] \end{aligned}$$

$$\mathrm{INCR}(x, y) \Leftrightarrow \mathrm{INSTR}(l(x), y) = 1$$

$$\mathrm{DECR}(x, y) \Leftrightarrow \mathrm{INSTR}(l(x), y) = 2 \;\&\; p_{\mathrm{VAR}(l(x), y)} \mid r(x)$$

$$\begin{aligned} \mathrm{BRANCH}(x, y) \Leftrightarrow {} & \mathrm{INSTR}(l(x), y) > 2 \;\&\; p_{\mathrm{VAR}(l(x), y)} \mid r(x) \\ & \& \; (\exists i)_{\le \mathrm{Lt}(y+1)} \mathrm{LABEL}(i, y) = \mathrm{LABEL}'(l(x), y) \end{aligned}$$

Now we can define SUCC(x, y), which, for program number y, gives the representative of the successor to the snapshot represented by x.

$$\mathrm{SUCC}(x,y) = \begin{cases} \langle l(x)+1, r(x)\rangle & \text{if SKIP}(x,y) \\ \langle l(x)+1, r(x)\cdot p_{\mathrm{VAR}(l(x),y)}\rangle & \text{if INCR}(x,y) \\ \langle l(x)+1, \lfloor r(x)/p_{\mathrm{VAR}(l(x),y)}\rfloor\rangle & \text{if DECR}(x,y) \\ \langle \min_{i \le \mathrm{Lt}(y+1)}[\mathrm{LABEL}(i,y) = \mathrm{LABEL}'(l(x),y)],\ r(x)\rangle & \\ & \text{if BRANCH}(x,y) \\ \langle \mathrm{Lt}(y+1)+1, r(x)\rangle & \text{otherwise.} \end{cases}$$

We also need

$$\mathrm{INIT}^{(n)}(x_1,\ldots,x_n) = \langle 1, \prod_{i=1}^{n}(p_{2i})^{x_i}\rangle,$$

which gives the representation of the initial snapshot for inputs $x_1, \ldots, x_n$, and

$$\mathrm{TERM}(x,y) \Leftrightarrow l(x) > \mathrm{Lt}(y+1),$$

which tests whether x represents a terminal snapshot for program y.

Putting these together we can define a primitive recursive function that gives the numbers of the successive snapshots produced by a given program.

$$\mathrm{SNAP}^{(n)}(x_1,\ldots,x_n,y,0) = \mathrm{INIT}^{(n)}(x_1,\ldots,x_n)$$

$$\mathrm{SNAP}^{(n)}(x_1,\ldots,x_n,y,i+1) = \mathrm{SUCC}(\mathrm{SNAP}^{(n)}(x_1,\ldots,x_n,y,i),y)$$

Thus,

$$\mathrm{STP}^{(n)}(x_1,\ldots,x_n,y,t) \Leftrightarrow \mathrm{TERM}(\mathrm{SNAP}^{(n)}(x_1,\ldots,x_n,y,t),y),$$

and it is clear that $\mathrm{STP}^{(n)}(x_1,\ldots,x_n,y,t)$ is primitive recursive. ■

By using the technique of the above proof, we can obtain the following important result.

Theorem 3.3 (Normal Form Theorem). Let $f(x_1, \ldots, x_n)$ be a partially computable function. Then there is a primitive recursive predicate $R(x_1, \ldots, x_n, y)$ such that

$$f(x_1,\ldots,x_n) = l\left(\min_z R(x_1,\ldots,x_n,z)\right).$$

Proof. Let y_0 be the number of a program that computes $f(x_1, \ldots, x_n)$. We shall prove the following equation, which clearly implies the desired result:

$$f(x_1, \ldots, x_n) = l\left(\min_z R(x_1, \ldots, x_n, z)\right) \tag{3.1}$$

where $R(x_1, \ldots, x_n, z)$ is the predicate

$$\begin{aligned} &\mathrm{STP}^{(n)}(x_1, \ldots, x_n, y_0, r(z)) \\ &\qquad \& \ (r(\mathrm{SNAP}^{(n)}(x_1, \ldots, x_n, y_0, r(z))))_1 = l(z). \end{aligned}$$

First consider the case when the righthand side of equation (3.1) is defined. Then, in particular, there exists a number z such that

$$\begin{aligned} &\mathrm{STP}^{(n)}(x_1, \ldots, x_n, y_0, r(z)) \\ &\qquad \text{and } (r(\mathrm{SNAP}^{(n)}(x_1, \ldots, x_n, y_0, r(z))))_1 = l(z). \end{aligned}$$

For any such z, the computation by the program with number y_0 has reached a terminal snapshot in $r(z)$ or fewer steps and $l(z)$ is the value held in the output variable Y, i.e., $l(z) = f(x_1, \ldots, x_n)$.

If, on the other hand, the right side is undefined, it must be the case that $\mathrm{STP}^{(n)}(x_1, \ldots, x_n, y_0, t)$ is false for all values of t, i.e., $f(x_1, \ldots, x_n)\uparrow$. ■

The normal form theorem leads to another characterization of the class of partially computable functions.

Theorem 3.4. A function is partially computable if and only if it can be obtained from the initial functions by a finite number of applications of composition, recursion, and minimalization.

Proof. That every function which can be so obtained is partially computable is an immediate consequence of Theorems 1.1, 2.1, 2.2, 3.1, and 7.2 in Chapter 3. Note that a partially computable predicate is necessarily computable, so Theorem 7.2 covers all applications of minimalization to a predicate obtained as described in the theorem.

Conversely, we can use the normal form theorem to write any given partially computable function in the form

$$l\left(\min_y R(x_1, \ldots, x_n, y)\right),$$

where R is a primitive recursive predicate and so is obtained from the initial functions by a finite number of applications of composition and recursion. Finally, our given function is obtained from R by one use of minimalization and then by composition with the primitive recursive function l. ∎

When $\min_y R(x_1, \ldots, x_n, y)$ is a total function [that is, when for each $x_1, \ldots, x_n$ there is at least one y for which $R(x_1, \ldots, x_n, y)$ is true], we say that we are applying the operation of *proper minimalization* to R. Now, if

$$l\left(\min_y R(x_1, \ldots, x_n, y)\right)$$

is total, then $\min_y R(x_1, \ldots, x_n, y)$ must be total. Hence we have

Theorem 3.5. A function is computable if and only if it can be obtained from the initial functions by a finite number of applications of composition, recursion, and *proper* minimalization.

Exercises

1. Show that for each u, there are infinitely many different numbers v such that for all x, $\Phi_u(x) = \Phi_v(x)$.

2. **(a)** Let

$$H_1(x) = \begin{cases} 1 & \text{if } \Phi(x, x)\downarrow \\ \uparrow & \text{otherwise.} \end{cases}$$

Show that $H_1(x)$ is partially computable.

(b) Let $A = \{a_1, \ldots, a_n\}$ be a finite set such that $\Phi(a_i, a_i)\uparrow$ for $1 \le i \le n$, and let

$$H_2(x) = \begin{cases} 1 & \text{if } \Phi(x, x)\downarrow \\ 0 & \text{if } x \in A \\ \uparrow & \text{otherwise.} \end{cases}$$

Show that $H_2(x)$ is partially computable.

(c) Give an infinite set B such that $\Phi(b, b)\uparrow$ for all $b \in B$ and such that

$$H_3(x) = \begin{cases} 1 & \text{if } \Phi(x, x)\downarrow \\ 0 & \text{if } x \in B \\ \uparrow & \text{otherwise} \end{cases}$$

is partially computable.

(d) Give an infinite set C such that $\Phi(c, c)\uparrow$ for all $c \in C$ and such that

$$H_4(x) = \begin{cases} 1 & \text{if } \Phi(x, x)\downarrow \\ 0 & \text{if } x \in C \\ \uparrow & \text{otherwise} \end{cases}$$

is not partially computable.

3. Give a program $\mathscr{P}$ such that $H_{\mathscr{P}}(x_1, x_2)$, defined

$$H_{\mathscr{P}}(x_1, x_2) \Leftrightarrow \text{program } \mathscr{P} \text{ eventually halts on inputs } x_1, x_2$$

is not computable.

4. Let $f(x_1, \ldots, x_n)$ be computed by program $\mathscr{P}$, and suppose that for some primitive recursive function $g(x_1, \ldots, x_n)$,

$$\text{STP}^{(n)}(x_1, \ldots, x_n, \#(\mathscr{P}), g(x_1, \ldots, x_n))$$

is true for all $x_1, \ldots, x_n$. Show that $f(x_1, \ldots, x_n)$ is primitive recursive.

5.* Give a primitive recursive function counter(x) such that if Φ_n is a computable predicate, then

$$\Phi_n(\text{counter}(n)) \Leftrightarrow \sim \text{HALT}(\text{counter}(n), \text{counter}(n)).$$

That is, counter(n) is a counterexample to the possibility that Φ_n computes HALT(x, x). [Compare this exercise with Exercise 2.5.]

6.* Give an upper bound on the length of the shortest $\mathscr{S}$ program that computes the function $\Phi_y(x)$.

4. Recursively Enumerable Sets

The close relation between predicates and sets, as described in Chapter 1, lets us use the language of sets in talking about solvable and unsolvable problems. For example, the predicate HALT(x, y) is the characteristic function of the set $\{(x, y) \in N^2 \mid \text{HALT}(x, y)\}$. To say that a set B, where $B \subseteq N^m$, belongs to some class of *functions* means that the characteristic function $P(x_1, \ldots, x_m)$ of B belongs to the class in question. Thus, in particular, to say that the set B is computable or recursive is just to say that $P(x_1, \ldots, x_m)$ is a computable function. Likewise, B is a primitive recursive set if $P(x_1, \ldots, x_m)$ is a primitive recursive predicate.

We have, for example,

Theorem 4.1. Let the sets B, C belong to some PRC class $\mathscr{C}$. Then so do the sets $B \cup C, B \cap C, \bar{B}$.

Proof. This is an immediate consequence of Theorem 5.1, Chapter 3. ■

As long as the Gödel numbering functions $[x_1, \ldots, x_n]$ and $(x)_i$ are available, we can restrict our attention to subsets of N. We have, for example,

Theorem 4.2. Let $\mathscr{C}$ be a PRC class, and let B be a subset of N^m, $m \geq 1$. Then B belongs to $\mathscr{C}$ if and only if

$$B' = \{[x_1, \ldots, x_m] \in N \mid (x_1, \ldots, x_m) \in B\}$$

belongs to $\mathscr{C}$.

Proof. If $P_B(x_1, \ldots, x_m)$ is the characteristic function of B, then

$$P_{B'}(x) \Leftrightarrow P_B((x)_1, \ldots, (x)_m) \ \& \ \mathrm{Lt}(x) \leq m \ \& \ x > 0$$

is the characteristic function of B', and $P_{B'}$ clearly belongs to $\mathscr{C}$ if P_B belongs to $\mathscr{C}$. On the other hand, if $P_{B'}(x)$ is the characteristic function of B', then

$$P_B(x_1, \ldots, x_m) \Leftrightarrow P_{B'}([x_1, \ldots, x_m])$$

is the characteristic function of B, and P_B clearly belongs to $\mathscr{C}$ if $P_{B'}$ belongs to $\mathscr{C}$. ■

It immediately follows, for example, that $\{[x, y] \in N \mid \mathrm{HALT}(x, y)\}$ is not a computable set.

Definition. The set $B \subseteq N$ is called *recursively enumerable* if there is a partially computable function $g(x)$ such that

$$B = \{x \in N \mid g(x)\downarrow\}. \tag{4.1}$$

The term *recursively enumerable* is usually abbreviated *r.e.* A set is recursively enumerable just when it is the domain of a partially computable function. If $\mathscr{P}$ is a program that computes the function g in (4.1), then B is simply the set of all inputs to $\mathscr{P}$ for which $\mathscr{P}$ eventually halts. If we think of $\mathscr{P}$ as providing an algorithm for testing for membership in B, we see that for numbers that do belong to B, the algorithm will provide a "yes" answer; but for numbers that do not, the algorithm will never terminate. If we invoke Church's thesis, r.e. sets B may be thought of intuitively as sets for which there exist algorithms related to B as in the previous sentence, but without stipulating that the algorithms be expressed by programs of the language $\mathscr{S}$. Such algorithms, sometimes called *semi-decision procedures*, provide a kind of "approximation" to solving the problem of testing membership in B.

We have

Theorem 4.3. If B is a recursive set, then B is r.e.

Proof. Consider the program $\mathscr{P}$:

$$[A] \quad \text{IF} \sim (X \in B) \text{ GOTO } A$$

Since B is recursive, the predicate $x \in B$ is computable and $\mathscr{P}$ can be expanded to a program of $\mathscr{S}$. Let $\mathscr{P}$ compute the function $h(x)$. Then, clearly,

$$B = \{x \in N \mid h(x)\downarrow\}. \qquad \blacksquare$$

If B and $\bar{B}$ are both r.e., we have a pair of algorithms that will terminate in case a given input is or is not in B, respectively. We can think of combining these two algorithms to obtain a single algorithm that will always terminate and that will tell us whether a given input belongs to B. This combined algorithm might work by "running" the two separate algorithms for longer and longer times until one of them terminates. This method of combining algorithms is called *dovetailing*, and the step-counter theorem enables us to use it in a rigorous manner.

Theorem 4.4. The set B is recursive if and only if B and $\bar{B}$ are both r.e.

Proof. If B is recursive, then by Theorem 4.1 so is $\bar{B}$, and hence by Theorem 4.3, they are both r.e.

Conversely, if B and $\bar{B}$ are both r.e., we may write

$$B = \{x \in N \mid g(x)\downarrow\},$$

$$\bar{B} = \{x \in N \mid h(x)\downarrow\},$$

where g and h are both partially computable. Let g be computed by program $\mathscr{P}$ and h be computed by program $\mathscr{Q}$, and let $p = \#(\mathscr{P})$, $q = \#(\mathscr{Q})$. Then the program that follows computes B. (That is, the program computes the characteristic function of B.)

```
[A]     IF STP^(1)(X, p, T) GOTO C
        IF STP^(1)(X, q, T) GOTO E
        T ← T + 1
        GOTO A
[C]     Y ← 1
```
■

Theorem 4.5. If B and C are r.e. sets so are $B \cup C$ and $B \cap C$.

Proof. Let

$$B = \{x \in N \mid g(x)\downarrow\},$$
$$C = \{x \in N \mid h(x)\downarrow\},$$

where g and h are both partially computable. Let $f(x)$ be the function computed by the program

$$Y \leftarrow g(X)$$
$$Y \leftarrow h(X)$$

Then $f(x)$ is defined if and only if $g(x)$ and $h(x)$ are both defined. Hence

$$B \cap C = \{x \in N \mid f(x)\downarrow\},$$

so that $B \cap C$ is also r.e.

To obtain the result for $B \cup C$ we must use dovetailing again. Let g and h be computed by programs $\mathscr{P}$ and $\mathscr{Q}$, respectively, and let $\#(\mathscr{P}) = p$, $\#(\mathscr{Q}) = q$. Let $k(x)$ be the function computed by the program

```
[A]     IF STP^(1)(X, p, T) GOTO E
        IF STP^(1)(X, q, T) GOTO E
        T ← T + 1
        GOTO A'
```

Then $k(x)$ is defined just in case *either* $g(x)$ *or* $h(x)$ is defined. That is,

$$B \cup C = \{x \in N \mid k(x)\downarrow\}.$$ ■

Definition. We write

$$W_n = \{x \in N \mid \Phi(x, n)\downarrow\}.$$

Then we have

Theorem 4.6 (Enumeration Theorem). A set B is r.e. if and only if there is an n for which $B = W_n$.

Proof. This is an immediate consequence of the definition of $\Phi(x, n)$. ■

The theorem gets its name from the fact that the sequence

$$W_0, W_1, W_2, \ldots$$

is an enumeration of all r.e. sets.

We define

$$K = \{n \in N \mid n \in W_n\}.$$

Now,

$$n \in W_n \Leftrightarrow \Phi(n, n)\downarrow \Leftrightarrow \text{HALT}(n, n).$$

Thus, K is the set of all numbers n such that program number n eventually halts on input n. We have

Theorem 4.7. K is r.e. but not recursive.

Proof. Since $K = \{n \in N \mid \Phi(n, n)\downarrow\}$ and (by the universality theorem—Theorem 3.1), $\Phi(n, n)$ is certainly partially computable, K is clearly r.e. If $\bar{K}$ were also r.e., by the enumeration theorem we would have

$$\bar{K} = W_i$$

for some i. Then

$$i \in K \Leftrightarrow i \in W_i \Leftrightarrow i \in \bar{K},$$

which is a contradiction. ■

Actually the *proof* of Theorem 2.1 already shows not only that HALT(x, z) is not computable, but also that HALT(x, x) is not computable, i.e., that K is not a recursive set. (This was Exercise 2.1.)

We conclude this section with some alternative ways of characterizing r.e. sets.

Theorem 4.8. Let B be an r.e. set. Then there is a primitive recursive predicate $R(x, t)$ such that $B = \{x \in N \mid (\exists t)R(x, t)\}$.

Proof. Let $B = W_n$. Then $B = \{x \in N \mid (\exists t)\text{STP}^{(1)}(x, n, t)\}$, and $\text{STP}^{(1)}$ is primitive recursive by Theorem 3.2. ■

Theorem 4.9. Let S be a nonempty r.e. set. Then there is a primitive recursive function $f(u)$ such that $S = \{f(n) \mid n \in N\} = \{f(0), f(1), f(2), \ldots\}$. That is, S is the range of f.

Proof. By Theorem 4.8

$$S = \{x \mid (\exists t) R(x, t)\},$$

where R is a primitive recursive predicate. Let x_0 be some fixed member of S (for example, the smallest). Let

$$f(u) = \begin{cases} l(u) & \text{if} \quad R(l(u), r(u)) \\ x_0 & \text{otherwise.} \end{cases}$$

Then by Theorem 5.4 in Chapter 3, f is primitive recursive. Each value $f(u)$ is in S, since x_0 is automatically in S, while if $R(l(u), r(u))$ is true, then certainly $(\exists t)R(l(u), t)$ is true, which implies that $f(u) = l(u) \in S$. Conversely, if $x \in S$, then $R(x, t_0)$ is true for some t_0. Then

$$f(\langle x, t_0 \rangle) = l(\langle x, t_0 \rangle) = x,$$

so that $x = f(u)$ for $u = \langle x, t_0 \rangle$. ■

Theorem 4.10. Let $f(x)$ be a partially computable function and let $S = \{f(x) \mid f(x)\downarrow\}$. (That is, S is the *range* of f.) Then S is r.e.

Proof. Let

$$g(x) = \begin{cases} 0 & \text{if} \quad x \in S \\ \uparrow & \text{otherwise.} \end{cases}$$

Since

$$S = \{x \mid g(x)\downarrow\},$$

it suffices to show that $g(x)$ is partially computable. Let $\mathscr{P}$ be a program that computes f and let $\#(\mathscr{P}) = p$. Then the following program computes $g(x)$:

```
[A]  IF ~ STP^(1)(Z, p, T) GOTO B
     V ← f(Z)
     IF V = X GOTO E
[B]  Z ← Z + 1
     IF Z ≤ T GOTO A
     T ← T + 1
     Z ← 0
     GOTO A
```

Note that in this program the macro expansion of $V \leftarrow f(Z)$ will be entered only when the step-counter test has already guaranteed that f is defined. ■

Combining Theorems 4.9 and 4.10, we have

Theorem 4.11. Suppose that $S \neq \emptyset$. Then the following statements are all equivalent:

1. S is r.e.;
2. S is the range of a primitive recursive function;
3. S is the range of a recursive function;
4. S is the range of a partial recursive function.

Proof. By Theorem 4.9, (1) implies (2). Obviously, (2) implies (3), and (3) implies (4). By Theorem 4.10, (4) implies (1). Hence all four statements are equivalent. ■

Theorem 4.11 provides the motivation for the term *recursively enumerable*. In fact, such a set (if it is nonempty) is enumerated by a recursive function.

Exercises

1. Let B be a subset of N^m, $m > 1$. We say that B is r.e. if $B = \{(x_1, \ldots, x_m) \in N^m \mid g(x_1, \ldots, x_m)\downarrow\}$ for some partially computable function $g(x_1, \ldots, x_m)$. Let

$$B' = \{[x_1, \ldots, x_m] \in N \mid (x_1, \ldots, x_m) \in B\}.$$

Show that B' is r.e. if and only if B is r.e.

2. Let $K_0 = \{\langle x, y\rangle \mid x \in W_y\}$. Show that K_0 is r.e.

3. Let f be an n-ary partial function. The graph of f, denoted gr(f), is the set $\{[x_1, \ldots, x_n, f(x_1, \ldots, x_n)] \mid f(x_1, \ldots, x_n)\downarrow\}$.

(**a**) Let $\mathscr{C}$ be a PRC class. Prove that if f belongs to $\mathscr{C}$ then gr(f) belongs to $\mathscr{C}$.

(**b**) Prove that if gr(f) is recursive then f is partially computable.

(**c**) Prove that the recursiveness of gr(f) does not necessarily imply that f is computable.

4. Let $B = \{f(n) \mid n \in N\}$, where f is a strictly increasing computable function [i.e., $f(n + 1) > f(n)$ for all n]. Prove that B is recursive.

5. Show that every infinite r.e. set has an infinite recursive subset.

6. Prove that an infinite set A is r.e. if and only if $A = \{f(n) \mid n \in N\}$ for some one–one computable function $f(x)$.

7. Let A, B be sets. Prove or disprove:
   (a) If $A \cup B$ is r.e., then A and B are both r.e.
   (b) If $A \subseteq B$ and B is r.e., then A is r.e.
8. Show that there is no computable function $f(x)$ such that $f(x) = \Phi(x, x) + 1$ whenever $\Phi(x, x)\downarrow$.
9. (a) Let $g(x), h(x)$ be partially computable functions. Show there is a partially computable function $f(x)$ such that $f(x)\downarrow$ for precisely those values of x for which either $g(x)\downarrow$ or $h(x)\downarrow$ (or both) and such that when $f(x)\downarrow$, either $f(x) = g(x)$ or $f(x) = h(x)$.
   (b) Can f be found fulfilling all the requirements of (a) but such that in addition $f(x) = g(x)$ whenever $g(x)\downarrow$? Proof?
10. (a) Let $A = \{y \mid (\exists t)P(t, y)\}$, where P is a computable predicate. Show that A is r.e.
    (b) Let $B = \{y \mid (\exists t_1) \cdots (\exists t_n)Q(t_1, \ldots, t_n, y)\}$, where Q is a computable predicate. Show that B is r.e.
11. Give a computable predicate $R(x, y)$ such that $\{y \mid (\forall t)R(t, y)\}$ is not r.e.

5. The Parameter Theorem

The parameter theorem (which has also been called the *iteration theorem* and the *s–m–n theorem*) is an important technical result that relates the various functions $\Phi^{(n)}(x_1, x_2, \ldots, x_n, y)$ for different values of n.

Theorem 5.1 (Parameter Theorem). For each $n, m > 0$, there is a primitive recursive function $S_m^n(u_1, u_2, \ldots, u_n, y)$ such that

$$\Phi^{(m+n)}(x_1, \ldots, x_m, u_1, \ldots, u_n, y) = \Phi^{(m)}(x_1, \ldots, x_m, S_m^n(u_1, \ldots, u_n, y)). \tag{5.1}$$

Suppose that values for variables $u_1, \ldots, u_n$ are fixed and we have in mind some particular value of y. Then the left side of (5.1) is a partially computable function of the m arguments $x_1, \ldots, x_m$. Letting q be the number of a program that computes this function of m variables, we have

$$\Phi^{(m+n)}(x_1, \ldots, x_m, u_1, \ldots, u_n, y) = \Phi^{(m)}(x_1, \ldots, x_m, q).$$

The parameter theorem tells us that not only does there exist such a number q, but that it can be obtained from $u_1, \ldots, u_n, y$ in a computable (in fact, primitive recursive) way.

Proof. The proof is by mathematical induction on n.

For $n = 1$, we need to show that there is a primitive recursive function $S_m^1(u, y)$ such that

$$\Phi^{(m+1)}(x_1, \ldots, x_m, u, y) = \Phi^{(m)}(x_1, \ldots, x_m, S_m^1(u, y)).$$

Here $S_m^1(u, y)$ must be the number of a program which, given m inputs $x_1, \ldots, x_m$, computes the same value as program number y does when given the $m + 1$ inputs $x_1, \ldots, x_m, u$. Let $\mathscr{P}$ be the program such that $\#(\mathscr{P}) = y$. Then $S_m^1(u, y)$ can be taken to be the number of a program which first gives the variable X_{m+1} the value u and then proceeds to carry out $\mathscr{P}$. X_{m+1} will be given the value u by the program

$$\left.\begin{array}{l} X_{m+1} \leftarrow X_{m+1} + 1 \\ \vdots \\ X_{m+1} \leftarrow X_{m+1} + 1 \end{array}\right\} u$$

The number of the unlabeled instruction

$$X_{m+1} \leftarrow X_{m+1} + 1$$

is

$$\langle 0, \langle 1, 2m + 1 \rangle \rangle = 16m + 10.$$

So we may take

$$S_m^1(u, y) = \left[\left(\prod_{i=1}^{u} p_i \right)^{16m+10} \cdot \prod_{j=1}^{\text{Lt}(y+1)} p_{u+j}^{(y+1)_j} \right] \dot{-} 1,$$

a primitive recursive function. Here the numbers of the instructions of $\mathscr{P}$ which appear as exponents in the prime power factorization of $y + 1$ have been shifted to the primes $p_{u+1}, p_{u+2}, \ldots, p_{u+\text{Lt}(y+1)}$.

To complete the proof, suppose the result known for $n = k$. Then we have

$$\begin{aligned} &\Phi^{(m+k+1)}(x_1, \ldots, x_m, u_1, \ldots, u_k, u_{k+1}, y) \\ &\quad = \Phi^{(m+k)}(x_1, \ldots, x_m, u_1, \ldots, u_k, S_{m+k}^1(u_{k+1}, y)) \\ &\quad = \Phi^{(m)}\big(x_1, \ldots, x_m, S_m^k(u_1, \ldots, u_k, S_{m+k}^1(u_{k+1}, y))\big), \end{aligned}$$

using first the result for $n = 1$ and then the induction hypothesis. But now, if we define

$$S_m^{k+1}(u_1, \ldots, u_k, u_{k+1}, y) = S_m^k(u_1, \ldots, u_k, S_{m+k}^1(u_{k+1}, y)),$$

we have the desired result. ■

We next give a sample application of the parameter theorem. It is desired to find a computable function $g(u, v)$ such that

$$\Phi_u(\Phi_v(x)) = \Phi_{g(u,v)}(x).$$

We have by the meaning of the notation that

$$\Phi_u(\Phi_v(x)) = \Phi(\Phi(x, v), u)$$

is a partially computable function of x, u, v. Hence, we have

$$\Phi_u(\Phi_v(x)) = \Phi^{(3)}(x, u, v, z_0)$$

for some number z_0. By the parameter theorem,

$$\Phi^{(3)}(x, u, v, z_0) = \Phi(x, S_1^2(u, v, z_0)) = \Phi_{S_1^2(u,v,z_0)}(x).$$

Exercises

1. Given a partially computable function $f(x, y)$, find a primitive recursive function $g(u, v)$ such that

$$\Phi_{g(u,v)}(x) = f(\Phi_u(x), \Phi_v(x)).$$

2. Show that there is a primitive recursive function $g(u, v, w)$ such that

$$\Phi^{(3)}(u, v, w, z) = \Phi_{g(u,v,w)}(z).$$

3. Let us call a partially computable function $g(x)$ *extendable* if there is a computable function $f(x)$ such that $f(x) = g(x)$ for all x for which $g(x)\downarrow$. Show that there is no algorithm for determining of a given z whether or not $\Phi_z(x)$ is extendable. [*Hint:* Exercise 8 of Section 4 shows that $\Phi(x, x) + 1$ is not extendable. Find an extendable function $k(x)$ such that the function

$$h(x, t) = \begin{cases} \Phi(x, x) + 1 & \text{if } \Phi(t, t)\downarrow \\ k(x) & \text{otherwise} \end{cases}$$

is partially computable.]

4.* A *programming system* is an enumeration $S = \{\phi_i^{(n)} \mid i \in N, n > 0\}$ of the partially computable functions. That is, for each partially computable function $f(x_1, \ldots, x_n)$ there is an i such that f is $\phi_i^{(n)}$.

(a) A programming system S is *universal* if for each $n > 0$, the function $\Psi^{(n)}$, defined

$$\Psi^{(n)}(x_1, \ldots, x_n, i) = \phi_i^{(n)}(x_1, \ldots, x_n),$$

is partially computable. That is, S is universal if a version of the universality theorem holds for S. Obviously,

$$\{\Phi_i^{(n)} \mid i \in N, n > 0\}$$

is a universal programming system. Prove that a programming system S is universal if and only if for each $n > 0$ there is a computable function f_n such that $\phi_i^{(n)} = \Phi_{f_n(i)}^{(n)}$ for all i.

(b) A universal programming system S is *acceptable* if for each $n, m > 0$ there is a computable function $s_m^n(u_1, \ldots, u_n, y)$ such that

$$\begin{aligned} &\Psi^{(m+n)}(x_1, \ldots, x_m, u_1, \ldots, u_n, y) \\ &\quad = \Psi^{(m)}(x_1, \ldots, x_m, s_m^n(u_1, \ldots, u_n, y)). \end{aligned}$$

That is, S is acceptable if a version of the parameter theorem holds for S. Again, $\{\Phi_i^{(n)} \mid i \in N, n > 0\}$ is obviously an acceptable programming system. Prove that S is acceptable if and only if for each $n > 0$ there is a computable function g_n such that $\Phi_i^{(n)} = \phi_{g_n(i)}^{(n)}$ for all i.

6. Diagonalization and Reducibility

So far we have seen very few examples of nonrecursive sets. We now discuss two general techniques for proving that given sets are not recursive or even that they are not r.e. The first method, *diagonalization*, turns on the demonstration of two assertions of the following sort:

1. A certain set A can be enumerated in a suitable fashion.
2. It is possible, with the help of the enumeration, to define an object b that is different from every object in the enumeration, i.e., $b \notin A$.

We sometimes say that b is defined by *diagonalizing over* A. In some diagonalization arguments the goal is simply to find some $b \notin A$. We will give an example of such an argument later in the chapter. The arguments we will consider in this section have an additional twist: the definition of b is such that *b must belong to A*, contradicting the assertion that we began with an enumeration of *all* of the elements in A. The end of the argument, then, is to draw some conclusion from this contradiction.

For example, the proof given for Theorem 2.1 is a diagonalization argument that the predicate HALT(x, y), or equivalently, the set

$$\{(x, y) \in N^2 \mid \text{HALT}(x, y)\},$$

is not computable. The set A in this case is the class of unary partially computable functions, and assertion 1 follows from the fact that $\mathscr{S}$ programs can be coded as numbers. For each n, let $\mathscr{P}_n$ be the program with number n. Then all unary partially computable functions occur among $\psi^{(1)}_{\mathscr{P}_0}, \psi^{(1)}_{\mathscr{P}_1}, \ldots$. We began by assuming that HALT(x, y) is computable, and we wrote a program $\mathscr{P}$ that computes $\psi^{(1)}_{\mathscr{P}}$. The heart of the proof consisted of showing that $\psi^{(1)}_{\mathscr{P}}$ does not appear among $\psi^{(1)}_{\mathscr{P}_0}, \psi^{(1)}_{\mathscr{P}_1}, \ldots$. In particular, we wrote $\mathscr{P}$ so that for every x, $\psi^{(1)}_{\mathscr{P}}(x)\downarrow$ if and only if $\psi^{(1)}_{\mathscr{P}_x}(x)\uparrow$, i.e.,

$$\text{HALT}(x, \#(\mathscr{P})) \Leftrightarrow \sim \text{HALT}(x, x),$$

so $\psi^{(1)}_{\mathscr{P}}$ differs from each function $\psi^{(1)}_{\mathscr{P}_0}, \psi^{(1)}_{\mathscr{P}_1}, \ldots$ on at least one input value. That is, n is a counterexample to the possibility that $\psi^{(1)}_{\mathscr{P}}$ is $\psi^{(1)}_{\mathscr{P}_n}$, since $\psi^{(1)}_{\mathscr{P}}(n)\downarrow$ if and only if $\psi^{(1)}_{\mathscr{P}_n}(n)\uparrow$. Now we have the unary partially computable function $\psi^{(1)}_{\mathscr{P}}$ that is not among $\psi^{(1)}_{\mathscr{P}_0}, \phi^{(1)}_{\mathscr{P}_1}, \ldots$, so assertion 2 is satisfied, giving us a contradiction. In the proof of Theorem 2.1 the contradiction was expressed a bit differently: Because $\psi^{(1)}_{\mathscr{P}}$ is partially computable, it *must* appear among $\psi^{(1)}_{\mathscr{P}_0}, \psi^{(1)}_{\mathscr{P}_1}, \ldots$, and, in particular, it must be $\psi^{(1)}_{\mathscr{P}_{\#(\mathscr{P})}}$, since $\mathscr{P}_{\#(\mathscr{P})}$ is $\mathscr{P}$ by definition, but we have the counterexample $\psi^{(1)}_{\mathscr{P}}(\#(\mathscr{P}))\downarrow$ if and only if $\psi^{(1)}_{\mathscr{P}_{\#(\mathscr{P})}}(\#(\mathscr{P}))\uparrow$, i.e.,

$$\text{HALT}(\#(\mathscr{P}), \#(\mathscr{P})) \Leftrightarrow \sim \text{HALT}(\#(\mathscr{P}), \#(\mathscr{P})).$$

Since we know assertion 1 to be true, and since assertion 2 depended on the assumption that HALT(x, y) is computable, HALT(x, y) cannot be computable.

To present the situation more graphically, we can represent the values of each function $\psi^{(1)}_{\mathscr{P}_0}, \psi^{(1)}_{\mathscr{P}_1}, \ldots$ by the infinite array

$$\begin{array}{cccc}
\boxed{\psi^{(1)}_{\mathscr{P}_0}(0)} & \psi^{(1)}_{\mathscr{P}_0}(1) & \psi^{(1)}_{\mathscr{P}_0}(2) & \cdots \\
\psi^{(1)}_{\mathscr{P}_1}(0) & \boxed{\psi^{(1)}_{\mathscr{P}_1}(1)} & \psi^{(1)}_{\mathscr{P}_1}(2) & \cdots \\
\psi^{(1)}_{\mathscr{P}_2}(0) & \psi^{(1)}_{\mathscr{P}_2}(1) & \boxed{\psi^{(1)}_{\mathscr{P}_2}(2)} & \cdots \\
\vdots & \vdots & \vdots &
\end{array}$$

Each row represents one function. It is along the *diagonal* of this array that we have arranged to find the counterexamples, which explains the origin of the term *diagonalization*.

We can use a similar argument to give an example of a non-r.e. set. Let TOT be the set of all numbers p such that p is the number of a program that computes a total function $f(x)$ of one variable. That is,

$$\text{TOT} = \{z \in N \mid (\forall x)\Phi(x, z)\downarrow\}.$$

Since

$$\Phi(x, z)\downarrow \Leftrightarrow x \in W_z,$$

TOT is simply the set of numbers z such that W_z is the set of all nonnegative integers.

We have

Theorem 6.1. TOT is not r.e.

Proof. Suppose that TOT were r.e. Since TOT $\neq \varnothing$, by Theorem 4.9 there is a computable function $g(x)$ such that TOT $= \{g(0), g(1), g(2), \ldots\}$. Let

$$h(x) = \Phi(x, g(x)) + 1.$$

Since each value $g(x)$ is the number of a program that computes a total function, $\Phi(u, g(x))\downarrow$ for all x, u and hence, in particular, $h(x)\downarrow$ for all x. Thus h is itself a computable function. Let h be computed by program $\mathscr{P}$, and let $p = \#(\mathscr{P})$. Then $p \in$ TOT, so that $p = g(i)$ for some i. Then

$$\begin{aligned} h(i) &= \Phi(i, g(i)) + 1 && \text{by definition of } h \\ &= \Phi(i, p) + 1 && \text{since } p = g(i) \\ &= h(i) + 1 && \text{since } h \text{ is computed by } \mathscr{P}, \end{aligned}$$

which is a contradiction. ■

Note that in the proof of Theorem 6.1, the set A is TOT itself, and this time assertion 1 was taken as an assumption, while assertion 2 is shown to be true. Theorem 6.1 helps to explain why we base the study of computability on partial functions rather than total functions. By Church's thesis, Theorem 6.1 implies that there is no algorithm to determine if an $\mathscr{S}$ program computes a total function.

Once some set such as K has been shown to be nonrecursive, we can use that set to give other examples of nonrecursive sets by way of the *reducibility* method.

Definition. Let A, B be sets. A is *many–one reducible to* B, written $A \leq_m B$, if there is a computable function f such that

$$A = \{x \in N \mid f(x) \in B\}.$$

That is, $x \in A$ if and only if $f(x) \in B$. (The word *many–one* simply refers to the fact that we do not require f to be one–one.)

If $A \leq_m B$, then in a sense testing membership in A is "no harder than" testing membership in B. In particular, to test $x \in A$, we can compute $f(x)$ and then test $f(x) \in B$.

Theorem 6.2. Suppose $A \leq_m B$.

1. If B is recursive, then A is recursive.
2. If B is r.e., then A is r.e.

Proof. Let $A = \{x \in N \mid f(x) \in B\}$, where f is computable, and let $P_B(x)$ be the characteristic function of B. Then

$$A = \{x \in N \mid P_B(f(x))\},$$

and if B is recursive then $P_B(f(x))$, the characteristic function of A, is computable.

Now suppose that B is r.e. Then $B = \{x \in N \mid g(x)\downarrow\}$ for some partially computable function g, and $A = \{x \in N \mid g(f(x))\downarrow\}$. But $g(f(x))$ is partially computable, so A is r.e. ∎

We generally use Theorem 6.2 in the form: If A is not recursive (r.e.), then B is not recursive (respectively, not r.e.). For example, let

$$K_0 = \{x \in N \mid \Phi_{r(x)}(l(x))\downarrow\} = \{\langle x, y\rangle \mid \Phi_y(x)\downarrow\}.$$

K_0 is clearly r.e. However, we can show by reducing K to K_0, that is, by showing that $K \leq_m K_0$, that K_0 is not recursive: $x \in K$ if and only if $\langle x, x\rangle \in K_0$, and the function $f(x) = \langle x, x\rangle$ is computable. In fact, it is easy to show that every r.e. set is many–one reducible to K_0: if A is r.e., then

$$\begin{aligned} A &= \{x \in N \mid g(x)\downarrow\} && \text{for some partially computable } g \\ &= \{x \in N \mid \Phi(x, z_0)\downarrow\} && \text{for some } z_0 \\ &= \{x \in N \mid \langle x, z_0\rangle \in K_0\}. \end{aligned}$$

Definition. A set A is *m-complete* if

1. A is r.e., and
2. for every r.e. set B, $B \leq_m A$.

So K_0 is m-complete. We can also show that K is m-complete. First we show that $K_0 \leq_m K$. This argument is somewhat more involved because K_0 seems, at first glance, to contain more information than K. K_0 represents the halting behavior of all partially computable functions on all inputs, while K represents only the halting behavior of partially computable functions on a single argument. We wish to take a pair $\langle n, q \rangle$ and transform it to a number $f(\langle n, q \rangle)$ of a single program such that

$$\Phi_q(n)\downarrow \quad \text{if and only if} \quad \Phi_{f(\langle n,q \rangle)}(f(\langle n, q \rangle))\downarrow,$$

i.e., such that $\langle n, q \rangle \in K_0$ if and only if $f(\langle n, q \rangle) \in K$. The parameter theorem turns out to be very useful here. Let $\mathscr{P}$ be the program

$$Y \leftarrow \Phi^{(1)}(l(X_2), r(X_2))$$

and let $p = \#(\mathscr{P})$. Then $\psi_{\mathscr{P}}(x_1, x_2) = \Phi^{(1)}(l(x_2), r(x_2))$, and

$$\psi_{\mathscr{P}}(x_1, x_2) = \Phi^{(2)}(x_1, x_2, p) = \Phi^{(1)}(x_1, S_1^1(x_2, p))$$

by the parameter theorem, so for any pair $\langle n, q \rangle$,

$$\Phi^{(1)}(n, q) = \psi_{\mathscr{P}}(x_1, \langle n, q \rangle) = \Phi^{(1)}_{S_1^1(\langle n,q \rangle, p)}(x_1). \tag{6.1}$$

Now, (6.1) holds for all values of x_1, so, in particular,

$$\Phi^{(1)}(n, q) = \Phi^{(1)}_{S_1^1(\langle n,q \rangle, p)}(S_1^1(\langle n, q \rangle, p)),$$

and therefore

$$\Phi^{(1)}(n, q)\downarrow \quad \text{if and only if} \quad \Phi^{(1)}_{S_1^1(\langle n,q \rangle, p)}(S_1^1(\langle n, q \rangle, p))\downarrow,$$

i.e.,

$$\langle n, q \rangle \in K_0 \quad \text{if and only if} \quad S_1^1(\langle n, q \rangle, p) \in K.$$

With p held constant $S_1^1(x, p)$ is a computable unary function, so $K_0 \leq_m K$.

To complete the argument that K is m-complete we need

Theorem 6.3. If $A \leq_m B$ and $B \leq_m C$, then $A \leq_m C$.

Proof. Let $A = \{x \in N \mid f(x) \in B\}$ and $B = \{x \in N \mid g(x) \in C\}$. Then $A = \{x \in N \mid g(f(x)) \in C\}$, and $g(f(x))$ is computable. ■

As an immediate consequence we have

Corollary 6.4. If A is m-complete, B is r.e., and $A \leq_m B$, then B is m-complete.

Proof. If C is r.e. then $C \leq_m A$, and $A \leq_m B$ by assumption, so $C \leq_m B$. ■

Thus, K is m-complete. Informally, testing membership in an m-complete set is "at least as difficult as" testing membership in any r.e. set. So an m-complete set is a good choice for showing by a reducibility argument that a given set is not computable. We expand on this subject in Chapter 8.

Actually, we have shown both $K \leq_m K_0$ and $K_0 \leq_m K$, so in a sense, testing membership in K and testing membership in K_0 are "equally difficult" problems.

Definition. $A \equiv_m B$ means that $A \leq_m B$ and $B \leq_m A$.

In general, for sets A and B, if $A \equiv_m B$ then testing membership in A has the "same difficulty as" testing membership in B.

To summarize, we have proved

Theorem 6.5.

1. K and K_0 are m-complete.
2. $K \equiv_m K_0$.

We can also use reducibility arguments to show that certain sets are not r.e. Let

$$\text{EMPTY} = \{x \in N \mid W_x = \varnothing\}.$$

Theorem 6.6. EMPTY is not r.e.

Proof. We will show that $\bar{K} \leq_m$ EMPTY. $\bar{K}$ is not r.e., so by Theorem 6.2, EMPTY is not r.e. Let $\mathscr{P}$ be the program

$$Y \leftarrow \Phi(X_2, X_2),$$

and let $p = \#(\mathscr{P})$. $\mathscr{P}$ ignores its first argument, so for a given z,

$$\psi_{\mathscr{P}}^{(2)}(x, z)\downarrow \text{ for all } x \quad \text{if and only if} \quad \Phi(z, z)\downarrow.$$

By the parameter theorem

$$\psi_{\mathscr{P}}^{(2)}(x_1, x_2) = \Phi^{(2)}(x_1, x_2, p) = \Phi^{(1)}(x_1, S_1^1(x_2, p)),$$

so, for any z,

$$
\begin{aligned}
z \in \bar{K} \quad & \text{if and only if} \quad \Phi(z, z)\uparrow \\
& \text{if and only if} \quad \psi_{\mathscr{P}}^{(2)}(x, z)\uparrow \text{ for all } x \\
& \text{if and only if} \quad \Phi^{(1)}(x, S_1^1(z, p))\uparrow \text{ for all } x \\
& \text{if and only if} \quad W_{S_1^1(z, p)} = \varnothing \\
& \text{if and only if} \quad S_1^1(z, p) \in \text{EMPTY}.
\end{aligned}
$$

$f(z) = S_1^1(z, p)$ is computable, so $\bar{K} \leq_m$ EMPTY. ■

Exercises

1. Show that the proof of Theorem 4.7 is a diagonalization argument.
2. Prove by diagonalization that there is no enumeration $f_0, f_1, f_2, \ldots$ of all total unary (not necessarily computable) functions on N.
3. Let $A = \{x \in N \mid \Phi_x(x)\downarrow \text{ and } \Phi_x(x) > x\}$.
   (a) Show that A is r.e.
   (b) Show by diagonalization that A is not recursive.
4. Show how the diagonalization argument in the proof of Theorem 6.1 fails for the set of all numbers p such that p is the number of a program that computes a partial function, i.e., the set N.
5. Let A, B be sets of numbers. Prove
   (a) $A \leq_m A$.
   (b) $A \leq_m B$ if and only if $\bar{A} \leq_m \bar{B}$.
6. Prove that no m-complete set is recursive.
7. Let A, B be m-complete. Show that $A \equiv_m B$.
8. Prove that $\bar{K} \not\leq_m K$, i.e., $\bar{K}$ is not many–one reducible to K.
9. For every number n, let $A_n = \{x \mid n \in W_x\}$.
   (a) Show that A_i is r.e. but not recursive, for all i.
   (b) Show that $A_i \equiv_m A_j$ for all i, j.
10. Define the predicate $P(x) \Leftrightarrow \Phi_x(x) = 1$. Show that $P(x)$ is not computable.
11. Define the predicate

$$Q(x) \Leftrightarrow \text{the variable } Y \text{ assumes the value 1 sometime during the computation of } \psi_{\mathscr{P}}(x), \text{ where } \#(\mathscr{P}) = x.$$

Show that $Q(x)$ is not computable. [*Hint:* Use the parameter theorem and a version of the universal program $\mathscr{U}_1$.]

12. Let INF $= \{x \in N \mid W_x$ is infinite$\}$. Show that INF $\equiv_m$ TOT.

13. Let FIN $= \{x \in N \mid W_x$ is finite$\}$. Show that $\bar{K} \leq_m$ FIN.

14.* Let

$$\text{MONOTONE} = \{y \in N \mid \Phi_y(x) \text{ is total and } \Phi_y(x) \leq \Phi_y(x+1) \text{ for all } x\}.$$

(a) Show by diagonalization that MONOTONE is not r.e.
(b) Show that MONOTONE $\equiv_m$ TOT.

7. Rice's Theorem

Using the reducibility method we can prove a theorem that gives us, at a single stroke, a wealth of interesting unsolvable problems concerning programs.

Let Γ be some collection of partially computable functions of one variable. We may associate with Γ the set (usually called an *index set*)

$$R_\Gamma = \{t \in N \mid \Phi_t \in \Gamma\}.$$

R_Γ is a recursive set just in case the predicate $g(t)$, defined $g(t) \Leftrightarrow \Phi_t \in \Gamma$, is computable. Consider the examples:

1. Γ is the set of computable functions;
2. Γ is the set of primitive recursive functions;
3. Γ is the set of partially computable functions that are defined for all but a finite number of values of x.

These examples make it plain that it would be interesting to be able to show that R_Γ is computable for various collections Γ. Invoking Church's thesis, we can say that R_Γ is a recursive set just in case there is an algorithm that accepts *programs* $\mathscr{P}$ as input and returns the value TRUE or FALSE depending on whether or not the function $\psi_{\mathscr{P}}^{(1)}$ does or does not belong to Γ. In fact, those who work with computer programs would be very pleased to possess algorithms that accept a program as input and which return as output some useful property of the partial function computed by that program. Alas, such algorithms are not to be found! This dismal conclusion follows from Rice's theorem.

Theorem 7.1 (Rice's Theorem). Let Γ be a collection of partially computable functions of one variable. Let there be partially computable functions $f(x)$, $g(x)$ such that $f(x)$ belongs to Γ but $g(x)$ does not. Then R_Γ is not recursive.

Proof. Let $h(x)$ be the function such that $h(x)\uparrow$ for all x. We assume first that $h(x)$ does not belong to Γ. Let q be the number of

$$Z \leftarrow \Phi(X_2, X_2)$$
$$Y \leftarrow f(X_1)$$

Then, for any i, $S_1^1(i, q)$ is the number of

$$X_2 \leftarrow i$$
$$Z \leftarrow \Phi(X_2, X_2)$$
$$Y \leftarrow f(X_1)$$

Now

$$\begin{aligned} i \in K \quad &\textit{implies} \quad \Phi(i,i)\downarrow \\ &\textit{implies} \quad \Phi_{S_1^1(i,q)}(x) = f(x) \text{ for all } x \\ &\textit{implies} \quad \Phi_{S_1^1(i,q)} \in \Gamma \\ &\textit{implies} \quad S_1^1(i,q) \in R_\Gamma, \end{aligned}$$

and

$$\begin{aligned} i \notin K \quad &\textit{implies} \quad \Phi(i,i)\uparrow \\ &\textit{implies} \quad \Phi_{S_1^1(i,q)}(x)\uparrow \text{ for all } x \\ &\textit{implies} \quad \Phi_{S_1^1(i,q)} = h \\ &\textit{implies} \quad \Phi_{S_1^1(i,q)} \notin \Gamma \\ &\textit{implies} \quad S_1^1(i,q) \notin R_\Gamma, \end{aligned}$$

so $K \leq_m R_\Gamma$. By Theorem 6.2, R_Γ is not recursive.

If $h(x)$ does belong to Γ, then the same argument with Γ and $f(x)$ replaced by $\overline{\Gamma}$ and $g(x)$ shows that $R_{\overline{\Gamma}}$ is not recursive. But $R_{\overline{\Gamma}} = \overline{R_\Gamma}$, so, by Theorem 4.1, R_Γ is not recursive in this case either. ■

Corollary 7.2. There are no algorithms for testing a given program $\mathscr{P}$ of the language $\mathscr{S}$ to determine whether $\psi_{\mathscr{P}}^{(1)}(x)$ belongs to any of the classes described in Examples 1–3.

Proof. In each case we only need find the required functions $f(x)$, $g(x)$ to show that R_Γ is not recursive. The corollary then follows by Church's thesis. For 1, 2, or 3 we can take, for example, $f(x) = u_1^1(x)$ and $g(x) = 1 - x$ [so that $g(x)$ is defined only for $x = 0, 1$]. ■

Exercises

1. Show that Rice's theorem is false if the requirement for functions $f(x), g(x)$ is omitted.
2. Show there is no algorithm to determine of a given program $\mathscr{P}$ in the language $\mathscr{S}$ whether $\psi_{\mathscr{P}}(x) = x^2$ for all x.
3. Show that there is no algorithm to determine of a pair of numbers u, v whether $\Phi_u(x) = \Phi_v(x)$ for all x.
4. Show that the set $A = \{x \mid \Phi_x$ is defined for at least one input$\}$ is r.e. but not recursive.
5. Use Rice's theorem to show that the following sets are not recursive. [See Section 6 for the definitions of the sets.]
   (a) TOT;
   (b) EMPTY;
   (c) INF;
   (d) FIN;
   (e) MONOTONE;
   (f) $\{y \in N \mid \Phi_y^{(1)}$ is a predicate$\}$.
6. Let Γ be a collection of partially computable functions of m variables, $m > 1$, and let $R_\Gamma^{(m)} = \{t \in N \mid \Phi_t^{(m)} \in \Gamma\}$. State and prove a version of Rice's theorem for collections of partially computable functions of m variables, $m > 1$.
7. Define the predicate

   PROPER(n) $\Leftrightarrow$ $\min_z \left[\Phi_n^{(2)}(x, z) = 3\right]$ is an application of proper minimalization to the predicate $\Phi_n^{(2)}(x, z) = 3$.

   Show that PROPER(x) is not computable.
8. Let Γ be a set of partially computable functions of one variable such that $\varnothing \subset R_\Gamma \subset N$. Show that R_Γ is r.e. if and only if it is m-complete.

*8. The Recursion Theorem

In the proof that HALT(x, y) is not computable, we gave (assuming HALT(x, y) to be computable) a program $\mathscr{P}$ such that

$$\text{HALT}(\#(\mathscr{P}), \#(\mathscr{P})) \Leftrightarrow \sim \text{HALT}(\#(\mathscr{P}), \#(\mathscr{P})).$$

We get a contradiction when we consider the behavior of the program $\mathscr{P}$ on input $\#(\mathscr{P})$. The phenomenon of a program acting on its own description is sometimes called *self-reference*, and it is the source of many fundamental results in computability theory. Indeed, the whole point of diagonalization in the proof of Theorem 2.1 is to get a contradictory self-reference. We turn now to a theorem which packages, so to speak, a general technique for obtaining self-referential behavior. It is one of the most important applications of the parameter theorem.

Theorem 8.1 (Recursion Theorem). Let $g(z, x_1, \ldots, x_m)$ be a partially computable function of $m + 1$ variables. Then there is a number e such that

$$\Phi_e^{(m)}(x_1, \ldots, x_m) = g(e, x_1, \ldots, x_m).$$

Discussion. Let $e = \#(\mathscr{P})$, so that $\psi_{\mathscr{P}}^{(m)}(x_1, \ldots, x_m) = \Phi_e^{(m)}(x_1, \ldots, x_m)$. The equality in the theorem says that the m-ary function $\psi_{\mathscr{P}}^{(m)}(x_1, \ldots, x_m)$ is equal to $g(z, x_1, \ldots, x_m)$ when the first argument of g is held constant at e. That is, $\mathscr{P}$ is a program that, in effect, gets access to its own number, e, and computes the m-ary function $g(e, x_1, \ldots, x_m)$. Note that since $x_1, \ldots, x_m$ can be arbitrary values, e generally does not appear among the inputs to $\psi_{\mathscr{P}}^{(m)}(x_1, \ldots, x_m)$, so $\mathscr{P}$ must somehow compute e. One might suppose that $\mathscr{P}$ might contain e copies of an instruction such as $Z \leftarrow Z + 1$, that is, an expansion of the macro $Z \leftarrow e$, but if $\mathscr{P}$ has at least e instructions, then certainly $\#(\mathscr{P}) > e$. The solution is to write $\mathscr{P}$ so that it computes e without having e "built in" to the program. In particular, we build into $\mathscr{P}$ a "partial description" of $\mathscr{P}$, and then have $\mathscr{P}$ compute e from the partial description. Let $\mathscr{Q}$ be the program

$$Z \leftarrow S_m^1(X_{m+1}, X_{m+1})$$
$$Y \leftarrow g(Z, X_1, \ldots, X_m)$$

We prefix $\#(\mathscr{Q})$ copies of the instruction $X_{m+1} \leftarrow X_{m+1} + 1$ to get the program $\mathscr{R}$:

$$X_{m+1} \leftarrow X_{m+1} + 1$$
$$\vdots \qquad \vdots$$
$$X_{m+1} \leftarrow X_{m+1} + 1$$
$$Z \leftarrow S_m^1(X_{m+1}, X_{m+1})$$
$$Y \leftarrow g(Z, X_1, \ldots, X_m)$$

After the first $\#(\mathscr{Q})$ instructions are executed, X_{m+1} holds the value $\#(\mathscr{Q})$, and $S_m^1(\#(\mathscr{Q}), \#(\mathscr{Q}))$, as defined in the proof of the parameter theorem, computes the number of the program consisting of $\#(\mathscr{Q})$ copies of $X_{m+1} \leftarrow X_{m+1} + 1$ followed by program $\mathscr{Q}$. *But that program is* $\mathscr{R}$. So $Z \leftarrow S_m^1(X_{m+1}, X_{m+1})$ gives Z the value $\#(\mathscr{R})$, and $Y \leftarrow g(Z, X_1, \ldots, X_m)$ causes $\mathscr{R}$ to output $g(\#(\mathscr{R}), x_1, \ldots, x_m)$. We take e to be $\#(\mathscr{R})$ and we have

$$\Phi_e^{(m)}(x_1, \ldots, x_m) = \psi_{\mathscr{R}}^{(m)}(x_1, \ldots, x_m) = g(e, x_1, \ldots, x_m).$$

We now formalize this argument.

Proof. Consider the partially computable function

$$g(S_m^1(v, v), x_1, \ldots, x_m)$$

where S_m^1 is the function that occurs in the parameter theorem. Then we have for some number z_0,

$$\begin{aligned} g(S_m^1(v, v), x_1, \ldots, x_m) &= \Phi^{(m+1)}(x_1, \ldots, x_m, v, z_0) \\ &= \Phi^{(m)}(x_1, \ldots, x_m, S_m^1(v, z_0)), \end{aligned}$$

where we have used the parameter theorem. Setting $v = z_0$ and $e = S_m^1(z_0, z_0)$, we have

$$g(e, x_1, \ldots, x_m) = \Phi^{(m)}(x_1, \ldots, x_m, e) = \Phi_e^{(m)}(x_1, \ldots, x_m). \qquad ■$$

We can use the recursion theorem to give another self-referential proof that HALT(x, y) is not computable. If HALT(x, y) were computable, then

$$f(x, y) = \begin{cases} \uparrow & \text{if HALT}(y, x) \\ 0 & \text{otherwise} \end{cases}$$

would be partially computable, so by the recursion theorem there would be a number e such that

$$\Phi_e(y) = f(e, y) = \begin{cases} \uparrow & \text{if HALT}(y, e) \\ 0 & \text{otherwise,} \end{cases}$$

that is,

$$\sim \text{HALT}(y, e) \Leftrightarrow \text{HALT}(y, e).$$

So HALT(x, y) is not computable. The self-reference occurs when Φ_e computes e, tests HALT(y, e), and then does the opposite of what HALT(y, e) says it does.

One of the many uses of the recursion theorem is to allow us to write down definitions of functions that involve the program used to compute the function as part of its definition. For a simple example we give

Corollary 8.2. There is a number e such that for all x

$$\Phi_e(x) = e.$$

Proof. We consider the computable function

$$g(z,x) = u_1^2(z,x) = z.$$

Applying the recursion theorem we obtain a number e such that

$$\Phi_e(x) = g(e,x) = e$$

and we are done. ■

It is tempting to be a little metaphorical about this result. The program with number e "consumes" its "environment" (i.e., the input x) and outputs a "copy" of itself. That is, it is, in miniature, a self-reproducing organism. This program has often been cited in considerations of the comparison between living organisms and machines.

For another example, let

$$f(x,t) = \begin{cases} k & \text{if } t = 0 \\ g(t \dot{-} 1, \Phi_x(t \dot{-} 1)) & \text{otherwise,} \end{cases}$$

where $g(x,y)$ is computable. It is clear that $f(x,t)$ is partially computable, so by the recursion theorem there is a number e such that

$$\Phi_e(t) = f(e,t) = \begin{cases} k & \text{if } t = 0 \\ g(t \dot{-} 1, \Phi_e(t \dot{-} 1)) & \text{otherwise.} \end{cases}$$

An easy induction argument on t shows that Φ_e is a total, and therefore computable, function. Now, Φ_e satisfies the equations

$$\Phi_e(0) = k$$

$$\Phi_e(t+1) = g(t, \Phi_e(t)),$$

that is, Φ_e is obtained from g by primitive recursion of the form (2.1) in Chapter 3, so the recursion theorem gives us another proof of Theorem 2.1 in Chapter 3. In fact, the recursion theorem can be used to justify definitions based on much more general forms of recursion, which explains how it came by its name.[1] We give one more example, in which we wish to

[1] For more on this subject, see Part 5.

know if there are partially computable functions f, g that satisfy the equations

$$\begin{aligned} f(0) &= 1 \\ f(t+1) &= g(2t) + 1 \\ g(0) &= 3 \\ g(2t+2) &= f(t) + 2. \end{aligned} \tag{8.1}$$

Let $F(z, t)$ be the partially computable function

$$F(z,x) = \begin{cases} 1 & \text{if } x = \langle 0,0\rangle \\ \Phi_z(\langle 1, 2(r(x) \dot{-} 1)\rangle) + 1 & \text{if } (\exists y)_{\le x}(x = \langle 0, y+1\rangle) \\ 3 & \text{if } x = \langle 1,0\rangle \\ \Phi_z(\langle 0, \lfloor (r(x) \dot{-} 2)/2 \rfloor\rangle) + 2 & \text{if } (\exists y)_{\le x}(x = \langle 1, 2y+2\rangle). \end{cases}$$

By the recursion theorem there is a number e such that

$$\begin{aligned} \Phi_e(x) &= F(e,x) \\ &= \begin{cases} 1 & \text{if } x = \langle 0,0\rangle \\ \Phi_e(\langle 1, 2(r(x) \dot{-} 1)\rangle) + 1 & \text{if } (\exists y)_{\le x}(x = \langle 0, y+1\rangle) \\ 3 & \text{if } x = \langle 1,0\rangle \\ \Phi_e(\langle 0, \lfloor (r(x) \dot{-} 2)/2 \rfloor\rangle) + 2 & \text{if } (\exists y)_{\le x}(x = \langle 1, 2y+2\rangle). \end{cases} \end{aligned}$$

Now, setting

$$f(x) = \Phi_e(\langle 0, x\rangle) \quad \text{and} \quad g(x) = \Phi_e(\langle 1, x\rangle)$$

we have

$$\begin{aligned} f(0) &= \Phi_e(\langle 0,0\rangle) = 1 \\ f(t+1) &= \Phi_e(\langle 0, t+1\rangle) = \Phi_e(\langle 1, 2t\rangle) + 1 = g(2t) + 1 \\ g(0) &= \Phi_e(\langle 1,0\rangle) = 3 \\ g(2t+2) &= \Phi_e(\langle 1, 2t+2\rangle) = \Phi_e(\langle 0, t\rangle) + 2 = f(t) + 2, \end{aligned}$$

so f, g satisfy (8.1).

Another application of the recursion theorem is

Theorem 8.3 (Fixed Point Theorem). Let $f(z)$ be a computable function. Then there is a number e such that

$$\Phi_{f(e)}(x) = \Phi_e(x)$$

for all x.

Proof. Let $g(z, x) = \Phi_{f(z)}(x)$, a partially computable function. By the recursion theorem, there is a number e such that

$$\Phi_e(x) = g(e, x) = \Phi_{f(e)}(x). \qquad \blacksquare$$

Usually a number n is considered to be a fixed point of a function $f(x)$ if $f(n) = n$. Clearly there are computable functions that have no fixed point in this sense, e.g., $s(x)$. The fixed point theorem says that for every computable function $f(x)$, there is a number e of a program that *computes the same function as* the program with number $f(e)$.

For example, let $P(x)$ be a computable predicate, let $g(x)$ be a computable function, and let while$(n) = \#(\mathscr{Q}_n)$, where $\mathscr{Q}_n$ is the program

$$\begin{array}{l} X_2 \leftarrow n \\ Y \leftarrow X \\ \text{IF } \sim P(Y) \text{ GOTO } E \\ Y \leftarrow \Phi_{X_2}(g(Y)) \end{array}$$

It should be clear that while(x) is a computable, in fact primitive recursive, function, so by the fixed point theorem there is a number e such that

$$\Phi_e(x) = \Phi_{\text{while}(e)}(x).$$

It follows from the construction of while(e) that

$$\Phi_e(x) = \Phi_{\text{while}(e)}(x) = \begin{cases} x & \text{if } \sim P(x) \\ \Phi_e(g(x)) & \text{otherwise.} \end{cases}$$

Moreover,

$$\Phi_e(g(x)) = \Phi_{\text{while}(e)}(g(x)) = \begin{cases} g(x) & \text{if } \sim P(g(x)) \\ \Phi_e(g(g(x))) & \text{otherwise,} \end{cases}$$

so

$$\Phi_e(x) = \Phi_{\text{while}(e)}(x) = \begin{cases} x & \text{if } \sim P(x) \\ g(x) & \text{if } P(x) \mathbin{\&} \sim P(g(x)) \\ \Phi_e(g(g(x))) & \text{otherwise,} \end{cases}$$

and continuing in this fashion we get

$$\Phi_e(x) = \Phi_{\mathrm{while}(e)}(x) = \begin{cases} x & \text{if } \sim P(x) \\ g(x) & \text{if } P(x) \,\&\, \sim P(g(x)) \\ g(g(x)) & \text{if } P(x) \,\&\, P(g(x)) \,\&\, \sim P(g(g(x))) \\ \vdots & \vdots \end{cases}$$

In other words, program e behaves like the pseudo-program

```
Y ← X
WHILE P(Y) DO
    Y ← g(Y)
END
```

We end this discussion of the recursion theorem by giving another proof of Rice's theorem. Let Γ, $f(x)$, $g(x)$ be as in the statement of Theorem 7.1.

Alternative Proof of Rice's Theorem.[2] Suppose that R_Γ were computable. Let

$$P_\Gamma(t) = \begin{cases} 1 & \text{if } t \in R_\Gamma \\ 0 & \text{otherwise.} \end{cases}$$

That is, P_Γ is the characteristic function of R_Γ. Let

$$h(t, x) = \begin{cases} g(x) & \text{if } t \in R_\Gamma \\ f(x) & \text{otherwise.} \end{cases}$$

Then, since (as in the proof of Theorem 5.4, Chapter 3)

$$h(t, x) = g(x) \cdot P_\Gamma(t) + f(x) \cdot \alpha(P_\Gamma(t)),$$

$h(t, x)$ is partially computable. Thus, by the recursion theorem, there is a number e such that

$$\Phi_e(x) = h(e, x) = \begin{cases} g(x) & \text{if } \Phi_e \text{ belongs to } \Gamma \\ f(x) & \text{otherwise.} \end{cases}$$

---

[2] This elegant proof was called to our attention by John Case.

Does e belong to R_Γ? Recalling that $f(x)$ belongs to Γ but $g(x)$ does not, we have

$$\begin{aligned} e \in R_\Gamma \quad & \textit{implies} \quad \Phi_e(x) = g(x) \\ & \textit{implies} \quad \Phi_e \text{ is not in } \Gamma \\ & \textit{implies} \quad e \notin R_\Gamma. \end{aligned}$$

But likewise,

$$\begin{aligned} e \notin R_\Gamma \quad & \textit{implies} \quad \Phi_e(x) = f(x) \\ & \textit{implies} \quad \Phi_e \text{ is in } \Gamma \\ & \textit{implies} \quad e \in R_\Gamma. \end{aligned}$$

This contradiction proves the theorem. ■

Exercises

1. Use the proof of Corollary 8.2 and the discussion preceding the proof of the recursion theorem to write a program $\mathscr{P}$ such that $\psi_{\mathscr{P}}(x) = \#(\mathscr{P})$.
2. Let $A = \{x \in N \mid \Phi_x(x)\downarrow \text{ and } \Phi_x(x) > x\}$. Use the recursion theorem to show that A is not recursive.
3. Show that there is a number e such that $W_e = \{e\}$.
4. Show that there is a program $\mathscr{P}$ such that $\psi_{\mathscr{P}}(x)\downarrow$ if and only if $x = \#(\mathscr{P})$.
5. **(a)** Show that there is a partially computable function f that satisfies the equations

$$\begin{aligned} f(x,0) &= x + 2 \\ f(x,1) &= 2 \cdot f(x,2x) \\ f(x,2t+2) &= 3 \cdot f(x,2t) \\ f(x,2t+3) &= 4 \cdot f(x,2t+1). \end{aligned}$$

   What is $f(2,5)$?

   (b) Prove that f is total.

   (c) Prove that f is unique. (That is, only one function satisfies the given equations.)
6. Give two distinct partially computable functions f, g that satisfy the equations

$$\begin{aligned} f(0) &= 2 & \qquad g(0) &= 2 \\ f(2t+2) &= 3 \cdot f(2t) & \qquad g(2t+2) &= 3 \cdot g(2t). \end{aligned}$$

   For the specific functions f, g that you give, what are $f(1)$ and $g(1)$?

7. Let $f(x) = x + 1$. Use the proof of the fixed point theorem and the discussion preceding the proof of the recursion theorem to give a program $\mathscr{P}$ such that $\Phi_{\#(\mathscr{P})}(x) = \Phi_{f(\#(\mathscr{P}))}(x)$. What unary function does $\mathscr{P}$ compute?

8. Give a function $f(y)$ such that, for all y, $f(y) > y$ and $\Phi_y(x) = \Phi_{f(y)}(x)$.

9. Give a function $f(y)$ such that, for all y, if $\Phi_y(x) = \Phi_{f(y)}(x)$, then $\Phi_y(x)$ is not total.

10. Show that the function while(x) defined following the fixed point theorem is primitive recursive. [*Hint:* Use the parameter theorem.]

11. (a) Prove that the recursion theorem can be strengthened to read: There are infinitely many numbers e such that

$$\Phi_e^{(m)}(x_1, \ldots, x_m) = g(e, x_1, \ldots, x_m).$$

(b) Prove that the fixed point theorem can be strengthened to read: There are infinitely many numbers e such that

$$\Phi_{f(e)}(x) = \Phi_e(x).$$

12. Prove the following version of the recursion theorem: There is a primitive recursive function self(x) such that for all z

$$\Phi_{\text{self}(z)}(x) = \Phi_z^{(2)}(\text{self}(z), x).$$

13. Prove the following version of the fixed point theorem: There is a primitive recursive function fix(u) such that for all x, u,

$$\Phi_{\text{fix}(u)}(x) = \Phi_{\Phi_u(\text{fix}(u))}(x).$$

14.* Let S be an acceptable programming system with universal functions $\Psi^{(m)}$. Prove the following: For every partially computable function $g(z, x_1, \ldots, x_m)$ there is a number e such that

$$\Psi_e^{(m)}(x_1, \ldots, x_m) = g(e, x_1, \ldots, x_m).$$

That is, a version of the recursion theorem holds for S. [See Exercise 5.4 for the definition of acceptable programming systems.]

*9. A Computable Function That Is Not Primitive Recursive

In Chapter 3 we showed that all primitive recursive functions are computable, but we did not settle the question of whether all computable functions are primitive recursive. We shall deal with this matter by showing how to obtain a function $h(x)$ that is computable but is not primitive recursive. Our method will be to construct a computable function $\phi(t, x)$ that enumerates all of the unary primitive recursive functions. That is, it will be the case that

1. for each fixed value $t = t_0$, the function $\phi(t_0, x)$ will be primitive recursive;
2. for each unary primitive recursive function $f(x)$, there will be a number t_0 such that $f(x) = \phi(t_0, x)$.

Once we have this function ϕ at our disposal, we can diagonalize, obtaining the unary computable function $\phi(x, x) + 1$ which must be different from all primitive recursive functions. (If it were primitive recursive, we would have

$$\phi(x, x) + 1 = \phi(t_0, x)$$

for some fixed t_0, and setting $x = t_0$ would lead to a contradiction.)

We will obtain our enumerating function by giving a new characterization of the unary primitive recursive functions. However, we begin by showing how to reduce the number of parameters needed in the operation of primitive recursion which, as defined in Chapter 3 (Eq. (2.2)), proceeds from the total n-ary function f and the total $n + 2$-ary function g to yield the $n + 1$-ary function h such that

$$h(x_1, \ldots, x_n, 0) = f(x_1, \ldots, x_n)$$

$$h(x_1, \ldots, x_n, t + 1) = g(t, h(x_1, \ldots, x_n, t), x_1, \ldots, x_n).$$

If $n > 1$ we can reduce the number of parameters needed from n to $n - 1$ by using the pairing functions. That is, let

$$\tilde{f}(x_1, \ldots, x_{n-1}) = f(x_1, \ldots, x_{n-2}, l(x_{n-1}), r(x_{n-1})),$$

$$\tilde{g}(t, u, x_1, \ldots, x_{n-1}) = g(t, u, x_1, \ldots, x_{n-2}, l(x_{n-1}), r(x_{n-1})),$$

$$\tilde{h}(x_1, \ldots, x_{n-1}, t) = h(x_1, \ldots, x_{n-2}, l(x_{n-1}), r(x_{n-1}), t).$$

Then, we have

$$\tilde{h}(x_1, \ldots, x_{n-1}, 0) = \tilde{f}(x_1, \ldots, x_{n-1})$$

$$\tilde{h}(x_1, \ldots, x_{n-1}, t + 1) = \tilde{g}\big(t, \tilde{h}(x_1, \ldots, x_{n-1}, t), x_1, \ldots, x_{n-1}\big).$$

Finally, we can retrieve the original function h from the equation

$$h(x_1, \ldots, x_n, t) = \tilde{h}(x_1, \ldots, x_{n-2}, \langle x_{n-1}, x_n \rangle, t).$$

By iterating this process we can reduce the number of parameters to 1, that is, to recursions of the form

$$\begin{aligned} h(x,0) &= f(x) \\ h(x,t+1) &= g(t,h(x,t),x) \end{aligned} \tag{9.1}$$

Recursions with no parameters, as in Eq. (2.1) in Chapter 3, can also readily be put into the form (9.1). Namely, to deal with

$$\begin{aligned} \psi(0) &= k \\ \psi(t+1) &= \theta(t,\psi(t)), \end{aligned}$$

we set $f(x) = k$ (which can be obtained by k compositions with $s(x)$ beginning with $n(x)$) and

$$g(x_1,x_2,x_3) = \theta(u_1^3(x_1,x_2,x_3),u_2^3(x_1,x_2,x_3))$$

in the recursion (9.1). Then, $\psi(t) = h(x,t)$ for all x. In particular, $\psi(t) = h(u_1^1(t),u_1^1(t))$.

We can simplify recursions of the form (9.1) even further by using the pairing functions to combine arguments. Namely, we set

$$\tilde{h}(x,t) = \langle h(x,t),\langle x,t\rangle\rangle.$$

Then, we have

$$\begin{aligned} \tilde{h}(x,0) &= \langle f(x),\langle x,0\rangle\rangle \\ \tilde{h}(x,t+1) &= \langle h(x,t+1),\langle x,t+1\rangle\rangle \\ &= \langle g(t,h(x,t),x),\langle x,t+1\rangle\rangle \\ &= \tilde{g}(\tilde{h}(x,t)), \end{aligned}$$

where

$$\tilde{g}(u) = \langle g(r(r(u)),l(u),l(r(u))),\langle l(r(u)),r(r(u))+1\rangle\rangle.$$

Once again, the original function h can be retrieved from $\tilde{h}$; we can use the equation

$$h(x,t) = l(\tilde{h}(x,t)).$$

Now this reduction in the complexity of recursions was only possible using the pairing functions. Nevertheless, we can use it to get a simplified characterization of the class of primitive recursive functions by adding the pairing functions to our initial functions. We may state the result as a theorem.

Theorem 9.1. The primitive recursive functions are precisely the functions obtainable from the initial functions

$$s(x), n(x), l(z), r(z), \langle x, y \rangle \quad \text{and} \quad u_i^n, \quad 1 \le i \le n$$

using the operations of composition and primitive recursion of the particular form

$$h(x,0) = f(x)$$
$$h(x, t+1) = g(h(x,t)).$$

The promised characterization of the unary primitive recursive functions is as follows.

Theorem 9.2. The unary primitive recursive functions are precisely those obtained from the initial functions $s(x) = x + 1$, $n(x) = 0$, $l(x)$, $r(x)$ by applying the following three operations on unary functions:

1. to go from $f(x)$ and $g(x)$ to $f(g(x))$;
2. to go from $f(x)$ and $g(x)$ to $\langle f(x), g(x) \rangle$;
3. to go from $f(x)$ and $g(x)$ to the function defined by the recursion

$$h(0) = 0$$
$$h(t+1) = \begin{cases} f\left(\dfrac{t}{2}\right) & \text{if } t+1 \text{ is odd,} \\ g\left(h\left(\dfrac{t+1}{2}\right)\right) & \text{if } t+1 \text{ is even.} \end{cases}$$

Proof. Let us write **PR** for the set of all functions obtained from the initial functions listed in the theorem using operations 1 through 3. We will show that **PR** is precisely the set of unary primitive recursive functions.

To see that all the functions in **PR** are primitive recursive, it is necessary only to consider operation 3. That is, we need to show that if f and g are primitive recursive, and h is obtained using operation 3, then h is also primitive recursive. What is different about operation 3 is that $h(t+1)$ is computed, not from $h(t)$ but rather from $h(t/2)$ or $h((t+1)/2)$, depending on whether t is even or odd. To deal with this we make use of Gödel numbering, setting

$$\tilde{h}(0) = 0,$$
$$\tilde{h}(n) = [h(0), \ldots, h(n-1)] \text{ if } n > 0.$$

We will show that $\tilde{h}$ is primitive recursive and then conclude that the same is true of h by using the equation[3]

$$h(n) = (\tilde{h}(n+1))_{n+1}.$$

Then (recalling that p_n is the nth prime number) we have

$$\begin{aligned}\tilde{h}(n+1) &= \tilde{h}(n) \cdot p_{n+1}^{h(n)} \\ &= \begin{cases} \tilde{h}(n) \cdot p_{n+1}^{f(\lfloor n/2 \rfloor)} & \text{if } n \text{ is odd,} \\ \tilde{h}(n) \cdot p_{n+1}^{g((\tilde{h}(n))_{\lfloor n/2 \rfloor})} & \text{if } n \text{ is even.} \end{cases}\end{aligned}$$

Here, we have used $\lfloor n/2 \rfloor$ because it gives the correct value whether n is even or odd and because we know from Chapter 3 that it is primitive recursive.

Next we will show that every unary primitive recursive function belongs to **PR**. For this purpose we will call a function $g(x_1, \ldots, x_n)$ *satisfactory* if it has the property that for any unary functions $h_1(t), \ldots, h_n(t)$ that belong to **PR**, the function $g(h_1(t), \ldots, h_n(t))$ also belongs to **PR**. Note that a *unary* function $g(t)$ that is satisfactory must belong to **PR** because $g(t) = g(u_1^1(t))$ and $u_1^1(t) = \langle l(t), r(t) \rangle$ belongs to **PR**. Thus, we can obtain our desired result by proving that all primitive recursive functions are satisfactory.[4]

We shall use the characterization of the primitive recursive functions of Theorem 9.1. Among the initial functions, we need consider only the pairing function $\langle x_1, x_2 \rangle$ and the projection functions u_i^n where $1 \leq i \leq n$. If $h_1(t)$ and $h_2(t)$ are in **PR**, then using operation 2 in the definition of **PR**, we see that $\langle h_1(t), h_2(t) \rangle$ is also in **PR**. Hence, $\langle x_1, x_2 \rangle$ is satisfactory. And evidently, if $h_1(t), \ldots, h_n(t)$ belong to **PR**, then $u_i^n(h_1(t), \ldots, h_n(t))$, which is simply equal to $h_i(t)$, certainly belongs to **PR**, so u_i^n is satisfactory.

To deal with composition, let

$$h(x_1, \ldots, x_n) = f(g_1(x_1, \ldots, x_n), \ldots, g_k(x_1, \ldots, x_n))$$

where $g_1, \ldots, g_k$ and f are satisfactory. Let $h_1(t), \ldots, h_n(t)$ be given functions that belong to **PR**. Then, setting

$$\tilde{g}_i(t) = g_i(h_1(t), \ldots, h_n(t))$$

---

[3] This is a general technique for dealing with recursive definitions for a given value in terms of smaller values, so-called *course-of-value recursions*. See Exercise 8.5 in Chapter 3.

[4] This is an example of what was called an *induction loading device* in Chapter 1.

for $1 \le i \le k$ we see that each $\tilde{g}_i$ belongs to **PR**. Hence

$$h(h_1(t), \ldots, h_n(t)) = f(\tilde{g}_1(t), \ldots, \tilde{g}_n(t))$$

belongs to **PR**, and so, h is satisfactory.

Finally, let

$$h(x, 0) = f(x)$$
$$h(x, t + 1) = g(h(x, t))$$

where f and g are satisfactory. Let $\psi(0) = 0$ and let $\psi(t + 1) = h(r(t), l(t))$. Recalling that

$$\langle a, b \rangle = 2^a(2b + 1) - 1,$$

we consider two cases according to whether $t + 1 = 2^a(2b + 1)$ is even or odd. If $t + 1$ is even, then $a > 0$ and

$$\begin{aligned} \psi(t + 1) &= h(b, a) \\ &= g(h(b, a - 1)) \\ &= g(\psi(2^{a-1}(2b + 1))) \\ &= g(\psi((t + 1)/2)). \end{aligned}$$

On the other hand, if $t + 1$ is odd, then $a = 0$ and

$$\begin{aligned} \psi(t + 1) &= h(b, 0) \\ &= f(b) \\ &= f(t/2). \end{aligned}$$

In other words,

$$\psi(0) = 0$$

$$\psi(t + 1) = \begin{cases} f\left(\dfrac{t}{2}\right) & \text{if } t + 1 \text{ is odd,} \\ g\left(\psi\left(\dfrac{t + 1}{2}\right)\right) & \text{if } t + 1 \text{ is even.} \end{cases}$$

Now f and g are satisfactory, and, being unary, they are therefore in **PR**. Since ψ is obtained from f and g using operation 3, ψ also belongs to **PR**. To retrieve h from ψ we can use $h(x, y) = \psi(\langle x, y \rangle + 1)$. So,

$$h(h_1(t), h_2(t)) = \psi(s(\langle h_1(t), h_2(t) \rangle))$$

from which we see that if h_1 and h_2 both belong to **PR**, then so does $h(h_1(t), h_2(t))$. Hence h is satisfactory. ■

Now we are ready to define the function $\phi(t, x)$, which we shall also write as $\phi_t(x)$, that will enumerate the unary primitive recursive functions:

$$\phi_t(x) = \begin{cases} x + 1 & \text{if } t = 0 \\ 0 & \text{if } t = 1 \\ l(x) & \text{if } t = 2 \\ r(x) & \text{if } t = 3 \\ \phi_{l(n)}(\phi_{r(n)}(x)) & \text{if } t = 4, n \geq 0 \\ \langle \phi_{l(n)}(x), \phi_{r(n)}(x) \rangle & \text{if } t = 5, n \geq 0 \\ 0 & \text{if } t = 6, n \geq 0 \text{ and } x = 0 \\ \phi_{l(n)}((x-1)/2) & \text{if } t = 6, n \geq 0 \text{ and } x \text{ is odd} \\ \phi_{r(n)}(\phi_t(x/2)) & \text{if } t = 6, n \geq 0 \text{ and } x \text{ is even} \end{cases}$$

Here $\phi_0(x)$, $\phi_1(x)$, $\phi_2(x)$, $\phi_3(x)$ are the four initial functions. For $t > 3$, t is represented as $3n + i$ where $n \geq 0$ and $i = 4$, 5, or 6, the three operations of Theorem 9.2 are then dealt with for values of t with the corresponding value of i. The pairing functions are used to guarantee all functions obtained for any value of t are eventually used in applying each of the operations. It should be clear from the definition that $\phi(t, x)$ is a total function and that it does enumerate all the unary primitive recursive functions. Although it is pretty clear that the definition provides an algorithm for computing the values of ϕ for any given inputs, for a rigorous proof more is needed. Fortunately, the recursion theorem makes it easy to provide such a proof. Namely, we set

$$g(z, t, x)$$

$$= \begin{cases} x + 1 & \text{if } t = 0 \\ 0 & \text{if } t = 1 \\ l(x) & \text{if } t = 2 \\ r(x) & \text{if } t = 3 \\ \Phi_z^{(2)}(l(n), \Phi_z^{(2)}(r(n), x)) & \text{if } t = 4, n \geq 0 \\ \langle \Phi_z^{(2)}(l(n), x), \Phi_z^{(2)}(r(n), x) \rangle & \text{if } t = 5, n \geq 0 \\ 0 & \text{if } t = 6, n \geq 0 \text{ and } x = 0 \\ \Phi_z^{(2)}(l(n), \lfloor x/2 \rfloor) & \text{if } t = 6, n \geq 0 \text{ and } x \text{ is odd} \\ \Phi_z^{(2)}(r(n), \Phi_z^{(2)}(t, \lfloor x/2 \rfloor)) & \text{if } t = 6, n \geq 0 \text{ and } x \text{ is even} \end{cases}$$

Then, $g(z, t, x)$ is partially computable, and by the recursion theorem, there is a number e such that

$$g(e, t, x) = \Phi_e^{(2)}(t, x).$$

Then, since $g(e, t, x)$ satisfies the definition of $\phi(t, x)$ and that definition determines ϕ uniquely as a total function, we must have

$$\phi(t, x) = g(e, t, x),$$

so that ϕ is computable.

The discussion at the beginning of this section now applies and we have our desired result.

Theorem 9.3. The function $\phi(x, x) + 1$ is a computable function that is not primitive recursive.

Exercises

1. Show that $\phi(t, x)$ is not primitive recursive.
2. Give a direct proof that $\phi(t, x)$ is computable by showing how to obtain an $\mathscr{S}$ program that computes ϕ. [*Hint:* Use the pairing functions to construct a stack for handling recursions.]

5

Calculations on Strings

1. Numerical Representation of Strings

So far we have been dealing exclusively with computations on numbers. Now we want to extend our point of view to include computations on strings of symbols on a given alphabet. In order to extend computability theory to strings on an alphabet A, we wish to associate numbers with elements of A^* in a one-one manner. We now describe one convenient way of doing this: Let A be some given alphabet. Since A is a set, there is no order implied among the symbols. However, *we will assume in this chapter that the elements of A have been placed in some definite order*. In particular, when we write $A = \{s_1, \dots, s_n\}$, we think of the sequence $s_1, \dots, s_n$ as corresponding to this given order. Now, let $w = s_{i_k} s_{i_{k-1}} \cdots s_{i_1} s_{i_0}$. Then we associate with w the integer

$$x = i_k \cdot n^k + i_{k-1} \cdot n^{k-1} + \cdots + i_1 \cdot n + i_0. \tag{1.1}$$

With $w = 0$, we associate the number 0. (It is for this reason that we use the same symbol for both.) For example, let A consist of the symbols a, b, c given in the order shown, and let $w = baacb$. Then, the corresponding integer is

$$x = 2 \cdot 3^4 + 1 \cdot 3^3 + 1 \cdot 3^2 + 3 \cdot 3^1 + 2 = 209.$$

In order to see that the representation (1.1) is unique, we show how to retrieve the subscripts $i_0, i_1, \ldots, i_k$ from x assuming that $x \neq 0$. We define the primitive recursive functions:

$$R^+(x, y) = \begin{cases} R(x, y) & \text{if } \sim(y \mid x) \\ y & \text{otherwise,} \end{cases}$$

$$Q^+(x, y) = \begin{cases} \lfloor x/y \rfloor & \text{if } \sim(y \mid x) \\ \lfloor x/y \rfloor \dot{-} 1 & \text{otherwise,} \end{cases}$$

where the functions $R(x, y)$ and $\lfloor x/y \rfloor$ are as defined in Chapter 3, Section 7. Then, as we shall easily show, for $y \neq 0$,

$$\frac{x}{y} = Q^+(x, y) + \frac{R^+(x, y)}{y}, \qquad 0 < R^+(x, y) \leq y.$$

This equation expresses ordinary division with quotient and remainder:

$$\frac{x}{y} = \lfloor x/y \rfloor + \frac{R(x, y)}{y},$$

as long as y is not a divisor of x. If y is a divisor of x we have

$$\frac{x}{y} = \lfloor x/y \rfloor = (\lfloor x/y \rfloor \dot{-} 1) + \frac{y}{y} = Q^+(x, y) + \frac{R^+(x, y)}{y}.$$

Thus, what we are doing differs from ordinary division with remainders in that "remainders" are permitted to take on values between 1 and y rather than between 0 and $y - 1$.

Now, let us set

$$u_0 = x, \qquad u_{m+1} = Q^+(u_m, n). \tag{1.2}$$

Thus, by (1.1)

$$\begin{aligned} u_0 &= i_k \cdot n^k + i_{k-1} \cdot n^{k-1} + \cdots + i_1 \cdot n + i_0, \\ u_1 &= i_k \cdot n^{k-1} + i_{k-1} \cdot n^{k-2} + \cdots + i_1, \\ &\vdots \\ u_k &= i_k. \end{aligned} \tag{1.3}$$

Therefore,

$$i_m = R^+(u_m, n), \qquad m = 0, 1, \ldots, k. \tag{1.4}$$

Hence, for any number x satisfying (1.1), the string w can be retrieved. It is worth noting that this can be accomplished using primitive recursive functions. If we write

$$g(0, n, x) = x,$$

$$g(m + 1, n, x) = Q^+(g(m, n, x), n),$$

then

$$g(m, n, x) = u_m \tag{1.5}$$

as defined by (1.2), where, of course, g is primitive recursive. Moreover, if we let $h(m, n, x) = R^+(g(m, n, x), n)$, then h is also primitive recursive, and by (1.4)

$$i_m = h(m, n, x), \qquad m = 0, 1, \ldots, k. \tag{1.6}$$

This method of representing strings by numbers is clearly related to the usual base n notation for numbers. To explore the connection, it is instructive to consider the alphabet

$$D = \{1, 2, 3, 4, 5, 6, 7, 8, 9, X\}$$

in the order shown. Then the number associated with the *string* 45 is

$$4 \cdot 10 + 5 = 45.$$

On the other hand, the number associated with $2X$ is

$$2 \cdot 10 + 10 = 30.$$

(Perhaps we should read $2X$ as "twenty-ten"!) Clearly a string on D that does not include X is simply the usual decimal notation for the number it represents. It is numbers whose decimal representation includes a 0 which now require an X.

Thus, in the general case of an alphabet A consisting of $s_1, \ldots, s_n$, ordered as shown, we see that we are simply using a base n representation in which the "digits" range from 1 to n instead of the usual 0 to $n - 1$. We are proceeding in this manner simply to avoid the lack of uniqueness of the usual base n representation:

$$79 = 079 = 0079 = 00079 = \text{etc.}$$

This lack of uniqueness is of course caused by the fact that leading zeros do not change the number being represented.

It is interesting to observe that the rules of elementary arithmetic (including the use of "carries") work perfectly well with our representation. Here are a few examples:

$$\begin{array}{r} 1\,7 \\ +\,1\,X\,3 \\ \hline 2\,1\,X \end{array} \qquad \text{which corresponds to} \qquad \begin{array}{r} 17 \\ +203 \\ \hline 220 \end{array}$$

$$\begin{array}{r} 2\,9 \\ -\,1\,X \\ \hline 9 \end{array} \qquad \text{which corresponds to} \qquad \begin{array}{r} 29 \\ -20 \\ \hline 9 \end{array}$$

$$\begin{array}{r} X\,5 \\ \times\,2\,X \\ \hline X\,4\,X \\ 1\,X\,X \\ \hline 3\,1\,4\,X \end{array} \qquad \text{which corresponds to} \qquad \begin{array}{r} 105 \\ \times 30 \\ \hline 3150 \end{array}$$

(Incidentally, this shows that the common belief that the modern rules of calculation required the introduction of a digit for 0 is unjustified.) Note in particular the following examples of adding 1:

$$\begin{array}{r} X\,1 \\ +\quad 1 \\ \hline X\,2 \end{array} \qquad \begin{array}{r} 3\,X \\ +\quad 1 \\ \hline 4\,1 \end{array} \qquad \begin{array}{r} 3\,X\,X \\ +\quad 1 \\ \hline 4\,1\,1 \end{array} \qquad \begin{array}{r} 7\,3\,X\,X \\ +\quad 1 \\ \hline 7\,4\,1\,1 \end{array} \qquad \begin{array}{r} 4\,9 \\ +\quad 1 \\ \hline 4\,X \end{array}$$

Adding 1 to X gives a result of 1 with a carry of 1. If the string ends in more than one X, the carry propagates. Subtracting 1 is similar, with a propagating carry produced by a string ending in 1:

$$\begin{array}{r} 1\,X \\ -\quad 1 \\ \hline 1\,9 \end{array} \qquad \begin{array}{r} X\,1 \\ -\quad 1 \\ \hline 9\,X \end{array} \qquad \begin{array}{r} 7\,1\,1 \\ -\quad 1 \\ \hline 6\,X\,X \end{array}$$

Now we return to the general case. Given the alphabet A consisting of $s_1, \ldots, s_n$ in the order shown, the string $w = s_{i_k} s_{i_{k-1}} \cdots s_{i_1} s_{i_0}$ is called the *base n notation for the number x* defined by (1.1). (0 is the base n notation for the null string 0 for every n.) Thus when n is fixed we can regard a partial function of one or more variables on A^* as a function of the corresponding numbers. (That is, the numbers are just those which the given strings represent in base n notation.) It now makes perfect sense to speak of an m-ary partial function on A^* with values in A^* as being *partially computable*, or when it is total, as being *computable*. Similarly we can say that an m-ary function on A^* is *primitive recursive*. Note that for any alphabet $A = \{s_1, \ldots, s_n\}$ with the symbols ordered as shown, s_1 denotes 1 in base n. Thus an m-ary predicate on A^* is simply a total m-ary function on A^* all of whose values are either s_1 or 0. And it now makes sense as well to speak of an m-ary predicate on A^* as being computable.

As was stated in Chapter 1, for a given alphabet A, any subset of A^* is called a *language* on A. Once again, by associating with the elements of A^* the corresponding numbers, we can speak of a language on A as being r.e., or recursive, or primitive recursive.

It is important to observe that whereas the usual base n notation using a 0 digit works only for $n \geq 2$, the representation (1.1) is valid even for $n = 1$. For an alphabet consisting of the single symbol 1, the string $1^{[x]}$ of length x is the base 1 notation for the number $\sum_{i=0}^{x-1} 1 \cdot (1)^i = \sum_{i=0}^{x-1} 1 = x$. That is, the base 1 (or *unary*) representation of the number x is simply a string of ones of length x.

In thinking of numbers (that is, elements of N) as inputs to and outputs from programs written in our language $\mathscr{S}$, no particular representation of these numbers was specified or required. Numbers occur in the theory as purely abstract entities, just as they do in ordinary mathematics. However, when we wish to refer to particular numbers, we do so in the manner familiar to all of us, by writing their decimal representations. These representations are, of course, really strings on the alphabet that consists of the decimal digits:

$$\{0, 1, 2, 3, 4, 5, 6, 7, 8, 9\}.$$

But it is essential to avoid confusing such strings with the numbers they represent. For this reason, for the remainder of this chapter we shall avoid the use of decimal digits as symbols in our alphabets. Thus, a string of decimal digits will *always be meant to refer to a number.*

Now, let A be some fixed alphabet containing exactly n symbols, say $A = \{s_1, s_2, \ldots, s_n\}$. For each $m \geq 1$, we define $\text{CONCAT}_n^{(m)}$ as follows:

$$\begin{aligned} \text{CONCAT}_n^{(1)}(u) &= u, \\ \text{CONCAT}_n^{(m+1)}(u_1, \ldots, u_m, u_{m+1}) &= zu_{m+1}, \end{aligned} \tag{1.7}$$

where

$$z = \text{CONCAT}_n^{(m)}(u_1, \ldots, u_m).$$

Thus, for given strings $u_1, \ldots, u_m \in A^*$, $\text{CONCAT}_n^{(m)}(u_1, \ldots, u_m)$ is simply the string obtained by placing the strings $u_1, \ldots, u_m$ one after the other, or, as is usually said, by *concatenating* them. We will usually omit the superscript, so that, for example we may write

$$\mathrm{CONCAT}_2(s_2 s_1, s_1 s_1 s_2) = s_2 s_1 s_1 s_1 s_2 .$$

Likewise,

$$\mathrm{CONCAT}_6(s_2 s_1, s_1 s_1 s_2) = s_2 s_1 s_1 s_1 s_2 .$$

However, the string $s_2 s_1$ represents the number 5 in base 2 and the number 13 in base 6. Also, the string $s_1 s_1 s_2$ represents the number 8 in base 2 and the number 44 in base 6. Finally, the string $s_2 s_1 s_1 s_1 s_2$ represents 48 in base 2 and 2852 in base 6. If we wish to think of CONCAT as defining functions on N (as will be necessary, for example, in showing that the functions (1.7) are primitive recursive), then the example we have been considering becomes

$$\mathrm{CONCAT}_2(5, 8) = 48 \qquad \text{and} \qquad \mathrm{CONCAT}_6(13, 44) = 2852.$$

The same example in base 10 gives

$$\mathrm{CONCAT}_{10}(21, 112) = 21112.$$

Bearing this discussion in mind, we now proceed to give a list of primitive recursive functions (on A^* or N, depending on one's point of view) that we will need later.

1. $f(u) = |u|$. This "length" function is most naturally understood as being defined on A^* and taking values in N. For each x, the number $\sum_{j=0}^{x} n^j$ has the base n representation $s_1^{[x+1]}$; hence this number is the smallest number whose base n representation contains $x + 1$ symbols. Thus,

$$|u| = \min_{x \le u} \left[\sum_{j=0}^{x} n^j > u \right].$$

2. $g(u, v) = \mathrm{CONCAT}_n(u, v)$. The primitive recursiveness of this function follows from the equation

$$\mathrm{CONCAT}_n(u, v) = u \cdot n^{|v|} + v.$$

3. $\mathrm{CONCAT}_n^{(m)}(u_1, \ldots, u_m)$, as defined in (1.7), is primitive recursive for each $m, n \ge 1$. This follows at once from the previous example using composition.
4. $\mathrm{RTEND}_n(w) = h(0, n, w)$, where h is as in (1.6). As a function of A^*, RTEND_n gives the rightmost symbol of a given word, as is clear from (1.3) and (1.6).

5. $\text{LTEND}_n(w) = h(|w| \dot{-} 1, n, w)$. LTEND_n gives the leftmost symbol of a given nonempty word.
6. $\text{RTRUNC}_n(w) = g(1, n, w)$. RTRUNC_n gives the result of removing the rightmost symbol from a given nonempty word, as is clear from (1.3) and (1.5). *When we can omit reference to the base n, we often write* w^- *for* $\text{RTRUNC}_n(w)$. *Note that* $0^- = 0$.
7. $\text{LTRUNC}_n(w) = w \dot{-} (\text{LTEND}_n(w) \cdot n^{|w| \dot{-} 1})$. In the notation of (1.3), for a given nonempty word w, $\text{LTRUNC}_n(w) = w - i_k \cdot n^k$, i.e., $\text{LTRUNC}_n(w)$ is the result of removing the leftmost symbol from w.

We will now use the list of primitive recursive functions that we have just given to prove the computability of a pair of functions that can be used in changing base. Thus, let $1 \leq n < l$. Let $A \subset \tilde{A}$, where A is an alphabet of n symbols and $\tilde{A}$ is an alphabet of l symbols. Thus a string that belongs to A^* also belongs to $\tilde{A}^*$. For any $x \in N$, let w be the word in A^* that represents x in base n. Then, we write $\text{UPCHANGE}_{n,l}(x)$ for the number which w represents in base l. For example, referring to our previous example, we have $\text{UPCHANGE}_{2,6}(5) = 13$, $\text{UPCHANGE}_{2,6}(8) = 44$, $\text{UPCHANGE}_{2,6}(48) = 2852$. Also $\text{UPCHANGE}_{2,10}(5) = 21$ and $\text{UPCHANGE}_{6,10}(13) = 21$.

Next, for $x \in N$, let w be the string in $\tilde{A}^*$ which represents x in base l, and let w' be obtained from w by crossing out all of the symbols that belong to $\tilde{A} - A$. Then, $w' \in A^*$, and we write $\text{DOWNCHANGE}_{n,l}(x)$ for the number which w' represents in base n. For example, the string $s_2 s_6 s_1$ represents the number 109 in base 6. To obtain $\text{DOWNCHANGE}_{2,6}(109)$ we cross out the s_6, obtaining the string $s_2 s_1$, which represents 5 in base 2; thus $\text{DOWNCHANGE}_{2,6}(109) = 5$.

Although $\text{UPCHANGE}_{n,l}$ and $\text{DOWNCHANGE}_{n,l}$ are actually primitive recursive functions, we will content ourselves with proving that they are computable:

Theorem 1.1. Let $0 < n < l$. Then the functions $\text{UPCHANGE}_{n,l}$ and $\text{DOWNCHANGE}_{n,l}$ are computable.

Proof. We begin with $\text{UPCHANGE}_{n,l}$. We write a program which extracts the successive symbols of the word that the given number represents in base n and uses them in computing the number that the given word represents in base l:

```
[A]    IF X = 0 GOTO E
       Z ← LTEND_n(X)
       X ← LTRUNC_n(X)
       Y ← l·Y + Z
       GOTO A
```

$\text{DOWNCHANGE}_{n,l}$ is handled similarly. Our program will extract the successive symbols of the word that the given number represents in base l. However, these symbols will only be used if they belong to the smaller alphabet, i.e., if as numbers they are $\le n$:

$$
\begin{array}{ll}
[A] & \text{IF } X = 0 \text{ GOTO } E \\
 & Z \leftarrow \text{LTEND}_l(X) \\
 & X \leftarrow \text{LTRUNC}_l(X) \\
 & \text{IF } Z > n \text{ GOTO } A \\
 & Y \leftarrow n \cdot Y + Z \\
 & \text{GOTO } A
\end{array}
$$

■

Exercises

1. (a) Write the numbers 40 and 12 in base 3 notation using the "digits" $\{1, 2, 3\}$.
   (b) Work out the multiplication $40 \cdot 12 = 480$ in base 3.
   (c) Compute $\text{CONCAT}_n(12, 15)$ for $n = 3$, 5, and 10. Why is no calculation required in the last case?
   (d) Compute the following: $\text{UPCHANGE}_{3,7}(15)$, $\text{UPCHANGE}_{2,7}(15)$, $\text{UPCHANGE}_{2,10}(15)$, $\text{DOWNCHANGE}_{3,7}(15)$, $\text{DOWNCHANGE}_{2,7}(15)$, $\text{DOWNCHANGE}_{2,10}(20)$.
2. Compute each of the following for $n = 3$.
   (a) $\text{CONCAT}_n^{(2)}(17, 32)$.
   (b) $\text{CONCAT}_n^{(3)}(17, 32, 11)$.
   (c) $\text{RTEND}_n(23)$.
   (d) $\text{LTEND}_n(29)$.
   (e) $\text{RTRUNC}_n(19)$.
   (f) $\text{LTRUNC}_n(18)$.
3. Do the previous exercise for $n = 4$.
4. Show that the function f whose value is the string formed of the symbols occurring in the odd-numbered places in the input [i.e., $f(a_1 a_2 a_3 \cdots a_n) = a_1 a_3 \cdots$] is computable.
5. Let $A = \{s_1, \ldots, s_n\}$, and let $P(x)$ be the predicate on N which is true just when the string in A^* that represents x has an even number of symbols. Show that $P(x)$ is primitive recursive.

6. If $u \neq 0$, let $\#(u, v)$ be the number of occurrences of u as a part of v [e.g., $\#(bab, ababab) = 2$]. Also, let $\#(0, v) = 0$. Prove that $\#(u, v)$ is primitive recursive.
7. Show that $\text{UPCHANGE}_{n,l}$ and $\text{DOWNCHANGE}_{n,l}$ are primitive recursive.
8. For $n \geq 2$, show that when $|u|$ is calculated with respect to base n notation, $|u| \leq \lfloor \log_n u \rfloor + 1$ for all $u \in N$.

2. A Programming Language for String Computations

From the point of view of string computations, the language $\mathscr{S}$ seems quite artificial. For example, the instruction

$$V \leftarrow V + 1$$

which is so basic for integers, seems entirely unnatural as a basic instruction for string calculations. Thus, for the alphabet $\{a, b, c\}$, applying this instruction to *bacc* produces *bbaa* because a carry is propagated. (This will perhaps seem more evident if, momentarily ignoring our promise to avoid the decimal digits as symbols in our alphabets, we use the alphabet $\{1, 2, 3\}$ and write

$$2133 + 1 = 2211.)$$

We are now going to introduce, for each $n > 0$, a programming language $\mathscr{S}_n$, which is specifically designed for string calculations on an alphabet of n symbols. The languages $\mathscr{S}_n$ will be supplied with the same input, output, and local variables as $\mathscr{S}$, except that we now think of them as having values in the set A^*, where A is an n symbol alphabet. Variables not otherwise initialized are to be initialized to 0. We use the same symbols as labels in $\mathscr{S}_n$ as in $\mathscr{S}$ and the same conventions regarding their use. The instruction types are shown in Table 2.1.

The formal rules of syntax in $\mathscr{S}_n$ are entirely analogous to those for $\mathscr{S}$, and we omit them. Similarly, we use macro expansions quite freely. An m-ary partial function on A^* which is computed by a program in $\mathscr{S}_n$ is said to be *partially computable* in $\mathscr{S}_n$. If the function is total and partially computable in $\mathscr{S}_n$, it is called *computable* in $\mathscr{S}_n$.

Although the instructions of $\mathscr{S}_n$ refer to strings, we can just as well think of them as referring to the numbers that the corresponding strings represent in base n. For example, the numerical effect of the instruction

$$X \leftarrow s_i X$$

Table 2.1

Instruction	Interpretation
$V \leftarrow \sigma V$ for each symbol σ in the alphabet A	Place the symbol σ to the left of the string which is the value of V.
$V \leftarrow V^-$	Delete the final symbol of the string which is the value of V. If the value of V is 0, leave it unchanged.
If V ENDS σ GOTO L for each symbol σ in the alphabet A and each label L	If the value of V ends in the symbol σ, execute next the first instruction labeled L; otherwise proceed to the next instruction.

in the n symbol alphabet $\{s_1, \ldots, s_n\}$ ordered as shown is to replace the numerical value x by $i \cdot n^{|x|} + x$. Just as the instructions of $\mathcal{S}$ are natural as basic numerical operations, but complex as string operations, so the instructions of $\mathcal{S}_n$ are natural as basic string operations, but complex as numerical operations.

We now give some macros for use in $\mathcal{S}_n$ with the corresponding expansions.

1. The macro IF $V \neq 0$ GOTO L has the expansion

$$
\begin{aligned}
&\text{IF } V \text{ ENDS } s_1 \text{ GOTO } L\\
&\text{IF } V \text{ ENDS } s_2 \text{ GOTO } L\\
&\quad\vdots\\
&\text{IF } V \text{ ENDS } s_n \text{ GOTO } L
\end{aligned}
$$

2. The macro $V \leftarrow 0$ has the expansion

$$
\begin{aligned}
[A]\quad & V \leftarrow V^-\\
& \text{IF } V \neq 0 \text{ GOTO } A
\end{aligned}
$$

3. The macro GOTO L has the expansion

$$
\begin{aligned}
&Z \leftarrow 0\\
&Z \leftarrow s_1 Z\\
&\text{IF } Z \text{ ENDS } s_1 \text{ GOTO } L
\end{aligned}
$$

4. The macro $V' \leftarrow V$ has the expansion shown in Fig. 2.1.

The macro expansion of $V' \leftarrow V$ in $\mathcal{S}_n$ is quite similar to that in $\mathcal{S}$.

$$
\begin{array}{ll}
 & Z \leftarrow 0 \\
 & V' \leftarrow 0 \\
[A] & \text{IF } V \text{ ENDS } s_1 \text{ GOTO } B_1 \\
 & \text{IF } V \text{ ENDS } s_2 \text{ GOTO } B_2 \\
 & \vdots \\
 & \text{IF } V \text{ ENDS } s_n \text{ GOTO } B_n \\
 & \text{GOTO } C \\
[B_i] & \left.\begin{array}{l} V \leftarrow V^- \\ V' \leftarrow s_i V' \\ Z \leftarrow s_i Z \\ \text{GOTO } A \end{array}\right\} i = 1, 2, \ldots, n \\
[C] & \text{IF } Z \text{ ENDS } s_1 \text{ GOTO } D_1 \\
 & \text{IF } Z \text{ ENDS } s_2 \text{ GOTO } D_2 \\
 & \vdots \\
 & \text{IF } Z \text{ ENDS } s_n \text{ GOTO } D_n \\
 & \text{GOTO } E \\
[D_i] & \left.\begin{array}{l} Z \leftarrow Z^- \\ V \leftarrow s_i V \\ \text{GOTO } C \end{array}\right\} i = 1, 2, \ldots, n
\end{array}
$$

Figure 2.1. Macro expansion of $V' \leftarrow V$ in $\mathscr{S}_n$.

The block of instructions

$$
\begin{array}{l}
\text{IF } V \text{ ENDS } s_1 \text{ GOTO } B_1 \\
\text{IF } V \text{ ENDS } s_2 \text{ GOTO } B_2 \\
\vdots \\
\text{IF } V \text{ ENDS } s_n \text{ GOTO } B_n
\end{array}
$$

is usually written simply

$$\text{IF } V \text{ ENDS } s_i \text{ GOTO } B_i \qquad (1 \le i \le n)$$

Such a block of instructions is referred to as a *filter* for obvious reasons. Note that at the point in the computation when the first "GOTO C" is executed, V' and Z will both have the original value of V, whereas V will have the value 0. On exiting, Z has the value 0, while V' retains the original value of V and V has been restored to its original value.

If $f(x_1, \ldots, x_m)$ is any function that is partially computable in $\mathscr{S}_n$, we permit the use in $\mathscr{S}_n$ of macros of the form

$$V \leftarrow f(V_1, \ldots, V_m)$$

The corresponding expansions are carried out in a manner entirely analogous to that discussed in Chapter 2, Section 5.

We conclude this section with two examples of functions that are computable in $\mathscr{S}_n$ for every n. The general results in the next section will

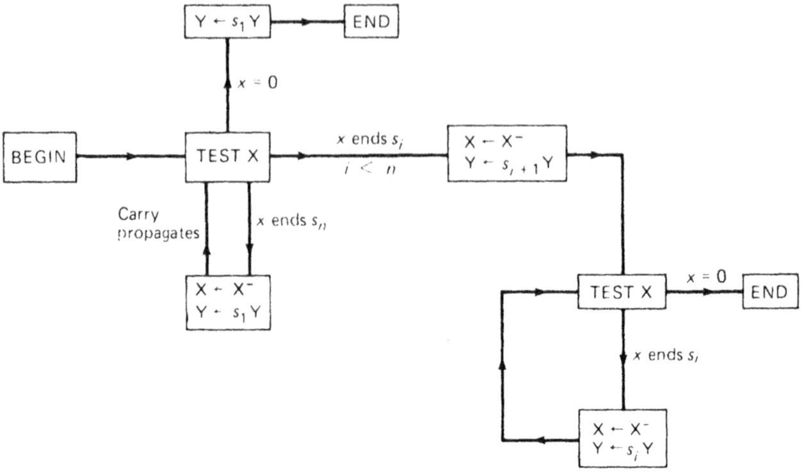

Figure 2.2. Flow chart for computing $x + 1$ in $\mathscr{S}_n$.

make it clear that these two examples are the only bit of programming in $\mathscr{S}_n$ that we shall need to carry out explicitly.

We want to show that the function $x + 1$ is computable in $\mathscr{S}_n$. We let our alphabet consist of the symbols $s_1, s_2, \ldots, s_n$ ordered as shown. The desired program is exhibited in Fig. 2.3; a flow chart that shows how the program works is shown in Fig. 2.2.

Our final example is a program that computes $x \dot{-} 1$ base n. A flow chart is given in Fig. 2.4 and the actual program in $\mathscr{S}_n$ is exhibited in Fig. 2.5. The reader should check both of these programs with some examples.

$[B]$	IF X ENDS s_i GOTO A_i	$(1 \le i \le n)$
	$Y \leftarrow s_1 Y$	
	GOTO E	
$[A_i]$	$X \leftarrow X^-$	
	$Y \leftarrow s_{i+1} Y$	$\Big\}\ 1 \le i < n$
	GOTO C	
$[A_n]$	$X \leftarrow X^-$	
	$Y \leftarrow s_1 Y$	
	GOTO B	
$[C]$	IF X ENDS s_i GOTO D_i	$(1 \le i \le n)$
	GOTO E	
$[D_i]$	$X \leftarrow X^-$	
	$Y \leftarrow s_i Y$	$\Big\}\ 1 \le i \le n$
	GOTO C	

Figure 2.3. Program that computes $x + 1$ in $\mathscr{S}_n$.

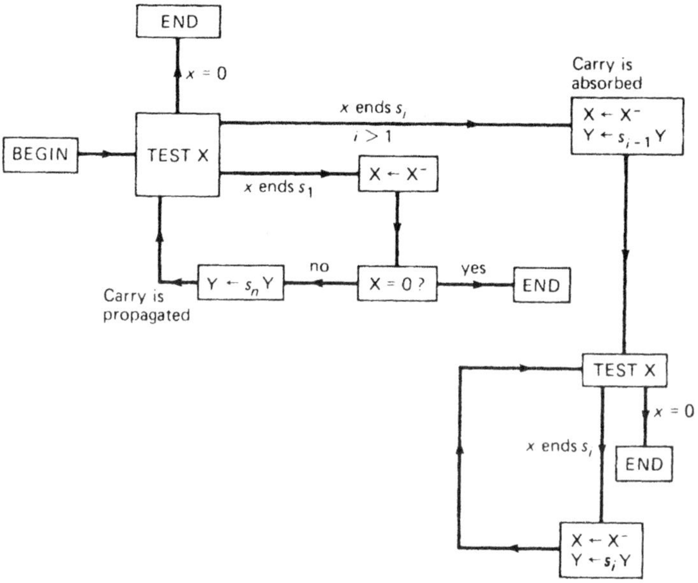

Figure 2.4. Flow chart for computing $x \dot{-} 1$ in $\mathscr{S}_n$.

$[B]$ IF X ENDS s_i GOTO A_i $(1 \leq i \leq n)$
GOTO E

$[A_i]$ $\left.\begin{array}{l} X \leftarrow X^- \\ Y \leftarrow s_{i-1}Y \\ \text{GOTO } C \end{array}\right\} 1 < i \leq n$

$[A_1]$ $X \leftarrow X^-$
IF $X \neq 0$ GOTO C_2
GOTO E

$[C_2]$ $Y \leftarrow s_nY$
GOTO B

$[C]$ IF X ENDS s_i GOTO D_i $(1 \leq i \leq n)$
GOTO E

$[D_i]$ $\left.\begin{array}{l} X \leftarrow X^- \\ Y \leftarrow s_iY \\ \text{GOTO } C \end{array}\right\} 1 \leq i \leq n$

Figure 2.5. Program that computes $x \dot{-} 1$ in $\mathscr{S}_n$.

Exercises

1. Let $A = \{s_1, s_2\}$. Write out the complete expansion of the macro $X \leftarrow Y$ in $\mathscr{S}_2$.

2. Write a program in $\mathscr{S}_n$ to compute the function f defined in Exercise 1.4.

3. Show that $f(u, v) = \widehat{uv}$ is computable in $\mathscr{S}_n$. ($\widehat{uv}$ is the concatenation of u and v, defined in Chapter 1.)

4. Let $A = \{s_1, \ldots, s_n\}$, and let $P(x)$ be the predicate on A^* which is true just when x has an even number of symbols. Show that $P(x)$ is computable in $\mathscr{S}_n$.

5. Write a program in $\mathscr{S}_n$ to compute $\#(u, v)$ as defined in Exercise 1.6.

6. Give an expansion in $\mathscr{S}_n$ for the macro $V \leftarrow V\sigma$, which means: Place the symbol σ to the right of the string that is the value of V.

7. Show that $f(x) = x^{\mathrm{R}}$ is computable in $\mathscr{S}_n$. (x^{R} is defined in Chapter 1, Section 3.)

8. Let $A = \{s_1, \ldots, s_n\}$, and let $g(u) = w$ for all strings u in A^*, where w is the base n notation for the number of symbols in u. Show that g is computable in $\mathscr{S}_n$.

9. Let $A = \{s_1, s_2\}$, and let $\mathscr{P}$ be the $\mathscr{S}_2$ program

$$Y \leftarrow X + 1$$

Write out the computation of $\mathscr{P}$ for input $x = s_2 s_2$.

10. Let $A = \{s_1, s_2, s_3\}$, and let $\mathscr{P}$ be the $\mathscr{S}_3$ program

$$Y \leftarrow X \dot{-} 1$$

Write out the computation of $\mathscr{P}$ for input $x = s_1 s_1$.

11. (a) Show that Theorem 1.1 in Chapter 3 holds if we substitute "computable in $\mathscr{S}_n$" for "computable."
(b) Show that Theorems 2.1 and 2.2 in Chapter 3 hold if we substitute "computable in $\mathscr{S}_n$" for "computable."
(c) Show that if $f(x_1, \ldots, x_n)$ is primitive recursive, then it is computable in $\mathscr{S}_n$.

3. The Languages $\mathscr{S}$ and $\mathscr{S}_n$

We now want to compare the functions that can be computed in the various languages we have been considering, namely, $\mathscr{S}$ and the different $\mathscr{S}_n$. For the purpose of making this comparison, we take the point of view that, in all of the languages, computations are "really" dealing with numbers, and that strings on an n letter alphabet are simply data objects being used to represent numbers (using base n of course).

We shall see that in fact all of these languages are equivalent. That is, a function f is partially computable if and only if it is partially computable in each $\mathscr{S}_n$ and therefore, also, f is partially computable in any one $\mathscr{S}_n$ if and only if it is partially computable in all of them.

To begin with we have

Theorem 3.1. A function is partially computable if and only if it is partially computable in $\mathscr{S}_1$.

Proof. It is easy to see that the languages $\mathscr{S}$ and $\mathscr{S}_1$ are really the same. That is, the numerical effect of the instructions

$$V \leftarrow s_1 V \qquad \text{and} \qquad V \leftarrow V^-$$

in $\mathscr{S}_1$ is the same as that of the corresponding instructions in $\mathscr{S}$:

$$V \leftarrow V + 1 \qquad \text{and} \qquad V \leftarrow V - 1.$$

Furthermore, the condition "V ENDS s_1" in $\mathscr{S}_1$ is equivalent to the condition $V \neq 0$ in $\mathscr{S}$. (Since s_1 is the only symbol, ending in s_1 is equivalent to being different from the null string.) ∎

This theorem shows that results we obtain about the languages $\mathscr{S}_n$ can always be specialized to give results about $\mathscr{S}$ by setting $n = 1$.

Next we shall prove

Theorem 3.2. If a function is partially computable, then it is also partially computable in $\mathscr{S}_n$ for each n.

Proof. Let the function f be computed by a program $\mathscr{P}$ in the language $\mathscr{S}$. We translate $\mathscr{P}$ into a program in $\mathscr{S}_n$ by replacing each instruction of $\mathscr{P}$ by a macro in $\mathscr{S}_n$ as follows.

We replace each instruction $V \leftarrow V + 1$ by the macro $V \leftarrow V + 1$, each instruction $V \leftarrow V - 1$ by the macro $V \leftarrow V \dot{-} 1$, and each instruction IF $V \neq 0$ GOTO L by the macro IF $V \neq 0$ GOTO L. Here we are using the fact, proved at the end of the preceding section, that $x + 1$ and $x \dot{-} 1$ are both computable in base n, and hence can each be used to define a macro in $\mathscr{S}_n$.

It is then obvious that the new program computes in $\mathscr{S}_n$ the same function f that $\mathscr{P}$ computes in $\mathscr{S}$. ∎

This is the first of many proofs by the method of *simulation*: a program in one language is "simulated" step by step by a corresponding program in a different language.

We could now prove directly that if a function is partially computable in $\mathscr{S}_n$ for any particular n, then it is in fact partially computable in our original sense. But it will be easier to delay doing so since the result will be an automatic consequence of our work on Post–Turing programs.

Exercises

1. Give a primitive recursive function $b_1(n, x)$ such that any partial function computed by an $\mathscr{S}$ program with x instructions is computed by some $\mathscr{S}_n$ program with no more than $b_1(n, x)$ instructions.
2. Give a primitive recursive function $b_2^{(m)}(n, x_1, \ldots, x_m, y)$ such that any partial function $f(x_1, \ldots, x_m)$ computed by an $\mathscr{S}$ program in y steps on inputs $x_1, \ldots, x_m$ is computed by some $\mathscr{S}_n$ program in no more than $b_2^{(m)}(n, x_1, \ldots, x_m, y)$ steps. [*Hint:* Note that after y steps no variable holds a value larger than $\max\{x_1, \ldots, x_m\} + y$.]
3. Let n be some fixed number > 0, and let $\#(\mathscr{P})$ be a numbering scheme for $\mathscr{S}_n$ programs defined exactly like the numbering scheme for $\mathscr{S}$ programs given in Chapter 4, except that $\#(I) = \langle a, \langle b, c\rangle\rangle$, where

$$b = \begin{cases} 0 & \text{if the statement in } I \text{ is } V \leftarrow V^- \\ i & \text{if the statement in } I \text{ is } V \leftarrow s_i V \\ \#(L') \cdot n + i & \\ & \text{if the statement in } I \text{ is IF } V \text{ ENDS } s_i \text{ GOTO } L'. \end{cases}$$

   (a) Define

$$\mathrm{HALT}_n(x, y) \Leftrightarrow \mathscr{S}_n \text{ program } y \text{ eventually halts on input } x.$$

   Show that the predicate $\mathrm{HALT}_n(x, y)$ is not computable in $\mathscr{S}_n$.

   (b) Define the universal function $\Phi_n^{(m)}$ for m-ary functions partially computable in $\mathscr{S}_n$ as follows:

$$\Phi_n^{(m)}(x_1, \ldots, x_m, y) = \psi_{\mathscr{P}}^{(m)}(x_1, \ldots, x_m), \quad \text{where } \#(\mathscr{P}) = y.$$

   (Of course, $\psi_{\mathscr{P}}^{(m)}$ is the m-ary partial function computed by the $\mathscr{S}_n$ program $\mathscr{P}$.) Show that for each $m > 0$, the function $\Phi_n^{(m)}(x_1, \ldots, x_m, y)$ is partially computable in $\mathscr{S}_n$.

   (c)* State and prove a version of the parameter theorem for $\mathscr{S}_n$.

   (d)* State and prove a version of the recursion theorem for $\mathscr{S}_n$.

   (e)* Show that $\mathscr{S}_n$ is an acceptable programming system. [See Exercise 5.4 in Chapter 4 for the definition of acceptable programming systems.]

4.* Give an upper bound on the length of the shortest $\mathscr{S}_1$ program which computes the function $\Phi_y(x)$ defined in Chapter 4. [See Exercise 3.6 in Chapter 4.]

4. Post–Turing Programs

In this section, we will study yet another programming language for string manipulation, the Post–Turing language $\mathscr{T}$. Unlike $\mathscr{S}_n$, the language $\mathscr{T}$ has no variables. All of the information being processed is placed on one linear tape. We can conveniently think of the tape as ruled into squares each of which can carry a single symbol (see Fig. 4.1). The tape is thought of as infinite in both directions. Each step of a computation is sensitive to just one symbol on the tape, the symbol on the square being "scanned." We can think of the tape passing through a device (like a tape recorder), or we can think of the computer as a tapehead that moves along the tape and is at each moment on one definite square (or we might say "tile"). With this simple scheme, there are not many steps we can imagine. The symbol being scanned can be altered. (That is, a new symbol can be "printed" in its place.) Or which instruction of a program is to be executed next can depend on which symbol is currently being scanned. Or, finally, the head can move one square to the left or right of the square presently scanned. We are led to the language shown in Table 4.1.

Although the formulation of $\mathscr{T}$ we have presented is closer in spirit to that originally given by Emil Post, it was Turing's analysis of the computation process that has made this formulation seem so appropriate. This language has played a fundamental role in theoretical computer science.

Turing's analysis was obtained by abstracting from the process carried out by a human being engaged in calculating according to a mechanical deterministic algorithm. Turing reasoned that there was no loss of generality in assuming that the person used a linear paper (like the paper tape in an old-fashioned adding machine or a printing calculator) instead of two-dimensional sheets of paper. Such a calculator is then engaged in observing symbols and writing symbols. Again without loss of generality, we can assume that only one symbol at a time is observed, since any finite group of symbols can be regarded as a single "megasymbol." Finally, we can assume that when the calculator shifts attention it is to an immediately adjacent symbol. For, to look, say, three symbols to the left is equivalent to moving one symbol to the left three successive times. And now we have arrived at precisely the Post–Turing language.

In order to speak of a function being computed by a Post–Turing program, we will need to deal with input and output. Let us suppose that

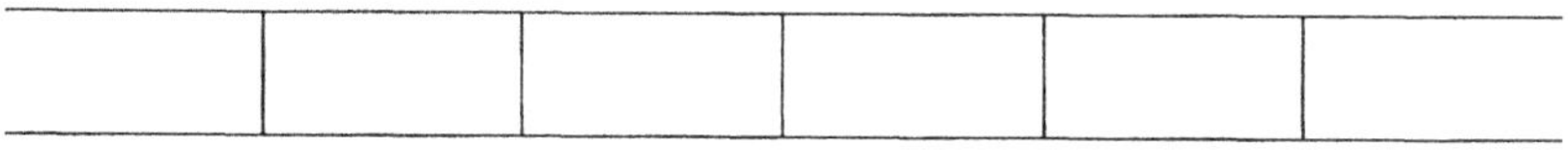

Figure 4.1

Table 4.1

Instruction	Interpretation
PRINT σ	Replace the symbol on the square being scanned by σ.
IF σ GOTO L	GOTO the first instruction labeled L *if* the symbol currently scanned is σ; otherwise, continue to the next instruction.
RIGHT	Scan the square immediately to the *right* of the square presently scanned.
LEFT	Scan the square immediately to the *left* of the square presently scanned.

we are dealing with string functions on the alphabet $A = \{s_1, s_2, \ldots, s_n\}$. We will use an additional symbol, written s_0, which we call the *blank* and use as a punctuation mark. *Often we write B for the blank instead of* s_0. All of our computations will be arranged so that *all but a finite number of squares on the tape are blank, i.e., contain the symbol B*. We show the contents of a tape by exhibiting a finite section containing all of the nonblank squares. We indicate the square currently being scanned by an arrow pointing up, just below the scanned square.

For example we can write

$$\begin{array}{ccccc} s_1 & s_2 & B & s_2 & s_1 \\ & & & \uparrow & \end{array}$$

to indicate that the tape consists of $s_1 s_2 B s_2 s_1$ with blank squares to the left and right, and that the square currently scanned contains the s_2 furthest to the right. We speak of a *tape configuration* as consisting of the tape contents together with a specification of one square as being currently *scanned*.

Now, to compute a partial function $f(x_1, \ldots, x_m)$ of m variables on A^*, we need to place the m strings $x_1, \ldots, x_m$ on the tape initially. We do this using the initial tape configuration:

$$\begin{array}{cccccccc} B & x_1 & B & x_2 & \ldots & B & x_m, \\ \uparrow & & & & & & \end{array}$$

That is, the inputs are separated by single blanks, and the symbol initially scanned is the blank immediately to the left of x_1. Here are a few examples:

1. $n = 1$, so the alphabet is $\{s_1\}$. We want to compute a function $f(x_1, x_2)$ and the initial values are $x_1 = s_1 s_1$, $x_2 = s_1$. Then the tape configuration initially will be

$$\begin{array}{ccccc} B & s_1 & s_1 & B & s_1. \\ \uparrow & & & & \end{array}$$

Of course, there are infinitely many blank squares to the left and right of the finite section we have shown:

$$\ldots\ B\ B\ B\ B\ s_1\ s_1\ B\ s_1\ B\ B\ B\ \ldots .$$
$$\uparrow$$

2. $n = 2$, $x_1 = s_1 s_2$, $x_2 = s_2 s_1$, $x_3 = s_2 s_2$. Then the tape configuration is initially

$$B\ s_1\ s_2\ B\ s_2\ s_1\ B\ s_2\ s_2 .$$
$$\uparrow$$

3. $n = 2$, $x_1 = 0$, $x_2 = s_2 s_1$, $x_3 = s_2$. Then the tape configuration is initially

$$B\ B\ s_2\ s_1\ B\ s_2 .$$
$$\uparrow$$

4. $n = 2$, $x_1 = s_1 s_2$, $x_2 = s_2 s_1$, $x_3 = 0$. Then the tape configuration is initially

$$B\ s_1\ s_2\ B\ s_2\ s_1\ B .$$
$$\uparrow$$

Note that there is no way to distinguish this initial tape configuration from that for which there are only two inputs $x_1 = s_1 s_2$ and $x_2 = s_2 s_1$. In other words, with this method of placing inputs on the tape, the number of arguments must be provided externally. It cannot be read from the tape.

A simple example of a Post–Turing program is given in Fig. 4.2.

Beginning with input x, this program outputs $s_2 s_1 x$. More explicitly, beginning with a tape configuration

$$B\ x$$
$$\uparrow$$

this program halts with the tape configuration

$$B\ s_2\ s_1\ x .$$
$$\uparrow$$

Figure 4.2

PRINT s_1
LEFT
PRINT s_2
LEFT

```
[A]   RIGHT
      IF s_1 GOTO A
      IF s_2 GOTO A
      IF s_3 GOTO A
      PRINT s_1
      RIGHT
      PRINT s_1
[C]   LEFT
      IF s_1 GOTO C
      IF s_2 GOTO C
      IF s_3 GOTO C
```

Figure 4.3

Next, for a slightly more complicated example, we consider Fig. 4.3. Here we are assuming that the alphabet is $\{s_1, s_2, s_3\}$. Let x be a string on this alphabet. Beginning with a tape configuration

$$\begin{array}{ll} B & x \\ \uparrow & \end{array}$$

this program halts with the tape configuration

$$\begin{array}{llll} B & x & s_1 & s_1. \\ \uparrow & & & \end{array}$$

The computation proceeds by first moving right until the blank to the right of x is located. The symbol s_1 is then printed twice and then the computation proceeds by moving left until the blank to the left of x is again located.

Figure 4.4 exhibits another example, this time with the alphabet $\{s_1, s_2\}$. The effect of this program is to "erase" all of the occurrences of s_2 in the input string, that is to replace each s_2 by B. For the purpose of reading output values off the tape, these additional Bs are ignored. Thus, if $f(x)$ is the function which this last program computes, we have, for example,

$$f(s_2 s_1 s_2) = s_1,$$
$$f(s_1 s_2 s_1) = s_1 s_1,$$
$$f(0) = 0.$$

Of course, the initial tape configuration

$$\begin{array}{llll} B & s_1 & s_2 & s_1 \\ \uparrow & & & \end{array}$$

```
[C]     RIGHT
        IF B GOTO E
        IF s2 GOTO A
        IF s1 GOTO C
[A]     PRINT B
        IF B GOTO C
```

Figure 4.4

```
[A]     RIGHT
        IF B GOTO E
        PRINT M
[B]     RIGHT
        IF s1 GOTO B
[C]     RIGHT
        IF s1 GOTO C
        PRINT s1
[D]     LEFT
        IF s1 GOTO D
        IF B GOTO D
        PRINT s1
        IF s1 GOTO A
```

Figure 4.5

leads to the final tape configuration

$$\begin{matrix} B & s_1 & B & s_1 & B \\ & & & & \uparrow \end{matrix}$$

but the blanks are ignored in reading the output.

For our final example we are computing a string function on the alphabet $\{s_1\}$. However, the program uses three symbols, B, s_1, and M. The symbol M is a *marker* to keep track of a symbol being copied. The program is given in Fig. 4.5. Beginning with the tape configuration

$$\begin{matrix} B & u \\ \uparrow & \end{matrix}$$

where u is a string in which only the symbol s_1 occurs, this program will terminate with the tape configuration

$$\begin{matrix} B & u & B & u. \\ & & \uparrow & \end{matrix}$$

(Thus we can say that this program computes the function $2x$ using unary notation.) The computation proceeds by replacing each successive s_1 (going from left to right) by the marker M and then copying the s_1 on the right.

We conclude this section with some definitions. Let $f(x_1, \ldots, x_m)$ be an m-ary partial function on the alphabet $\{s_1, \ldots, s_n\}$. Then the program $\mathscr{P}$ in the Post–Turing language $\mathscr{T}$ is said to *compute* f if when started in the tape configuration

$$\begin{array}{l} B\ x_1\ B\ \ldots\ B\ x_m \\ \uparrow \end{array}$$

it eventually halts if and only if $f(x_1, \ldots, x_m)$ is defined and if, on halting, the string $f(x_1, \ldots, x_m)$ can be read off the tape by ignoring all symbols other than $s_1, \ldots, s_n$. (That is, any "markers" left on the tape as well as blanks are to be ignored.) Note that we are thus permitting $\mathscr{P}$ to contain instructions that mention symbols other than $s_1, \ldots, s_n$.

The program $\mathscr{P}$ will be said to compute f *strictly* if two additional conditions are met:

1. no instruction in $\mathscr{P}$ mentions any symbol other than $s_0, s_1, \ldots, s_n$;
2. whenever $\mathscr{P}$ halts, the tape configuration is of the form

$$\begin{array}{l} \ldots\ B\ B\ B\ B\ y\ B\ B\ \ldots, \\ \qquad\qquad\quad \uparrow \end{array}$$

where the string y contains no blanks.

Thus when $\mathscr{P}$ computes f strictly, the output is available in a consecutive block of squares on the tape.

Exercises

1. Write out the computation performed by the Post–Turing program in Fig. 4.4 on input string $s_1 s_2 s_2 s_1$. Do the same for input $s_1 s_2 s_3 s_1$.

2. Write out the computation performed by the Post–Turing program in Fig. 4.5 on input string $s_1 s_1 B s_1 s_1 s_1$. Do the same for input $s_1 s_1 B s_1 B s_1 s_1$.

3. For each of the following functions, construct a Post–Turing program that computes the function strictly.

(a) $f(u, v) = \widehat{uv}$.

(b) the predicate $P(x)$ given in Exercise 2.4.

(c) the function $f(x) = x^{\mathrm{R}}$ (see Exercise 2.7).

(d) the function $\#(u, v)$ given in Exercise 1.6.

4. For each of the following functions, construct a Post–Turing program using only the symbols s_0, s_1 that computes the function in base 1 strictly.
   (a) $f(x, y) = x + y$.
   (b) $f(x) = 2x$.
   (c) $f(x, y) = x \dot{-} y$.
   (d) $f(x, y) = 2x + y \dot{-} 1$.
5. Construct a Post–Turing program using only the symbols s_0, s_1, s_2 that computes the function $s(x) = x + 1$ in base 2 strictly.

5. Simulation of $\mathscr{S}_n$ in $\mathscr{T}$

In this section we will prove

Theorem 5.1. If $f(x_1, \ldots, x_m)$ is partially computable in $\mathscr{S}_n$, then there is a Post–Turing program that computes f strictly.

Let $\mathscr{P}$ be a program in $\mathscr{S}_n$ which computes f. We assume that in addition to the input variables $X_1, \ldots, X_m$ and the output variable Y, $\mathscr{P}$ uses the local variables $Z_1, \ldots, Z_k$. Thus, altogether $\mathscr{P}$ uses $m + k + 1$ variables:

$$X_1, \ldots, X_m, Z_1, \ldots, Z_k, Y.$$

We set $l = m + k + 1$ and write these variables, in the same order, as

$$V_1, \ldots, V_l.$$

We shall construct a Post–Turing program $\mathscr{Q}$ that simulates $\mathscr{P}$ step by step. Since all of the information available to $\mathscr{Q}$ will be on the tape, we must allocate space on the tape to contain the values of the variables $V_1, \ldots, V_l$. Our scheme is simply that at the beginning of each simulated step, the tape configuration will be as follows:

$$\begin{array}{l} B\ x_1\ B\ x_2\ B\ \ldots\ B\ x_m\ B\ z_1\ B\ \ldots\ B\ z_k\ B\ y, \\ \uparrow \end{array}$$

where $x_1, x_2, \ldots, x_m, z_1, \ldots, z_k, y$ are the current values computed for the variables $X_1, X_2, \ldots, X_m, Z_1, \ldots, Z_k, Y$. This scheme is especially convenient in that the initial tape configuration

$$\begin{array}{l} B\ x_1\ B\ x_2\ B\ \ldots\ B\ x_m \\ \uparrow \end{array}$$

is already in the correct form, since the remaining variables are initialized to be 0. So we must show how to program the effect of each instruction type of $\mathscr{S}_n$ in the language $\mathscr{T}$. Various macros in $\mathscr{T}$ will be useful in doing this, and we now present them.

The macro

GOTO L

has the expansion

```
IF s_0 GOTO L
IF s_1 GOTO L
   ⋮
IF s_n GOTO L
```

The macro

RIGHT TO NEXT BLANK

has the expansion

```
[A]     RIGHT
        IF B GOTO E
        GOTO A
```

Similarly the macro

LEFT TO NEXT BLANK

has the expansion

```
[A]     LEFT
        IF B GOTO E
        GOTO A
```

The macro

MOVE BLOCK RIGHT

has the expansion

```
[C]      LEFT
         IF s_0 GOTO A_0
         IF s_1 GOTO A_1
           ⋮
         IF s_n GOTO A_n
[A_i]    RIGHT      }
         PRINT s_i  }  i = 1, 2, ..., n
         LEFT       }
         GOTO C     }
[A_0]    RIGHT
         PRINT B
         LEFT
```

The effect of the macro MOVE BLOCK RIGHT beginning with a tape configuration

in which the string in the rectangular box contains no blanks, is to terminate with the tape configuration

Finally we will use the macro

ERASE A BLOCK

whose expansion is

```
[A]   RIGHT
      IF B GOTO E
      PRINT B
      GOTO A
```

This program causes the head to move to the right, with everything erased between the square at which it begins and the first blank to its right.

We adopt the convention that a number ≥ 0 in square brackets after the name of a macro indicates that the macro is to be repeated that number of times. For example,

RIGHT TO NEXT BLANK [3]

is short for

RIGHT TO NEXT BLANK
RIGHT TO NEXT BLANK
RIGHT TO NEXT BLANK

We are now ready to show how to simulate the three instruction types in the language $\mathscr{S}_n$ by Post–Turing programs. We begin with

$$V_j \leftarrow s_i V_j$$

In order to place the symbol s_i to the left of the jth variable on the tape, the values of the variables $V_j, \ldots, V_l$ must all be moved over one square to the right to make room. After the s_i has been inserted, we must remember to go back to the blank at the left of the value of V_1 in order to be ready for the next simulated instruction. The program is

RIGHT TO NEXT BLANK [l]
MOVE BLOCK RIGHT [$l - j + 1$]
RIGHT
PRINT s_i
LEFT TO NEXT BLANK [j]

Next we must show how to simulate

$$V_j \leftarrow V_j^-$$

The complication is that if the value of V_j is the null word, we want it left unchanged. So we move to the blank immediately to the right of the value of V_j. By moving one square to the left we can detect whether the value of V_j is null (if it is, there are two consecutive blanks). Here is the program:

```
      RIGHT TO NEXT BLANK [j]
      LEFT
      IF B GOTO C
      MOVE BLOCK RIGHT [j]
      RIGHT
      GOTO E
[C]   LEFT TO NEXT BLANK [j - 1]
```

The final instruction type in $\mathscr{S}_n$ is

$$\text{IF } V_j \text{ ENDS } s_i \text{ GOTO } L$$

and the corresponding Post–Turing program is

```
      RIGHT TO NEXT BLANK [j]
      LEFT
      IF s_i GOTO C
      GOTO D
[C]   LEFT TO NEXT BLANK [j]
      GOTO L
[D]   RIGHT
      LEFT TO NEXT BLANK [j]
```

This completes the simulation of the three instruction types of $\mathscr{S}_n$. Thus, given our program $\mathscr{P}$ in the language $\mathscr{S}_n$, we can compile a corresponding program of $\mathscr{T}$. When this corresponding program terminates, the tape configuration will be

$$\ldots \; B \; B \; B \; x_1 \; B \; \ldots \; B \; x_m \; B \; z_1 \; B \; \ldots \; B \; z_k \; B \; y \; B \; B \; B \; \ldots,$$
$$\uparrow$$

where the values between blanks are those of the variables of $\mathscr{P}$ on its termination. However, we wish only y to remain as output. Hence to obtain our program $\mathscr{Q}$ in the language $\mathscr{T}$ we put at the end of the compiled Post–Turing program the following:

ERASE A BLOCK $[l - 1]$

After this last has been executed, all but the last block will have been erased and the tape configuration will be

$$\ldots \; B \; B \; B \; B \; y \; B \; B \; B \; \ldots.$$
$$\uparrow$$

Thus, the output is in precisely the form required for us to be able to assert that our Post–Turing program computes f *strictly*.

Exercises

1. **(a)** Use the construction in the proof of Theorem 5.1 to give a Post–Turing program that computes the function $f(x)$ computed by the $\mathscr{S}_2$ program

```
[A]   IF X ENDS s1 GOTO B
      X ← X⁻
      IF X ≠ 0 GOTO A
      GOTO E
[B]   Y ← s1Y
      X ← X⁻
      GOTO A
```

   (b) Do the same as (a) for $f(x_1, x_2)$.

2. Answer question 1(a) with the instruction $[B]Y \leftarrow s_1Y$ replaced by $[B]Y \leftarrow Y + 1$.

3. Give a primitive recursive function $b_1(n, x, z)$ such that any partial function computed by an $\mathscr{S}_n$ program that has x instructions and that uses only variables among $X_1, \ldots, X_l, Z_1, \ldots, Z_k, Y$ is computed strictly by a Post–Turing program with no more than $b_1(n, x, l + k + 1)$ instructions.

4. Give a primitive recursive function $b_2^{(m)}(n, x_1, \ldots, x_m, y, z)$ such that any partial function computed by an $\mathscr{S}_n$ program in y steps on input $x_1, \ldots, x_m$, using only variables among $X_1, \ldots, X_l, Z_1, \ldots, Z_k, Y$, is computed strictly by some Post–Turing program in no more than $b_2^{(m)}(n, x_1, \ldots, x_m, y, l + k + 1)$ steps. [*Hint:* Note that after y steps no variable holds a value larger than $\max\{x_1, \ldots, x_m\} + y$.]

5.* Give an upper bound on the length of the shortest Post–Turing program that computes $\Phi_y(x)$. [See Exercise 3.4.]

6. Simulation of $\mathscr{T}$ in $\mathscr{S}$

In this section we will prove

Theorem 6.1. If there is a Post–Turing program that computes the partial function $f(x_1, \ldots, x_m)$, then f is partially computable.

What this theorem asserts is that if the m-ary partial function f on A^* is computed by a program of $\mathscr{T}$, then there is a program of $\mathscr{S}$ that computes f (regarded as an m-ary partial function on the base n numerical values of the strings). Before giving the proof we observe some of the consequences of this theorem. As shown in Fig. 6.1, the theorem completes a "circle" of implications. Thus all of the conditions in the figure are equivalent. To summarize:

Theorem 6.2. Let f be an m-ary partial function on A^*, where A is an alphabet of n symbols. Then the following conditions are all equivalent:

1. f is partially computable;
2. f is partially computable in $\mathscr{S}_n$;

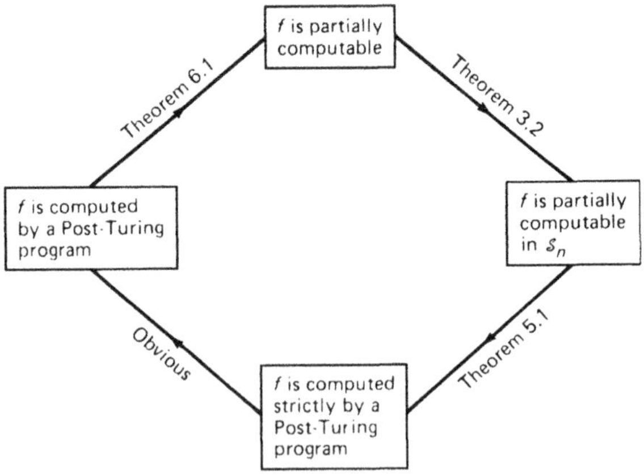

Figure 6.1

3. f is computed strictly by a Post–Turing program;
4. f is computed by a Post–Turing program.

The equivalence of so many different notions of computability constitutes important evidence for the correctness of our identification of intuitive computability with these notions, i.e., for the correctness of Church's thesis.

Shifting our point of view to that of an m-ary partial function on N, we have

Corollary 6.3. For any $n, l \geq 1$, an m-ary partial function f on N is partially computable in $\mathscr{S}_n$ if and only if it is also partially computable in $\mathscr{S}_l$.

Proof. Each of these conditions is equivalent to the function f being partially computable. ■

By considering the language $\mathscr{S}_1$ we have

Corollary 6.4. Every partially computable function is computed strictly by some Post–Turing program that uses only the symbols s_0, s_1.

Now we return to the proof of Theorem 6.1. Let $\mathscr{P}$ be a Post–Turing program that computes f. We want to construct a program $\mathscr{Q}$ in the language $\mathscr{S}$ that computes f. $\mathscr{Q}$ will consist of three sections:

BEGINNING
MIDDLE
END

The MIDDLE section will simulate $\mathscr{P}$ in a step-by-step "interpretive" manner. The task of BEGINNING is to arrange the input to $\mathscr{Q}$ in the appropriate format for MIDDLE, and the task of END is to extract the output.

Let us suppose that f is an m-ary partial function on A^*, where $A = \{s_1, \ldots, s_n\}$. The Post–Turing program $\mathscr{P}$ will also use the blank B and perhaps additional symbols (we are not assuming that the computation is *strict*!) $s_{n+1}, \ldots, s_r$. We write the symbols that $\mathscr{P}$ uses in the order

$$s_1, \ldots, s_n, s_{n+1}, \ldots, s_r, B.$$

The program $\mathscr{Q}$ will simulate $\mathscr{P}$ by using the numbers that strings on this alphabet represent in base $r + 1$ as "codes" for the corresponding strings. Note that as we have arranged the symbols, the blank B represents the number $r + 1$. For this reason *we will write the blank as* s_{r+1} *instead of* s_0. The tape configuration at a given stage in the computation by $\mathscr{P}$ will be kept track of by $\mathscr{Q}$ using three numbers stored in the variables L, H, and R. The value of H will be the numerical value of the symbol currently being scanned by the *head*. The value of L will be a number which represents in base $r + 1$ a string of symbols w such that the tape contents *to the left of the head* consists of infinitely many blanks followed by w. The value of R represents in a similar manner the string of symbols to the *right* of the head. For example, consider the tape configuration

$$\ldots \ B \ B \ B \ B \ s_2 \ s_1 \ B \ \underset{\uparrow}{s_3} \ s_1 \ s_2 \ B \ B \ B \ldots .$$

Here $r = 3$, so we will use the base 4. Then we would have

$$H = 3.$$

We might have

$$L = 2 \cdot 4^2 + 1 \cdot 4 + 4 = 40,$$
$$R = 1 \cdot 4 + 2 = 6.$$

An alternative representation could show some of the blanks on the left or right explicitly. For example, recalling that B represents $r + 1 = 4$,

$$L = 4 \cdot 4^3 + 2 \cdot 4^2 + 1 \cdot 4 + 4 = 296,$$
$$R = 1 \cdot 4^3 + 2 \cdot 4^2 + 4 \cdot 4 + 4 = 116.$$

Now it is easy to simulate the instruction types of $\mathscr{T}$ by programs of $\mathscr{S}$.

An instruction PRINT s_i is simulated by

$$H \leftarrow i$$

An instruction IF s_i GOTO L is simulated by

$$\text{IF } H = i \text{ GOTO } L$$

An instruction RIGHT is simulated by

$$\begin{aligned}
&L \leftarrow \text{CONCAT}_{r+1}(L, H)\\
&H \leftarrow \text{LTEND}_{r+1}(R)\\
&R \leftarrow \text{LTRUNC}_{r+1}(R)\\
&\text{IF } R \neq 0 \text{ GOTO } E\\
&R \leftarrow r + 1
\end{aligned}$$

Similarly an instruction LEFT is simulated by

$$\begin{aligned}
&R \leftarrow \text{CONCAT}_{r+1}(H, R)\\
&H \leftarrow \text{RTEND}_{r+1}(L)\\
&L \leftarrow \text{RTRUNC}_{r+1}(L)\\
&\text{IF } L \neq 0 \text{ GOTO } E\\
&L \leftarrow r + 1
\end{aligned}$$

Now the section MIDDLE of $\mathscr{Q}$ can be assembled simply by replacing each instruction of $\mathscr{P}$ by its simulation.

In writing BEGINNING and END we must deal with the fact that f is an m-ary function on $\{s_1, \ldots, s_n\}^*$. Thus the initial values of $X_1, \ldots, X_m$ for $\mathscr{Q}$ will be numbers that represent the input strings in base n. Theorem 1.1 will enable us to change base as required. The section BEGINNING has the task of calculating the initial values of L, H, R, that is, the values corresponding to the tape configuration

$$\begin{matrix} B & x_1 & B & x_2 & B & \ldots & B & x_m, \\ \uparrow & & & & & & & \end{matrix}$$

where the numbers $x_1, \ldots, x_m$ are represented in base n notation. Thus the section BEGINNING of $\mathscr{Q}$ can simply be taken to be

$$\begin{aligned}
&L \leftarrow r + 1\\
&H \leftarrow r + 1\\
&Z_1 \leftarrow \text{UPCHANGE}_{n, r+1}(X_1)\\
&Z_2 \leftarrow \text{UPCHANGE}_{n, r+1}(X_2)\\
&\vdots\\
&Z_m \leftarrow \text{UPCHANGE}_{n, r+1}(X_m)\\
&R \leftarrow \text{CONCAT}_{r+1}(Z_1, r+1, Z_2, r+1, \ldots, r+1, Z_m)
\end{aligned}$$

Finally, the section END of $\mathscr{Q}$ can be taken simply to be

$$Z \leftarrow \text{CONCAT}_{r+1}(L, H, R)$$

$$Y \leftarrow \text{DOWNCHANGE}_{n, r+1}(Z)$$

We have now completed the description of the program $\mathscr{Q}$ that simulates $\mathscr{P}$, and our proof is complete. ■

Exercises

1. Use the construction in the proof of Theorem 6.1 to give an $\mathscr{S}$ program that computes the same unary function as the Post-Turing program in Fig. 4.5. You may use macros.

2. For any Post–Turing program $\mathscr{P}$, let $\#(\mathscr{P})$ be $\#(\mathscr{Q})$, where $\mathscr{Q}$ is the $\mathscr{S}$ program obtained for $\mathscr{P}$ in the proof of Theorem 6.1, and let $\mathrm{HALT}_{\mathscr{T}}(x, y)$ be defined

$$\mathrm{HALT}_{\mathscr{T}}(x, y) \Leftrightarrow y \text{ is the number of a Post–Turing program that eventually halts on input } x.$$

Show that $\mathrm{HALT}_{\mathscr{T}}(x, y)$ is not a computable predicate.

3.* Show that the Post–Turing programs, under an appropriate ordering $\mathscr{P}_0, \mathscr{P}_1, \ldots,$ are an acceptable programming system. [See Exercise 5.4 in Chapter 4 for the definition of acceptable programming systems.]

6

Turing Machines

1. Internal States

Now we turn to a variant of the Post–Turing language that is closer to Turing's original formulation. Instead of thinking of a list of instructions, we imagine a device capable of various internal *states*. The device is, at any particular instant, scanning a square on a linear tape just like the one used by Post–Turing programs. The combination of the current internal state with the symbol on the square currently scanned is then supposed to determine the next "action" of the device. As suggested by Turing's analysis of the computation process (see Chapter 5, Section 4), we can take the next action to be either "printing" a symbol on the scanned square or moving one square to the right or left. Finally, the device must be permitted to enter a new state.

We use the symbols $q_1, q_2, q_3, \ldots$ to represent states and we write $s_0, s_1, s_2, \ldots$ to represent symbols that can appear on the tape, where as usual $s_0 = B$ is the "blank." By a *quadruple* we mean an expression of one of the following forms consisting of four symbols:

1. $q_i \quad s_j \quad s_k \quad q_l,$
2. $q_i \quad s_j \quad R \quad q_l,$
3. $q_i \quad s_j \quad L \quad q_l.$

We intend a quadruple of type 1 to signify that in state q_i scanning symbol s_j, the device will print s_k and go into state q_l. Similarly, a quadruple of type 2 signifies that in state q_i scanning s_j the device will move one square to the right and then go into state q_l. Finally, a quadruple of type 3 is like one of type 2 except that the motion is to the left.

We now define a *Turing machine* to be a finite set of quadruples, no two of which begin with the same pair $q_i s_j$. Actually, any finite set of quadruples is called a *nondeterministic Turing machine*. But for the present we will deal only with *deterministic Turing machines*, which satisfy the additional "consistency" condition forbidding two quadruples of a given machine to begin with the same pair $q_i s_j$, thereby guaranteeing that at any stage a Turing machine is capable of only one action. Nondeterministic Turing machines are discussed in Section 5.

The *alphabet* of a given Turing machine $\mathscr{M}$ consists of all of the symbols s_i which occur in quadruples of $\mathscr{M}$ *except* s_0.

We stipulate that a Turing machine always begins in state q_1. Moreover, a Turing machine will *halt* if it is in state q_i scanning s_j and *there is no quadruple of the machine which begins* $q_i s_j$. With these understandings, and using the same conventions concerning input and output that were employed in connection with Post–Turing programs, it should be clear what it means to say that some given Turing machine $\mathscr{M}$ *computes* a partial function f on A^* for a given alphabet A.

Just as for Post–Turing programs, we may speak of a Turing machine $\mathscr{M}$ that computes a function *strictly*, namely: assuming that $\mathscr{M}$ computes f where f is a partial function on A^*, we say that $\mathscr{M}$ *computes f strictly* if

1. the alphabet of $\mathscr{M}$ is a subset of A;
2. whenever $\mathscr{M}$ halts, the final configuration has the form

$$\begin{array}{c} By \\ \uparrow \\ q_i \end{array}$$

where y contains no blanks.

Writing $s_0 = B$, $s_1 = 1$ consider the Turing machine with alphabet $\{1\}$:

$$\begin{array}{cccc} q_1 & B & R & q_2 \\ q_2 & 1 & R & q_2 \\ q_2 & B & 1 & q_3 \\ q_3 & 1 & R & q_3 \\ q_3 & B & 1 & q_1 . \end{array}$$

Table 1.1

	State		
Symbol	q_1	q_2	q_3
B	$R\ q_2$	$1\ q_3$	$1\ q_1$
1		$R\ q_2$	$R\ q_3$

We can check the computation:

$$\begin{array}{llllll} B111, & B111,\ldots, & B111B, & B1111, & B1111B, & B11111 \\ \uparrow & \uparrow & \quad\uparrow & \quad\uparrow & \quad\quad\uparrow & \quad\quad\uparrow \\ q_1 & q_2 & \quad q_2 & \quad q_3 & \quad\quad q_3 & \quad\quad q_1 \end{array}$$

The computation halts because there is no quadruple beginning $q_1 1$. Clearly, this Turing machine computes (but not strictly) the function $f(x) = x + 2$, where we are using unary (base 1) notation. The steps of the computation, which explicitly exhibit the state of the machine, the string of symbols on the tape, as well as the individual square on the tape being scanned, are called *configurations*.

It is sometimes helpful to exhibit a Turing machine by giving a state versus symbol table. Thus, for example the preceding Turing machine could be represented as shown in Table 1.1.

Another useful representation is by a state transition diagram. The Turing machine being discussed thus could be represented by the diagram shown in Fig. 1.1.

We now prove

Theorem 1.1. Any partial function that can be computed by a Post–Turing program can be computed by a Turing machine using the same alphabet.

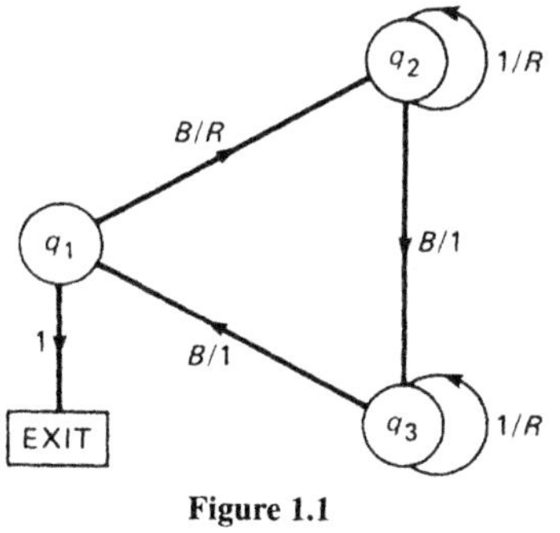

Figure 1.1

Proof. Let $\mathscr{P}$ be a given Post–Turing program consisting of the instructions $I_1, \ldots, I_K$, and let $s_0, s_1, \ldots, s_n$ be a list that includes all of the symbols mentioned in $\mathscr{P}$. We shall construct a Turing machine $\mathscr{M}$ that simulates $\mathscr{P}$.

The idea is that $\mathscr{M}$ will be in state q_i precisely when $\mathscr{P}$ is about to execute instruction I_i. Thus, if I_i is "PRINT s_k," then we place in $\mathscr{M}$ all of the quadruples

$$q_i \;\; s_j \;\; s_k \;\; q_{i+1}, \qquad j = 0, 1, \ldots, n.$$

If I_i is "RIGHT," then we place in $\mathscr{M}$ all of the quadruples

$$q_i \;\; s_j \;\; R \;\; q_{i+1}, \qquad j = 0, 1, \ldots, n.$$

If I_i is "LEFT," then we place in $\mathscr{M}$ all of the quadruples

$$q_i \;\; s_j \;\; L \;\; q_{i+1}, \qquad j = 0, 1, \ldots, n.$$

Finally, if I_i is "IF s_k GOTO L," let m be the least number such that I_m is labeled L if there is an instruction of $\mathscr{P}$ labeled L; otherwise let $m = K + 1$. We place in $\mathscr{M}$ the quadruple

$$q_i \;\; s_k \;\; s_k \;\; q_m$$

as well as all of the quadruples:

$$q_i \;\; s_j \;\; s_j \;\; q_{i+1}, \qquad j = 0, 1, \ldots, n; \quad j \neq k.$$

It is clear that the actions of $\mathscr{M}$ correspond precisely to the instructions of $\mathscr{P}$, so we are done. ■

Using Corollary 6.4 from Chapter 5 and the *proof* of Theorem 1.1, we have

Theorem 1.2. Let f be an m-ary partially computable function on A^* for a given alphabet A. Then there is a Turing machine $\mathscr{M}$ that computes f strictly.

It is particularly interesting to apply this theorem to the case $A = \{1\}$. Thus, if $f(x_1, \ldots, x_m)$ is any partially computable function on N, there is a Turing machine that computes f using only the symbols B and 1. The initial configuration corresponding to inputs $x_1, \ldots, x_m$ is

$$\begin{array}{l} B \;\; 1^{[x_1]} \;\; B \;\; \ldots \;\; B \;\; 1^{[x_m]} \\ \uparrow \\ q_1 \end{array}$$

and the final configuration when $f(x_1, \ldots, x_m)\downarrow$ will be

$$\begin{array}{l} B \quad 1^{[f(x_1, \ldots, x_m)]}. \\ \uparrow \\ q_{K+1} \end{array}$$

Next we shall consider a variant notion of Turing machines: machines that consist of quintuples instead of quadruples. There are two kinds of quintuples:

$$q_i \quad s_j \quad s_k \quad R \quad q_l,$$
$$q_i \quad s_j \quad s_k \quad L \quad q_l.$$

The first quintuple signifies that when the machine is in state q_i scanning s_j it will print s_k and *then* move one square to the right and go into state q_l. And naturally, the second quintuple is the same, except that the motion is to the left. A finite set of quintuples no two of which begin with the same pair $q_i s_j$ is called a *quintuple Turing machine*. We can easily prove

Theorem 1.3. Any partial function that can be computed by a Turing machine can be computed by a quintuple Turing machine using the same alphabet.

Proof. Let $\mathscr{M}$ be a Turing machine with states $q_1, \ldots, q_K$ and alphabet $\{s_1, \ldots, s_n\}$. We construct a quintuple Turing machine $\bar{\mathscr{M}}$ to simulate $\mathscr{M}$. The states of $\bar{\mathscr{M}}$ will be $q_1, \ldots, q_K, q_{K+1}, \ldots, q_{2K}$.

For each quadruple of $\mathscr{M}$ of the form $q_i\, s_j\, R\, q_l$ we place the corresponding quintuple $q_i\, s_j\, s_j\, R\, q_l$ in $\bar{\mathscr{M}}$. Similarly, for each quadruple $q_i\, s_j\, L\, q_l$ in $\mathscr{M}$, we place the quintuple $q_i\, s_j\, s_j\, L\, q_l$ in $\bar{\mathscr{M}}$. And, for each quadruple $q_i\, s_j\, s_k\, q_l$ in $\mathscr{M}$, we place in $\bar{\mathscr{M}}$ the quintuple $q_i\, s_j\, s_k\, R\, q_{K+l}$. Finally we place in $\bar{\mathscr{M}}$ all quintuples of the form

$$q_{K+i} \quad s_j \quad s_j \quad L \quad q_i, \qquad i = 1, \ldots, K; \quad j = 0, 1, \ldots, n.$$

Quadruples requiring motion are simulated easily by quintuples. But a quadruple requiring a "print" necessitates using a quintuple which causes a motion after the "print" has taken place. The final list of quintuples undoes the effect of this unwanted motion. The extra states $q_{K+1}, \ldots, q_{2K}$ serve to "remember" that we have gone a square too far to the right. ■

Finally, we will complete another circle by proving

Theorem 1.4. Any partial function that can be computed by a quintuple Turing machine can be computed by a Post–Turing program using the same alphabet.

Combining Theorems 1.1, 1.3, and 1.4, we will have

Corollary 1.5. For a given partial function f, the following are equivalent:

1. f can be computed by a Post–Turing program;
2. f can be computed by a Turing machine;
3. f can be computed by a quintuple Turing machine.

Proof of Theorem 1.4. Let $\mathscr{M}$ be a given quintuple Turing machine with states $q_1, \ldots, q_K$ and alphabet $\{s_1, \ldots, s_n\}$. We associate with each state q_i a label A_i and with each pair $q_i s_j$ a label B_{ij}. Each label A_i is to be placed next to the first instruction in the filter:

```
[A_i]     IF s_0 GOTO B_i0
          IF s_1 GOTO B_i1
          .
          .
          .
          IF s_n GOTO B_in
```

If $\mathscr{M}$ contains the quintuple $q_i\, s_j\, s_k\, R\, q_l$, then we introduce the block of instructions

```
[B_ij]     PRINT s_k
           RIGHT
           GOTO A_l
```

Similarly, if $\mathscr{M}$ contains the quintuple $q_i\, s_j\, s_k\, L\, q_l$, then we introduce the block of instructions:

```
[B_ij]     PRINT s_k
           LEFT
           GOTO A_l
```

Finally, if there is no quintuple in $\mathscr{M}$ beginning $q_i s_j$, we introduce the block

```
[B_ij]  GOTO E
```

Then we can easily construct a Post–Turing program that simulates $\mathscr{M}$ simply by putting all of these blocks and filters one under the other. The order is irrelevant except for one restriction: The filter labeled A_1 must begin the program. The entire program is listed in Figure 1.2. ∎

$$
\begin{array}{ll}
[A_1] & \text{IF } s_0 \text{ GOTO } B_{10} \\
 & \vdots \\
 & \text{IF } s_n \text{ GOTO } B_{1n} \\
[A_2] & \text{IF } s_0 \text{ GOTO } B_{20} \\
 & \vdots \\
 & \text{IF } s_n \text{ GOTO } B_{2n} \\
 & \vdots \\
[A_K] & \text{IF } s_0 \text{ GOTO } B_{K0} \\
 & \vdots \\
 & \text{IF } s_n \text{ GOTO } B_{Kn} \\
[B_{i_1 j_1}] & \text{PRINT } s_{k_1} \\
 & \text{RIGHT} \\
 & \text{GOTO } A_{l_1} \\
[B_{i_2 j_2}] & \text{PRINT } s_{k_2} \\
 & \vdots
\end{array}
$$

Figure 1.2

Exercises

1. Let T be the Turing machine consisting of the quadruples

$$
\begin{array}{llll}
q_1 & B & R & q_2 \\
q_2 & 1 & R & q_3 \\
q_3 & B & R & q_4 \\
q_4 & 1 & B & q_1 \\
q_4 & B & R & q_4 .
\end{array}
$$

For each integer x, let $g(x)$ be the number of occurrences of 1 on the tape when and if T halts when started with the read–write head one square to the left of the initial 1, with input $1^{[x]}$. What is the function $g(x)$?

2. Write out the quadruples constituting a Turing machine that computes the function

$$
f(x) = \begin{cases} 1 & \text{if } x \text{ is a perfect square} \\ 0 & \text{otherwise} \end{cases}
$$

in base 1. Exhibit the state transition diagram for your machine.

3. Give precise definitions of *configuration*, *computation*, and *Turing machine $\mathscr{M}$ computes the function f*. (Compare Chapter 2, Section 3.)

4. For each of the following functions, construct a Turing machine that computes the function strictly.

(a) $f(u, v) = \widehat{uv}$.

(b) $P(x) \Leftrightarrow x$ has an even number of symbols.

(c) $f(x)$ given in Exercise 1.4 in Chapter 5.

(d) $f(x) = x^R$. [x^R is defined in Chapter 1, Section 3.]

(e) $\#(u, v)$ given in Exercise 1.6 in Chapter 5.

5. Construct Turing machines for Exercise 4.4 in Chapter 5.

6. Construct a Turing machine for Exercise 4.5 in Chapter 5.

7. Using the construction in the proof of Theorem 1.1, transform the Post–Turing program in Figure 4.4 of Chapter 5 into an equivalent Turing machine.

8. Using the construction in the proof of Theorem 1.3, transform the Turing machine in Table 1.1 into an equivalent quintuple Turing machine.

9. Construct a quintuple Turing machine that computes $f(x, y) = x \dot{-} y$ in base 1 strictly.

10.* Show that any partially computable function can be computed by a quintuple Turing machine with two states. [*Hint:* A quintuple Turing machine $\mathscr{M}$ with n states and m symbols (including s_0) can be simulated by a quintuple Turing machine $\mathscr{M}'$ with two states and $4mn + m$ symbols. The $4mn$ new symbols represent the current state and currently scanned symbol of $\mathscr{M}$, as well as additional bookkeeping information. Transferring this stored information to an adjacent square can be done by a "loop" that moves the tape head back and forth.]

2. A Universal Turing Machine

Let us now recall the partially computable function $\Phi(x, z)$ from Chapter 4. For fixed z, $\Phi(x, z)$ is the unary partial function computed by the program whose number is z. Let $\mathscr{M}$ be a Turing machine (in either quadruple or quintuple form) that computes this function with alphabet $\{1\}$. For reasons that we will explain, it is appropriate to call this machine $\mathscr{M}$ *universal.*

Let $g(x)$ be any partially computable function of one variable and let z_0 be the number of some program in the language $\mathscr{S}$ that computes g. Then if we begin with a configuration

$$\begin{array}{l} B \ x \ B \ z_0 \\ \uparrow \\ q_1 \end{array}$$

(where x and z_0 are written as blocks of ones, i.e., in unary notation), and let $\mathscr{M}$ proceed to compute, $\mathscr{M}$ will compute $\Phi(x, z_0)$, i.e., $g(x)$. Thus, $\mathscr{M}$ can be used to compute any partially computable function of one variable.

$\mathscr{M}$ provides a suggestive model of an all-purpose computer, in which data and programs are stored together in a single "memory." We can think of z_0 as a coded version of the program for computing g and x as the input to that program. Turing's construction of a universal computer in 1936 provided reason to believe that, at least in principle, an all-purpose computer would be possible, and was thus an anticipation of the modern digital computer.

Exercises

1.* **(a)** Define a numbering $\#(\mathscr{M})$ of Turing machines like the numbering $\#(\mathscr{P})$ of $\mathscr{S}$ programs given in Chapter 4.

(b) Prove a version of the parameter theorem for Turing machines.

(c) Prove a version of the recursion theorem for Turing machines.

(d) Show that there is a Turing machine $\mathscr{M}$ that prints $\#(\mathscr{M})$ when started with any input tape.

(e) Show that Turing machines are an acceptable programming system. [Acceptable programming systems are defined in Exercise 5.4 in Chapter 4.]

2.* Give an upper bound on the size of the smallest universal Turing machine. [See Exercise 5.5 in Chapter 5.]

3. The Languages Accepted by Turing Machines

Given a Turing machine $\mathscr{M}$ with alphabet $A = \{s_1, \ldots, s_n\}$, a word $u \in A^*$ is said to be *accepted* by $\mathscr{M}$ if when $\mathscr{M}$ begins with the configuration

$$\begin{array}{l} s_0 \ u \\ \uparrow \\ q_1 \end{array}$$

it will eventually halt. The set of all words $u \in A^*$ that $\mathcal{M}$ accepts is called the *language accepted by* $\mathcal{M}$. An important problem in the theory of computation involves characterizing the languages accepted by various kinds of computing devices. It is easy for us to solve this problem for Turing machines.

Theorem 3.1. A language is accepted by some Turing machine if and only if the language is r.e.

Proof. Let L be the language accepted by a Turing machine $\mathcal{M}$ with alphabet A. Let $g(x)$ be the unary function on A^* that $\mathcal{M}$ computes. Then g is a partially computable function (by Corollary 1.5 and by Theorem 6.2 in Chapter 5). Now,

$$L = \{x \in A^* \mid g(x)\downarrow\}. \tag{3.1}$$

Hence L is r.e.

Conversely, let L be r.e. Then there is a partially computable function $g(x)$ such that (3.1) holds. Using Theorem 1.2, let $\mathcal{M}$ be a Turing machine with alphabet $\{s_1, \ldots, s_n\}$ that computes $g(x)$ strictly. Then $\mathcal{M}$ accepts L. ∎

Naturally Theorem 3.1 is also true for quintuple Turing machines.

Let us consider the special case $A = \{1\}$. Then we have

Theorem 3.2. A set U of numbers is r.e. if and only if there is a Turing machine $\mathcal{M}$ with alphabet $\{1\}$ that accepts $1^{[x]}$ if and only if $x \in U$.

Proof. This follows immediately from Theorem 3.1 and the fact that the base 1 representation of the number x is the string $1^{[x]}$. ∎

This is an appropriate place to consider some annoying ambiguities in our notation of r.e. language. Thus, for example, consider the language

$$L_0 = \{a^{[n]} \mid n > 0\},$$

on the alphabet $\{a, b\}$. According to our definitions, to say that L_0 is an r.e. language is to say that the set of numbers which the strings in L_0 represent in base 2 is an r.e. set of numbers. But, *this set of numbers is not determined until an order is specified for the letters of the alphabet.* If we take a, b in the order shown, then the set of numbers which represent strings in L_0 is clearly

$$Q_1 = \{2^n - 1 \mid n > 0\},$$

while if we take the letters in the order b, a, the set of numbers which represent strings in L_0 is

$$Q_2 = \{2x \mid x \in Q_1\} = \{2^{n+1} - 2 \mid n > 0\}.$$

Now, although there is no difficulty whatever in showing that Q_1 and Q_2 are both r.e. sets, it is nevertheless a thoroughly unsatisfactory state of affairs to be forced to be concerned with such matters in asserting that L_0 is an r.e. language. Here Theorem 3.1 comes to our rescue. The notion of a given string being accepted by a Turing machine does not involve imposing any order on the symbols of the alphabet. Hence, Theorem 3.1 implies immediately that whether a particular language on a given alphabet is r.e. is *independent of how the symbols of the alphabet are ordered*. The same is clearly true of a language L on a given alphabet A being *recursive* since this is equivalent to L and $A^* - L$ both being r.e.

Another ambiguity arises from the fact that a particular language may be considered with respect to more than one alphabet. Thus, let A be an n-letter alphabet and let $\tilde{A}$ be an m-letter alphabet containing A, so that $m > n$. Then a language L on the alphabet A is simply some subset of A^*, so that L is also a language on the larger alphabet $\tilde{A}$. Thus, depending on whether we are thinking of L as a language on A or as a language on $\tilde{A}$, we will have to read the strings in L as being the notation for integers in *base n or in base m, respectively*. Hence, we are led to the unpleasant possibility that whether L is r.e. might actually depend on which alphabet we are considering. As an example, we may take $A = \{a\}$ and $\tilde{A} = \{a, b\}$, and consider the language L_0 above, where

$$L_0 \subseteq A^* \subseteq \tilde{A}^*.$$

We have already seen that our original definition of L_0's being r.e. as a language on the alphabet $\tilde{A}$ amounts to requiring that the set of numbers Q_1 or Q_2 (depending on the order of the symbols a, b) be r.e. However, if we take our alphabet to be A, then the relevant set of numbers is

$$Q_3 = \{n \in N \mid n > 0\}.$$

We remove all such ambiguities by proving

Theorem 3.3. Let $A \subseteq \tilde{A}$ where A and $\tilde{A}$ are alphabets and let $L \subseteq A^*$. Then L is an r.e. language on the alphabet A if and only if L is an r.e. language on $\tilde{A}$.

Proof. Let L be r.e. on A and let $\mathcal{M}$ be a Turing machine with alphabet A that accepts L. Without loss of generality, we can assume that $\mathcal{M}$ begins by moving right until it finds a blank and then returns to its original position. Let $\tilde{\mathcal{M}}$ be obtained from $\mathcal{M}$ by adjoining to it the quadruples $q\,s\,s\,q$ for each symbol $s \in \tilde{A} - A$, and each state q of $\mathcal{M}$. Thus $\tilde{\mathcal{M}}$ will enter an "infinite loop" if it ever encounters a symbol in $\tilde{A} - A$. Since $\tilde{\mathcal{M}}$ has alphabet $\tilde{A}$ and accepts the language L, we conclude from Theorem 3.1 that L is an r.e. language on $\tilde{A}$.

Conversely, let L be r.e. as a language on $\tilde{A}$, and let $\mathcal{M}$ be a Turing machine with alphabet $\tilde{A}$ that accepts L. Let $g(x)$ be the function on A^* that $\mathcal{M}$ computes. (The symbols belonging to $\tilde{A} - A$ thus serve as "markers.") Since $L \subseteq A^*$, we have

$$L = \{x \in A^* \mid g(x)\downarrow\}.$$

Since $g(x)$ is partially computable, it follows that L is an r.e. language on A. ■

Corollary 3.4. Let $A, \tilde{A}, L$ be as in Theorem 3.3. Then L is a recursive language on A if and only if L is a recursive language on $\tilde{A}$.

Proof. First let L be a recursive language on A. Then L and $A^* - L$ are r.e. languages on A and therefore on $\tilde{A}$. Moreover, since

$$\tilde{A}^* - L = (\tilde{A}^* - A^*) \cup (A^* - L),$$

and since $\tilde{A}^* - A^*$ is r.e., as the reader can easily show (see Exercise 6), it follows from Theorem 4.5 in Chapter 4 that $\tilde{A}^* - L$ is r.e. Hence, L is a recursive language on $\tilde{A}$.

Conversely, if L is a recursive language on $\tilde{A}$, then L and $\tilde{A}^* - L$ are r.e. languages on $\tilde{A}$ and therefore L is an r.e. language on A. Moreover, since

$$A^* - L = (\tilde{A}^* - L) \cap A^*,$$

and since A^* is obviously r.e. (as a language on A and therefore on $\tilde{A}$), it follows from Theorem 4.5 in Chapter 4 that $A^* - L$ is an r.e. language on $\tilde{A}$ and hence on A. Thus, L is a recursive language on A. ■

Exercises

1. Write out the quadruples constituting a Turing machine that accepts the language consisting of all words on the alphabet $\{a, b\}$ of the form $a^{[i]}ba^{[i]}$.
2. Give a Turing machine that accepts $\{1^{[i]}B1^{[j]}B1^{[i+j]} \mid i, j \in N\}$.
3. Give a Turing machine that accepts $\{w \in \{a, b\}^* \mid w = w^R\}$.

4. Show that there is a Turing machine that accepts the language $\{1^{[x]}B1^{[y]} \mid \Phi_y(x)\downarrow\}$.
5. Show that there is no Turing machine that accepts the language $\{1^{[y]} \mid \Phi_y(x)\downarrow \text{ for all } x \in N\}$.
6. Complete the proof of Corollary 3.4 by showing that $\tilde{A}^* - A^*$ is an r.e. language.

4. The Halting Problem for Turing Machines

We can use the results of the previous section to obtain a sharpened form of the unsolvability of the halting problem.

By the halting problem for a *fixed* given Turing machine $\mathcal{M}$ we mean the problem of finding an algorithm *to determine whether* $\mathcal{M}$ *will eventually halt starting with a given configuration.* We have

Theorem 4.1. There is a Turing machine $\mathcal{K}$ with alphabet {1} that has an unsolvable halting problem.

Proof. Take for the set U in Theorem 3.2, some r.e. set that is not recursive (e.g., the set K from Chapter 4). Let $\mathcal{K}$ be the corresponding Turing machine. Thus $\mathcal{K}$ accepts a string of ones if and only if its length belongs to U. Hence, $x \in U$ if and only if $\mathcal{K}$ eventually halts when started with the configuration

$$\begin{array}{l} B \;\; 1^{[x]} \\ \uparrow \\ q_1 \end{array}$$

Thus, if there were an algorithm for solving the halting problem for $\mathcal{K}$, it could be used to test a given number x for membership in U. Since U is not recursive, such an algorithm is impossible. ■

This is really a stronger result than was obtained in Chapter 4. What we can prove about Turing machines just using Theorem 2.1 in Chapter 4 is that there is no algorithm that can be used, given a Turing machine and an initial configuration, to determine whether the Turing machine will ever halt. Our present result gives a *fixed* Turing machine whose halting problem is unsolvable. Actually, this result could also have been easily obtained from the earlier one by using a universal Turing machine.

Next, we show how the unsolvability of the halting problem can be used to obtain another unsolvable problem concerning Turing machines. We begin with a Turing machine $\mathcal{K}$ with alphabet {1} that has an unsolvable halting problem. Let the states of $\mathscr{K}$ be $q_1, \ldots, q_k$. We will construct a Turing machine $\tilde{\mathscr{K}}$ by adjoining to the quadruples of $\mathscr{K}$ the following quadruples:

$$q_i \ B \ B \ q_{k+1}$$

for $i = 1, 2, \ldots, k$ for which no quadruple of $\mathscr{K}$ begins $q_i B$, and

$$q_i \ 1 \ 1 \ q_{k+1}$$

for $i = 1, 2, \ldots, k$ when no quadruple of $\mathscr{K}$ begins $q_i 1$. Thus, $\mathscr{K}$ eventually halts beginning with a given configuration if and only if $\tilde{\mathscr{K}}$ eventually is in state q_{k+1}. We conclude

Theorem 4.2. There is a Turing machine $\tilde{\mathscr{K}}$ with alphabet $\{1\}$ and a state q_m such that there is no algorithm that can determine whether $\tilde{\mathscr{K}}$ will ever arrive at state q_m when it begins at a given configuration.

Exercises

1. Prove that there is a Turing machine $\mathscr{M}$ such that there is no algorithm that can determine of a given configuration whether $\mathscr{M}$ will eventually halt with a completely blank tape when started with the given tape configuration.
2. Prove that there is a Turing machine $\mathscr{M}$ with alphabet $\{s_1, s_2\}$ such that there is no algorithm that can determine whether $\mathscr{M}$ starting with a given configuration will ever print the symbol s_2.
3. Let $\mathscr{M}_0, \mathscr{M}_1, \ldots$ be a list of all Turing machines, and let f_i be the unary partial function computed by $\mathscr{M}_i$, $i = 0, 1, \ldots$. Suppose $g(x)$ is a total function such that for all $x \geq 0$ and all $0 \leq i \leq x$, if $f_i(x)\downarrow$ then $f_i(x) < g(x)$. Show that $g(x)$ is not computable.
4. Jill and Jack have been working as programmers for a year. They are discussing their work. We listen in:

   JACK: We are working on a wonderful program, AUTOCHECK. AUTOCHECK will accept Pascal programs as inputs and will return the values OK or LOOPS depending on whether the given program is or is not free of infinite loops.

   JILL: Big deal! We have a mad mathematician in our firm who has developed an algorithm so complicated that no program can be written to execute it no matter how much space and time is allowed.

   Comment on and criticize Jack and Jill's statements.

5. Nondeterministic Turing Machines

As already mentioned, a *nondeterministic Turing machine* is simply an arbitrary finite set of quadruples. Thus, what we have been calling a Turing machine is simply a special kind of nondeterministic Turing machine. For emphasis, we will sometimes refer to ordinary Turing machines as *deterministic*.

A configuration

$$\begin{array}{ccc} \ldots & s_j & \ldots \\ & \uparrow & \\ & q_i & \end{array}$$

is called *terminal* with respect to a given nondeterministic Turing machine (and the machine is said to *halt*) if it contains no quadruple beginning $q_i\, s_j$. (This, of course, is exactly the same as for deterministic Turing machines.) We use the symbol $\vdash$ (borrowed from logic) placed between a pair of configurations to indicate that the transition from the configuration on the left to the one on the right is permitted by one of the quadruples of the machine under consideration.

As an example, consider the nondeterministic Turing machine given by the quadruples

$$\begin{array}{cccc} q_1 & B & R & q_2 \\ q_2 & 1 & R & q_3 \\ q_2 & B & B & q_4 \\ q_3 & 1 & R & q_2 \\ q_3 & B & B & q_3 \\ q_4 & B & R & q_4 \\ q_4 & B & B & q_5 \end{array}$$

Then we have

$$\begin{array}{ccccccccccccccccccccccccccccc} B & 1 & 1 & 1 & 1 & \vdash & B & 1 & 1 & 1 & 1 & \vdash & B & 1 & 1 & 1 & 1 & \vdash & B & 1 & 1 & 1 & 1 & \vdash & B & 1 & 1 & 1 & 1 \\ \uparrow & & & & & & & \uparrow & & & & & & & \uparrow & & & & & & & & \uparrow & & & & & & \uparrow \\ q_1 & & & & & & & q_2 & & & & & & & q_3 & & & & & & & & q_2 & & & & & & q_3 \end{array}$$

$$\begin{array}{cccccccccccccc} \vdash & B & 1 & 1 & 1 & 1 & B & B & \vdash & B & 1 & 1 & 1 & 1 & B. \\ & & & & & & \uparrow & & & & & & & & \uparrow \\ & & & & & & q_2 & & & & & & & & q_4 \end{array}$$

So far the computation has been entirely determined; however, at this point the nondeterminism plays a role. We have

$$\begin{array}{ccccccccccccc} B&1&1&1&1&B&\vdash&B&1&1&1&1&B,\\ &&&&&\uparrow&&&&&&&\uparrow\\ &&&&&q_4&&&&&&&q_5\end{array}$$

at which the machine halts. But we also have

$$\begin{array}{ccccccccccccccccccccccccc} B&1&1&1&1&B&\vdash&B&1&1&1&1&B&B&\vdash&B&1&1&1&1&B&B&B&\vdash&\cdots .\\ &&&&&\uparrow&&&&&&&&\uparrow&&&&&&&&&\uparrow&&\\ &&&&&q_4&&&&&&&&q_4&&&&&&&&&q_4&&\end{array}$$

Let $A = \{s_1, \ldots, s_n\}$ be a given alphabet and let $u \in A^*$. Then the nondeterministic Turing machine $\mathscr{M}$ is said to *accept* u if there exists a sequence of configurations $\gamma_1, \gamma_2, \ldots, \gamma_m$ such that γ_1 is the configuration

$$\begin{array}{cc} s_0 & u\\ \uparrow & \\ q_1 & \end{array}$$

γ_m is terminal with respect to $\mathscr{M}$, and $\gamma_1 \vdash \gamma_2 \vdash \gamma_3 \vdash \cdots \vdash \gamma_m$. In this case, the sequence $\gamma_1, \gamma_2, \ldots, \gamma_m$ is called an *accepting computation by* $\mathscr{M}$ *for* u. If A is the alphabet of $\mathscr{M}$, then the *language accepted by* $\mathscr{M}$ is the set of all $u \in A^*$ that are accepted by $\mathscr{M}$.

Of course, for deterministic Turing machines, this definition gives nothing new. However, it is important to keep in mind the distinctive feature of acceptance by nondeterministic Turing machines. It is perfectly possible to have an infinite sequence

$$\gamma_1 \vdash \gamma_2 \vdash \gamma_3 \vdash \cdots$$

of configurations, where γ_1 is

$$\begin{array}{cc} s_0 & u\\ \uparrow & \\ q_1 & \end{array}$$

even though u is accepted by $\mathscr{M}$. It is only necessary that there be *some* sequence of transitions leading to a terminal configuration. One sometimes expresses this by saying, "The machine is always permitted to guess the correct next step."

Thus in the example given above, taking the alphabet $A = \{1\}$, we have that $\mathscr{M}$ accepts 1111. In fact the language accepted by $\mathscr{M}$ is $\{1^{[2^n]}\}$. (See Exercise 3.)

Since a Turing machine is also a nondeterministic Turing machine, Theorem 3.1 can be weakened to give

Theorem 5.1. For every r.e. language L, there is a nondeterministic Turing machine $\mathscr{M}$ that accepts L.

The converse is also true: the language accepted by a nondeterministic Turing machine must be r.e. By Church's thesis, it is clear that this should be true. It is only necessary to "run" a nondeterministic Turing machine $\mathscr{M}$ on a given input u, *following all alternatives at each step*, and giving the value (say) 0, if termination is reached along any branch. This defines a function that is *intuitively* partially computable and whose domain is the language accepted by $\mathscr{M}$. However, a detailed proof along these lines would be rather messy.

Fortunately the converse of Theorem 5.1 will be an easy consequence of the methods we will develop in the next chapter.

Exercises

1. Explain why nondeterministic Turing machines are unsuitable for defining functions.
2. Let L be the set of all words on the alphabet $\{a, b\}$ that contain at least two consecutive occurrences of b. Construct a nondeterministic Turing machine that *never moves left* and accepts L.
3. Show that the nondeterministic Turing machine $\mathscr{M}$ used as an example in this section accepts the set $\{1^{[2n]}\}$.
4. Let

$$L_1 = \{w \in \{a, b\}^* \mid w \text{ has an even number of } a\text{'s}\},$$

$$L_2 = \{w \in \{a, b\}^* \mid w \text{ has an odd number of } b\text{'s}\}.$$

   (a) Give deterministic Turing machines $\mathscr{M}_1, \mathscr{M}_2$ that accept L_1, L_2, respectively, and combine them to get a nondeterministic Turing machine that accepts $L_1 \cup L_2$.

   (b) Give a deterministic Turing machine that accepts $L_1 \cup L_2$.

5. Give a nondeterministic Turing machine that accepts $\{1^{[n]} \mid n \text{ is prime}\}$.
6. If we replace "the first instruction labeled L" by "some instruction labeled L" in the interpretation of Post–Turing instructions of the form IF σ GOTO L, then we get *nondeterministic Post–Turing programs*. Show that a language is accepted by a nondeterministic Post–

Turing program if and only if it is accepted by a nondeterministic Turing machine (where acceptance of a language by a Post–Turing program is defined just like acceptance by a Turing machine).

6. Variations on the Turing Machine Theme

So far we have three somewhat different formulations of Turing's conception of computation: the Post–Turing programming language, Turing machines as made up of quadruples, and quintuple Turing machines. The proof that these formulations are equivalent was quite simple. This is true in part because all three involved a single tapehead on a single two-way infinite tape. But it is easy to imagine other arrangements. In fact, Turing's original formulation was in terms of a tape that was infinite in only one direction, that is, with a first or leftmost square (see Fig. 6.1). We can also think of permitting several tapes, each of which can be one-way or two-way infinite and each with its own tapehead. There might even be several tapeheads per tape. As one would expect, programs can be shorter when several tapes are available. But, if we believe Church's thesis, we certainly would expect all of these formulations to be equivalent. In this section we will indicate briefly how this equivalence can be demonstrated.

Let us begin by considering one-way infinite tapes. To make matters definite, we assume that we are representing a Turing machine as a set of quadruples. It is necessary to make a decision about the effect of a quadruple $q_i\, s_j\, L\, q_k$ in case the tapehead is already at the left end of the tape. There are various possibilities, and it really does not matter very much which we adopt. For definiteness we assume that an instruction to move left will be interpreted as a *halt* in case the tapehead is already at the leftmost square. Now it is pretty obvious that anything that a Turing machine could do on a one-way infinite tape could also be done on a two-way infinite tape, and we leave details to the reader.

How can we see that any partially computable function can be computed by a Turing machine on a one-way infinite tape? One way is by simply examining the proof of Theorem 5.1 in Chapter 5, which shows how a

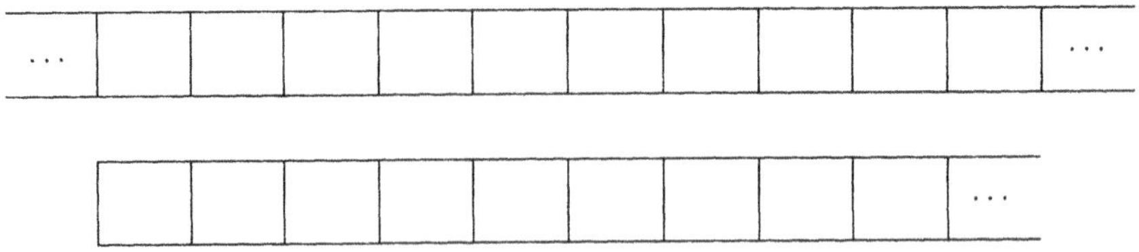

Figure 6.1. Two-way infinite versus one-way infinite tape.

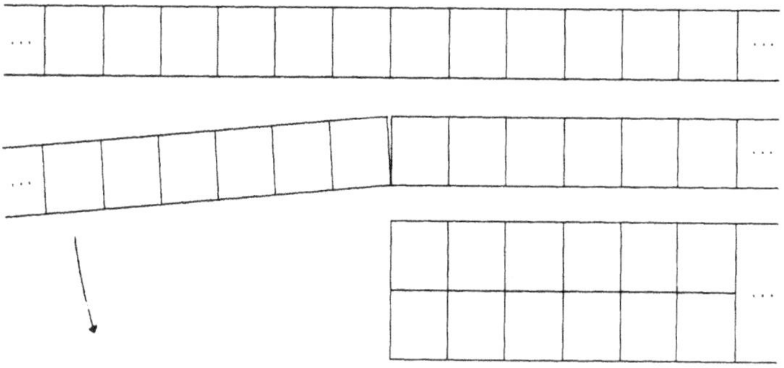

Figure 6.2

computation in any of the languages $\mathscr{S}_n$ can be simulated by a program in the Post–Turing language $\mathscr{T}$. In fact, the program $\mathscr{Q}$ in the language $\mathscr{T}$ which is constructed to simulate a given program $\mathscr{P}$ in the language $\mathscr{S}_n$ has the particular property that when $\mathscr{Q}$ is executed, the tapehead never moves to the left of the square initially scanned. Hence, the program $\mathscr{Q}$ would work exactly as well on a one-way infinite tape whose leftmost square is initially scanned. And, it is an easy matter, as in the proof of Theorem 1.1, to convert $\mathscr{Q}$ into a Turing machine.

Although this is an entirely convincing argument, we would like to mention another approach which is interesting in its own right, namely, we directly face the question, how can the information contained in a two-way infinite tape be handled by a Turing machine with one tapehead on a one-way infinite tape? The intuitive idea is to think of a two-way infinite tape as being "hinged" so it can be folded as in Fig. 6.2. Thus our two-way infinite tape can be represented by a one-way infinite tape with two "tracks," an "upper" and a "lower." Moreover, by adding enough symbols to the alphabet, we can code each pair consisting of an upper and a lower symbol by a single symbol.

Thus, let us begin with a Turing machine $\mathscr{M}$ with alphabet $A = \{s_1, \ldots, s_n\}$ and states $q_1, \ldots, q_K$. Let $\mathscr{M}$ compute a unary[1] partial function g on A_0^*, where $A_0 \subseteq A$. Thus the input configuration when $\mathscr{M}$ is computing $g(x)$ for $x \in A_0^*$ will be

$$\begin{array}{c} B\ x \\ \uparrow \\ q_1 \end{array}$$

[1] The restriction to unary functions is, of course, not essential.

We will construct a Turing machine $\bar{\mathscr{M}}$ that computes g on a one-way infinite tape. The initial configuration for $\bar{\mathscr{M}}$ will be

$$\begin{array}{c} \# \; B \; x \\ \uparrow \\ q_1 \end{array}$$

where # is a special symbol that will occupy the leftmost square on the tape for most of the computation. The alphabet of $\bar{\mathscr{M}}$ will be

$$A \cup \{\#\} \cup \left\{b_j^i \,|\, 0 \le i, j \le n\right\},$$

where we think of the symbol b_j^i as indicating that s_i is on the upper track and s_j is on the lower track. The states of $\bar{\mathscr{M}}$ are q_1, q_2, q_3, q_4, q_5, and

$$\{\bar{q}_i, \tilde{q}_i \,|\, i = 1, 2, \ldots, K\}$$

as well as certain additional states.

We can think of the quadruples constituting $\bar{\mathscr{M}}$ as made up of three sections: BEGINNING, MIDDLE, and END. BEGINNING serves to copy the input on the upper track putting blanks on the corresponding lower track of each square. BEGINNING consists of the quadruples

$$\begin{array}{lllll} q_1 & B & R & q_2 & \\ q_2 & s_i & R & q_2 & i = 1, 2, \ldots, n, \\ q_2 & B & L & q_3 & \\ q_3 & s_i & b_0^i & q_3 & i = 0, 1, 2, \ldots, n, \\ q_3 & b_0^i & L & q_3 & i = 0, 1, 2, \ldots, n, \\ q_3 & \# & R & \bar{q}_1 . & \end{array}$$

Thus, starting with the configuration

$$\begin{array}{c} \# \; B \; s_2 \; s_1 \; s_3 \\ \uparrow \\ q_1 \end{array}$$

BEGINNING will halt in the configuration

$$\begin{array}{c} \# \; b_0^0 \; b_0^2 \; b_0^1 \; b_0^3 \; B. \\ \uparrow \\ \bar{q}_1 \end{array}$$

Note that b_0^0 is different from $s_0 = B$. MIDDLE will consist of quadruples corresponding to those of $\mathscr{M}$ as well as additional quadruples as indicated

Table 6.1

	Quadruple of $\mathscr{M}$	Quadruple of $\bar{\mathscr{M}}$	
(a)	$q_i \quad s_j \quad s_k \quad q_l$	$\bar{q}_i \quad b^j_m \quad b^k_m \quad \bar{q}_l$	$m = 0, 1, \ldots, n$
		$\tilde{q}_i \quad b^m_j \quad b^m_k \quad \tilde{q}_l$	$m = 0, 1, \ldots, n$
(b)	$q_i \quad s_j \quad R \quad q_l$	$\bar{q}_i \quad b^j_m \quad R \quad \bar{q}_l$	$m = 0, 1, \ldots, n$
		$\tilde{q}_i \quad b^m_j \quad L \quad \tilde{q}_l$	$m = 0, 1, \ldots, n$
(c)	$q_i \quad s_j \quad L \quad q_l$	$\bar{q}_i \quad b^j_m \quad L \quad \bar{q}_l$	$m = 0, 1, \ldots, n$
		$\tilde{q}_i \quad b^m_j \quad R \quad \tilde{q}_l$	$m = 0, 1, \ldots, n$
(d)	———	$\bar{q}_i \quad B \quad b^0_0 \quad \bar{q}_i$	$i = 1, 2, \ldots, K$
		$\tilde{q}_i \quad B \quad b^0_0 \quad \tilde{q}_i$	$i = 1, 2, \ldots, K$
(e)	———	$\bar{q}_i \quad \# \quad R \quad \tilde{q}_i$	$i = 1, 2, \ldots, K$
		$\tilde{q}_i \quad \# \quad R \quad \bar{q}_i$	$i = 1, 2, \ldots, K$

in Table 6.1. The states $\bar{q}_i, \tilde{q}_i$ correspond to actions on the upper track and lower track, respectively. Note in (b) and (c) that on the lower track left and right are reversed. The quadruples in (d) replace single blanks B by double blanks b^0_0 as needed. The quadruples (e) arrange for switchover from the upper to the lower track. It should be clear that MIDDLE simulates $\mathscr{M}$.

END has the task of translating the output into a word on the original alphabet A. This task is complicated by the fact that the output is split between the two tracks. To begin with, END contains the following quadruples:

$$\left.\begin{matrix} \bar{q}_i & b^j_m & b^j_m & q_4 \\ \tilde{q}_i & b^m_j & b^m_j & q_4 \end{matrix}\right\} \quad \begin{matrix} \text{whenever } \mathscr{M} \text{ contains no quadruple} \\ \text{beginning } q_i s_j \text{, for } m = 0, 1, \ldots, n; \ 0 \le i, j \le n, \end{matrix}$$

$$\begin{matrix} q_4 & b^i_j & L & q_4, \\ q_4 & \# & B & q_5. \end{matrix}$$

For each initial configuration for which $\mathscr{M}$ halts, the effect of BEGINNING, MIDDLE, and this part of END is to ultimately produce a configuration of the form

$$\begin{matrix} B & b^{i_1}_{j_1} & b^{i_2}_{j_2} & \cdots & b^{i_k}_{j_k}. \\ \uparrow & & & & \\ q_5 & & & & \end{matrix}$$

The remaining task of END is to convert the tape contents into

$$s_{j_k} \ s_{j_{k-1}} \ \cdots \ s_{j_1} \ s_{i_1} \ s_{i_2} \ \cdots \ s_{i_k}.$$

$[D]$	RIGHT TO NEXT BLANK
	MOVE BLOCK RIGHT
	RIGHT
$[C]$	RIGHT
	IF b_j^i GOTO A_j^i $\quad (0 \le i, j \le n)$
	IF B GOTO F
	GOTO C
$[A_j^i]$	PRINT s_i $\quad (0 < i \le n, 0 \le j \le n)$
	GOTO B_j
$[A_j^0]$	PRINT # $\quad (0 \le j \le n)$
	GOTO B_j
$[B_j]$	LEFT TO NEXT BLANK $\quad (0 < j \le n)$
	PRINT s_j
	GOTO D
$[B_0]$	LEFT TO NEXT BLANK
	PRINT #
	GOTO D
$[F]$	LEFT
	IF s_j GOTO F $\quad (0 < j \le n)$
	IF # GOTO G
	IF B GOTO E
$[G]$	PRINT B
	GOTO F

Figure 6.3

Instead of giving quadruples for accomplishing this, we exhibit a program in the Post–Turing language $\mathscr{T}$, so that we can make use of some of the macros available in that language. Of course, this program can easily be translated into a set of quadruples using the method of proof of Theorem 1.1. Because our macros for $\mathscr{T}$ were designed for use with "blocks" of symbols containing no blanks, we will use # instead of $s_0 = B$ in carrying out our translation. One final pass will be needed to replace each # by B. The program is given in Fig. 6.3.

Each b_j^i is processed going from left to right. b_j^i is replaced by s_i (or by # if $i = 0$) and s_j (or # if $j = 0$) is printed on the left. The "MOVE BLOCK RIGHT" macro is used to make room on the tape for printing the successive symbols from the "lower" track. As an example let us apply the program of Fig. 6.3 to the configuration

$$\begin{array}{l} B \; b_1^2 \; b_1^0 \; b_0^1 . \\ \uparrow \end{array}$$

B	b_1^2	b_1^0	b_0^1						D
↑									
B	B	b_1^2	b_1^0	b_0^1	B				C
	↑								
B	B	b_1^2	b_1^0	b_0^1	B				A_1^2
		↑							
B	B	s_2	b_1^0	b_0^1	B				B_1
		↑							
B	s_1	s_2	b_1^0	b_0^1	B				D
	↑								
B	B	s_1	s_2	b_1^0	b_0^1	B			A_1^0
				↑					
B	s_1	s_1	s_2	#	b_0^1	B			D
	↑								
B	#	s_1	s_1	s_2	#	s_1	B		D
	↑								
B	B	#	s_1	s_1	s_2	#	s_1	B	F
								↑	
B	B	B	s_1	s_1	s_2	B	s_1	B	E
	↑								

Figure 6.4

In Fig. 6.4 we show various stages in the computation; in each case the tape configuration is followed by the label on the next instruction to be executed.

The technique of thinking of the tape of a Turing machine as decomposed into a number of parallel tracks has numerous uses. (It will appear again in Chapter 11.) For the moment we note that it can be used to simulate the behavior of a multitape Turing machine by an ordinary Turing machine. For, in the first place a second track can be used to show the position of a tapehead on a one-tape machine as in the example shown in Fig. 6.5; the 1 under the s_3 shows the position of the head. In an entirely similar manner the contents of k tapes and the position of the tapehead on each can be represented as a single tape with $2k$ tracks. Using this representation, it is easy to see how to simulate any computation by a k-tape Turing machine using only one tape. The same result can also be obtained indirectly using the method of proof of Theorem 6.1 in Chapter 5 to show that any function computed by a k-tape Turing machine is partially computable.

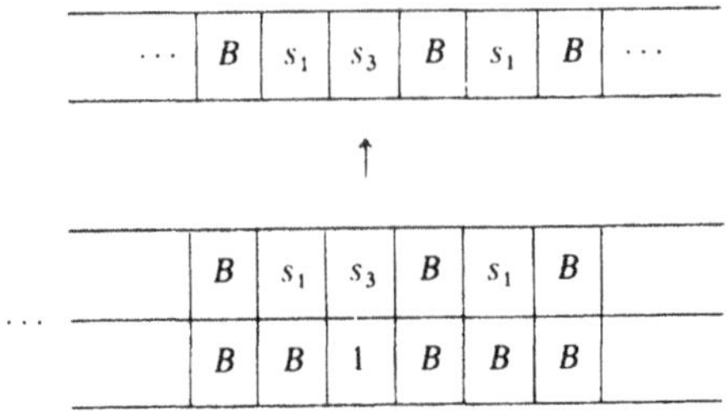

Figure 6.5

Exercises

1. Give a formal description of a Turing machine that uses three tapes: one with a "read only" head for input, one with a "write only" head for output, and one for "working." Give an appropriate definition of computability by such machines and prove the equivalence with computability by ordinary Turing machines.
2. Do the same for a Turing machine with input tape, output tape, and k working tapes for any $k \geq 1$.
3. Let the Post–Turing language be augmented by the instructions UP, DOWN so that it can deal with computations on a two-dimensional "tape" infinite in all four directions. Supply an appropriate definition of what it means to compute a function by a program in this language, and prove that any function computed by such a program is partially computable.
4. Adapt the construction in this section so that it works for binary functions.

7

Processes and Grammars

1. Semi-Thue Processes

In this chapter we will see how the methods of computability theory can be used to deal with combinatorial problems involving substitution of one substring for another in a string.

Definition. Given a pair of words $g, \bar{g}$ on some alphabet, the expression

$$g \to \bar{g}$$

is called a *semi-Thue production* or simply a *production*. The term *rewrite rule* is also used.

Thue is from Axel Thue, a Norwegian mathematician, and is pronounced too-ay.

If P is the semi-Thue production $g \to \bar{g}$, then we write

$$u \underset{P}{\Rightarrow} v$$

to mean that there are (possibly null) words r, s such that

$$u = rgs \qquad \textit{and} \qquad v = r\bar{g}s.$$

(In other words, v is obtained from u by a replacement of g by $\bar{g}$.)

Definition. A *semi-Thue process* is a finite set of semi-Thue productions.

If Π is a semi-Thue process, we write

$$u \underset{\Pi}{\Rightarrow} v$$

to mean that

$$u \underset{P}{\Rightarrow} v$$

for some production P which belongs to Π. Finally, we write

$$u \overset{*}{\underset{\Pi}{\Rightarrow}} v$$

if there is a sequence

$$u = u_1 \underset{\Pi}{\Rightarrow} u_2 \underset{\Pi}{\Rightarrow} \cdots \underset{\Pi}{\Rightarrow} u_n = v.$$

The sequence $u_1, u_2, \ldots, u_n$ is then called a *derivation of v from u*. In particular (taking $n = 1$)

$$u \overset{*}{\underset{\Pi}{\Rightarrow}} u.$$

When no ambiguity results we often omit the explicit reference to Π, writing simply $u \Rightarrow v$ and $u \overset{*}{\Rightarrow} v$.

Here is a simple example: We let $\Pi = \{ab \to aa, ba \to bb\}$. Then we have

$$aba \Rightarrow abb \Rightarrow aab \Rightarrow aaa.$$

Thus,

$$aba \overset{*}{\Rightarrow} aaa,$$

and the sequence of words aba, abb, aab, aaa is a derivation of aaa from aba.

Exercises

1. Let Π be the semi-Thue process with the production $ba \to ab$.
   (a) Give two different derivations of $aaabbb$ from $abbaba$.
   (b) Give the set of all words in $\{a, b\}^*$ from which $aabb$ can be derived.
   (c) Give the set of all words which can be derived from $bbaa$.
2. Let Π be the semi-Thue process with productions $ba \to ab$ and $ab \to ba$. Show that for all words $u, v \in \{a, b\}^*$, $u \overset{*}{\underset{\Pi}{\Rightarrow}} v$ if and only if $v \overset{*}{\underset{\Pi}{\Rightarrow}} u$.

3. Give a semi-Thue process Π such that $1^{[x]} \underset{\Pi}{\overset{*}{\Rightarrow}} 1^{[y]}$ if and only if $|x - y|$ is even.

4. Let $A = \{1, 2, b_1^1, b_2^1, b_1^2, b_2^2, c_1, c_2, d_1, d_2\#\}$. Give a semi-Thue process Π such that $\#b_{j_1}^{i_1} \cdots b_{j_n}^{i_n}\# \underset{\Pi}{\overset{*}{\Rightarrow}} w \in \{1, 2\}^*$, for all words $b_{j_1}^{i_1} \cdots b_{j_n}^{i_n}$, where $i_1 \cdots i_n, j_1 \cdots j_n$ are binary representations of numbers and $i_1 \cdots i_n + j_1 \cdots j_n = w$. [*Hint:* The symbols c_1, c_2 are used to remember the need to carry 1, and d_1, d_2 are used to remember the need to carry 2.]

2. Simulation of Nondeterministic Turing Machines by Semi-Thue Processes

Let us begin with a nondeterministic Turing machine $\mathcal{M}$ with alphabet $\{s_1, \ldots, s_K\}$, and states $q_1, q_2, \ldots, q_n$. We shall show how to simulate $\mathcal{M}$ by a semi-Thue process $\Sigma(\mathcal{M})$ on the alphabet

$$s_0, s_1, \ldots, s_K, q_0, q_1, q_2, \ldots, q_n, q_{n+1}, h.$$

Each stage in a computation by $\mathcal{M}$ is specified completely by the current configuration. We shall code each such stage by a word on the alphabet of $\Sigma(\mathcal{M})$. For example, the configuration

$$\begin{array}{ccccccc} s_1 & s_1 & s_3 & s_2 & s_0 & s_1 & s_2 \\ & & & \uparrow & & & \\ & & & q_4 & & & \end{array}$$

will be represented by the single word

$$hs_1s_1s_3q_4s_2s_0s_1s_2h. \tag{2.1}$$

Note that h is used as a *beginning* and *end* marker, and the symbol q_4 indicates the state of $\mathcal{M}$ and is placed immediately to the left of the scanned square. A word like (2.1) will be called a *Post word.* Of course, the same configuration can be represented by infinitely many Post words because any number of additional blanks may be shown on the left or right. For example,

$$hs_0s_0s_1s_1s_3q_4s_2s_0s_1s_2s_0h$$

is a Post word representing the same configuration that (2.1) does.

In general, a word huq_ivh, where $0 \le i \le n + 1$, is called a *Post word* if u and v are words on the subalphabet $\{s_0, s_1, \ldots, s_K\}$. We shall show how to associate suitable semi-Thue productions with each quadruple of $\mathcal{M}$; the productions simulate the effect of that quadruple on Post words.

1. For each quadruple of $\mathcal{M}$ of the form $q_i\ s_j\ s_k\ q_l$ we place in $\Sigma(\mathcal{M})$ the production
$$q_i s_j \to q_l s_k .$$

2. For each quadruple of $\mathcal{M}$ of the form $q_i\ s_j\ R\, q_l$ we place in $\Sigma(\mathcal{M})$ the productions
$$q_i s_j s_k \to s_j q_l s_k , \qquad k = 0, 1, \dots, K,$$
$$q_i s_j h \to s_j q_l s_0 h .$$

3. For each quadruple of $\mathcal{M}$ of the form $q_i\ s_j\ L\, q_l$ we place in $\Sigma(\mathcal{M})$ the productions
$$s_k q_i s_j \to q_l s_k s_j , \qquad k = 0, 1, \dots, K,$$
$$h q_i s_j \to h q_l s_0 s_j .$$

To see how these productions simulate the behavior of $\mathcal{M}$, suppose $\mathcal{M}$ is in configuration

$$\begin{array}{cccc} s_2 & s_1 & s_0 & s_3 . \\ & \uparrow & & \\ & q_4 & & \end{array}$$

This configuration is represented by the Post word
$$h s_2 q_4 s_1 s_0 s_3 h .$$
Now suppose $\mathcal{M}$ contains the quadruple
$$q_4\ s_1\ s_3\ q_5 .$$
Then $\Sigma(\mathcal{M})$ contains the production
$$q_4 s_1 \to q_5 s_3 ,$$
so that
$$h s_2 q_4 s_1 s_0 s_3 h \underset{\Sigma(\mathcal{M})}{\Rightarrow} h s_2 q_5 s_3 s_0 s_3 h .$$
The Post word on the right then corresponds to the configuration immediately following application of the above quadruple. Now suppose that $\mathcal{M}$ contains the quadruple
$$q_4\ s_1\ R\ q_3 .$$

(Of course, if $\mathcal{M}$ is a *deterministic* Turing machine, it cannot contain both of these quadruples.) Then $\Sigma(\mathcal{M})$ contains the production

$$q_4 s_1 s_0 \to s_1 q_3 s_0,$$

so that

$$h s_2 q_4 s_1 s_0 s_3 h \underset{\Sigma(\mathcal{M})}{\Rightarrow} h s_2 s_1 q_3 s_0 s_3 h.$$

Finally if $\mathcal{M}$ contains the quadruple

$$q_4 \ \ s_1 \ \ L \ q_2,$$

then $\Sigma(\mathcal{M})$ contains the production

$$s_2 q_4 s_1 \to q_2 s_2 s_1,$$

so that

$$h s_2 q_4 s_1 s_0 s_3 h \underset{\Sigma(\mathcal{M})}{\Rightarrow} h q_2 s_2 s_1 s_0 s_3 h.$$

The productions involving h are to take care of cases where motion to the right or left would go past the part of the tape included in the Post word, so that an additional blank must be added. For example, if the configuration is

$$\begin{array}{ccc} s_2 & s_3 & s_1 \\ & & \uparrow \\ & & q_4 \end{array}$$

and $\mathcal{M}$ contains the quadruple

$$q_4 \ \ s_1 \ \ R \ q_3,$$

then $\Sigma(\mathcal{M})$ contains the production

$$q_4 s_1 h \to s_1 q_3 s_0 h$$

and we have

$$h s_2 s_3 q_4 s_1 h \underset{\Sigma(\mathcal{M})}{\Rightarrow} h s_2 s_3 s_1 q_3 s_0 h,$$

so that the needed blank on the right has been inserted. The reader will readily verify that blanks on the left are similarly supplied when needed.

We now complete the specification of $\Sigma(\mathcal{M})$:

4. Whenever $q_i s_j (i = 1, \ldots, n;\ j = 0, 1, \ldots, K)$ are *not* the first two symbols of a quadruple of $\mathcal{M}$, we place in $\Sigma(\mathcal{M})$ the production

$$q_i s_j \to q_{n+1} s_j.$$

Thus, q_{n+1} serves as a "halt" state.

5. Finally, we place in $\Sigma(\mathscr{M})$ the productions

$$q_{n+1}s_i \to q_{n+1}, \qquad i = 0, 1, \ldots, K,$$
$$q_{n+1}h \to q_0 h$$
$$s_i q_0 \to q_0, \qquad i = 0, 1, \ldots, K.$$

We have

Theorem 2.1. Let $\mathscr{M}$ be a *deterministic* Turing machine, and let w be a Post word on the alphabet of $\Sigma(\mathscr{M})$. Then

1. there is at most one word z such that $w \underset{\Sigma(\mathscr{M})}{\Rightarrow} z$, and
2. if there is a word z satisfying (1), then z is a Post word.

Proof. We have $w = huq_ivh$.

If $1 \le i \le n$, then

a. if $v = 0$ no production of $\Sigma(\mathscr{M})$ applies to w;
b. if v begins with the symbol s_j and there is a (necessarily unique) quadruple of $\mathscr{M}$ which begins $q_i s_j$, then there is a uniquely applicable production of $\Sigma(\mathscr{M})$ and the result of applying it will be a Post word;
c. if v begins with the symbol s_j and there is no quadruple of $\mathscr{M}$ which begins $q_i s_j$, then the one applicable production of $\Sigma(\mathscr{M})$ is

$$q_i s_j \to q_{n+1}s_j,$$

which yields another Post word when applied to w.

If $i = n + 1$, then

a. if $v = 0$, the only applicable production of $\Sigma(\mathscr{M})$ is

$$q_{n+1}h \to q_0 h,$$

which yields a Post word;
b. if v begins with the symbol s_j, the only applicable production of $\Sigma(\mathscr{M})$ is

$$q_{n+1}s_j \to q_{n+1},$$

which again yields a Post word.

Finally, if $i = 0$, then

a. if $u = 0$, no production of $\Sigma(\mathscr{M})$ can be applied;
b. if u ends with s_j, the only applicable production of $\Sigma(\mathscr{M})$ is

$$s_j q_0 \to q_0,$$

which yields a Post word. ■

Our next result makes precise the sense in which $\Sigma(\mathcal{M})$ simulates $\mathcal{M}$.

Theorem 2.2. Let $\mathcal{M}$ be a nondeterministic Turing machine. Then, for each string u on the alphabet of $\mathcal{M}$, $\mathcal{M}$ accepts u if and only if

$$hq_1 s_0 uh \underset{\Sigma(\mathcal{M})}{\overset{*}{\Rightarrow}} hq_0 h.$$

Proof. Let the alphabet of $\mathcal{M}$ be $s_1, \ldots, s_K$. First let us suppose that $\mathcal{M}$ accepts u. Then, if $\mathcal{M}$ begins in the configuration

$$\begin{array}{cc} s_0 & u \\ \uparrow & \\ q_1 & \end{array}$$

it will eventually reach a state q_i scanning a symbol s_k where no quadruple of $\mathcal{M}$ begins $q_i\, s_k$. Then we will have (for appropriate words v, w on the alphabet of $\mathcal{M}$)

$$hq_1 s_0 uh \underset{\Sigma(\mathcal{M})}{\overset{*}{\Rightarrow}} hvq_i s_k wh \underset{\Sigma(\mathcal{M})}{\Rightarrow} hvq_{n+1} s_k wh$$

$$\underset{\Sigma(\mathcal{M})}{\overset{*}{\Rightarrow}} hvq_0 h \underset{\Sigma(\mathcal{M})}{\overset{*}{\Rightarrow}} hq_0 h.$$

Next suppose that $\mathcal{M}$ does not accept u. Then, beginning with configuration

$$\begin{array}{c} s_0 u \\ \uparrow \\ q_1 \end{array}$$

$\mathcal{M}$ will never halt. Let

$$w_1 = hq_1 s_0 uh,$$

and suppose that

$$w_1 \underset{\Sigma(\mathcal{M})}{\Rightarrow} w_2 \underset{\Sigma(\mathcal{M})}{\Rightarrow} w_3 \underset{\Sigma(\mathcal{M})}{\Rightarrow} \cdots \underset{\Sigma(\mathcal{M})}{\Rightarrow} w_m.$$

Then each w_j, $1 \le j \le m$, must contain a symbol q_i with $1 \le i \le n$. Hence there can be no derivation of a Post word containing q_0 from w_1, and so, in particular, there is no derivation of $hq_0 h$ from w_1. ■

Definition. The *inverse* of the production $g \to \bar{g}$ is the production $\bar{g} \to g$.

For example, the inverse of the production $ab \to aa$ is the production $aa \to ab$.

Let us write $\Omega(\mathscr{M})$ for the semi-Thue process which consists of the inverses of all the productions of $\Sigma(\mathscr{M})$. Then an immediate consequence of Theorem 2.2 is

Theorem 2.3. Let $\mathscr{M}$ be a nondeterministic Turing machine. Then for each string u in the alphabet of $\mathscr{M}$, $\mathscr{M}$ accepts u if and only if

$$hq_0h \underset{\Omega(\mathscr{M})}{\overset{*}{\Rightarrow}} hq_1s_0uh.$$

Exercises

1. **(a)** Give $\Sigma(\mathscr{M})$, where $\mathscr{M}$ is the Turing machine in Table 1.1 of Chapter 6.

   (b) Give a derivation that shows that $hq_1s_0111h \underset{\Sigma(\mathscr{M})}{\Rightarrow} hq_0h$.

   (c) Describe $\{u \mid hq_0h \underset{\Omega(\mathscr{M})}{\overset{*}{\Rightarrow}} hq_1s_0uh\}$.
2. Give a semi-Thue process Π such that, for all words $u, v \in \{1,2\}^*$, $hq_1s_0us_0vh \underset{\Pi}{\overset{*}{\Rightarrow}} w \in \{1,2\}^*$, where $u + v = w$ in binary notation.
3. Show that for any partially computable function $f(x)$, there is a semi-Thue process Π such that for all $x \in N$, $1^{[x]} \underset{\Pi}{\overset{*}{\Rightarrow}} 1^{[y]}$ if and only if $y = f(x)$.

3. Unsolvable Word Problems

Definition. The *word problem* for a semi-Thue process Π is the problem of determining for any given pair u, v of words on the alphabet of Π whether $u \underset{\Pi}{\overset{*}{\Rightarrow}} v$.

We shall prove

Theorem 3.1. There is a Turing machine $\mathscr{M}$ such that the word problem is unsolvable for both the semi-Thue processes $\Sigma(\mathscr{M})$ and $\Omega(\mathscr{M})$.

Proof. By Theorem 3.1 in Chapter 6, there is a Turing machine $\mathscr{M}$ (in fact, deterministic) that accepts a nonrecursive language. Suppose first that the word problem for $\Sigma(\mathscr{M})$ were solvable. Then there would be an algorithm for testing given words v, w to determine whether $v \underset{\Sigma(\mathscr{M})}{\overset{*}{\Rightarrow}} w$. By Theorem 2.2, we could use this algorithm to determine whether $\mathscr{M}$ will accept a given word u by testing whether

$$hq_1s_0uh \underset{\Sigma(\mathscr{M})}{\overset{*}{\Rightarrow}} hq_0h.$$

We would thus have an algorithm for testing a given word u to see whether $\mathscr{M}$ will accept it. But such an algorithm cannot exist since the language accepted by $\mathscr{M}$ is not a recursive set.

Finally, an algorithm that solved the word problem for $\Omega(\mathscr{M})$ would also solve the word problem for $\Sigma(\mathscr{M})$, since

$$v \underset{\Sigma(\mathscr{M})}{\overset{*}{\Rightarrow}} w \qquad \textit{if and only if} \qquad w \underset{\Omega(\mathscr{M})}{\overset{*}{\Rightarrow}} v. \qquad \blacksquare$$

Definition. A semi-Thue process is called a *Thue process* if the inverse of each production in the process is also in the process.

The fact that Thue processes are in fact "two-way" processes is a curious coincidence.

We write $g \leftrightarrow \bar{g}$ to combine the production $g \to \bar{g}$ and its inverse $\bar{g} \to g$.

For each Turing machine $\mathscr{M}$, we write

$$\Theta(\mathscr{M}) = \Sigma(\mathscr{M}) \cup \Omega(\mathscr{M}),$$

so that $\Theta(\mathscr{M})$ is a Thue process. We have

Theorem 3.2 (Post's Lemma). Let $\mathscr{M}$ be a deterministic Turing machine. Let u be a word on the alphabet of $\mathscr{M}$ such that

$$hq_1s_0uh \underset{\Theta(\mathscr{M})}{\overset{*}{\Rightarrow}} hq_0h.$$

Then

$$hq_1s_0uh \underset{\Sigma(\mathscr{M})}{\overset{*}{\Rightarrow}} hq_0h.$$

Proof. Let the sequence

$$hq_1s_0uh = w_1, w_2, \ldots, w_l = hq_0h$$

be a derivation in $\Theta(\mathscr{M})$. Since w_1 is a Post word, and each production of $\Theta(\mathscr{M})$ transforms Post words into Post words, we can conclude that the entire derivation consists of Post words. We need to show how to eliminate use of productions belonging to $\Omega(\mathscr{M})$ from this derivation. So let us assume that the last time in the derivation that a production of $\Omega(\mathscr{M})$ was used was in getting from w_i to w_{i+1}. That is, we assume

$$w_i \underset{\Omega(\mathscr{M})}{\overset{*}{\Rightarrow}} w_{i+1}; \qquad w_{i+1} \underset{\Sigma(\mathscr{M})}{\Rightarrow} w_{i+2} \underset{\Sigma(\mathscr{M})}{\overset{*}{\Rightarrow}} w_l = hq_0h.$$

Now, $\Omega(\mathcal{M})$ consists of inverses of productions of $\Sigma(\mathcal{M})$; hence we must have

$$w_{i+1} \underset{\Sigma(\mathcal{M})}{\Rightarrow} w_i .$$

Moreover, we must have $i + 1 < l$ because no production of $\Sigma(\mathcal{M})$ can be applied to $w_l = hq_0h$. Now, w_{i+1} is a Post word and

$$w_{i+1} \underset{\Sigma(\mathcal{M})}{\Rightarrow} w_i , \qquad w_{i+1} \underset{\Sigma(\mathcal{M})}{\Rightarrow} w_{i+2} .$$

By Theorem 2.1, we conclude that $w_{i+2} = w_i$. Thus the transition from w_i to w_{i+1} and then back to $w_{i+2} = w_i$ is clearly an unnecessary detour. That is, the sequence

$$w_1, w_2, \ldots, w_i, w_{i+3}, \ldots, w_l$$

from which w_{i+1}, w_{i+2} have been omitted is a derivation in $\Theta(\mathcal{M})$.

We have shown that any derivation that uses a production belonging to $\Omega(\mathcal{M})$ can be shortened. Continuing this procedure, we eventually obtain a derivation using only productions of $\Sigma(\mathcal{M})$. ∎

Theorem 3.3 (Post–Markov). If the deterministic Turing machine $\mathcal{M}$ accepts a nonrecursive set, then the word problem for the Thue process $\Theta(\mathcal{M})$ is unsolvable.

Proof. Let u be a word on the alphabet of $\mathcal{M}$. Then we have, using Theorems 2.2 and 3.2,

$$\mathcal{M} \text{ accepts } u$$

if and only if

$$hq_1s_0uh \overset{*}{\underset{\Sigma(\mathcal{M})}{\Rightarrow}} hq_0h$$

if and only if

$$hq_1s_0uh \overset{*}{\underset{\Theta(\mathcal{M})}{\Rightarrow}} hq_0h .$$

Hence, an algorithm for solving the word problem for $\Theta(\mathcal{M})$ could be used to determine whether or not $\mathcal{M}$ will accept u, which is impossible. ∎

Now we consider semi-Thue processes on an alphabet of two symbols.

Theorem 3.4. There is a semi-Thue process on the alphabet $\{a, b\}$ whose word problem is unsolvable. Moreover, for each production $g \to h$ of this semi-Thue process, $g, h \neq 0$.

Proof. Let us begin with a semi-Thue process Π on the alphabet $A = \{a_1, \ldots, a_n\}$ and with productions

$$g_i \to \bar{g}_i, \qquad i = 1, 2, \ldots, m,$$

whose word problem is unsolvable. We also assume that for each $i = 1, 2, \ldots, m$, $g_i \neq 0$ and $\bar{g}_i \neq 0$. This is legitimate because this condition is satisfied by the productions of $\Sigma(\mathscr{M})$.

We write

$$a'_j = ba^{[j]}b, \qquad j = 1, 2, \ldots, n,$$

where there is a string of a's of length j between the two b's. Finally, for any word $w \neq 0$ in A^*,

$$w = a_{j_1} a_{j_2} \cdots a_{j_k},$$

we write

$$w' = a'_{j_1} a'_{j_2} \cdots a'_{j_k}.$$

In addition we let $0' = 0$. Then, we consider the semi-Thue process Π' on the alphabet $\{a, b\}$ whose productions are

$$g'_i \to \bar{g}'_i, \qquad i = 1, 2, \ldots, m.$$

We have

Lemma 1. If $u \underset{\Pi}{\Rightarrow} v$, then $u' \underset{\Pi'}{\Rightarrow} v'$.

Proof. We have $u = rg_is$, $v = r\bar{g}_is$. Hence $u' = r'g'_is'$, $v' = r'\bar{g}'_is'$, so that $u' \underset{\Pi'}{\Rightarrow} v'$. ∎

Lemma 2. If $u' \underset{\Pi'}{\Rightarrow} w$, then for some $v \in A^*$ we have $w = v'$ and $u \underset{\Pi}{\Rightarrow} v$.

Proof. We have $u' = pg'_iq$, $w = p\bar{g}'_iq$. Now, since $g_i \neq 0$, g'_i begins and ends with the letter b. Hence each of p and q either begins and ends with b or is 0, so that $p = r'$, $q = s'$. Then, $u = rg_is$. Let $v = r\bar{g}_is$. Then $w = v'$ and $u \underset{\Pi}{\Rightarrow} v$. ∎

Lemma 3. $u \overset{*}{\underset{\Pi}{\Rightarrow}} v$ if and only if $u' \overset{*}{\underset{\Pi'}{\Rightarrow}} v'$.

Proof. If $u = u_1 \underset{\Pi}{\Rightarrow} u_2 \underset{\Pi}{\Rightarrow} \cdots \underset{\Pi}{\Rightarrow} u_n = v$, then by Lemma 1

$$u' = u'_1 \underset{\Pi'}{\Rightarrow} u'_2 \underset{\Pi'}{\Rightarrow} \cdots \underset{\Pi'}{\Rightarrow} u'_n = v'.$$

Conversely, if

$$u' = w_1 \underset{\Pi'}{\Rightarrow} w_2 \underset{\Pi'}{\Rightarrow} \cdots \underset{\Pi'}{\Rightarrow} w_n = v',$$

then by Lemma 2, for each w_i there is a string $u_i \in A^*$ such that $w_i = u_i'$. Thus,

$$u' = u_1' \underset{\Pi'}{\Rightarrow} u_2' \underset{\Pi'}{\Rightarrow} \cdots \underset{\Pi'}{\Rightarrow} u_n' = v'.$$

By Lemma 2 once again,

$$u = u_1 \underset{\Pi}{\Rightarrow} u_2 \underset{\Pi}{\Rightarrow} \cdots \underset{\Pi}{\Rightarrow} u_n = v,$$

so that $u \overset{*}{\underset{\Pi}{\Rightarrow}} v$. ■

Proof of Theorem 3.4 Concluded. By Lemma 3, if the word problem for Π' were solvable, the word problem for Π would also be solvable. Hence, the word problem for Π' is unsolvable. ■

In the preceding proof it is clear that if the semi-Thue process Π with which we begin is actually a Thue process, then Π' will be a Thue process on $\{a, b\}$. We conclude

Theorem 3.5. There is a Thue process on the alphabet $\{a, b\}$ whose word problem is unsolvable. Moreover, for each production $g \to h$ of this Thue process, $g, h \neq 0$.

Exercises

1. Let Π be the semi-Thue process with productions $cde \to ce$, $d \to cde$. Use the construction in the proof of Theorem 3.4 to get a semi-Thue process Π' with productions on $\{a, b\}$ such that $u \overset{*}{\underset{\Pi}{\Rightarrow}} v$ if and only if $u' \overset{*}{\underset{\Pi'}{\Rightarrow}} v'$ for all words $u, v \in \{c, d, e\}^*$.
2. A semi-Thue system is defined to be a pair (u_0, Π), where Π is a semi-Thue process and u_0 is a given word on the alphabet of Π. A word w is called a *theorem* of (u_0, Π) if $u_0 \overset{*}{\underset{\Pi}{\Rightarrow}} w$. Show that there is a semi-Thue system for which no algorithm exists to determine whether a given string is a theorem of the system.
3. Let Π be a semi-Thue process containing only one production. Show that Π has a solvable word problem.

4.* Give an upper bound on the size of the smallest semi-Thue process with an undecidable word problem. [See Exercise 2.2 in Chapter 6.]

4. Post's Correspondence Problem

The Post correspondence problem first appeared in a paper by Emil Post in 1946. It was only much later that this problem was seen to have important applications in the theory of formal languages.

Our treatment of the Post correspondence problem is a simplification of a proof due to Floyd, itself much simpler than Post's original work.

The correspondence problem may conveniently be thought of as a solitaire game played with special sets of dominoes. Each domino has a word (on some given alphabet) appearing on each half. A typical domino is shown in Fig. 4.1. A *Post correspondence system* is simply a finite set of dominoes of this kind. Figure 4.2 gives a simple example of a Post correspondence system using three dominoes and the alphabet $\{a, b\}$. Each move in the solitaire game defined by a particular Post correspondence system consists of simply placing one of the dominoes of the system to the right of the dominoes laid down on previous moves. The key fact is that the *dominoes are not used up by being played, so that each one can be used any number of times*. The way to "win" the game is to reach a situation where the very same word appears on the top halves as on the bottom halves of the dominoes when we read across from left to right. Figure 4.3 shows how to win the game defined by the dominoes of Fig. 4.2. (Note that one of the dominoes is used twice.) The word *aabbbb* which appears across both the top halves and bottom halves is called a *solution* of the given Post correspondence system. Thus a Post correspondence system possesses a *solution* if and only if it is possible to win the game defined by that system.

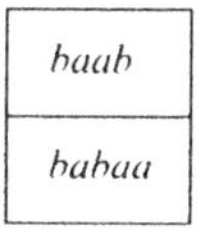

Figure 4.1

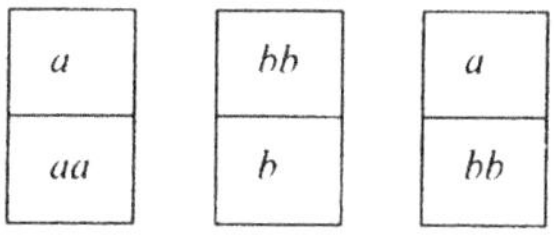

Figure 4.2

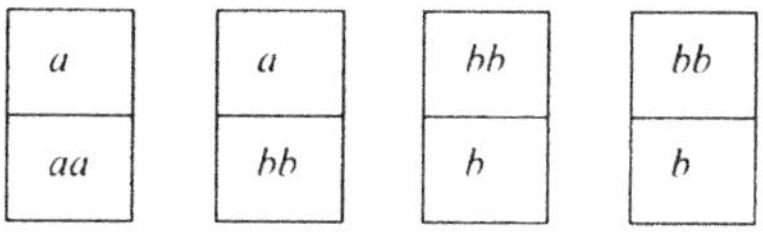

Figure 4.3

We shall prove

Theorem 4.1. There is no algorithm that can test a given Post correspondence system to determine whether it has a solution.

Proof. Using Theorem 3.4, we begin with a semi-Thue process Π on the alphabet $\{a, b\}$ whose word problem is unsolvable. We modify Π in the following trivial way: we add to the productions of Π the two productions

$$a \to a, \qquad b \to b.$$

Naturally this addition has no effect on whether

$$u \overset{*}{\underset{\Pi}{\Rightarrow}} v$$

for given words u, v. However, it does guarantee that whenever $u \overset{*}{\underset{\Pi}{\Rightarrow}} v$, there is a derivation

$$u = u_1 \underset{\Pi}{\Rightarrow} u_2 \underset{\Pi}{\Rightarrow} \cdots \underset{\Pi}{\Rightarrow} u_m = v,$$

where m is an odd number. This is because with the added productions we have

$$u_i \underset{\Pi}{\Rightarrow} u_i$$

for each i, so that any step in a derivation (e.g., the first) can be repeated if necessary to change the length of the derivation from an even to an odd number.

Let u and v be any given words on the alphabet $\{a, b\}$. We shall construct a Post correspondence system $P_{u,v}$ (which depends on Π as well as on the words u and v) such that $P_{u,v}$ has a solution if and only if $u \overset{*}{\underset{\Pi}{\Rightarrow}} v$. Once we have obtained this $P_{u,v}$ we are through. For, if there were an algorithm for testing given Post correspondence systems for possessing a solution, this algorithm could be applied in particular to $P_{u,v}$ and therefore to determine whether $u \overset{*}{\underset{\Pi}{\Rightarrow}} v$; since Π has an unsolvable word problem, this is impossible.

We proceed to show how to construct $P_{u,v}$. Let the productions of Π (including the two we have just added) be $g_i \to h_i$, $i = 1, 2, \ldots, n$. The alphabet of $P_{u,v}$ consists of the eight symbols

$$a \; b \; \tilde{a} \; \tilde{b} \; [\;] \; * \; \tilde{*}.$$

For any word w on $\{a, b\}$, we write $\tilde{w}$ for the word on $\{\tilde{a}, \tilde{b}\}$ obtained by placing " $\sim$ " on top of each symbol of w. $P_{u,v}$ is then to consist of the $2n + 4$ dominoes shown in Fig. 4.4. Note that because Π contains the productions $a \to a$ and $b \to b$, $P_{u,v}$ contains the four dominoes

a	$\tilde{a}$	b	$\tilde{b}$
$\tilde{a}$	a	$\tilde{b}$	b

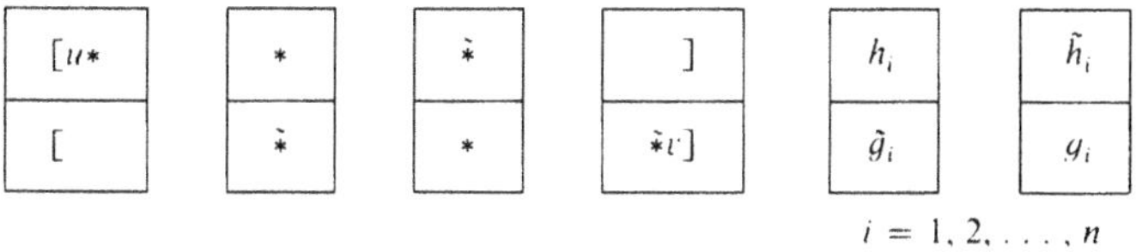

Figure 4.4

Therefore, it is clear that in our play it is legitimate to use dominoes of the form

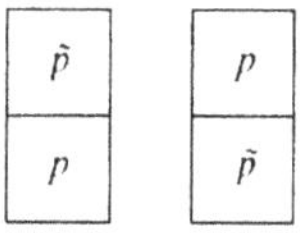

where p is any word on $\{a, b\}$, since any such dominoes can be assembled by lining up single dominoes selected appropriately from the previous four.

We proceed to show that $P_{u,v}$ has a solution if and only if $u \underset{\Pi}{\overset{*}{\Rightarrow}} v$.

First suppose that $u \underset{\Pi}{\overset{*}{\Rightarrow}} v$. Let

$$u = u_1 \underset{\Pi}{\Rightarrow} u_2 \underset{\Pi}{\Rightarrow} \cdots \underset{\Pi}{\Rightarrow} u_m = v,$$

where m is an odd number. Thus, for each i, $1 \le i < m$, we can write

$$u_i = p_i g_{j_i} q_i, \qquad u_{i+1} = p_i h_{j_i} q_i,$$

where the transition from u_i to u_{i+1} is via the j_ith production of Π. Then we claim that the word

$$\left[u_1 * \tilde{u}_2 \tilde{*} u_3 * \cdots * \tilde{u}_{m-1} \tilde{*} u_m\right] \tag{4.1}$$

is a solution of $P_{u,v}$. To see this, let us begin to play by laying down the dominoes:

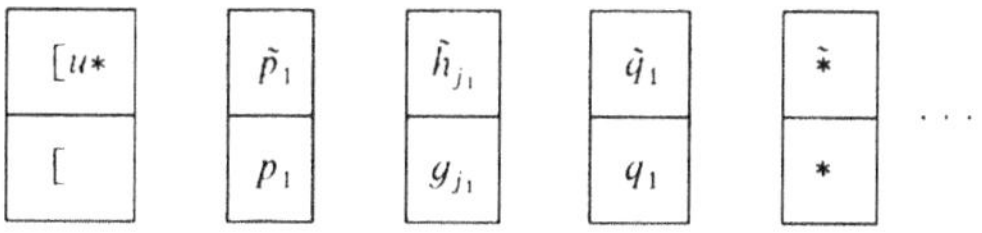

At this stage, the word on top is

$$[u_1 * \tilde{u}_2 \tilde{*}$$

while the word on the bottom is

$$[u_1 * .$$

We can continue to play as follows:

$[u*$	$\tilde{p}_1$	$\tilde{h}_{j_1}$	$\tilde{q}_1$	$\tilde{*}$	p_2	h_{j_2}	q_2	$*$
$[$	p_1	g_{j_1}	q_1	$*$	$\tilde{p}_2$	$\tilde{g}_{j_2}$	$\tilde{q}_2$	$\tilde{*}$

Now the word on top is

$$[u_1 * \tilde{u}_2 \tilde{*} u_3 *$$

and the word on the bottom is

$$[u_1 * \tilde{u}_2 \tilde{*}$$

Recalling that m is an odd number we see that we can win by continuing as follows:

$[u*$	$\tilde{p}_1$	$\cdots$	$\tilde{*}$	p_{m-1}	$h_{j_{m-1}}$	q_{m-1}	$]$
$[$	p_1	$\cdots$	$*$	$\tilde{p}_{m-1}$	$\tilde{g}_{j_{m-1}}$	$\tilde{q}_{m-1}$	$\tilde{*}v]$

;

for, at this point the word both on top and on bottom is (4.1).

Conversely suppose that $P_{u,v}$ has a solution w. Examining Fig. 4.4, we see that the only possible way to win involves playing

$[u*$	and	$]$
$[$		$\tilde{*}v]$

first and last, respectively. This is because none of the other dominoes in $P_{u,v}$ have tops and bottoms which begin (or end) with the same symbol. Thus, w must begin with [and end with]. Let us write $w = [z]y$, where z contains no]. (Of course it is quite possible that $y = 0$.) Since the only domino containing] contains it on the far right on top and on bottom, we see that $[z]$ itself is already a solution to $P_{u,v}$. We work with this solution. So far we know that the game looks like this:

$[u*$	$\cdots$	$]$
$[$		$\tilde{*}v]$

so that the solution $[z]$ looks like this:

$$[u * \cdots \tilde{*} v].$$

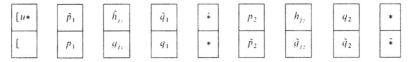

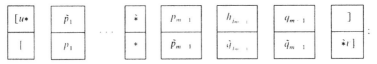

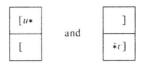

Continuing from the left we see that the play must go

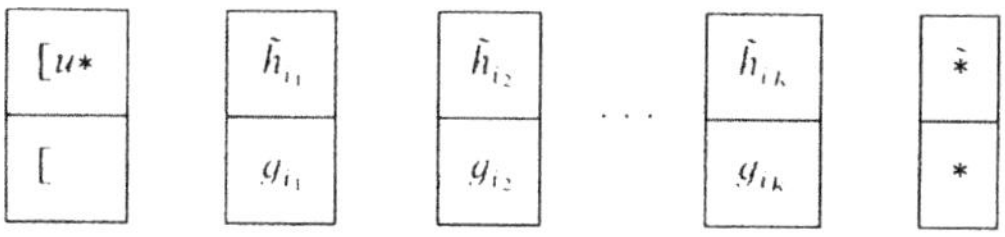

where $g_{i_1} g_{i_2} \cdots g_{i_k} = u$. (This is necessary in order for the bottom to "catch up" with the $u*$ which is already on top.) Writing $u = u_1$ and $u_2 = h_{i_1} h_{i_2} \cdots h_{i_k}$ we see that $u_1 \overset{*}{\underset{\Pi}{\Rightarrow}} u_2$ and that the solution has the form

$$\left[u_1 * \tilde{u}_2 \tilde{*} \cdots \tilde{*}\, v \right].$$

Now we see how the play must continue:

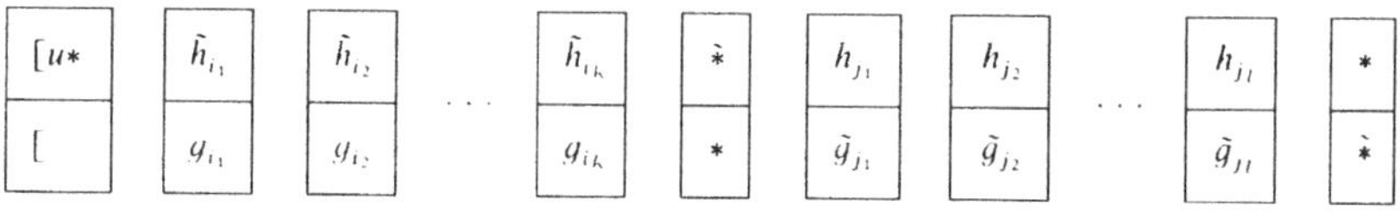

where of course $u_2 = g_{j_1} g_{j_2} \cdots g_{j_l}$. Again, writing $u_3 = h_{j_1} h_{j_2} \cdots h_{j_l}$ we have that $u_2 \overset{*}{\underset{\Pi}{\Rightarrow}} u_3$ and that the solution has the form

$$\left[u_1 * \tilde{u}_2 \tilde{*} u_3 * \cdots \tilde{*}\, v \right].$$

Continuing, it is clear that the solution can be written

$$\left[u_1 * \tilde{u}_2 \tilde{*} u_3 * \cdots * \tilde{u}_{m-1} \tilde{*} u_m \right],$$

where

$$u = u_1 \overset{*}{\underset{\Pi}{\Rightarrow}} u_2 \overset{*}{\underset{\Pi}{\Rightarrow}} u_3 \overset{*}{\underset{\Pi}{\Rightarrow}} \cdots \overset{*}{\underset{\Pi}{\Rightarrow}} u_{m-1} \overset{*}{\underset{\Pi}{\Rightarrow}} u_m = v,$$

so that $u \overset{*}{\underset{\Pi}{\Rightarrow}} v$. ■

Exercises

1. Let Π be the semi-Thue process with productions $aba \to a$, $b \to aba$, and let $u = bb$, $v = aaaaaa$. Describe the Post correspondence system $P_{u,v}$ and give a solution to $P_{u,v}$.
2. Find a solution to the Post correspondence problem defined by the dominoes

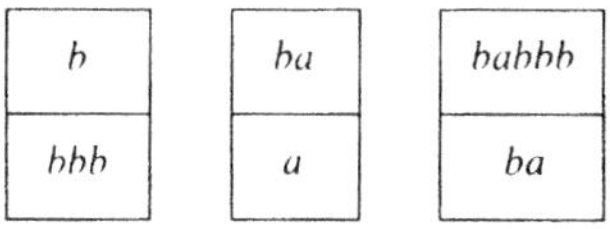

3. Find an algorithm for Post correspondence problems whose alphabet consists of just one symbol.

5. Grammars

A *phrase-structure grammar* or simply a *grammar* is just a semi-Thue process in which the letters of the alphabet are separated into two disjoint sets called the *variables* and the *terminals*, with one of the variables singled out as the *start symbol*. It is customary (but, of course, not necessary) to use lower case letters for terminals, capital letters for variables, and in particular the letter S for the start symbol.

Let Γ be a grammar with start symbol S and let $\mathscr{V}, T$ be the sets of variables and terminals of Γ, respectively. Then we define

$$L(\Gamma) = \{u \in T^* \mid S \overset{*}{\Rightarrow} u\},$$

and call $L(\Gamma)$ the *language generated by* Γ. Our purpose in this section is to characterize languages which can be generated by grammars.

We first prove

Theorem 5.1. Let U be a language accepted by a nondeterministic Turing machine. Then there is a grammar Γ such that $U = L(\Gamma)$.

Proof. Let $U \subseteq T^*$ and let $\mathscr{M}$ be a nondeterministic Turing machine that accepts U. We will construct Γ by modifying the semi-Thue process $\Omega(\mathscr{M})$ from Section 2. Let $\mathscr{M}$ have the states $q_1, \ldots, q_n$. Then we recall that the alphabet of $\Omega(\mathscr{M})$ [which is the same as that of $\Sigma(\mathscr{M})$] consists of $s_0, q_0, q_1, q_2, \ldots, q_n, q_{n+1}, h$ in addition to the letters of the alphabet of $\mathscr{M}$. We let the *terminals* of Γ be just the letters of T, and the variables of Γ be the symbols from the alphabet of $\Omega(\mathscr{M})$ not in T, together with the two additional symbols S and q. S is to be the start symbol of Γ. The productions of Γ are then the productions of $\Omega(\mathscr{M})$ together with the productions

$$S \to hq_0h$$

$$hq_1s_0 \to q$$

$$qs \to sq \qquad \text{for each} \quad s \in T$$

$$qh \to 0.$$

Now, let $\mathscr{M}$ accept $u \in T^*$. Then, using Theorem 2.3, we have

$$S \underset{\Gamma}{\Rightarrow} hq_0h \overset{*}{\underset{\Gamma}{\Rightarrow}} hq_1s_0uh \underset{\Gamma}{\Rightarrow} quh \overset{*}{\underset{\Gamma}{\Rightarrow}} uqh \underset{\Gamma}{\Rightarrow} u,$$

so that $u \in L(\Gamma)$.

Conversely, let $u \in L(\Gamma)$. Then $u \in T^*$ and $S \overset{*}{\underset{\Gamma}{\Rightarrow}} u$. Examining the list of productions of Γ, we see that we must in fact have

$$S \underset{\Gamma}{\Rightarrow} hq_0h \overset{*}{\underset{\Gamma}{\Rightarrow}} vqhz \underset{\Gamma}{\Rightarrow} vz = u.$$

Proceeding further, we see that the symbol q could only be introduced using the production

$$hq_1s_0 \to q.$$

Hence, our derivation must have the form

$$S \underset{\Gamma}{\Rightarrow} hq_0h \overset{*}{\underset{\Gamma}{\Rightarrow}} xhq_1s_0yhz \underset{\Gamma}{\Rightarrow} xqyhz \overset{*}{\underset{\Gamma}{\Rightarrow}} xyqhz \underset{\Gamma}{\Rightarrow} xyz = u,$$

where of course $xy = v$. Thus, there is a derivation of xhq_1s_0yhz from hq_0h in Γ. Moreover, this must actually be a derivation in $\Omega(\mathscr{M})$ since the added productions are clearly inapplicable. Moreover, the productions of $\Omega(\mathscr{M})$ always lead from Post words to Post words. Hence, xhq_1s_0yhz must be a Post word. That is, $x = z = 0$ and $u = xyz = y$. We conclude that

$$hq_0h \overset{*}{\underset{\Omega(\mathscr{M})}{\Rightarrow}} hq_1s_0uh.$$

Thus by Theorem 2.3, $\mathscr{M}$ accepts u. ■

Now, let us begin with a grammar Γ and see what we can say about $L(\Gamma)$. Thus, let the alphabet of Γ be

$$\{s_1, \ldots, s_n, V_1, \ldots, V_k\},$$

where $T = \{s_1, \ldots, s_n\}$ is the set of terminals, $V_1, \ldots, V_k$ are the variables, and $S = V_1$ is the start symbol. Let us order the alphabet of Γ as shown. Thus strings on this alphabet are notations for integers in the base $n + k$. We have

Lemma 1. The predicate $u \underset{\Gamma}{\Rightarrow} v$ is primitive recursive.

Proof. Let the productions Γ be $g_i \to h_i$, $i = 1, 2, \ldots, l$. We write, for $i = 1, 2, \ldots, l$,

$$\text{PROD}_i(u, v) \Leftrightarrow (\exists r, s)_{\leq u}[u = \text{CONCAT}(r, g_i, s) \;\&\; v = \text{CONCAT}(r, h_i, s)].$$

Since, by Chapter 5, Section 1, CONCAT is primitive recursive, each of the predicates PROD_i is primitive recursive. But

$$u \underset{\Gamma}{\Rightarrow} v \Leftrightarrow \mathrm{PROD}_1(u,v) \vee \mathrm{PROD}_2(u,v) \vee \cdots \vee \mathrm{PROD}_l(u,v),$$

and the result follows. ■

We write DERIV(u, y) to mean that for some m, $y = [u_1, \ldots, u_m, 1]$, where the sequence $u_1, \ldots, u_m$ is a derivation of u from S in Γ. (The "1" has been added to avoid complications in case $u_m = u = 0$.) Then, since the value of S in base $n + k$ is $n + 1$ [because $S = V_1$ is the $(n + 1)$th symbol in our alphabet], we have

$$\begin{aligned} \mathrm{DERIV}(u,y) \Leftrightarrow (\exists m)_{\le y}\Big(m + 1 = \mathrm{Lt}(y) \;\&\; (y)_1 = n + 1 \\ \&\; (y)_m = u \;\&\; (y)_{m+1} = 1 \\ \&\; (\forall j)_{<m}\Big\{j = 0 \vee \Big[(y)_j \underset{\Gamma}{\Rightarrow} (y)_{j+1}\Big]\Big\}\Big). \end{aligned}$$

Using Lemma 1, we have proved

Lemma 2. DERIV(u, y) is primitive recursive.

Also, by definition of DERIV(u, y), we have for every word u on the alphabet of Γ

$$S \overset{*}{\underset{\Gamma}{\Rightarrow}} u \Leftrightarrow (\exists y)\mathrm{DERIV}(u,y). \tag{5.1}$$

Finally, (5.1) shows that

$$S \overset{*}{\underset{\Gamma}{\Rightarrow}} u \Leftrightarrow \min_y \mathrm{DERIV}(u,y)\downarrow.$$

Hence, by Lemma 2 and Theorem 7.2 in Chapter 3, we see that $\{u \mid S \overset{*}{\underset{\Gamma}{\Rightarrow}} u\}$ is r.e. But

$$L(\Gamma) = T^* \cap \left\{u \mid S \overset{*}{\underset{\Gamma}{\Rightarrow}} u\right\} \tag{5.2}$$

(where T is the alphabet of terminals of Γ), so that $L(\Gamma)$ is the intersection of two r.e. sets and hence is r.e. Combining this result with Theorem 5.1 in Chapter 6 and Theorem 5.1 in this chapter, we have

Theorem 5.2. A language U is r.e. if and only if there is a grammar Γ such that $U = L(\Gamma)$.

We now are able to obtain easily the promised converse to Theorem 5.1 in Chapter 6. In fact putting Theorem 3.1 in Chapter 6 and Theorems 5.1 and 5.2 in this chapter all together, we have

Theorem 5.3. Let L be a given language. Then the following conditions are all equivalent:

1. L is r.e.;
2. L is accepted by a deterministic Turing machine;
3. L is accepted by a nondeterministic Turing machine;
4. there is a grammar Γ such that $L = L(\Gamma)$.

Theorem 5.3 involves some of the main concerns of theoretical computer science: on the one hand, the relation between grammars, the languages they generate, and the devices that accept them; on the other hand, the relation, for various devices, between determinism and nondeterminism.

We will conclude this section by obtaining a result that will be needed in Chapter 11, but can easily be proved at this point.

Definition. A grammar Γ is called *context-sensitive* if for each production $g \to h$ of Γ we have $|g| \le |h|$.

Lemma 3. If Γ is context-sensitive, then

$$\{u \mid S \overset{*}{\underset{\Gamma}{\Rightarrow}} u\}$$

is recursive.

Proof. It will suffice to obtain a recursive bound for y in formula (5.1). Since

$$1 = |u_1| \le |u_2| \le \cdots \le |u_m| = |u|$$

for any derivation $u_1, \ldots, u_m$ of u from S in the context-sensitive grammar Γ, we must have

$$u_1, u_2, \ldots, u_m \le g(u),$$

where $g(u)$ is the smallest number which represents a string of length $|u| + 1$ in base $n + k$. Now, since $g(u)$ is simply the value in base $n + k$ of a string consisting of $|u| + 1$ repetitions of 1, we have

$$g(u) = \sum_{i=0}^{|u|} (n + k)^i,$$

which is primitive recursive because $|u|$ is primitive recursive. Next, note that we may assume that the derivation

$$S = u_1 \Rightarrow u_2 \Rightarrow \cdots \Rightarrow u_m = u$$

contains no repetitions. This is because given a sequence of steps

$$z = u_i \Rightarrow u_{i+1} \Rightarrow \cdots \Rightarrow u_{i+l} = z,$$

we could simply eliminate the steps $u_{i+1}, \ldots, u_{i+l}$. Hence the length m of the derivation is bounded by the total number of distinct strings of length $\le |u|$ on our alphabet of $n + k$ symbols. But this number is just $g(u)$. Hence,

$$[u_1, \ldots, u_m, 1] = \prod_{i=1}^{m} p_i^{u_i} \cdot p_{m+1} \le h(u),$$

where we have written $h(u)$ for the primitive recursive function defined by

$$h(u) = \prod_{i=1}^{g(u)} p_i^{g(u)} \cdot p_{g(u)+1}.$$

Finally, we have

$$S \overset{*}{\underset{\Gamma}{\Rightarrow}} u \Leftrightarrow (\exists y)_{\le h(u)} \mathrm{DERIV}(u, y),$$

which gives the result. ∎

Theorem 5.4. If Γ is a context-sensitive grammar, then $L(\Gamma)$ is recursive.

Proof. We will use Lemma 3 and Eq. (5.2). Since T^* is a recursive set, the result follows at once. ∎

Exercises

1. For each of the following languages L, give a grammar Γ such that $L = L(\Gamma)$.
   (a) $L = \{a^{[n]}b^{[n]} \mid n \in N\}$.
   (b) $L = \{a^{[n]}b^{[m]} \mid n \le m\}$.
   (c) $L = \{ww^R \mid w \in \{a, b\}^*\}$.
2. Use the construction in the proof of Theorem 5.1 to give a grammar Γ such that $L(\Gamma) = \{1^{[m]}B1^{[n]}B1^{[m+n]} \mid m, n \ge 0\}$.
3. Write down the proof of Theorem 5.2.

4. (a) Let Γ have the variables S, B, C, the terminals a, b, c and the productions

$$S \to aSBC, \qquad S \to aBC,$$
$$CB \to BC, \qquad bB \to bb,$$
$$aB \to ab, \qquad bC \to bc,$$
$$cC \to cc.$$

Prove that for each $n \neq 0$, $a^{[n]}b^{[n]}c^{[n]} \in L(\Gamma)$.

(b)* Prove that $L(\Gamma) = \{a^{[n]}b^{[n]}c^{[n]} \mid n \neq 0\}$.

6. Some Unsolvable Problems Concerning Grammars

How much information can we hope to obtain about $L(\Gamma)$ by a computation that uses the grammar Γ as input? Not much at all, as we shall see.

Let $\mathscr{M}$ be a Turing machine and let u be some given word on the alphabet of $\mathscr{M}$. We shall construct a grammar Γ_u as follows:

The variables of Γ_u are the *entire* alphabet of $\Sigma(\mathscr{M})$ together with S (the start symbol) and V. There is just one terminal, namely, a. The productions of Γ_u are all of the productions of $\Sigma(\mathscr{M})$ together with

$$S \to hq_1 s_0 uh$$
$$hq_0 h \to V$$
$$V \to aV$$
$$V \to a.$$

Then it follows at once from Theorems 2.1 and 2.2 that $S \overset{*}{\underset{\Gamma_u}{\Rightarrow}} V$ if and only if $\mathscr{M}$ accepts u. Thus we have

Lemma. If $\mathscr{M}$ accepts u, then $L(\Gamma_u) = \{a^{[i]} \mid i \neq 0\}$. If $\mathscr{M}$ does not accept u, then $L(\Gamma_u) = \varnothing$.

Now we can select $\mathscr{M}$ so that the language it accepts is not recursive. Then there is no algorithm for determining for given u whether $\mathscr{M}$ accepts u. But the lemma obviously implies the equivalences

$$\mathscr{M} \text{ accepts } u \Leftrightarrow L(\Gamma_u) \neq \varnothing$$
$$\Leftrightarrow L(\Gamma_u) \text{ is infinite}$$
$$\Leftrightarrow a \in L(\Gamma_u).$$

We have obtained

Theorem 6.1. There is no algorithm to determine of a given grammar Γ whether

1. $L(\Gamma) = \varnothing$,
2. $L(\Gamma)$ is infinite, or
3. $v_0 \in L(\Gamma)$ for a fixed word v_0.

We can also prove

Theorem 6.2. There is no algorithm for determining of a given pair Γ, Δ of grammars whether

1. $L(\Delta) \subseteq L(\Gamma)$,
2. $L(\Delta) = L(\Gamma)$.

Proof. Let Δ be the grammar with the single variable S, the single terminal a, and the productions

$$S \to aS$$
$$S \to a.$$

Then $L(\Delta) = \{a^{[i]} \mid i \neq 0\}$. Thus we have by the previous lemma

$$\mathscr{M} \text{ accepts } u \Leftrightarrow L(\Delta) = L(\Gamma_u) \Leftrightarrow L(\Delta) \subseteq L(\Gamma_u).$$

The result follows at once. ■

Exercise

1. Show that there is no algorithm to determine of a given grammar Γ whether
   (a) $L(\Gamma)$ contains at least one word with exactly three symbols;
   (b) v_0 is the shortest word in $L(\Gamma)$ for some given word v_0;
   (c) $L(\Gamma) = A^*$ for some given alphabet A.

*7. Normal Processes

Given a pair of words g and $\bar{g}$ we write

$$gz \to z\bar{g}$$

to indicate a kind of transformation on strings called a *normal production*. If P is the normal production $gz \to z\bar{g}$ we write

$$u \underset{P}{\Rightarrow} v$$

if for some string z we have

$$u = gz, \qquad v = z\bar{g}.$$

That is, v can be obtained from u by crossing off g from the left of u and adjoining $\bar{g}$ to the right. A *normal process* is simply a finite set of normal productions. If ν is a normal process, we write

$$u \underset{\nu}{\Rightarrow} v$$

to mean that

$$u \underset{P}{\Rightarrow} v$$

for some production P in ν. Finally, we write

$$u \overset{*}{\underset{\nu}{\Rightarrow}} v$$

to mean that there is a sequence (called a *derivation*)

$$u = u_1 \underset{\nu}{\Rightarrow} u_2 \underset{\nu}{\Rightarrow} \cdots \underset{\nu}{\Rightarrow} u_m = v.$$

The *word problem* for ν is the problem of determining of two given words u, v whether $u \overset{*}{\underset{\nu}{\Rightarrow}} v$.

Let Π be a semi-Thue process on the alphabet $\{a, b\}$ with an unsolvable word problem. We shall show how to simulate Π by a normal process ν on the alphabet $\{a, b, \tilde{a}, \tilde{b}\}$. As earlier, if $u \in \{a, b\}^*$, we write $\tilde{u}$ for the word on $\{\tilde{a}, \tilde{b}\}$ obtained by placing $\sim$ above each letter in u. Let the productions of Π be

$$g_i \to h_i, \qquad i = 1, 2, \ldots, n.$$

Then the productions of ν will be

$$\begin{aligned} g_i z &\to z\tilde{h}_i \qquad i = 1, 2, \ldots, n \\ az &\to z\tilde{a} \\ bz &\to z\tilde{b} \\ \tilde{a}z &\to za \\ \tilde{b}z &\to zb. \end{aligned}$$

A word on $\{a, b, \tilde{a}, \tilde{b}\}$ is called *proper* if it can be written in one of the forms $u\tilde{v}$ or $\tilde{u}v$, where u, v are words on $\{a, b\}$. We say that two words are *associates* if there is a derivation of one from the other *using only the last four productions* of ν. A word on $\{a, b\}$ of length n has $2n$ associates, all of which are proper. For example, the associates of *baab* are as follows:

$$baab \Rightarrow aab\tilde{b} \Rightarrow ab\tilde{b}\tilde{a} \Rightarrow b\tilde{b}\tilde{a}\tilde{a} \Rightarrow \tilde{b}\tilde{a}\tilde{a}\tilde{b} \Rightarrow \tilde{a}\tilde{a}\tilde{b}b \Rightarrow \tilde{a}\tilde{b}ba \Rightarrow \tilde{b}baa \Rightarrow baab.$$

Generally for $u, v \in \{a, b\}^*$, the proper words $u\tilde{v}$ and $\tilde{u}v$ are associates of each other and also of the word vu. In fact, vu is the unique word on $\{a, b\}$ which is an associate of $u\tilde{v}$. Thus, a word is proper just in case it is an associate of a word on $\{a, b\}$.

Lemma 1. If $u \underset{\Pi}{\Rightarrow} v$, then $u \overset{*}{\underset{\nu}{\Rightarrow}} v$.

Proof. We have $u = pg_iq$, $v = ph_iq$ for some i. Then

$$u \overset{*}{\underset{\nu}{\Rightarrow}} g_iq\tilde{p} \underset{\nu}{\Rightarrow} q\tilde{p}\tilde{h}_i \overset{*}{\underset{\nu}{\Rightarrow}} ph_iq. \qquad \blacksquare$$

Lemma 2. If $u \overset{*}{\underset{\Pi}{\Rightarrow}} v$, then $u \overset{*}{\underset{\nu}{\Rightarrow}} v$.

Proof. Immediate from Lemma 1. ■

Lemma 3. Let u be proper and let $u \underset{\nu}{\Rightarrow} v$. Then there are words r, s on $\{a, b\}$ that are associates of u, v, respectively, such that $r \overset{*}{\underset{\Pi}{\Rightarrow}} s$.

Proof. If v is an associate of u, then u and v are both associates of some word r on $\{a, b\}$, and the result follows because $r \overset{*}{\underset{\Pi}{\Rightarrow}} r$.

If v is not an associate of u, the production used to obtain v from u must be one of the $g_iz \to z\tilde{h}_i$. Since u is proper, we have $u = g_iq\tilde{p}$, where p, q are words on $\{a, b\}$. Then $v = q\tilde{p}\tilde{h}_i$. Thus, setting

$$r = pg_iq, \qquad s = ph_iq,$$

the result follows because $r \underset{\Pi}{\Rightarrow} s$. ■

Lemma 4. Let u be proper and let $u \overset{*}{\underset{\nu}{\Rightarrow}} v$. Then there are words r, s on $\{a, b\}$ that are associates of u, v, respectively, such that $r \overset{*}{\underset{\Pi}{\Rightarrow}} s$.

Proof. By induction on the length of the derivation in ν of v from u. The result is obvious if the derivation has length 1. Suppose the result is known for derivations of length m, and let

$$u = u_1 \underset{\nu}{\Rightarrow} u_2 \underset{\nu}{\Rightarrow} \cdots \underset{\nu}{\Rightarrow} u_m \underset{\nu}{\Rightarrow} u_{m+1} = v.$$

By the induction hypothesis, there are words r, z on $\{a, b\}$ that are associates of u, u_m, respectively, such that $r \overset{*}{\underset{\Pi}{\Rightarrow}} z$. By Lemma 3, u_{m+1} is an associate of a word s on $\{a, b\}$ such that $z \overset{*}{\underset{\Pi}{\Rightarrow}} s$. Thus, $r \overset{*}{\underset{\Pi}{\Rightarrow}} s$. ■

Lemma 5. Let u, v be words on $\{a, b\}$. Then $u \overset{*}{\underset{\nu}{\Rightarrow}} v$ if and only if $u \overset{*}{\underset{\Pi}{\Rightarrow}} v$.

Proof. By Lemma 2 we know that $u \overset{*}{\underset{\Pi}{\Rightarrow}} v$ implies $u \overset{*}{\underset{\nu}{\Rightarrow}} v$. Conversely, if $u \overset{*}{\underset{\nu}{\Rightarrow}} v$, by Lemma 4, $r \overset{*}{\underset{\Pi}{\Rightarrow}} s$, where r, s are words on $\{a, b\}$ that are associates of u, v, respectively. But since u, v are already words on $\{a, b\}$, we have $r = u$, $s = v$, so that $u \overset{*}{\underset{\Pi}{\Rightarrow}} v$. ■

Since Π was chosen to have an unsolvable word problem, it is now clear that ν has an unsolvable word problem. For, by Lemma 5, if we had an algorithm for deciding whether $u \overset{*}{\underset{\nu}{\Rightarrow}} v$, we could use it to decide whether $u \overset{*}{\underset{\Pi}{\Rightarrow}} v$.

We have proved

Theorem 7.1. There is a normal process on a four-letter alphabet with an unsolvable word problem.

Exercise

1. Show that there is a normal process with an unsolvable word problem whose alphabet contains only two letters.

8

Classifying Unsolvable Problems

1. Using Oracles

Once one gets used to the fact that there are explicit problems, such as the halting problem, that have no algorithmic solution, one is led to consider questions such as the following.

Suppose we were given a "black box" or, as one says, an *oracle*, which somehow can tell us whether a given Turing machine with given input eventually halts. (Of course, by Church's thesis, the behavior of such an "oracle" cannot be characterized by an algorithm.) Then it is natural to consider a kind of program that is allowed to ask questions of our oracle and to use the answers in its further computation. Which noncomputable functions will now become computable?

In this chapter we will see how to give a precise answer to such questions. To begin with, we shall have to modify the programming language $\mathscr{S}$ introduced in Chapter 2, to permit the use of "oracles." Specifically, we change the definition of "statement" (in Chapter 2, Section 3) to allow statements of the form $V \leftarrow O(V)$ instead of $V \leftarrow V$. The modified version of $\mathscr{S}$ thus contains four kinds of statement: increment, decrement, conditional branch, and this new kind of statement which we call an *oracle statement*. The definitions of *instruction*, *program*, *state*, *snapshot*, and *terminal snapshot* remain exactly as in Chapter 2.

We now let G be some partial function on N with values in N, and we shall think of G as an oracle. Let $\mathscr{P}$ be a program of length n and let (i, σ) be a nonterminal snapshot of $\mathscr{P}$, i.e., $i \leq n$. We define the snapshot (j, τ) to be the *G-successor* of (i, σ) exactly as in the definition of *successor* in Chapter 2, Section 3, except that Case 3 is now replaced by

Case 3. *The ith instruction of $\mathscr{P}$ is $V \leftarrow O(V)$ and σ contains the equation* $V = m$. If $G(m)\downarrow$, then $j = i + 1$ and τ is obtained from σ by replacing the equation $V = m$ by $V = G(m)$. If $G(m)\uparrow$, then (i, σ) has no successor.

Thus, when $G(m)\downarrow$, execution of this oracle statement has the intuitive effect of answering the computer's question "$G(m) = ?$". When $G(m)\uparrow$, an "out-of-bounds" condition is recognized, and the computer halts without reaching a terminal snapshot. Of course, when G is total, every nonterminal snapshot has a successor.

A *G-computation* is defined just like *computation* except that the word *successor* is replaced by *G-successor*. A number m that is replaced by $G(m)$ in the course of a G-computation (under Case 3) is called an *oracle query of the G-computation*. We define $\psi_{\mathscr{P},G}^{(m)}(r_1, r_2, \ldots, r_m)$ exactly as we defined $\psi_{\mathscr{P}}^{(m)}(r_1, r_2, \ldots, r_m)$ in Chapter 2, Section 4, except that the word *computation* is replaced by *G-computation*.

Now, let G be a total function. Then, the partial function $\psi_{\mathscr{P},G}^{(m)}(x_1, \ldots, x_m)$ is said to be G-computed by $\mathscr{P}$. A partial function f is said to be *partially G-computable* or *G-partial recursive* if there is a program that G-computes it. A partially G-computable function that is total is called *G-computable* or *G-recursive*. Note that we have not defined *partially G-computable* unless G is a total function.

We have a few almost obvious theorems.

Theorem 1.1. If f is partially computable, then f is partially G-computable for all total functions G.

Proof. Clearly, we can assume that f is computed by a program $\mathscr{P}$ containing no statements of the form[1] $V \leftarrow V$. Now this program $\mathscr{P}$ is also

---

[1] Unlabeled statements $V \leftarrow V$ can just be deleted, and

$$[L] \quad V \leftarrow V$$

can be replaced by

$$\begin{aligned} [L] \quad & V \leftarrow V + 1 \\ & V \leftarrow V - 1. \end{aligned}$$

a program in the new revised sense; moreover, a computation of $\mathscr{P}$ is the same thing as a G-computation of $\mathscr{P}$ since $\mathscr{P}$ contains no oracle statements. Hence $\psi_{\mathscr{P},G}^{(m)} = \psi_{\mathscr{P}}^{(m)}$ for all G. ■

We write I for the identity function $I(x) = x$. (Thus, $I = u_1^1$.)

Theorem 1.2. f is partially computable if and only if f is partially I-computable.

Proof. If f is partially computable, then by Theorem 1.1 it is certainly partially I-computable. Conversely, let $\mathscr{P}$ I-compute f. Let $\mathscr{P}'$ be obtained from $\mathscr{P}$ by replacing each oracle statement $V \leftarrow O(V)$ by $V \leftarrow V$. Then, $\mathscr{P}'$ is a program in the original sense and $\mathscr{P}'$ computes f. ■

Theorem 1.3. Let G be a total function. Then G is G-computable.

Proof. The following program[2] clearly G-computes G:

$$X \leftarrow O(X)$$
$$Y \leftarrow X$$
■

Theorem 1.4. The class of G-computable functions is a PRC class.

Proof. Exactly like the proof of Theorem 3.1 in Chapter 3. ■

This last proof illustrates a situation, which turns out to be quite typical, in which the proof of an earlier theorem can be used virtually intact to prove a theorem relative to an "oracle" G. One speaks of a *relativized* theorem and of *relativizing* a proof. It is a matter of taste how much detail to provide in such a case.

Theorem 1.5. Let F be partially G-computable and let G be H-computable. Then F is partially H-computable.

Proof. Let $\mathscr{P}$ be a program which G-computes F. Let $\mathscr{P}'$ be obtained from $\mathscr{P}$ by replacing each oracle statement $V \leftarrow O(V)$ by a macro expansion obtained from some program which H-computes G. Then clearly, $\mathscr{P}'$ H-computes F. ■

Theorem 1.6. Let G be any computable function. Then a function F is partially computable if and only if it is partially G-computable.

---

[2] Of course, we can freely use macro expansions, as explained in Chapter 2.

Proof. Theorem 1.1 gives the result in one direction. For the converse, let F be partially G-computable. By Theorem 1.2, G is I-computable. Hence, by Theorem 1.5, F is partially I-computable and so, by Theorem 1.2 again, F is partially computable. ■

It is useful to be able to work with "oracles" that are functions of more than one variable. We introduce this notion by using a familiar coding device from Chapter 3, Section 8.

Definition. Let f be a total n-ary function on N, $n > 1$. Then we say that g is (*partially*) *f-computable* to mean that g is (*partially*) *G-computable*, where

$$G(x) = f((x)_1, \ldots, (x)_n). \tag{1.1}$$

Theorem 1.7. Let f be a total n-ary function on N. Then f is f-computable.

Proof. Let G be defined by (1.1). Then

$$f(x_1, \ldots, x_n) = G([x_1, \ldots, x_n]).$$

Hence the following program G-computes f:

$$Z \leftarrow [X_1, \ldots, X_n]$$
$$Z \leftarrow O(Z)$$
$$Y \leftarrow Z$$
■

Since predicates are also total functions we can speak of a function being (partially) P-computable, where P is a predicate. Also, we speak of a function being (partially) A-computable when $A \subseteq N$; as usual, we simply identify A with the predicate that is its characteristic function.

Exercises

1. Provide a suitable definition of computability by a Post–Turing program relative to an oracle and prove an appropriate equivalence theorem.
2. For a given total function G from N to N, define the class $\text{Rec}(G)$ to be the class of functions obtained from G and the initial functions of Chapter 3 using composition, recursion, and minimalization. Prove that every function in $\text{Rec}(G)$ is partially G-computable.

2. Relativization of Universality

We now proceed to relativize the development in Chapter 4. As in Chapter 4, Section 1, we define an instruction number $\#(I) = \langle a, \langle b, c \rangle \rangle$ for all instructions I. The only difference is that $b = 0$ now indicates an oracle statement instead of one of the form $V \leftarrow V$. For $\mathscr{P}$ a program, we now define $\#(\mathscr{P})$ as before. As indicated in Chapter 4, in order to avoid ambiguity we must not permit a program ending in the instruction whose number is 0. This instruction is now the unlabeled statement $Y \leftarrow O(Y)$. Hence, for complete rigor, if we wish to end a program with $Y \leftarrow O(Y)$, we will have to provide the statement with a spurious label.

We define $\Phi_G^{(n)}(x_1, \ldots, x_n, y)$ to be $\psi_{\mathscr{P}, G}^{(n)}(x_1, \ldots, x_n)$ where $\mathscr{P}$ is the *unique* program such that $\#(\mathscr{P}) = y$. We also write $\Phi_G(x, y)$ for $\Phi_G^{(1)}(x, y)$. We have

Theorem 2.1 (Relativized Universality Theorem). Let G be total. Then the function $\Phi_G^{(n)}(x_1, \ldots, x_n, y)$ is partially G-computable.

Proof. The proof of this theorem is essentially contained in the program of Fig. 2.1. The daggers (‡) indicate the changes from the unrelativized universal program in Fig. 3.1 in Chapter 4. As in that case, what we have is essentially an interpretative program. The new element is of course the interpretation of oracle statements. This occurs in the following program segment which, not surprisingly, itself contains an oracle statement:

$$
\begin{array}{ll}
[O] & W \leftarrow (S)_{r(U)+1} \\
 & B \leftarrow W \\
 & B \leftarrow O(B) \\
 & S \leftarrow \lfloor S/P^W \rfloor \cdot P^B
\end{array}
$$

The program segment works as follows. First, W and B are both set to the current value of the variable in the oracle statement being interpreted. Then an oracle statement gives B a new value which is G of the old value. Finally, this new value is stored as an exponent on the appropriate prime in the number S. The remainder of the program works exactly as in the unrelativized case. ■

Let G be any *partial* function on N with values in N. Then we define the relativized step-counter predicate by

$\mathrm{STP}_G^{(n)}(x_1, \ldots, x_n, y, t) \Leftrightarrow$ there is a G-computation of program number y of length $\leq t + 1$ beginning with inputs $x_1, \ldots, x_n$.

	$Z \leftarrow X_{n+1} + 1$	
	$S \leftarrow \prod_{i=1}^{n} (p_{2i})^{X_i}$	
	$K \leftarrow 1$	
[C]	IF $K = \mathrm{Lt}(Z) + 1 \vee K = 0$ GOTO F	
	$U \leftarrow r((Z)_K)$	
	$P \leftarrow p_{r(U)+1}$	
	IF $l(U) = 0$ GOTO O	(‡)
	IF $l(U) = 1$ GOTO A	
	IF $\sim(P \mid S)$ GOTO N	
	IF $l(U) = 2$ GOTO M	
	$K \leftarrow \min_{i \le \mathrm{Lt}(Z)} [l((Z)_i) + 2 = l(U)]$	
	GOTO C	
[O]	$W \leftarrow (S)_{r(U)+1}$	(‡)
	$B \leftarrow W$	(‡)
	$B \leftarrow O(B)$	(‡)
	$S \leftarrow \lfloor S/P^W \rfloor \cdot P^B$	(‡)
	GOTO N	(‡)
[M]	$S \leftarrow \lfloor S/P \rfloor$	
	GOTO N	
[A]	$S \leftarrow S \cdot P$	
[N]	$K \leftarrow K + 1$	
	GOTO C	
[F]	$Y \leftarrow (S)_1$	

Figure 2.1. Program that G-computes $\Phi_G^{(n)}(X_1, \ldots, X_n, X_{n+1})$.

As in the unrelativized case, we have

Theorem 2.2 (Relativized Step-Counter Theorem). For any total function G, the predicates $\mathrm{STP}_G^{(n)}(x_1, \ldots, x_n, y, t)$ are G-computable.

In Chapter 4 we proved that the unrelativized predicates $\mathrm{STP}^{(n)}$ are primitive recursive, but we do not need such a sharp result here. Instead, we modify the program in Fig. 2.1 by adding a variable Q that functions as a step counter. Then each time through the main loop, Q is increased by 1, so that the program will "know" when a given number of steps has been

```
        Z ← X_{n+1} + 1
        S ← ∏_{i=1}^{n} (p_{2i})^{X_i}
        K ← 1
[C]     Q = Q + 1                                   (*)
        IF Q > X_{n+2} + 1 GOTO E                   (*)
        IF K = Lt(Z) + 1 ∨ K = 0 GOTO F
        U ← r((Z)_K)
        P ← p_{r(U)+1}
        IF l(U) = 0 GOTO O                          (‡)
        IF l(U) = 1 GOTO A
        IF ~(P | S) GOTO N
        IF l(U) = 2 GOTO M
        K ← min_{i≤Lt(Z)} [l((Z)_i) + 2 = l(U)]
        GOTO C
[O]     W ← (S)_{r(U)+1}                            (‡)
        B ← W                                       (‡)
        B ← O(B)                                    (‡)
        S ← ⌊S/P^W⌋ · P^B                           (‡)
        GOTO N                                      (‡)
[M]     S ← ⌊S/P⌋
        GOTO N
[A]     S ← S · P
[N]     K ← K + 1
        GOTO C
[F]     Y ← 1                                       (*)
```

Figure 2.2. Program that G-computes $\mathrm{STP}_G^{(n)}(X_1, \ldots, X_n, X_{n+1}, X_{n+2})$.

exceeded. The program is given in Fig. 2.2. The asterisks (*) indicate changes from the relativized universal program and the daggers (‡), as before, indicate the changes made in relativizing.

We shall now consider certain partial functions with finite domains, and use numbers as codes for them. For every $u \in N$ we define

$$\{u\}(i) = \begin{cases} (r(u))_{i+1} & \text{for } i < l(u) \\ \uparrow & \text{for } i \geq l(u). \end{cases} \tag{2.1}$$

Thus, if $l(u) = 0$, then $\{u\} = \varnothing$, the nowhere defined function. Also, if

$$u = \langle k, [a_0, a_1, \ldots, a_{k-1}]\rangle,$$

then $\{u\}(x) = a_x$ for $x = 0, 1, \ldots, k - 1$ and $\{u\}(x)\uparrow$ for $x \geq k$.

Theorem 2.3. The predicate

$$P(x_1, \ldots, x_n, y, t, u) \Leftrightarrow \mathrm{STP}^{(n)}_{\{u\}}(x_1, \ldots, x_n, y, t)$$

is computable.

Proof. We will transform the program in Fig. 2.2 into one that computes $P(x_1, \ldots, x_n, x_{n+1}, x_{n+2}, x_{n+3})$. We need only replace the single oracle statement $B \leftarrow O(B)$ by instructions that operate on x_{n+3} to obtain the required information about $\{x_{n+3}\}$. This involves first testing for $\{x_{n+3}\}(b)\downarrow$, where b is the value of the variable B. If $\{x_{n+3}\}(b)\uparrow$, computation should halt with output 0, because there is no computation in this case. Otherwise B should be given the value $\{x_{n+3}\}(b)$. Thus, by (2.1), it suffices to replace the oracle statement $B \leftarrow O(B)$ in the program in Fig. 2.2 by the following pair of instructions:

$$\text{IF } l(X_{n+3}) \leq B \text{ GOTO } E$$

$$B \leftarrow (r(X_{n+3}))_{B+1}$$ ■

Now, let G be a total function. Then, we define

$$u \prec G$$

to mean that $\{u\}(i) = G(i)$ for $0 \leq i < l(u)$. [Of course, by (2.1), $\{u\}(i)\uparrow$ for $i \geq l(u)$.] For a number u such that $u \prec G$, values of G can be retrieved by using the equations

$$G(i) = (r(u))_{i+1}, \qquad i = 0, 1, \ldots, l(u) - 1.$$

We can use the predicate $\mathrm{STP}^{(n)}_{\{u\}}(x_1, \ldots, x_n, y, t)$ to obtain an important result that isolates the noncomputability of the relativized step-counter predicate in a way that will prove helpful. The simple observation on which this result capitalizes is that any G-computation can contain only finitely many oracle queries.

Theorem 2.4 (Finiteness Theorem). Let G be a total function. Then, we have

$$\mathrm{STP}^{(n)}_G(x_1, \ldots, x_n, y, t) \Leftrightarrow (\exists u)\left[u \prec G \;\&\; \mathrm{STP}^{(n)}_{\{u\}}(x_1, \ldots, x_n, y, t)\right].$$

Proof. First suppose that $\text{STP}_G^{(n)}(x_1, \ldots, x_n, y, t)$ is true for some given values of $x_1, \ldots, x_n, y, t$, and let $\mathscr{P}$ be the program with $\#(\mathscr{P}) = y$. Let $s_1, s_2, \ldots, s_k$ be a G-computation of $\mathscr{P}$ where s_1 is the initial snapshot corresponding to the input values $x_1, x_2, \ldots, x_n$ and where $k \leq t + 1$. Let M be the largest value of an oracle query of this G-computation, and let $u = \langle M + 1, [G(0), G(1), \ldots, G(M)] \rangle$. Thus, $u \prec G$ and $\{u\}(m) = G(m)$ for all $m \leq M$. Hence, $s_1, s_2, \ldots, s_k$ is likewise a $\{u\}$-computation of $\mathscr{P}$. Since $k \leq t + 1$, $\text{STP}_{\{u\}}^{(n)}(x_1, \ldots, x_n, y, t)$ is true.

Conversely, let us be given $u \prec G$ such that $\text{STP}_{\{u\}}^{(n)}(x_1, \ldots, x_n, y, t)$ is true, and let $\#(\mathscr{P}) = y$. Let $s_1, s_2, \ldots, s_k$ be a $\{u\}$-computation of $\mathscr{P}$ where s_1 is the initial snapshot corresponding to the input values $x_1, x_2, \ldots, x_n$ and where $k \leq t + 1$. For each m that is an oracle query of this $\{u\}$-computation, we must have $\{u\}(m)\downarrow$, since otherwise one of the snapshots in this $\{u\}$-computation would be nonterminal and yet not have a successor. Since $u \prec G$, we must have $\{u\}(m) = G(m)$ for all such m. Hence $s_1, s_2, \ldots, s_k$ is likewise a G-computation of $\mathscr{P}$. Since $k \leq t + 1$, $\text{STP}_G^{(n)}(x_1, \ldots, x_n, y, t)$ is true. ■

To conclude this section we turn to the parameter theorem (Theorem 5.1 in Chapter 4).

Theorem 2.5 (Relativized and Strengthened Parameter Theorem). For each $n, m > 0$, there is a primitive recursive function $S_m^n(u_1, \ldots, u_n, y)$ such that for every total function G:

$$\Phi_G^{(m+n)}(x_1, \ldots, x_m, u_1, \ldots, u_n, y) = \Phi_G^{(m)}(x_1, \ldots, x_m, S_m^n(u_1, \ldots, u_n, y)). \tag{2.2}$$

Moreover, the functions S_m^n have the property:

$$S_m^n(u_1, \ldots, u_n, y) = S_m^n(\bar{u}_1, \ldots, \bar{u}_n, y) \quad \text{implies} \quad u_1 = \bar{u}_1, \ldots, u_n = \bar{u}_n.$$

Proof. The functions S_m^n are defined exactly as in the proof of Theorem 5.1 in Chapter 4. We briefly give the proof again in a slightly different way. Thus, let $\#(\mathscr{P}) = y$; then the function $S_m^1(u, y)$ is defined to be the number of the program $\overline{\mathscr{P}}$ obtained from $\mathscr{P}$ by preceding it by the statement

$$X_{m+1} \leftarrow X_{m+1} + 1$$

repeated u times. Since $\overline{\mathscr{P}}$ on inputs $x_1, \ldots, x_m$ will do exactly what $\mathscr{P}$ would have done on inputs $x_1, \ldots, x_m, u$ we have

$$\Phi_G^{(m+1)}(x_1, \ldots, x_m, u, y) = \Phi_G^{(m)}(x_1, \ldots, x_m, S_m^1(u, y)),$$

the desired result for $n = 1$. To complete the proof, we define S_m^n for $n > 1$ by the recursion

$$S_m^{k+1}(u_1, \ldots, u_k, u_{k+1}, y) = S_m^k(u_1, \ldots, u_k, S_{m+k}^1(u_{k+1}, y)).$$

It is now easy to prove by induction on n that if $\#(\mathscr{P}) = y$, then $S_m^n(u_1, \ldots, u_n, y) = \#(\overline{\mathscr{P}})$, where $\overline{\mathscr{P}}$ is obtained from $\mathscr{P}$ by preceding it by the following program consisting of $u_n + \cdots + u_1$ statements.

$$\left.\begin{array}{l} X_{m+1} \leftarrow X_{m+1} + 1 \\ \vdots \\ X_{m+1} \leftarrow X_{m+1} + 1 \end{array}\right\} u_1$$
$$\vdots$$
$$\left.\begin{array}{l} X_{m+n} \leftarrow X_{m+n} + 1 \\ \vdots \\ X_{m+n} \leftarrow X_{m+n} + 1 \end{array}\right\} u_n$$

Hence, $\overline{\mathscr{P}}$ on inputs $x_1, \ldots, x_m$ will do exactly what $\mathscr{P}$ would have done on inputs $x_1, \ldots, x_m, u_1, \ldots, u_n$. Thus, we obtain (2.2).

Finally, let

$$S_m^n(u_1, \ldots, u_n, y) = S_m^n(\bar{u}_1, \ldots, \bar{u}_n, y) = \#(\overline{\mathscr{P}}),$$

and let $y = \#(\mathscr{P})$. Then, $\overline{\mathscr{P}}$ consists of a list of increment statements followed by $\mathscr{P}$, and for $1 \le i \le n$, u_i and $\bar{u}_i$ are both simply the number of times the statement

$$X_{m+i} \leftarrow X_{m+i} + 1$$

occurs in $\overline{\mathscr{P}}$ preceding $\mathscr{P}$. Thus, $u_i = \bar{u}_i$. ■

Exercises

1. (a) Show that the functions S_m^n do *not* have the property:

$$S_m^n(u_1, \ldots, u_n, y) = S_m^n(\bar{u}_1, \ldots, \bar{u}_n, \bar{y})$$
$$\text{implies } u_1 = \bar{u}_1, \ldots, u_n = \bar{u}_n, y = \bar{y}.$$

   (b) Can the definition of S_m^n be modified so the parameter theorem continues to hold, but so the condition of (a) holds as well? How?

2. Prove the converse of Exercise 1.2.

3. Reducibility

If A and B are *sets* such that A is B-recursive, we also say that A is *Turing-reducible to B* and we write $A \leq_t B$. We have

Theorem 3.1. $A \leq_t A$. If $A \leq_t B$ and $B \leq_t C$, then $A \leq_t C$.

Proof. The first statement follows at once from Theorem 1.3 and the second from Theorem 1.5. ■

Any relation on the subsets of N for which Theorem 3.1 is true is called a *reducibility.* Many reducibilities have been studied. For example, we introduced many–one reducibility in Chapter 4. We can also define a restricted form of many–one reducibility.

Definition. We write $A \leq_1 B$ and say that A is *one–one reducible to B* if there is a one–one recursive function f (i.e., $f(x) = f(y)$ implies $x = y$) such that

$$A = \{x \in N \mid f(x) \in B\}.$$

Theorem 3.2. $A \leq_1 B$ implies $A \leq_m B$ implies $A \leq_t B$.

Proof. The first implication is immediate. For the second implication, let $A = \{x \in N \mid f(x) \in B\}$, where f is recursive. Then the following program B-computes A:

$$X \leftarrow f(X)$$
$$X \leftarrow O(X)$$
$$Y \leftarrow X$$
■

Theorem 3.3. $\leq_1$ and $\leq_m$ are both reducibilities.

Proof. Clearly $A = \{x \in N \mid I(x) \in A\}$, where I is the identity function. Hence $A \leq_1 A$ and therefore $A \leq_m A$.

Let $A \leq_m B$ and $B \leq_m C$, and let

$$A = \{x \in N \mid f(x) \in B\}, \qquad B = \{x \in N \mid g(x) \in C\},$$

where f, g are recursive. Then

$$A = \{x \in N \mid g(f(x)) \in C\},$$

so that $A \leq_m C$. If, moreover, f and g are one–one and $h(x) = g(f(x))$, then h is also one–one, because

$$\begin{aligned} h(x) = h(y) \quad &\text{implies} \quad g(f(x)) = g(f(y)) \\ &\text{implies} \quad f(x) = f(y) \\ &\text{implies} \quad x = y. \end{aligned}$$
■

Thus, we have three examples, $\leq_1$, $\leq_m$, and $\leq_t$, of reducibilities. Polynomial–time reducibility, $\leq_p$, which we will study in Chapter 15, is another example. (In fact, historically, polynomial–time reducibility was suggested by many–one reducibility.) There are a number of simple properties that all reducibilities share. To work some of these out, let us write $\leq_Q$ to represent an arbitrary reducibility. By replacing Q by 1, m, t (or even p) we specialize to the particular reducibilities we have been studying. We write $A \nleq_Q B$ to indicate that it is not the case that $A \leq_Q B$.

Definition. $A \equiv_Q B$ means that $A \leq_Q B$ and $B \leq_Q A$.

Theorem 3.4. For any reducibility $\leq_Q$:

$$A \equiv_Q A,$$

$$A \equiv_Q B \quad \text{implies} \quad B \equiv_Q A,$$

$$A \equiv_Q B \quad \text{and} \quad B \equiv_Q C \quad \text{implies} \quad A \equiv_Q C.$$

Proof. Immediate from the definition. ■

Definition. Let **W** be a collection of subsets of N and let $\leq_Q$ be a reducibility. **W** is called *Q-closed* if it has the property

$$A \in \mathbf{W} \text{ and } B \leq_Q A \quad \text{implies} \quad B \in \mathbf{W}.$$

Also, a set $A \in \mathbf{W}$ is called *Q-complete for* **W** if for every $B \in \mathbf{W}$ we have $B \leq_Q A$.

NP-completeness, which will be studied in Chapter 15, is, in the present terminology, *polynomial–time completeness for* **NP**. Completeness of a set A is often proved by showing that a set already known to be complete can be reduced to A.

Theorem 3.5. Let A be Q-complete for **W**, let $B \in \mathbf{W}$, and let $A \leq_Q B$. Then B is Q-complete for **W**.

Proof. Let $C \in \mathbf{W}$. Then $C \leq_Q A$. Hence $C \leq_Q B$. ■

If **W** is a collection of subsets of N, we write

$$\text{co-}\mathbf{W} = \{A \subseteq N \mid \bar{A} \in \mathbf{W}\}.$$

Theorem 3.6. Let co-**W** be Q-closed, let A be Q-complete for **W**, and let $A \in$ co-**W**. Then we have **W** = co-**W**.

Proof. Let $B \in \mathbf{W}$. Then, since A is Q-complete for $\mathbf{W}$, $B \leq_Q A$. Since $A \in$ co-$\mathbf{W}$ and co-$\mathbf{W}$ is Q-closed, $B \in$ co-$\mathbf{W}$. This proves that $\mathbf{W} \subseteq$ co-$\mathbf{W}$.

Next let $B \in$ co-$\mathbf{W}$. Then $\bar{B} \in \mathbf{W}$. By what has already been shown, $\bar{B} \in$ co-$\mathbf{W}$. Hence $B \in \mathbf{W}$. This proves that co-$\mathbf{W} \subseteq \mathbf{W}$. ■

As we shall see, Theorem 3.6 is quite useful. Our applications will be to the case of one–one and many–one reducibility. For this purpose, it is useful to note

Theorem 3.7. If $A \leq_{\mathrm{m}} B$, then $\bar{A} \leq_{\mathrm{m}} \bar{B}$. Likewise if $A \leq_1 B$, then $\bar{A} \leq_1 \bar{B}$.

Proof. If $A = \{x \in N \mid f(x) \in B\}$, then clearly $\bar{A} = \{x \in N \mid f(x) \in \bar{B}\}$. ■

Corollary 3.8. If $\mathbf{W}$ is m-closed or 1-closed, then so is co-$\mathbf{W}$.

Proof. Let $B \in$ co-$\mathbf{W}$, $A \leq_{\mathrm{m}} B$. By the theorem, $\bar{A} \leq_{\mathrm{m}} \bar{B}$. Since $\bar{B} \in \mathbf{W}$ and $\mathbf{W}$ is m-closed, $\bar{A} \in \mathbf{W}$. Hence $A \in$ co-$\mathbf{W}$. Similarly for $\leq_1$. ■

For a concrete example, we may take $\mathbf{W}$ to be the collection of r.e. subsets of N. (For notation, the reader should review Chapter 4, Section 4.) We have

Theorem 3.9. K is 1-complete for the class of r.e. sets.

Proof. Let A be any r.e. set. We must show that $A \leq_1 K$. Since A is r.e., we have

$$A = \{x \in N \mid f(x)\downarrow\},$$

where f is a partially computable function. Let $g(t, x) = f(x)$ for all t, x. Thus, g is also partially computable. Using the (unrelativized) universality and parameter theorems, we have for a suitable number e:

$$g(t, x) = \Phi^{(2)}(t, x, e) = \Phi(t, S_1^1(x, e)).$$

Hence,

$$\begin{aligned} A &= \{x \in N \mid f(x)\downarrow\} \\ &= \left\{x \in N \mid g(S_1^1(x, e), x)\downarrow\right\} \\ &= \left\{x \in N \mid \Phi(S_1^1(x, e), S_1^1(x, e))\downarrow\right\} \\ &= \{x \in N \mid S_1^1(x, e) \in K\}. \end{aligned}$$

Thus, $A \leq_{\mathrm{m}} K$. But, by the strengthened version of the parameter theorem (Theorem 2.5), $S_1^1(x, e)$ is actually one–one. Hence, $A \leq_1 K$. ■

The class of r.e. sets is easily seen to be m-closed. Thus, let f be partially computable, let $A = \{x \in N \mid f(x)\downarrow\}$, and let $B = \{x \in N \mid g(x) \in A\}$, where g is computable. Then

$$B = \{x \in N \mid f(g(x))\downarrow\},$$

so that B is r.e. Applying Theorems 3.2, 3.6, and 3.9 and Corollary 3.8, we obtain the not very interesting conclusion:

If $\bar{K}$ is r.e., then the complement of every r.e. set is r.e.

Since we know that $\bar{K}$ is in fact not r.e., this does us no good. However, Corollary 3.8 and Theorem 3.6 together with the fact that there is an r.e. set (e.g., K) whose complement is not r.e. permits us to conclude

Theorem 3.10. If A is m-complete for the class of r.e. sets, then $\bar{A}$ is not r.e., so that A is not recursive.

We conclude this section with a simple but important construction. For $A, B \subseteq N$ we write

$$A \oplus B = \{2x \mid x \in A\} \cup \{2x + 1 \mid x \in B\}.$$

Intuitively, $A \oplus B$ contains the information in both A and B and nothing else. This suggests the truth of the following simple result.

Theorem 3.11. $A \leq_t A \oplus B$, $B \leq_t A \oplus B$. If $A \leq_t C$ and $B \leq_t C$, then $A \oplus B \leq_t C$.

Proof. The following program $(A \oplus B)$-computes A:

$$X \leftarrow 2X$$
$$X \leftarrow O(X)$$
$$Y \leftarrow X$$

If the first instruction is replaced by $X \leftarrow 2X + 1$, the program $(A \oplus B)$-computes B.

Finally, let C_A, C_B be the characteristic functions of A and B, respectively. Assuming that A and B are both C-computable, there must be programs that C-compute the functions C_A and C_B, respectively. Hence, we may use macros

$$Y \leftarrow C_A(X) \quad \text{and} \quad Y \leftarrow C_B(X)$$

in programs that have C available as oracle. Thus, the following program C-computes $A \oplus B$:

```
        IF 2 | X GOTO D
        X ← ⌊(X ∸ 1)/2⌋
        Y ← C_B(X)
        GOTO E
[D]     X ← ⌊X/2⌋
        Y ← C_A(X)
```

■

Exercises

1. Let $U = \{x \in N \mid l(x) \in W_{r(x)}\}$. Show that U is 1-complete for the class of r.e. sets.
2. Let $K \leq_t A$ and let

$$C = \left\{x \in K \mid \Phi_x(x) \notin A \oplus \bar{A}\right\}.$$

Prove that $A \leq_1 C$, $C \leq_t A$, but $C \not\leq_m A$.
3. Prove that Theorem 3.11 holds with $\leq_t$ replaced by $\leq_m$.
4. Let FIN $= \{x \in N \mid W_x$ is finite$\}$. Prove that $K \leq_1$ FIN.
5. Prove that if $B, \bar{B} \neq \varnothing$, then for every recursive set A, $A \leq_m B$.

4. Sets r.e. Relative to an Oracle

If G is a total function (of one or more arguments) we say that a set $B \subseteq N$ is *G-recursively enumerable* (abbreviated *G-r.e.*) if there is a partially G-computable function g such that

$$B = \{x \in N \mid g(x)\downarrow\}.$$

By Theorem 1.6, r.e. sets are then simply sets that are G-r.e. for some computable function G.

It is easy to relativize the proofs in Chapter 4, Section 4, using, in particular, the relativized step-counter theorem. We give some of the results and leave the details to the reader.

Theorem 4.1. If B is a G-recursive set, then B is G-r.e.

Theorem 4.2. The set B is G-recursive if and only if B and $\bar{B}$ are both G-r.e.

Theorem 4.3. If B and C are G-r.e. sets, so are $B \cup C$ and $B \cap C$.

Next, we obtain

Theorem 4.4. The set A is G-r.e. if and only if there is a G-computable predicate $Q(x,t)$ such that

$$A = \{x \in N \mid (\exists t)Q(x,t)\}. \tag{4.1}$$

Proof. First let A be G-r.e. Then, there is a partially G-computable function h such that

$$A = \{x \in N \mid h(x)\downarrow\}.$$

Writing $h(x) = \Phi_G(x, z_0)$, we have

$$A = \{x \in N \mid (\exists t)\text{STP}_G^{(1)}(x, z_0, t)\},$$

which gives the result in one direction.

Conversely, let (4.1) hold, where Q is a G-computable predicate. Let $h(x)$ be the partial function which is G-computed by the following program:

```
[B]    Z ← Q(X,Y)
       Y ← Y + 1
       IF Z = 0 GOTO B
```

Then clearly,

$$A = \{x \in N \mid h(x)\downarrow\},$$

so that A is G-r.e. ■

Corollary 4.5. The set A is G-r.e. if and only if there is a G-recursive set B such that

$$A = \{x \in N \mid (\exists y)(\langle x, y\rangle \in B)\}.$$

Proof. If B is G-recursive, then the predicate $\langle x, y\rangle \in B$ is G-computable (by Theorem 1.4) and hence, by the theorem, A is G-r.e.

Conversely, if A is G-r.e., we have a G-computable predicate Q such that (4.1) holds. Letting $B = \{z \in N \mid Q(l(z), r(z))\}$, B is (again by Theorem 1.4) G-recursive and

$$A = \{x \in N \mid (\exists y)(\langle x, y\rangle \in B)\}.$$ ■

For any unary function G, we write

$$W_n^G = \{x \in N \mid \Phi_G(x,n)\downarrow\}.$$

(Thus $W_n = W_n^I$.) For the remainder of this section, *G will be a unary total function.* We have at once

Theorem 4.6 (Relativized Enumeration Theorem). A set B is G-r.e. if and only if there is an n for which $B = W_n^G$.

We define

$$G' = \{n \in N \mid n \in W_n^G\}.$$

(Thus, $K = I'$.) G' is called the *jump* of G. We have

Theorem 4.7. G' is G-r.e. but not G-recursive.

This is just the relativization of Theorem 4.7, in Chapter 4, and the proof of that theorem relativizes easily. However, we include the details because of the importance of the result.

Proof of Theorem 4.7. Since

$$G' = \{n \in N \mid \Phi_G(n,n)\downarrow\},$$

the relativized universality theorem shows that G' is G-r.e. If $\overline{G'}$ were also G-r.e., we would have $\overline{G'} = W_i^G$ for some $i \in N$. Then

$$i \in G' \Leftrightarrow i \in W_i^G \Leftrightarrow i \in \overline{G'},$$

a contradiction. ■

Our next result is essentially a relativization of Theorem 3.9.

Theorem 4.8. The following assertions are all equivalent:

a. $A \leq_1 G'$;
b. $A \leq_m G'$;
c. A is G-r.e.

Proof. It is obvious that assertion a implies b. To see that b implies c, let h be a recursive function such that

$$x \in A \quad \text{if and only if} \quad h(x) \in G'.$$

Then

$$x \in A \quad \text{if and only if} \quad \Phi_G(h(x), h(x))\downarrow,$$

so that A is G-r.e.

Finally, to see that c implies a, let A be G-r.e., so that we can write

$$A = \{x \in N \mid f(x)\downarrow\},$$

where f is partially G-computable. Let $g(t, x) = f(x)$ for all t, x. By the relativized universality and parameter theorems, we have, for some number e,

$$g(t, x) = \Phi_G^{(2)}(t, x, e) = \Phi_G(t, S_1^1(x, e)).$$

Hence,

$$\begin{aligned} A &= \{x \in N \mid f(x)\downarrow\} \\ &= \{x \in N \mid g(S_1^1(x, e), x)\downarrow\} \\ &= \{x \in N \mid \Phi_G(S_1^1(x, e), S_1^1(x, e))\downarrow\} \\ &= \{x \in N \mid S_1^1(x, e) \in G'\}. \end{aligned}$$

Since, by Theorem 2.5, $S_1^1(x, e)$ is one–one, we have $A \leq_1 G'$. ■

Theorem 4.9. If F and G are total unary functions and F is G-recursive, then $F' \leq_1 G'$.

Proof. By Theorem 4.7, F' is F-r.e. That is, we can write

$$F' = \{x \in N \mid f(x)\downarrow\},$$

where f is partially F-computable. By Theorem 1.5, f is also partially G-computable. Hence F' is G-r.e. By Theorem 4.8, $F' \leq_1 G'$. ■

By iterating the jump operation, we can obtain a hierarchy of problems each of which is "more unsolvable" than the preceding one.

We write $G^{(n)}$ for the jump iterated n times. That is, we define

$$\begin{aligned} G^{(0)} &= G, \\ G^{(n+1)} &= (G^{(n)})'. \end{aligned}$$

We have

Theorem 4.10. $\varnothing^{(n+1)}$ is $\varnothing^{(n)}$-r.e. but not $\varnothing^{(n)}$-recursive.

Proof. Immediate from Theorem 4.7. ■

It should be noted that, by Theorem 4.9, $K \equiv_1 \varnothing'$, since I and $\varnothing$ are both recursive and $K = I'$. Later we shall see that much more can be said along these lines.

Exercise

1. Show that there are sets A, B, C such that A is B-r.e. and B is C-r.e., but A is not C-r.e.

5. The Arithmetic Hierarchy

The arithmetic hierarchy, which we will study in this section, is one of the principal tools used in classifying unsolvable problems.

Definition. Σ_0 is the class of recursive sets. For each $n \in N$, Σ_{n+1} is the class of sets which are A-r.e. for some set A that belongs to Σ_n. For all n, $\Pi_n = \text{co-}\Sigma_n$, $\Delta_n = \Sigma_n \cap \Pi_n$.

Note that Σ_1 is the class of r.e. sets and that $\Sigma_0 = \Pi_0 = \Delta_0 = \Delta_1$ is the class of recursive sets.

Theorem 5.1. $\Sigma_n \subseteq \Sigma_{n+1}$, $\Pi_n \subseteq \Pi_{n+1}$.

Proof. For any set $A \in \Sigma_n$, A is A-r.e. and hence $A \in \Sigma_{n+1}$. The rest follows by taking complements. ∎

Theorem 5.2. $\varnothing^{(n)} \in \Sigma_n$.

Proof. By induction. For $n = 0$ the result is obvious. The inductive step follows at once from Theorem 4.10. ∎

Theorem 5.3. $A \in \Sigma_{n+1}$ if and only if A is $\varnothing^{(n)}$-r.e.

Proof. If A is $\varnothing^{(n)}$-r.e., it follows at once from Theorem 5.2 that $A \in \Sigma_{n+1}$.

We prove the converse by induction. If $A \in \Sigma_1$, then A is r.e., so, of course, A is $\varnothing$-r.e. Assume the result known for $n = k$ and let $A \in \Sigma_{k+2}$. Then A is B-r.e. for some $B \in \Sigma_{k+1}$. By the induction hypothesis, B is $\varnothing^{(k)}$-r.e. By Theorem 4.8, $A \leq_1 B'$ and $B \leq_1 \varnothing^{(k+1)}$. By Theorem 4.9, $B' \leq_1 \varnothing^{(k+2)}$. Hence $A \leq_1 \varnothing^{(k+2)}$, and by Theorem 4.8 again, A is $\varnothing^{(k+1)}$-r.e. ∎

Corollary 5.4. For $n \geq 1$ the following are all equivalent:

$$A \leq_1 \varnothing^{(n)};$$

$$A \leq_m \varnothing^{(n)};$$

$$A \in \Sigma_n.$$

Proof. This follows at once from Theorems 4.8 and 5.3. ∎

Corollary 5.5. For $n \geq 1$, $\varnothing^{(n)}$ is 1-complete for Σ_n.

Proof. Immediate from Theorem 5.2 and Corollary 5.4. ∎

Corollary 5.6. For $n \geq 1$, Σ_n and Π_n are both m-closed and hence 1-closed.

Proof. Let $A \in \Sigma_n$, $B \leq_m A$. Then using Corollary 5.4 twice, $B \leq_m \varnothing^{(n)}$, and hence $B \in \Sigma_n$. This proves that Σ_n is m-closed. The result for Π_n is now immediate from Corollary 3.8. ∎

Theorem 5.7. $A \in \Delta_{n+1}$ if and only if $A \leq_t \varnothing^{(n)}$.

Proof. Immediate from Theorems 4.2 and 5.3. ∎

In particular, since $K \equiv_t \varnothing'$ (actually $K \equiv_1 \varnothing'$), Δ_2 consists of all sets that are K-recursive, that is, sets for which there are algorithms that can decide membership by making use of an oracle for the halting problem.

Theorem 5.8. $\Sigma_n \cup \Pi_n \subseteq \Delta_{n+1}$.

Proof. For $n = 0$, the inclusion becomes an equality, so we assume $n \geq 1$. If $A \in \Sigma_n$, then by Corollary 5.4, $A \leq_1 \varnothing^{(n)}$, so by Theorem 5.7, $A \in \Delta_{n+1}$. If $A \in \Pi_n$, then $\bar{A} \leq_1 \varnothing^{(n)}$. But clearly $A \leq_t \bar{A}$ (for example, by Theorem 1.4). Hence $A \leq_t \varnothing^{(n)}$ and by Theorem 5.7, $A \in \Delta_{n+1}$. ∎

Theorem 5.9. For $n \geq 1$, $\varnothing^{(n)} \in \Sigma_n - \Delta_n$.

Proof. By Theorem 4.10, $\varnothing^{(n)}$ is not $\varnothing^{(n-1)}$-recursive. ∎

Theorem 5.10 (Kleene's Hierarchy Theorem). We have for $n \geq 1$

1. $\Delta_n \subset \Sigma_n$, $\Delta_n \subset \Pi_n$;
2. $\Sigma_n \subset \Sigma_{n+1}$, $\Pi_n \subset \Pi_{n+1}$;
3. $\Sigma_n \cup \Pi_n \subset \Delta_{n+1}$.

Proof.

1. By definition $\Delta_n \subseteq \Sigma_n$, $\Delta_n \subseteq \Pi_n$. By Theorem 5.9, $\varnothing^{(n)} \in \Sigma_n - \Delta_n$, and so $\overline{\varnothing^{(n)}} \in \Pi_n - \Delta_n$. Thus the inclusions are proper.
2. By Theorem 5.1 we need show only that the inclusions are proper. But $\varnothing^{(n+1)} \in \Sigma_{n+1}$. If $\varnothing^{(n+1)} \in \Sigma_n$, by Theorem 5.8, $\varnothing^{(n+1)} \in \Delta_{n+1}$, contradicting Theorem 5.9. Likewise $\overline{\varnothing^{(n+1)}} \in \Pi_{n+1} - \Pi_n$.

3. By Theorem 5.8, we need show only that the inclusion is proper. Let $A_n = \emptyset^{(n)} \oplus \overline{\emptyset^{(n)}}$. We shall show that $A_n \in \Delta_{n+1} - (\Sigma_n \cup \Pi_n)$. By Theorem 3.11 (with $C = \emptyset^{(n)}$), we have $A_n \leq_t \emptyset^{(n)}$. Hence $A_n \in \Delta_{n+1}$. Also,

$$\emptyset^{(n)} = \{x \in N \mid 2x \in A_n\},$$

$$\overline{\emptyset^{(n)}} = \{x \in N \mid 2x + 1 \in A_n\}.$$

Hence $\emptyset^{(n)} \leq_1 A_n$, $\overline{\emptyset^{(n)}} \leq_1 A_n$. Suppose that $A_n \in \Sigma_n$. Then, by Corollary 5.6, $\overline{\emptyset^{(n)}} \in \Sigma_n$, so that $\emptyset^{(n)} \in \Delta_n$, contradicting Theorem 5.9. Likewise if $A_n \in \Pi_n$, then $\emptyset^{(n)} \in \Pi_n$ and hence $\emptyset^{(n)} \in \Delta_n$. ■

Since we have now seen that for all $n \geq 1$, $\Sigma_n \neq \text{co-}\Sigma_n$, and since we know that for $n \geq 1$, Σ_n and Π_n are each m-closed, we may apply Theorem 3.6 to obtain the following extremely useful result.

Theorem 5.11. If A is m-complete for Σ_n, then $A \notin \Pi_n$. Likewise, if A is m-complete for Π_n, then $A \notin \Sigma_n$.

6. Post's Theorem

In order to make use of the arithmetic hierarchy, we will employ an alternative characterization of the classes Σ_n, Π_n involving strings of quantifiers. This alternative formulation is most naturally expressed in terms of predicates rather than sets. Hence we will use the following terminology.

We first associate with each predicate $P(x_1, \ldots, x_s)$ the set

$$A = \{x \in N \mid P((x)_1, \ldots, (x)_s)\}.$$

Then we say that P *is* Σ_n or that P *is a* Σ_n *predicate* to mean that $A \in \Sigma_n$. Likewise, we say that P *is* Π_n *or* Δ_n if $A \in \Pi_n$ or $A \in \Delta_n$, respectively. Notice that we continue to regard Σ_n and Π_n as consisting of subsets of N, and we will not speak of a predicate as being a *member* of Σ_n or Π_n.

Our terminology involves a slight anomaly for unary predicates. We have just defined $P(x)$ to be Σ_n (or Π_n) if the set $A = \{x \in N \mid P((x)_1)\}$ belongs to Σ_n (or Π_n), whereas it would be more natural to speak of $P(x)$ as being Σ_n (or Π_n) depending on whether $B = \{x \in N \mid P(x)\}$ belongs to Σ_n (or Π_n). Fortunately, there is really no conflict, for we have

Theorem 6.1. Let $B = \{x \in N \mid P(x)\}$. Then $P(x)$ is Σ_n if and only if $B \in \Sigma_n$. Likewise for Π_n, Δ_n.

Proof. For $n = 0$, the result is obvious, so assume that $n \geq 1$. $P(x)$ is Σ_n (or Π_n, or Δ_n) if and only if the set $A = \{x \in N \mid P((x)_1)\}$ belongs to Σ_n (or Π_n or Δ_n). Now,

$$A = \{x \in N \mid (x)_1 \in B\},$$

and

$$B = \{x \in N \mid 2^x \in A\}.$$

Thus $A \equiv_{\mathrm{m}} B$. By Corollary 5.6, this gives the result. ■

Theorem 6.2. Let $P(x_1, \ldots, x_s)$ be a Σ_n predicate and let

$$Q(t_1, \ldots, t_k) \Leftrightarrow P(f_1(t_1, \ldots, t_k), \ldots, f_s(t_1, \ldots, t_k)),$$

where $f_1, \ldots, f_s$ are computable functions. Then Q is also Σ_n. Likewise for Π_n.

Proof. Let

$$A = \{x \in N \mid P((x)_1, \ldots, (x)_s)\},$$

$$B = \{t \in N \mid Q((t)_1, \ldots, (t)_k)\}.$$

We shall prove that $B \leq_{\mathrm{m}} A$. It will thus follow that if $A \in \Sigma_n$ (or Π_n), then $B \in \Sigma_n$ (or Π_n), giving the desired result.

We have

$$\begin{aligned} t \in B &\Leftrightarrow Q((t)_1, \ldots, (t)_k) \\ &\Leftrightarrow P(f_1((t)_1, \ldots, (t)_k), \ldots, f_s((t)_1, \ldots, (t)_k)) \\ &\Leftrightarrow [f_1((t)_1, \ldots, (t)_k), \ldots, f_s((t)_1, \ldots, (t)_k)] \in A, \end{aligned}$$

so that $B \leq_{\mathrm{m}} A$. ■

Theorem 6.3. A predicate P is Σ_n (or Π_n) if and only if $\sim P$ is Π_n (or Σ_n).

Proof. $A = \{x \in N \mid P((x)_1, \ldots, (x)_s)\}$ implies

$$\bar{A} = \{x \in N \mid \sim P((x)_1, \ldots, (x)_s)\}.$$ ■

Theorem 6.4. Let $P(x_1, \ldots, x_s), Q(x_1, \ldots, x_s)$ be Σ_n (or Π_n). Then the predicates P & Q and $P \vee Q$ are likewise Σ_n (or Π_n).

Proof. For $n = 0$, the result is obvious. Assume that $n \geq 1$ and let

$$A = \{x \in N \mid P((x)_1, \ldots, (x)_s)\},$$
$$B = \{x \in N \mid Q((x)_1, \ldots, (x)_s)\},$$
$$C = \{x \in N \mid P((x)_1, \ldots, (x)_s) \mathbin{\&} Q((x)_1, \ldots, (x)_s)\},$$
$$D = \{x \in N \mid P((x)_1, \ldots, (x)_s) \vee Q((x)_1, \ldots, (x)_s)\}.$$

Thus, $C = A \cap B$ and $D = A \cup B$. If P and Q are Σ_n, then $A, B \in \Sigma_n$. Thus, by Theorem 5.3, A and B are both $\varnothing^{(n-1)}$-r.e. By Theorem 4.3, C and D are likewise $\varnothing^{(n-1)}$-r.e., and so $P \mathbin{\&} Q$ and $P \vee Q$ are Σ_n.

If P and Q are Π_n, then $A, B \in \Pi_n$ so that $\bar{A}, \bar{B} \in \Sigma_n$. By Theorems 4.3 and 5.3, $\bar{A} \cap \bar{B} = (\overline{A \cup B}) \in \Sigma_n$ and $\bar{A} \cup \bar{B} = (\overline{A \cap B}) \in \Sigma_n$. Hence $D, C \in \Pi_n$, so that both $P \vee Q$ and $P \mathbin{\&} Q$ are Π_n. ■

Theorem 6.5. Let $Q(x_1, \ldots, x_s, y)$ be Σ_n, $n \geq 1$, and let

$$P(x_1, \ldots, x_s) \Leftrightarrow (\exists y) Q(x_1, \ldots, x_s, y).$$

Then P is also Σ_n.

Proof. Let

$$A = \{x \in N \mid Q((x)_1, \ldots, (x)_s, (x)_{s+1})\},$$
$$B = \{x \in N \mid P((x)_1, \ldots, (x)_s)\}.$$

We are given that $A \in \Sigma_n$, i.e., that A is $\varnothing^{(n-1)}$-r.e., and we must show that B is likewise $\varnothing^{(n-1)}$-r.e.

By Theorem 4.4, we may write

$$A = \{x \in N \mid (\exists t) R(x, t)\},$$

where R is $\varnothing^{(n-1)}$-recursive. Hence,

$$\begin{aligned} Q(x_1, \ldots, x_s, y) &\Leftrightarrow [x_1, \ldots, x_s, y] \in A \\ &\Leftrightarrow (\exists t) R([x_1, \ldots, x_s, y], t). \end{aligned}$$

Thus,

$$\begin{aligned} x \in B &\Leftrightarrow P((x)_1, \ldots, (x)_s) \\ &\Leftrightarrow (\exists y) Q((x)_1, \ldots, (x)_s, y) \\ &\Leftrightarrow (\exists y)(\exists t) R([(x)_1, \ldots, (x)_s, y], t) \\ &\Leftrightarrow (\exists z) R([(x)_1, \ldots, (x)_s, l(z)], r(z)). \end{aligned}$$

By Theorems 1.4 and 4.4, B is $\varnothing^{(n-1)}$-r.e. ■

Theorem 6.6. Let $Q(x_1, \ldots, x_s, y)$ be Π_n, $n \geq 1$, and let

$$P(x_1, \ldots, x_s) \Leftrightarrow (\forall y)Q(x_1, \ldots, x_s, y).$$

Then P is also Π_n.

Proof. $\sim P(x_1, \ldots, x_s) \Leftrightarrow (\exists y) \sim Q(x_1, \ldots, x_s, y)$. The result follows from Theorems 6.3 and 6.5. ■

The main result of this section is

Theorem 6.7 (Post's Theorem). A predicate $P(x_1, \ldots, x_s)$ is Σ_{n+1} if and only if there is a Π_n predicate $Q(x_1, \ldots, x_s, y)$ such that

$$P(x_1, \ldots, x_s) \Leftrightarrow (\exists y)Q(x_1, \ldots, x_s, y). \tag{6.1}$$

Proof. If (6.1) holds, with Q a Π_n predicate, it is easy to see that P must be Σ_{n+1}. By Theorem 5.8, Q is certainly itself Σ_{n+1}, and therefore, by Theorem 6.5, P is Σ_{n+1}.

The converse is somewhat more difficult. Let us temporarily introduce the following terminology: we will say that a predicate $P(x_1, \ldots, x_s)$ is $\exists^{n+1}$ if it can be expressed in the form (6.1), where Q is Π_n. Then Post's theorem just says that the Σ_{n+1} and the $\exists^{n+1}$ predicates are the same. We have already seen that all $\exists^{n+1}$ predicates are Σ_{n+1}.

Lemma 1. If a predicate is Σ_n, then it is $\exists^{n+1}$.

Proof. For $n = 0$, the result is obvious. Let $n \geq 1$, and let $P(x_1, \ldots, x_s)$ be Σ_n. Let

$$A = \{x \in N \mid P((x)_1, \ldots, (x)_s)\}.$$

Then A is $\varnothing^{(n-1)}$-r.e., so by Theorem 4.4,

$$A = \{x \in N \mid (\exists t)R(x, t)\},$$

where R is $\varnothing^{(n-1)}$-recursive. Thus

$$P(x_1, \ldots, x_s) \Leftrightarrow (\exists t)R([x_1, \ldots, x_s], t).$$

It remains to show that $R([x_1, \ldots, x_s], t)$ is Π_n. But in fact, by Theorem 1.4, $R([x_1, \ldots, x_s], t)$ is $\varnothing^{n-1}$-recursive, so that it is actually Δ_n and hence certainly Π_n. ■

Lemma 2. If a predicate is Π_n, then it is $\exists^{n+1}$.

Proof. If $P(x_1, \ldots, x_s)$ is Π_n, we need only set

$$Q(x_1, \ldots, x_s, y) \Leftrightarrow P(x_1, \ldots, x_s),$$

so that, of course,

$$P(x_1, \ldots, x_s) \Leftrightarrow (\exists y)Q(x_1, \ldots, x_s, y).$$

Since

$$\{x \in N \mid Q((x)_1, \ldots, (x)_s, (x)_{s+1})\} = \{x \in N \mid P((x)_1, \ldots, (x)_s)\},$$

the predicate Q is also Π_n, which gives the result. ■

Lemma 3. If $P(x_1, \ldots, x_s, z)$ is $\exists^{n+1}$ and

$$Q(x_1, \ldots, x_s) \Leftrightarrow (\exists z)P(x_1, \ldots, x_s, z),$$

then Q is $\exists^{n+1}$.

Proof. We may write

$$P(x_1, \ldots, x_s, z) \Leftrightarrow (\exists y)R(x_1, \ldots, x_s, z, y),$$

where R is Π_n. Then

$$\begin{aligned} Q(x_1, \ldots, x_s) &\Leftrightarrow (\exists z)(\exists y)R(x_1, \ldots, x_s, z, y) \\ &\Leftrightarrow (\exists t)R(x_1, \ldots, x_s, l(t), r(t)), \end{aligned}$$

which is $\exists^{n+1}$ by Theorem 6.2. ■

Lemma 4. If P and Q are $\exists^{n+1}$, then so are P & Q and $P \vee Q$.

Proof. Let us write

$$P(x_1, \ldots, x_s) \Leftrightarrow (\exists y)R(x_1, \ldots, x_s, y),$$
$$Q(x_1, \ldots, x_s) \Leftrightarrow (\exists z)S(x_1, \ldots, x_s, z),$$

where R and S are Π_n. Then

$$P(x_1, \ldots, x_s) \,\&\, Q(x_1, \ldots, x_s) \Leftrightarrow (\exists y)(\exists z)[R(x_1, \ldots, x_s, y) \,\&\, S(x_1, \ldots, x_s, z)]$$

and

$$P(x_1, \ldots, x_s) \vee Q(x_1, \ldots, x_s) \Leftrightarrow (\exists y)(\exists z)[R(x_1, \ldots, x_s, y) \vee S(x_1, \ldots, x_s, z)].$$

The result follows from Theorem 6.4 and Lemmas 2 and 3. ■

Lemma 5. If $P(x_1, \ldots, x_s, t)$ is $\exists^{n+1}$ and

$$Q(x_1, \ldots, x_s, y) \Leftrightarrow (\forall t)_{\leq y}P(x_1, \ldots, x_s, t),$$

then Q is $\exists^{n+1}$.

Proof. Let

$$P(x_1, \ldots, x_s, t) \Leftrightarrow (\exists z) R(x_1, \ldots, x_s, t, z),$$

where R is Π_n. Thus,

$$\begin{aligned} Q(x_1, \ldots, x_s, y) &\Leftrightarrow (\forall t)_{\leq y} (\exists z) R(x_1, \ldots, x_s, t, z) \\ &\Leftrightarrow (\exists u)(\forall t)_{\leq y} R(x_1, \ldots, x_s, t, (u)_{t+1}), \end{aligned}$$

where we are using the Gödel number $u = [z_0, z_1, \ldots, z_y]$ to encode the sequence of values of z corresponding to $t = 0, 1, \ldots, y$. Thus,

$$\begin{aligned} Q(x_1, \ldots, x_s, y) &\Leftrightarrow (\exists u)(\forall t)[t > y \vee R(x_1, \ldots, x_s, t, (u)_{t+1})] \\ &\Leftrightarrow (\exists u) S(x_1, \ldots, x_s, y, u), \end{aligned}$$

where S is Π_n. For $n = 0$, we have used Theorem 6.3 from Chapter 3; and for $n > 0$, we have used the fact that the predicate $t > y$ is recursive (and hence certainly Π_n), and Theorems 6.2, 6.4, and 6.6. ■

We now recall from Section 2 that $u \prec G$ means that

$$\{u\}(i) = G(i) \quad \text{for} \quad 0 \leq i < l(u).$$

Lemma 6. Let $R(x)$ be Σ_n. Then the predicate $u \prec R$ is $\exists^{n+1}$.

Proof. We have

$$\begin{aligned} u \prec R \Leftrightarrow{}& (\forall i)_{< l(u)} \{[(r(u))_{i+1} = 1 \,\&\, R(i)] \vee [(r(u))_{i+1} = 0 \,\&\, \sim R(i)]\} \\ \Leftrightarrow{}& l(u) = 0 \vee (\exists z)(z + 1 = l(u) \,\&\, (\forall i)_{\leq z} \{[(r(u))_{i+1} = 1 \,\&\, R(i)] \\ & \vee [(r(u))_{i+1} = 0 \,\&\, \sim R(i)]\}). \end{aligned}$$

Thus, using Lemmas 1–5 and the fact that the predicate $\sim R(i)$ is Π_n, we have the result. ■

Proof of Theorem 6.7 (Post's Theorem) Concluded. Let $P(x_1, \ldots, x_s)$ be any Σ_{n+1} predicate. Let

$$A = \{x \in N \mid P((x)_1, \ldots, (x)_s)\}.$$

Then $A \in \Sigma_{n+1}$, which means that A is B-r.e. for some set $B \in \Sigma_n$. Let $R(x)$ be the characteristic function of B, so that by Theorem 6.1, R is Σ_n. Since A is B-r.e., we are able to write

$$A = \{x \in N \mid f(x)\downarrow\},$$

where f is partially B-computable. Let f be B-computed by a program with number y_0. Then, using Theorem 2.4 (the finiteness theorem), we have

$$\begin{aligned} x \in A &\Leftrightarrow (\exists t)\mathrm{STP}_R^{(1)}(x, y_0, t) \\ &\Leftrightarrow (\exists t)(\exists u)\{u \prec R \ \& \ \mathrm{STP}_{\{u\}}^{(1)}(x, y_0, t)\}. \end{aligned}$$

Thus,

$$P(x_1, \ldots, x_s) \Leftrightarrow (\exists t)(\exists u)\{u \prec R \ \& \ \mathrm{STP}_{\{u\}}^{(1)}([x_1, \ldots, x_s], y_0, t)\}.$$

Therefore by Theorem 2.3 and Lemmas 3, 4, and 6, P is $\exists^{n+1}$. ■

Now that we know that being Σ_{n+1} and $\exists^{n+1}$ are the same, we may rewrite Lemma 5 as

Corollary 6.8. If $P(x_1, \ldots, x_s, t)$ is Σ_n and

$$Q(x_1, \ldots, x_s, y) \Leftrightarrow (\forall t)_{\leq y} P(x_1, \ldots, x_s, t),$$

then Q is also Σ_n.

Also, we can easily obtain the following results.

Corollary 6.9. A predicate $P(x_1, \ldots, x_s)$ is Π_{n+1} if and only if there is a Σ_n predicate $Q(x_1, \ldots, x_s, y)$ such that

$$P(x_1, \ldots, x_s) \Leftrightarrow (\forall y)Q(x_1, \ldots, x_s, y).$$

Proof. Immediate from Post's theorem and Theorem 6.3. ■

Corollary 6.10. If $P(x_1, \ldots, x_s, t)$ is Π_n, and

$$Q(x_1, \ldots, x_s, y) \Leftrightarrow (\exists t)_{\leq y} P(x_1, \ldots, x_s, t),$$

then Q is also Π_n.

Proof. Immediate from Corollary 6.8 and Theorem 6.3. ■

We are now in a position to survey the situation. We call a predicate $P(x_1, \ldots, x_s)$ *arithmetic* if there is a recursive predicate $R(x_1, \ldots, x_s, y_1, \ldots, y_n)$ such that

$$P(x_1, \ldots, x_s) \Leftrightarrow (Q_1 y_1)(Q_2 y_2) \cdots (Q_n y_n) R(x_1, \ldots, x_s, y_1, \ldots, y_n), \tag{6.2}$$

where each of $Q_1, \ldots, Q_n$ is either the symbol $\exists$ or the symbol $\forall$. We say that the Q_i are *alternating* if for $1 \leq i < n$ when Q_i is $\exists$, then Q_{i+1} is $\forall$ and vice versa. Then we have

Theorem 6.11.

a. Every predicate that is Σ_n or Π_n for any n is arithmetic.
b. Every arithmetic predicate is Σ_n for some n (and also Π_n for some n).
c. A predicate is Σ_n (or Π_n) if and only if it can be represented in the form (6.2) with $Q_1 = \exists$ (or $Q_1 = \forall$) and the Q_i alternating.

Proof. Since Σ_0 and Π_0 predicates are just recursive, they are arithmetic. Proceeding by induction, if we know, for some particular n, that all Σ_n and Π_n predicates are arithmetic, then Theorem 6.7 and Corollary 6.9 show that the same is true for Σ_{n+1} and Π_{n+1} predicates. This proves a.

For b we proceed by induction on n, the number of quantifiers. For $n = 0$, we have a Σ_0 (and a Π_0) predicate. If the result is known for $n = k$, then it follows for $n = k + 1$ using Theorems 6.5–6.7 and Corollary 6.9.

Finally, c is easily proved by mathematical induction using Theorem 6.7 and Corollary 6.9. ■

7. Classifying Some Unsolvable Problems

We will now see how to apply the arithmetic hierarchy. We begin with the set

$$\text{TOT} = \{z \in N \mid (\forall x)\Phi(x, z)\downarrow\},$$

which consists of all numbers of programs which compute total functions. This set was discussed in Chapter 4, Section 6, where it was shown that TOT is not r.e. Without relying on this previous discussion, we shall obtain much sharper information about TOT.

We begin by observing that

$$\text{TOT} = \{z \in N \mid (\forall x)(\exists t)\text{STP}^{(1)}(x, z, t)\},$$

so that TOT $\in \Pi_2$. We shall prove

Theorem 7.1. TOT is 1-complete for Π_2. Therefore, TOT $\notin \Sigma_2$.

Proof. The second assertion follows from the first by Theorem 5.11.

Since we know that TOT $\in \Pi_2$, it remains to show that for any $A \in \Pi_2$, we have $A \leq_1$ TOT. For $A \in \Pi_2$, we can write

$$A = \{w \in N \mid (\forall x)(\exists y)R(x, y, w)\},$$

where R is recursive. Let

$$h(x, w) = \min_y R(x, y, w),$$

so that h is partially computable. Let h be computed by a program with number e. Thus,

$$(\exists y)R(x,y,w) \Leftrightarrow h(x,w)\downarrow \Leftrightarrow \Phi^{(2)}(x,w,e)\downarrow \Leftrightarrow \Phi(x,S_1^1(w,e))\downarrow,$$

where we have used the parameter theorem. Hence,

$$\begin{aligned} w \in A &\Leftrightarrow (\forall x)(\exists y)R(x,y,w) \\ &\Leftrightarrow (\forall x)\left[\Phi(x,S_1^1(w,e))\downarrow\right] \\ &\Leftrightarrow S_1^1(w,e) \in \text{TOT}. \end{aligned}$$

Since, by Theorem 2.5, $S_1^1(w,e)$ is one–one, we can conclude that

$$A \leq_1 \text{TOT}. \qquad \blacksquare$$

As a second simple example we consider

$$\text{INF} = \{z \in N \mid W_z \text{ is infinite}\}.$$

We have

$$z \in \text{INF} \Leftrightarrow (\forall x)(\exists y)(y > x \ \&\ y \in W_z).$$

Now

$$y \in W_z \Leftrightarrow (\exists t)\text{STP}^{(1)}(y,z,t),$$

and hence the predicate $y \in W_z$ is Σ_1. Using Theorems 6.4 and 6.5, $(\exists y)(y > x \ \&\ y \in W_z)$ is also Σ_1, and finally $\text{INF} \in \Pi_2$. We shall show that INF is also 1-complete for Π_2. By Theorem 3.5, it suffices to show that $\text{TOT} \leq_1 \text{INF}$ since we already know that TOT is 1-complete for Π_2.

To do this we shall obtain a recursive one–one function $f(x)$ such that

$$W_x = N \quad \textit{implies} \quad W_{f(x)} = N$$

and (7.1)

$$W_x \neq N \quad \textit{implies} \quad W_{f(x)} \text{ is finite}.$$

Having done this we will be through since we will have

$$x \in \text{TOT} \Leftrightarrow f(x) \in \text{INF},$$

and therefore,

$$\text{TOT} \leq_1 \text{INF}.$$

The intuitive idea behind the construction of f is that program number $f(x)$ will "accept" a given input z if and only if program number x

"accepts" successively inputs $0, 1, \ldots, z$. We can write this intuitive idea in the form of an equation as follows:

$$W_{f(x)} = \{z \in N \mid (\forall k)_{\leq z}(k \in W_x)\}.$$

Now it is a routine matter to use the parameter theorem to obtain f. We first note that, by Corollary 6.8, the predicate $(\forall k)_{\leq z}(k \in W_x)$ is Σ_1. Hence, as earlier, there is a number e such that

$$\begin{aligned}(\forall k)_{\leq z}(k \in W_x) &\Leftrightarrow \Phi^{(2)}(z, x, e)\downarrow \\ &\Leftrightarrow \Phi(z, S_1^1(x, e))\downarrow \\ &\Leftrightarrow z \in W_{S_1^1(x,e)}.\end{aligned}$$

Thus the desired function $f(x)$ is simply $S_1^1(x, e)$, which is one–one, as we know from Theorem 2.5.

This completes the proof that INF is 1-complete for Π_2. Hence also, INF $\notin \Sigma_2$.

The following notation will be useful.

Definition. Let $A, B, C \subseteq N$. Then we write $A \leq_m (B, C)$ to mean that there is a recursive function f such that

$$x \in A \quad \textit{implies} \quad f(x) \in B$$

and

$$x \in \bar{A} \quad \textit{implies} \quad f(x) \in C.$$

If f is one–one we write $A \leq_1 (B, C)$.

Thus $A \leq_1 B$ is simply the assertion: $A \leq_1 (B, \bar{B})$.

It will be useful to note that by (7.1), we have actually proved

$$\text{TOT} \leq_1 \left(\text{TOT}, \overline{\text{INF}}\right). \tag{7.2}$$

Now, we have

Theorem 7.2. If $A \leq_1 (B, C)$, $B \subseteq D$, and $C \cap D = \varnothing$, then $A \leq_1 D$.

Proof. We have a recursive one–one function f such that

$$x \in A \quad \textit{implies} \quad f(x) \in B \quad \textit{implies} \quad f(x) \in D$$

and

$$x \in \bar{A} \quad \textit{implies} \quad f(x) \in C \quad \textit{implies} \quad f(x) \in \bar{D}. \qquad \blacksquare$$

Our final example will classify a Σ_3 set, and is considerably more difficult than either of those considered so far.

Theorem 7.3. Let

$$\mathrm{COF} = \{x \in N \mid \overline{W}_x \text{ is finite}\}.$$

Then COF is 1-complete for Σ_3.

Lemma 1. $\mathrm{COF} \in \Sigma_3$.

Proof.

$$\mathrm{COF} = \{x \in N \mid (\exists n)(\forall k)(k \le n \vee k \in W_x)\}.$$

Since the predicate in parentheses is Σ_1, the result follows from Theorem 6.11. ■

We introduce the notation

$${}_nW_x = \{m \in N \mid \mathrm{STP}^{(1)}(m, x, n)\}.$$

Intuitively, ${}_nW_x$ is the set of numbers that program number x "accepts" in $\le n$ steps. Clearly,

$$W_x = \bigcup_{n \in N} ({}_nW_x).$$

We also define

$${}_nW_x^r = \{m < r \mid m \in {}_nW_x\}.$$

We write $L(n, x)$ to mean that

$${}_{n+1}W_x^n = {}_nW_x^n.$$

Clearly $L(n, x)$ is a recursive predicate. We write

$$R(x, n) \Leftrightarrow (\forall r)_{\le n}(r \in W_x) \vee [L(n, x) \,\&\, (\exists k)_{< n}(k \notin {}_nW_x)].$$

Since $R(x, n)$ is Σ_1 we can use the parameter theorem, as in the previous example, to find a recursive one–one function $g(x)$ such that

$$W_{g(x)} = \{n \mid R(x, n)\}.$$

Lemma 2. If $x \in \mathrm{TOT}$, then $g(x) \in \mathrm{TOT}$. If $x \notin \mathrm{INF}$, then $g(x) \in \mathrm{COF} - \mathrm{TOT}$.

Proof. If $x \in \mathrm{TOT}$, then $W_x = N$, so that $(\forall r)_{\le n}(r \in W_x)$ is true for all n. Hence $R(x, n)$ is true for all n, i.e., $W_{g(x)} = N$ and $g(x) \in \mathrm{TOT}$.

Now let $x \notin \text{INF}$, i.e., W_x is finite. Therefore, there is a number n_0 such that for all $n > n_0$, we have

$$W_x = {}_nW_x^n$$

and

$$(\exists k)_{<n}(k \notin W_x).$$

Thus, for $n > n_0$,

$${}_{n+1}W_x^n = {}_nW_x^n,$$

i.e., $L(n, x)$ is true. Thus, $n > n_0$ implies that $R(x, n)$ is true, i.e., that $n \in W_{g(x)}$. We have shown that all sufficiently large integers belong to $W_{g(x)}$. Hence $g(x) \in \text{COF}$. It remains to show that $g(x) \notin \text{TOT}$.

Let s be the least number not in W_x. We consider two cases.

Case 1. $s \notin W_{g(x)}$. Then surely $g(x) \notin \text{TOT}$.

Case 2. $s \in W_{g(x)}$. That is, $R(x, s)$ is true. But $(\forall r)_{\leq s}(r \in W_x)$ must be false because $s \notin W_x$. Hence $L(s,x)$ must be true and $(\exists k)_{<s}(k \notin_s W_x)$. Now this number k is less than s, which is the least number not in W_x. Hence $k \in W_x$. Since $k \notin_s W_x$,

$$(\exists n)_{n \geq s}[k \notin_n W_x \;\&\; k \in_{n+1} W_x]. \qquad (7.3)$$

Now we claim that this number $n \notin W_{g(x)}$, which will show that in this case also $g(x) \notin \text{TOT}$. Thus, suppose that $n \in W_{g(x)}$, i.e., that $R(x, n)$ is true. Since $s \notin W_x$ and $n \geq s$, the condition $(\forall r)_{\leq n}(r \in W_x)$ must be false. Thus we would have to have $L(n, x)$, i.e., ${}_{n+1}W_x^n = {}_nW_x^n$. But by (7.3), $k < s \leq n$, $k \in_{n+1} W_x$, and $k \notin_n W_x$. This is a contradiction. ∎

Lemma 3. $\text{TOT} \leq_1 (\text{TOT}, \text{COF} - \text{TOT})$.

Proof. Let f be the recursive one–one function satisfying (7.1) and let g be as above. Let $h(x) = g(f(x))$. Then using Lemma 2 and (7.1), we have

$x \in \text{TOT}$ *implies* $f(x) \in \text{TOT}$ *implies* $h(x) \in \text{TOT}$,
$x \notin \text{TOT}$ *implies* $f(x) \notin \text{INF}$ *implies* $h(x) \in \text{COF} - \text{TOT}$. ∎

Now let $A \in \Sigma_3$. We wish to show that $A \leq_1 \text{COF}$. By Post's theorem, we can write

$$x \in A \Leftrightarrow (\exists n)B(x, n),$$

where B is Π_2. Using the pairing functions, let

$$C = \{t \in N \mid (\exists n)_{\leq l(t)} B(r(t), n)\}.$$

Thus, $C \in \Pi_2$. Theorem 7.1, $C \leq_1$ TOT. Hence, using Lemma 3, $C \leq_1$ (TOT, COF − TOT). Let θ be a recursive one–one function such that

$$\begin{aligned} t \in C &\quad \textit{implies} \quad \theta(t) \in \text{TOT}, \\ t \notin C &\quad \textit{implies} \quad \theta(t) \in \text{COF} - \text{TOT}. \end{aligned} \tag{7.4}$$

Consider the Σ_1 predicate $r(z) \in W_{\theta(\langle l(z), x \rangle)}$. Using the parameter theorem as usual, we can write this in the form $z \in W_{\psi(x)}$, where ψ is a one–one recursive function. Thus,

$$W_{\psi(x)} = \{\langle k, m \rangle \mid m \in W_{\theta(\langle k, x \rangle)}\}. \tag{7.5}$$

The theorem then follows at once from

Lemma 4. $x \in A$ if and only if $\psi(x) \in$ COF.

Proof. Let $x \in A$. Then $B(x, n)$ is true for some least value of n. Hence, for all $k \geq n$, we have $\langle k, x \rangle \in C$. By (7.4), $\theta(\langle k, x \rangle) \in$ TOT for all $k \geq n$. Since n is the least value for which $B(x, n)$ is true, $B(x, k)$ is false for $k < n$. Hence, for $k < n$, $\langle k, x \rangle \notin C$. Thus, by (7.4), $\theta(\langle k, x \rangle) \in$ COF − TOT. To recapitulate,

$$k \geq n \quad \textit{implies} \quad \theta(\langle k, x \rangle) \in \text{TOT},$$

and (7.6)

$$k < n \quad \textit{implies} \quad \theta(\langle k, x \rangle) \in \text{COF} - \text{TOT}.$$

Thus, by (7.5) we see that for $k \geq n$, $\langle k, m \rangle \in W_{\psi(x)}$ for all m. For each $k < n$, $W_{\theta(\langle k, x \rangle)}$ contains all but a finite set of m. Thus, altogether, $W_{\psi(x)}$ can omit at most finitely many integers, i.e., $\psi(x) \in$ COF.

Now, let $x \notin A$. Then, $B(x, n)$ is false for all n. Therefore, $\langle k, x \rangle \notin C$ for all k. By (7.4),

$$\theta(\langle k, x \rangle) \in \text{COF} - \text{TOT} \quad \textit{for all} \quad k \in N,$$

and thus certainly,

$$\theta(\langle k, x \rangle) \notin \text{TOT} \quad \textit{for all} \quad k \in N.$$

That is, for every $k \in N$, there exists m such that $m \notin W_{\theta(\langle k, x \rangle)}$, i.e., by (7.5), such that $\langle k, m \rangle \notin W_{\psi(x)}$. Thus, $\overline{W}_{\psi(x)}$ is infinite, and hence $\psi(x) \notin$ COF. ■

Exercises

1. Show that the following sets belong to Σ_3.
   (a) $\{x \in N \mid$ there is a recursive function f such that $\Phi_x \subseteq f\}$.
   (b) $\{\langle x, y \rangle \mid x \in N \;\&\; y \in N \;\&\; W_x - W_y$ is finite$\}$.

2. (a) Prove that for each m, n there is a predicate $U(x_1, \ldots, x_m, y)$ which is Σ_n, such that for every Σ_n predicate $P(x_1, \ldots, x_m)$ there is a number y_0 with

$$P(x_1, \ldots, x_m) \Leftrightarrow U(x_1, \ldots, x_m, y_0).$$

(b) State and prove a similar result for Π_n.

3. Use the previous exercise to prove that for each n, $\Pi_n - \Sigma_n \neq \varnothing$.

8. Rice's Theorem Revisited

In Chapter 4, we gave a proof of Rice's theorem (Theorem 7.1) using the original parameter theorem. We get a somewhat stronger result using the strengthened form of the parameter theorem.

Definition. Let Γ be a set of partially computable functions of one variable. As in Chapter 4, Section 7, we write

$$R_\Gamma = \{t \in N \mid \Phi_t \in \Gamma\}.$$

We call Γ *nontrivial* if $\Gamma \neq \varnothing$ and there is at least one partially computable function $g(x)$ such that $g \notin \Gamma$.

Theorem 8.1 (Strengthened Form of Rice's Theorem). Let Γ be a nontrivial collection of partially computable functions of one variable. Then, $K \leq_1 R_\Gamma$ or $\bar{K} \leq_1 R_\Gamma$, so that R_Γ is not recursive.

Thus not only is R_Γ nonrecursive, but the halting problem can be "solved" using R_Γ as an oracle. Actually, the first proof of Rice's theorem already shows that either $K \leq_m R_\Gamma$ or $\bar{K} \leq_m R_\Gamma$. We give essentially the same proof here, using the strengthened form of the parameter theorem to upgrade the result to one–one reducibility.

Proof. We recall (Chapter 1, Section 2) that $\varnothing$ is a partially computable function, namely, the nowhere defined function.

Case 1. $\varnothing \notin \Gamma$. Since Γ is nontrivial, it contains at least one function, say f. Since $f \in \Gamma$ and $\varnothing \notin \Gamma$, $f \neq \varnothing$; f must be defined for at least one value. Let

$$\Omega(x, t) = \begin{cases} f(t) & \text{if } x \in K \\ \uparrow & \text{if } x \notin K. \end{cases}$$

Since

$$x \in K \Leftrightarrow \Phi(x, x)\downarrow,$$

it is clear that Ω is partially computable. Using the parameter theorem in its strengthened form, we can write

$$\Omega(x,t) = \Phi_{g(x)}(t),$$

where g is a one–one recursive function. Then we have

$$x \in K \textit{ implies } \Phi_{g(x)} = f \textit{ implies } g(x) \in R_\Gamma;$$

$$x \notin K \textit{ implies } \Phi_{g(x)} = \varnothing \textit{ implies } g(x) \notin R_\Gamma.$$

Thus, $K \leq_1 R_\Gamma$.

Case 2. $\varnothing \in \Gamma$. Now let Δ be the class of all partially computable functions not in Γ. Thus, $R_\Gamma = \overline{R}_\Delta$ and $\varnothing \notin \Delta$. By Case 1, $K \leq_1 R_\Delta$, and hence by Theorem 3.7, $\overline{K} \leq_1 R_\Gamma$. ■

Exercises

1. State and prove a relativized version of Rice's theorem.
2. **(a)** Develop a code for partial functions from N to N with finite domains, writing f_n for the nth such function.
   (b) Prove the Rice–Shapiro theorem: R_Γ is r.e. if and only if $\Gamma = \varnothing$ or there is a recursive function $t(x)$ such that

$$\Gamma = \{g \mid (\exists x)(g \supseteq f_{t(x)})\}.$$

9. Recursive Permutations

Definition. A one–one recursive function f whose domain and range are both N is called a *recursive permutation.*

With each recursive permutation f we may associate its *inverse* f^{-1}:

$$f^{-1}(t) = \min_x (t = f(x)).$$

Then, f^{-1} is clearly likewise a recursive permutation.

Definition. Let $A, B \subseteq N$. Then A and B are said to be *recursively isomorphic*, written $A \equiv B$, if there is a recursive permutation f such that $x \in A$ if and only if $f(x) \in B$.

Since a recursive permutation provides what is essentially a mere change of notation, recursively isomorphic sets may be thought of as containing the same "information" presented in different notation.

It is obvious that $A \equiv B$ implies $A \equiv_1 B$. Remarkably, the converse statement is also true.

Theorem 9.1 (Myhill). If $A \equiv_1 B$, then $A \equiv B$.

In our proof of this theorem we shall need to code sequences of ordered pairs of numbers. We shall speak of *the code* of the sequence

$$(a_1, b_1), \dots, (a_n, b_n) \tag{9.1}$$

of pairs of elements of N meaning the number

$$u = \langle n, [\langle a_1, b_1 \rangle, \dots, \langle a_n, b_n \rangle] \rangle.$$

Thus, the numbers a_i, b_i can be retrieved from the code u by using the relations

$$\left.\begin{aligned} a_i &= l((r(u))_i) \\ b_i &= r((r(u))_i) \end{aligned}\right\} \qquad i = 1, 2, \dots, l(u).$$

Note that every natural number is the code of a unique finite (possibly empty) sequence of ordered pairs. (E.g., $3 = \langle 2,0 \rangle$ codes (0,0), (0,0).)

We say that the finite sequence (9.1) *associates A and B*, where $A, B \subseteq N$, if

1. $a_i \neq a_j$ for $1 \leq i < j \leq n$;
2. $b_i \neq b_j$ for $1 \leq i < j \leq n$;
3. for each i, $1 \leq i \leq n$, either $a_i \in A$ and $b_i \in B$ or $a_i \notin A$ and $b_i \notin B$.

We shall prove the

Lemma. Let $A \leq_1 B$. Then there is a computable function $k(u, v)$ such that if u codes the sequence (9.1) that associates A and B and $a \notin \{a_1, a_2, \dots, a_n\}$, then there is a b such that $k(u, a)$ codes the sequence

$$(a_1, b_1), \dots, (a_n, b_n), (a, b) \tag{9.2}$$

that also associates A and B.

Proof. Let f be a recursive one–one function such that

$$x \in A \quad \text{if and only if} \quad f(x) \in B. \tag{9.3}$$

We provide an algorithm for computing b from u and a. $k(u, a)$ can then be set equal to the code of (9.2), i.e.

$$k(u, a) = \langle l(u) + 1, r(u) \cdot p_{l(u)+1}^{\langle a, b \rangle} \rangle.$$

The numbers $f(a_1), f(a_2), \ldots, f(a_n), f(a)$ are all distinct, because f is one–one. Hence, at least one of these $n + 1$ numbers does not belong to the set $\{b_1, b_2, \ldots, b_n\}$. Our algorithm for obtaining b begins by computing $f(a)$. If $f(a) \notin \{b_1, b_2, \ldots, b_n\}$, we set $b = f(a)$. Otherwise, $f(a) = b_i$ for some i and we try $f(a_i)$, because

$$a \in A \Leftrightarrow f(a) = b_i \in B \Leftrightarrow a_i \in A \Leftrightarrow f(a_i) \in B.$$

If $f(a_i) \notin \{b_1, b_2, \ldots, b_n\}$, we set $b = f(a_i)$. Otherwise, if $f(a_i) = b_j$, we continue the process, trying $f(a_j)$. By 1 and 2, none of the a_i and b_i obtained in this way duplicate previous ones. Thus, by our earlier remark the process must terminate in a value b. Using (9.3), we see that either $a \in A$ and $b \in B$ or $a \notin A$ and $b \notin B$. ■

Proof of Theorem 9.1. Since $A \leq_1 B$, by the Lemma there is a computable function $k(u, v)$ such that if u codes (9.1) that associates A and B and $a \notin \{a_1, a_2, \ldots, a_n\}$, then for some b, $k(u, a)$ codes the sequence (9.2) that also associates A and B. But since $B \leq_1 A$, we can also apply the Lemma to obtain a computable function $\bar{k}(u, v)$ such that if u codes (9.1) that associates A and B and $b \notin \{b_1, b_2, \ldots, b_n\}$, then for some a, $\bar{k}(u, b)$ codes the sequence (9.2) that likewise associates A and B.

We let $\nu(0) = 0$, which codes the empty sequence. (Note that the empty sequence does associate A and B.) We let

$$\nu(2x + 1) = \begin{cases} \nu(2x) & \text{if } x \text{ is one of the left components} \\ & \quad \text{of the sequence coded by } \nu(2x) \\ k(\nu(2x), x) & \text{otherwise;} \end{cases}$$

$$\nu(2x + 2) = \begin{cases} \nu(2x + 1) & \text{if } x \text{ is one of the right components} \\ & \quad \text{of the sequence coded by } \nu(2x + 1) \\ \bar{k}(\nu(2x + 1), x) & \text{otherwise.} \end{cases}$$

Thus, we have

1. ν is a computable function.
2. For each x, $\nu(x)$ codes a sequence that associates A and B.
3. The sequence coded by $\nu(x + 1)$ is identical to, or is an extension of, the sequence coded by $\nu(x)$.
4. For each $a \in N$, there is an x such that a pair (a, b) occurs in the sequence coded by $\nu(x)$. (In fact, we can take $x = 2a + 1$.)
5. For each $b \in N$, there is an x such that a pair (a, b) occurs in the sequence coded by $\nu(x)$. (In fact, we can take $x = 2b + 2$.)

We now define the function f by setting $f(a)$ to be the number b such that the pair (a, b) appears in the sequence coded by some $\nu(x)$. b is uniquely determined because all the $\nu(x)$ code sequences that associate A and B. f is clearly computable. In fact,

$$f(a) = \min_b (\exists i)_{\leq l(\nu(2a+1))}[(r(\nu(2a + 1)))_i = \langle a, b\rangle].$$

By 5, the range of f is N; thus f is a recursive permutation and hence, $A \equiv B$. ■

Exercises

1. Prove that $K \equiv U$, where U is defined in Exercise 3.1.
2. Prove that

$$A \oplus \bar{A} \equiv \overline{A \oplus \bar{A}}.$$

Part 2

Grammars and Automata

9

Regular Languages

1. Finite Automata

Computability theory, discussed in Part 1, is the theory of computation obtained when limitations of space and time are deliberately ignored. In automata theory, which we study in this chapter, computation is studied in a context in which bounds on space and time are entirely relevant. The point of view of computability theory is exemplified in the behavior of a Turing machine (Chapter 6) in which a read–write head moves back and forth on an infinite tape, with no preset limit on the number of steps required to reach termination.[1] At the opposite pole, one can imagine a device which moves from left to right on a finite input tape, and it is just such devices, the so-called *finite automata*, that we will now study. Since a finite automaton will have only one opportunity to scan each square in its motion from left to right, nothing is to be gained by permitting the device to "print" new symbols on its tape.

Unlike modern computers, whose action is controlled in part by an internally stored list of instructions called a *program*, the computing

[1] The present chapter does not depend on familiarity with the material in Chapters 2–8. Any exercises that refer to earlier material are marked with an *.

Table 1.1

δ	a	b
q_1	q_2	q_4
q_2	q_2	q_3
q_3	q_4	q_3
q_4	q_4	q_4

devices we will consider in this chapter have no such programs and no internal memory for storing either programs or partial results. In addition, since, as we just indicated, a finite automaton is permitted only a single pass over the tape, there is no external memory available. Instead, there are internal *states* that control the automaton's behavior and also function as memory in the sense of being able to retain some information about what has been read from the input tape up to a given point.

Thus, a finite automaton can be thought of as a very limited computing device which, after reading a string of symbols on the input tape, either accepts the input or rejects it, depending upon the state the machine is in when it has finished reading the tape.

The machine begins by reading the leftmost symbol on the tape, in a specified state called the *initial state* (the automaton is in this state whenever it is initially "turned on"). If at a given time, the machine is in a state q_i reading a given symbol s_j on the input tape, the device moves one square to the right on the tape and enters a state q_k. The current state of the automaton plus the symbol on the tape being read completely determine the automaton's next state.

Definition. A *finite automaton* $\mathcal{M}$ on the alphabet[2] $A = \{s_1, \ldots, s_n\}$ with states $Q = \{q_1, \ldots, q_m\}$ is given by a function δ that maps each pair (q_i, s_j), $1 \leq i \leq m$, $1 \leq j \leq n$, into a state q_k, together with a set $F \subseteq Q$. One of the states, usually q_1, is singled out and called the *initial* state. The states belonging to the set F are called the *final* or *accepting* states. δ is called the *transition function*.

We can represent the function δ using a state versus symbol table. An example is given in Table 1.1, where the alphabet is $\{a, b\}$, $F = \{q_3\}$, and q_1

[2] For an introduction to alphabets and strings, see Chapter 1, Section 3.

is the initial state. It is easy to check that for the tapes

a	a	b	b	b

b	a	b	a

a	a	b	a

a	b	b	b

the automaton will terminate in states q_3, q_4, q_4, and q_3, respectively. We shall say that the automaton *accepts* the strings *aabbb* and *abbb* (because $q_3 \in F$), while it *rejects* the strings *baba* and *aaba* (because $q_4 \notin F$), i.e., that it *accepts* the first and fourth of the preceding tapes and *rejects* the second and third.

To proceed more formally, let $\mathcal{M}$ be a finite automaton with transition function δ, initial state q_1, and accepting states F. If q_i is any state of $\mathcal{M}$ and $u \in A^*$, where A is the alphabet of $\mathcal{M}$, we shall write $\delta^*(q_i, u)$ for the state which $\mathcal{M}$ will enter if it begins in state q_i at the left end of the string u and moves across u until the entire string has been processed. A formal definition by recursion is

$$\delta^*(q_i, 0) = q_i,$$

$$\delta^*(q_i, us_j) = \delta\big(\delta^*(q_i, u), s_j\big).$$

Obviously, $\delta^*(q_i, s_j) = \delta(q_i, s_j)$. Then we say that $\mathcal{M}$ *accepts* a word u provided that $\delta^*(q_1, u) \in F$. $\mathcal{M}$ *rejects* u means that $\delta^*(q_1, u) \in Q - F$. Finally, the *language accepted by* $\mathcal{M}$, written $L(\mathcal{M})$, is the set of all $u \in A^*$ accepted by $\mathcal{M}$:

$$L(\mathcal{M}) = \{u \in A^* \mid \delta^*(q_1, u) \in F\}.$$

A language is called *regular* if there exists a finite automaton that accepts it.

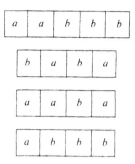

It is important to realize that the notion of regular language does not depend on the particular alphabet. That is, if $L \subseteq A^*$ and $A \subseteq B$, then there is an automaton on the alphabet A that accepts L if and only if there is one on the alphabet B that accepts L. That is, an automaton with alphabet B can be contracted to one on the alphabet A by simply restricting the transition function δ to A; clearly this will have no effect on which elements of A^* are accepted. Likewise, an automaton $\mathcal{M}$ with alphabet A can be expanded to one with alphabet B by introducing a new "trap" state q and decreeing

$$\delta(q_i, b) = q \text{ for all states } q_i \text{ of } \mathcal{M} \text{ and all } b \in B - A,$$

$$\delta(q, b) = q \text{ for all } b \in B.$$

Leaving the set of accepting states unchanged (so that q is not an accepting state), we see that the expanded automaton accepts the same language as $\mathcal{M}$.

Returning to the automaton given by Table 1.1 with $F = \{q_3\}$, it is easy to see that the language it accepts is

$$\{a^{[n]}b^{[m]} \mid n, m > 0\}. \tag{1.1}$$

Thus we have shown that (1.1) is a regular language.

We conclude this section by mentioning another way to represent the transition function δ. We can draw a graph in which each state is represented by a *vertex*. Then, the fact that $\delta(q_i, s_j) = q_k$ is represented by drawing an arrow from vertex q_i to vertex q_k and labeling it s_j. The diagram thus obtained is called the *state transition diagram* for the given automaton. The state transition diagram for the transition function of Table 1.1 is shown in Fig. 1.1.

Exercises

1. In each of the following examples, an alphabet A and a language L are indicated with $L \subseteq A^*$. In each case show that L is regular by constructing a finite automaton $\mathcal{M}$ that accepts L.

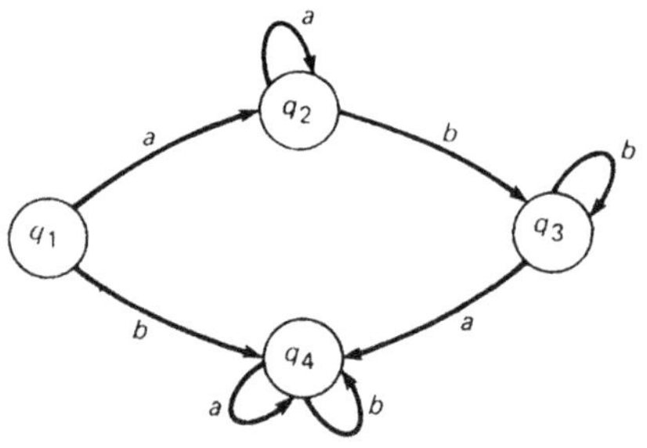

Figure 1.1

(a) $A = \{1\}$; $L = \{1^{[6k]}1 \mid k \geq 0\}$.

(b) $A = \{a, b\}$; L consists of all words whose final four symbols form the string *bbab*.

(c) $A = \{a, b\}$; L consists of all words whose final five symbols include two a's and three b's.

(d) $A = \{0, 1\}$; L consists of all strings that, when considered as binary numbers, have a value which is an integral multiple of 5.

(e) $$A = \left\{ \begin{pmatrix} 0 \\ 0 \\ 0 \end{pmatrix}, \begin{pmatrix} 0 \\ 0 \\ 1 \end{pmatrix}, \begin{pmatrix} 0 \\ 1 \\ 0 \end{pmatrix}, \begin{pmatrix} 0 \\ 1 \\ 1 \end{pmatrix}, \begin{pmatrix} 1 \\ 0 \\ 0 \end{pmatrix}, \begin{pmatrix} 1 \\ 0 \\ 1 \end{pmatrix}, \begin{pmatrix} 1 \\ 1 \\ 0 \end{pmatrix}, \begin{pmatrix} 1 \\ 1 \\ 1 \end{pmatrix} \right\};$$

$\mathcal{M}$ is to be a binary addition checker in the sense that it accepts strings of binary triples

$$\begin{pmatrix} a_1 \\ b_1 \\ c_1 \end{pmatrix} \begin{pmatrix} a_2 \\ b_2 \\ c_2 \end{pmatrix} \cdots \begin{pmatrix} a_n \\ b_n \\ c_n \end{pmatrix}$$

such that $c_1 c_2 \cdots c_n$ is the sum of $a_1 a_2 \cdots a_n$ and $b_1 b_2 \cdots b_n$ when each is treated as a binary number.

(f) $A = \{a, b, c\}$. A *palindrome* is a word such that $w = w^R$. That is, it reads the same backward and forward. L consists of all palindromes of length less than or equal to 6.

(g) $A = \{a, b\}$; L consists of all strings $s_1 s_2 \cdots s_n$ such that $s_{n-2} = b$. (Note that L contains no strings of length less than 3.)

(h) $A = \{a, b\}$; L consists of all words in which three a's occur consecutively.

(i) $A = \{a, b\}$; L consists of all words in which three a's do not occur consecutively.

2. (a) Suppose that the variable names in your favorite programming language are words w on the alphabet $\{A, \ldots, Z, 0, \ldots, 9\}$ such that $1 \leq |w| \leq 8$ and such that the first symbol of w belongs to $\{A, \ldots, Z\}$. Give a finite automaton that accepts the language consisting of these variable names.

(b) Now, remove the restriction $|w| \leq 8$ and give a finite automaton that accepts this extended language.

3. Describe the language accepted by each of the following finite automata. In each case the initial state is q_1.

(a)

δ_1	a	b	c
q_1	q_2	q_3	q_4
q_2	q_2	q_4	q_5
q_3	q_4	q_3	q_5
q_4	q_4	q_4	q_4
q_5	q_4	q_4	q_5

$F_1 = \{q_5\}$.

(b) $\delta_2 = \delta_1$, $F_2 = \{q_4\}$.

(c)

δ_3	a	b	c
q_1	q_2	q_2	q_1
q_2	q_3	q_2	q_1
q_3	q_1	q_3	q_2

$F_3 = \{q_2\}$.

4. Let $A = \{s_1, \ldots, s_n\}$. How many finite automata are there on A with exactly m states, $m > 0$?

5. Show that there is a regular language that is not accepted by any finite automaton with just one accepting state.

6. For any regular language L, define rank(L) = the least number n such that L is accepted by some finite automaton with n states. Prove that for every $n > 0$ there is a regular language L with rank(L) = n.

7. Prove or disprove the following: If L_1, L_2 are regular languages such that $L_1 \subseteq L_2$, then rank(L_1) $\leq$ rank(L_2).

8.* Let $\mathcal{M}$ be a finite automaton on the alphabet $A = \{s_1, \ldots, s_n\}$ with states $Q = \{q_1, \ldots, q_m\}$, transition function δ, initial state q_1, and accepting states F. Give a Turing machine $\mathcal{M}'$ that accepts $L(\mathcal{M})$.

2. Nondeterministic Finite Automata

Next we modify the definition of a finite automaton to permit transitions at each stage to either zero, one, or more than one states. Formally, we accomplish this by altering the definition of a finite automaton in the previous section by making the values of the transition function δ be *sets of states, i.e., sets of elements of Q* (rather than members of Q). The devices

Table 2.1

δ	a	b
q_1	$\{q_1, q_2\}$	$\{q_1, q_3\}$
q_2	$\{q_4\}$	$\varnothing$
q_3	$\varnothing$	$\{q_4\}$
q_4	$\{q_4\}$	$\{q_4\}$

so obtained are called *nondeterministic finite automata* (*ndfa*), and sometimes ordinary finite automata are then called *deterministic finite automata* (*dfa*). An ndfa on a given alphabet A with set of states Q is specified by giving such a transition function δ [which maps each pair (q_i, s_j) into a possibly empty subset of Q] and a fixed subset F of Q. For an ndfa, we define

$$\delta^*(q_i, 0) = \{q_i\},$$

$$\delta^*(q_i, us_j) = \bigcup_{q \in \delta^*(q_i, u)} \delta(q, s_j).$$

Thus, in calculating $\delta^*(q_i, u)$, one accumulates *all* states that the automaton can enter when it reaches the right end of u, beginning at the left end of u in state q_i. An ndfa $\mathscr{M}$ with initial state q_1 *accepts* $u \in A^*$ if $\delta^*(q_1, u) \cap F \neq \varnothing$, i.e., if at least one of the states at which $\mathscr{M}$ ultimately arrives belongs to F. Finally, $L(\mathscr{M})$, *the language accepted by* $\mathscr{M}$, is the set of all strings accepted by $\mathscr{M}$.

An example is given in Table 2.1 and Figure 2.1. Here $F = \{q_4\}$. It is not difficult to see that this ndfa accepts a string on the alphabet $\{a, b\}$ just in case at least one of the symbols has two successive occurrences in the string.

In state q_1, if the next character read is an a, then there are two possibilities. It might be that this a is the first of the desired pair of a's. In that case we would want to remember that we had found one a and hence

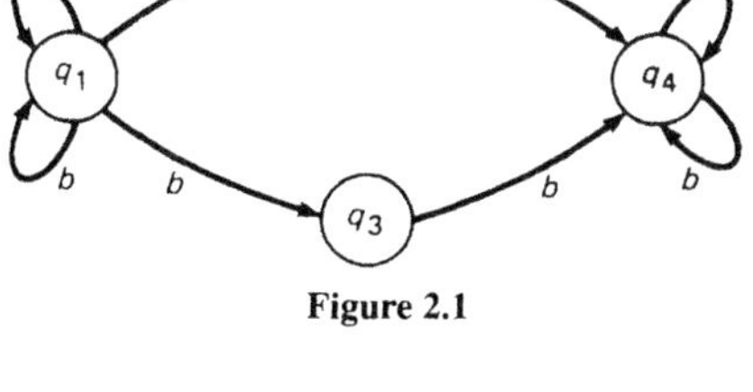

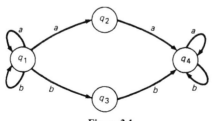

Figure 2.1

enter state q_2 to record that fact. On the other hand, it might be that the symbol following this a will be a b. Then this a is of no help in attaining the desired goal and hence we would remain in q_1. Since we are not able to look ahead in the string, we cannot at this point determine which role the current a is playing and so the automaton "simultaneously" hypothesizes both possibilities. If the next character read is b, then since there is no transition from q_2 reading b, the choice has been resolved and the automaton will be in state q_1. If instead, the character following the first a is another a, then since $q_2 \in \delta(q_1, a)$ and $q_4 \in \delta(q_2, a)$, and on any input the automaton once in state q_4 remains in q_4, the input string will be accepted because q_4 is an accepting state. A similar analysis can be made if a b is read when the automaton is in state q_1.

Strictly speaking, a dfa is *not* just a special kind of ndfa, although it is frequently thought of as such. This is because for a dfa, $\delta(q, s)$ is a state, whereas for an ndfa it is a set of states. But it is natural to identify the dfa $\mathscr{M}$ with transition function δ, with the closely related ndfa $\bar{\mathscr{M}}$ whose transition function $\bar{\delta}$ is given by

$$\bar{\delta}(q, s) = \{\delta(q, s)\},$$

and which has the same final states as $\mathscr{M}$. Obviously $L(\mathscr{M}) = L(\bar{\mathscr{M}})$.

The main theorem on nondeterministic finite automata is

Theorem 2.1. A language is accepted by an ndfa if and only if it is regular. Equivalently, a language is accepted by an ndfa if and only if it is accepted by a dfa.

Proof. As we have just seen, a language accepted by a dfa is also accepted by an ndfa. Conversely, let $L = L(\mathscr{M})$, where $\mathscr{M}$ is an ndfa with transition function δ, set of states $Q = \{q_1, \ldots, q_m\}$, and set of final states F. We will construct a dfa $\tilde{\mathscr{M}}$ such that $L(\tilde{\mathscr{M}}) = L(\mathscr{M}) = L$. The idea of the construction is that the individual states of $\tilde{\mathscr{M}}$ will be sets of states of $\mathscr{M}$.

Thus, we proceed to specify the dfa $\tilde{\mathscr{M}}$ on the same alphabet as $\mathscr{M}$. The states of $\tilde{\mathscr{M}}$ are just the 2^m sets of states (including $\emptyset$) of $\mathscr{M}$. We write these as $\tilde{Q} = \{Q_1, Q_2, \ldots, Q_{2^m}\}$, where in particular $Q_1 = \{q_1\}$ is to be the initial state of $\tilde{\mathscr{M}}$. The set $\mathscr{F}$ of final states of $\tilde{\mathscr{M}}$ is given by

$$\mathscr{F} = \{Q_i \mid Q_i \cap F \neq \emptyset\}.$$

The transition function $\tilde{\delta}$ of $\tilde{\mathscr{M}}$ is then defined by

$$\tilde{\delta}(Q_i, s) = \bigcup_{q \in Q_i} \delta(q, s).$$

Now, we have

Lemma 1. Let $R \subseteq \tilde{Q}$. Then

$$\tilde{\delta}\left(\bigcup_{Q_i \in R} Q_i, s\right) = \bigcup_{Q_i \in R} \tilde{\delta}(Q_i, s).$$

Proof. Let $\bigcup_{Q_i \in R} Q_i = Q$. Then by definition,

$$\begin{aligned}\tilde{\delta}(Q, s) &= \bigcup_{q \in Q} \delta(q, s)\\ &= \bigcup_{Q_i \in R} \bigcup_{q \in Q_i} \delta(q, s)\\ &= \bigcup_{Q_i \in R} \tilde{\delta}(Q_i, s).\end{aligned}$$ ■

Lemma 2. For any string u,

$$\tilde{\delta}^*(Q_i, u) = \bigcup_{q \in Q_i} \delta^*(q, u).$$

Proof. The proof is by induction on $|u|$. If $|u| = 0$, then $u = 0$ and

$$\tilde{\delta}^*(Q_i, 0) = Q_i = \bigcup_{q \in Q_i} \{q\} = \bigcup_{q \in Q_i} \delta^*(q, 0).$$

If $|u| = l + 1$ and the result is known for $|u| = l$, we write $u = vs$, where $|v| = l$, and observe that, using Lemma 1 and the induction hypothesis,

$$\begin{aligned}\tilde{\delta}^*(Q_i, u) &= \tilde{\delta}^*(Q_i, vs)\\ &= \tilde{\delta}(\tilde{\delta}^*(Q_i, v), s)\\ &= \tilde{\delta}\left(\bigcup_{q \in Q_i} \delta^*(q, v), s\right)\\ &= \bigcup_{q \in Q_i} \tilde{\delta}(\delta^*(q, v), s)\\ &= \bigcup_{q \in Q_i} \bigcup_{r \in \delta^*(q, v)} \delta(r, s)\\ &= \bigcup_{q \in Q_i} \delta^*(q, vs)\\ &= \bigcup_{q \in Q_i} \delta^*(q, u).\end{aligned}$$ ■

Lemma 3. $L(\mathcal{M}) = L(\tilde{\mathcal{M}})$.

Proof. $u \in L(\tilde{\mathcal{M}})$ if and only if $\tilde{\delta}^*(Q_1, u) \in \mathcal{F}$. But, by Lemma 2,

$$\tilde{\delta}^*(Q_1, u) = \tilde{\delta}^*(\{q_1\}, u) = \delta^*(q_1, u).$$

Hence,

$$\begin{aligned} u \in L(\tilde{\mathcal{M}}) \quad & \text{if and only if} \quad \delta^*(q_1, u) \in \mathcal{F} \\ & \text{if and only if} \quad \delta^*(q_1, u) \cap F \neq \varnothing \\ & \text{if and only if} \quad u \in L(\mathcal{M}). \end{aligned}$$ ■

Proof of Theorem 2.1 Concluded. Theorem 2.1 is an immediate consequence of Lemma 3. ■

Note that this proof is constructive. Not only have we shown that if a language is accepted by some ndfa, it is also accepted by some dfa, but we have also provided, within the proof, an algorithm for carrying out the conversion. This is important because, although it is frequently easier to design an ndfa than a dfa to accept a particular language, actual machines that are built are deterministic.

Exercises

1. Describe the language accepted by each of the following ndfas. In each case the initial state is q_1.

(a)

δ_1	a	b	c
q_1	$\{q_1, q_2, q_3\}$	$\varnothing$	$\varnothing$
q_2	$\varnothing$	$\{q_4\}$	$\varnothing$
q_3	$\varnothing$	$\varnothing$	$\{q_4\}$
q_4	$\varnothing$	$\varnothing$	$\varnothing$

$F_1 = \{q_4\}$.

(b) $\delta_2 = \delta_1$, $F_2 = \{q_1, q_2, q_3\}$.

(c)

δ_3	a	b
q_1	$\{q_2\}$	$\varnothing$
q_2	$\varnothing$	$\{q_1, q_3\}$
q_3	$\{q_1, q_3\}$	$\varnothing$

$F_3 = \{q_2\}$.

2. For each dfa $\mathcal{M}$ in Exercise 1.3, transform $\mathcal{M}$ into an ndfa $\mathcal{M}'$ which accepts $L(\mathcal{M})$. Then transform $\mathcal{M}'$ into a dfa $\mathcal{M}''$ by way of the construction in the proof of Theorem 2.1.
3. Let $\mathcal{M}$ be a dfa with a single accepting state. Consider the ndfa $\mathcal{M}'$ formed by reversing the roles of the initial and accepting states and reversing the direction of the arrows of all transitions in the transition diagram. Describe $L(\mathcal{M}')$ in terms of $L(\mathcal{M})$.
4. Prove that, given any ndfa $\mathcal{M}_1$, there exists an ndfa $\mathcal{M}_2$ with exactly one accepting state such that

$$L(\mathcal{M}_2) = L(\mathcal{M}_1) \quad \text{or} \quad L(\mathcal{M}_2) = L(\mathcal{M}_1) - \{0\}.$$

5. (a) The construction in the proof of Theorem 2.1 shows that any regular language accepted by an ndfa with n states is accepted by some dfa with 2^n states. Show that there is a regular language that is accepted by an ndfa with two states, not accepted by any ndfa with fewer than two states, and accepted by a dfa with two states.
   (b) Show that there is a regular language that is accepted by an ndfa with two states and not accepted by any dfa with fewer than four states.
   (c) Show that there is a regular language that is accepted by an ndfa with three states and not accepted by any dfa with fewer than eight states.

3. Additional Examples

We first give two simple examples of finite automata and their associated regular languages.

For our first example we consider a unary even parity checker. That is, we want to design a finite automaton over the alphabet {1} such that the machine terminates in an accepting state if and only if the input string contains an even number of ones. Intuitively then, the machine must contain two states which "remember" whether an even or an odd number of ones have been encountered so far. When the automaton begins, no ones, and hence an even number of ones, have been read; hence the initial state q_1 will represent the even parity state, and q_2, the odd parity state. Furthermore, since we want to accept words containing an even number of ones, q_1 will be an accepting state.

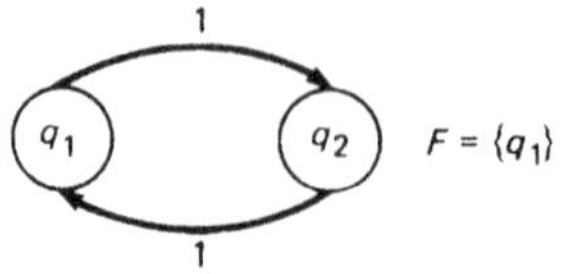

Figure 3.1

Thus the finite automaton to perform the required task is as shown in Fig. 3.1, and the language it accepts is

$$\{(11)^{[n]} \mid n \geq 0\}.$$

We next consider a slightly more complicated example. Suppose we wish to design a finite automaton that will function as a 25¢ candy vending machine. The alphabet consists of the three symbols n, d, and q (representing nickel, dime, and quarter, respectively—no pennies, please!). If more than 25¢ is deposited, no change is returned and no credit is given for the overage. Intuitively, the states keep track of the amount of money deposited so far. The automaton is exhibited in Fig. 3.2, with each state labeled to indicate its role. The state labeled 0 is the initial state. Note that the state labeled d is a "dead" state; i.e., once that state is entered it may never be left. Whenever sufficient money has been inserted so that the automaton has entered the 25¢ (accepting) state, any additional coins will send the machine into this dead state, which may be thought of as a coin return state. Presumably when in the accepting state, a button can be pressed to select your candy and the machine is reset to 0.

Unlike the previous example, the language accepted by this finite automaton is a finite set. It consists of the following combinations of nickels, dimes, and quarters: {*nnnnn*, *nnnnd*, *nnnnq*, *nnnd*, *nnnq*, *nndn*, *nndd*, *nndq*, *nnq*, *ndnn*, *ndnd*, *ndnq*, *ndd*, *ndq*, *nq*, *dnnn*, *dnnd*, *dnnq*, *dnd*, *dnq*, *ddn*, *ddd*, *ddq*, *dq*, *q*}.

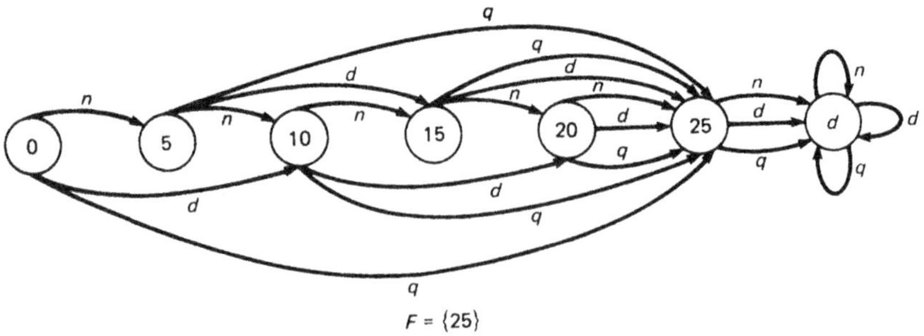

Figure 3.2

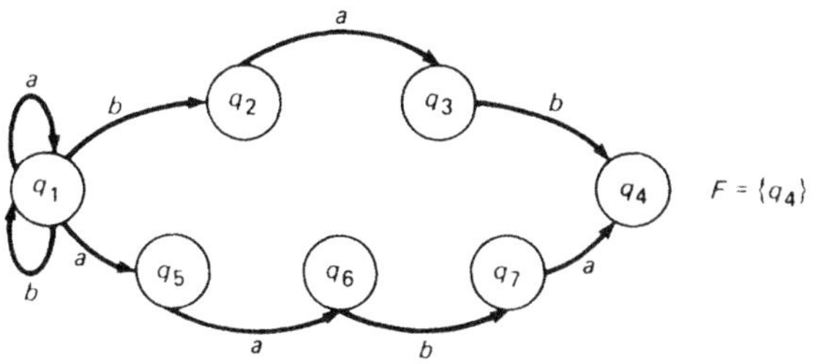

Figure 3.3

Suppose we wish to design an automaton on the alphabet $\{a, b\}$ that accepts all and only strings which end in *bab* or *aaba*. A real-world analog of this problem might arise in a demographic study in which people of certain ethnic groups are to be identified by checking to see if their family name ends in certain strings of letters.

It is easy to design the desired ndfa: see Fig. 3.3.

As our final example, we discuss a slightly more complicated version of the first example considered in Section 1:

$$L = \{a^{[n_1]}b^{[m_1]} \cdots a^{[n_k]}b^{[m_k]} \mid n_1, m_1, \ldots, n_k, m_k > 0\}.$$

An ndfa $\mathscr{M}$ such that $L(\mathscr{M}) = L$ is shown in Fig. 3.4.

These two examples of ndfas illustrate an important characteristic of such machines: not only is it permissible to have many alternative transitions for a given state–symbol pair, but frequently there are no transitions for a given pair. In a sense, this means that whereas for a dfa one has to describe what happens for *any* string whether or not that string is a word in the language, for an ndfa one need only describe the behavior of the automaton for words in the language.

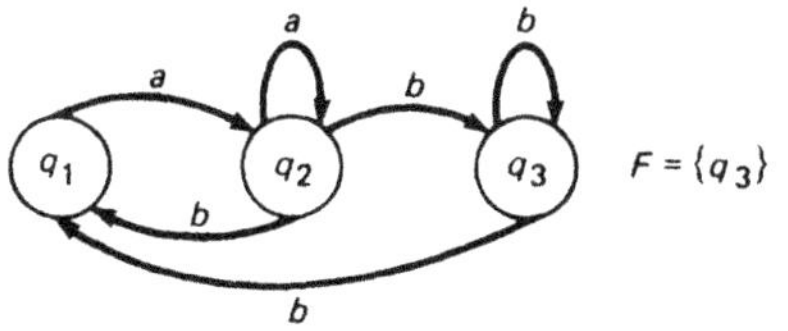

Figure 3.4

4. Closure Properties

We will be able to prove that the class of regular languages is closed under a large number of operations. It will be helpful that, by the equivalence theorems of the previous two sections, we can use deterministic or nondeterministic finite automata to suit our convenience.

Definition. A dfa is called *nonrestarting* if there is no pair q, s for which

$$\delta(q, s) = q_1,$$

where q_1 is the initial state.

Theorem 4.1. There is an algorithm that will transform a given dfa $\mathscr{M}$ into a nonrestarting dfa $\tilde{\mathscr{M}}$ such that $L(\tilde{\mathscr{M}}) = L(\mathscr{M})$.

Proof. Let $Q = \{q_1, q_2, \ldots, q_n\}$ be the set of states of $\mathscr{M}$, q_1 the initial state, F the set of accepting states, and δ the transition function. We construct $\tilde{\mathscr{M}}$ with the set of states $\tilde{Q} = Q \cup \{q_{n+1}\}$, initial state q_1, and transition function $\tilde{\delta}$ defined by

$$\tilde{\delta}(q, s) = \begin{cases} \delta(q, s) & \text{if} \quad q \in Q \text{ and } \delta(q, s) \neq q_1 \\ q_{n+1} & \text{if} \quad q \in Q \text{ and } \delta(q, s) = q_1, \end{cases}$$

$$\tilde{\delta}(q_{n+1}, s) = \tilde{\delta}(q_1, s).$$

Thus, there is no transition into state q_1 for $\tilde{\mathscr{M}}$. The set of accepting states $\tilde{F}$ of $\tilde{\mathscr{M}}$ is defined by

$$\tilde{F} = \begin{cases} F & \text{if} \quad q_1 \notin F \\ F \cup \{q_{n+1}\} & \text{if} \quad q_1 \in F. \end{cases}$$

To see that $L(\mathscr{M}) = L(\tilde{\mathscr{M}})$ as required, one need only observe that $\tilde{\mathscr{M}}$ follows the same transitions as $\mathscr{M}$ except that whenever $\mathscr{M}$ reenters q_1, $\tilde{\mathscr{M}}$ enters q_{n+1}. ■

Theorem 4.2. If L and $\tilde{L}$ are regular languages, then so is $L \cup \tilde{L}$.

Proof. Without loss of generality, by Theorem 4.1, let $\mathscr{M}, \tilde{\mathscr{M}}$ be nonrestarting dfas that accept L and $\tilde{L}$, respectively, with Q, q_1, F, δ and $\tilde{Q}, \tilde{q}_1, \tilde{F}, \tilde{\delta}$ the set of states, initial state, set of accepting states, and transition function of $\mathscr{M}$ and $\tilde{\mathscr{M}}$, respectively. We also assume that $\mathscr{M}$ and $\tilde{\mathscr{M}}$ have no states in common, i.e., $Q \cap \tilde{Q} = \varnothing$. Furthermore, by the discussion in Section 1, we can assume that the alphabets of L and $\tilde{L}$ are the same, say, A. We define the *ndfa* $\check{\mathscr{M}}$ with states $\check{Q}$, initial state $\check{q}_1$, set of accepting states $\check{F}$, and transition function $\check{\delta}$ as follows:

$$\check{Q} = Q \cup \tilde{Q} \cup \{\check{q}_1\} - \{q_1, \tilde{q}_1\}.$$

(That is, $\check{\mathcal{M}}$ contains a new initial state $\check{q}_1$ and all states of $\mathcal{M}$ and $\tilde{\mathcal{M}}$ except their initial states.)

$$\check{F} = \begin{cases} F \cup \tilde{F} \cup \{\check{q}_1\} - \{q_1, \tilde{q}_1\} & \text{if} \quad q_1 \in F \text{ or } \tilde{q}_1 \in \tilde{F} \\ F \cup \tilde{F} & \text{otherwise.} \end{cases}$$

The transition function of $\check{\mathcal{M}}$ is defined as follows for $s \in A$:

$$\check{\delta}(q, s) = \begin{cases} \{\delta(q, s)\} & \text{if} \quad q \in Q - \{q_1\} \\ \{\tilde{\delta}(q, s)\} & \text{if} \quad q \in \tilde{Q} - \{\tilde{q}_1\} \end{cases}$$

$$\check{\delta}(\check{q}_1, s) = \{\delta(q_1, s)\} \cup \{\tilde{\delta}(\tilde{q}_1, s)\}.$$

Thus, since $Q \cap \tilde{Q} = \varnothing$ and $\mathcal{M}$ and $\tilde{\mathcal{M}}$ are nonrestarting, once a first transition has been selected, the automaton $\check{\mathcal{M}}$ is locked into one of the two automata $\mathcal{M}$ and $\tilde{\mathcal{M}}$. Hence $L(\check{\mathcal{M}}) = L \cup \tilde{L}$. ■

Theorem 4.3. Let $L \subseteq A^*$ be a regular language. Then $A^* - L$ is regular.

Proof. Let $\mathcal{M}$ be a dfa that accepts L. Let $\mathcal{M}$ have alphabet A, set of states Q, and set of accepting states F. Let $\bar{\mathcal{M}}$ be exactly like $\mathcal{M}$ except that it accepts precisely when $\mathcal{M}$ rejects. That is, the set of accepting states of $\bar{\mathcal{M}}$ is $Q - F$. Then $\bar{\mathcal{M}}$ clearly accepts $A^* - L$. ■

Theorem 4.4. If L_1 and L_2 are regular languages, then so is $L_1 \cap L_2$.

Proof. Let $L_1, L_2 \subseteq A^*$. Then we have the De Morgan identity:

$$L_1 \cap L_2 = A^* - ((A^* - L_1) \cup (A^* - L_2)).$$

Theorems 4.2 and 4.3 then give the result. ■

Theorem 4.5. $\varnothing$ and $\{0\}$ are regular languages.

Proof. $\varnothing$ is clearly the language accepted by any automaton whose set of accepting states is empty. Next, the automaton with states q_1, q_2, alphabet $\{a\}$, accepting states $F = \{q_1\}$, and transition function $\delta(q_1, a) = \delta(q_2, a) = q_2$ clearly accepts $\{0\}$, as does any nonrestarting dfa on any alphabet provided $F = \{q_1\}$. ■

Theorem 4.6. Let $u \in A^*$. Then $\{u\}$ is a regular language.

Proof. For $u = 0$, we already know this from Theorem 4.5. Otherwise let $u = a_1 a_2 \cdots a_l a_{l+1}$, where $a_1, a_2, \ldots, a_l, a_{l+1} \in A$. Let $\mathcal{M}$ be the ndfa with states $q_1, q_2, \ldots, q_{l+2}$, initial state q_1, accepting state q_{l+2}, and transition function δ given by

$$\delta(q_i, a_i) = \{q_{i+1}\}, \qquad i = 1, \ldots, l+1,$$

$$\delta(q_i, a) = \emptyset \qquad \text{for} \quad a \in A - \{a_i\}.$$

Then $L(\mathscr{M}) = \{u\}$. ■

Corollary 4.7. Every finite subset of A^* is regular.

Proof. We have already seen that $\emptyset$ is regular. If $L = \{u_1, \ldots, u_n\}$, where $u_1, \ldots, u_n \in A^*$, we note that

$$L = \{u_1\} \cup \{u_2\} \cup \cdots \cup \{u_n\},$$

and apply Theorems 4.2 and 4.6. ■

Exercises

1. Let $A = \{a, b\}$, let $L_1 \subseteq A^*$ consist of all words with at least two occurrences of a, and let $L_2 \subseteq A^*$ consist of all words with at least two occurrences of b. For each of the following languages L, give an ndfa that accepts L.

(a) $L = L_1 \cup L_2$.

(b) $L = A^* - L_1$.

(c) $L = A^* - L_2$.

(d) $L = L_1 \cap L_2$.

2. Use the constructions in the proofs of Theorem 4.6 and Corollary 4.7 to give an ndfa that accepts the language $\{ab, ac, ad\}$.

3. **(a)** Let L, L' be regular languages. Prove that $L - L'$ is regular.

(b) Let L, L' be languages such that L is regular, $L \cup L'$ is regular, and $L \cap L' = \emptyset$. Prove that L' is regular.

4. Let L_1, L_2 be regular languages with $\text{rank}(L_1) = n_1$ and $\text{rank}(L_2) = n_2$. [See Exercise 1.6 for the definition of *rank*.]

(a) Use Theorems 4.1, 4.2, and 2.1 to give an upper bound on $\text{rank}(L_1 \cup L_2)$.

(b) Use Theorems 4.1, 4.2, 4.3, 4.4, and 2.1 to give an upper bound on $\text{rank}(L_1 \cap L_2)$.

5.* Let A_1, A_2 be alphabets, and let f be a function from A_1^* to subsets of A_2^*. f is a *substitution* on A_1 if $f(0) = \{0\}$ and, for all nonnull words

$a_1 \cdots a_n \in A_1^*$, where $a_1, \ldots, a_n \in A_1$, $f(a_1 \cdots a_n) = f(a_1) \cdots f(a_n) = \{ u_1 \cdots u_n \,|u_i \in f(a_i), 1 \le i \le n\}$. For $L \subseteq A_1^*$, $f(L) = \bigcup_{w \in L} f(w)$.

(a) Let $A_1 = \{a, b\}$, $A_2 = \{c, d, e\}$, let f be the substitution on A_1 such that $f(a) = \{cc, 0\}$ and $f(b) = \{w \in A_2^* \mid w$ ends in $e\}$, and let $L = \{a^{[m]}b^{[n]} \mid m, n \ge 0\}$. What is $f(L)$?

(b) Let A_1, A_2 be alphabets, let f be a substitution on A_1 such that $f(a) \subseteq A_2^*$ is a regular language for all $a \in A_1$, and let L be a regular language on A_1. Prove that $f(L)$ is a regular language on A_2.

(c) Let A_1, A_2 be alphabets, and let g be a function from A_1^* to A_2^*. g is a *homomorphism* on A_1 if $g(0) = 0$ and, for all nonnull words $a_1 \cdots a_n \in A_1^*$, where $a_1, \ldots, a_n \in A_1$, $g(a_1 \cdots a_n) = g(a_1) \cdots g(a_n)$. For $L \subseteq A_1^*$, $g(L) = \{g(w) \mid w \in L\}$. Use (b) to show that if g is a homomorphism on A_1 and $L \subseteq A_1^*$ is regular, then $g(L)$ is regular.

5. Kleene's Theorem

In this section we will see how the class of regular languages can be characterized as the class of all languages obtained from finite languages using a few operations.

Definition. Let $L_1, L_2 \subseteq A^*$. Then, we write

$$L_1 \cdot L_2 = L_1 L_2 = \{uv \mid u \in L_1 \text{ and } v \in L_2\}.$$

Definition. Let $L \subseteq A^*$. Then we write

$$L^* = \{u_1 u_2 \cdots u_n \mid n \ge 0, u_1, \ldots, u_n \in L\}.$$

With respect to this last definition, note that

1. $0 \in L^*$ automatically because $n = 0$ is allowed;
2. for A^* the present notation is consistent with what we have been using.

Theorem 5.1. If $L, \tilde{L}$ are regular languages, then $L \cdot \tilde{L}$ is a regular language.

Proof. Let $\mathscr{M}$ and $\tilde{\mathscr{M}}$ be dfas that accept L and $\tilde{L}$, respectively, with Q, q_1, F, δ and $\tilde{Q}, \tilde{q}_1, \tilde{F}, \tilde{\delta}$ the set of states, initial state, set of accepting states, and transition function, respectively. Assume that $\mathscr{M}$ and $\tilde{\mathscr{M}}$ have no states in common, i.e., $Q \cap \tilde{Q} = \varnothing$. By our discussion in Section 1, we can assume without loss of generality that the alphabets of L and $\tilde{L}$ are the same. Consider the ndfa $\dot{\mathscr{M}}$ formed by "gluing together" $\mathscr{M}$ and $\tilde{\mathscr{M}}$ in the following way. The set $\dot{Q}$ of states of $\dot{\mathscr{M}}$ is $Q \cup \tilde{Q}$, and the initial state is q_1. We will define the transition function $\dot{\delta}$ of $\dot{\mathscr{M}}$ in such a way that the transitions of $\dot{\mathscr{M}}$ will contain all transitions of $\mathscr{M}$ and $\tilde{\mathscr{M}}$. In addition $\dot{\delta}(q, s)$ will contain $\tilde{\delta}(\tilde{q}_1, s)$ for every $q \in F$. Thus, any time a symbol of the input string causes $\mathscr{M}$ to enter an accepting state, $\dot{\mathscr{M}}$ can either continue by treating the next symbol of the input as being from the word of L or as the first symbol of the word of $\tilde{L}$. Formally we define $\dot{\delta}$ as follows:

$$\dot{\delta}(q, s) = \begin{cases} \{\delta(q, s)\} & \text{for} \quad q \in Q - F \\ \{\delta(q, s)\} \cup \{\tilde{\delta}(\tilde{q}_1, s)\} & \text{for} \quad q \in F \\ \{\tilde{\delta}(q, s)\} & \text{for} \quad q \in \tilde{Q}. \end{cases}$$

Thus, $\dot{\mathscr{M}}$ begins by behaving exactly like $\mathscr{M}$. However, just when $\mathscr{M}$ has accepted a word and would make a transition from an accepting state, $\dot{\mathscr{M}}$ may proceed as if it were $\tilde{\mathscr{M}}$ making a transition from $\tilde{q}_1$.

Finally, if $0 \in \tilde{L}$ we set $\dot{F} = F \cup \tilde{F}$, and if $0 \notin \tilde{L}$ we set $\dot{F} = \tilde{F}$. Clearly, $L \cdot \tilde{L} = L(\dot{\mathscr{M}})$, so that $L \cdot \tilde{L}$ is a regular language. ■

Theorem 5.2. If L is a regular language, then so is L^*.

Proof. Let $\mathscr{M}$ be a nonrestarting dfa that accepts L with alphabet A, set of states Q, initial state q_1, accepting states F, and transition function δ. We construct the ndfa $\overset{*}{\mathscr{M}}$ with the same states and initial state as $\mathscr{M}$, and accepting state q_1. The transition function $\bar{\delta}$ is defined as follows:

$$\bar{\delta}(q, s) = \begin{cases} \{\delta(q, s)\} & \text{if} \quad \delta(q, s) \notin F \\ \{\delta(q, s)\} \cup \{q_1\} & \text{if} \quad \delta(q, s) \in F. \end{cases}$$

That is, whenever $\mathscr{M}$ would enter an accepting state, $\overset{*}{\mathscr{M}}$ will enter either the corresponding accepting state or the initial state. Clearly $L^* = L(\overset{*}{\mathscr{M}})$, so that L^* is a regular language. ■

Theorem 5.3 (Kleene's Theorem). A language is regular if and only if it can be obtained from finite languages by applying the three operators $\cup$, $\cdot$, $*$ a finite number of times.

The characterization of regular languages that Kleene's theorem gives resembles the definition of the primitive recursive functions and the characterization of the partially computable functions of Theorem 3.5 in

Chapter 4. In each case one begins with some initial objects and applies certain operations a finite number of times.

Proof. Every finite language is regular by Corollary 4.7, and if $L = L_1 \cup L_2$ or $L = L_1 \cdot L_2$ or $L = L_1^*$, where L_1 and L_2 are regular, then L is regular by Theorems 4.2, 5.1, and 5.2, respectively. Therefore, by induction on the number of applications of $\cup$, $\cdot$, and $*$, any language obtained from finite languages by applying these operators a finite number of times is regular.

On the other hand, let L be a regular language, $L = L(\mathscr{M})$, where $\mathscr{M}$ is a dfa with states $q_1, \ldots, q_n$. As usual, q_1 is the initial state, F is the set of accepting states, δ is the transition function, and $A = \{s_1, \ldots, s_K\}$ is the alphabet. We define the sets $R_{i,j}^k$, $i, j > 0, k \geq 0$, as follows:

$$R_{i,j}^k = \{x \in A^* \mid \delta^*(q_i, x) = q_j \text{ and } \mathscr{M} \text{ passes through no state } q_l \text{ with } l > k \text{ as it moves across } x\}.$$

More formally, $R_{i,j}^k$ is the set of words $x = s_{i_1} s_{i_2} \cdots s_{i_r} s_{i_{r+1}}$ such that we can write

$$\delta(q_i, s_{i_1}) = q_{j_1},$$
$$\delta(q_{j_1}, s_{i_2}) = q_{j_2},$$
$$\vdots$$
$$\delta(q_{j_{r-1}}, s_{i_r}) = q_{j_r},$$
$$\delta(q_{j_r}, s_{i_{r+1}}) = q_j,$$

where $j_1, j_2, \ldots, j_r \leq k$. Now, we observe that $R_{i,i}^0 = \{0\}$, and, for $i \neq j$,

$$R_{i,j}^0 = \{a \in A \mid \delta(q_i, a) = q_j\},$$

since for a word of length 1, $\mathscr{M}$ passes directly from state q_i into state q_j while in processing any word of length > 1, $\mathscr{M}$ will pass through some intermediate state $q_l, l \geq 1$. Thus $R_{i,j}^0$ is a finite set. Furthermore, we have

$$R_{i,j}^{k+1} = R_{i,j}^k \cup \left[R_{i,k+1}^k \cdot (R_{k+1,k+1}^k)^* \cdot R_{k+1,j}^k\right]. \tag{5.1}$$

This rather imposing formula really states something quite simple: The set $R_{i,j}^{k+1}$ contains all the elements of $R_{i,j}^k$ and in addition contains strings x, such that $\mathscr{M}$ in scanning x passes through the state q_{k+1} (but through none with larger subscript) some finite number of times. Such a string can be decomposed into a left end, which $\mathscr{M}$ enters in state q_i and leaves in state q_{k+1} (passing only through states with subscripts less than $k + 1$ in the process), followed by some finite number of pieces each of which $\mathscr{M}$ enters and leaves in state q_{k+1} (passing only through q_l with $l \leq k$), and a right end which $\mathscr{M}$ enters in state q_{k+1} and leaves in state q_j (again passing only through states with subscript $\leq k$ in between). Now we have

Lemma. Each $R_{i,j}^k$ can be obtained from finite languages by a finite number of applications of the operations $\cup$, $\cdot$, *.

Proof. We prove by induction on k that for all i, j, the set $R_{i,j}^k$ has the desired property. For $k = 0$ this is obvious, since $R_{i,j}^0$ is finite.
Assuming the result known for k, (5.1) yields the result for $k + 1$. ■

Proof of Kleene's Theorem Concluded. We note that

$$L(\mathscr{M}) = \bigcup_{q_j \in F} R_{1,j}^n;$$

thus, the result follows at once from the lemma. ■

Kleene's theorem makes it possible to give names to regular languages in a particularly simple way. Let us begin with an alphabet $A = \{s_1, s_2, \ldots, s_k\}$. Then we define the corresponding alphabet:

$$\tilde{A} = \{\mathbf{s_1}, \mathbf{s_2}, \ldots, \mathbf{s_k}, \mathbf{0}, \boldsymbol{\varnothing}, \boldsymbol{\cup}, \boldsymbol{\cdot}, \boldsymbol{*}, \mathbf{(}, \mathbf{)}\}.$$

The class of *regular expressions* on the alphabet A is then defined to be the subset of $\tilde{A}^*$ determined by the following:

1. $\boldsymbol{\varnothing}, \mathbf{0}, \mathbf{s_1}, \ldots, \mathbf{s_k}$ are regular expressions.
2. If α and β are regular expressions, then so is $\mathbf{(}\alpha \boldsymbol{\cup} \beta\mathbf{)}$.
3. If α and β are regular expressions, then so is $\mathbf{(}\alpha \boldsymbol{\cdot} \beta\mathbf{)}$.
4. If α is a regular expression, then so is $\alpha\boldsymbol{*}$.
5. No expression is regular unless it can be generated using a finite number of applications of 1–4.

Here are a few examples of regular expressions on the alphabet $A = \{a, b, c\}$:

$$\mathbf{(a \cdot (b^* \cup c^*))}$$
$$\mathbf{(0 \cup (a \cdot b)^*)}$$
$$\mathbf{(c^* \cdot b^*)}.$$

For each regular expression γ, we define a corresponding regular language $\langle \gamma \rangle$ by recursion according to the following "semantic" rules:[3]

$$\langle s_i \rangle = \{s_i\},$$
$$\langle \mathbf{0} \rangle = \{0\},$$
$$\langle \varnothing \rangle = \varnothing,$$
$$\langle (\alpha \cup \beta) \rangle = \langle \alpha \rangle \cup \langle \beta \rangle,$$
$$\langle (\alpha \cdot \beta) \rangle = \langle \alpha \rangle \cdot \langle \beta \rangle,$$
$$\langle \alpha^* \rangle = \langle \alpha \rangle^*.$$

When $\langle \gamma \rangle = L$, we say that the regular expression γ *represents* L. Thus,

$$\langle (\mathbf{a} \cdot (\mathbf{b}^* \cup \mathbf{c}^*)) \rangle = \{ab^{[n]} \mid n \geq 0\} \cup \{ac^{[m]} \mid m \geq 0\},$$
$$\langle (\mathbf{0} \cup (\mathbf{a} \cdot \mathbf{b})^*) \rangle = \langle (\mathbf{a} \cdot \mathbf{b})^* \rangle = \{(ab)^{[n]} \mid n \geq 0\},$$
$$\langle (\mathbf{c}^* \cdot \mathbf{b}^*) \rangle = \{c^{[m]}b^{[n]} \mid m, n \geq 0\}.$$

We have

Theorem 5.4. For every finite subset L of A^*, there is a regular expression γ on A such that $\langle \gamma \rangle = L$.

Proof. If $L = \varnothing$, then $L = \langle \varnothing \rangle$. If $L = \{0\}$, then $L = \langle \mathbf{0} \rangle$. If $L = \{x\}$, where $x = s_{i_1} s_{i_2} \cdots s_{i_l}$, then

$$L = \langle (\mathbf{s}_{i_1} \cdot (\mathbf{s}_{i_2} \cdot (\mathbf{s}_{i_3} \cdots \mathbf{s}_{i_l}) \cdots)) \rangle.$$

This gives the result for languages L consisting of 0 or 1 element. Assuming the result known for languages of k elements, let L have $k + 1$ elements. Then we can write

$$L = L_1 \cup \{x\},$$

where $x \in A^*$ and L_1 contains k elements. By the induction hypothesis, there is a regular expression α such that $\langle \alpha \rangle = L_1$. By the one-element case already considered, there is a regular expression β such that $\langle \beta \rangle = \{x\}$. Then we have

$$\langle (\alpha \cup \beta) \rangle = \langle \alpha \rangle \cup \langle \beta \rangle = L_1 \cup \{x\} = L. \qquad \blacksquare$$

---

[3] For more on this subject see Part 5.

Theorem 5.5 (Kleene's Theorem—Second Version). A language $L \subseteq A^*$ is regular if and only if there is a regular expression γ on A such that $\langle \gamma \rangle = L$.

Proof. For any regular expression γ, the regular language $\langle \gamma \rangle$ is built up from finite languages by applying $\cup$, $\cdot$, * a finite number of times, so $\langle \gamma \rangle$ is regular by Kleene's theorem.

On the other hand, let L be a regular language. If L is finite then, by Theorem 5.4, there is a regular expression γ such that $\langle \gamma \rangle = L$. Otherwise, by Kleene's theorem, L can be obtained from certain finite languages by a finite number of applications of the operations $\cup, \cdot, *$. By beginning with regular expressions representing these finite languages, we can build up a regular expression representing L by simply indicating each use of the operations $\cup$, $\cdot$, * by writing $\cup$, $\cdot$, $*$, respectively, and punctuating with (and). ■

Exercises

1. (a) For each language L described in Exercise 1.1, give a regular expression α such that $L = \langle \alpha \rangle$.
   (b) For each dfa $\mathscr{M}$ described in Exercise 1.3, give a regular expression α such that $L(\mathscr{M}) = \langle \alpha \rangle$.
   (c) For each ndfa $\mathscr{M}$ described in Exercise 2.1, give a regular expression α such that $L(\mathscr{M}) = \langle \alpha \rangle$.
2. For regular expressions α, β, let us write $\alpha \equiv \beta$ to mean that $\langle \alpha \rangle = \langle \beta \rangle$. For α, β, γ given regular expressions, prove the following identities.
   (a) $(\alpha \cup \alpha) \equiv \alpha$.
   (b) $((\alpha \cdot \beta) \cup (\alpha \cdot \gamma)) \equiv (\alpha \cdot (\beta \cup \gamma))$.
   (c) $((\beta \cdot \alpha) \cup (\gamma \cdot \alpha)) \equiv ((\beta \cup \gamma) \cdot \alpha)$.
   (d) $(\alpha^* \cdot \alpha^*) \equiv \alpha^*$.
   (e) $(\alpha \cdot \alpha^*) \equiv (\alpha^* \cdot \alpha)$.
   (f) $\alpha^{**} \equiv \alpha^*$.
   (g) $(\mathbf{0} \cup (\alpha \cdot \alpha^*)) \equiv \alpha^*$.
   (h) $((\alpha \cdot \beta)^* \cdot \alpha) \equiv (\alpha \cdot (\beta \cdot \alpha)^*)$.
   (i) $(\alpha \cup \beta)^* \equiv (\alpha^* \cdot \beta^*)^* \equiv (\alpha^* \cup \beta^*)^*$.
3. Using the identities of Exercise 2 prove that

$$((\mathbf{abb})^*(\mathbf{ba})^*(\mathbf{b} \cup \mathbf{aa})) \equiv (\mathbf{abb})^*((\mathbf{0} \cup (\mathbf{b}(\mathbf{ab})^*\mathbf{a}))\mathbf{b} \cup (\mathbf{ba})^*(\mathbf{aa})).$$

(Note that parentheses and the symbol "·" have been omitted to facilitate reading.)

4. Let α, β be given regular expressions such that $0 \notin \langle \alpha \rangle$. Consider the equation in the "unknown" regular expression ξ:

$$\xi \equiv (\beta \cup (\xi \cdot \alpha)).$$

Prove that this equation has the solution

$$\xi \equiv (\beta \cdot \alpha^*)$$

and that the solution is unique in the sense that if ξ' also satisfies the equation, then $\xi \equiv \xi'$.

5. Let $L = \{x \in \{a, b\}^* \mid x \neq 0 \text{ and } bb \text{ is not a substring of } x\}$.

(a) Show that L is regular by constructing a dfa $\mathcal{M}$ such that $L = L(\mathcal{M})$.

(b) Find a regular expression γ such that $L = \langle \gamma \rangle$.

6. Let $L = \langle((\mathbf{a} \cdot \mathbf{a}) \cup (\mathbf{a} \cdot \mathbf{a} \cdot \mathbf{a}))^*\rangle$. Find a dfa $\mathcal{M}$ that accepts L.

7. Describe an algorithm that, given any regular expression α, produces an ndfa $\mathcal{M}$ that accepts $\langle \alpha \rangle$.

8. Let L_1, L_2 be regular languages with $\text{rank}(L_1) = n_1$ and $\text{rank}(L_2) = n_2$. [See Exercise 1.6 for the definition of rank.]

(a) Use Theorem 5.1 to give an upper bound on $\text{rank}(L_1 \cdot L_2)$.

(b) Use Theorem 5.2 to give an upper bound on $\text{rank}(L_1^*)$.

9. Let $A = \{s_1, \ldots, s_n\}$.

(a) Give a function b_1 such that $\text{rank}(\langle \alpha \rangle) \leq b_1(\alpha)$ for all regular expressions α on A.

(b) Define the *size* of a regular expression on A as follows.

$$\begin{aligned}
&\text{size}(\varnothing) &&= 1\\
&\text{size}(\mathbf{0}) &&= 1\\
&\text{size}(\mathbf{s}_i) &&= 1 \quad i = 1, \ldots, n\\
&\text{size}((\alpha \cup \beta)) &&= \text{size}(\alpha) + \text{size}(\beta) + 1\\
&\text{size}((\alpha \cdot \beta)) &&= \text{size}(\alpha) + \text{size}(\beta) + 1\\
&\text{size}(\alpha^*) &&= \text{size}(\alpha) + 1
\end{aligned}$$

Give a numeric function b_2 such that $\text{rank}(\langle \alpha \rangle) \leq b_2(\text{size}(\alpha))$ for all regular expressions α on A.

(c)* Verify that b_2 is primitive recursive.

10.* Let $A = \{s_1, \ldots, s_n\}$, let α, β be regular expressions on A, and let P_α, P_β be primitive recursive predicates such that for all $w \in A^*$,

$P_\alpha(w) = 1$ if and only if $w \in \langle \alpha \rangle$ and $P_\beta(w) = 1$ if and only if $w \in \langle \beta \rangle$.

(a) Give a primitive recursive predicate $P_{(\alpha \cup \beta)}$ such that $P_{(\alpha \cup \beta)}(w) = 1$ if and only if $w \in \langle (\alpha \cup \beta) \rangle$.

(b) Give a primitive recursive predicate $P_{(\alpha \cdot \beta)}$ such that $P_{(\alpha \cdot \beta)}(w) = 1$ if and only if $w \in \langle (\alpha \cdot \beta) \rangle$.

(c) Give a primitive recursive predicate P_{α^*} such that $P_{\alpha^*}(w) = 1$ if and only if $w \in \langle \alpha^* \rangle$.

(d) Use parts (a), (b), and (c) to show that for all regular expressions γ on A, there is a primitive recursive predicate P_γ such that $P_\gamma(w) = 1$ if and only if $w \in \langle \gamma \rangle$.

6. The Pumping Lemma and Its Applications

We will make use of the following basic combinatorial fact:

> *Pigeon-Hole Principle.* If $(n + 1)$ objects are distributed among n sets, then at least one of the sets must contain at least two objects.

We will use this pigeon-hole principle to prove the following result.

Theorem 6.1 (Pumping Lemma). Let $L = L(\mathcal{M})$, where $\mathcal{M}$ is a dfa with n states. Let $x \in L$, where $|x| \geq n$. Then we can write $x = uvw$, where $v \neq 0$ and $uv^{[i]}w \in L$ for all $i = 0, 1, 2, 3, \ldots$.

Proof. Since x consists of at least n symbols, $\mathcal{M}$ must go through at least n state transitions as it scans x. Including the initial state, this requires at least $n + 1$ (not necessarily distinct) states. But since there are only n states in all, we conclude (here is the pigeon-hole principle!) that $\mathcal{M}$ must be in at least one state more than once. Let q be a state in which $\mathcal{M}$ finds itself at least twice. Then we can write $x = uvw$, where

$$\delta^*(q_1, u) = q,$$
$$\delta^*(q, v) = q,$$
$$\delta^*(q, w) \in F.$$

That is, $\mathcal{M}$ arrives in state q for the first time after scanning the last (right-hand) symbol of u and then again after scanning the last symbol of v. Since this "loop" can be repeated any number of times, it is clear that

$$\delta^*(q_1, uv^{[i]}w) = \delta^*(q_1, uvw) \in F.$$

Hence $uv^{[i]}w \in L$. ■

Theorem 6.2. Let $\mathcal{M}$ be a dfa with n states. Then, if $L(\mathcal{M}) \neq \emptyset$, there is a string $x \in L(\mathcal{M})$ such that $|x| < n$.

Proof. Let x be a string in $L(\mathcal{M})$ of the shortest possible length. Suppose $|x| \geq n$. By the pumping lemma, $x = uvw$, where $v \neq 0$ and $uw \in L(\mathcal{M})$. Since $|uw| < |x|$, this is a contradiction. Thus $|x| < n$. ■

This theorem furnishes an algorithm for testing a given dfa $\mathcal{M}$ to see whether the language it accepts is empty. We need only "run" $\mathcal{M}$ on all strings of length less than the number of states of $\mathcal{M}$. If none is accepted, we will be able to conclude that $L(\mathcal{M}) = \emptyset$.

Next we turn to infinite regular languages. If $L = L(\mathcal{M})$ is infinite, then L must surely contain words having length greater than the number of states of $\mathcal{M}$. Hence from the pumping lemma, we can conclude

Theorem 6.3. If L is an infinite regular language, then there are words u, v, w, such that $v \neq 0$ and $uv^{[i]}w \in L$ for $i = 0, 1, 2, 3, \ldots$.

This theorem is useful in showing that certain languages are *not* regular. However, for infinite regular languages we can say even more.

Theorem 6.4. Let $\mathcal{M}$ be a dfa with n states. Then $L(\mathcal{M})$ is infinite if and only if $L(\mathcal{M})$ contains a string x such that $n \leq |x| < 2n$.

Proof. First let $x \in L(\mathcal{M})$ with $n \leq |x| < 2n$. By the pumping lemma, we can write $x = uvw$, where $v \neq 0$ and $uv^{[i]}w \in L(\mathcal{M})$ for all i. But then $L(\mathcal{M})$ is infinite.

Conversely, let $L(\mathcal{M})$ be infinite. Then $L(\mathcal{M})$ must contain strings of length $\geq 2n$. Let $x \in L(\mathcal{M})$, where x has the shortest possible length $\geq 2n$. We write $x = x_1 x_2$, where $|x_1| = n$. Thus $|x_2| \geq n$. Then using the pigeon-hole principle as in the proof of the pumping lemma, we can write $x_1 = uvw$, where

$$\delta^*(q_1, u) = q,$$

$$\delta^*(q, v) = q \qquad \text{with} \quad 1 \leq |v| \leq n,$$

$$\delta^*(q, wx_2) \in F.$$

Thus $uwx_2 \in L(\mathcal{M})$. But

$$|uwx_2| \geq |x_2| \geq n,$$

and $|uwx_2| < |x|$, and since x was a shortest word of $L(\mathcal{M})$ with length at least $2n$, we have

$$n \leq |uwx_2| < 2n. \qquad ■$$

This theorem furnishes an algorithm for testing a given dfa $\mathcal{M}$ to determine whether $L(\mathcal{M})$ is finite. We need only run $\mathcal{M}$ on all strings x such that $n \leq |x| < 2n$, where $\mathcal{M}$ has n states. $L(\mathcal{M})$ is infinite just in case $\mathcal{M}$ accepts at least one of these strings.

For another example of an algorithm, let $\mathcal{M}_1, \mathcal{M}_2$ be dfas on the alphabet A and let us seek to determine whether $L(\mathcal{M}_1) \subseteq L(\mathcal{M}_2)$. Using the methods of proof of Theorems 4.2–4.4, we can obtain a dfa $\mathcal{M}$ such that

$$L(\mathcal{M}) = L(\mathcal{M}_1) \cap [A^* - L(\mathcal{M}_2)].$$

Then $L(\mathcal{M}_1) \subseteq L(\mathcal{M}_2)$ if and only if $L(\mathcal{M}) = \emptyset$. Since Theorem 6.2 enables us to test algorithmically whether $L(\mathcal{M}) = \emptyset$, we have an algorithm by means of which we can determine whether $L(\mathcal{M}_1) \subseteq L(\mathcal{M}_2)$. Moreover, since $L(\mathcal{M}_1) = L(\mathcal{M}_2)$ just when $L(\mathcal{M}_1) \subseteq L(\mathcal{M}_2)$ and $L(\mathcal{M}_2) \subseteq L(\mathcal{M}_1)$, we also have an algorithm for testing whether $L(\mathcal{M}_1) = L(\mathcal{M}_2)$.

The pumping lemma also furnishes a technique for showing that given languages are not regular. For example, let $L = \{a^{[n]}b^{[n]} \mid n > 0\}$, and suppose that $L = L(\mathcal{M})$, where $\mathcal{M}$ is a dfa with m states. We get a contradiction by showing that there is a word $x \in L$, with $|x| \geq m$, such that there is no way of writing $x = uvw$, with $v \neq 0$, so that $\{uv^{[i]}w \mid i \geq 0\} \subseteq L$. Let $x = a^{[l]}b^{[l]}$, where $2l \geq m$, and let $a^{[l]}b^{[l]} = uvw$. Then either $v = a^{[l_1]}$ or $v = a^{[l_1]}b^{[l_2]}$ or $v = b^{[l_2]}$, with $l_1, l_2 \leq l$, and in each case $uvvw \notin L$, contradicting the pumping lemma, so there can be no such dfa $\mathcal{M}$, and L is not regular.

This example and the exercises at the end of Section 7 show that finite automata are incapable of doing more than a limited amount of counting.

Exercises

1. Given a word w and a dfa $\mathcal{M}$, a test to determine if $w \in L(\mathcal{M})$ is a *membership test*.
   - **(a)** Let $\mathcal{M}_1, \mathcal{M}_2$ be arbitrary dfas on alphabet $A = \{s_1, \ldots, s_n\}$, where $\mathcal{M}_1$ has m_1 states and $\mathcal{M}_2$ has m_2 states. Give an upper bound $f(m_1, m_2)$ on the number of membership tests necessary to determine if $L(\mathcal{M}_1) = L(\mathcal{M}_2)$.
   - **(b)*** Verify that f is primitive recursive.
2. **(a)** Describe an algorithm that, for any regular expressions α and β, determines if $\langle \alpha \rangle = \langle \beta \rangle$.
   - **(b)** Give a function $g(x, y)$ such that the algorithm in part (a) requires at most $g(\text{size}(\alpha), \text{size}(\beta))$ membership tests. [See Exercise 5.9 for the definition of size(α).]
   - **(c)*** Verify that g is primitive recursive.

7. The Myhill – Nerode Theorem

We conclude this chapter by giving another characterization of the regular languages on an alphabet A. We begin with a pair of definitions.

Definition. Let $L \subseteq A^*$, where A is an alphabet. For strings $x, y \in A^*$, we write $x \equiv_L y$ to mean that for every $w \in A^*$ we have $xw \in L$ if and only if $yw \in L$.

It is obvious that $\equiv_L$ has the following properties.

$$x \equiv_L x.$$

$$\text{If } x \equiv_L y, \text{ then } y \equiv_L x.$$

$$\text{If } x \equiv_L y \text{ and } y \equiv_L z, \text{ then } x \equiv_L z.$$

(Relations having these three properties are known as *equivalence relations*.)

It is also obvious that

$$\text{If } x \equiv_L y, \text{ then for all } w \in A^*, \ xw \equiv_L yw.$$

Definition. Let $L \subseteq A^*$, where A is an alphabet. Let $S \subseteq A^*$. Then S is called a *spanning set for L* if

1. S is finite, and
2. for every $x \in A^*$, there is a $y \in S$ such that $x \equiv_L y$.

Then we can prove

Theorem 7.1 (Myhill–Nerode). A language is regular if and only if it has a spanning set.

Proof. First let L be regular. Then $L = L(\mathcal{M})$, where $\mathcal{M}$ is a dfa with set of states Q, initial state q_1, and transition function δ. Let us call a state $q \in Q$ *reachable* if there exists $y \in A^*$ such that

$$\delta^*(q_1, y) = q. \tag{7.1}$$

For each reachable state q, we select one particular string y that satisfies (7.1) and we write it as y_q. Thus,

$$\delta^*(q_1, y_q) = q$$

for every reachable state q. We set

$$S = \{y_q \mid q \text{ is reachable}\}.$$

S is clearly finite. To show that S is a spanning set for L, we let $x \in A^*$ and show how to find $y \in S$ such that $x \equiv_L y$. In fact, let $\delta^*(q_1, x) = q$, and set $y = y_q$. Thus, $y \in S$ and $\delta^*(q_1, y) = q$. Now for every $w \in A^*$,

$$\delta^*(q_1, xw) = \delta^*(q, w) = \delta^*(q_1, yw).$$

Hence, $\delta^*(q_1, xw) \in F$ if and only if $\delta^*(q_1, yw) \in F$; i.e., $xw \in L$ if and only if $yw \in L$. Thus, $x \equiv_L y$.

Conversely, let $L \subseteq A^*$ and let $S \subseteq A^*$ be a spanning set for L. We show how to construct a dfa $\mathcal{M}$ such that $L(\mathcal{M}) = L$. We define the set of states of $\mathcal{M}$ to be $Q = \{q_x \mid x \in S\}$, where we have associated a state q_x with each element $x \in S$. Since S is a spanning set for L, there is an $x_0 \in S$ such that $0 \equiv_L x_0$; we take q_{x_0} to be the initial state of $\mathcal{M}$. We let the final states of $\mathcal{M}$ be

$$F = \{q_y \mid y \in L\}.$$

Finally, for $a \in A$, we set $\delta(q_x, a) = q_y$, where $y \in S$ and $xa \equiv_L y$. Then we claim that for all $w \in A^*$,

$$\delta^*(q_x, w) = q_y, \qquad \text{where} \quad xw \equiv_L y.$$

We prove this claim by induction on $|w|$. For $|w| = 0$, we have $w = 0$. Moreover, $\delta^*(q_x, 0) = q_x$ and $x0 = x \equiv_L x$. Suppose our claim is known for all words w such that $|w| = k$, and consider $w \in A^*$ with $|w| = k + 1$. Then $w = ua$, where $|u| = k$ and $a \in A$. We have

$$\delta^*(q_x, w) = \delta(\delta^*(q_x, u), a) = \delta(q_y, a) = q_z,$$

where, using the induction hypothesis, $xu \equiv_L y$ and, by definition of δ, $ya \equiv_L z$. Then $xw = xua \equiv_L ya \equiv_L z$, which proves the claim. Now, we have

$$L(\mathcal{M}) = \{w \in A^* \mid \delta^*(q_{x_0}, w) \in F\}.$$

Let $\delta^*(q_{x_0}, w) = q_y$. Then by the way x_0 was defined and our claim,

$$w \equiv_L x_0 w \equiv_L y.$$

Thus, $w \in L$ if and only if $y \in L$, which in turn is true if and only if $q_y \in F$, i.e., if and only if $w \in L(\mathcal{M})$. Hence $L = L(\mathcal{M})$. ∎

Like the pumping lemma, the Myhill–Nerode theorem furnishes a technique for showing that a given language is not regular. For example, let $L = \{a^{[n]}b^{[n]} \mid n > 0\}$ again, and let n_1, n_2 be distinct numbers > 0. Then $a^{[n_1]}b^{[n_1]} \in L$ and $a^{[n_2]}b^{[n_1]} \notin L$, so $a^{[n_1]} \not\equiv_L a^{[n_2]}$, and since $\equiv_L$ is an equivalence relation, there can be no word w such that $a^{[n_1]} \equiv_L w$ and $a^{[n_2]} \equiv_L w$. But if there were a spanning set $S = \{w_1, \ldots, w_m\}$ for L, then by the pigeon-hole principle, there would have to be at least two distinct words among $\{a, aa, \ldots, a^{[m+1]}\}$, say $a^{[i]}$ and $a^{[j]}$, and some $w_k \in S$ such that $a^{[i]} \equiv_L w_k$ and $a^{[j]} \equiv_L w_k$, which is impossible. Therefore L has no spanning set, and by the Myhill–Nerode theorem, L is not regular.

Exercises

1. (a) For each language L described in Exercise 1.1, give a spanning set for L.

   (b) For each dfa $\mathcal{M}$ described in Exercise 1.3, give a spanning set for $L(\mathcal{M})$.

   (c) For each ndfa $\mathcal{M}$ described in Exercise 2.1, give a spanning set for $L(\mathcal{M})$.

2. Prove that there is no dfa that accepts exactly the set of all words that are palindromes over a given alphabet containing at least two symbols. (For a definition of palindrome, see Exercise 1.1f.)

3. u is called an *initial segment* of a word w if there is a word v such that $w = uv$. Let L be a regular language. Prove that the language consisting of all initial segments of words of L is a regular language.

4. Let L be a regular language and L' the language consisting of all words w such that both w and $w \cdot w$ are words in L. Prove that L' is regular.

5. Prove the following statement, if it is true, or give a counterexample: Every language that is a subset of a regular language is regular.

6. Prove that each of the following is *not* a regular language.

   (a) The language on the alphabet $\{a, b\}$ consisting of all strings in which the number of occurrences of b is greater than the number of occurrences of a.

   (b) The language L over the alphabet $\{., 0, 1, \ldots, 9\}$, consisting of all strings that are initial segments of the infinite decimal expansion of π. [$L = \{3, 3., 3.1, 3.14, 3.141, 3.1415, \ldots\}$.]

   (c) The language L over the alphabet $\{a, b\}$ consisting of all strings that are initial segments of the infinite string

   $$babaabaaabaaaab\ldots$$

7. Let $L = \{a^{[i]}b^{[j]} \mid i \neq j\}$. Show that L is not regular.

8. Let $L = \{a^{[n]}b^{[2n]} \mid n > 0\}$. Show that L is not regular.

9. Let $L = \{a^{[n]}b^{[m]} \mid 0 < n \leq m\}$. Show that L is not regular.

10. Let $L = \{a^{[p]} \mid p \text{ is a prime number}\}$. Show that L is not regular.

11. Let $\mathcal{M}$ be a finite automaton with alphabet A, set of states $Q = \{q_1, \ldots, q_n\}$, initial state q_1, and transition function δ. Let $a_1, a_2, a_3, \ldots$ be an *infinite sequence* of symbols of A. We can think of these symbols as being "fed" to $\mathcal{M}$ in the given order producing a sequence of states $r_1, r_2, r_3, \ldots$, where r_1 is just the initial state q_1 and $r_{i+1} = \delta(r_i, a_i)$, $i = 1, 2, 3, \ldots$. Suppose there are integers p, k such that

$$a_{i+p} = a_i \qquad \text{for all} \quad i \geq k.$$

Prove that there are integers l, s such that $s \leq np$ and

$$r_{i+s} = r_i \qquad \text{for all} \quad i \geq l.$$

[*Hint:* Use the pigeon-hole principle.]

12. (a) Let L be a regular language, and let S be a spanning set for L. S is a *minimal* spanning set for L if there is no spanning set for L that has fewer elements than S, and S is *independent* if there is no pair s, s' of distinct elements of S such that $s \equiv_L s'$. Prove that S is minimal if and only if it is independent.

(b) Let L be a regular language, and let S, S' be spanning sets for L. S and S' are *isomorphic* if there is a one–one function f from S onto S' such that $s \equiv_L f(s)$ for all $s \in S$. Prove that if S and S' are both minimal, then they are isomorphic.

(c) A dfa $\mathcal{M}$ is a *minimal* dfa for a regular language L if $L = L(\mathcal{M})$ and if there is no dfa $\mathcal{M}'$ with fewer states than $\mathcal{M}$ such that $L(\mathcal{M}') = L(\mathcal{M})$. Let L be a regular language, let $\mathcal{M}$ be a dfa that accepts L, and let S be a spanning set for L constructed from $\mathcal{M}$ as in the proof of Theorem 7.1. Prove that if $\mathcal{M}$ is a minimal dfa for L then S is a minimal spanning set for L. Why is the converse to this statement false?

(d) Let L be a regular language, let S be a spanning set for L, and let $\mathcal{M}$ be the dfa constructed from S as in the proof of Theorem 7.1. Prove that S is a minimal spanning set for L if and only if $\mathcal{M}$ is a minimal dfa for L.

(e) Let $\mathcal{M}$ and $\mathcal{M}'$ be dfas on alphabet A with states Q and Q', initial states q_1 and q_1', accepting states F and F', and transition functions δ and δ'. $\mathcal{M}$ and $\mathcal{M}'$ are *isomorphic* if there is a one–one function g from Q onto Q' such that $g(q_1) = q_1'$, $q \in F$ if and only if $g(q) \in F'$, and $\delta'(g(q), s) = g(\delta(q, s))$ for all $q \in Q$ and $s \in A$. (Informally, $\mathcal{M}$ and $\mathcal{M}'$ are identical but for a renaming of the states.) Prove that, if $\mathcal{M}$ and $\mathcal{M}'$ are both minimal dfas for some regular language L, then they are isomorphic.

13. Let L_1, L_2 be languages on some alphabet A. The *right quotient* of L_1 by L_2, denoted L_1/L_2, is $\{x \mid xy \in L_1 \text{ for some } y \in L_2\}$. Prove that if L_1 and L_2 are regular, then L_1/L_2 is regular.

14. Let $L = \{a^{[p]}b^{[m]} \mid p \text{ is a prime number}, m > 0\} \cup \{a^{[n]} \mid n \geq 0\}$.

(a) Show that L is not regular. [*Hint:* See Exercise 4.3 and Exercises 10 and 13 above.]

(b) Explain why the pumping lemma alone is not sufficient to show that L is not regular.

(c) State and prove a stronger version of the pumping lemma which is sufficient to show that L is not regular.

10

Context-Free Languages

1. Context-Free Grammars and Their Derivation Trees

Let $\mathscr{V}, T$ be a pair of disjoint alphabets. A *context-free production* on $\mathscr{V}, T$ is an expression

$$X \to h$$

where $X \in \mathscr{V}$ and $h \in (\mathscr{V} \cup T)^*$. The elements of $\mathscr{V}$ are called *variables*, and the elements of T are called *terminals*. If P stands for the production $X \to h$ and $u, v \in (\mathscr{V} \cup T)^*$, we write

$$u \underset{P}{\Rightarrow} v$$

to mean that there are words $p, q \in (\mathscr{V} \cup T)^*$ such that $u = pXq$ and $v = phq$. In other words, v results from u by replacing the variable X by the word h. Productions $X \to 0$ are called *null productions*. A *context-free grammar* Γ *with variables* $\mathscr{V}$ *and terminals* T consists of a finite set of context-free productions on $\mathscr{V}, T$ together with a designated symbol $S \in \mathscr{V}$ called the *start symbol*. Collectively, the set $\mathscr{V} \cup T$ is called the *alphabet* of Γ. If none of the productions of Γ is a null production, Γ is called a *positive context-free grammar*.[1]

[1] Those who have read Chapter 7 should note that every positive context-free grammar is a context-sensitive grammar in the sense defined there. For the moment we are not assuming familiarity with Chapter 7. However, the threads will all be brought together in the next chapter.

If Γ is a context-free grammar with variables $\mathscr{V}$ and terminals T, and if $u, v \in (\mathscr{V} \cup T)^*$, we write

$$u \underset{\Gamma}{\Rightarrow} v$$

to mean that $u \underset{P}{\Rightarrow} v$ for some production P of Γ. We write

$$u \overset{*}{\underset{\Gamma}{\Rightarrow}} v$$

to mean there is a sequence $u_1, \ldots, u_m$ where $u = u_1, u_m = v$, and

$$u_i \underset{\Gamma}{\Rightarrow} u_{i+1} \qquad \text{for} \quad 1 \leq i < m.$$

The sequence $u_1, \ldots, u_m$ is called a *derivation of v from u in* Γ. The number m is called the *length* of the derivation.[2] The symbol Γ below the $\Rightarrow$ may be omitted when no ambiguity results. Finally, we define

$$L(\Gamma) = \{u \in T^* \mid S \overset{*}{\Rightarrow} u\}.$$

$L(\Gamma)$ is called the language *generated* by Γ. A language $L \subseteq T^*$ is called *context-free* if there is a context-free grammar Γ such that $L = L(\Gamma)$.

A simple example of a context-free grammar Γ is given by $\mathscr{V} = \{S\}$, $T = \{a, b\}$, and the productions

$$S \to aSb, \qquad S \to ab.$$

Here we clearly have

$$L(\Gamma) = \{a^{[n]}b^{[n]} \mid n > 0\};$$

thus, this language is context-free. We showed in Chapter 9, Section 6, that $L(\Gamma)$ is not regular. Later we shall see that every regular language is context-free. For the meanwhile we have proved

Theorem 1.1. The language $L = \{a^{[n]}b^{[n]} \mid n > 0\}$ is context-free but not regular.

We now wish to discuss the relation between context-free grammars in general and positive context-free grammars. It is obvious that if Γ is a positive context-free grammar, then $0 \notin L(\Gamma)$. We shall show that except for this limitation, everything that can be done using context-free grammars can be done with positive context-free grammars. This will require some messy technicalities, but working out the details now will simplify matters later.

[2] Some authors use the number $m - 1$ as the length of the derivation.

Definition. We define the *kernel* of a given context-free grammar Γ, written $\ker(\Gamma)$, to be the set of variables V of Γ such that $V \overset{*}{\underset{\Gamma}{\Rightarrow}} 0$.

As an example consider the context-free grammar Γ_0 with productions

$$S \to XYYX, \qquad S \to aX, \qquad X \to 0, \qquad Y \to 0.$$

Then $\ker(\Gamma_0) = \{X, Y, S\}$. This example suggests an algorithm for locating the elements of $\ker(\Gamma)$ for a given context-free grammar Γ. We let

$$\mathscr{V}_0 = \{V \mid V \to 0 \text{ is a production of } \Gamma\},$$

$$\mathscr{V}_{i+1} = \mathscr{V}_i \cup \{V \mid V \to \alpha \text{ is a production of } \Gamma, \text{ where } \alpha \in \mathscr{V}_i^*\}.$$

Thus for Γ_0, $\mathscr{V}_0 = \{X, Y\}$, $\mathscr{V}_1 = \{X, Y, S\}$, and $\mathscr{V}_i = \mathscr{V}_1$ for all $i > 1$. S is in $\mathscr{V}_1$ because $XYYX \in \mathscr{V}_0^*$. In the general case it is clear, because Γ has only finitely many variables, that a stage k will eventually be reached for which $\mathscr{V}_{k+1} = \mathscr{V}_k$ and that then $\mathscr{V}_i = \mathscr{V}_k$ for all $i > k$. We have

Lemma 1. If $\mathscr{V}_k = \mathscr{V}_{k+1}$, then $\ker(\Gamma) = \mathscr{V}_k$.

Proof. It is clear that $\mathscr{V}_i \subseteq \ker(\Gamma)$ for all i. Conversely, we show that if $V \in \ker(\Gamma)$, then $V \in \mathscr{V}_k$. We prove this by induction on the length of a derivation of 0 from V in Γ. If there is such a derivation of length 2, then $V \underset{\Gamma}{\Rightarrow} 0$, so that $V \to 0$ is a production of Γ and $V \in \mathscr{V}_0$. Let us assume the result for all derivations of length $< r$ and let $V = \alpha_1 \Rightarrow \alpha_2 \Rightarrow \cdots \Rightarrow \alpha_{r-1} \Rightarrow \alpha_r = 0$ be a derivation of length r in Γ. The words $\alpha_1, \alpha_2, \ldots, \alpha_{r-1}$ must consist entirely of variables, since terminals cannot be eliminated by context-free productions. Let $\alpha_2 = V_1 V_2 \cdots V_s$. Then we have $V_i \overset{*}{\underset{\Gamma}{\Rightarrow}} 0$, $i = 1, 2, \ldots, s$, by derivations of length $< r$. By the induction hypothesis, each $V_i \in \mathscr{V}_k$. Since Γ contains the production $V \to V_1 V_2 \cdots V_s$, and $\alpha_2 \in \mathscr{V}_k^*$, we have $V \in \mathscr{V}_{k+1} = \mathscr{V}_k$. ■

Lemma 2. There is an algorithm that will transform a given context-free grammar Γ into a positive context-free grammar $\bar{\Gamma}$ such that $L(\Gamma) = L(\bar{\Gamma})$ or $L(\Gamma) = L(\bar{\Gamma}) \cup \{0\}$.

Proof. We begin by computing $\ker(\Gamma)$. Then we obtain $\bar{\Gamma}$ by first adding all productions that can be obtained from the productions of Γ by deleting from their righthand sides one or more variables belonging to $\ker(\Gamma)$ and by then deleting all productions (old and new) of the form $V \to 0$. (In our example, $\bar{\Gamma}_0$ would have the productions $S \to XYYX$, $S \to aX$, $S \to a$, $S \to YYX$, $S \to XYX$, $S \to XYY$, $S \to XY$, $S \to YY$, $S \to YX$, $S \to XX$, $S \to X$, $S \to Y$.) We shall show that $L(\Gamma) = L(\bar{\Gamma})$ or $L(\Gamma) = L(\bar{\Gamma}) \cup \{0\}$.

Let $V \to \beta_1 \beta_2 \cdots \beta_s$ be a production of $\bar{\Gamma}$ that is not a production of Γ, where $\beta_1, \beta_2, \ldots, \beta_s \in (\mathscr{V} \cup T)$, and where this production was obtained from a production of Γ of the form

$$V \to u_0 \beta_1 u_1 \beta_2 \cdots \beta_s u_s,$$

with $u_0, u_1, u_2, \ldots, u_s \in (\ker(\Gamma))^*$. [Of course, u_0, u_s might be 0. But since $0 \in (\ker(\Gamma))^*$, this creates no difficulty.] Now,

$$u_i \overset{*}{\underset{\Gamma}{\Rightarrow}} 0, \qquad i = 0, 1, 2, \ldots, s,$$

so that

$$V \underset{\Gamma}{\Rightarrow} u_0 \beta_1 u_1 \beta_2 \cdots u_{s-1} \beta_s u_s \overset{*}{\underset{\Gamma}{\Rightarrow}} \beta_1 \beta_2 \cdots \beta_s .$$

Thus, the effect of this new production of $\bar{\Gamma}$ can be simulated in Γ. This proves that $L(\bar{\Gamma}) \subseteq L(\Gamma)$.

It remains to show that if $v \in L(\Gamma)$ and $v \neq 0$, then $v \in L(\bar{\Gamma})$. Let T be the set of terminals of Γ (and also of $\bar{\Gamma}$). We shall prove by induction the stronger assertion:

For any variable V, if $V \overset{*}{\underset{\Gamma}{\Rightarrow}} w \neq 0$ for $w \in T^*$, then $V \overset{*}{\underset{\bar{\Gamma}}{\Rightarrow}} w$.

If in fact $V \underset{\Gamma}{\Rightarrow} w$, then Γ contains the production $V \to w$ which is also a production of $\bar{\Gamma}$. Otherwise we may write

$$V \underset{\Gamma}{\Rightarrow} w_0 V_1 w_1 V_2 w_2 \cdots V_s w_s \overset{*}{\underset{\Gamma}{\Rightarrow}} w,$$

where $V_1, \ldots, V_s$ are variables and $w_0, w_1, w_2, \ldots, w_s$ are (possible null) words on the terminals. Then w can be written

$$w = w_0 v_1 w_1 v_2 w_2 \cdots v_s w_s,$$

where

$$V_i \overset{*}{\underset{\Gamma}{\Rightarrow}} v_i, \qquad i = 1, 2, \ldots, s.$$

Since each v_i must have a shorter derivation from V_i than w has from V, we may proceed inductively by assuming that for each v_i which is not 0, $V_i \overset{*}{\underset{\bar{\Gamma}}{\Rightarrow}} v_i$. On the other hand, if $v_i = 0$, then $V_i \in \ker(\Gamma)$. We set

$$V_i^0 = \begin{cases} 0 & \text{if } \ v_i = 0 \\ V_i & \text{otherwise.} \end{cases}$$

Then $V \to w_0 V_1^0 w_1 V_2^0 w_2 \cdots V_s^0 w_s$ is one of the productions of $\bar{\Gamma}$. Hence we have

$$V \underset{\bar{\Gamma}}{\Rightarrow} w_0 V_1^0 w_1 V_2^0 w_2 \cdots V_s^0 w_s \overset{*}{\underset{\bar{\Gamma}}{\Rightarrow}} w_0 v_1 w_1 v_2 w_2 \cdots v_s w_s = w. \qquad \blacksquare$$

We can now easily prove

Theorem 1.2. A language L is context-free if and only if there is a positive context-free grammar Γ such that

$$L = L(\Gamma) \text{ or } L = L(\Gamma) \cup \{0\}. \tag{1.1}$$

Moreover, there is an algorithm that will transform a context-free grammar Δ for which $L = L(\Delta)$ into a positive context-free grammar Γ that satisfies (1.1).

Proof. If L is context-free with $L = L(\Delta)$ for a context-free grammar Δ, then we can use the algorithm of Lemma 2 to construct a positive context-free grammar Γ such that $L = L(\Gamma)$ or $L = L(\Gamma) \cup \{0\}$.

Conversely, if Γ is a positive context-free grammar and $L = L(\Gamma)$, there is nothing to prove since a positive context-free grammar is already a context-free grammar. If $L = L(\Gamma) \cup \{0\}$, let S be the start symbol of Γ and let $\tilde{\Gamma}$ be the context-free grammar obtained from Γ by introducing $\tilde{S}$ as a new start symbol and adding the productions

$$\tilde{S} \to S, \qquad \tilde{S} \to 0.$$

Clearly, $L(\tilde{\Gamma}) = \mathrm{L}(\Gamma) \cup \{0\}$. ■

Now, let Γ be a *positive* context-free grammar with alphabet $T \cup \mathscr{V}$, where T consists of the terminals and $\mathscr{V}$ is the set of variables. We will make use of *trees* consisting of a finite number of points called *nodes* or *vertices*, each of which is labeled by a letter of the alphabet, i.e., an element of $T \cup \mathscr{V}$. Certain vertices will have other nodes as *immediate successors*, and the immediate successors of a given node are to be in some definite order. It is helpful (though of course not part of the formal development) to think of the immediate successors of a given node as being physically *below* the given node and arranged from left to right in their given order. Nodes are to be connected by line segments to their immediate successors. There is to be exactly one node which is not an immediate successor; this node is called the *root*. Each node other than the root is to be the immediate successor of precisely one node, its *predecessor*. Nodes which have no immediate successors are called *leaves*.

A tree is called a Γ-*tree* if it satisfies the following conditions:

1. the root is labeled by a variable;
2. each vertex which is not a leaf is labeled by a variable;
3. if a vertex is labeled X and its immediate successors are labeled $\alpha_1, \alpha_2, \ldots, \alpha_k$ (reading from left to right), then $X \to \alpha_1 \alpha_2 \cdots \alpha_k$ is a production of Γ.

Let $\mathscr{T}$ be a Γ-tree, and let ν be a vertex of $\mathscr{T}$ which is labeled by the variable X. Then we shall speak of the *subtree* $\mathscr{T}^{\nu}$ *of* $\mathscr{T}$ *determined by* ν. The vertices of $\mathscr{T}^{\nu}$ are ν, its immediate successors in $\mathscr{T}$, their immediate successors, and so on. The vertices of $\mathscr{T}^{\nu}$ are labeled exactly as they are in $\mathscr{T}$. (In particular, the root of $\mathscr{T}^{\nu}$ is ν which is labeled X.) Clearly, $\mathscr{T}^{\nu}$ is itself a Γ-tree.

If $\mathscr{T}$ is a Γ-tree, we write $\langle \mathscr{T} \rangle$ for the word that consists of *the labels of the leaves of* $\mathscr{T}$ reading from left to right (a vertex to the left of a given node is regarded as also being to the left of each of its immediate successors). If the root of $\mathscr{T}$ is labeled by the start symbol S of Γ and if $w = \langle \mathscr{T} \rangle$, then $\mathscr{T}$ is called a *derivation tree for* w *in* Γ. Thus the tree shown in Fig. 1.1 is a derivation tree for $a^{[4]}b^{[3]}$ in the grammar shown in the same figure.

Theorem 1.3. If Γ is a positive context-free grammar, and $S \overset{*}{\underset{\Gamma}{\Rightarrow}} w$, then there is a derivation tree for w in Γ.

Proof. Our proof is by induction on the length of a derivation of w from S in Γ. If this length is 1, then $w = S$ and the required derivation tree consists of a single vertex labeled S (being both root and leaf).

Now let w have a derivation from S of length $r + 1$, where the result is known for derivations of length r. Then we have $S \overset{*}{\Rightarrow} v \Rightarrow w$ with $v, w \in (\mathscr{V} \cup T)^*$, where the induction hypothesis applies to the derivation $S \overset{*}{\Rightarrow} v$. Thus, we may assume that we have a derivation tree for v. Now since $v \Rightarrow w$, we must have $v = xXy$ and $w = x\alpha_1 \cdots \alpha_k y$, where Γ contains the production $X \to \alpha_1 \cdots \alpha_k$. Then the derivation tree for v can be extended to yield a derivation tree for w simply by giving k immediate successors to the node labeled X, labeled $\alpha_1, \ldots, \alpha_k$ from left to right. ■

Before considering the converse of Theorem 1.3, it will be helpful to consider the following derivations of $a^{[4]}b^{[3]}$ from S with respect to the grammar indicated in Fig. 1.1:

1. $S \Rightarrow aXbY \Rightarrow a^{[2]}XbY \Rightarrow a^{[3]}XbY \Rightarrow a^{[4]}bY \Rightarrow a^{[4]}b^{[2]}Y \Rightarrow a^{[4]}b^{[3]}$
2. $S \Rightarrow aXbY \Rightarrow a^{[2]}XbY \Rightarrow a^{[2]}Xb^{[2]}Y \Rightarrow a^{[3]}Xb^{[2]}Y \Rightarrow a^{[3]}Xb^{[3]} \Rightarrow a^{[4]}b^{[3]}$
3. $S \Rightarrow aXbY \Rightarrow aXb^{[2]}Y \Rightarrow aXb^{[3]} \Rightarrow a^{[2]}Xb^{[3]} \Rightarrow a^{[3]}Xb^{[3]} \Rightarrow a^{[4]}b^{[3]}$.

Now, if the proof of Theorem 1.3 is applied to these three derivations, the very same derivation tree is obtained—namely, the one shown in Fig. 1.1. This shows that there does not exist a one–one correspondence between derivations and derivation trees, but that rather, several derivations may give rise to the same tree. Hence, there is no unique derivation which we can hope to be able to read off a given derivation tree.

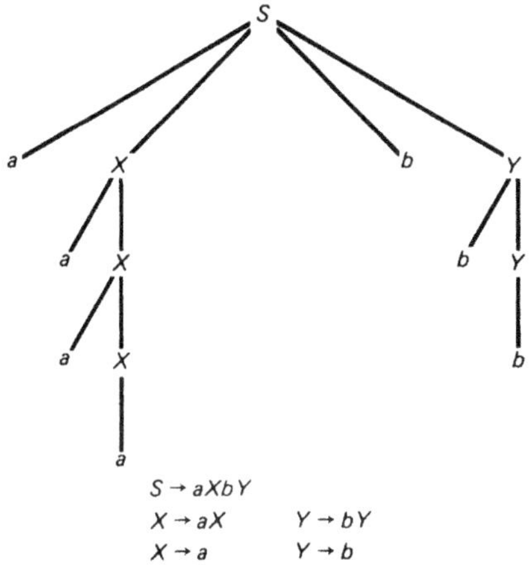

Figure 1.1. A derivation tree for $a^{[4]}b^{[3]}$ in the indicated grammar.

Definition. We write $u \Rightarrow_l v$ (in Γ) if $u = xXy$ and $v = xzy$, where $X \to z$ is a production of Γ and $x \in T^*$. If, instead, $x \in (T \cup \mathscr{V})^*$ but $y \in T^*$, we write $u \Rightarrow_r v$.

Thus, when $u \Rightarrow_l v$, it is the *leftmost* variable in u for which a substitution is made, whereas when $u \Rightarrow_r v$, it is the *rightmost* variable in u. A derivation

$$u_1 \Rightarrow_l u_2 \Rightarrow_l u_3 \Rightarrow_l \cdots \Rightarrow_l u_n$$

is called a *leftmost* derivation, and then we write $u_1 \overset{*}{\Rightarrow}_l u_n$. Similarly, a derivation

$$u_1 \Rightarrow_r u_2 \Rightarrow_r u_3 \Rightarrow_r \cdots \Rightarrow_r u_n$$

is called a *rightmost* derivation, and we write $u_1 \overset{*}{\Rightarrow}_r u_n$. In the preceding examples of derivations of $a^{[4]}b^{[3]}$ from S in the grammar of Fig. 1.1, 1 is leftmost, 3 is rightmost, and 2 is neither.

Now we shall see how, given a derivation tree $\mathscr{T}$ for a word $w \in T^*$, we can obtain a *leftmost derivation of w from S* and a *rightmost derivation of w from S*. Let the word which consists of the labels of the immediate successors of the root of $\mathscr{T}$ (reading from left to right) be $v_0X_1v_1X_2 \cdots X_rv_r$, where $v_0, v_1, \ldots, v_r \in T^*$, $X_1, X_2, \ldots, X_r \in \mathscr{V}$, and $X_1, X_2, \ldots, X_r$ label the vertices $\nu_1, \ldots, \nu_r$, which are immediate successors of the root of $\mathscr{T}$. (Of course, some of the v_i may be 0.) Then $S \to v_0X_1v_1X_2 \cdots X_rv_r$ is one of the productions of Γ. Now it is possible that the immediate successors of the root of $\mathscr{T}$ are all leaves; this is precisely the case where $w = v_0$ and $r = 0$. If this is the case, then we have $S \Rightarrow_l w$ and $S \Rightarrow_r w$, so that we do have a leftmost as well as a rightmost derivation of w from S.

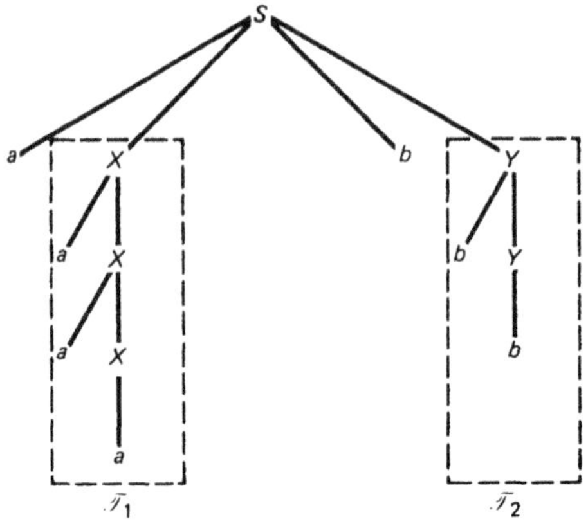

Figure 1.2. Decomposition of the tree of Fig. 1.1 as in the proof of the existence of leftmost and rightmost derivations.

Otherwise, i.e., for $r > 0$, we consider the trees $\mathcal{T}_i = \mathcal{T}^{\nu_i}$, $i = 1, 2, \ldots, r$. Here $\mathcal{T}_i$ has its root ν_i labeled X_i and is made up of the part of $\mathcal{T}$ consisting of ν_i, its immediate successors, their immediate successors, etc. (see Fig. 1.2). Let Γ_i be the grammar whose productions and alphabet are the same as for Γ but which has start symbol X_i. Then $\mathcal{T}_i$ is a derivation tree in Γ_i. Let $\mathcal{T}_i$ be a derivation tree for w_i in Γ_i. Then, clearly,

$$w = v_0 w_1 v_1 w_2 v_2 \cdots w_r v_r .$$

Moreover, since each $\mathcal{T}_i$ contains fewer vertices than $\mathcal{T}$, we may assume inductively that for $i = 1, 2, \ldots, r$

$$X_i \overset{*}{\Rightarrow}_l w_i \qquad \text{and} \qquad X_i \overset{*}{\Rightarrow}_r w_i .$$

Hence we have

$$\begin{aligned} S &\Rightarrow_l v_0 X_1 v_1 X_2 \cdots X_r v_r \\ &\overset{*}{\Rightarrow}_l v_0 w_1 v_1 X_2 \cdots X_r v_r \\ &\overset{*}{\Rightarrow}_l v_0 w_1 v_1 w_2 \cdots X_r v_r \\ &\vdots \\ &\overset{*}{\Rightarrow}_l v_0 w_1 v_1 w_2 \cdots w_r v_r = w \end{aligned}$$

and

$$\begin{aligned} S &\Rightarrow_r v_0 X_1 v_1 X_2 \cdots X_r v_r \\ &\overset{*}{\Rightarrow}_r v_0 X_1 v_1 X_2 \cdots w_r v_r \\ &\vdots \\ &\overset{*}{\Rightarrow}_r v_0 X_1 v_1 w_2 \cdots w_r v_r \\ &\overset{*}{\Rightarrow}_r v_0 w_1 v_1 w_2 \cdots w_r v_r = w. \end{aligned}$$

So we have shown how to obtain a leftmost and a rightmost derivation of w from S in Γ.

Now, Theorem 1.3 tells us that if $w \in L(\Gamma)$, there is a derivation tree for w in Γ. And we have just seen that if there is a derivation tree for w in Γ, then there are both leftmost and rightmost derivations of w from S in Γ [so that, in particular, $w \in L(\Gamma)$]. Putting all of this information together we have

Theorem 1.4. Let Γ be a positive context-free grammar with start symbol S and terminals T. Let $w \in T^*$. Then the following conditions are equivalent:

1. $w \in L(\Gamma)$;
2. there is a derivation tree for w in Γ;
3. there is a leftmost derivation of w from S in Γ;
4. there is a rightmost derivation of w from S in Γ.

Definition. A positive context-free grammar is called *branching* if it has no productions of the form $X \to Y$, where X and Y are variables.

For a derivation tree in a branching grammar Γ, each vertex that is not a leaf cannot be the *only* immediate successor of its predecessor. Since we shall find it useful to work with branching grammars, we prove

Theorem 1.5. There is an algorithm that transforms a given positive context-free grammar Γ into a branching context-free grammar Δ such that $L(\Delta) = L(\Gamma)$.

Proof. Let $\mathscr{V}$ be the set of variables of Γ. First suppose that Γ contains productions

$$X_1 \to X_2, \qquad X_2 \to X_3, \qquad \ldots, \qquad X_k \to X_1, \tag{1.2}$$

where $k \geq 1$ and $X_1, X_2, \ldots, X_k \in \mathscr{V}$. Then, we can eliminate the productions (1.2) and replace each variable X_i in the remaining productions of Γ by the new variable X. (If one of $X_1, \ldots, X_k$ is the start symbol, then X must now be the start symbol.) Obviously the language generated is not changed by this transformation.

Thus, we need consider only the case where no "cycles" like (1.2) occur in Γ. If Γ is not branching, it must contain a production $X \to Y$ such that Γ contains no productions of the form $Y \to Z$. We eliminate the production $X \to Y$, but add to Γ productions $X \to x$ for each word $x \in (\mathscr{V} \cup T)^*$ for which $Y \to x$ is a production of Γ. Again the language generated is unchanged, but the number of productions that Γ contains of the form

$U \to V$ has been decreased. Iterating this process we arrive at a grammar Δ containing no productions of the form $U \to V$, which is therefore of the required form. ■

A *path* in a Γ-tree $\mathcal{T}$ is a sequence $\alpha_1, \alpha_2, \ldots, \alpha_k$ of vertices of $\mathcal{T}$ such that α_{i+1} is an immediate successor of α_i for $i = 1, 2, \ldots, k-1$. All of the vertices on the path are called *descendants* of α_1.

A particularly interesting situation arises when two different vertices α, β lie on the same path in the derivation tree $\mathcal{T}$ and are labeled by the same variable X. In such a case one of the vertices is a descendant of the other, say, β is a descendant of α. $\mathcal{T}^\beta$ is then not only a subtree of $\mathcal{T}$ but also of $\mathcal{T}^\alpha$. [In fact, $(\mathcal{T}^\alpha)^\beta = \mathcal{T}^\beta$.] We wish to consider two important

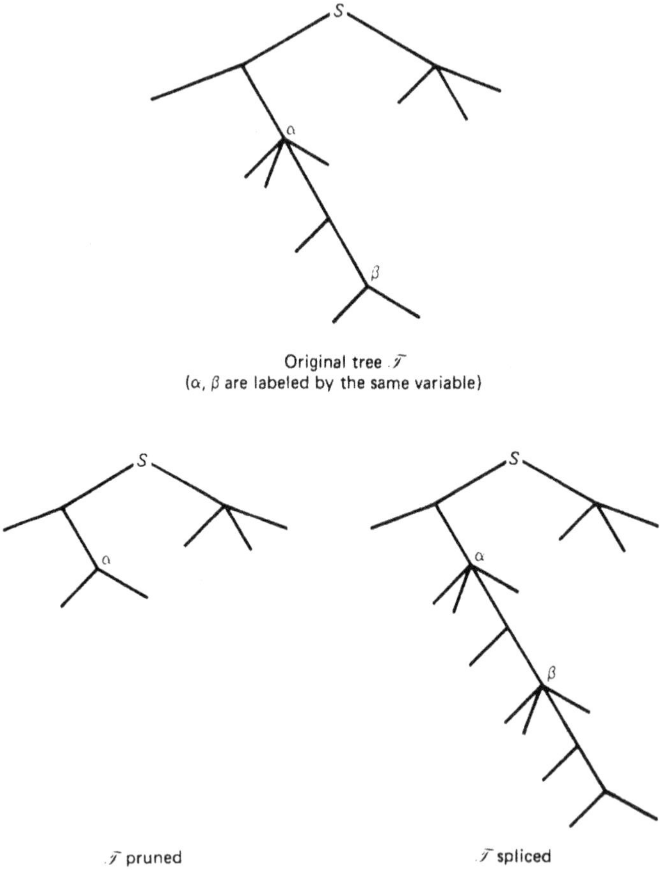

Figure 1.3

operations on the derivation tree $\mathcal{T}$ which can be performed in this case. The first operation, which we call *pruning*, is to remove the subtree $\mathcal{T}^\alpha$ from the vertex α and to *graft* the subtree $\mathcal{T}^\beta$ in its place. The second operation, which we call *splicing*, is to remove the subtree $\mathcal{T}^\beta$ from the vertex β and to *graft* an exact copy of $\mathcal{T}^\alpha$ in its place. (See Fig. 1.3.) *Because α and β are labeled by the same variable, the trees obtained by pruning and splicing are themselves derivation trees.*

Let $\mathcal{T}_\text{p}$ and $\mathcal{T}_\text{s}$ be trees obtained from a derivation tree $\mathcal{T}$ in a *branching* grammar by pruning and splicing, respectively, where α and β are as before. We have $\langle\mathcal{T}\rangle = r_1\langle\mathcal{T}^\alpha\rangle r_2$ for words r_1, r_2 and $\langle\mathcal{T}^\alpha\rangle = q_1\langle\mathcal{T}^\beta\rangle q_2$ for words q_1, q_2. Since α, β are distinct vertices, and since the grammar is branching, q_1 and q_2 cannot both be 0. (That is, $q_1 q_2 \neq 0$.) Also,

$$\langle\mathcal{T}_\text{p}\rangle = r_1\langle\mathcal{T}^\beta\rangle r_2 \quad \text{and} \quad \langle\mathcal{T}_\text{s}\rangle = r_1 q_1^{[2]}\langle\mathcal{T}^\beta\rangle q_2^{[2]} r_2. \tag{1.3}$$

Since $q_1 q_2 \neq 0$, we have $|\langle\mathcal{T}^\beta\rangle| < |\langle\mathcal{T}^\alpha\rangle|$ and hence $|\langle\mathcal{T}_\text{p}\rangle| < |\langle\mathcal{T}\rangle|$. From this last inequality and Theorem 1.4, we can easily infer

Theorem 1.6. Let Γ be a branching context-free grammar, let $u \in L(\Gamma)$, and let u have a derivation tree $\mathcal{T}$ in Γ that has two different vertices on the same path labeled by the same variable. Then there is a word $v \in L(\Gamma)$ such that $|v| < |u|$.

Proof. Since $u = \langle\mathcal{T}\rangle$, we need only take $v = \langle\mathcal{T}_\text{p}\rangle$. ■

Exercises

1. Find a context-free grammar generating the set of arithmetic statements of Pascal (or FORTRAN).
2. Consider the grammar Γ with start symbol S and productions

$$S \to XXYY \qquad X \to XX \qquad Y \to YY$$
$$X \to a \qquad Y \to b.$$

Show that Γ generates the same language as the grammar of Fig. 1.1.

3. Show that $\varnothing$ is a context-free language.
4. Give three languages that are context-free but not regular. Justify your answer.
5. Give a context-free grammar Γ such that $\ker(\Gamma) = \mathscr{V}_4$ and $\mathscr{V}_i \neq \ker(\Gamma)$, $i = 1, 2, 3$.

6. Let Γ be a context-free grammar with productions $X_1 \to \alpha_1, \ldots, X_n \to \alpha_n$. We define the *width* of Γ as $\Sigma_{i=1}^{n} |\alpha_i|$.
   (a) Give a function $f(w)$ such that for any context-free grammar Γ with width w, there is a positive context-free grammar $\bar{\Gamma}$ such that $L(\Gamma) = L(\bar{\Gamma})$ or $L(\Gamma) = L(\bar{\Gamma}) \cup \{0\}$ and $\bar{\Gamma}$ has no more than $f(w)$ productions.
   (b) Give a grammar Γ with width w for which any such $\bar{\Gamma}$ has at least $f(w)/2$ productions.
7. (a) Let Γ be the grammar in Exercise 2. Give two different derivation trees for *aaabb*. From each tree obtain a leftmost and a rightmost derivation of *aaabb* from S.
   (b) Let Γ' be the grammar in Fig. 1.1. Prove that for every $w \in L(\Gamma')$, there is a unique derivation tree for w in Γ'.
8. Let Γ be the grammar with productions

$$\begin{array}{llll} S \to VW & W \to aW & X \to S & Y \to X \\ S \to W & W \to X & X \to W & Y \to c \\ V \to bX & W \to Y & X \to Z & Z \to V \\ V \to b & & & \end{array}$$

   and start symbol S. Use the construction in the proof of Theorem 1.5 to give a branching context-free grammar Δ such that $L(\Delta) = L(\Gamma)$. Can any of the resulting productions be eliminated from Δ?

2. Regular Grammars

We shall now see that regular languages are generated by context-free grammars of an especially simple form.

Definition. A context-free grammar is called *regular* if each of its productions has one of the two forms

$$U \to aV \quad \text{or} \quad U \to a,$$

where U, V are variables and a is a terminal.

Then we have

Theorem 2.1. If L is a regular language, then there is a regular grammar Γ such that either $L = L(\Gamma)$ or $L = L(\Gamma) \cup \{0\}$.

Proof. Let $L = L(\mathcal{M})$, where $\mathcal{M}$ is a dfa with states $q_1, \ldots, q_m$, alphabet $\{s_1, \ldots, s_n\}$, transition function δ, and set of accepting states F. We construct a grammar Γ with variables $q_1, \ldots, q_m$, terminals $s_1, \ldots, s_n$, and start symbol q_1. The productions are

1. $q_i \to s_r q_j$ whenever $\delta(q_i, s_r) = q_j$, and
2. $q_i \to s_r$ whenever $\delta(q_i, s_r) \in F$.

Clearly the grammar Γ is regular. We shall show that $L(\Gamma)$ is just $L - \{0\}$.

First, suppose $u \in L$, $u \neq 0$; let $u = s_{i_1} s_{i_2} \cdots s_{i_l} s_{i_{l+1}}$. Thus, $\delta^*(q_1, u) \in F$, so that we have

$$\delta(q_1, s_{i_1}) = q_{j_1}, \quad \delta(q_{j_1}, s_{i_2}) = q_{j_2}, \quad \ldots, \quad \delta(q_{j_l}, s_{i_{l+1}}) = q_{j_{l+1}} \in F. \tag{2.1}$$

Hence, the grammar Γ contains the productions

$$q_1 \to s_{i_1} q_{j_1}, \quad q_{j_1} \to s_{i_2} q_{j_2}, \quad \ldots, \quad q_{j_{l-1}} \to s_{i_l} q_{j_l}, \quad q_{j_l} \to s_{i_{l+1}}. \tag{2.2}$$

Thus, we have in Γ

$$\begin{aligned} q_1 &\Rightarrow s_{i_1} q_{j_1} \\ &\Rightarrow s_{i_1} s_{i_2} q_{j_2} \\ &\vdots \\ &\Rightarrow s_{i_1} s_{i_2} \cdots s_{i_l} q_{j_l} \\ &\Rightarrow s_{i_1} s_{i_2} \cdots s_{i_l} s_{i_{l+1}} = u, \end{aligned} \tag{2.3}$$

so that $u \in L(\Gamma)$.

Conversely, suppose that $u \in L(\Gamma)$, $u = s_{i_1} s_{i_2} \cdots s_{i_l} s_{i_{l+1}}$. Then there is a derivation of u from q_1 in Γ, which must be of the form (2.3). Hence, the productions listed in (2.2) must belong to Γ, and finally, the transitions (2.1) must hold in $\mathscr{M}$. Thus, $u \in L(\mathscr{M})$. ∎

Theorem 2.2. Let Γ be a regular grammar. Then $L(\Gamma)$ is a regular language.

Proof. Let Γ have the *variables* $V_1, V_2, \ldots, V_K$, where $S = V_1$ is the start symbol, and the *terminals* $s_1, \ldots, s_n$. Since Γ is assumed to be regular, its productions are of the form $V_i \to s_r V_j$ and $V_i \to s_r$. We shall construct an ndfa $\mathscr{M}$ which accepts precisely $L(\Gamma)$.

The states of $\mathscr{M}$ will be $V_1, V_2, \ldots, V_K$ and an additional state W. V_1 will be the initial state and W will be the only accepting state, i.e., $F = \{W\}$. Let

$$\delta_1(V_i, s_r) = \{V_j \mid V_i \to s_r V_j \text{ is a production of } \Gamma\},$$

$$\delta_2(V_i, s_r) = \begin{cases} \{W\} & \text{if } V_i \to s_r \text{ is a production of } \Gamma \\ \varnothing & \text{otherwise.} \end{cases}$$

Then we take as the transition function δ of $\mathscr{M}$

$$\delta(V_i, s_r) = \delta_1(V_i, s_r) \cup \delta_2(V_i, s_r).$$

This completes the specification of $\mathscr{M}$.

Now let $u = s_{i_1} s_{i_2} \cdots s_{i_l} s_{i_{l+1}} \in L(\Gamma)$. Thus, we must have

$$V_1 \Rightarrow s_{i_1} V_{j_1} \Rightarrow s_{i_1} s_{i_2} V_{j_2} \overset{*}{\Rightarrow} s_{i_1} s_{i_2} \cdots s_{i_l} V_{j_l} \Rightarrow s_{i_1} s_{i_2} \cdots s_{i_l} s_{i_{l+1}}, \tag{2.4}$$

where Γ contains the productions

$$\begin{aligned} V_1 &\to s_{i_1} V_{j_1}, \\ V_{j_1} &\to s_{i_2} V_{j_2}, \\ &\vdots \\ V_{j_{l-1}} &\to s_{i_l} V_{j_l}, \\ V_{j_l} &\to s_{i_{l+1}}. \end{aligned} \tag{2.5}$$

Thus,

$$\begin{aligned} V_{j_1} &\in \delta(V_1, s_{i_1}), \\ V_{j_2} &\in \delta(V_{j_1}, s_{i_2}), \\ &\vdots \\ V_{j_l} &\in \delta(V_{j_{l-1}}, s_{i_l}), \\ W &\in \delta(V_{j_l}, s_{i_{l+1}}). \end{aligned} \tag{2.6}$$

Thus, $W \in \delta^*(V_1, u)$ and $u \in L(\mathscr{M})$.

Conversely, if $u = s_{i_1} s_{i_2} \cdots s_{i_l} s_{i_{l+1}}$ is accepted by $\mathscr{M}$, then there must be a sequence of transitions of the form (2.6). Hence, the productions of (2.5) must all belong to Γ, so that there is a derivation of the form (2.4) of u from V_1. ■

In order to combine Theorems 2.1 and 2.2 in a single equivalence, it is necessary to show only that if L is a regular language, then so is $L \cup \{0\}$. But this follows at once from Theorems 4.2 and 4.5 in Chapter 9.

Combining Theorems 2.1 and 2.2 with this discussion, we have

Theorem 2.3. A language L is regular if and only if there is a regular grammar Γ such that either $L = L(\Gamma)$ or $L = L(\Gamma) \cup \{0\}$.

Since regular grammars are context-free grammars, we have

Corollary 2.4. Every regular language is context-free.

The converse of Corollary 2.4 is not true, however, as we have already observed in Theorem 1.1.

There are more extensive classes of context-free grammars which can be shown to generate only regular languages. A particularly important example for us (see Section 7) is the class of *right-linear* grammars.

Definition. A context-free grammar is called *right-linear* if each of its productions has one of the two forms

$$U \to xV \quad \text{or} \quad U \to x, \tag{2.7}$$

where U, V are variables and $x \neq 0$ is a word consisting entirely of terminals.

Thus a regular grammar is just a right-linear grammar in which $|x| = 1$ for each string x in (2.7). We have

Theorem 2.5. Let Γ be a right-linear grammar. Then $L(\Gamma)$ is regular.

Proof. Given a right-linear grammar Γ, we construct a regular grammar $\bar{\Gamma}$ as follows.

We replace each production of Γ of the form

$$U \to a_1 a_2 \cdots a_n V, \qquad n > 1,$$

by the productions

$$\begin{aligned} U &\to a_1 Z_1, \\ Z_1 &\to a_2 Z_2, \\ &\vdots \\ Z_{n-2} &\to a_{n-1} Z_{n-1}, \\ Z_{n-1} &\to a_n V, \end{aligned}$$

where $Z_1, \ldots, Z_{n-1}$ are new variables. Also, we replace each production

$$U \to a_1 a_2 \cdots a_n, \qquad n > 1,$$

by a list of productions similar to the preceding list except that instead of the last production we have

$$Z_{n-1} \to a_n.$$

It is obvious that $\bar{\Gamma}$ is regular and that $L(\bar{\Gamma}) = L(\Gamma)$. ■

Exercises

1. (a) For each regular language L described in Exercise 1.1 of Chapter 9, give a regular grammar Γ such that $L(\Gamma) = L - \{0\}$.

(b) For each dfa $\mathcal{M}$ in Exercise 1.3 of Chapter 9, give a regular grammar Γ such that $L(\Gamma) = L(\mathcal{M}) - \{0\}$.

(c) For each ndfa $\mathcal{M}$ in Exercise 2.1 of Chapter 9, give a regular grammar Γ such that $L(\Gamma) = L(\mathcal{M}) - \{0\}$.

2. Let Γ be the grammar with productions

$$\begin{array}{llll} S \to aS & X \to bX & Z \to aZ & Z \to a \\ S \to aX & X \to bZ & Z \to bZ & Z \to b \\ S \to aY & Y \to cX & Z \to cZ & Z \to c \\ & Y \to cZ & & \end{array}$$

and start symbol S. Give an ndfa $\mathcal{M}$ such that $L(\mathcal{M}) = L(\Gamma)$.

3. Let Γ be the grammar with productions

$$\begin{array}{lll} S \to aX & Y \to aY & Z \to aS \\ S \to bY & Y \to bZ & Z \to bS \\ X \to aZ & & Z \to a \\ X \to bX & & Z \to b \end{array}$$

and start symbol S.

(a) Use the construction in the proof of Theorem 2.2 to give an ndfa $\mathcal{M}$ with five states such that $L(\mathcal{M}) = L(\Gamma)$.

(b) Transform $\mathcal{M}$ into a dfa $\mathcal{M}'$ with four states such that $L(\mathcal{M}') = L(\Gamma)$.

4. Prove that for every regular language L, there is a regular grammar Γ with start symbol S such that $L = L(\Gamma)$ or $L = L(\Gamma) \cup \{0\}$ and such that every $w \in L(\Gamma)$ has exactly one derivation from S in Γ.

5. Prove that for every $n \geq 1$, there is a regular language generated by no regular grammar with fewer than n variables.

6. (a) Write a context-free grammar to generate all and only regular expressions over the alphabet $\{a, b\}$.

(b) Can a regular grammar generate this language? Support your answer.

7. A grammar Γ is *self-embedding* if there is a variable X such that

$$X \overset{*}{\underset{\Gamma}{\Rightarrow}} vXw, \qquad \text{where} \quad v, w \in (\mathscr{V} \cup T)^* - \{0\}.$$

Let L be a context-free language. Prove that L is regular if and only if there is a non-self-embedding context-free grammar Γ such that $L(\Gamma) = L$.

8. For a language L, the *reverse* of L, denoted L^R, is $\{w^R \mid w \in L\}$.

(a) Let Γ be a regular grammar and let $L = L(\Gamma)$. Show that there is an ndfa which accepts L^R.

(b) Conclude from (a) that a language L is regular if and only if L^R is regular.

(c) Let Γ be a grammar such that all productions are of the form $U \to Vs$ or $U \to s$, where U, V are variables and s is a terminal. Show that $L(\Gamma)$ is regular.

(d) A grammar is *left-linear* if each of its productions is of the form $U \to Vx$ or $U \to x$, where U, V are variables and x is a word consisting entirely of terminals. Prove that a language L is regular if and only if there is a left-linear grammar Γ such that $L = L(\Gamma)$ or $L = L(\Gamma) \cup \{0\}$.

3. Chomsky Normal Form

Although context-free grammars are extremely simply, there are even simpler special classes of context-free grammars that suffice to give all context-free languages. Such classes are called *normal forms*.

Definition. A context-free grammar Γ with variables $\mathscr{V}$ and terminals T is in *Chomsky normal form* if each of its productions has one of the two forms

$$X \to YZ \quad \text{or} \quad X \to a,$$

where $X, Y, Z \in \mathscr{V}$ and $a \in T$.

Then we can prove

Theorem 3.1. There is an algorithm that transforms a given positive context-free grammar Γ into a Chomsky normal form grammar Δ such that $L(\Gamma) = L(\Delta)$.

Proof. Using Theorem 1.5, we begin with a branching context-free grammar Γ with variables $\mathscr{V}$ and terminals T. We continue by "disguising" the terminals as variables. That is, for each $a \in T$ we introduce a new variable X_a. Then we modify Γ by replacing each production $X \to x$ *for which x is not a single terminal* by $X \to x'$, where x' is obtained from x by replacing each terminal a by the corresponding new variable X_a. In addition all of the productions $X_a \to a$ are added. Clearly the grammar thus obtained generates the same language as Γ and has all of its productions in one of the two forms

$$X \to X_1 X_2 \ldots X_k, \qquad k \geq 2, \tag{3.1}$$

$$X \to a, \tag{3.2}$$

where $X, X_1, \ldots, X_k$ are variables and a is a terminal. To obtain a Chomsky normal form grammar we need to eliminate all of the productions of type (3.1) for which $k > 2$. We can do this by introducing the new variables $Z_1, Z_2, \ldots, Z_{k-2}$ and replacing (3.1) by the productions

$$\begin{aligned} X &\to X_1 Z_1 \\ Z_1 &\to X_2 Z_2 \\ &\vdots \\ Z_{k-3} &\to X_{k-2} Z_{k-2} \\ Z_{k-2} &\to X_{k-1} X_k. \end{aligned}$$

Thus we obtain a grammar in Chomsky normal form that generates $L(\Gamma)$. ■

As an example, let us convert the grammar of Fig. 1.1 to Chomsky normal form. ,

Step 1. Eliminate productions of the form $X_1 \to X_2$: there are no such productions so we skip this step.

Step 2. Disguise the terminals as variables: the grammar now consists of the productions

$$\begin{array}{lll} S \to X_a X X_b Y & X \to X_a X & Y \to X_b Y \\ X \to a & Y \to b & \\ X_a \to a & X_b \to b. & \end{array}$$

Step 3. Obtain Chomsky normal form by replacing the production $S \to X_a X X_b Y$ by the productions

$$\begin{aligned} S &\to X_a Z_1, \\ Z_1 &\to X Z_2, \\ Z_2 &\to X_b Y. \end{aligned}$$

The final Chomsky normal form grammar thus obtained consists of the productions

$$\begin{array}{llll} S \to X_a Z_1 & Z_1 \to X Z_2 & Z_2 \to X_b Y & \\ X \to X_a X & X \to a & Y \to X_b Y & X_b \to b \\ X_a \to a & & Y \to b & \end{array}$$

Exercises

1. (a) Find context-free grammars Γ_1, Γ_2 such that

$$L(\Gamma_1) = \{a^{[i]}b^{[j]} \mid i \geq j > 0\}$$

$$L(\Gamma_2) = \{a^{[2i]}b^{[i]} \mid i > 0\}.$$

   (b) Find Chomsky normal form grammars that generate the same languages.

2. Let $T = \{\downarrow, p, q\}$ be the set of terminals for the grammar Γ:

$$S \to p, \qquad S \to q, \qquad S \to \downarrow SS.$$

   Find a Chomsky normal form grammar that generates $L(\Gamma)$.

3.* A context-free grammar is said to be in *Greibach normal form* if every production of the grammar is of the form

$$X \to aY_1Y_2 \cdots Y_k, \qquad k \geq 0,$$

   where $a \in T$ and $X, Y_1, Y_2, \ldots, Y_k \in \mathscr{V}$. Show that there is an algorithm that transforms any positive context-free grammar into one in Greibach normal form that generates the same language.

4.* Show that there is an algorithm that transforms any positive context-free grammar into a grammar that generates the same language for which every production is of the form

$$A \to a,$$

$$A \to aB,$$

   or

$$A \to aBC,$$

   $A, B, C \in \mathscr{V}$, $a \in T$.

4. Bar-Hillel's Pumping Lemma

An important application of Chomsky normal form is in the proof of the following key theorem, which is an analog for context-free languages of the pumping lemma for regular languages.

Theorem 4.1 (Bar-Hillel's Pumping Lemma). Let Γ be a Chomsky normal form grammar with exactly n variables, and let $L = L(\Gamma)$. Then, for every $x \in L$ for which $|x| > 2^n$, we have $x = r_1 q_1 r q_2 r_2$, where

1. $|q_1 r q_2| \le 2^n$;
2. $q_1 q_2 \ne 0$;
3. for all $i \ge 0$, $r_1 q_1^{[i]} r q_2^{[i]} r_2 \in L$.

Lemma. Let $S \overset{*}{\underset{\Gamma}{\Rightarrow}} u$, where Γ is a Chomsky normal form grammar. Suppose that $\mathscr{T}$ is a derivation tree for u in Γ and that no path in $\mathscr{T}$ contains more than k nodes. Then $|u| \le 2^{k-2}$.

Proof. First, suppose that $\mathscr{T}$ has just one leaf labeled by a terminal a. Then $u = a$, and $\mathscr{T}$ has just two nodes, which are labeled S and a, respectively. Thus, no path in $\mathscr{T}$ contains more than two nodes and $|u| = 1 \le 2^{2-2}$.

Otherwise, since Γ is in Chomsky normal form, the root of $\mathscr{T}$ must have exactly two immediate successors α, β in $\mathscr{T}$ labeled by variables, say, X and Y, respectively. (In this case, Γ contains the production $S \to XY$.) Now we will consider the two trees $\mathscr{T}_1 = \mathscr{T}^\alpha$ and $\mathscr{T}_2 = \mathscr{T}^\beta$ whose roots are labeled X and Y, respectively. (See Fig. 4.1.)

In each of $\mathscr{T}_1$ and $\mathscr{T}_2$ the longest path must contain $\le k - 1$ nodes. Proceeding inductively, we may assume that each of $\mathscr{T}_1, \mathscr{T}_2$ have $\le 2^{k-3}$ leaves. Hence,

$$|u| \le 2^{k-3} + 2^{k-3} = 2^{k-2}. \qquad ■$$

Proof of Theorem 4.1. Let $x \in L$, where $|x| > 2^n$, and let $\mathscr{T}$ be a derivation tree for x in Γ. Let $\alpha_1, \alpha_2, \ldots, \alpha_m$ be a path in $\mathscr{T}$ where m is *as large as possible*. Then $m \ge n + 2$. (For, if $m \le n + 1$, by the lemma, $|x| \le 2^{n-1}$.)

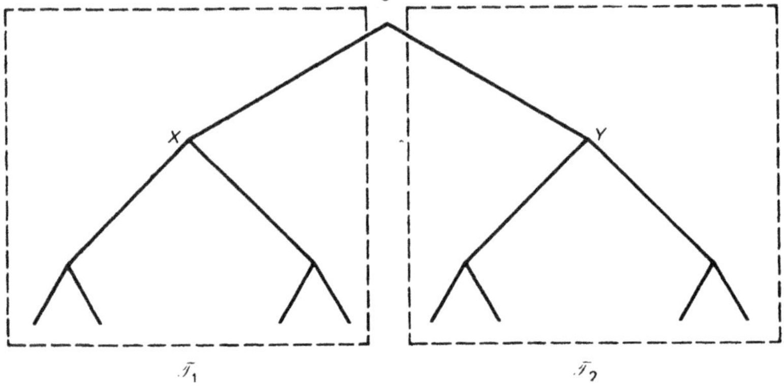

Figure 4.1

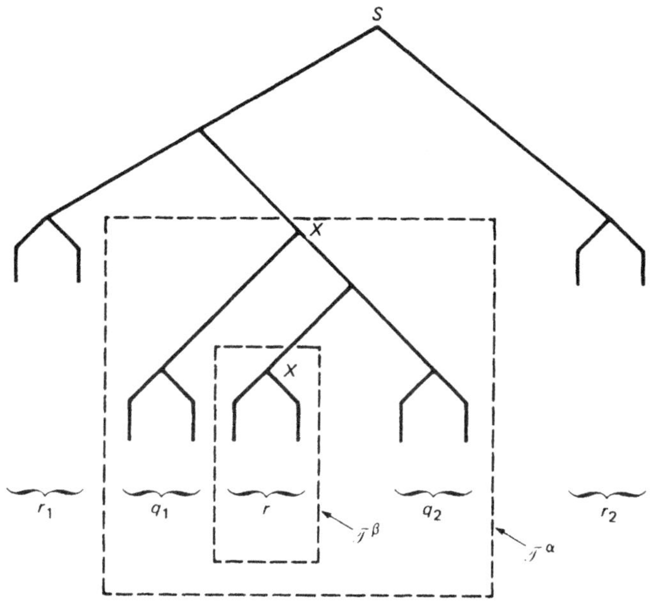

Figure 4.2

α_m is a leaf (otherwise we could get a longer path) and so is labeled by a terminal. $\alpha_1, \alpha_2, \ldots, \alpha_{m-1}$ are all labeled by variables. Let us write

$$\gamma_i = \alpha_{m+i-n-2}, \qquad i = 1, 2, \ldots, n + 2$$

so that the sequence of vertices $\gamma_1, \gamma_2, \ldots, \gamma_{n+2}$ is simply the path consisting of the vertices

$$\alpha_{m-n-1}, \alpha_{m-n}, \alpha_{m-n+1}, \ldots, \alpha_{m-1}, \alpha_m,$$

where $\gamma_{n+2} = \alpha_m$ is labeled by a terminal, and $\gamma_1, \ldots, \gamma_{n+1}$ are labeled by variables. Since there are only n variables in the alphabet of Γ, the pigeon-hole principle guarantees that there is a variable X that labels two different vertices: $\alpha = \gamma_i$ and $\beta = \gamma_j$, $i < j$. (See Fig. 4.2.) Hence, the discussion of *pruning* and *splicing* at the end of Section 1 can be applied. We let the words $q_1 q_2, r_1, r_2$ be defined as in that discussion and set $r = \langle \mathscr{T}^\beta \rangle$. Then [recalling (1.3)] we have

$$\langle \mathscr{T}_p \rangle = r_1 r r_2,$$

$$\langle \mathscr{T}_s \rangle = r_1 q_1^{[2]} r q_2^{[2]} r_2,$$

$$\langle (\mathscr{T}_s)_s \rangle = r_1 q_1^{[3]} r q_2^{[3]} r_2.$$

Since pruning and splicing a derivation tree in Γ yields a new derivation tree in Γ, we see that all of these words belong to $L(\Gamma)$. If, in addition, we iterate the splicing operation, we see that all of the words $r_1 q_1^{[i]} r q_2^{[i]} r_2$, $i \geq 0$, belong to $L(\Gamma)$.

Finally, we note that the path $\gamma_i, \ldots, \gamma_{n+2}$ in $\mathscr{T}^\alpha$ consists of $\leq n + 2$ nodes and that no path in $\mathscr{T}^\alpha$ can be longer. (This is true simply because if there were a path in $\mathscr{T}^\alpha$ consisting of more than $n + 3 - i$ vertices, it could be extended backward through $\alpha = \gamma_i$ to yield a path in $\mathscr{T}$ consisting of more than m vertices.) Hence by the lemma

$$|q_1 r q_2| = |q_1 \langle \mathscr{T}^\beta \rangle q_2| = |\langle \mathscr{T}^\alpha \rangle| \leq 2^n. \qquad \blacksquare$$

As an example of the uses of Bar-Hillel's pumping lemma, we show that the language $L = \{a^{[n]} b^{[n]} c^{[n]} \mid n > 0\}$ is *not* context-free.

Suppose that L is context-free with $L = L(\Gamma)$, where Γ is a Chomsky normal form grammar with n variables. Choose k so large that $|a^{[k]} b^{[k]} c^{[k]}| > 2^n$ (i.e., choose $k > 2^n/3$). Then we would have $a^{[k]} b^{[k]} c^{[k]} = r_1 q_1 r q_2 r_2$, where, setting

$$x_i = r_1 q_1^{[i]} r q_2^{[i]} r_2,$$

we have $x_i \in L$ for $i = 0, 1, 2, 3, \ldots$. In particular,

$$x_2 = r_1 q_1 q_1 r q_2 q_2 r_2 \in L.$$

Since the elements of L consist of a block of a's, followed by a block of b's, followed by a block of c's, we see that q_1 and q_2 must each contain only one of these letters. Thus, one of the three letters occurs neither in q_1 nor in q_2. But since as $i = 2, 3, 4, 5, \ldots$, x_i contains more and more copies of q_1 and q_2 and since $q_1 q_2 \neq 0$, it is impossible for x_i to have the same number of occurrences of a, b, and c. This contradiction shows that L is not context-free.

We have proved

Theorem 4.2. The language $L = \{a^{[n]} b^{[n]} c^{[n]} \mid n > 0\}$ is not context-free.

Exercises

1. Show that $\{a^{[i]} \mid i$ is a prime number$\}$ is not context-free.
2. Show that $\{a^{[i^2]} \mid i > 0\}$ is not context-free.
3. Show that a context-free language on a one-letter alphabet is regular.

5. Closure Properties

We now consider for context-free languages, some of the closure properties previously discussed for regular languages.

Theorem 5.1. If L_1, L_2 are context-free languages, then so is $L_1 \cup L_2$.

Proof. Let $L_1 = L(\Gamma_1)$, $L_2 = L(\Gamma_2)$, where Γ_1, Γ_2 are context-free grammars with disjoint sets of variables $\mathscr{V}_1$ and $\mathscr{V}_2$, and start symbols S_1, S_2, respectively. Let Γ be the context-free grammar with variables $\mathscr{V}_1 \cup \mathscr{V}_2 \cup \{S\}$ and start symbol S. The productions of Γ are those of Γ_1 and Γ_2, together with the two additional productions $S \to S_1, S \to S_2$. Then obviously $L(\Gamma) = L(\Gamma_1) \cup L(\Gamma_2)$, so that $L_1 \cup L_2 = L(\Gamma)$. ■

Surprisingly enough, the class of context-free languages is not closed under intersection. In fact, let Γ_1 be the context-free grammar whose productions are

$$S \to Sc, \qquad S \to Xc, \qquad X \to aXb, \qquad X \to ab.$$

Then clearly,

$$L_1 = L(\Gamma_1) = \{a^{[n]}b^{[n]}c^{[m]} \mid n, m > 0\}.$$

Now, let Γ_2 be the grammar whose productions are

$$S \to aS, \qquad S \to aX, \qquad X \to bXc, \qquad X \to bc.$$

Then

$$L_2 = L(\Gamma_2) = \{a^{[m]}b^{[n]}c^{[n]} \mid n, m > 0\}.$$

Thus, L_1 and L_2 are context-free languages. But

$$L_1 \cap L_2 = \{a^{[n]}b^{[n]}c^{[n]} \mid n > 0\},$$

which, by Theorem 4.2, is not context-free. We have proved

Theorem 5.2. There are context-free languages L_1 and L_2 such that $L_1 \cap L_2$ is not context-free.

Corollary 5.3. There is a context-free language $L \subseteq A^*$ such that $A^* - L$ is not context-free.

Proof. Suppose otherwise, i.e., for every context-free language $L \subseteq A^*$, $A^* - L$ is also context-free. Then the De Morgan identity

$$L_1 \cap L_2 = A^* - ((A^* - L_1) \cap (A^* - L_2))$$

together with Theorem 5.1 would contradict Theorem 5.2. ■

Although, as we have just seen, the intersection of context-free languages need not be context-free, the situation is different if one of the two languages is regular.

Theorem 5.4. If R is a regular language and L is a context-free language, then $R \cap L$ is context-free.

Proof. Let A be an alphabet such that $L, R \subseteq A^*$. Let $L = L(\Gamma)$ or $L(\Gamma) \cup \{0\}$, where Γ is a positive context-free grammar with variables $\mathscr{V}$, terminals A and start symbol S. Finally, let $\mathscr{M}$ be a dfa that accepts R with states Q, initial state $q_1 \in Q$, accepting states $F \subseteq Q$, and transition function δ. Now, for each symbol $\sigma \in A \cup \mathscr{V}$ and each ordered pair $p, q \in Q$, we introduce a new symbol σ^{pq}. We shall construct a positive context-free grammar $\tilde{\Gamma}$ whose terminals are just the elements of A (i.e., the terminals of Γ) and whose set of variables consists of a start symbol $\tilde{S}$ together with all of the new symbols σ^{pq} for $\sigma \in A \cup \mathscr{V}$ and $p, q \in Q$. (Note that for $a \in A$, a is a terminal, but a^{pq} is a variable for each $p, q \in Q$.) The productions of $\tilde{\Gamma}$ are as follows:

1. $\tilde{S} \to S^{q_1 q}$ for all $q \in F$.
2. $X^{pq} \to \sigma_1^{pr_1}\sigma_2^{r_1 r_2} \cdots \sigma_n^{r_{n-1}q}$ for all productions $X \to \sigma_1\sigma_2 \cdots \sigma_n$ of Γ and all $p, r_1, r_2, \ldots, r_{n-1}, q \in Q$.
3. $a^{pq} \to a$ for all $a \in A$ and all $p, q \in Q$ such that $\delta(p, a) = q$.

We shall now prove that $L(\tilde{\Gamma}) = L(\Gamma) \cap R$. Since $\tilde{\Gamma}$ is clearly a positive context-free grammar, and since $R \cap L = L(\tilde{\Gamma})$ or $R \cap L = L(\tilde{\Gamma}) \cup \{0\}$, the theorem follows from Theorem 1.2.

First let $u = a_1 a_2 \cdots a_n \in L(\Gamma) \cap R$. Since $u \in L(\Gamma)$, we have

$$S \overset{*}{\underset{\Gamma}{\Rightarrow}} a_1 a_2 \cdots a_n .$$

Using productions 1 and 2 of $\tilde{\Gamma}$, we have

$$\tilde{S} \underset{\tilde{\Gamma}}{\Rightarrow} S^{q_1, q_{n+1}} \overset{*}{\underset{\tilde{\Gamma}}{\Rightarrow}} a_1^{q_1 q_2} a_2^{q_2 q_3} \cdots a_n^{q_n q_{n+1}}, \tag{5.1}$$

where $q_2, q_3, \ldots, q_n$ are arbitrary states of $\mathscr{M}$ and q_{n+1} is any state in F. (q_1 is of course the initial state.) But since $u \in L(\mathscr{M})$, we can choose the states $q_2, q_3, \ldots, q_{n+1}$ so that

$$\delta(q_i, a_i) = q_{i+1}, \qquad i = 1, 2, \ldots, n, \tag{5.2}$$

and $q_{n+1} \in F$. In this case, not only does (5.1) hold, but also the productions

$$a_i^{q_i q_{i+1}} \to a_i, \qquad i = 1, 2, \ldots, n, \tag{5.3}$$

all belong to $\tilde{\Gamma}$. Hence, finally,

$$\tilde{S} \underset{\tilde{\Gamma}}{\overset{*}{\Rightarrow}} a_1 a_2 \cdots a_n .$$

Conversely, let $\tilde{S} \underset{\tilde{\Gamma}}{\overset{*}{\Rightarrow}} u \in A^*$. We shall need the following

Lemma. Let $\sigma^{pq} \underset{\tilde{\Gamma}}{\overset{*}{\Rightarrow}} u \in A^*$. Then, $\delta^*(p,u) = q$. Moreover, if σ is a variable, then $\sigma \underset{\Gamma}{\overset{*}{\Rightarrow}} u$.

Since $\tilde{S} \underset{\tilde{\Gamma}}{\Rightarrow} S^{q_1 q} \underset{\tilde{\Gamma}}{\overset{*}{\Rightarrow}} u$ where $q \in F$, we can use the Lemma to conclude that $\delta^*(q_1, u) = q$, and $S \underset{\Gamma}{\overset{*}{\Rightarrow}} u$. Hence, $u \in R \cap L(\Gamma)$. Theorem 5.4 then follows immediately. ■

It remains to prove the Lemma.

Proof of Lemma. The proof is by induction on the length of a derivation of u from σ^{pq} in $\tilde{\Gamma}$. If that length is 2, we must have $\sigma \in A$, $u = \sigma$. Then, $\delta^*(p,u) = \delta(p,u) = q$. Otherwise we can write

$$\sigma^{pq} \underset{\tilde{\Gamma}}{\Rightarrow} \sigma_1^{r_0 r_1} \sigma_2^{r_1 r_2} \cdots \sigma_n^{r_{n-1} r_n} \underset{\tilde{\Gamma}}{\overset{*}{\Rightarrow}} u$$

where we have written $r_0 = p$ and $r_n = q$. Thus, we have

$$\sigma_i^{r_{i-1} r_i} \underset{\tilde{\Gamma}}{\overset{*}{\Rightarrow}} u_i, \qquad i = 1, 2, \ldots, n, \tag{5.4}$$

where $u = u_1 u_2 \cdots u_n$. Clearly, the induction hypothesis can be applied to the derivations in (5.4) so that $\delta^*(r_{i-1}, u_i) = r_i$, $i = 1, 2, \ldots, n$. Hence $\delta^*(p,u) = r_n = q$. Also, if σ_i is a variable, the induction hypothesis will give $\sigma_i \underset{\Gamma}{\overset{*}{\Rightarrow}} u_i$, while otherwise $\sigma_i \in A$ and $\sigma_i = u_i$. Finally,

$$\sigma \to \sigma_1 \sigma_2 \cdots \sigma_n$$

must be a production of Γ. Hence, we have

$$\begin{aligned} \sigma &\underset{\Gamma}{\Rightarrow} \sigma_1 \sigma_2 \cdots \sigma_n \\ &\underset{\Gamma}{\overset{*}{\Rightarrow}} u_1 u_2 \cdots u_n = u. \end{aligned}$$ ■

Let A, P be alphabets such that $P \subseteq A$. For each letter $a \in A$, let us write

$$a^0 = \begin{cases} 0 & \text{if } a \in P \\ a & \text{if } a \in A - P. \end{cases}$$

If $x = a_1 a_2 \cdots a_n \in A^*$, we write

$$\mathrm{Er}_P(x) = a_1^0 a_2^0 \cdots a_n^0 .$$

In other words, $\mathrm{Er}_P(x)$ is the word that results from x when all the symbols in it that are part of the alphabet P are "erased." If $L \subseteq A^*$, we also write

$$\mathrm{Er}_P(L) = \{\mathrm{Er}_P(x) \mid x \in L\}.$$

Finally, if Γ is any context-free grammar with terminals T and if $P \subseteq T$, we write $\mathrm{Er}_P(\Gamma)$ for the context-free grammar with terminals $T - P$, the same variables and start symbol as Γ, and productions

$$X \to \mathrm{Er}_P(v)$$

for each production $X \to v$ of Γ. [Note that even if Γ is a positive context-free grammar, $\mathrm{Er}_P(\Gamma)$ may not be positive; that is, it is possible that $\mathrm{Er}_P(v) = 0$ even if $v \neq 0$.] We have

Theorem 5.5. If Γ is a context-free grammar and $\tilde{\Gamma} = \mathrm{Er}_P(\Gamma)$, then $L(\tilde{\Gamma}) = \mathrm{Er}_P(L(\Gamma))$.[3]

Proof. Let S be the start symbol of Γ and $\tilde{\Gamma}$. Suppose that $w \in L(\Gamma)$. We have

$$S = w_1 \underset{\Gamma}{\Rightarrow} w_2 \cdots \underset{\Gamma}{\Rightarrow} w_m = w.$$

Let $v_i = \mathrm{Er}_P(w_i)$, $i = 1, 2, \ldots, m$. Then clearly,

$$S = v_1 \underset{\tilde{\Gamma}}{\Rightarrow} v_2 \cdots \underset{\tilde{\Gamma}}{\Rightarrow} v_m = \mathrm{Er}_P(w),$$

so that $\mathrm{Er}_P(w) \in L(\tilde{\Gamma})$. This proves that $L(\tilde{\Gamma}) \supseteq \mathrm{Er}_P(L(\Gamma))$.

To complete the proof it will suffice to show that whenever $X \overset{*}{\underset{\tilde{\Gamma}}{\Rightarrow}} v \in (T - P)^*$, there is a word $w \in T^*$ such that $X \overset{*}{\underset{\Gamma}{\Rightarrow}} w$ and $v = \mathrm{Er}_P(w)$. We do this by induction on the length of a derivation of v from X in $\tilde{\Gamma}$. If $X \underset{\tilde{\Gamma}}{\Rightarrow} v$, then $X \to v$ is a production of $\tilde{\Gamma}$, so that $X \to w$ is a production of Γ for some w with $\mathrm{Er}_P(w) = v$. Proceeding inductively, let there be a derivation of v from X in $\tilde{\Gamma}$ of length $k > 2$, where the result is known

[3] Readers familiar with the terminology may enjoy noting that this theorem states that the "operators" L and Er_P commute.

for all derivations of length $< k$. Then, we can write

$$X \underset{\Gamma}{\Rightarrow} u_0 V_1 u_1 V_2 u_2 \cdots V_s u_s \overset{*}{\underset{\Gamma}{\Rightarrow}} v,$$

where $u_0, u_1, \ldots, u_s \in (T - P)^*$ and $V_1, V_2, \ldots, V_s$ are variables. Thus, there are words $\bar{u}_0, \bar{u}_1, \ldots, \bar{u}_s \in T^*$ such that $u_i = \mathrm{Er}_P(\bar{u}_i)$, $i = 0, 1, \ldots, s$, and

$$X \to \bar{u}_0 V_1 \bar{u}_1 V_2 \bar{u}_2 \cdots V_s \bar{u}_s$$

is a production of Γ. Also we can write

$$v = u_0 v_1 u_1 v_2 u_2 \cdots v_s u_s,$$

where

$$V_i \overset{*}{\underset{\Gamma}{\Rightarrow}} v_i, \qquad i = 1, \ldots, s. \tag{5.5}$$

Since (5.5) clearly involves derivations of length $< k$, the induction hypothesis applies, and we can conclude that there are words $\bar{v}_i \in T^*$, $i = 1, 2, \ldots, s$, such that $v_i = \mathrm{Er}_P(\bar{v}_i)$ and $V_i \overset{*}{\underset{\Gamma}{\Rightarrow}} \bar{v}_i$, $i = 1, 2, \ldots, s$. Hence, we have

$$X \underset{\Gamma}{\Rightarrow} \bar{u}_0 V_1 \bar{u}_1 V_2 \bar{u}_2 \cdots V_s \bar{u}_s \overset{*}{\underset{\Gamma}{\Rightarrow}} \bar{u}_0 \bar{v}_1 \bar{u}_1 \bar{v}_2 \bar{u}_2 \cdots \bar{v}_s \bar{u}_s.$$

But

$$\mathrm{Er}_P\left(\bar{u}_0 \bar{v}_1 \bar{u}_1 \bar{v}_2 \cdots \bar{v}_s \bar{u}_s\right) = u_0 v_1 u_1 v_2 u_2 \cdots v_s u_s = v,$$

which completes the proof. ■

Corollary 5.6. If $L \subseteq A^*$ is a context-free language and $P \subseteq A$, then $\mathrm{Er}_P(L)$ is also a context-free language.

Proof. Let $L = L(\Gamma)$, where Γ is a context-free grammar, and let $\tilde{\Gamma} = \mathrm{Er}_P(\Gamma)$. Then, by Theorem 5.5, $\mathrm{Er}_P(L) = L(\tilde{\Gamma})$, so that $\mathrm{Er}_P(L)$ is context-free. ■

Exercises

1. For each of the following, give languages L_1, L_2 on alphabet $\{a, b\}$ such that
   (a) L_1, L_2 are context-free but not regular, and $L_1 \cup L_2$ is regular;
   (b) L_1, L_2 are context-free, $L_1 \neq L_2$, and $L_1 \cup L_2$ is not regular;

(c) L_1, L_2 are context-free but not regular, $L_1 \cap L_2 \neq \varnothing$, and $L_1 \cap L_2$ is regular;

(d) L_1, L_2 are context-free but not regular, $L_1 \neq L_2$, and $L_1 \cap L_2$ is context-free but not regular.

2. Let L, L' be context-free languages. Prove the following.

(a) $L \cdot L'$ is context-free.

(b) L^* is context-free.

(c) $L^R = \{w^R \mid w \in L\}$ is context-free.

3. Give languages R, L_1, L_2 on alphabet $\{a, b\}$ such that R is regular, L_1, L_2 are context-free but not regular, and

(a) $R \cap L_1$ is regular;

(b) $R \cap L_2$ is not regular.

4. Give a context-free language L on alphabet $A = \{a, b\}$ such that L is not regular and $A^* - L$ is context-free.

5. Let $R = \{a^{[m]}b^{[n]} \mid m \geq 0,\ n > 0\}$, $L = \{a^{[n]}b^{[n]} \mid n > 0\}$. Use the construction in the proof of Theorem 5.4 to give a grammar Γ such that $L(\Gamma) = R \cap L$.

6. Give alphabets A, P such that $P \neq \varnothing$, $P \subseteq A$, $P \neq A$, and give languages $L_1, L_2 \subseteq A^*$ such that

(a) L_1 is not context-free and $\mathrm{Er}_P(L_1)$ is regular;

(b) L_2 is context-free and $\mathrm{Er}_P(L_2)$ is not regular.

7. Prove that if $L \subseteq A^*$ is regular and $P \subseteq A$, then $\mathrm{Er}_P(L)$ is also regular.

8. Let A_1, A_2 be alphabets, let $L \subseteq A_1^*$ be context-free, let f be a substitution on A_1 such that $f(a) \subseteq A_2^*$ is context-free for all $a \in A_1$, and let g be a homomorphism from A_1^* to A_2^*. [See Exercise 4.5 in Chapter 9 for the definitions of *substitution* and *homomorphism*.]

(a) Prove that $f(L)$ is context-free.

(b) Prove that $g(L)$ is context-free.

9. Let $A_1 = \{a_1, \ldots, a_n\}$, let $L \subseteq A_1^*$ be context-free, and let $R \subseteq A_1^*$ be regular.

(a) Let $A_2 = \{a'_1, \ldots, a'_n\}$, where $A_1 \cap A_2 = \varnothing$, and let f be a substitution on A_1 such that $f(a_i) = \{a_i, a'_i\}$, $1 \leq i \leq n$. Show that $A_2^* \cdot R \cap f(L)$ is context-free. [See Exercise 8.]

(b) Let g be the homomorphism on $A_1 \cup A_2$ such that $g(a_i) = 0$ and $g(a'_i) = a_i$, $1 \leq i \leq n$. Show that $g(A_2^* \cdot R \cap f(L))$ is context-free.

(c) Show that $g(A_2^* \cdot R \cap f(L)) = L/R$, the right quotient of L by R. [See Exercise 7.13 in Chapter 9 for the definition of *right quotient*.]

(d) Conclude that if L is context-free and R is regular, then L/R is context-free.

*6. Solvable and Unsolvable Problems[4]

Let Γ be a context-free grammar with terminals T and start symbol S, let $u \in T^*$, and let us consider the problem of determining whether $u \in L(\Gamma)$. First let $u = 0$. Then we can use the algorithms provided in Section 1 to compute $\ker(\Gamma)$. Since $0 \in L(\Gamma)$ if and only if $S \in \ker(\Gamma)$, we can answer the question in this case. For $u \neq 0$, we use Theorems 1.2 and 3.1 to obtain a Chomsky normal form grammar Δ such that $u \in L(\Gamma)$ if and only if $u \in L(\Delta)$. To test whether $u \in L(\Delta)$, we use the following:

Lemma. Let Δ be a Chomsky normal form grammar with terminals T. Let V be a variable of Δ and let

$$V \overset{*}{\underset{\Delta}{\Rightarrow}} u \in T^*.$$

Then there is a derivation of u from V in Δ of length $2|u|$.

Proof. The proof is by induction on $|u|$. If $|u| = 1$, then u is a terminal and Δ must contain a production $V \to u$, so that we have a derivation of u from V of length 2.

Now, let $V \overset{*}{\underset{\Delta}{\Rightarrow}} u$, where $|u| > 1$, and let us assume the result known for all strings of length $< |u|$. Recalling the definition of a Chomsky normal form grammar, we see that

$$V \Rightarrow XY \overset{*}{\Rightarrow} u.$$

Thus, we must have $X \overset{*}{\Rightarrow} v$, $Y \overset{*}{\Rightarrow} w$, $u = vw$ where $|v|, |w| < |u|$. By the induction hypothesis we have derivations

$$X = \alpha_1 \Rightarrow \alpha_2 \Rightarrow \cdots \Rightarrow \alpha_{2|v|} = v,$$

$$Y = \beta_1 \Rightarrow \beta_2 \Rightarrow \cdots \Rightarrow \beta_{2|w|} = w.$$

Hence, we can write the derivation

$$V \Rightarrow XY = \alpha_1 Y \Rightarrow \alpha_2 Y \Rightarrow \cdots \Rightarrow \alpha_{2|v|} Y = v\beta_1 \Rightarrow v\beta_2 \Rightarrow \cdots \Rightarrow v\beta_{2|w|},$$

[4] The * does *not* refer to the material through Theorem 6.4.

where $v\beta_{2|w|} = vw = u$. But this derivation is of length $2|v| + 2|w| = 2|u|$, which completes the proof. ■

Now to test $u \in L(\Delta)$, we simply write out all derivations from S of length $2|u|$. We have $u \in L(\Delta)$ if and only if at least one of these derivations terminates in the string u.

We have proved

Theorem 6.1.[5] There is an algorithm that will test a given context-free grammar Γ and a given word u to determine whether $u \in L(\Gamma)$.

Next we wish to consider the question of whether a given context-free grammar generates the empty language $\emptyset$. Let Γ be a given context-free grammar. We first check as previously to decide whether $0 \in L(\Gamma)$. If $0 \in L(\Gamma)$, we know that $L(\Gamma) \neq \emptyset$. Otherwise we use Theorems 1.2 and 1.5 to obtain a *branching* context-free grammar $\bar{\Gamma}$ such that $L(\Gamma) = L(\bar{\Gamma})$. Let $\bar{\Gamma}$ have n variables and set of terminals T. Suppose that $L(\bar{\Gamma}) \neq \emptyset$. Let $u \in L(\bar{\Gamma})$, where u has the shortest possible length of any word in $L(\bar{\Gamma})$. Then *in any derivation tree for u in* $\bar{\Gamma}$, *each path contains fewer than* $n + 2$ *nodes*. This is because, if there were a path containing at least $n + 2$ nodes, at least $n + 1$ of them would be labeled by variables, and by the pigeon-hole principle, Theorem 1.6 would apply and yield a word $v \in L(\bar{\Gamma})$ with $|v| < |u|$. Thus, we conclude that

$L(\bar{\Gamma}) \neq \emptyset$ *if and only if there is a derivation tree* $\mathcal{T}$ *in* Γ *of a word* $u \in T^*$ *such that each path in* $\mathcal{T}$ *contains fewer than* $n + 2$ *nodes.*

It is a straightforward matter (at least in principle) to write out explicitly all derivation trees in Γ in which no path has length $\geq n + 2$. To test whether $L(\bar{\Gamma}) \neq \emptyset$, it suffices to note whether there is such a tree $\mathcal{T}$ for which $\langle \mathcal{T} \rangle \in T^*$. Thus we have

Theorem 6.2. There is an algorithm to test a given context-free grammar Γ to determine whether $L(\Gamma) = \emptyset$.

Next we seek an algorithm to test whether $L(\Gamma)$ is finite or infinite for a given context-free grammar Γ. Such an algorithm can easily be obtained from the following.

---

[5] This result follows at once from Theorem 5.4 in Chapter 7; but the algorithm given here is of some independent interest.

Theorem 6.3. Let Γ be a Chomsky normal form grammar with exactly n variables. Then $L(\Gamma)$ is infinite if and only if there is a word $x \in L(\Gamma)$ such that

$$2^n < |x| \le 2^{n+1}.$$

Proof. If there is a word $x \in L(\Gamma)$ with $|x| > 2^n$, then by Bar-Hillel's pumping lemma, $L(\Gamma)$ is infinite.

Conversely, let $L(\Gamma)$ be infinite. Let u be a word of shortest possible length such that $u \in L(\Gamma)$ and $|u| > 2^{n+1}$. By Bar-Hillel's pumping lemma, we have $u = r_1 q_1 r q_2 r_2$ where $q_1 q_2 \neq 0$, $|q_1 r q_2| \le 2^n$ and $x = r_1 r r_2 \in L(\Gamma)$. Now,

$$|x| \ge |r_1 r_2| = |u| - |q_1 r q_2| > 2^n.$$

Since $|x| < |u|$, the manner in which we chose u guarantees that $|x| \le 2^{n+1}$. ■

Theorem 6.4. There is an algorithm to test a given context-free grammar Γ to determine whether $L(\Gamma)$ is finite or infinite.

Proof. Given context-free grammar Γ with terminals T, we use the algorithms of Theorems 1.2 and 3.1 to construct a Chomsky normal form grammar Δ with $L(\Gamma) = L(\Delta)$ or $L(\Delta) \cup \{0\}$. Let Δ have n variables and let $l = 2^n$. Then we simply use Theorem 6.1 to test each word $u \in T^*$ for which $l < |u| \le 2l$ to see whether $u \in L(\Gamma)$. $L(\Gamma)$ is infinite if and only if at least one of these words u does belong to $L(\Gamma)$. ■

Remarkably enough, there are also some very simple unsolvable problems related to context-free grammars.[6] The easiest way to obtain these results is to associate a pair of context-free grammars with each Post correspondence system.

Thus, suppose we are given the finite set of dominoes:

$$\begin{array}{|c|}\hline u_i \\ \hline v_i \\ \hline\end{array}$$

$i = 1, 2, \ldots, n$, where $u_i, v_i \in A^*$ for some given alphabet A. We introduce n new symbols $c_1, c_2, \ldots, c_n$ and define two context-free grammars Γ_1, Γ_2, both of which have as their terminals $A \cup \{c_1, c_2, \ldots, c_n\}$. Γ_1 has the

[6] The remainder of this section depends on Chapter 7. Readers who have not covered this material should move on to Section 7.

single variable S_1, its start symbol, and Γ_2 has S_2 as its only variable and start symbol. The productions of Γ_1 are

$$S_1 \to u_i S_1 c_i, \qquad S_1 \to u_i c_i, \qquad i = 1, 2, \ldots, n,$$

and those of Γ_2 are

$$S_2 \to v_i S_2 c_i, \qquad S_2 \to v_i c_i, \qquad i = 1, 2, \ldots, n.$$

Now, the given Post correspondence system has a solution if and only if we can have

$$u_{i_1} u_{i_2} \cdots u_{i_m} = v_{i_1} v_{i_2} \cdots v_{i_m}$$

for some sequence $i_1, i_2, \ldots, i_m$. Moreover,

$$L(\Gamma_1) = \{u_{i_1} u_{i_2} \cdots u_{i_m} c_{i_m} \cdots c_{i_2} c_{i_1}\}$$

and

$$L(\Gamma_2) = \{v_{i_1} v_{i_2} \cdots v_{i_m} c_{i_m} \cdots c_{i_2} c_{i_1}\}.$$

Thus, we have

Theorem 6.5. $L(\Gamma_1) \cap L(\Gamma_2) \neq \varnothing$ if and only if the given Post correspondence problem has a solution.

Using Theorem 4.1 in Chapter 7, we conclude

Theorem 6.6. There is no algorithm to test a given pair of context-free grammars Γ_1, Γ_2 to determine whether $L(\Gamma_1) \cap L(\Gamma_2) = \varnothing$.

Another important unsolvability result about context-free grammars concerns ambiguity.

Definition. A context-free grammar Γ is called *ambiguous* if there is a word $u \in L(\Gamma)$ that has two different leftmost derivations in Γ. If Γ is not ambiguous, it is said to be *unambiguous*.

Theorem 6.7. There is no algorithm to test a given context-free grammar to determine whether it is ambiguous.

Proof. Once again we begin with a Post correspondence system, and form the two context-free grammars Γ_1, Γ_2 used in proving Theorem 6.5. Γ_1 and Γ_2 are obviously both unambiguous. Now let Γ have start symbol S and all of the productions of Γ_1 and Γ_2, together with $S \to S_1$ and $S \to S_2$. Then, since the first step of a derivation from S in Γ involves an irreversible commitment to either Γ_1 or Γ_2, Γ will be ambiguous just in case $L(\Gamma_1) \cap L(\Gamma_2) \neq \varnothing$. By Theorem 6.5 this will be the case if and only if the given Post correspondence system has a solution. The result now follows again from Theorem 4.1 in Chapter 7. ■

Another unsolvability result is given in Exercise 8.16.

Exercises

1. Let Γ_1 be the grammar with productions $S \to aS$, $S \to a$, and let Γ_2 be the grammar with productions $S \to SS$, $S \to a$.
   (a) How many derivation trees are there for $a^{[6]}$ in Γ_1? In Γ_2?
   (b) How many derivations of $a^{[4]}$ from S are there in Γ_1? In Γ_2?
   (c) How many leftmost derivations of $a^{[6]}$ from S are there in Γ_1? In Γ_2?
2. Write a context-free grammar Γ such that
$$L(\Gamma) = \{a^{[i]}b^{[j]}c^{[k]} \mid i = j \vee j = k\}.$$
   This language is an example of an *inherently ambiguous language*, i.e., a language such that every grammar that generates it is ambiguous. Explain why this language is inherently ambiguous.
3. Give an unambiguous context-free grammar that generates the same language as the ambiguous grammar
$$\begin{aligned} S &\to aB \\ S &\to Ab \\ A &\to aAB \\ B &\to ABb \\ A &\to a \\ B &\to b. \end{aligned}$$

7. Bracket Languages

Let A be some finite set. Although we think of A as an alphabet, we will also wish to permit $A = \varnothing$. Let B be the alphabet we get from A by adjoining the $2n$ *new* symbols $(_i,)_i$, $i = 1, 2, \ldots, n$, where n is some given positive integer. We will write $\text{PAR}_n(A)$ for the language consisting of all strings in B^* that are correctly "paired," thinking of each pair $(_i,)_i$ as matching left and right brackets. More precisely, $\text{PAR}_n(A) = L(\Gamma_0)$, where

Γ_0 is the context-free grammar with the single variable S, terminals B, and the productions

1. $S \to a$ for all $a \in A$,
2. $S \to \underset{i}{(} S \underset{i}{)}$, $i = 1, 2, \ldots, n$.
3. $S \to SS$, $S \to 0$.

The languages $\mathrm{PAR}_n(A)$ are called *bracket languages*.

Let us consider the example $A = \{a, b, c\}$, $n = 2$. For ease of reading we will use the symbol (for $\underset{1}{(}$,) for $\underset{1}{)}$, [for $\underset{2}{(}$, and] for $\underset{2}{)}$. Then $cb[(ab)c](a[b]c) \in \mathrm{PAR}_2(A)$, as the reader should easily verify. Also, $()[\,] \in \mathrm{PAR}_2(A)$, since we have

$$S \Rightarrow SS \overset{*}{\Rightarrow} (S)[S] \overset{*}{\Rightarrow} ()[\,].$$

Bracket languages have the following properties.

Theorem 7.1. $\mathrm{PAR}_n(A)$ is a context-free language such that

a. $A^* \subseteq \mathrm{PAR}_n(A)$;
b. if $x, y \in \mathrm{PAR}_n(A)$, so is xy;
c. if $x \in \mathrm{PAR}_n(A)$, so is $\underset{i}{(} x \underset{i}{)}$, for $i = 1, 2, \ldots, n$;
d. if $x \in \mathrm{PAR}_n(A)$ and $x \notin A^*$, then we can write $x = u \underset{i}{(} v \underset{i}{)} w$, for some $i = 1, 2, \ldots, n$, where $u \in A^*$ and $v, w \in \mathrm{PAR}_n(A)$.

Proof. Since $\mathrm{PAR}_n(A) = L(\Gamma_0)$ where Γ_0 is a context-free grammar, $\mathrm{PAR}_n(A)$ must be context-free. Property a follows at once on considering the productions 1 and 3. For b, let $S \overset{*}{\Rightarrow} x$, $S \overset{*}{\Rightarrow} y$. Then using the productions 3, we have

$$S \Rightarrow SS \overset{*}{\Rightarrow} xy.$$

For c, let $S \overset{*}{\Rightarrow} x$. Then using the productions 2, we have

$$S \Rightarrow \underset{i}{(} S \underset{i}{)} \overset{*}{\Rightarrow} \underset{i}{(} x \underset{i}{)}.$$

To prove d, note first that we can assume $|x| > 1$ because otherwise $x \in A^*$. Then, a derivation of x from S must begin by using a production containing S on the right. We proceed by induction assuming the result for all strings of length $< |x|$. There are two cases.

Case 1. $S \Rightarrow \underset{i}{(} S \underset{i}{)} \overset{*}{\Rightarrow} \underset{i}{(} v \underset{i}{)} = x$, where $S \overset{*}{\Rightarrow} v$; the result then follows (without using the induction hypothesis) with $u = w = 0$.

Case 2. $S \Rightarrow SS \overset{*}{\Rightarrow} rs = x$ where $S \overset{*}{\Rightarrow} r$, $S \overset{*}{\Rightarrow} s$, and $r \neq 0$, $s \neq 0$. Clearly, $|r|, |s| < |x|$. If $r \in A^*$, then $|s| > 1$ and we can use the induction hypothesis to write $s = u \underset{i}{(} v \underset{i}{)} w$, where $u \in A^*$ and $v, w \in$

$\text{PAR}_n(A)$, and the desired result follows since $ru \in A^*$. Otherwise, we can use the induction hypothesis to write $r = u\,{}_i^{(}v{}_i^{)}w$ where $u \in A^*$ and $v, w \in \text{PAR}_n(A)$, so that the result follows since $ws \in \text{PAR}_n(A)$ by b. ■

Historically, the special case $A = \varnothing$ has played an important role in studying context-free languages. The language $\text{PAR}_n(\varnothing)$ is called the *Dyck language* of order n and is usually written D_n.

Now let us begin with a Chomsky normal form grammar Γ, with terminals T and productions

$$X_i \to Y_i Z_i, \qquad i = 1, 2, \ldots, n, \tag{7.1}$$

in addition to certain productions of the form $V \to a$ with $a \in T$. We will construct a new grammar Γ_s which we call the *separator* of Γ. The terminals of Γ_s are the symbols of T together with $2n$ new symbols ${}_i^{(}, {}_i^{)}$, $i = 1, 2, \ldots, n$. Thus a pair of "brackets" has been added for each of the productions (7.1).

The productions of Γ_s are

$$X_i \to {}_i^{(} Y_i {}_i^{)} Z_i, \qquad i = 1, 2, \ldots, n,$$

as well as all of the productions of Γ of the form $V \to a$ with $a \in T$.

As an example, let Γ have the productions

$$S \to XY, \qquad S \to YX, \qquad Y \to ZZ,$$
$$X \to a, \qquad Z \to a.$$

Then Γ is ambiguous as we can see from the leftmost derivations:

$$S \Rightarrow XY \Rightarrow aY \Rightarrow aZZ \Rightarrow aaZ \Rightarrow aaa,$$
$$S \Rightarrow YX \Rightarrow ZZX \Rightarrow aZX \Rightarrow aaX \Rightarrow aaa.$$

The productions of Γ_s can be written

$$S \to (X)Y, \qquad S \to [Y]X, \qquad Y \to \{Z\}Z,$$
$$X \to a, \qquad Z \to a,$$

using ( ), [ ], and { } in place of the numbered brackets. The two derivations just given then become

$$S \Rightarrow (X)Y \Rightarrow (a)Y \Rightarrow (a)\{Z\}Z \Rightarrow (a)\{a\}Z \Rightarrow (a)\{a\}a,$$
$$S \Rightarrow [Y]X \Rightarrow [\{Z\}Z]X \Rightarrow [\{a\}Z]X \Rightarrow [\{a\}a]X \Rightarrow [\{a\}a]a.$$

Γ_s thus *separates* the two derivations in Γ. The bracketing in the words $(a)\{a\}a, [\{a\}a]a$ enables their respective derivation trees to be recovered.

If we write P for the set of brackets $\underset{i}{(}, \underset{i}{)}$, $i = 1, 2, \ldots, n$, then clearly $\Gamma = \mathrm{Er}_P(\Gamma_s)$. Hence by Theorem 5.5,

Theorem 7.2. $\mathrm{Er}_P(L(\Gamma_s)) = L(\Gamma)$.

We also will prove

Lemma 1. $L(\Gamma_s) \subseteq \mathrm{PAR}_n(T)$.

Proof. We show that if $X \underset{\Gamma_s}{\overset{*}{\Rightarrow}} w \in (T \cup P)^*$ for any variable X, then $w \in \mathrm{PAR}_n(T)$. The proof is by induction on the length of a derivation of w from X in Γ_s. If this length is 2, then w is a single terminal and the result is clear. Otherwise we can write

$$X = X_i \underset{\Gamma_s}{\Rightarrow} \underset{i}{(}Y_i\underset{i}{)}Z_i \underset{\Gamma_s}{\overset{*}{\Rightarrow}} \underset{i}{(}u\underset{i}{)}v = w,$$

where $Y_i \underset{\Gamma_s}{\overset{*}{\Rightarrow}} u$ and $Z_i \underset{\Gamma_s}{\overset{*}{\Rightarrow}} v$. By the induction hypothesis, $u, v \in \mathrm{PAR}_n(T)$. By b and c of Theorem 7.1, so is w. ■

Now let Δ be the grammar whose variables, start symbol, and terminals are those of Γ_s and whose productions are as follows:

1. all productions $V \to a$ from Γ (or equivalently Γ_s) with $a \in T$,
2. all productions $X_i \to \underset{i}{(}Y_i$, $i = 1, 2, \ldots, n$,
3. all productions $V \to a\underset{i}{)}Z_i$, $i = 1, 2, \ldots, n$, for which $V \to a$ is a production of Γ with $a \in T$.

We have

Lemma 2. $L(\Delta)$ is regular.

Proof. Since Δ is obviously right-linear, the result follows at once from Theorem 2.5. ■

Lemma 3. $L(\Gamma_s) \subseteq L(\Delta)$.

Proof. We show that if $X \underset{\Gamma_s}{\overset{*}{\Rightarrow}} u \in (T \cup P)^*$ then $X \underset{\Delta}{\overset{*}{\Rightarrow}} u$. If u has a derivation of length 2, then $u \in T$, and $X \to u$ is a production of Γ_s and of Γ and therefore also of Δ. Thus $X \underset{\Delta}{\overset{*}{\Rightarrow}} u$.

Proceeding by induction, let

$$X = X_i \underset{\Gamma_s}{\Rightarrow} \underset{i}{(}Y_i\underset{i}{)}Z_i \underset{\Gamma_s}{\overset{*}{\Rightarrow}} \underset{i}{(}v\underset{i}{)}w = u,$$

where the induction hypothesis applies to $Y_i \overset{*}{\underset{\Gamma_s}{\Rightarrow}} v$ and to $Z_i \overset{*}{\underset{\Gamma_s}{\Rightarrow}} w$. Thus, $Y_i \overset{*}{\underset{\Delta}{\Rightarrow}} v$ and $Z_i \overset{*}{\underset{\Delta}{\Rightarrow}} w$. Let $v = za$, $a \in T$. (See Exercise 3.) Then, examining the productions of the grammar Δ, we see that we must have

$$Y_i \overset{*}{\underset{\Delta}{\Rightarrow}} zV \underset{\Delta}{\Rightarrow} za = v,$$

where $V \to a$ is a production of Γ. But then we have

$$X_i \underset{\Delta}{\Rightarrow} {}^{(}_{i}Y_i \overset{*}{\underset{\Delta}{\Rightarrow}} {}^{(}_{i}zV \underset{\Delta}{\Rightarrow} {}^{(}_{i}za{}^{)}_{i}Z_i \overset{*}{\underset{\Delta}{\Rightarrow}} {}^{(}_{i}v{}^{)}_{i}w = u. \qquad \blacksquare$$

Lemma 4. $L(\Delta) \cap \mathrm{PAR}_n(T) \subseteq L(\Gamma_s)$.

Proof. Let $X \overset{*}{\underset{\Delta}{\Rightarrow}} u$, where $u \in \mathrm{PAR}_n(T)$. We shall prove that $X \overset{*}{\underset{\Gamma_s}{\Rightarrow}} u$. The proof is by induction on the total number of occurrences of the symbols ${}^{(}_{i}\,, {}^{)}_{i}$ in u. If this number is 0, then, examining the productions of Δ, we see that $u \in T$ and the production $X \to u$ is in Δ and hence in Γ_s. Thus $X \overset{*}{\underset{\Gamma_s}{\Rightarrow}} u$.

Now let $X \overset{*}{\underset{\Delta}{\Rightarrow}} u$, where u contains occurrences of the bracket symbols ${}^{(}_{i}\,, {}^{)}_{i}$ and where the result is known for words v containing fewer occurrences of these symbols than u. Examining the productions of Δ, we see that our derivation of u from X must begin with one of the productions 2. (If the derivation began with a production of the form 1, then u would be a terminal. If the derivation began with a production of the form 3, then $u = a{}^{)}_{i}w$ for some word w, which is impossible by Theorem 7.1d.) Therefore $u = {}^{(}_{i}z$, for some word z and some $i = 1, 2, \ldots, n$. By Theorem 7.1d, $u = {}^{(}_{i}v{}^{)}_{i}w$, where $v, w \in \mathrm{PAR}_n(T)$. In our derivation of u in Δ, the symbol ${}^{)}_{i}$ can only arise from the use of one of the productions of the form 3, say, $V \to a{}^{)}_{i}Z_i$, where $a \in T$ and $V \to a$ is a production of Γ. Then v must end in a, so that we can write $v = \bar{v}a$, where

$$X = X_i \underset{\Delta}{\Rightarrow} {}^{(}_{i}Y_i \overset{*}{\underset{\Delta}{\Rightarrow}} {}^{(}_{i}\bar{v}V \underset{\Delta}{\Rightarrow} {}^{(}_{i}\bar{v}a{}^{)}_{i}Z_i \overset{*}{\underset{\Delta}{\Rightarrow}} {}^{(}_{i}v{}^{)}_{i}w$$

and $Z_i \overset{*}{\underset{\Delta}{\Rightarrow}} w$. Moreover, since $V \to a$ is a production of Γ, it is also one of the productions of Δ of the form 1. Therefore, we have in Δ

$$Y_i \overset{*}{\underset{\Delta}{\Rightarrow}} \bar{v}V \underset{\Delta}{\Rightarrow} \bar{v}a = v.$$

Since v and w must each contain fewer occurrences of ${}^{(}_{i}\,, {}^{)}_{i}$ than u, we have by the induction hypothesis

$$Y_i \overset{*}{\underset{\Gamma_s}{\Rightarrow}} v, \qquad Z_i \overset{*}{\underset{\Gamma_s}{\Rightarrow}} w.$$

Hence,

$$X_i \underset{\Gamma_s}{\Rightarrow} \overset{(}{_i} Y \overset{)}{_i} {}_i Z_i \underset{\Gamma_s}{\overset{*}{\Rightarrow}} \overset{(}{_i} v \overset{)}{_i} w = u. \qquad \blacksquare$$

We are now ready to state

Theorem 7.3. Let Γ be a grammar in Chomsky normal form with terminals T. Then there is a regular language R such that

$$L(\Gamma_s) = R \cap \text{PAR}_n(T).$$

Proof. Let Δ be defined as above and let $R = L(\Delta)$. The result then follows at once from Lemmas 1–4. ■

Theorem 7.4 (Chomsky–Schützenberger Representation Theorem). A language $L \subseteq T^*$ is context-free if and only if there is a regular language R and a number n such that

$$L = \text{Er}_P(R \cap \text{PAR}_n(T)), \tag{7.2}$$

where $P = \{\overset{(}{_i}, \overset{)}{_i} \mid i = 1, 2, \ldots, n\}$.

Proof. It is clear by Theorems 7.2 and 7.3 that for every grammar Γ in Chomsky normal form, $L = L(\Gamma)$ satisfies (7.2). For an arbitrary context-free language L, by Theorems 1.2 and 3.1, there is a Chomsky normal form grammar Γ such that

$$L = L(\Gamma) \text{ or } L = L(\Gamma) \cup \{0\}.$$

If

$$L(\Gamma) = \text{Er}_P(R \cap \text{PAR}_n(T)),$$

then

$$L(\Gamma) \cup \{0\} = \text{Er}_P((R \cup \{0\}) \cap \text{PAR}_n(T))$$

since, by Theorem 7.1a, $0 \in \text{PAR}_n(T)$. But, by Theorems 4.2 and 4.5 in Chapter 9, $R \cup \{0\}$ is a regular language.

It remains only to show that any language L that satisfies (7.2) must be context-free. But since, by Theorem 7.1, $\text{PAR}_n(T)$ is context-free, this result follows at once from Theorem 5.4 and Corollary 5.6. ■

The Chomsky–Schützenberger theorem is usually expressed in terms of the Dyck languages $D_n = \text{PAR}_n(\varnothing)$. Since our form of the theorem is equivalent to the more usual form, we will give only a very brief sketch of the proof of the usual form. It is necessary to go back to the construction of Γ_s. Each element a of T is now thought of as a "left bracket" and is supplied with a "twin" a' to act as its corresponding right bracket. A new grammar Γ_t is then defined to have the same productions $X_i \to {}^{(}_{i} Y^{)}_{i}$, $i = 1, 2, \ldots, n$, as Γ_s but to have productions

$$V \to aa'$$

for each production $V \to a$ of Γ. Then clearly, $L(\Gamma_t)$ can be obtained from $L(\Gamma_s)$ by simply replacing all occurrences of letters $a \in T$ in words of $L(\Gamma_s)$ by aa'. By replacing a by aa' in productions of the forms 1 and 3 of Δ, we obtain a right linear grammar Δ' such that

$$L(\Gamma_t) = L(\Delta') \cap \mathrm{PAR}_n(T'),$$

where $T' = \{a, a' \mid a \in T\}$. But in fact $L(\Gamma_t) \subseteq D_m$, where $m = n + k$ and there are k letters in T. Thus,

$$L(\Gamma_t) = L(\Delta') \cap D_m .$$

Finally letting $Q = \{ {}^{(}_{i}, {}^{)}_{i} \mid i = 1, 2, \ldots, n\} \cup \{a' \mid a \in T\}$, we have

$$L(\Gamma) = \mathrm{Er}_Q(L(\Gamma_t)) = \mathrm{Er}_Q(L(\Delta') \cap D_m).$$

Thus, we get

Theorem 7.5. A language L is context-free if and only if there is a regular language R, an alphabet Q, and an integer m such that

$$L = \mathrm{Er}_Q(R \cap D_m).$$

Exercises

1. Let A be a finite set of symbols, n a positive integer, and $\mathrm{PAR}_n(A) = L(\Gamma_0)$, where Γ_0 is the grammar given in the definition of $\mathrm{PAR}_n(A)$. Show that Γ_0 is ambiguous. [See Section 6 for the definition of *ambiguous grammars*.]
2. Let Γ be the grammar with productions

$$\begin{array}{lll} S \to XZ & X \to a & \\ S \to XY & Y \to b & Z \to SY \end{array}$$

and start symbol S.

(a) Give Γ_s.

(b) Give Δ, as defined following Lemma 1, for Γ.

(c) Show that $L(\Gamma_s) \neq \text{PAR}_3(\{a, b\})$.

(d) Show that $L(\Gamma_s) \neq L(\Delta)$.

3. Let Γ be a grammar in Chomsky normal form with variables $\mathscr{V}$ and terminals T, and let $T \cup P$ be the terminals of Γ_s. Prove that for all $V \in \mathscr{V}$ and all w such that $V \overset{*}{\underset{\Gamma_s}{\Rightarrow}} w$, $w = vs$ for some $v \in (\mathscr{V} \cup T \cup P)^*$ and some $s \in \mathscr{V} \cup T$.

4. Let Γ be a regular grammar, and let Γ' be the Chomsky normal form grammar derived from Γ by the construction in the proof of Theorem 3.1. Prove that $L(\Gamma'_s)$ is regular.

8. Pushdown Automata

We are now ready to discuss the question of what kind of automaton is needed for accepting context-free languages. We take our cue from Theorem 7.2, and begin by trying to construct an appropriate automaton for recognizing $L(\Gamma_s)$, where Γ is a given Chomsky normal form grammar. We know that $L(\Gamma_s) = R \cap \text{PAR}_n(T)$, where R is a regular language. Thus R is accepted by a finite automaton. The problem we need to solve is this: what additional facilities does this finite automaton require in order to check that some given word belongs to $\text{PAR}_n(T)$? Those familiar with "stacks" and their uses will see at once that what is needed is a "pushdown stack" as an auxiliary storage device. Such a device behaves in a last-in–first-out manner. At each step in a computation with a pushdown stack one or both of a pair of operations can be performed:

1. The symbol at the "top" of the stack may be read and discarded. (This operation is called *popping the stack*.)
2. A new symbol may be "*pushed*" onto the stack.

A stack can be used to identify a string as belonging to $\text{PAR}_n(T)$ as follows: For each pair $\overset{(}{_i}, \overset{)}{_i}$, $i = 1, 2, \ldots, n$, a special symbol J_i is introduced. Now, as our automaton moves from left to right over a string, it pushes J_i onto the stack whenever it sees $\overset{(}{_i}$, and it pops the stack, eliminating a J_i, whenever it sees $\overset{)}{_i}$. Such an automaton will successfully scan the entire string and terminate with an empty stack just in case the string belongs to $\text{PAR}_n(T)$.

To move toward making these ideas precise, let T be a given alphabet and let $P = \{\overset{(}{_i}, \overset{)}{_i} \mid i = 1, 2, \ldots, n\}$. Let $\Omega = \{J_1, J_2, \ldots, J_n\}$, where we have introduced a single symbol J_i for each pair $\overset{(}{_i}, \overset{)}{_i}$, $1 \leq i \leq n$. Let $u \in (T \cup P)^*$, say, $u = c_1 c_2 \cdots c_k$, where $c_1, c_2, \ldots, c_k \in T \cup P$. We define a sequence $\gamma_j(u)$ of elements of Ω^* as follows:

$$\gamma_1(u) = 0$$

$$\gamma_{j+1}(u) = \begin{cases} \gamma_j(u) & \text{if} \quad c_j \in T \\ J_i \gamma_j(u) & \text{if} \quad c_j = {}_i^(\\ \alpha & \text{if} \quad c_j = {}_i^) \text{ and } \gamma_j(u) = J_i \alpha, \end{cases}$$

for $j = 1, 2, \ldots, k$. Note that if $c_j = {}_i^)$, $\gamma_{j+1}(u)$ will be *undefined unless* γ_j *begins with the symbol* J_i *for the very same value of* i. Of course, if a particular $\gamma_r(u)$ is undefined, all $\gamma_j(u)$ with $j > r$ will also be undefined.

Definition. We say that the word $u \in (T \cup P)^*$ is *balanced* if $\gamma_j(u)$ is defined for $1 \leq j \leq |u| + 1$ and $\gamma_{|u|+1}(u) = 0$.

The heuristic considerations with which we began suggest

Theorem 8.1. Let T be an alphabet and let

$$P = \{{}_i^(, {}_i^) \mid i = 1, 2, \ldots, n\}, \qquad T \cap P = \varnothing.$$

Let $u \in (T \cup P)^*$, let $\Omega = \{J_1, J_2, \ldots, J_n\}$. Then $u \in \text{PAR}_n(T)$ if and only if u is balanced.

The proof is via a series of easy lemmas.

Lemma 1. If $u \in T^*$, then u is balanced.

Proof. Clearly $\gamma_j(u) = 0$ for $1 \leq j \leq |u| + 1$ in this case. ■

Lemma 2. If u and v are balanced, so is uv.

Proof. Clearly $\gamma_j(uv) = \gamma_j(u)$ for $1 \leq j \leq |u| + 1$. Since $\gamma_{|u|+1}(u) = 0 = \gamma_{|u|+1}(uv) = \gamma_1(v)$, we have $\gamma_{|u|+j}(uv) = \gamma_j(v)$ for $1 \leq j \leq |v| + 1$. Hence, $\gamma_{|uv|+1}(uv) = \gamma_{|u|+|v|+1}(uv) = \gamma_{|v|+1}(v) = 0$. ■

Lemma 3. Let $v = {}_i^(u {}_i^)$. Then u is balanced if and only if v is balanced.

Proof. We have $\gamma_1(v) = 0$, $\gamma_2(v) = J_i$, $\gamma_{j+1}(v) = \gamma_j(u)J_i$, $j = 1, 2, \ldots, |v| - 1$. In particular, $\gamma_{|v|}(v) = \gamma_{|u|+2}(v) = \gamma_{|u|+1}(u)J_i$. Thus, if u is balanced, then $\gamma_{|u|+1}(u) = 0$, so that $\gamma_{|v|}(v) = J_i$ and $\gamma_{|v|+1}(v) = 0$. Conversely, if v is balanced, $\gamma_{|v|+1}(v) = 0$, so that $\gamma_{|v|}(v)$ must be J_i and $\gamma_{|u|+1}(u) = 0$. ■

Lemma 4. If u is balanced and uv is balanced, then v is balanced.

Proof. $\gamma_j(uv) = \gamma_j(u)$ for $1 \leq j \leq |u| + 1$. Since $\gamma_{|u|+1}(u) = 0$, we have $\gamma_{|u|+j}(uv) = \gamma_j(v)$ for $1 \leq j \leq |v| + 1$. Finally,

$$0 = \gamma_{|uv|+1}(uv) = \gamma_{|u|+|v|+1}(uv) = \gamma_{|v|+1}(v). \qquad \blacksquare$$

Lemma 5. If $u \in \text{PAR}_n(T)$, then u is balanced.

Proof. The proof is by induction on the total number of occurrences of the symbols $(_i,)_i$ in u. If this number is 0, then $u \in T^*$, so by Lemma 1, u is balanced.

Proceeding by induction, let u have $k > 0$ occurrences of the symbols $(_i,)_i$, where the result is known for all strings with fewer than k occurrences of these symbols. Then, by Theorem 7.1d, we can write $u = v(_i w)_i z$, where $v, w, z \in \text{PAR}_n(T)$. By the induction hypothesis, v, w, z are all balanced, and by Lemmas 2 and 3, u is therefore balanced. ■

Lemma 6. If u is balanced, then $u \in \text{PAR}_n(T)$.

Proof. If $u \in T^*$, the result follows from Theorem 7.1a. Otherwise, we can write $u = xy$, where $x \in T^*$ and the initial symbol of y is in P. By the definition of $\gamma_j(u)$, we will have $\gamma_j(u) = 0$ for $1 \leq j \leq |x| + 1$. Therefore, the initial symbol of y cannot be one of the $)_i$. Thus we can write $u = x(_i z$, and $\gamma_{|x|+2}(u) = J_i$. Since u is balanced, $\gamma_{|u|+1}(u) = 0$, and we can let k be the least integer $> |x| + 1$ for which $\gamma_k(u) = 0$. Then $\gamma_{k-1}(u) = J_i$ and the $(k - 1)$th symbol of u must be $)_i$. Thus $u = x(_i v)_i w$, where $k = |x| + |v| + 3$. Thus $0 = \gamma_{|x|+|v|+3}(u) = \gamma_{|x|+|v|+3}(x(_i v)_i)$. Hence $x(_i v)_i$ is balanced. By Lemma 4, w is balanced. Since $x \in T^*$, x is balanced, and by Lemma 4 again, $(_i v)_i$ is balanced. By Lemma 3, v is balanced. Since $x \in T^*$, $x \in \text{PAR}_n(T)$. Since $|v|, |w| < |u|$, we can assume by mathematical induction that it is already known that $v, w \in \text{PAR}_n(T)$. By b and c of Theorem 7.1, we conclude that $u \in \text{PAR}_n(T)$. ■

Theorem 8.1 is an immediate consequence of Lemmas 5 and 6.

We now give a precise definition of *pushdown automata*. We begin with a finite set of *states* $Q = \{q_1, \ldots, q_m\}$, q_1 being the *initial state*, a subset $F \subseteq Q$ of *final*, or *accepting*, *states*, a *tape alphabet* A, and a *pushdown alphabet* Ω. (We usually use lowercase letters for elements of A and capital letters for elements of Ω.) We assume that the symbol **0** does not belong to either A or Ω and write $\bar{A} = A \cup \{0\}, \bar{\Omega} = \Omega \cup \{\mathbf{0}\}$. A *transition* is a quintuple of the form

$$q_i a U \colon V q_j$$

where $a \in \bar{A}$ and $U, V \in \bar{\Omega}$. Intuitively, if $a \in A$ and $U, V \in \Omega$, this is to read: "In state q_i scanning a, with U on top of the stack, move one square to the right, 'pop' the stack removing U, 'push' V onto the stack, and enter state q_j." If $a = \mathbf{0}$, motion to the right does not take place and the stack action can occur regardless of what symbol is actually being scanned. Similarly, $U = \mathbf{0}$ indicates that nothing is to be popped and $V = \mathbf{0}$ that nothing is to be pushed. A *pushdown automaton* is specified by a finite set of transitions. The *distinct* transitions $q_i aU{:}Vq_j$, $q_i bW{:}Xq_k$ are called *incompatible* if one of the following is the case:

1. $a = b$ and $U = W$;
2. $a = b$ and U or W is $\mathbf{0}$;
3. $U = W$ and a or b is $\mathbf{0}$;
4. a or b is $\mathbf{0}$ and U or W is $\mathbf{0}$.

A pushdown automaton is *deterministic* if it has no pair of incompatible transitions.

Let $u \in A^*$ and let $\mathcal{M}$ be a pushdown automaton. Then a *u-configuration for* $\mathcal{M}$ is a triple $\Delta = (k, q_i, \alpha)$, where $1 \le k \le |u| + 1$, q_i is a state of $\mathcal{M}$, and $\alpha \in \Omega^*$. [Intuitively, the u-configuration (k, q_i, α) stands for the situation in which u is written on $\mathcal{M}$'s tape, $\mathcal{M}$ is scanning the kth symbol of u—or, if $k = |u| + 1$, has completed scanning u—and α is the string of symbols on the pushdown stack.] We speak of q_i as the *state at configuration* Δ and of α as the *stack contents at configuration* Δ. If $\alpha = 0$, we say the *stack is empty at* Δ. For a pair of u-configurations, we write

$$u{:}\ (k, q_i, \alpha) \vdash_{\mathcal{M}} (l, q_j, \beta)$$

if $\mathcal{M}$ contains a transition $q_i aU{:}Vq_j$, where $\alpha = U\gamma$, $\beta = V\gamma$ for some $\gamma \in \Omega^*$, and either

1. $l = k$ and $a = \mathbf{0}$, or
2. $l = k + 1$ and the kth symbol of u is a.

Note that the equation $\alpha = U\gamma$ is to be read simply $\alpha = \gamma$ in case $U = \mathbf{0}$; likewise for $\beta = V\gamma$.

A sequence $\Delta_1, \Delta_2, \ldots, \Delta_m$ of u-configurations is called a *u-computation by* $\mathcal{M}$ if

1. $\Delta_1 = (1, q, 0)$ for some $q \in Q$,
2. $\Delta_m = (|u| + 1, p, \gamma)$ for some $p \in Q$ and $\gamma \in \Omega^*$, and
3. $u{:}\ \Delta_i \vdash_{\mathcal{M}} \Delta_{i+1}$ for $1 \le i < m$.

This u-computation is called *accepting* if the state at Δ_1 is the initial state q_1, the state p at Δ_m is in F, and the stack at Δ_m is empty. We say that $\mathcal{M}$ *accepts* the string $u \in A^*$ if there is an accepting u-computation by $\mathcal{M}$. We write $L(\mathcal{M})$ for the set of strings accepted by $\mathcal{M}$, and we call $L(\mathcal{M})$ the *language accepted by* $\mathcal{M}$.

Acceptance can alternatively be defined either by requiring only that the state at Δ_m is in F or only that $\gamma = 0$. It is not difficult to prove that the class of languages accepted by pushdown automata is not changed by either of these alternatives. (See Exercise 8.)

A few examples should provide readers with some intuition for working with pushdown automata.

Example $\mathscr{M}_1$ Tape alphabet = $\{a, b\}$, pushdown alphabet = $\{A\}$, $Q = \{q_1, q_2\}$, $F = \{q_2\}$. The transitions are

$$q_1 a \mathbf{0}: Aq_1$$
$$q_1 bA: \mathbf{0} q_2$$
$$q_2 bA: \mathbf{0} q_2.$$

The reader should verify that $L(\mathscr{M}_1) = \{a^{[n]}b^{[n]} \mid n > 0\}$.

Example $\mathscr{M}_2$ Tape alphabet = $\{a, b, c\}$, pushdown alphabet = $\{A, B\}$, $Q = \{q_1, q_2\}$, $F = \{q_2\}$. The transitions are

$$q_1 a \mathbf{0}: Aq_1$$
$$q_1 b \mathbf{0}: Bq_1$$
$$q_1 c \mathbf{0}: \mathbf{0} q_2$$
$$q_2 aA: \mathbf{0} q_2$$
$$q_2 bB: \mathbf{0} q_2.$$

Here, $L(\mathscr{M}_2) = \{ucu^R \mid u \in \{a, b\}^*\}$.

Example $\mathscr{M}_3$ Tape alphabet = $\{a, b\}$, pushdown alphabet = $\{A, B\}$, $Q = \{q_1, q_2\}$, $F = \{q_2\}$,

$$q_1 a \mathbf{0}: Aq_1$$
$$q_1 b \mathbf{0}: Bq_1$$
$$q_1 aA: \mathbf{0} q_2$$
$$q_1 bB: \mathbf{0} q_2$$
$$q_2 aA: \mathbf{0} q_2$$
$$q_2 bB: \mathbf{0} q_2.$$

In this case, $L(\mathscr{M}_3) = \{uu^R \mid u \in \{a, b\}^*, u \neq 0\}$. Note that while $\mathscr{M}_1, \mathscr{M}_2$ are deterministic, $\mathscr{M}_3$ is a nondeterministic pushdown automaton. Does there exist a deterministic pushdown automaton that accepts $L(\mathscr{M}_3)$? Why not?

$L(\mathcal{M}_1)$, $L(\mathcal{M}_2)$, and $L(\mathcal{M}_3)$ are all context-free languages. We begin our investigation of the relationship between context-free languages and pushdown automata with the following theorem.

Theorem 8.2. Let Γ be a Chomsky normal form grammar with separator Γ_s. Then there is a deterministic pushdown automaton $\mathcal{M}$ such that $L(\mathcal{M}) = L(\Gamma_s)$.

Proof. Let T be the set of terminals of Γ. By Theorem 7.3, for suitable n,

$$L(\Gamma_s) = R \cap \mathrm{PAR}_n(T),$$

where R is a regular language. Let $P = \{\overset{(}{_i}, \overset{)}{_i} \mid i = 1, 2, \ldots, n\}$. Let $\mathcal{M}_0$ be a dfa with alphabet $T \cup P$ that accepts R. Let $Q = \{q_1, \ldots, q_m\}$ be the states of $\mathcal{M}_0$, q_1 the initial state, $F \subseteq Q$ the accepting states, and δ the transition function. We construct a pushdown automaton $\mathcal{M}$ with tape alphabet $T \cup P$ and the same states, initial state, and accepting states as $\mathcal{M}_0$, $\mathcal{M}$ is to have the pushdown alphabet $\Omega = \{J_1, \ldots, J_n\}$. The transitions of $\mathcal{M}$ are as follows for all $q \in Q$:

a. for each $a \in T$, $qa\mathbf{0}$: $\mathbf{0}p$, where $p = \delta(q, a)$;
b. for $i = 1, 2, \ldots, n$, $q\overset{(}{_i}\mathbf{0}$: $J_i p_i$, where $p_i = \delta(q, \overset{(}{_i})$;
c. for $i = 1, 2, \ldots, n$, $q\overset{)}{_i} J_i$: $\mathbf{0}\bar{p}_i$, where $\bar{p}_i = \delta(q, \overset{)}{_i})$.

Since the second entry in these transitions is never $\mathbf{0}$, we see that for any $u \in (T \cup P)^*$, a u-computation must be of length $|u| + 1$. It is also clear that no two of the transitions in a–c are incompatible; thus, $\mathcal{M}$ is deterministic.

Now, let $u \in L(\Gamma_s)$, $u = c_1 c_2 \cdots c_K$, where $c_1, c_2, \ldots, c_K \in (T \cup P)$. Since $u \in R$, the dfa $\mathcal{M}_0$ accepts u. Thus, there is a sequence $p_1, p_2, \ldots, p_{K+1} \in Q$ such that $p_1 = q_1$, $p_{K+1} \in F$, and $\delta(p_i, c_i) = p_{i+1}$, $i = 1, 2, \ldots, K$. Since $u \in \mathrm{PAR}_n(T)$, by Theorem 8.1, u is balanced, so that $\gamma_j(u)$ is defined for $j = 1, 2, \ldots, K+1$ and $\gamma_{K+1}(u) = 0$. We let

$$\Delta_j = (j, p_j, \gamma_j(u)), \qquad j = 1, 2, \ldots, K+1.$$

To see that the sequence $\Delta_1, \Delta_2, \ldots, \Delta_{K+1}$ is an accepting u-computation by $\mathcal{M}$, it remains only to check that

$$u\colon \Delta_j \vdash_{\mathcal{M}} \Delta_{j+1}, \qquad j = 1, 2, \ldots, K.$$

But this clear from the definition of $\gamma_j(u)$.

Conversely, let $\mathcal{M}$ accept $u = c_1 c_2 \cdots c_K$. Thus, let $\Delta_1, \Delta_2, \ldots, \Delta_{K+1}$ be an accepting u-computation by $\mathcal{M}$. Let

$$\Delta_j = (j, p_j, \gamma_j), \qquad j = 1, 2, \ldots, K+1.$$

Since

$$u\colon \Delta_j \vdash_{\mathcal{M}} \Delta_{j+1}, \qquad j = 1, 2, \ldots, K,$$

and $\gamma_1 = 0$, we see that γ_j satisfies the defining recursion for $\gamma_j(u)$ and hence, $\gamma_j = \gamma_j(u)$ for $j = 1, 2, \ldots, K+1$. Since $\gamma_{K+1} = 0$, u is balanced and hence $u \in \mathrm{PAR}_n(T)$. Finally, we have $p_1 = q_1$, $p_{K+1} \in F$, and $\delta(p_j, c_j) = p_{j+1}$. Therefore the dfa $\mathcal{M}_0$ accepts u, and $u \in R$. ■

We call a pushdown automaton *atomic* (whether or not it is deterministic) if all of its transitions are of one of the forms

i. $pa\mathbf{0}\colon \mathbf{0}q$,
ii. $p\mathbf{0}U\colon \mathbf{0}q$,
iii. $p\mathbf{00}\colon Vq$.

Thus, at each step in a computation an atomic pushdown automaton can read the tape and move right, or pop a symbol off the stack or push a symbol on the stack. But, unlike pushdown automata in general, it cannot perform more than one of these actions in a single step.

Let $\mathcal{M}$ be a given *atomic* pushdown automaton with tape alphabet T and pushdown alphabet $\Omega = \{J_1, J_2, \ldots, J_n\}$. We set

$$P = \left\{\overset{(}{_i}, \overset{)}{_i} \mid i = 1, 2, \ldots, n\right\}$$

and show how to use the "brackets" to define a kind of "record" of a computation by $\mathcal{M}$. Let $\Delta_1, \Delta_2, \ldots, \Delta_m$ be a (not necessarily accepting) v-computation by $\mathcal{M}$, where $v = c_1 c_2 \cdots c_K$ and $c_k \in T$, $k = 1, 2, \ldots, K$, and where $\Delta_i = (l_i, p_i, \gamma_i)$, $i = 1, 2, \ldots, m$. We set

$$w_1 = 0$$

$$w_{i+1} = \begin{cases} w_i c_{l_i} & \text{if } \gamma_{i+1} = \gamma_i \\ w_i \overset{(}{_j} & \text{if } \gamma_{i+1} = J_j \gamma_i \\ w_i \overset{)}{_j} & \text{if } \gamma_i = J_j \gamma_{i+1} \end{cases} \qquad 1 \le i < m.$$

[Note that $\gamma_{i+1} = \gamma_i$ is equivalent to $l_{i+1} = l_i + 1$ and is the case when a transition of form i is used in getting from Δ_i to Δ_{i+1}; the remaining two cases occur when transitions of the form iii or ii, respectively, are used.] Now let $w = w_m$, so that $\mathrm{Er}_P(w) = v$ and $m = |w| + 1$. This word w is called the *record* of the given v-computation $\Delta_1, \ldots, \Delta_m$ by $\mathcal{M}$. From w we can read off not only the word v but also the sequence of "pushes" and "pops" as they occur. In particular, $w_i, 1 < i \le m$, indicates how $\mathcal{M}$ goes from Δ_{i-1} to Δ_i.

Now we want to modify the pushdown automaton $\mathscr{M}$ of Theorem 8.2 so that it will accept $L(\Gamma)$ instead of $L(\Gamma_s)$. In doing so we will have to give up determinism. The intuitive idea is to use nondeterminism by permitting our modified pushdown automaton to "guess" the location of the "brackets" $\overset{(}{_i}, \overset{)}{_i}$. Thus, continuing to use the notation of the proof of Theorem 8.2, we define a pushdown automaton $\bar{\mathscr{M}}$ with the same states, initial state, accepting states, and pushdown alphabet as $\mathscr{M}$. However, the tape alphabet of $\bar{\mathscr{M}}$ will be T (rather than $T \cup P$). The transitions of $\bar{\mathscr{M}}$ are, for all $q \in Q$:

a. for each $a \in T$, $qa\mathbf{0}$: $\mathbf{0}p$, where $p = \delta(q, a)$ [i.e., the same as the transitions a of $\mathscr{M}$];
b. for $i = 1, 2, \ldots, n$, $q\mathbf{00}$: $J_i p_i$, where $p_i = \delta(q, \overset{(}{_i})$;
c. for $i = 1, 2, \ldots, n$, $q\mathbf{0}J_i$: $\mathbf{0}\bar{p}_i$ where $\bar{p}_i = \delta(q, \overset{)}{_i})$.

Depending on the transition function δ, $\bar{\mathscr{M}}$ can certainly be nondeterministic. We shall prove that $L(\bar{\mathscr{M}}) = L(\Gamma)$. Note that $\bar{\mathscr{M}}$ is atomic (although $\mathscr{M}$ is not).

First, let $v \in L(\Gamma)$. Then, since $\mathrm{Er}_p(L(\Gamma_s)) = L(\Gamma)$, there is a word $w \in L(\Gamma_s)$ such that $\mathrm{Er}_p(w) = v$. By Theorem 8.2, $w \in L(\mathscr{M})$. Let $\Delta_1, \Delta_2, \ldots, \Delta_m$ be an accepting w-computation by $\mathscr{M}$ (where in fact $m = |w| + 1$). Let

$$\Delta_i = (i, p_i, \gamma_i), \qquad i = 1, 2, \ldots, m.$$

Let $n_i = 1$ if w: $\Delta_i \vdash_{\mathscr{M}} \Delta_{i+1}$ via a transition belonging to group a; otherwise $n_i = 0, 1 \leq i < m$. Let

$$l_1 = 1,$$

$$l_{i+1} = l_i + n_i, \qquad 1 \leq i < m.$$

Finally let

$$\bar{\Delta}_i = (l_i, p_i, \gamma_i), \qquad i = 1, 2, \ldots, m.$$

Then, as is easily checked,

$$v: \bar{\Delta}_i \vdash_{\bar{\mathscr{M}}} \bar{\Delta}_{i+1}, \qquad 1 \leq i < m.$$

Since $\bar{\Delta}_m = (|v| + 1, q, 0)$ with $q \in F$, we have $v \in L(\bar{\mathscr{M}})$.

Conversely, let $v \in L(\bar{\mathscr{M}})$. Let $\bar{\Delta}_1, \bar{\Delta}_2, \ldots, \bar{\Delta}_m$ be an accepting v-computation by $\bar{\mathscr{M}}$, where we may write

$$\bar{\Delta}_i = (l_i, p_i, \gamma_i), \qquad i = 1, 2, \ldots, m.$$

Using the fact that $\bar{\mathcal{M}}$ is atomic, we can let w be the *record* of this computation in the sense defined earlier so that $\mathrm{Er}_P(w) = v$ and $m = |w| + 1$. We write

$$\Delta_i = (i, p_i, \gamma_i), \qquad i = 1, 2, \ldots, m,$$

and easily observe that

$$w: \Delta_i \vdash_{\mathcal{M}} \Delta_{i+1}, \qquad i = 1, 2, \ldots, m.$$

[In effect, whenever $\bar{\mathcal{M}}$ pushes J_i onto its stack, $\overset{(}{_i}$ is inserted into w; and whenever $\bar{\mathcal{M}}$ pops J_i, $\overset{)}{_i}$ is inserted into w. This makes the transitions b, c of $\mathcal{M}$ behave on w just the way the corresponding transitions of $\bar{\mathcal{M}}$ behave on v.] Since $p_m \in F$ and $\gamma_m = 0$, $\Delta_1, \Delta_2, \ldots, \Delta_m$ is an accepting w-computation by $\mathcal{M}$. Thus, by Theorem 8.2, $w \in L(\Gamma_s)$. Hence $v \in L(\Gamma)$.

We have shown that $L(\Gamma) = L(\bar{\mathcal{M}})$. Hence we have proved

Theorem 8.3. Let Γ be a Chomsky normal form grammar. Then there is a pushdown automaton $\bar{\mathcal{M}}$ such that $L(\bar{\mathcal{M}}) = L(\Gamma)$.

Now let L be any context-free language. By Theorems 1.2 and 3.1 there is a Chomsky normal form grammar Γ such that $L = L(\Gamma)$ or $L(\Gamma) \cup \{0\}$. In the former case, we have shown how to obtain a pushdown automaton $\bar{\mathcal{M}}$ such that $L = L(\bar{\mathcal{M}})$. For the latter case we first modify the dfa $\mathcal{M}_0$ used in the *proof of Theorem 8.2 so that it is nonrestarting*. We know that this can be done without changing the regular language that $\mathcal{M}_0$ accepts by Theorem 4.1 in Chapter 9. By carrying out the construction of a pushdown automaton $\bar{\mathcal{M}}$ for which $L(\bar{\mathcal{M}}) = L(\Gamma)$ using the modified version of $\mathcal{M}_0$, $\bar{\mathcal{M}}$ will have the property that none of its transitions has q_1 as its final symbol. That is, $\bar{\mathcal{M}}$ will never return to its initial state. Thus, if we define $\mathcal{M}'$ to be exactly like $\bar{\mathcal{M}}$ except for having as its set of accepting states

$$F' = F \cup \{q_1\},$$

we see that $L(\mathcal{M}') = L(\bar{\mathcal{M}}) \cup \{0\} = L(\Gamma) \cup \{0\}$. Thus we have proved

Theorem 8.4. For every context-free language L, there is a pushdown automaton $\mathcal{M}$ such that $L = L(\mathcal{M})$.

We will end this section by proving the converse of this result. Thus we must begin with a pushdown automaton and prove that the language it accepts is context-free. As a first step toward this goal, we will show that we can limit our considerations to atomic pushdown automata.

Theorem 8.5. Let $\mathcal{M}$ be a pushdown automaton. Then there is an atomic pushdown automaton $\bar{\mathcal{M}}$ such that $L(\mathcal{M}) = L(\bar{\mathcal{M}})$.

Proof. For each transition

$$paU: Vq$$

of $\mathscr{M}$ for which $a, U, V \neq \mathbf{0}$, we introduce two new states r, s and let $\bar{\mathscr{M}}$ have the transitions

$$pa\mathbf{0}: \mathbf{0}r,$$
$$r\mathbf{0}U: \mathbf{0}s,$$
$$s\mathbf{0}\mathbf{0}: Vq.$$

If exactly one of a, U, V is $\mathbf{0}$, then only two transitions are needed for $\bar{\mathscr{M}}$. Finally, for each transition $p\mathbf{0}\mathbf{0}: \mathbf{0}q$, we introduce a new state t and replace $p\mathbf{0}\mathbf{0}: \mathbf{0}q$ with the transitions $p\mathbf{0}\mathbf{0}: Jt$, $t\mathbf{0}J: \mathbf{0}q$, where J is an arbitrary symbol of the pushdown alphabet (or a new symbol if the pushdown alphabet of $\mathscr{M}$ is empty). Otherwise, $\bar{\mathscr{M}}$ is exactly like $\mathscr{M}$. Clearly, $L(\bar{\mathscr{M}}) = L(\mathscr{M})$. ■

Theorem 8.6. For every pushdown automaton $\mathscr{M}$, $L(\mathscr{M})$ is a context-free language.

Proof. Without loss of generality, by using Theorem 8.5 we can assume that $\mathscr{M}$ is atomic. Let $\mathscr{M}$ have states $Q = \{q_1, \ldots, q_m\}$, initial state q_1, final states F, tape alphabet T, and pushdown alphabet $\Omega = \{J_1, \ldots, J_n\}$. Let $P = \{ (_i,)_i \mid i = 1, \ldots, n\}$. Let $L \subseteq (T \cup P)^*$ consist of the records of every accepting u-computation by $\mathscr{M}$, and let $R = L(\mathscr{M}_0)$, where $\mathscr{M}_0$ is the ndfa with alphabet $T \cup P$, the same states, initial state, and accepting states as $\mathscr{M}$, and transition function δ defined as follows. For each $q \in Q$,

$$\delta(q, a) = \{p \in Q \mid \mathscr{M} \text{ has the transition } qa\mathbf{0}: \mathbf{0}p\} \text{ for } a \in T,$$

$$\delta(q, (_i) = \{p \in Q \mid \mathscr{M} \text{ has the transition } q\mathbf{0}\mathbf{0}: J_i p\},\ i = 1, \ldots, n,$$

$$\delta(q,)_i) = \{p \in Q \mid \mathscr{M} \text{ has the transition } q\mathbf{0}J_i: \mathbf{0}p\},\ i = 1, \ldots, n.$$

Let $w \in L$ be the record of an accepting u-computation $\Delta_1, \ldots, \Delta_m$, where $\Delta_i = (l_i, p_i, \gamma_i)$, $i = 1, \ldots, m$. An easy induction on i shows that $p_i \in \delta^*(q_1, w_i)$, $i = 1, \ldots, m$, so, in particular, $p_m \in \delta^*(q_1, w)$, which implies $w \in R$, since p_m must be an accepting state. Moreover, another easy induction on i shows that $\gamma_i(w) = \gamma_i$, $i = 1, \ldots, m$, which implies that $\gamma_i(w)$ is defined for $1 \leq i \leq |w| + 1$ and $\gamma_{|w|+1}(w) = \gamma_{|w|+1} = 0$ (since $\Delta_1, \ldots, \Delta_m$ is accepting), i.e., w is balanced. Therefore, by Theorem 8.1, $w \in R \cap \mathrm{PAR}_n(T)$, and so $L \subseteq R \cap \mathrm{PAR}_n(T)$.

On the other hand, let $w = c_1 \cdots c_r$ be a balanced word in R, i.e., $w \in R \cap \mathrm{PAR}_n(T)$, let $u = d_1 \cdots d_s$ be $\mathrm{Er}_P(w)$, and let $p_1, \ldots, p_{r+1}$ be some sequence of states such that $p_1 = q_1, p_{r+1} \in F$, and $p_{i+1} \in \delta(p_i, c_i)$ for $i = 1, \ldots, r$. We claim that

$$(l_1, p_1, \gamma_1(w)), (l_2, p_2, \gamma_2(w)), \ldots, (l_{r+1}, p_{r+1}, \gamma_{r+1}(w)), \qquad (8.1)$$

where

$$l_1 = 1$$

$$l_{i+1} = \begin{cases} l_i + 1 & \text{if} \quad c_i \in T \\ l_i & \text{otherwise,} \end{cases}$$

is an accepting u-computation by $\mathscr{M}$ and that w is its record. Clearly, we have $(l_1, p_1, \gamma_1(w)) = (1, q_1, 0), l_{r+1} = |u| + 1, p_{r+1} \in F$, and $\gamma_{r+1}(w) = 0$ (since w is balanced), so we just need to show that

$$u: (l_i, p_i, \gamma_i(w)) \vdash_{\mathscr{M}} (l_{i+1}, p_{i+1}, \gamma_{i+1}(w)) \qquad (8.2)$$

for $i = 1, \ldots, r$. For arbitrary $i = 1, \ldots, r$, if $\gamma_{i+1}(w) = \gamma_i(w)$, then $c_i \in T$, so $p_{i+1} \in \delta(p_i, c_i)$, and $\mathscr{M}$ has the transition $p_i c_i \mathbf{0}: \mathbf{0} p_{i+1}$. Now, a simple induction on i shows that $\mathrm{Er}_P(c_1 \cdots c_{i-1}) = d_1 \ldots d_{l_i - 1}$, $i = 1, \ldots, r + 1$ (where $c_1 \cdots c_0$ represents 0), from which we can show

$$\text{if} \quad c_i \in T \quad \text{then} \quad d_{l_i} = c_i, \qquad i = 1, \ldots, r.$$

In particular, for any $i = 1, \ldots, r$, if $c_i \in T$, then

$$\mathrm{Er}_P(c_1 \cdots c_i) = \mathrm{Er}_P(c_1 \cdots c_{i-1}) c_i = d_1 \cdots d_{l_i - 1} c_i,$$

so c_i must be d_{l_i} since c_i is not deleted when Er_P is applied to w. Therefore, $\mathscr{M}$ has the transition $p_i d_{l_i} \mathbf{0}: \mathbf{0} p_{i+1}$ and $l_{i+1} = l_i + 1$, so (8.2) is satisfied. If, instead, $\gamma_{i+1}(w) = J_j \gamma_i(w)$ for some $j = 1, \ldots, n$, then $c_i = {}^{(}_{j}$, so $p_{i+1} \in \delta(p_i, {}^{(}_{j})$ and $\mathscr{M}$ has the transition $p_i \mathbf{00}: J_j p_{i+1}$. Moreover, $c_j \notin T$, so $l_{i+1} = l_i$, and (8.2) is satisfied in this case as well. Finally, if $\gamma_i(w) = J_j \gamma_{i+1}(w)$ for some $j = 1, \ldots, n$, then $c_i = {}^{)}_{j}$, so $p_{i+1} \in \delta(p_i, {}^{)}_{j})$ and $\mathscr{M}$ has the transition $p_i \mathbf{0} J_j: \mathbf{0} p_{i+1}$. Moreover, $l_{i+1} = l_i$, so again (8.2) is satisfied. Therefore, (8.1) is an accepting u-computation by $\mathscr{M}$. If we set $w_i = c_1 \cdots c_{i-1}$, $i = 1, \ldots, r + 1$, then an induction on i shows that w is indeed the record of (8.1), so $w \in L$, and we have $R \cap \mathrm{PAR}_n(T) \subseteq L$. Therefore, $L = R \cap \mathrm{PAR}_n(T)$, and

$$L(\mathscr{M}) = \mathrm{Er}_P(R \cap \mathrm{PAR}_n(T)).$$

Finally, by Theorems 5.4 and 7.1 and Corollary 5.6, $L(\mathscr{M})$ is context-free. ■

Exercises

1. Let T be an alphabet, $P = \{(_i,)_i \mid i = 1, \ldots, n\}$, $w = a_1 \cdots a_m \in \text{PAR}_n(T)$. Integers j, k, where $1 \le j < k \le m$, are *matched in* w if $w = a_1 \cdots a_{j-1} (_i a_{j+1} \cdots a_{k-1})_i a_{k+1} \cdots a_m$, for some $1 \le i \le n$, and if $a_{j+1} \cdots a_{k-1}$ is balanced. Let Γ be a Chomsky normal form grammar.
   - (a) Let $w = (_i x \in L(\Gamma_s)$. Prove that there is exactly one k, $1 < k \le |w|$, such that 1 and k are matched in w.
   - (b) Show that Γ_s is unambiguous. [See Section 6 for the definition of *ambiguous grammars*.]
2. (a) For pushdown automaton $\mathcal{M}_1$ in the examples, give the accepting u_1-computation for $u_1 = aabb$.
   - (b) For pushdown automaton $\mathcal{M}_2$ in the examples and $u_2 = abcbba$, give the longest sequence $\Delta_1 = (1, q_1, 0), \Delta_2, \ldots, \Delta_m$ of u_2-configurations that satisfies condition 3 in the definition of u-computations.
   - (c) For pushdown automaton $\mathcal{M}_3$ in the examples, give all possible u_3-computations, accepting or not, for $u_3 = aaaa$.
3. For each of the following languages L, give a pushdown automaton that accepts L.
   - (a) $\{a^{[n]}b^{[2n]} \mid n > 0\}$.
   - (b) $\{a^{[n]}b^{[m]} \mid 0 < n \le m\}$.
   - (c) $\{a^{[n]}b^{[m]} \mid n \ne m\}$.
   - (d) $\{a^{[n]}b^{[m]}a^{[n]} \mid m, n > 0\} \cup \{a^{[n]}c^{[n]} \mid n > 0\}$.
4. Let $\mathcal{M}$ be the pushdown automaton with $Q = \{q_1\}$, $F = \{q_1\}$, and transitions

$$q_1 a \mathbf{0}\colon A q_1 \quad q_1 a B\colon \mathbf{0} q_1$$
$$q_1 b \mathbf{0}\colon B q_1 \quad q_1 b A\colon \mathbf{0} q_1 .$$

   What is $L(\mathcal{M})$?

5. Let $\mathcal{M}$ be the pushdown automaton with $Q = \{q_1, q_2, q_3, q_4, q_5\}$, $F = \{q_5\}$, and transitions

$$\begin{array}{lll} q_1\mathbf{00}\colon Zq_2 & q_3bA\colon \mathbf{0}q_3 & q_4a\mathbf{0}\colon \mathbf{0}q_4 \\ q_2a\mathbf{0}\colon Aq_2 & q_3bZ\colon \mathbf{0}q_4 & q_4b\mathbf{0}\colon \mathbf{0}q_4 \\ q_2bA\colon \mathbf{0}q_3 & q_3a\mathbf{0}\colon \mathbf{0}q_4 & q_4\mathbf{0}A\colon \mathbf{0}q_4 \\ q_2bZ\colon \mathbf{0}q_4 & q_3\mathbf{0}Z\colon \mathbf{0}q_5 & q_4\mathbf{0}Z\colon \mathbf{0}q_4 . \end{array}$$

(a) What is $L(\mathcal{M})$?

(b) Prove that for every $u \in \{a, b\}^*$, there is a u-computation by $\mathcal{M}$.

6. Show that every regular language accepted by a (deterministic) finite automaton with n states is accepted by a (deterministic) pushdown automaton with n states and an empty pushdown alphabet.

7. Show that every regular language R is accepted by a pushdown automaton with at most two states, and if $0 \in R$ then R is accepted by a pushdown automaton with one state.

8. Let $\mathcal{M}$ be a pushdown automaton with initial state q_1, accepting states F, and tape alphabet A, let $u \in A^*$, and let $\Delta_1 = (1, q_1, 0), \ldots, \Delta_m = (|u| + 1, p, \gamma)$ be a u-computation by $\mathcal{M}$. We say that $\mathcal{M}$ *accepts u by final state* if $p \in F$, and that $\mathcal{M}$ *accepts u by empty stack* if $\gamma = 0$. $T(\mathcal{M}) = \{u \in A^* \mid \mathcal{M}$ accepts u by final state$\}$, and $N(\mathcal{M}) = \{u \in A^* \mid \mathcal{M}$ accepts u by empty stack$\}$.

   (a) Let $\mathcal{M}_1, \mathcal{M}_2, \mathcal{M}_3$ be the pushdown automata from the examples. Give $T(\mathcal{M}_i)$, $N(\mathcal{M}_i)$, $i = 1, 2, 3$.

   (b) Prove that a language L is context-free if and only if $L = T(\mathcal{M})$ for some pushdown automaton $\mathcal{M}$.

   (c) Prove that a language L is context-free if and only if $L \cup \{0\} = N(\mathcal{M})$ for some pushdown automation $\mathcal{M}$.

9. Let $\mathcal{M}$ be a pushdown automaton with tape alphabet A, and let $u \in A^*$. An infinite sequence $\Delta_1, \Delta_2, \ldots$ of u-configurations for $\mathcal{M}$ is an *infinite u-computation by* $\mathcal{M}$ if for some n and some x such that $u = xy$ for some y, each finite sequence $\Delta_1, \ldots, \Delta_n, \ldots, \Delta_{n+m}$, $m \geq 0$, is an x-computation by $\mathcal{M}$. It is an *accepting* infinite u-computation if $\Delta_1, \ldots, \Delta_k$ is an accepting u-computation by $\mathcal{M}$ for some k.

   (a) Give a pushdown automaton $\mathcal{M}_1$ and word u_1 such that there is a nonaccepting infinite u-computation by $\mathcal{M}_1$.

   (b) Give a pushdown automaton $\mathcal{M}_2$ and word u_2 such that there is an accepting infinite u_2-computation $(l_1, p_1, \gamma_1), (l_2, p_2, \gamma_2), \ldots$ by $\mathcal{M}_2$ where, for some k, p_l is an accepting state for all $l \geq k$.

   (c) Give a pushdown automaton $\mathcal{M}_3$ and word u_3 such that there is an accepting infinite u_3-computation $(l_1, p_1, \gamma_1), (l_2, p_2, \gamma_2), \ldots$ by $\mathcal{M}_3$ where there is no k such that p_l is an accepting state for all $l \geq k$.

10. Give the incompatible pairs among the following transitions. In each case, give the condition(s) 1, 2, 3, or 4 by which the pair is incompatible.

$$\begin{array}{lll} q_1 a J_1\colon \mathbf{0} q_1 & q_1 a \mathbf{0}\colon J_2 q_1 & q_1 b J_2\colon J_1 q_1 \\ q_1 b J_1\colon \mathbf{0} q_1 & q_1 \mathbf{0} J_1\colon \mathbf{0} q_1 & q_1 \mathbf{0}\mathbf{0}\colon J_1 q_1 \\ q_1 a J_1\colon \mathbf{0} q_2 & & \end{array}$$

11. Let $T = \{a, b\}$, $P = \{(_1,)_1, (_2,)_2\}$, $\Omega = \{J_1, J_2\}$. We will write (,), [,] for $(_1,)_1, (_2,)_2$, respectively. Give $\gamma_i(w)$, $1 \le i \le |w| + 1$, for each of the following.

(a) $w = a(b[ba]a)b[a]$.

(b) $w = (ab[ab)a]$.

(c) $w = a[b]]a$.

(d) $w = (a([b]a)$.

12. Let Γ be the grammar with productions $S \to SS$, $S \to a$.

(a) Use the construction in the proof of Theorem 8.2 to give a deterministic pushdown automaton that accepts $L(\Gamma_s)$.

(b) Use the construction in the proof of Theorem 8.3 to give a pushdown automaton that accepts $L(\Gamma)$.

13. **(a)** For pushdown automata $\mathcal{M}_1, \mathcal{M}_2, \mathcal{M}_3$ in the examples, use the construction in the proof of Theorem 8.5 to give atomic pushdown automata $\bar{\mathcal{M}}_1, \bar{\mathcal{M}}_2, \bar{\mathcal{M}}_3$.

(b) Answer Exercise 2 for $\bar{\mathcal{M}}_1, \bar{\mathcal{M}}_2, \bar{\mathcal{M}}_3$.

14. Let $\mathcal{M}$ be the pushdown automaton with $Q = \{q_1, q_2\}$, initial state q_1, $F = \{q_2\}$, tape alphabet $\{a, b\}$, pushdown alphabet $\{A\}$, and transitions

$$\begin{array}{llll} q_1 a \mathbf{0}\colon \mathbf{0} q_1 & q_1 \mathbf{0}\mathbf{0}\colon A q_1 & q_2 a \mathbf{0}\colon \mathbf{0} q_1 & q_2 \mathbf{0} A\colon \mathbf{0} q_2 \\ q_1 b \mathbf{0}\colon \mathbf{0} q_2 & q_1 \mathbf{0} A\colon \mathbf{0} q_1 & q_2 b \mathbf{0}\colon \mathbf{0} q_2 & q_2 \mathbf{0}\mathbf{0}\colon A q_2 . \end{array}$$

Use the constructions in Theorems 8.6 and 5.4 to give a context-free grammar Γ such that $L(\Gamma) = L(\mathcal{M})$.

15. Let us call a *generalized pushdown automaton* a device that functions just like a pushdown automaton except that it can write any finite sequence of symbols on the stack in a single step. Show that, for every generalized pushdown automaton $\mathcal{M}$, there is a pushdown automaton $\bar{\mathcal{M}}$ such that $L(\mathcal{M}) = L(\bar{\mathcal{M}})$.

16.* Let

$$P = \left\{ \begin{array}{|c|} \hline u_i \\ \hline s v_i \\ \hline \end{array} \;\middle|\; i = 1, 2, \ldots, k \right\}$$

be a set of dominoes on the alphabet A. Let $B = \{c_1, \ldots, c_k\}$ be an alphabet such that $A \cap B = \varnothing$. Let $c \notin A \cup B$. Let

$$R = \{ycy^R \mid y \in A^*B^*\},$$

$$L_1 = \{u_{i_1}u_{i_2} \cdots u_{i_n}c_{i_n}c_{i_{n-1}} \cdots c_{i_2}c_{i_1}\},$$

$$L_2 = \{v_{i_1}v_{i_2} \cdots v_{i_n}c_{i_n}c_{i_{n-1}} \cdots c_{i_2}c_{i_1}\},$$

$$S_p = \{ycz^R \mid y \in L_1, z \in L_2\}.$$

Recall that by Theorem 6.5, the Post correspondence problem P has a solution if and only if $L_1 \cap L_2 \neq \varnothing$.

(a) Show that the Post correspondence problem P has *no* solution if and only if $R \cap S_p = \varnothing$.

(b) Show that $(A \cup B \cup \{c\})^* - R$ and $(A \cup B \cup \{c\})^* - S_P$ are both context-free. [*Hint:* Construct pushdown automata.]

(c) From (a) and (b) show how to conclude that there is no algorithm that can determine for a given context-free grammar Γ with terminals T whether $L(\Gamma) \cup \{0\} = T^*$.

(d) Now show that there is no algorithm that can determine for a given context-free grammar Γ_1 and regular grammar Γ_2 whether

(i) $L(\Gamma_1) = L(\Gamma_2)$,

(ii) $L(\Gamma_1) \supseteq L(\Gamma_2)$.

17.* Let $\mathcal{M}$ be a pushdown automaton with $Q = \{q_1, \ldots, q_m\}$, tape alphabet $A = \{a_1, \ldots, a_n\}$, and pushdown alphabet $\Omega = \{J_1, \ldots, J_l\}$, and let $p, p' \in Q, J, J' \in \Omega$. A sequence $(1, p_1, \gamma_1), \ldots, (1, p_k, \gamma_k)$ of 0-configurations for $\mathcal{M}$ is a *reaching sequence by* $\mathcal{M}$ *from* (p, J) *to* (p', J') if $p_1 = p$, $\gamma_1 = J$, $p_k = p'$, $\gamma_k = J'\delta$ for some $\delta \in \Omega^*$, $|\gamma_i| > 0$ for $1 \le i \le k$, and $0{:}(1, p_i, \gamma_i) \vdash_{\mathcal{M}} (1, p_{i+1}, \gamma_{i+1})$ for $1 \le i < k$. (p, J) is a *looping pair of* $\mathcal{M}$ if there is a reaching sequence by $\mathcal{M}$ from (p, J) to (p, J).

(a) Prove that if $\mathcal{M}$ has a u-computation $\Delta_1, \ldots, \Delta_k = (|u| + 1,$ $p, J\gamma)$ for some looping pair (p, J) of $\mathcal{M}$, then $\mathcal{M}$ has an infinite uw-computation for every $w \in A^*$. [See Exercise 9 for the definition of infinite u-computations.]

(b) Prove that if (p, J) is a looping pair for $\mathcal{M}$, then there is a reaching sequence $\Delta_1 = (1, p, J), \ldots, \Delta_k = (1, p, J\delta)$ by $\mathcal{M}$ from (p, J) to (p, J) such that $|\delta| \le lm$ [*Hint:* Consider the pigeonhole principle and the proofs of the pumping lemmas.]

(c) Prove that if (p, J) is a looping pair of $\mathcal{M}$, then there is a reaching sequence $\Delta_1, \ldots, \Delta_k$ by $\mathcal{M}$ from (p, J) to (p, J) with $k \leq m(l + 1)^{lm} + 1$.

(d) Give an algorithm that will determine, for a pushdown automaton $\mathcal{M}$ and pair (p, J), whether or not (p, J) is a looping pair of $\mathcal{M}$.

(e) Prove that if $\mathcal{M}$ has an infinite u-computation, for some $u \in A^*$, then $\mathcal{M}$ has a looping pair.

(f) Suppose now that $\mathcal{M}$ is deterministic. Prove that there is a deterministic pushdown automaton $\mathcal{M}'$ such that

(i) there is no infinite u-computation by $\mathcal{M}'$ for any $u \in A^*$;

(ii) there is a u-computation by $\mathcal{M}'$ for every $u \in A^*$, and

(iii) $T(\mathcal{M}') = T(\mathcal{M})$. [See Exercise 8 for the definition of $T(\mathcal{M})$.]

(g) A language L is a *deterministic context-free language* if $L = T(\mathcal{M})$ for some deterministic pushdown automaton $\mathcal{M}$. Prove that if $L \subseteq A^*$ is a deterministic context-free language, then $A^* - L$ is also a deterministic context-free language.

(h) Show that $\{a^{[i]}b^{[j]}c^{[k]} \mid i \neq j \text{ or } j \neq k\}$ is a context-free language which is not deterministic.

(i) Show that there is an algorithm that can determine for a given deterministic pushdown automaton $\mathcal{M}$ and dfa $\mathcal{M}'$ whether $T(\mathcal{M}) = L(\mathcal{M}')$.

9. Compilers and Formal Languages

A *compiler* is a program that takes as input a program (known as the *source program*) written in a high-level language such as COBOL, FORTRAN, or Pascal and translates it into an equivalent program (known as the *object program*) in a low-level language such as an assembly language or a machine language. Just as in Chapters 2 and 5 we found it easier to write programs with the aid of macros, most programmers find programming in a high-level language faster, easier, and less tedious than in a low-level language. Thus the need for compilers.

The translation process is divided into a sequence of phases, of which the first two are of particular interest to us. *Lexical analysis*, which is the first phase of the compilation process, consists of dividing the characters of the source program into groups called *tokens*. Tokens are the logical units of an instruction and include keywords such as **IF, THEN,** and **DO,** operators such as + and $*$, predicates such as $>$, variable names, labels, constants, and punctuation symbols such as (and ;.

The reason that the lexical analysis phase of compilation is of interest to us is that it represents an application of the theory of finite automata and regular expressions. The lexical analyzer must identify tokens, determine types, and store this information into a *symbol table* for later use. Typically, compiler writers use nondeterministic finite automata to design these token recognizers. For example, the following is an ndfa that recognizes unsigned integer constants.

Similarly, a nondeterministic finite automaton that recognizes variable names might look like this:

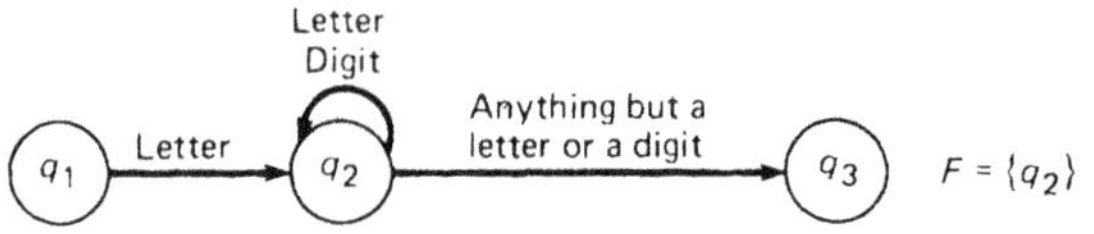

We end our brief discussion of lexical analysis by noting that it is not always a simple task to properly determine the division into tokens. For example, in FORTRAN, the statements

DO 10 I = 1.11

and

DO 10 I = 1,11

look very similar but are in fact totally unrelated instructions. The first is an assignment statement that assigns to a variable named DO10I (embedded blanks are ignored in FORTRAN) the value 1.11. The second is a **DO** loop that indicates that the body is to be performed 11 times. It is not until the "." or "," is encountered that the statement type can be determined.

At the completion of the lexical analysis phase of compilation, tokens have been identified, their types determined, and when appropriate, the value entered in the symbol table. At this point, the second phase of compilation, known as *syntactic analysis* or *parsing*, begins. It is in this second phase that context-free grammars play a central role.

For programming languages that are context-free, the parsing problem amounts to determining for a given context-free grammar Γ and word w

1. whether $w \in L(\Gamma)$, and
2. if $w \in L(\Gamma)$, how w could have been generated.

Intuitively, the parsing phase of the compilation process consists of the construction of derivation or parse trees whose leaves are the tokens identified by the lexical analyzer.

Thus, for example, if our grammar included the productions

$$
\begin{aligned}
S &\to \text{while-statement} \\
S &\to \text{assignment-statement} \\
\text{while-statement} &\to \textbf{while} \text{ cond } \textbf{do } S \\
\text{cond} &\to \text{cond} \vee \text{cond} \\
\text{cond} &\to \text{rel} \\
\text{rel} &\to \text{exp pred exp} \\
\text{exp} &\to \text{exp} + \text{exp} \\
\text{exp} &\to \text{var} \\
\text{exp} &\to \text{const} \\
\text{pred} &\to \, > \\
\text{pred} &\to \, = \\
\text{assignment-statement} &\to \text{var} \leftarrow \text{exp}
\end{aligned}
$$

then the parse tree for the statement

$$\textbf{while } x > y \vee z = 2 \textbf{ do } w \leftarrow x + 4$$

is given by Fig. 9.1.

The parsing is usually accomplished by simulating the behavior of a pushdown automaton that accepts $L(\Gamma)$ either starting from the root of the tree or the leaves of the tree. In the former case, this is known as *top-down parsing* and in the latter case, *bottom-up parsing*.

Most programming languages are for the most part context-free. (A major exception is the coordination of declarations and uses.) A common technique involves the definition of a superset of the programming language which can be accepted by a deterministic pushdown automaton. This is desirable since there are particularly fast algorithms for parsing grammars associated with deterministic pushdown automata.

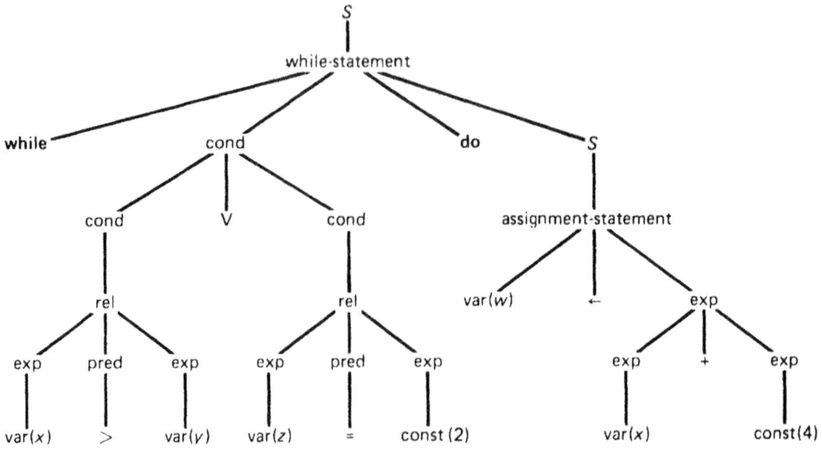

Figure 9.1

Exercise

1. Give a context-free grammar for generating valid Pascal arithmetic expressions over the alphabet $\{a, b, +, -, *, /, \uparrow, (,)\}$, where variable names are elements of $\{a, b\}^*$ of length at least 1. Is the grammar ambiguous? What are the implications of this?

11

Context-Sensitive Languages

1. The Chomsky Hierarchy

We are now going to place our work in the context of Noam Chomsky's hierarchy of grammars and languages. An arbitrary (phrase structure) grammar (recall Chapter 7, Section 5) is called a type 0 grammar. A *context-sensitive grammar* (recall Chapter 7, Section 5) is called a type 1 grammar. A *positive context-free grammar* (recall Chapter 10, Section 1) is called a type 2 grammar, and a *regular grammar* (recall Chapter 10, Section 2) is called a type 3 grammar. The inclusions suggested by the numbering obviously hold: every regular grammar is context-free, and every positive context-free grammar is context-sensitive. (Of course, grammars containing productions of the form $V \to 0$ cannot be context-sensitive.)

For each type of grammar, there is a corresponding class of languages:

$$\text{A language } L \text{ is } \begin{bmatrix} \textit{regular} \\ \textit{context-free} \\ \textit{context-sensitive} \\ \textit{r.e.} \end{bmatrix} \text{ or of type } \begin{bmatrix} 3 \\ 2 \\ 1 \\ 0 \end{bmatrix}$$

$$\text{if and only if there is a } \begin{bmatrix} \textit{regular} \\ \textit{positive context-free} \\ \textit{context-sensitive} \\ \textit{phrase structure} \end{bmatrix} \text{ grammar } \Gamma$$

such that

$$L = L(\Gamma) \quad \text{or} \quad L = L(\Gamma) \cup \{0\}.$$

For regular languages this statement is just Theorem 2.3 in Chapter 10. For context-free languages, it is Theorem 1.2 in Chapter 10. For context-sensitive languages we take it as a definition. For r.e. languages it is Theorem 5.2 in Chapter 7, and the special reference to {0} is not needed. We have

Theorem 1.1. Every regular language is context-free. Every context-free language is context-sensitive. Every context-sensitive language is recursive.

Proof. The first two statements follow simply from the corresponding inclusions among the types of grammar. The third follows at once from Theorem 5.4 in Chapter 7. ■

We would like to show that the inclusions of Theorem 1.1 are proper, that is, that none of the four classes mentioned in the theorem is identical to any of the others. We have seen in Theorem 1.1 in Chapter 10, that the language $L = \{a^{[n]}b^{[n]} | n > 0\}$ is context-free but not regular. Similarly, we saw in Theorem 4.2 in Chapter 10 that the language $\{a^{[n]}b^{[n]}c^{[n]} | n > 0\}$ is not context-free, while Exercise 5.4 in Chapter 7 shows that it is context-sensitive. This takes care of the first two inclusions of Theorem 1.1. The following theorem takes care of the remaining one.

Theorem 1.2. There is a recursive language on the alphabet {1} that is not context-sensitive.

Proof. We first code each context-sensitive grammar Γ with terminal alphabet {1} by a string on the five-letter alphabet $A = \{1, V, b, \to, /\}$. We do this simply by replacing each variable by a distinct string of the form $Vb^{[j]}$, using the arrow "$\to$" as usual between the left and right sides of productions, and using the slash "/" to separate productions. (Of course, not every string on this alphabet is actually the code for a context-sensitive grammar.) Now, the strings that code context-sensitive grammars may be placed in alphabetic order (or equivalently, in numerical order, regarding each string on A as the base 5 notation for an integer, as in Chapter 5). We let L_i be the context-sensitive language generated by the ith context-sensitive grammar in this enumeration, $i = 1, 2, 3, \ldots$. Then we set

$$L = \{1^{[i]} \mid 1^{[i]} \notin L_i, i \neq 0\}.$$

This is, of course, a typical diagonal construction, and we easily show that

L is not context-sensitive. For, if $L = L_{i_0}$, then

$$1^{[i_0]} \in L \qquad \text{if and only if} \quad 1^{[i_0]} \notin L_{i_0}$$
$$\qquad\qquad\quad \text{if and only if} \quad 1^{[i_0]} \notin L.$$

To see that L is recursive we note that there is an algorithm which given i will return a context-sensitive grammar Γ_i that generates L_i. Then $1^{[i]}$ can be tested for membership in L_i using the algorithm developed in the *proof* of Theorem 5.4 in Chapter 7. ■

For each class of languages corresponding to types 0, 1, 2, 3, we are concerned with questions of the following kinds: What can we determine algorithmically about a language from a grammar which generates it? What kinds of device will accept precisely the languages belonging to the class? Under what operations are the classes closed? We have been dealing with these questions for languages of types 0, 2, and 3. Now, we will see what can be said about languages of type 1, i.e., context-sensitive languages. We begin by considering the question of closure under union. We will need the

Lemma. There is an algorithm that will transform a given context-sensitive grammar Γ into a context-sensitive grammar Δ such that the left sides of the productions of Δ contain no terminals and $L(\Gamma) = L(\Delta)$.

Proof. We "disguise" the terminals as variables as in the proof of Theorem 3.1 in Chapter 10, except that now we need to replace the terminals on both the left and right sides of the productions. The resulting grammar, Δ, consists of productions of the form $X_1 \cdots X_m \to Y_1 \cdots Y_n$, $m \le n$, and $X_a \to a$, where $X_1, \ldots, X_m, Y_1, \ldots, Y_n, X_a$ are variables and a is a terminal. Clearly, $L(\Delta) = L(\Gamma)$. ■

Theorem 1.3. If L_1, L_2 are context-sensitive languages, then so is $L_1 \cup L_2$.

Proof. Assume $L_1 = L(\Gamma_1)$ or $L(\Gamma_1) \cup \{0\}$, $L_2 = L(\Gamma_2)$ or $L(\Gamma_2) \cup \{0\}$, where Γ_1 and Γ_2 are context-sensitive grammars with disjoint sets of variables of the form obtained in the Lemma. We construct Γ from Γ_1 and Γ_2 exactly as in the proof of Theorem 5.1 in Chapter 10, so that Γ is also context-sensitive and $L(\Gamma) = L(\Gamma_1) \cup L(\Gamma_2)$. Clearly, $L_1 \cup L_2 = L(\Gamma)$ or $L(\Gamma) \cup \{0\}$. ■

Exercises

1. Show that $\{w \in \{a, b, c\}^* | w$ has an equal number of a's, b's, c's$\}$ is context-sensitive.

2. Let Γ be the grammar with productions

$$\begin{array}{lll} S \to \lambda XY\rho & Aa \to aA & AY \to XYa \\ \lambda X \to \lambda aA & Ab \to bA & BY \to XYb \\ \lambda X \to \lambda bB & Ba \to aB & CY \to cc \\ \lambda X \to \lambda ccC & Bb \to bB & aX \to Xa \\ & Ca \to aC & bX \to Xb, \\ & Cb \to bC & \end{array}$$

where $\mathscr{V} = \{S, X, Y, A, B, C\}$ and $T = \{\lambda, \rho, a, b, c\}$. What is $L(\Gamma)$?

3. Show that $\{ww | w \in \{a, b\}^*\}$ is context-sensitive.

4. Apply the construction in the proof of the Lemma to the grammar in Exercise 2.

5. Show that the proof of Theorem 1.3 fails if we do not assume that Γ_1, Γ_2 conform to the conditions of the Lemma.

6. **(a)** Let Γ be a context-sensitive grammar. Show that there is a context-sensitive grammar Γ' such that $L(\Gamma') = L(\Gamma)$ and such that, for every production $u \to v$ in Γ, $|u| \leq 2$ and $|v| \leq 2$.

(b) Prove that a language L is context-sensitive if and only if it is generated by a grammar Γ, with variables $\mathscr{V}$ and terminals T, such that every production in Γ has the form $uVw \to uvw$, where $u, w \in (\mathscr{V} \cup T)^*$, $V \in \mathscr{V}$, and $v \in (\mathscr{V} \cup T)^* - \{0\}$. [*Note:* This explains the origin of the term *context-sensitive*.]

2. Linear Bounded Automata

We are now going to deal with the question: which devices accept context-sensitive languages? We define a *linear bounded automaton* on the alphabet $C = \{s_1, s_2, \ldots, s_n\}$ to be a *nondeterministic Turing machine* $\mathscr{M}$ on the alphabet $C \cup \{\lambda, \rho\}$ such that the only quadruples $\mathscr{M}$ contains beginning $q\lambda$ or $q\ \rho$ are of the forms $q\ \lambda\ R p$ and $q\ \rho\ L p$, respectively, such that $\mathscr{M}$ has a *final state*, written $\tilde{q}$, where no quadruple of $\mathscr{M}$ begins $\tilde{q}$, and finally such that for every quadruple $q\, a\, b\, p$ in $\mathscr{M}$, we have $b \neq \lambda, \rho$. Thus, when scanning λ, $\mathscr{M}$ can move only right, and when scanning ρ, $\mathscr{M}$ can move only left, and the symbols λ, ρ can never be printed in the course of a computation. Thus, the effect of the additional symbols λ and ρ is simply to prevent the machine from moving beyond the confines of the given string on the tape. Because of this we can code a configuration of $\mathscr{M}$ by a triple $(i, q, \lambda w \rho)$, where $0 \leq i \leq |w| + 1$; i gives the position of the tape-head (i.e., of the scanned square), q is the current state; and $\lambda w \rho$ is the

Table 2.1

Quadruple in $\mathscr{M}$	Corresponding transition
$q\ a\ b\ p$	$(\lvert u\rvert, q, uav) \vdash_{\mathscr{M}} (\lvert u\rvert, p, ubv)$
$q\ a\ R\ p$	$(\lvert u\rvert, q, uav) \vdash_{\mathscr{M}} (\lvert u\rvert + 1, p, uav)$
$q\ a\ L\ p$	$(\lvert u\rvert, q, uav) \vdash_{\mathscr{M}} (\lvert u\rvert - 1, p, uav)$

tape contents, $w \in (C \cup \{s_0\})^*$. (Recall that s_0 is the blank.) As usual, for configurations γ, δ we write $\gamma \vdash_{\mathscr{M}} \delta$ to mean that one of the quadruples of $\mathscr{M}$ permits the transition from γ to δ, and write $\gamma \overset{*}{\vdash}_{\mathscr{M}} \delta$ to mean that there is a sequence of configurations $\gamma = \gamma_1, \gamma_2, \ldots, \gamma_k = \delta$ such that $\gamma_i \vdash_{\mathscr{M}} \gamma_{i+1}$ for $1 \leq i < k$. Table 2.1 shows which transitions are permitted by each quadruple in $\mathscr{M}$ (here $a \in C \cup \{s_0, \lambda, \rho\}, b \in C \cup \{s_0\}$). (Of course, for $a = \lambda, \rho$, only quadruples of the second and third kind, respectively, can occur in $\mathscr{M}$.)

$\mathscr{M}$ is said to *accept* a string $w \in C^*$ if

$$(1, q_1, \lambda w \rho) \overset{*}{\vdash}_{\mathscr{M}} (i, \tilde{q}, \lambda w' \rho),$$

where q_1 is the initial state of $\mathscr{M}$ and, of course, $\tilde{q}$ is the final state. (Note carefully that unlike the situation for Turing machines, a configuration will be regarded as "accepting" only if $\mathscr{M}$ is in its final state $\tilde{q}$.) If $A \subseteq C$, we write $L_A(\mathscr{M})$ for the set of all $w \in A^*$ that are accepted by $\mathscr{M}$. The main theorem is

Theorem 2.1 (Landweber–Kuroda). The language $L \subseteq A^*$ is context-sensitive if and only if there is a linear bounded automaton $\mathscr{M}$ such that $L = L_A(\mathscr{M})$.

We begin with

Lemma 1. There is an algorithm that transforms any given context-sensitive grammar Γ with terminals T into a linear bounded automaton $\mathscr{M}$ such that $L(\Gamma) = L_T(\mathscr{M})$.

Proof. Let $\mathscr{V}$ be the set of variables of Γ, and let $S \in \mathscr{V}$ be the start symbol. The alphabet of $\mathscr{M}$ will be $T \cup \mathscr{V}$. Let the productions of Γ be $u_i \to v_i$, $i = 1, 2, \ldots, m$, where

$$u_i = \alpha_1^{(i)}\alpha_2^{(i)} \cdots \alpha_{k_i}^{(i)} \quad \text{and} \quad v_i = \beta_1^{(i)}\beta_2^{(i)} \cdots \beta_{l_i}^{(i)}; \tag{2.1}$$

$$\alpha_1^{(i)}, \alpha_2^{(i)}, \ldots, \alpha_{k_i}^{(i)}, \beta_1^{(i)}, \beta_2^{(i)}, \ldots, \beta_{l_i}^{(i)} \in T \cup \mathscr{V},$$

and $k_i \leq l_i$. Then we set

$$\alpha_{k_i+1}^{(i)} = \alpha_{k_i+2}^{(i)} = \cdots = \alpha_{l_i}^{(i)} = s_0.$$

That is, we fill out the left side of each production with blanks. Since $\mathscr{M}$ is operating nondeterministically, it can seek the word v_i on the tape and replace it by u_i, thus undoing the work of the production. It will help in following the construction of the automaton $\mathscr{M}$ if we think of it as operating in one of these four phases: *initialization*, *searching*, *production undoing*, and *termination*. The states of $\mathscr{M}$ will be the *initial* state q_1, the *search* state σ, the *return* state $\bar{\sigma}$, the *undoing* states $p_j^{(i)}, q_j^{(i)}$ for $1 \leq i \leq m$ and $1 \leq j \leq l_i$ [l_i is as defined in Eqs. (2.1)], and the *termination* states $\tau, \bar{\tau}$.

Phase 1 (Initialization) We place in $\mathscr{M}$ the quadruples

$$\left.\begin{matrix} q_1 & a & a & \sigma \\ q_1 & a & a & \tau \end{matrix}\right\} \qquad a \in \mathscr{V} \cup T \cup \{s_0\}.$$

Thus in Phase 1, $\mathscr{M}$ operating nondeterministically "decides" to enter either the search or the termination phase.

Phase 2 (Search) We place in $\mathscr{M}$ the quadruples

$$\begin{matrix} \sigma & a & R & \sigma & & a \neq \rho \\ \sigma & \rho & L & \bar{\sigma} & & \\ \sigma & \beta_1^{(i)} & \beta_1^{(i)} & p_1^{(i)} & & 1 \leq i \leq m \\ \bar{\sigma} & a & L & \bar{\sigma} & & a \neq \lambda \\ \bar{\sigma} & \lambda & R & q_1. & & \end{matrix}$$

In Phase 2, $\mathscr{M}$ moves right along the tape searching for one of the initial symbols $\beta_1^{(i)}$ of the right side of a production. Finding one, $\mathscr{M}$ may enter an *undoing* state. If $\mathscr{M}$ encounters the right end marker ρ while still in state σ, it enters the *return* state $\bar{\sigma}$ and goes back to the beginning.

Phase 3 (Production Undoing) We place in $\mathscr{M}$ the quadruples, for $1 \leq j < l_i$, $1 \leq i \leq m$,

$$\begin{matrix} p_j^{(i)} & \beta_j^{(i)} & \alpha_j^{(i)} & q_j^{(i)} \\ q_j^{(i)} & \alpha_j^{(i)} & R & p_{j+1}^{(i)} \\ p_{l_i}^{(i)} & \beta_{l_i}^{(i)} & \alpha_{l_i}^{(i)} & \bar{\sigma} \end{matrix}$$

together with the quadruples

$$p_j^{(i)} \quad s_0 \quad R \quad p_j^{(i)}.$$

When operating in Phase 3, $\mathscr{M}$ has the opportunity to replace the right side of one of the productions on the tape by the left side (ignoring any blanks that might have been introduced by previous replacements). If $\mathcal{M}$ succeeds, it can enter the return state $\bar{\sigma}$, return to the left, and begin again.

Phase 4 (Termination) We place in $\mathcal{M}$ the quadruples

$$\begin{array}{cccc} \tau & s_0 & R & \tau \\ \tau & S & R & \bar{\tau} \\ \bar{\tau} & s_0 & R & \bar{\tau} \\ \bar{\tau} & \rho & L & \tilde{q}. \end{array}$$

Thus if $\mathcal{M}$ ever returns to state q_1 with the tape contents

$$\lambda s_0^{[i]} S s_0^{[j]} \rho \qquad i, j \geq 0$$

(where, of course, S is the start symbol of Γ), then $\mathcal{M}$ will have the opportunity to move all the way to the right in this phase and to enter the final state $\tilde{q}$.

Thus, $\mathcal{M}$ will accept a word $w \in T^*$ just in case there is a derivation of w from S in Γ. ■

Lemma 2. If $L \subseteq A^*$ is a context-sensitive language, then there is a linear bounded automaton $\mathcal{M}$ such that $L = L_A(\mathcal{M})$.

Proof. We have $L = L(\Gamma)$ or $L(\Gamma) \cup \{0\}$ for a context-sensitive grammar Γ. In the first case, $\mathcal{M}$ can be obtained as in Lemma 1. In the second case, we modify the automaton $\mathcal{M}$ of Lemma 1 by adding the quadruple $q_1 \;\; \rho \;\; L \;\; \tilde{q}$. The modified automaton accepts 0 as well as the strings that $\mathcal{M}$ accepts. ■

Now, we wish to discuss the converse situation: we are given a linear bounded automaton $\mathcal{M}$ and alphabet A and wish to obtain a context-sensitive grammar Γ such that $L(\Gamma) = L_A(\mathcal{M}) - \{0\}$. The construction will be similar to the simulation, in Chapter 7, of a Turing machine by a semi-Thue process. However, the coding must be tighter because all the productions need to be non-length-decreasing.

Let $\mathcal{M}$ be the given linear bounded automaton with alphabet C where $A \subseteq C$, initial state q_1, and final state $\tilde{q}$. To begin with, we will only consider words $u \in C^*$ for which $|u| \geq 2$; such words can be written awb, where $w \in C^*$, $a, b \in C$. We wish to code a configuration $(i, q, \lambda awb\rho)$ of $\mathcal{M}$ by a word of length $|awb| = |w| + 2$. To help us in doing this, we will use five variants on each letter $a \in C$:

$$a \;\; {}^{\lceil}a \;\; a^{\rceil} \;\; \overleftarrow{a} \;\; \vec{a}.$$

The interpretation of these markings is

$^\ulcorner a$: a on the left end of the word;
$a^\urcorner$: a on the right end of the word;
$\overleftarrow{a}$: a on the left end, but the symbol being scanned is λ, one square to the left of a;
$\overrightarrow{a}$: a on the right end, but the symbol being scanned is ρ, one square to the right of a.

Finally, the current state will ordinarily be indicated by a subscript on the scanned symbol. If however, the scanned symbol is λ or ρ, the subscript will be on the adjacent symbol, marked, as just indicated, by an arrow. Thus, if $\mathscr{M}$ has n states we introduce $3(n + 1) + 2n$ symbols for each $a \in C$. (Note that $\overleftarrow{a}$ and $\overrightarrow{a}$ always have a subscript.) The examples in Table 2.2 should make matters plain. Of course, this encoding only works for words $\lambda w \rho$ for which $|w| \geq 2$.

Now we will construct a semi-Thue process Σ such that given configurations γ, δ of $\mathscr{M}$ and their codes $\overline{\gamma}$, $\overline{\delta}$, respectively, we shall have

$$\gamma \vdash_{\mathscr{M}} \delta \quad \text{if and only if} \quad \overline{\gamma} \underset{\Sigma}{\Rightarrow} \overline{\delta}.$$

As for Turing machines, we define Σ by introducing suitable productions corresponding to each quadruple of $\mathscr{M}$. The correspondence is shown in Table 2.3, where we have written $\overline{C}$ for $C \cup \{s_0\}$.

Now, since these productions simulate the behavior of $\mathscr{M}$ in an obvious and direct manner, we see that $\mathscr{M}$ will accept the string $aub, u \in C^*$, $a, b \in C$, just in case there is a derivation, from the initial word $^\ulcorner a_{q_1} u b^\urcorner$ using these productions, of a word containing $\tilde{q}$ as a subscript. To put this result in a more manageable form, we add to the alphabet of Σ the symbol S and add to Σ the "cleanup" productions

$$\alpha_{\tilde{q}} \to S, \qquad \alpha S \to S, \qquad S\alpha \to S, \tag{2.2}$$

where α can be any one of a, $^\ulcorner a$, $a^\urcorner$, $\overleftarrow{a}$, or $\overrightarrow{a}$, for any $a \in C$. Since these productions will transform the codes for configurations with the final state

Table 2.2

Configuration	Code
$(3, q, \lambda ababc\rho)$	$^\ulcorner aba_q bc^\urcorner$
$(1, q, \lambda ababc\rho)$	$^\ulcorner a_q babc^\urcorner$
$(5, q, \lambda ababc\rho)$	$^\ulcorner ababc^\urcorner_q$
$(0, q, \lambda ababc\rho)$	$\overleftarrow{a}_q babc^\urcorner$
$(6, q, \lambda ababc\rho)$	$^\ulcorner abab\overrightarrow{c}_q$

Table 2.3

Quadruple of $\mathcal{M}$		Productions of Σ	
$q\ a\ b\ p,$	$a, b \in \bar{C}$	$a_q \to b_p$ ${}^{\lceil}a_q \to {}^{\lceil}b_p$ $a_q^{\rceil} \to b_p^{\rceil}$	
$q\ a\ R\ p,$	$a \in \bar{C}$	$a_q b \to ab_p$ ${}^{\lceil}a_q b \to {}^{\lceil}ab_p$ $a_q b^{\rceil} \to ab_p^{\rceil}$ ${}^{\lceil}a_q b^{\rceil} \to {}^{\lceil}ab_p^{\rceil}$	all $b \in \bar{C}$
		$a_q^{\rceil} \to \vec{a}_p$	
$q\ \lambda\ R\ p$		$\overleftarrow{a}_q \to {}^{\lceil}a_p,$	all $a \in \bar{C}$
$q\ a\ L\ p,$	$a \in \bar{C}$	$ba_q \to b_p a$ $ba_q^{\rceil} \to b_p a^{\rceil}$ ${}^{\lceil}ba_q \to {}^{\lceil}b_p a$ ${}^{\lceil}ba_q^{\rceil} \to {}^{\lceil}b_p a^{\rceil}$	all $b \in \bar{C}$
		${}^{\lceil}a_q \to \overleftarrow{a}_p$	
$q\ \rho\ L\ p$		$\vec{a}_q \to a_p^{\rceil},$	all $a \in \bar{C}$

$\tilde{q}$ into the single symbol S, and since there is no other way to obtain the single symbol S using the productions of Σ, we have

Lemma 3. $\mathcal{M}$ accepts the string $aub, a, b \in C, u \in C^*$, if and only if

$$ {}^{\lceil}a_{q_1}ub^{\rceil} \underset{\Sigma}{\overset{*}{\Rightarrow}} S. $$

Now let Ω be the semi-Thue process whose productions are the *inverses* of the productions of Σ. (See Chapter 7, Section 2.) Then we have

Lemma 4. $\mathcal{M}$ accepts the string $aub, a, b \in C, u \in C^*$, if and only if

$$ S \underset{\Omega}{\overset{*}{\Rightarrow}} {}^{\lceil}a_{q_1}ub^{\rceil}. $$

Now we are ready to define a context-sensitive grammar Γ. Let the terminals of Γ be the members of A, let the variables of Γ be

1. the symbols from the alphabet of Ω that do not belong to A, and
2. symbols a^0 for each $a \in A$.

Finally, the productions of Γ are the productions of Ω together with

$$\left.\begin{array}{l} {}^{[}a_{q_1} \to a^0 \\ a^0 b \to ab^0 \\ a^0 b^{]} \to ab \end{array}\right\} \quad \text{for all} \quad a, b \in A. \tag{2.3}$$

It is easy to check that Γ is in fact context-sensitive. [Of course, the productions (2.2) must be read from right to left, since it is the inverses of (2.2) that appear in Γ.] Moreover, using Lemma 4 and (2.3), we have

Lemma 5. Let $w \in A^*$. Then $w \in L(\Gamma)$ if and only if $|w| \geq 2$ and $w \in L_A(\mathscr{M})$.

Now let $\mathscr{M}$ be a given linear bounded automaton, A a given alphabet, and let Γ be the context-sensitive grammar just constructed. Then, by Lemma 5, we have

$$L_A(\mathscr{M}) = L(\Gamma) \cup L_0,$$

where L_0 is the set of words $w \in A^*$ accepted by $\mathscr{M}$ such that $|w| < 2$. But L_0 is finite, hence (Corollary 4.7 in Chapter 9) L_0 is a regular language, and so is certainly context-sensitive. Finally, using Theorem 1.3, we see that $L_A(\mathscr{M})$ is context-sensitive. This, together with Lemma 2, completes the proof of Theorem 2.1. ∎

Exercises

1. Let $\mathscr{M}$ be the linear bounded automaton with initial state q_1, final state $\tilde{q}$, and quadruples

$$\begin{array}{llll} q_1 & a & R & q_2 \\ q_1 & b & R & q_3 \\ q_1 & c & R & q_1 \\ q_1 & \rho & L & \tilde{q} \end{array} \qquad \begin{array}{llll} q_2 & b & R & q_1 \\ q_2 & c & R & q_1 \\ q_3 & a & R & q_1 \\ q_3 & c & R & q_1. \end{array}$$

   What is $L(\mathscr{M})$?

2. Give a deterministic linear bounded automaton $\mathscr{M}$ that accepts $\{w \in \{a, b, c\}^* \mid w$ has an equal number of a's, b's, c's$\}$.

3. Give a linear bounded automaton $\mathscr{M}$ that accepts $\{ww \mid w \in \{a, b\}^*\}$.

4. Let $\mathscr{M}$ be the linear bounded automaton with initial state q_1, final state $\tilde{q}$, and quadruples

$$\begin{array}{llll} q_1 & a & R & q_1 \\ q_1 & b & R & q_2 \end{array} \qquad \begin{array}{llll} q_2 & b & R & q_2 \\ q_2 & \rho & L & \tilde{q}. \end{array}$$

(a) Use the construction in the proof of Theorem 2.1 to give a grammar Γ such that $L(\Gamma) = L(\mathscr{M})$.

(b) Give a derivation of *aabb* in Γ.

5. Let Γ be the grammar with start symbol S and productions $S \to aSb$, $S \to ab$.

   (a) Use the construction in the proof of Theorem 2.1 to give a linear bounded automaton $\mathscr{M}$ such that $L(\mathscr{M}) = L(\Gamma)$.

   (b) Give an accepting computation by $\mathscr{M}$ for input *aabb*.

6. Prove that every context-free language is accepted by a deterministic linear bounded automaton.

7. Show that there is an algorithm to test a given linear bounded automaton $\mathscr{M}$ and word w to determine whether or not $\mathscr{M}$ will eventually halt on input w. That is, the halting problem is solvable for linear bounded automata. [*Hint:* Consider the pigeon-hole principle.]

3. Closure Properties

We have already seen that the context-sensitive languages are closed under union (Theorem 1.3), and now we consider *intersection*. Here, although the context-free languages are *not* closed under intersection (Theorem 5.2 in Chapter 10), we can prove

Theorem 3.1. If L_1 and L_2 are context-sensitive languages, then so is $L_1 \cap L_2$.

Proof. Let $L_1 = L_A(\mathscr{M}_1)$, $L_2 = L_A(\mathscr{M}_2)$, where $\mathscr{M}_1, \mathscr{M}_2$ are linear bounded automata. The idea of the proof is to test a string w for membership in $L_1 \cap L_2$ by first seeing whether $\mathscr{M}_1$ will also accept w and then, if $\mathscr{M}_1$ does, to see whether $\mathscr{M}_2$ will also accept w. The difficulty is that $\mathscr{M}_1$ may destroy the input w in the process of testing it. If we were working with Turing machines, we would be able to deal with this kind of problem by saving a copy of the input on a part of the tape that remained undisturbed. Since linear bounded automata have no extra space, the problem must be solved another way. The solution uses an important idea: we think of our tape as consisting of a number of separate "tracks," in this case two tracks. We will construct a linear bounded automaton $\mathscr{M}$ that will work as follows:

1. $\mathscr{M}$ will copy the input so it appears on both the upper and the lower track of the tape;

2. $\mathcal{M}$ will simulate $\mathcal{M}_1$ working on the upper track only;
3. if $\mathcal{M}_1$ has accepted, $\mathcal{M}$ will then simulate $\mathcal{M}_2$ working on the lower track (on which the original input remains undisturbed).

Thus, let us assume that $\mathcal{M}_1$ and $\mathcal{M}_2$ both have the alphabet $C = \{s_1, s_2, \ldots, s_n\}$. (Of course, in addition they may use the symbols λ, ρ, s_0.) $\mathcal{M}$ will be a linear bounded automaton using, in addition, the symbols $b_j^i, 0 \le i, j \le n$. We think of the presence of the symbol b_j^i as indicating that s_i is on the "upper track" while s_j is on the "lower track" at the indicated position. Finally we assume that q_1 is the initial state of $\mathcal{M}_1$, that $\tilde{q}$ is its final state, and that q_2 is the initial state of $\mathcal{M}_2$. We also assume that the sets of states of $\mathcal{M}_1$ and $\mathcal{M}_2$ are disjoint. $\mathcal{M}$ is to have initial state q_0 and have the same final state as $\mathcal{M}_2$. $\mathcal{M}$ is to contain the following quadruples (for $0 \le i \le n$):

(1) Initialization:

$$\begin{array}{cccc} q_0 & s_i & b_i^i & \bar{q} \\ \bar{q} & b_i^i & R & q_0 \\ q_0 & \rho & L & \bar{\bar{q}} \\ \bar{\bar{q}} & b_i^i & L & \bar{\bar{q}} \\ \bar{\bar{q}} & \lambda & R & q_1 . \end{array}$$

Here $\bar{q}, \bar{\bar{q}}$ are not among the states of $\mathcal{M}_1$ and $\mathcal{M}_2$. These quadruples cause $\mathcal{M}$ to copy the input on both "tracks" and then to return to the leftmost symbol of the input.

(2) For each quadruple of $\mathcal{M}_1$, the corresponding quadruples, obtained by replacing each s_i by $b_j^i, j = 0, 1, \ldots, n$, are to be in $\mathcal{M}$. These quadruples cause $\mathcal{M}$ to simulate $\mathcal{M}_1$ operating on the "upper" track. In addition, $\mathcal{M}$ is to have the quadruples for $0 \le i, j \le n$:

$$\begin{array}{cccc} \tilde{q} & b_j^i & R & \tilde{q} \\ \tilde{q} & \rho & L & \tilde{p} \\ \tilde{p} & b_j^i & s_j & \tilde{p} \\ \tilde{p} & s_j & L & \tilde{p} \\ \tilde{p} & \lambda & R & q_2 . \end{array}$$

Here again $\tilde{p}$ does not occur among the states of $\mathcal{M}_1, \mathcal{M}_2$. These quadruples cause $\mathcal{M}$ to restore the "lower" track and then to enter the initial state of $\mathcal{M}_2$ scanning the leftmost input symbol.

(3) Finally, $\mathcal{M}$ is to contain all the quadruples of $\mathcal{M}_2$.

Since it is plain that

$$L_A(\mathcal{M}) = L_A(\mathcal{M}_1) \cap L_A(\mathcal{M}_2),$$

the proof is complete. ■

As an application, we obtain an unsolvability result about context-sensitive grammars.

Theorem 3.2. There is no algorithm for determining of a given context-sensitive grammar Γ whether $L(\Gamma) = \varnothing$.

Proof. Suppose there were such an algorithm. We can show that there would then be an algorithm for determining of two given *context-free* grammars Γ_1, Γ_2 whether $L(\Gamma_1) \cap L(\Gamma_2) = \varnothing$, thus contradicting Theorem 6.6 in Chapter 10. For, since Γ_1, Γ_2 are context-sensitive, the constructive nature of the proofs of Theorems 2.1 and 3.1 will enable us to obtain a context-sensitive grammar Γ with $L(\Gamma) = L(\Gamma_1) \cap L(\Gamma_2)$. ■

We turn now to a question about context-sensitive languages that was one of the outstanding open problems in theoretical computer science for over two decades. In 1964 Kuroda raised the question: Are the context-sensitive languages closed under complementation? It remained unsettled until 1987, when Neil Immerman showed that the answer is yes. What is particularly interesting is that, after more than twenty years, the solution turned out to be surprisingly straightforward.

We will show that if $L \subseteq A^*$ is accepted by a linear bounded automaton, then so is $A^* - L$. Suppose that L is accepted by the linear bounded automaton $\mathcal{M}$ with alphabet $\{s_1, \ldots, s_{n-1}\}$ and states $\{q_1, \ldots, q_k\}$. (We take q_k to be $\tilde{q}$, and we will sometimes write λ, ρ as s_n, s_{n+1}, respectively.) We want to find another linear bounded automaton $\mathcal{N}$ which accepts when $\mathcal{M}$ rejects and vice versa. This would be easy if $\mathcal{M}$ were deterministic, but suppose $\mathcal{M}$ is nondeterministic. If $w \notin L$ then *every* computation by $\mathcal{M}$ on input $\lambda w \rho$ is nonaccepting, so if we constructed $\mathcal{N}$ to simulate $\mathcal{M}$ and enter the final state $\tilde{q}$ precisely when $\mathcal{M}$ halts in a state other than $\tilde{q}$, then every halting computation by $\mathcal{N}$ would enter $\tilde{q}$ and $\mathcal{N}$ would accept w (if it has at least one halting computation). However, if $w \in L$, then $\mathcal{M}$ could still have some computations which halt in some state other than $\tilde{q}$, in which case $\mathcal{N}$ *would still accept* w. Thus, we need $\mathcal{N}$ to accept only when *every* computation of $\mathcal{M}$ fails to end in state $\tilde{q}$.

The problem is that it is not at all clear how to construct $\mathcal{N}$ so that a single computation by $\mathcal{N}$ can correctly gather information about every computation by $\mathcal{M}$. We could deterministically simulate $\mathcal{M}$, using a stack to remember "branch points." However, $\mathcal{M}$ has $|\lambda w\rho| \cdot k \cdot n^{|w|}$ distinct configurations with $|\lambda w\rho|$ tape squares, so a nonlooping computation by $\mathcal{M}$ on input $\lambda w\rho$ could run for as many as $|\lambda w\rho| \cdot k \cdot n^{|w|} - 1$ steps, and each step could require adding more information to the stack. There is no way, then, that such a stack can be stored in $|w|$ tape squares, even using multiple tracks as in the proof of Theorem 3.1. Actually, there are simulation techniques that are much more efficient in terms of space, but none are known that are sufficiently parsimonious for our purposes here.

The solution discovered by Immerman is to store sufficient information about the possible computations by *counting* configurations. The largest value that needs to be stored is $|\lambda w\rho| \cdot k \cdot n^{|w|}$, which for any $w \neq 0$ can be represented in base n notation by a string of length

$$\leq \log_n |\lambda w\rho| + \log_n k + |w| + 1 \leq c \cdot |w|$$

for some constant c. (We can ignore the case $w = 0$ since the decision to accept or reject 0 can be built explicitly into the quadruples of $\mathcal{N}$.) The important thing is that c does not depend on w, so we can construct $\mathcal{N}$ to maintain each such counter on c tracks, regardless of the length of the input. In fact, it will be convenient to consider the c tracks holding a counter as a single track with c "subtracks."

The other objects we need to represent are configurations. If the initial configuration is $(1, q_1, \lambda w\rho)$, then it is clear that we can represent on a single track any configuration $(i, q, \lambda x\rho)$ where $|x| = |w|$. For example, we could add to the alphabet of some track new symbols $s_i^j, 0 \leq i \leq n + 1$ and $1 \leq j \leq k$. Then s_i^j in square l on this track represents $\mathcal{M}$ in state q_j scanning square l (on its own tape) holding symbol s_i. Not every string on the alphabet

$$\{s_0, \ldots, s_{n+1}\} \cup \{s_i^j \,|\, 0 \leq i \leq n + 1; 1 \leq j \leq k\}$$

represents a configuration of $\mathcal{M}$, but it is clear that the representations of all configurations of $\mathcal{M}$ with $|\lambda w\rho|$ tape squares can be written one after another, say, in ascending numerical order, on some track. We will call the ith configuration in this enumeration C_i. Of these configurations, some may never occur in any computation by $\mathcal{M}$ on input $\lambda w\rho$. We say that a configuration $(i, q, \lambda x\rho)$ is *reachable* from w if $(1, q_1, \lambda w\rho) \vdash_{\mathcal{M}}^{*} (i, q, \lambda x\rho)$.

We describe the behavior of $\mathcal{N}$ by means of two nondeterministic procedures, the COUNT phase and the TEST phase. Although these are written in an informal high-level notation, it should be clear that $\mathcal{N}$ can be constructed to carry them out, using no more than $|\lambda w\rho|$ tape squares. We begin with the TEST phase, described in Figure 3.1, where we will see the importance of being able to count the reachable configurations. Suppose

> COUNTER ← 0
> *for* $i = 1$ *to* $|\lambda w\rho| \cdot k \cdot n^{|w|}$
>   CONFIG ← C_i
>   *nondeterministically simulate some computation by* $\mathcal{M}$
>   *on* $\lambda w\rho$ *until it reaches* CONFIG *or terminates*
>   *if* CONFIG *has been reached then*
>   *if* CONFIG *is accepting*
>   *then enter* q' *and halt*
>   *else* COUNTER ← COUNTER + 1
> *end for*
> *if* COUNTER $= r$ *then enter* $\tilde{q}$ *and halt*
>   *else enter* q' *and halt*

Figure 3.1. The TEST phase of $\mathcal{N}$.

we have a tape with w on track 1 and r on track 2, where r is the number of configurations reachable from w. We will write this tape as $\lambda w/r\rho$. The TEST phase needs four tracks in addition to tracks 1 and 2. Two are needed for variables i and COUNTER, which hold numbers $\leq |\lambda w\rho| \cdot k \cdot n^{|w|}$, one is needed for CONFIG, which holds representations of configurations with $|\lambda w\rho|$ tape squares, and a fourth is needed to simulate computations by $\mathcal{M}$ on input $\lambda w\rho$. It is clear that each track is large enough for its purpose. Let q' be some nonfinal state of $\mathcal{N}$.

Claim 1. Executing the TEST phase, $\mathcal{N}$ accepts w/r if and only if $\mathcal{M}$ rejects w.

If $\mathcal{M}$ accepts w, there are at most $r - 1$ nonaccepting reachable configurations, so any computation by $\mathcal{N}$ will either

- run forever simulating some computation by $\mathcal{M}$;
- simulate some computation by $\mathcal{M}$ that halts in state $\tilde{q}$; or
- end with COUNTER $< r$.

Therefore, no computation by $\mathcal{N}$ ends in state $\tilde{q}$, and $\mathcal{N}$ rejects w/r. If $\mathcal{M}$ rejects w then $\mathcal{N}$ can "guess" computations by $\mathcal{M}$ that reach every reachable configuration. None of these is accepting, so $\mathcal{N}$ finishes with COUNTER $= r$ and accepts w/r. This proves Claim 1.

Finally, we need to show that $\mathcal{N}$ can correctly compute r prior to entering the TEST phase. It might seem that $\mathcal{N}$ could simply guess r and then continue with the TEST phase. The problem is that, if $\mathcal{N}$ incorrectly guesses some $r' < r$, then some computation by $\mathcal{N}$ in the TEST phase might end with COUNTER $= r'$ and accept w when it should reject it. Therefore, it is not enough that *some* computation by $\mathcal{N}$ reach the TEST phase with the correct value of r. We must ensure that *every* computation

```
        i ← 0
        COUNTER ← 0
        NEW-COUNTER ← 1
[Main]  if NEW_COUNTER = COUNTER then
            delete all but tracks 1 and 2 from tape
            goto TEST phase
        i ← i + 1
        COUNTER ← NEW_COUNTER
        NEW_COUNTER ← 0
        for j = 1 to |λwρ| · k · n^|w|
            t ← 0
            CONFIG1 ← C_j
            for l = 1 to |λwρ| · k · n^|w|
                CONFIG2 ← C_l
                nondeterministically simulate some computation
                    by M on λwρ until it reaches CONFIG2 or
                    until i steps have been executed
                if CONFIG2 has been reached then
                    t ← t + 1
                    if CONFIG2 = CONFIG1 or
                            CONFIG2 ⊢_M CONFIG1 then
                        NEW_COUNTER ← NEW_COUNTER + 1
                        leave inner loop
            end for
            if l > |λwρ| · k · n^|w| and t < COUNTER then
                enter q' and halt
        end for
        goto Main
```

Figure 3.2. The COUNT phase of $\mathcal{N}$.

by $\mathcal{N}$ that gets as far as the TEST phase must do so with the correct value of r. We will now show that this can be done.

For all $i \geq 0$, let r_i be the number of configurations of $\mathcal{M}$ that can be reached from $(1, q_1, \lambda w \rho)$ in no more than i steps. Then there is some i_0 such that $r_{i_0} = r_{i_0+1} = r$. We will argue by induction that each r_i, for $1 \leq i \leq i_0$, is correctly computed by the COUNT phase, given in Figure 3.2. The input is the initial tape $\lambda w \rho$. We also need tracks to hold variables NEW_COUNTER, COUNTER, i, j, l, t, CONFIG1, and CONFIG2, and a track to use in simulating computations by $\mathcal{M}$. Again it is clear that sufficient space is available. We stipulate that NEW_COUNTER, which will eventually hold r, should be stored on track 2.

Claim 2. For $i \geq 0$, any computation by $\mathcal{N}$ on $\lambda w \rho$ that completes i executions of the main loop has the correct value of r_i in NEW_COUNTER.

The claim is obvious for $i = 0$, so we assume it is true for some $i \geq 0$ and show that it is true for $i + 1$. Suppose some computation completes $i + 1$ executions of the main loop. Then throughout the $i + 1$st execution of the main loop, COUNTER $= r_i$ by the induction hypothesis (since COUNTER is set to NEW_COUNTER at the beginning of the loop). CONFIG1 ranges over all configurations with $|\lambda w \rho|$ tape squares, and for each value of CONFIG1 we want NEW_COUNTER to be incremented just in case CONFIG1 is reachable within $i + 1$ steps. Now, for each value of CONFIG1, the inner *for* loop[1] ends either with $l \leq |\lambda w \rho| \cdot k \cdot n^{|w|}$, meaning that the current CONFIG1 has been found to be reachable within $i + 1$ steps, or with $l > |\lambda w \rho| \cdot k \cdot n^{|w|}$ and $t =$ COUNTER, meaning that all r_i of the configurations reachable within i steps have been found and none of them leads to CONFIG1 in 0 or 1 steps, i.e., CONFIG1 is not reachable within $i + 1$ steps. In the first case NEW_COUNTER is incremented and in the second case it is not, so the claim is true for $i + 1$.

To conclude we simply note that at least one computation by $\mathcal{N}$ on $\lambda w \rho$ will correctly guess the appropriate computations by $\mathcal{M}$ to simulate and will execute the main loop $i_0 + 1$ times, leaving r on track 2. Any such computation will then go on to execute the TEST phase, and, by Claim 1, $\mathcal{N}$ will accept w if and only if $\mathcal{M}$ rejects w. Therefore, by Theorem 2.1 we have proved

Theorem 3.3. If $L \subseteq A^*$ is context-sensitive, then so is $A^* - L$.

We conclude this chapter by mentioning another major problem concerning context-sensitive languages that remains open: is every context-sensitive language accepted by a *deterministic* linear bounded automaton?

Exercises

1. Let L, L' be context-sensitive languages. Prove the following.
 (a) $L \cdot L'$ is context-sensitive.
 (b) L^* is context-sensitive.
 (c) $L^R = \{w^R \mid w \in L\}$ is context-sensitive.

2. Let $L \subseteq A^*$ be an r.e. language. Show that there is a context-sensitive language $L' \subseteq (A \cup \{c\})^*$ such that for all $w \in A^*$, we have

$$w \in L \qquad \text{if and only if} \quad wc^{[i]} \in L' \text{ for some } i \geq 0.$$

[1] We are assuming here that when a loop of the form *for* $i = 1$ *to* n runs to completion, it leaves $i = n + 1$.

3. Show that for every r.e. language L there is a context-sensitive grammar Γ such that the grammar obtained from Γ by adding a single production of the form $V \to 0$ generates L. [*Hint:* Use Exercise 2 and take c to be the variable V.]

4. Give alphabets A, P and a context-sensitive language $L \subseteq A^*$ such that $\mathrm{Er}_P(L)$ is not context-sensitive.

5. Let A_1, A_2 be alphabets and let $L \subseteq A_1^*$ be context-sensitive. Let f be a substitution on A_1 such that for each $a \in A, f(a) \subseteq A_2^*$ is context-sensitive and $0 \notin f(a)$. Let g be a homomorphism from A_1^* to A_2^* such that $g(a) \neq 0$ for all $a \in A$. [See Exercise 4.5 in Chapter 9 for the definitions of *substitution* and *homomorphism*.]

   (a) Prove that $f(L)$ is context-sensitive.

   (b) Prove that $g(L)$ is context-sensitive.

   (c) Give a context-sensitive language L' and homomorphism h such that $h(L')$ is not context-sensitive.

Part 3

Logic

12

Propositional Calculus

1. Formulas and Assignments

Let A be some given alphabet and let $\mathscr{A} \subseteq A^*$. Let $B = A \cup \{\neg, \wedge, \vee, \supset, \leftrightarrow, (,)\}$, where we assume that these additional symbols are not already in A. $\neg$, $\wedge$, $\vee$, $\supset$, $\leftrightarrow$ are called *(propositional) connectives*. Then by a *propositional formula over* $\mathscr{A}$ we mean any element of B^* which either belongs to $\mathscr{A}$ or is obtainable from elements of $\mathscr{A}$ by repeated applications of the following operations on B^*:

1. transform α into $\neg\alpha$;
2. transform α and β into $(\alpha \wedge \beta)$;
3. transform α and β into $(\alpha \vee \beta)$;
4. transform α and β into $(\alpha \supset \beta)$;
5. transform α and β into $(\alpha \leftrightarrow \beta)$.

When the meaning is clear from the context, *propositional formulas over* $\mathscr{A}$ will be called *$\mathscr{A}$-formulas* or even just *formulas* for short. In this context the elements of $\mathscr{A}$ (which are automatically $\mathscr{A}$-formulas) are called *atoms*.

To make matters concrete we can take $A = \{\boldsymbol{p}, \boldsymbol{q}, \boldsymbol{r}, \boldsymbol{s}, \mathbf{I}\}$, and let

$$\mathscr{A} = \{\boldsymbol{p}\mathbf{I}^{[i]}, \boldsymbol{q}\mathbf{I}^{[i]}, \boldsymbol{r}\mathbf{I}^{[i]}, \boldsymbol{s}\mathbf{I}^{[i]} | i \in N\}.$$

In this case the atoms are called propositional variables. We can think of the suffix $\mathbf{I}^{[i]}$ as a subscript and write $p_i = \boldsymbol{p}\mathbf{I}^{[i]}$, $q_i = \boldsymbol{q}\mathbf{I}^{[i]}$, etc. Here are a few examples of formulas:

$$((\neg p \supset q) \supset p),$$
$$((((p \wedge q) \supset r) \wedge ((p_1 \wedge q_1) \supset r_1)) \supset \neg s),$$
$$(((p_1 \vee \neg p_2) \vee p_3) \wedge (\neg p_1 \vee p_3)).$$

Although the special case of propositional variables really suffices for studying propositional formulas, it is useful in order to include later applications, to allow the more general case of an arbitrary language of atoms. (In fact our assumption that the atoms form a language is not really necessary.)

By an *assignment* on a given set of atoms $\mathscr{A}$ we mean a function v which maps each atom into the set {FALSE, TRUE} = {0, 1}, where (recall Chapter 1, Section 4), as usual, we are identifying FALSE with 0 and TRUE with 1. Thus for each atom α we will have $v(\alpha) = 0$ or $v(\alpha) = 1$. Given an assignment v on a set of atoms $\mathscr{A}$, we now show how to define a value $\gamma^v \in \{0, 1\}$ for each $\mathscr{A}$-formula γ. The definition is by recursion and proceeds as follows:

1. if α is an atom, then $\alpha^v = v(\alpha)$;
2. if $\gamma = \neg \beta$, then $\gamma^v = \begin{cases} 1 & \text{if } \beta^v = 0 \\ 0 & \text{if } \beta^v = 1; \end{cases}$
3. $(\alpha \wedge \beta)^v = \begin{cases} 1 & \text{if } \alpha^v = \beta^v = 1 \\ 0 & \text{otherwise}; \end{cases}$
4. $(\alpha \vee \beta)^v = \begin{cases} 0 & \text{if } \alpha^v = \beta^v = 0 \\ 1 & \text{otherwise}; \end{cases}$
5. $(\alpha \supset \beta)^v = \begin{cases} 0 & \text{if } \alpha^v = 1 \text{ and } \beta^v = 0 \\ 1 & \text{otherwise}; \end{cases}$
6. $(\alpha \leftrightarrow \beta)^v = \begin{cases} 1 & \text{if } \alpha^v = \beta^v \\ 0 & \text{otherwise}. \end{cases}$

A set Ω of $\mathscr{A}$-formulas is said to be *truth-functionally satisfiable*, or just *satisfiable* for short, if there is an assignment v on $\mathscr{A}$ such that $\alpha^v = 1$ for all $\alpha \in \Omega$; otherwise Ω is said to be (truth-functionally) *unsatisfiable*. If $\Omega = \{\gamma\}$ consists of a single formula, then we say that γ is (truth-functionally) *satisfiable* if Ω is; γ is (truth-functionally) *unsatisfiable* if Ω is unsatisfiable. γ is called a *tautology* if $\gamma^v = 1$ for all assignments v. It is obvious that

Theorem 1.1. γ is tautology if and only if $\neg \gamma$ is unsatisfiable.

We agree to write $\alpha = \beta$ for $\mathscr{A}$-formulas α, β to mean that for every assignment v on $\mathscr{A}$, $\alpha^v = \beta^v$. This convention amounts to thinking of an $\mathscr{A}$-formula as naming a mapping from $\{0, 1\}^n$ into $\{0, 1\}$ for some $n \in N$, so that two $\mathscr{A}$-formulas are regarded as the same if they determine the same

Table 1.1

α	β	$\neg\alpha$	$(\neg\alpha \vee \beta)$	$(\alpha \supset \beta)$	$(\beta \supset \alpha)$	$(\alpha \leftrightarrow \beta)$
1	1	0	1	1	1	1
0	1	1	1	1	0	0
1	0	0	0	0	1	0
0	0	1	1	1	1	1

mappings. [Thus, in high school algebra one writes $x^2 - 1 = (x-1)(x+1)$, although $x^2 - 1$ and $(x-1)(x+1)$ are quite different as *expressions*, because they determine the same mappings on numbers.] With this understanding, we are able to eliminate some of the connectives in favor of others in a systematic manner. In particular, the equations

$$(\alpha \supset \beta) = (\neg\alpha \vee \beta), \tag{1.1}$$

$$(\alpha \leftrightarrow \beta) = ((\alpha \supset \beta) \wedge (\beta \supset \alpha)) \tag{1.2}$$

enable us to limit ourselves to the connectives $\neg$, $\wedge$, $\vee$. The truth of these two equations is easily verified by examining the "truth" tables in Table 1.1, which show all four possibilities for the pair α^v, β^v.

With our use of the equal sign, all tautologies are equal to one another and likewise all unsatisfiable formulas are equal to one another. Since the equations

$$\alpha^v = 1 \quad \text{for all} \quad v, \qquad \beta^v = 0 \quad \text{for all} \quad v$$

determine α to be a tautology and β to be unsatisfiable, it is natural to write 1 for any $\mathscr{A}$-formula which is a tautology and 0 for any $\mathscr{A}$-formula which is unsatisfiable. Thus $\alpha = 1$ means that α is a tautology, and $\alpha = 0$ means that α is unsatisfiable.

The system of $\mathscr{A}$-formulas, under the operations $\neg$, $\wedge$, $\vee$ and involving the "constants" 0, 1 obeys algebraic laws, some of which are analogous to laws satisfied by the real numbers under the operations $-$, $\cdot$, $+$; but there are some striking differences as well. Specifically, we have, for all $\mathscr{A}$-formulas α, β, γ

absorption:

$$(\alpha \wedge 1) = \alpha \qquad\qquad (\alpha \vee 0) = \alpha$$

contradiction; excluded middle:

$$(\alpha \wedge \neg\alpha) = 0 \qquad\qquad (\alpha \vee \neg\alpha) = 1$$

$$(\alpha \wedge 0) = 0 \qquad\qquad (\alpha \vee 1) = 1$$

idempotency:

$$(\alpha \wedge \alpha) = \alpha \qquad\qquad (\alpha \vee \alpha) = \alpha$$

commutativity:

$$(\alpha \wedge \beta) = (\beta \wedge \alpha) \qquad (\alpha \vee \beta) = (\beta \vee \alpha)$$

associativity:

$$(\alpha \wedge (\beta \wedge \gamma)) = ((\alpha \wedge \beta) \wedge \gamma) \qquad (\alpha \vee (\beta \vee \gamma)) = ((\alpha \vee \beta) \vee \gamma)$$

distributivity:

$$(\alpha \wedge (\beta \vee \gamma)) = ((\alpha \wedge \beta) \vee (\alpha \wedge \gamma)) \qquad (\alpha \vee (\beta \wedge \gamma)) = ((\alpha \vee \beta) \wedge (\alpha \vee \gamma))$$

De Morgan laws:

$$\neg(\alpha \wedge \beta) = (\neg \alpha \vee \neg \beta) \qquad \neg(\alpha \vee \beta) = (\neg \alpha \wedge \neg \beta)$$

double negation:

$$\neg \neg \alpha = \alpha$$

These equations, which are easily checked using truth tables, are the basis of the so-called Boolean algebra. In each row, the equations on the left and right can be obtained from one another by simply interchanging all occurrences of "$\vee$" with "$\wedge$" and of "0" with "1." This is a special case of a general principle. The truth tables in Table 1.2 show that if we think of 0 as representing "TRUE," and 1, "FALSE" (instead of the other way around), the tables for "$\wedge$" and "$\vee$" will simply be interchanged. Thus a being from another planet watching us doing propositional calculus might be able to guess that that was in fact what we were doing. But this being would have no way to tell which truth value we were representing by 0 and which by 1, and therefore could not say which of the two connectives represents "and" and which "or." Therefore we have the

> *General Principle of Duality*: Any correct statement involving $\wedge$, $\vee$ and 0, 1, can be translated into another correct statement in which 0 and 1 have been interchanged and $\wedge$ and $\vee$ have been interchanged.

Of course, in carrying out the translation, notions defined in terms of 0, 1, $\wedge$, and $\vee$ must be replaced by their *duals*. For example, the dual of "*α is a tautology*" is "*α is unsatisfiable*." (The first is "$\alpha^{v} = 1$ for all v"; the

Table 1.2

α	β	$(\alpha \wedge \beta)$	$(\alpha \vee \beta)$
1	1	1	1
0	1	0	1
1	0	0	1
0	0	0	0

second is "$\alpha^v = 0$ for all v".) Thus the dual of the correct statement

$$\textit{if } \alpha \textit{ is a tautology, so is } (\alpha \vee \beta)$$

is the equally correct statement

$$\textit{if } \alpha \textit{ is unsatisfiable, so is } (\alpha \wedge \beta).$$

Returning to our list of algebraic laws, we note that in particular the operations $\wedge$ and $\vee$ are commutative and associative. We take advantage of this associativity to write simply

$$\bigwedge_{i \le k} \alpha_i = (\alpha_1 \wedge \alpha_2 \wedge \cdots \wedge \alpha_k)$$

$$\bigvee_{i \le k} \alpha_i = (\alpha_1 \vee \alpha_2 \vee \cdots \vee \alpha_k)$$

without bothering to specify any particular grouping of the indicated formulas. We freely omit parentheses that are not necessary to avoid ambiguity.

Exercises

1. For each of the following formulas tell whether it is (i) satisfiable, (ii) a tautology, (iii) unsatisfiable.
   (a) $((p \supset (q \supset r)) \supset ((p \supset q) \supset (p \supset r)))$.
   (b) $((p \supset (q \supset r)) \leftrightarrow ((p \wedge q) \supset r))$.
   (c) $(p \wedge \neg q)$.
   (d) $((p \vee q) \supset p)$.
   (e) $((\neg(p \supset q) \supset (p \wedge \neg q))$.
2. Apply the general principle of duality to each of the following true statements:
   (a) $(p \vee \neg p)$ is a tautology.
   (b) $(p \supset (q \supset p))$ is a tautology.
3. Prove that if α and β are formulas, then $\alpha = \beta$ if and only if the formula $(\alpha \leftrightarrow \beta)$ is a tautology.
4. Verify the laws of absorption, contradiction, etc. given in this section.
5. Let $\mathscr{A}$ be a set of atoms, and define

$$\mathscr{A}_0 = \mathscr{A}$$

$$\mathscr{A}_{n+1} = \mathscr{A} \cup \{\neg \alpha, (\alpha \wedge \beta), (\alpha \vee \beta), (\alpha \supset \beta), (\alpha \leftrightarrow \beta) \mid \alpha, \beta \in \mathscr{A}_n\}.$$

   Show by induction on n that for all $\alpha \in \mathscr{A}_n$, the number of left parentheses equals the number of right parentheses. Conclude that any propositional formula over $\mathscr{A}$ has an equal number of left and right parentheses.

6. Let $\mathscr{A}, \mathscr{A}'$ be sets of atoms such that $\mathscr{A} \subseteq \mathscr{A}'$, and let v, v' be assignments on $\mathscr{A}, \mathscr{A}'$, respectively, such that $v(\alpha) = v'(\alpha)$ for all atoms α in $\mathscr{A}$. Define $\mathscr{A}_n$, $n \geq 0$, as in Exercise 5, and show by induction on n that $v(\alpha) = v'(\alpha)$ for all formulas $\alpha \in \mathscr{A}_n$. Conclude that $v(\alpha) = v'(\alpha)$ for all propositional formulas over $\mathscr{A}$.

2. Tautological Inference

Let $\gamma_1, \gamma_2, \ldots, \gamma_n, \gamma$ be $\mathscr{A}$-formulas. Then we write

$$\gamma_1, \gamma_2, \ldots, \gamma_n \vDash \gamma$$

and call γ a *tautological consequence* of the *premises* $\gamma_1, \ldots, \gamma_n$ if for every assignment v on $\mathscr{A}$ for which $\gamma_1^v = \gamma_2^v = \cdots = \gamma_n^v = 1$, we have also $\gamma^v = 1$. This relation of tautological consequence is the most important concept in the propositional calculus. However, we can easily prove

Theorem 2.1. The relation $\gamma_1, \gamma_2, \ldots, \gamma_n \vDash \gamma$ is equivalent to each of the following:

1. the formula $((\gamma_1 \wedge \gamma_2 \wedge \cdots \wedge \gamma_n) \supset \gamma)$ is a tautology;
2. the formula $(\gamma_1 \wedge \gamma_2 \wedge \cdots \wedge \gamma_n \wedge \neg\gamma)$ is unsatisfiable.

Proof. $((\gamma_1 \wedge \gamma_2 \wedge \cdots \wedge \gamma_n) \supset \gamma)$ is *not* a tautology just in case for some assignment v, $(\gamma_1 \wedge \gamma_2 \wedge \cdots \wedge \gamma_n)^v = 1$ but $\gamma^v = 0$. That is, just in case for some assignment v, $\gamma_1^v = \gamma_2^v = \cdots = \gamma_n^v = 1$ but $\gamma^v = 0$, which means simply that it is not the case that $\gamma_1, \gamma_2, \ldots, \gamma_n \vDash \gamma$. Likewise

$$(\gamma_1 \wedge \gamma_2 \wedge \cdots \wedge \gamma_n \wedge \neg\gamma)$$

is *satisfiable* if and only if for some assignment v, $\gamma_1^v = \gamma_2^v = \cdots = \gamma_n^v = (\neg\gamma)^v = 1$, i.e., $\gamma_1^v = \gamma_2^v = \cdots = \gamma_n^v = 1$, but $\gamma^v = 0$. ■

Thus the problem of tautological inference is reduced to testing a formula for satisfiability, or for being a tautology. Of course, in principle, such a test can be carried out by simply constructing a truth table. However, a truth table for a formula containing n different atoms will require 2^n rows. Hence, truth table construction may be quite unfeasible even for formulas of modest size.

Consider the example

$$((p \wedge q) \supset (r \wedge s)), ((p_1 \wedge q_1) \supset r_1), ((r_1 \wedge s) \supset s_1), p, q, q_1, p_1 \models s_1 . \tag{2.1}$$

Since there are eight atoms, a truth table would contain $2^8 = 256$ rows. In this example we can reason directly. If v makes all the premises TRUE, then $(p \wedge q)^v = (p_1 \wedge q_1)^v = 1$. Therefore, $(r \wedge s)^v = r_1^v = 1$, and in particular $s^v = 1$. Thus, $(r_1 \wedge s)^v = 1$ and finally, $s_1^v = 1$. We will use Theorem 2.1 to develop more systematic methods for doing such problems.

Exercises

1. Which of the following are correct?
   (a) $(p \supset q), p \models q$.
   (b) $(p \supset q), q \models p$.
   (c) $(p \supset q), \neg q \models \neg p$.
   (d) $(p \supset (q \supset r)), (\neg s \vee p), q \models (s \supset r)$.
2. Apply Theorem 2.1 to Exercise 1.
3. Prove or disprove each of the following.
   (a) $\alpha, \beta \models \gamma$ if and only if $\alpha \models (\beta \supset \gamma)$.
   (b) $\alpha \models \beta$ and $\beta \models \alpha$ if and only if $\alpha = \beta$.
   (c) if $\alpha \models \beta$ or $\alpha \models \gamma$ then $\alpha \models (\beta \vee \gamma)$.
   (d) if $\alpha \models \beta$ or $\alpha \models \gamma$ then $\alpha \models (\beta \wedge \gamma)$.
   (e) if $\alpha \models \beta$ and $\alpha \models \gamma$ then $\alpha \models (\beta \wedge \gamma)$.
   (f) if $\alpha \models \beta$ and $\alpha \models \gamma$ then $\alpha \models (\beta \vee \gamma)$.
   (g) if $\alpha \models \neg \alpha$ then $\neg \alpha$ is a tautology.
   (h) if $\alpha, \beta \models \gamma$ then $\alpha \models \gamma$ or $\beta \models \gamma$.
   (i) if $\alpha \models \gamma$ then $\alpha, \beta \models \gamma$.
   (j) if $\alpha \models (\beta \vee \gamma)$ then $\alpha \models \beta$ or $\alpha \models \gamma$.
4. (a) Show that if α is unsatisfiable then $\alpha \models \beta$ for any formula β.
   (b) Show that if β is a tautology then $\alpha \models \beta$ for any formula α.

3. Normal Forms

We will now describe some algebraic procedures for simplifying $\mathscr{A}$-formulas:

(I) ELIMINATE $\supset$ AND $\leftrightarrow$.

Simply use Eq. (1.2) for each occurrence of $\leftrightarrow$. After all such occurrences have been eliminated, use Eq. (1.1) for each occurrence of $\supset$.

Assuming (I) accomplished, we move on to

(II) MOVE $\neg$ INWARD.

For any occurrence of $\neg$ that is not immediately to the left of an atom either

1. the occurrence immediately precedes another $\neg$, in which case the pair $\neg\neg$ can be eliminated using the law of double negation; or
2. the occurrence immediately precedes an $\mathscr{A}$-formula of the form $(\alpha \wedge \beta)$ or $(\alpha \vee \beta)$, in which case one of the De Morgan laws can be applied to move the $\neg$ inside the parentheses.

After (II) has been applied some finite number of times, a formula will be obtained to which (II) can no longer be applied. Such a formula must have each $\neg$ immediately preceding an atom.

As an example of the use of (I) and (II) consider the formula

$$(((p \leftrightarrow q) \supset (r \supset s)) \wedge (q \supset \neg (p \wedge r))). \qquad (3.1)$$

Eliminating $\leftrightarrow$ gives

$$((((p \supset q) \wedge (q \supset p)) \supset (r \supset s)) \wedge (q \supset \neg (p \wedge r))).$$

Eliminate $\supset$:

$$(\neg ((\neg p \vee q) \wedge (\neg q \vee p)) \vee (\neg r \vee s)) \wedge (\neg q \vee \neg (p \wedge r)). \quad (3.2)$$

Move $\neg$ inward:

$$(\neg (\neg p \vee q) \vee \neg (\neg q \vee p) \vee (\neg r \vee s)) \wedge (\neg q \vee \neg p \vee \neg r).$$

Move $\neg$ inward:

$$((p \wedge \neg q) \vee (q \wedge \neg p) \vee \neg r \vee s) \wedge (\neg q \vee \neg p \vee \neg r). \quad (3.3)$$

A formula λ is called a *literal* if either λ is an atom or λ is $\neg \alpha$, where α is an atom. Note that if $\lambda = \neg \alpha$, for α an atom, then $\neg \lambda = \neg\neg \alpha = \alpha$. For α an atom it is convenient to write $\bar{\alpha}$ for $\neg \alpha$.

With this notation (3.3) becomes

$$((p \wedge \bar{q}) \vee (q \wedge \bar{p}) \vee \bar{r} \vee s) \wedge (\bar{q} \vee \bar{p} \vee \bar{r}). \qquad (3.4)$$

The distributive laws can be used to carry out further simplification, analogous to "multiplying out" in elementary algebra. However, the fact that there are two distributive laws available is a complication because the "multiplying out" can proceed in two directions. As we shall see, each direction gives rise to a specific so-called *normal form.*

A handy technique that makes use of the reader's facility with elementary algebra is to actually replace the symbols $\wedge$, $\vee$ by $+$, $\cdot$ and then calculate as in ordinary algebra. Since there are two distributive laws available, correct results will be obtained either by replacing $\wedge$ by $+$ and $\vee$ by $\cdot$ or vice versa. Thus, writing $\cdot$ for $\wedge$ (and even omitting the $\cdot$ as in elementary algebra) and $+$ for $\vee$, (3.4) can be written

$$
\begin{aligned}
&(p\bar{q} + q\bar{p} + \bar{r} + s) \cdot (\bar{q} + \bar{p} + \bar{r}) \\
&\quad = p\bar{q}\bar{q} + p\bar{q}\bar{p} + p\bar{q}\bar{r} + q\bar{p}\bar{q} + q\bar{p}\bar{p} + q\bar{p}\bar{r} + \bar{r}\bar{q} + \bar{r}\bar{p} + \bar{r}\bar{r} \\
&\qquad + s\bar{q} + s\bar{p} + s\bar{r} \\
&\quad = p\bar{q} + 0 + p\bar{q}\bar{r} + 0 + q\bar{p} + q\bar{p}\bar{r} + \bar{r}\bar{q} \\
&\qquad + \bar{r}\bar{p} + \bar{r} + s\bar{q} + s\bar{p} + s\bar{r} \\
&\quad = (p \wedge \bar{q}) \vee (p \wedge \bar{q} \wedge \bar{r}) \vee (q \wedge \bar{p}) \vee (q \wedge \bar{p} \wedge \bar{r}) \\
&\qquad \vee (\bar{r} \wedge \bar{q}) \vee (\bar{r} \wedge \bar{p}) \vee \bar{r} \vee (s \wedge \bar{q}) \vee (s \wedge \bar{p}) \vee (s \wedge \bar{r}), \qquad (3.5)
\end{aligned}
$$

where we have used the principles of contradiction and absorption. Alternatively, writing $+$ for $\wedge$ and $\cdot$ for $\vee$, (3.4) can be written

$$
\begin{aligned}
&(p + \bar{q})(q + \bar{p})\bar{r}s + \bar{q}\bar{p}\bar{r} \\
&\quad = (pq + p\bar{p} + \bar{q}q + \bar{q}\bar{p})\bar{r}s + \bar{q}\bar{p}\bar{r} \\
&\quad = (pq + 1 + 1 + \bar{q}\bar{p})\bar{r}s + \bar{q}\bar{p}\bar{r} \\
&\quad = pq\bar{r}s + \bar{q}\bar{p}\bar{r}s + \bar{q}\bar{p}\bar{r} \\
&\quad = (p \vee q \vee \bar{r} \vee s) \wedge (\bar{q} \vee \bar{p} \vee \bar{r} \vee s) \wedge (\bar{q} \vee \bar{p} \vee \bar{r}). \qquad (3.6)
\end{aligned}
$$

Let λ_i be a sequence of *distinct* literals, $1 \le i \le n$. Then the formula $\bigvee_{i \le n} \lambda_i$ is called an $\vee$-clause and the formula $\bigwedge_{i \le n} \lambda_i$ is called an $\wedge$-clause. A pair of literals λ, λ' are called *mates* if $\lambda' = \neg \lambda$. We have

Theorem 3.1. Let λ_i be a literal for $1 \le i \le n$. Then the following are equivalent:

1. $\bigvee_{i \le n} \lambda_i$ is a tautology;
2. $\bigwedge_{i \le n} \lambda_i$ is unsatisfiable;
3. some pair $\lambda_i, \lambda_j, 1 \le i, j \le n$, is a pair of mates.

Proof. If $\lambda_j = \neg \lambda_i$, then obviously, $\bigvee_{i \le n} \lambda_i$ is a tautology and $\bigwedge_{i \le n} \lambda_i$ is unsatisfiable. If, on the other hand, the λ_i contain no pair of mates, then there are assignments v, w such that $v(\lambda_i) = 1$, $w(\lambda_i) = 0$ for $1 \le i \le n$. Then $(\bigvee_{i \le n} \lambda_i)^w = 0$, $(\bigwedge_{i \le n} \lambda_i)^v = 1$, so that $\bigvee_{i \le n} \lambda_i$ is not a tautology and $\bigwedge_{i \le n} \lambda_i$ is satisfiable. ∎

Let κ_i, $1 \le i \le n$, be a sequence of *distinct* $\vee$*-clauses*. Then the $\mathscr{A}$-formula $\bigwedge_{i \le n} \kappa_i$ is said to be in *conjunctive normal form* (CNF). Dually, if κ_i, $1 \le i \le n$, is a sequence of *distinct* $\wedge$*-clauses*, then the $\mathscr{A}$-formula $\bigvee_{i \le n} \kappa_i$ is in *disjunctive normal form* (DNF). Note that (3.6) is in CNF and (3.5) is in DNF. We say that (3.6) is *a CNF of* (3.1) and that (3.5) is *a DNF of* (3.1). It should be clear that the procedures we have been describing will yield a CNF and a DNF for each $\mathscr{A}$-formula. Thus we have

Theorem 3.2. There is an algorithm which will transform any given $\mathscr{A}$-formula α into a formula β in CNF such that $\beta = \alpha$. There is a similar (in fact, dual) algorithm for DNF.

Because of Theorem 2.1, the following result is of particular importance.

Theorem 3.3. A formula in CNF is a tautology if and only if each of its $\vee$-clauses is a tautology. Dually, a formula in DNF is unsatisfiable if and only if each of its $\wedge$-clauses is unsatisfiable.

Proof. Let $\alpha = \bigwedge_{i \le n} \kappa_i$, where each κ_i is an $\vee$-clause. If each κ_i is a tautology, then for any assignment v we have $\kappa_i^v = 1$ for $1 \le i \le n$, so that $\alpha^v = 1$; hence α is a tautology. If some κ_i is not a tautology, then there is an assignment v such that $\kappa_i^v = 0$; hence $\alpha^v = 0$ and α is not a tautology.

The proof for DNF is similar. Alternatively, we can invoke the general principle of duality. ∎

Let us try to use these methods in applying Theorem 2.1 to example (2.1). First, using Theorem 2.1(1), we wish to know whether the following formula is a tautology:

$$((((p \wedge q) \supset (r \wedge s)) \wedge ((p_1 \wedge q_1) \supset r_1) \\ \wedge ((r_1 \wedge s) \supset s_1) \wedge p \wedge q \wedge q_1 \wedge p_1) \supset s_1).$$

Use of (I) yields

$$(\neg((\neg(p \wedge q) \vee (r \wedge s)) \wedge (\neg(p_1 \wedge q_1) \vee r_1) \\ \wedge(\neg(r_1 \wedge s) \vee s_1) \wedge p \wedge q \wedge q_1 \wedge p_1) \vee s_1).$$

Use of (II) gives

$$(\neg(\neg(p \wedge q) \vee (r \wedge s)) \vee \neg(\neg(p_1 \wedge q_1) \vee r_1) \\ \vee \neg(\neg(r_1 \wedge s) \vee s_1) \vee \neg p \vee \neg q \vee \neg q_1 \vee \neg p_1 \vee s_1).$$

Use of (II) again yields

$$((p \wedge q \wedge \neg(r \wedge s)) \vee (p_1 \wedge q_1 \wedge \neg r_1) \vee (r_1 \wedge s \wedge \neg s_1)$$
$$\vee \neg p \vee \neg q \vee \neg q_1 \vee \neg p_1 \vee s_1).$$

One final use of (II) gives

$$((p \wedge q \wedge (\neg r \vee \neg s)) \vee (p_1 \wedge q_1 \wedge \neg r_1)$$
$$\vee (r_1 \wedge s \wedge \neg s_1) \vee \neg p \vee \neg q \vee \neg q_1 \vee \neg p_1 \vee s_1). \tag{3.7}$$

To apply Theorem 3.3, it is necessary to find a CNF of (3.7). So we replace $\wedge$ by $+$ and $\vee$ by $\cdot$:

$$(p + q + \bar{r}\bar{s})(p_1 + q_1 + \bar{r}_1)(r_1 + s + \bar{s}_1)\bar{p}\bar{q}\bar{q}_1\bar{p}_1 s_1 \tag{3.8}$$

and see that the CNF of (3.7) will consist of 27 clauses. Here are three "typical" clauses from this CNF:

$$(p \vee p_1 \vee r_1 \vee \bar{p} \vee \bar{q} \vee \bar{q}_1 \vee \bar{p}_1 \vee s_1)$$
$$(\bar{r} \vee \bar{s} \vee q_1 \vee r_1 \vee \bar{p} \vee \bar{q} \vee \bar{q}_1 \vee \bar{p}_1 \vee s_1)$$
$$(q \vee \bar{r}_1 \vee \bar{s}_1 \vee \bar{p} \vee \bar{q} \vee \bar{q}_1 \vee \bar{p}_1 \vee s_1).$$

Each of these clauses contains a pair of literals that are mates: $p, \bar{p}$ in the first (and also $p_1, \bar{p}_1$); $q_1, \bar{q}_1$ in the second; and $q, \bar{q}$ in the third (also $s_1, \bar{s}_1$). The same will be true for the remaining 24 clauses. But this is clearly not the basis for a very efficient algorithm. What if we try Theorem 2.1(2) on the same example? Then we need to show that the following formula is unsatisfiable:

$$((p \wedge q) \supset (r \wedge s)) \wedge ((p_1 \wedge q_1) \supset r_1) \wedge ((r_1 \wedge s) \supset s_1)$$
$$\wedge p \wedge q \wedge q_1 \wedge p_1 \wedge \neg s_1. \tag{3.9}$$

Using (I) we obtain

$$((\neg(p \wedge q) \vee (r \wedge s)) \wedge (\neg(p_1 \wedge q_1) \vee r_1)$$
$$\wedge (\neg(r_1 \wedge s) \vee s_1) \wedge p \wedge q \wedge q_1 \wedge p_1 \wedge \neg s_1).$$

Using (II) we obtain

$$((\neg p \vee \neg q \vee (r \wedge s)) \wedge (\neg p_1 \vee \neg q_1 \vee r_1) \\ \wedge (\neg r_1 \vee \neg s \vee s_1) \wedge p \wedge q \wedge q_1 \wedge p_1 \wedge \neg s_1). \tag{3.10}$$

To find a DNF formula equal to this we replace $\wedge$ by $\cdot$ and $\vee$ by $+$, obtaining

$$(\bar{p} + \bar{q} + rs)(\bar{p}_1 + \bar{q}_1 + r_1)(\bar{r}_1 + \bar{s} + s_1)pqq_1p_1\bar{s}_1 .$$

But this is exactly the same as (3.8) except that each literal has been replaced by its mate! Once again we face essentially the same 27 clauses.

Suppose we seek a formula in CNF equal to (3.10) instead of a formula in DNF. We need only replace $\wedge$ by $+$ and $\vee$ by $\cdot$:

$$\bar{p}\bar{q}(r + s) + \bar{p}_1\bar{q}_1 r_1 + \bar{r}_1\bar{s}s_1 + p + q + q_1 + p_1 + \bar{s}_1 .$$

In this manner, we get a formula in which almost all "multiplying out" has already occurred. The CNF is simply

$$(\bar{p} \vee \bar{q} \vee r) \wedge (\bar{p} \vee \bar{q} \vee s) \wedge (\bar{p}_1 \vee \bar{q}_1 \vee r_1) \\ \wedge (\bar{r}_1 \vee \bar{s} \vee s_1) \wedge p \wedge q \wedge q_1 \wedge p_1 \wedge \bar{s}_1 . \tag{3.11}$$

It consists of nine short, easily obtained clauses.

A moment's reflection will show that this situation is entirely typical. Because the formula of Theorem 2.1(2) has the form

$$(\gamma_1 \wedge \gamma_2 \wedge \cdots \wedge \gamma_n \wedge \neg \gamma),$$

we can get a CNF formula simply by obtaining a CNF for each of the (ordinarily short) formulas $\gamma_1, \gamma_2, \ldots, \gamma_n, \neg \gamma$. However, to obtain a DNF, which according to Theorem 3.3 is what we really want, we will have to multiply out $(n + 1)$ polynomials. If, say, each of $\gamma_1, \ldots, \gamma_n, \neg \gamma$ is an $\vee$-clause consisting of k literals, then the DNF will consist of k^{n+1} $\wedge$-clauses. And the general principle of duality guarantees (as we have already seen in our particular example) that the same discouraging arithmetic will emerge should we attempt instead to use Theorem 2.1(1). In this case a DNF will generally be easy to get, whereas a CNF (which is what we really want) will require a good deal of computing time.

These considerations lead to the following problem:

Satisfiability Problem. Find an efficient algorithm for testing an $\mathscr{A}$-formula in CNF to determine whether it is truth-functionally satisfiable.

This problem has been of central importance in theoretical computer science, not only for the reasons already given, but also for others that will emerge in Chapter 15.

Exercises

1. Find CNF and DNF formulas equal to each of the following.
   (a) $((p \wedge (q \vee r)) \vee (q \wedge (p \vee r)))$.
   (b) $((\neg p \vee (p \wedge \neg q)) \wedge (r \vee (\neg p \wedge q)))$.
   (c) $(p \supset (q \leftrightarrow r))$.
2. Find a DNF formula that has the truth table

p	q	r	
1	1	1	0
0	1	1	1
1	0	1	1
0	0	1	1
1	1	0	0
0	1	0	1
1	0	0	0
0	0	0	0

   [*Hint:* The second row of the table corresponds to the $\wedge$-clause $(\neg p \wedge q \wedge r)$. Each row for which the value is 1 similarly determines an $\wedge$-clause.]
3. Show how to generalize Exercise 2 to obtain a DNF formula corresponding to any given truth table.
4. Describe a dual of the method of Exercise 3 which, for any formula α, gives a DNF formula β such that $\alpha = \neg \beta$. Then show how to turn $\neg \beta$ into a CNF formula γ such that $\alpha = \gamma$. Apply the method to the truth table in Exercise 2. [*Hint:* Each row in the truth table for which the value is 0 corresponds to an $\wedge$-clause which should *not* be true.]
5. Let $\mathscr{A} = \{p, q, r\}$.
   (a) Give a DNF formula α over $\mathscr{A}$ such that $\alpha^v = 1$ for exactly three assignments v on $\mathscr{A}$.
   (b) Give a CNF formula β over $\mathscr{A}$ such that $\beta^v = 1$ for exactly three assignments v on $\mathscr{A}$.

6. **(a)** Let α be

$$(p \wedge q \wedge r) \vee (p \wedge q \wedge \neg r) \vee (p \wedge \neg q \wedge r) \vee (p \wedge \neg q \wedge \neg r).$$

Give DNF formulas β, γ, δ with $3, 2, 1$ $\wedge$-clauses, respectively, such that $\alpha = \beta = \gamma = \delta$.

(b) Let α be

$$(p \vee q \vee r) \wedge (p \vee q \vee \neg r) \wedge (p \vee \neg q \vee r) \wedge (p \vee \neg q \vee \neg r).$$

Give CNF formulas β, γ, δ with $3, 2, 1$ $\vee$-clauses, respectively, such that $\alpha = \beta = \gamma = \delta$.

7. Give a CNF formula α with two $\vee$-clauses such that $\alpha \neq \beta$ for all CNF formulas β with one $\vee$-clause.

8. Use a normal form to show the correctness of the inference

$$(p \supset q), (r \vee \neg q), \neg (p \wedge r) \vDash \neg p.$$

4. The Davis–Putnam Rules

In order to make it easier to state algorithms for manipulating formulas in CNF, it will be helpful to give a simple representation of such formulas as sets. From now on we use the word *clause* to mean $\vee$-clause. We represent the clause $\kappa = \bigvee_{j \le m} \lambda_j$ as the *set* $\kappa = \{\lambda_j | j \le m\}$, and we represent the formula $\alpha = \bigwedge_{i \le n} \kappa_i$, where each κ_i is a clause, as the *set* $\alpha = \{\kappa_i | i \le n\}$. In so doing we lose the order of the clauses and the order of the literals in each clause; however, by the commutative laws, this does not matter.

It is helpful to speak of the empty set of literals as the *empty clause*, written $\square$, and of the empty set of clauses as the *empty formula*, written simply $\varnothing$. Since it is certainly true, although vacuously so, that there is an assignment (in fact any assignment will do) which makes every clause belonging to the empty formula true, it is natural and appropriate to agree that *the empty formula* $\varnothing$ is *satisfiable* (in fact, it is a tautology). On the other hand, there is no assignment which makes some literal belonging to the empty clause $\square$ true (because there are no such literals). Thus, we should regard the empty clause $\square$ as being *unsatisfiable*. Hence any formula α such that $\square \in \alpha$ will be unsatisfiable as well.

We will give some rules for manipulating formulas in CNF that are helpful in designing algorithms for testing such formulas for satisfiability. By Theorem 3.1, a clause κ is a tautology if and only if $\lambda, \neg\lambda \in \kappa$ for some literal λ. Now, if $\kappa \in \alpha$ and κ is a tautologous clause, then α is satisfiable if and only if $\alpha - \{\kappa\}$ is. Hence, we can assume that the sets of clauses with which we deal contain no clauses which are tautologies. The following terminology is helpful: a clause $\kappa = \{\lambda\}$, consisting of a single literal, is called a *unit*. If α is a set of clauses and λ is a literal, then a clause κ is called *λ-positive* if $\lambda \in \kappa$, κ is called *λ-negative* if $\neg\lambda \in \kappa$, and κ is called *λ-neutral* if κ is neither λ-positive nor λ-negative. Since tautologous clauses have been excluded, no clause can be both λ-positive and λ-negative. We write α_λ^+ for the set of λ-positive clauses of α, α_λ^- for the set of λ-negative clauses of α, and α_λ^0 for the set of λ-neutral clauses of α. Thus for every literal λ, we have the decomposition $\alpha = \alpha_\lambda^+ \cup \alpha_\lambda^- \cup \alpha_\lambda^0$. Finally, we write

$$\mathrm{POS}_\lambda(\alpha) = \alpha_\lambda^0 \cup \{\kappa - \{\lambda\} \mid \kappa \in \alpha_\lambda^+\},$$

$$\mathrm{NEG}_\lambda(\alpha) = \alpha_\lambda^0 \cup \{\kappa - \{\neg\lambda\} \mid \kappa \in \alpha_\lambda^-\}.$$

Our main result is

Theorem 4.1 (Splitting Rule). Let α be a formula in CNF, and let λ be a literal. Then α is satisfiable if and only if at least one of the pair $\mathrm{POS}_\lambda(\alpha)$ and $\mathrm{NEG}_\lambda(\alpha)$ is satisfiable.

Proof. First let α be satisfiable, say $\alpha^v = 1$. Thus $\kappa^v = 1$ for all $\kappa \in \alpha$. That is, for each $\kappa \in \alpha$, there is a literal $\mu \in \kappa$ such that $\mu^v = 1$. Now, we must have either $\lambda^v = 1$ or $\lambda^v = 0$. Suppose first that $\lambda^v = 0$. We know that for each $\kappa \in \alpha_\lambda^+$, there is a literal $\mu \in \kappa$ such that $\mu^v = 1$. Thus this μ is not λ. Thus, for $\kappa \in \alpha_\lambda^+$, $(\kappa - \{\lambda\})^v = 1$. Hence, in this case, $\mathrm{POS}_\lambda(\alpha)^v = 1$. If, instead, $\lambda^v = 1$, we can argue similarly that for each $\kappa \in \alpha_\lambda^-$, $(\kappa - \{\neg\lambda\})^v = 1$ and hence that $\mathrm{NEG}_\lambda(\alpha)^v = 1$.

Conversely, let $\mathrm{POS}_\lambda(\alpha)^v = 1$ for some assignment v. Then we define the assignment w by stipulating that

$$\lambda^w = 0; \qquad \mu^w = \mu^v \qquad \text{for all literals} \quad \mu \neq \lambda, \neg\lambda.$$

Now, if $\kappa \in \alpha_\lambda^0$, then $\kappa^w = \kappa^v = 1$; if $\kappa \in \alpha_\lambda^+$, then $\kappa^w = (\kappa - \{\lambda\})^w = (\kappa - \{\lambda\})^v = 1$; finally, if $\kappa \in \alpha_\lambda^-$, then $\kappa^w = 1$ because $(\neg\lambda)^w = 1$. Thus, $\alpha^w = 1$.

If, instead, $\mathrm{NEG}_\lambda(\alpha)^v = 1$ for some assignment v, we define w by

$$\lambda^w = 1; \qquad \mu^w = \mu^v \qquad \text{for all literals } \mu \neq \lambda, \neg\lambda.$$

Then if $\kappa \in \alpha_\lambda^0$, we have $\kappa^w = \kappa^v = 1$; if $\kappa \in \alpha_\lambda^-$, then $\kappa^w = (\kappa - \{\neg \lambda\})^w = (\kappa - \{\neg \lambda\})^v = 1$; finally, if $\kappa \in \alpha_\lambda^+$, then $\kappa^w = 1$ because $\lambda^w = 1$. Thus again $\alpha^w = 1$. ■

This theorem has the virtue of eliminating one literal, but at the price of considering two formulas instead of one. For this reason, it is of particular interest to find special cases in which we do not need to consider both $\text{POS}_\lambda(\alpha)$ and $\text{NEG}_\lambda(\alpha)$.

Thus, suppose that $\alpha_\lambda^- = \varnothing$. Then $\text{NEG}_\lambda(\alpha) = \alpha_\lambda^0 \subseteq \text{POS}_\lambda(\alpha)$. Hence, in this case, for any assignment v we have $\text{POS}_\lambda(\alpha)^v = 1$ *implies* $\text{NEG}_\lambda(\alpha)^v = 1$. Therefore, we conclude

Corollary 4.2 (Pure Literal Rule). If $\alpha_\lambda^- = \varnothing$, then α is satisfiable if and only if $\text{NEG}_\lambda(\alpha) = \alpha_\lambda^0$ is satisfiable.

For another useful special case, suppose that the unit clause $\{\lambda\} \in \alpha$. Then, since $\{\lambda\} - \{\lambda\} = \square$, we conclude that $\square \in \text{POS}_\lambda(\alpha)$. Hence, $\text{POS}_\lambda(\alpha)$ is unsatisfiable, and we have

Corollary 4.3 (Unit Rule). If $\{\lambda\} \in \alpha$, then α is satisfiable if and only if $\text{NEG}_\lambda(\alpha)$ is satisfiable.

To illustrate this last corollary by an example, let α be (3.11), which is a CNF of (3.9). Using the set representation,

$$\alpha = \{\{\bar{p}, \bar{q}, r\}, \{\bar{p}, \bar{q}, s\}, \{\bar{p}_1, \bar{q}_1, r_1\}, \{\bar{r}_1, \bar{s}, s_1\}, \{p\}, \{q\}, \{q_1\}, \{p_1\}, \{\bar{s}_1\}\}. \tag{4.1}$$

Thus, there are nine clauses, of which five are units. Using the unit clause $\{p\}$, Corollary 4.3 tells us that α is satisfiable if and only if $\text{NEG}_p(\alpha)$ is. That is, we need to test for satisfiability the set of clauses

$$\{\{\bar{q}, r\}, \{\bar{q}, s\}, \{\bar{p}_1, \bar{q}_1, r_1\}, \{\bar{r}_1, \bar{s}, s_1\}, \{q\}, \{q_1\}, \{p_1\}, \{\bar{s}_1\}\}.$$

Using the unit rule again, this time choosing the unit clause $\{q\}$, we reduce to

$$\{\{r\}, \{s\}, \{\bar{p}_1, \bar{q}_1, r_1\}, \{\bar{r}_1, \bar{s}, s_1\}, \{q_1\}, \{p_1\}, \{\bar{s}_1\}\}.$$

Using the unit clause $\{s\}$, we get

$$\{\{r\}, \{\bar{p}_1, \bar{q}_1, r_1\}, \{\bar{r}_1, s_1\}, \{q_1\}, \{p_1\}, \{\bar{s}_1\}\}.$$

Successive uses of the unit clauses $\{q_1\}, \{p_1\}, \{\bar{s}_1\}$ yield

$$\{\{r\}, \{\bar{p}_1, r_1\}, \{\bar{r}_1, s_1\}, \{p_1\}, \{\bar{s}_1\}\};$$
$$\{\{r\}, \{r_1\}, \{\bar{r}_1, s_1\}, \{\bar{s}_1\}\};$$
$$\{\{r\}, \{r_1\}, \{\bar{r}_1\}\}.$$

This last, containing the unit clauses $\{r_1\}$ and $\{\bar{r}_1\}$, is clearly unsatisfiable. Or, alternatively, applying the unit rule one last time, we obtain

$$\text{NEG}_{r_1}(\{\{r\}, \{r_1\}, \{\bar{r}_1\}\}) = \{\{r\}, \square\},$$

which is unsatisfiable because it contains the empty clause $\square$.

So we have shown by this computation that (4.1), and therefore (3.9), is unsatisfiable. And by Theorem 2.1, we then can conclude (once again) that the tautological inference (2.1) is valid.

A slight variant of this computation would begin by applying Corollary 4.2, the pure literal rule, to (4.1), using the literal r. This has the effect of simply deleting the first clause. The rest of the computation might then go as previously, but with the initial clause deleted at each stage.

For another example, recall (3.6), which was obtained as a CNF of (3.1). Written as a set of clauses this becomes

$$\beta = \{\{p, q, \bar{r}, s\}, \{\bar{q}, \bar{p}, \bar{r}, s\}, \{\bar{q}, \bar{p}, \bar{r}\}\}. \tag{4.2}$$

Here the pure literal rule can be applied using either of the literals $\bar{r}, s$. Thus, we have that β is satisfiable if and only if $\beta_{\bar{r}}^0$ is satisfiable, if and only if β_s^0 is satisfiable. And we have

$$\beta_{\bar{r}}^0 = \varnothing, \qquad \beta_s^0 = \{\{\bar{q}, \bar{p}, \bar{r}\}\}.$$

From the first we see at once that β is satisfiable; if we wish to use the second, we can note by inspection that $(\beta_s^0)^v = 1$, where $v(q) = v(p) = v(r) = 0$, or we can use the pure literal rule a second time (using any of the three available literals) and once again arrive at the empty formula $\varnothing$.

We next turn to an example that has no unit clauses and to which the pure literal rule is not applicable:

$$\alpha = \{\{\bar{q}, p\}, \{r, p\}, \{\bar{p}, \bar{q}\}, \{\bar{p}, s\}, \{q, \bar{r}\}, \{q, \bar{s}\}\}.$$

Thus we are led to use the splitting rule forming, say,

$$\text{POS}_p(\alpha) = \{\{\bar{q}\}, \{r\}, \{q, \bar{r}\}, \{q, \bar{s}\}\},$$

$$\text{NEG}_p(\alpha) = \{\{\bar{q}\}, \{s\}, \{q, \bar{r}\}, \{q, \bar{s}\}\}.$$

Applying the pure literal rule once and then the unit rule twice to $\text{POS}_p(\alpha)$, we obtain successively

$$\{\{\bar{q}\},\{r\},\{q,\bar{r}\}\}, \qquad \{\{r\},\{\bar{r}\}\}, \qquad \{\square\},$$

so that $\text{POS}_p(\alpha)$ is unsatisfiable. Doing the same to $\text{NEG}_p(\alpha)$ we obtain successively

$$\{\{\bar{q}\},\{s\},\{q,\bar{s}\}\}, \qquad \{\{s\},\{\bar{s}\}\}, \qquad \{\square\},$$

so that $\text{NEG}_p(\alpha)$ is likewise unsatisfiable. By Theorem 4.1, we can thus conclude that α is unsatisfiable.

These examples suggest a rather systematic recursive procedure (sometimes known as the Davis–Putnam procedure) for testing a given formula α in CNF for satisfiability. The procedure as we shall describe it will not be completely deterministic; there will be situations in which one of a number of literals is to be selected. We will write the recursive procedure using two variables, γ for a set of clauses and $\mathscr{S}$ for a stack of sets of clauses. We write TOP($\mathscr{S}$) for the set of clauses at the top of the stack $\mathscr{S}$, POP($\mathscr{S}$) for $\mathscr{S}$ after TOP($\mathscr{S}$) has been removed, PUSH($\beta, \mathscr{S}$) for the stack obtained by putting β on the top of $\mathscr{S}$, and $\varnothing$ for the empty stack. The procedure is as follows:

$\gamma \leftarrow \alpha$; $\mathscr{S} \leftarrow \varnothing$;
while $\gamma \neq \varnothing$ *and* ($\square \notin \gamma$ *or* $\mathscr{S} \neq \varnothing$)
    if $\square \in \gamma$
    then $\gamma \leftarrow$ TOP($\mathscr{S}$); $\mathscr{S} \leftarrow$ POP($\mathscr{S}$);
    else *if* $\gamma_\lambda^- = \varnothing$
        then $\gamma \leftarrow \gamma_\lambda^0$;
        else *if* $\{\lambda\} \in \gamma$
            then $\gamma \leftarrow \text{NEG}_\lambda(\gamma)$;
            else $\mathscr{S} \leftarrow \text{PUSH}(\text{NEG}_\lambda(\gamma), \mathscr{S})$; $\gamma \leftarrow \text{POS}_\lambda(\gamma)$;
end while
if $\gamma = \varnothing$ *then return* SATISFIABLE
        else return UNSATISFIABLE

Thus, this procedure will terminate returning SATISFIABLE whenever γ is the empty formula $\varnothing$, whether or not the stack $\mathscr{S}$ is empty. (This is all right because the original formula will be satisfiable if any one of the formulas obtained by repeated uses of the splitting rule is satisfiable, and, of course, $\varnothing$ is satisfiable.) The procedure will terminate returning UNSATISFIABLE if $\square \in \gamma$ and $\mathscr{S} = \varnothing$. (Here, γ is unsatisfiable, and no formulas remain in $\mathscr{S}$ as the result of uses of the splitting rule.) If neither of these termination conditions is satisfied, the algorithm will first test for $\square \in \gamma$. If $\square \in \gamma$, it rejects (since γ is unsatisfiable) and "pops" the stack. Otherwise it attempts to apply first the pure literal rule and then the unit rule. If both attempts fail, it chooses (nondeterministically) some literal λ, takes $\text{POS}_\lambda(\gamma)$ as the new formula to work on, and "pushes" $\text{NEG}_\lambda(\gamma)$ onto the stack for future reference.

It is not difficult to see that the algorithm just given must always terminate. Let us say that a set of clauses α *reduces* to a set of clauses β if for each clause κ in β there is a clause $\bar{\kappa}$ in α such that $\kappa \subseteq \bar{\kappa}$. Then, at the beginning of each pass through the *while* loop, γ is a set of clauses to which α reduces and the stack consists of a list of sets of clauses to each of which α reduces. Since, for a given α, there are only a finite number of distinct configurations of this kind, and none can be repeated, the algorithm must eventually terminate.

Exercises

1. Let α be $\{\{p,q,r\},\{p,\bar{q}\},\{\bar{p},\bar{r}\}\}$. For $\lambda = p, q, r, \bar{p}, \bar{q}, \bar{r}$, give α_λ^+, α_λ^-, α_λ^0, $\text{POS}_\lambda(\alpha)$, $\text{NEG}_\lambda(\alpha)$. Which of these sets are necessarily equal?

2. Use the Davis–Putnam rules to show the correctness of the inference in Exercise 3.8.

3. Use the Davis–Putnam rules to show the correctness of the following inference.

   If John went swimming, then he lost his glasses and did not go to the movies. If John ate too much meat and did not go to the movies, then he will suffer indigestion. Therefore, if John ate too much meat and went swimming, then he will suffer indigestion.

4. Test the following set of clauses for satisfiability:

$$\{p,\bar{q},r,s\}$$
$$\{\bar{p},q,\bar{r}\}$$
$$\{\bar{r},s\}$$
$$\{\bar{q},r\}$$
$$\{p,\bar{s}\}.$$

5. Modify the Davis–Putnam procedure so that when the answer is SATISFIABLE on input α, it returns an assignment v such that $\alpha^v = 1$.

6. How many distinct computations can be performed by the Davis–Putnam procedure on input $\{\{p, q, r\}, \{p, \bar{q}\}, \{\bar{p}, \bar{r}\}\}$?
7. Let α be a CNF formula with n distinct atoms.
   (a) What is the maximum number of formulas that can be on the stack in the Davis–Putnam procedure at any given time?
   (b) Suppose that α is satisfiable. Show that if the Davis–Putnam procedure always makes the right choice of λ at each stage, the *while* loop executes no more than n times.
   (c) How many times must the *while* loop execute on input

$$\{\{p, q\}, \{p, \bar{q}\}, \{\bar{p}, q\}, \{\bar{p}, \bar{q}\}\}?$$

   On input

$$\{\{p, q, r\}, \{p, q, \bar{r}\}, \{p, \bar{q}, r\}, \{p, \bar{q}, \bar{r}\},$$
$$\{\bar{p}, q, r\}, \{\bar{p}, q, \bar{r}\}, \{\bar{p}, \bar{q}, r\}, \{\bar{p}, \bar{q}, \bar{r}\}\}?$$

5. Minimal Unsatisfiability and Subsumption

We begin with

Theorem 5.1. Let the clauses κ_1, κ_2 satisfy $\kappa_1 \subset \kappa_2$. Then if α is a formula in CNF such that $\kappa_1, \kappa_2 \in \alpha$, then α is satisfiable if and only if $\alpha - \{\kappa_2\}$ is satisfiable.

Proof. Clearly, if α is satisfiable, so is $\alpha - \{\kappa_2\}$.

Conversely, if $(\alpha - \{\kappa_2\})^v = 1$, then $\kappa_1^v = 1$, so that also $\kappa_2^v = 1$. Hence, $\alpha^v = 1$. ■

Thus, if in fact $\kappa_1, \kappa_2 \in \alpha$ and $\kappa_1 \subset \kappa_2$, we may simply drop κ_2 and test $\alpha - \{\kappa_2\}$ for satisfiability. The operation of dropping κ_2 in such a case is called *subsumption*. Unfortunately, there is no efficient algorithm known for testing a large set of clauses for the possibility of applying *subsumption*.

Definition. A finite set of clauses α is called *minimally unsatisfiable* if

1. α is unsatisfiable, and
2. for all $\beta \subset \alpha$, β is satisfiable.

Definition. A finite set of clauses α is said to be *linked* if whenever $\lambda \in \kappa_1$ and $\kappa_1 \in \alpha$, there is a clause $\kappa_2 \in \alpha$ such that $\neg \lambda \in \kappa_2$. That is, each literal in a clause of α has a *mate* in another clause of α.

Then it is very easy to prove

Theorem 5.2. Let α be minimally unsatisfiable. Then

1. for no $\kappa_1, \kappa_2 \in \alpha$ can we have $\kappa_1 \subset \kappa_2$, and
2. α is linked.

Proof. Condition 1 is an immediate consequence of Theorem 5.1. To verify 2, suppose that α is minimally unsatisfiable but not linked. Then, there is a literal λ in a clause $\kappa \in \alpha$ such that the literal $\neg \lambda$ occurs in none of the clauses of α, i.e., $\alpha_\lambda^- = \emptyset$. Thus, by the pure literal rule, α_λ^0 is unsatisfiable. But since $\alpha_\lambda^0 \subset \alpha$, this is a contradiction. ∎

Exercise

1. Give a minimally unsatisfiable CNF formula with four clauses.

6. Resolution

Let κ_1, κ_2 be clauses such that $\lambda \in \kappa_1$ and $\neg \lambda \in \kappa_2$. Then we write

$$\text{res}_\lambda(\kappa_1, \kappa_2) = (\kappa_1 - \{\lambda\}) \cup (\kappa_2 - \{\neg \lambda\}).$$

The clause $\text{res}_\lambda(\kappa_1, \kappa_2)$ is then called the *resolvent of* κ_1, κ_2 *with respect to the literal* λ. The operation of forming resolvents has been the basis of a very large number of computer programs designed to perform logical deductions. We have

Theorem 6.1. Let λ be an atom and let κ_1, κ_2 be clauses such that $\lambda \in \kappa_1$, $\neg \lambda \in \kappa_2$. Then

$$\kappa_1, \kappa_2 \models \text{res}_\lambda(\kappa_1, \kappa_2).$$

Proof. Let v be an assignment such that $\kappa_1^v = \kappa_2^v = 1$. Now if $\lambda^v = 1$, then $(\kappa_2 - \{\neg \lambda\})^v = 1$, while if $\lambda^v = 0$, then $(\kappa_1 - \{\lambda\})^v = 1$. In either case, therefore, $\text{res}_\lambda(\kappa_1, \kappa_2)^v = 1$. ∎

Let α be a finite set of clauses. A sequence of clauses $\kappa_1, \kappa_2, \ldots, \kappa_n = \kappa$ is called a *resolution derivation of* κ *from* α if for each i, $1 \leq i \leq n$, either $\kappa_i \in \alpha$ or there are $j, k < i$ and a literal λ such that $\kappa_i = \text{res}_\lambda(\kappa_j, \kappa_k)$. A resolution derivation of $\square$ from α is called a *resolution refutation of* α. We define

$$\text{RES}_\lambda(\alpha) = \alpha_\lambda^0 \cup \{\text{res}_\lambda(\kappa_1, \kappa_2) | \kappa_1 \in \alpha_\lambda^+, \kappa_2 \in \alpha_\lambda^-\}.$$

We have

Theorem 6.2. Let α be a formula in CNF and let λ be a literal. Then α is satisfiable if and only if $\text{RES}_\lambda(\alpha)$ is satisfiable.

Proof. First let $\alpha^v = 1$. Then if $\kappa \in \alpha_\lambda^0$, we have also $\kappa \in \alpha$, so that $\kappa^v = 1$. Furthermore, if $\kappa = \text{res}_\lambda(\kappa_1, \kappa_2)$, with $\kappa_1 \in \alpha_\lambda^+$, $\kappa_2 \in \alpha_\lambda^-$, then $\kappa_1^v = 1$, $\kappa_2^v = 1$, so that, by Theorem 6.1, $\kappa^v = 1$. Since for all $\kappa \in \text{RES}_\lambda(\alpha)$, we have $\kappa^v = 1$, it follows that $\text{RES}_\lambda(\alpha)^v = 1$.

Conversely, let $\text{RES}_\lambda(\alpha)^v = 1$. We claim that either $\text{POS}_\lambda(\alpha)^v = 1$ or $\text{NEG}_\lambda(\alpha)^v = 1$. For, suppose that $\text{POS}_\lambda(\alpha)^v = 0$. Since $\alpha_\lambda^0 \subseteq \text{RES}_\lambda(\alpha)$, we have $(\alpha_\lambda^0)^v = 1$. So for some $\kappa_1 \in \alpha_\lambda^+$, we must have $(\kappa_1 - \{\lambda\})^v = 0$. However, *for all* $\kappa_2 \in \alpha_\lambda^-$ and this κ_1, we must have $\text{res}_\lambda(\kappa_1, \kappa_2)^v = [(\kappa_1 - \{\lambda\}) \cup (\kappa_2 - \{\neg \lambda\})]^v = 1$. Thus, for all $\kappa_2 \in \alpha_\lambda^-$ we have $(\kappa_2 - \{\neg \lambda\})^v = 1$, i.e., $\text{NEG}_\lambda(\alpha)^v = 1$. This proves our claim that either $\text{POS}_\lambda(\alpha)$ or $\text{NEG}_\lambda(\alpha)$ must be satisfiable. By Theorem 4.1, i.e., the splitting rule, α is satisfiable. ■

Theorem 6.2 suggests another procedure for testing a formula α in CNF for satisfiability. As with the Davis–Putnam rules, seek a literal of α to which the pure literal or unit rule can be applied. If none is to be found, choose a literal λ of α and compute $\text{RES}_\lambda(\alpha)$. Continue recursively.

As with the Davis–Putnam procedure, this procedure must eventually terminate in $\{\square\}$ or $\varnothing$; this is because the number of literals is successively diminished. This procedure has the advantage of not requiring a stack of formulas, but the disadvantage that the problem may get considerably larger because of the use of the $\text{RES}_\lambda(\alpha)$ operation. Unfortunately, the present procedure is also called the Davis–Putnam procedure in the literature. To add to the confusion, it seems that computer implementations of the "Davis–Putnam procedure" have been almost exclusively of the procedure introduced in Section 4, whereas theoretical analyses of the computational complexity of the "Davis–Putnam procedure" have tended to deal with the procedure we have just introduced.

Theorem 6.3. Let α be a formula in CNF and suppose that there is a resolution derivation of the clause κ from α. Then $\alpha^v = 1$ implies $\kappa^v = 1$.

Proof. Let $\kappa_1, \kappa_2, \ldots, \kappa_n = \kappa$ be a resolution derivation of κ from α. We shall prove that $\kappa_i^v = 1$ for $1 \leq i \leq n$, which will prove the result. To prove this by induction, we assume that $\kappa_j^v = 1$ for all $j < i$. (Of course for the case $i = 1$, this induction hypothesis is true vacuously.) Now, there are two cases. If $\kappa_i \in \alpha$, then $\kappa_i^v = 1$. Otherwise $\kappa_i = \text{res}_\lambda(\kappa_j, \kappa_k)$, where

$j, k < i$. Hence, by the induction hypothesis, $\kappa_j^v = \kappa_k^v = 1$. So by Theorem 6.1, $\kappa_i^v = 1$. ■

Theorem 6.4 (Ground Resolution Theorem). The formula α in CNF is unsatisfiable if and only if there is a resolution refutation of α.

Proof. First let there be a resolution refutation of α, but suppose that nevertheless $\alpha^v = 1$. Then, by Theorem 6.3, $\square^v = 1$, which is impossible.

Conversely, let α be unsatisfiable. Let $\lambda_1, \lambda_2, \ldots, \lambda_k$ be a list of all the *atoms* that occur in α. Let

$$\alpha_0 = \alpha, \qquad \alpha_i = \mathrm{RES}_{\lambda_i}(\alpha_{i-1}), \qquad i = 1, 2, \ldots, k.$$

Clearly each α_i contains only the atoms λ_j for which $i < j \leq k$. Hence α_k contains no atoms at all, and must be either $\varnothing$ or $\{\square\}$. On the other hand, by Theorem 6.2, we have that α_i is unsatisfiable for $0 \leq i \leq k$. Hence $\alpha_k = \{\square\}$. Now, let the sequence $\kappa_1, \kappa_2, \ldots, \kappa_m$ of clauses consist, first, of all of the clauses of $\alpha_0 = \alpha$, then, all of the clauses of α_1, and so on through all of the clauses of α_k. But this last means that $\kappa_m = \square$. Moreover, it is clear from the definition of the RES_λ operation that $\kappa_1, \kappa_2, \ldots, \kappa_m$ is a resolution derivation. ■

To illustrate the ground resolution theorem, we apply it to (4.1) to show, once again, that (3.9) is unsatisfiable. Here then is a resolution refutation of the formula α of (4.1):

$$\{\bar{p}, \bar{q}, s\}, \{\bar{r}_1, \bar{s}, s_1\}, \{\bar{p}, \bar{q}, \bar{r}_1, s_1\}, \{p\}, \{\bar{q}, \bar{r}_1, s_1\}, \{q\}, \{\bar{r}_1, s_1\},$$
$$\{\bar{s}_1\}, \{\bar{r}_1\}, \{\bar{p}_1, \bar{q}_1, r_1\}, \{\bar{p}_1, \bar{q}_1\}, \{q_1\}, \{\bar{p}_1\}, \{p_1\}, \square.$$

Exercises

1. **(a)** Use the resolution method to answer Exercise 1.1.
   (b) Do the same for Exercise 2.1.
2. Give a resolution refutation that shows the correctness of the inference of Exercise 3.8.
3. Do the same for the inference of Exercise 4.3.
4. Let $\alpha_0, \ldots, \alpha_n$ be CNF formulas, and let β be a DNF formula $\bigvee_{i \leq m} \beta_i$. Show that $\alpha_0, \ldots, \alpha_n \vDash \beta$ if and only if there is a resolution refutation of $\bigcup_{i \leq n} \alpha_i \cup \{\neg \beta_i \mid i \leq m\}$.

5. Let λ be an atom and let κ_1, κ_2 be clauses such that $\lambda \in \kappa_1$, $\neg \lambda \in \kappa_2$. Prove or disprove the following.
   (a) $\text{res}_\lambda(\kappa_1, \kappa_2) \models (\kappa_1 \vee \kappa_2)$.
   (b) $\text{res}_\lambda(\kappa_1, \kappa_2) \models \kappa_1$ or $\text{res}_\lambda(\kappa_1, \kappa_2) \models \kappa_2$.
   (c) $\text{res}_\lambda(\kappa_1, \kappa_2) \models (\kappa_1 \wedge \kappa_2)$.
   (d) if $\text{res}_\lambda(\kappa_1, \kappa_2)$ is valid, then $(\kappa_1 \wedge \kappa_2)$ is valid.
6. Let α be a formula in CNF and let λ be a literal. Prove or disprove that α is valid if and only if $\text{RES}_\lambda(\alpha)$ is valid.

7. The Compactness Theorem

Now, we will prove a theorem relating infinite sets of $\mathscr{A}$-formulas to their finite subsets.

Definition. A set Ω of $\mathscr{A}$-formulas is called *finitely satisfiable* if for every finite set $\Delta \subseteq \Omega$, the set Δ is truth-functionally satisfiable.

We have

Theorem 7.1. Let Ω be finitely satisfiable and let α be an $\mathscr{A}$-formula. Then either $\Omega \cup \{\alpha\}$ or $\Omega \cup \{\neg \alpha\}$ is finitely satisfiable.

Proof. Suppose to the contrary that Ω is finitely satisfiable but that neither $\Omega \cup \{\alpha\}$ nor $\Omega \cup \{\neg \alpha\}$ is finitely satisfiable. Then there are finite sets $\Delta_1, \Delta_2 \subseteq \Omega$ such that $\Delta_1 \cup \{\alpha\}$ and $\Delta_2 \cup \{\neg \alpha\}$ are both truth-functionally unsatisfiable. But $\Delta_1 \cup \Delta_2$ is a finite subset of Ω and hence there must be an assignment v such that for each $\beta \in \Delta_1 \cup \Delta_2$, we have $\beta^v = 1$. Now, either $\alpha^v = 1$ or $\alpha^v = 0$. In the first case $\Delta_1 \cup \{\alpha\}$ is satisfiable, and in the second case $\Delta_2 \cup \{\neg \alpha\}$ is satisfiable. This is a contradiction. ■

Now we will need to use a general property of infinite languages.

Enumeration Principle. Let L be an infinite subset of A^*, where A is an alphabet (and therefore is finite). Then there is an infinite sequence or *enumeration* $w_0, w_1, w_2, \ldots$ which consists of all the words in L each listed exactly once.

The truth of this enumeration principle can be seen in many ways. One is simply to imagine the elements of L written in order of increasing length, and to order words of the same length among themselves like the entries in a dictionary. Alternatively, one can regard the strings on A as notations for numbers in some base (as in Chapter 5) and arrange the elements of L in numerical order. (Actually, as it is not difficult to see, these two methods yield the same enumeration.) Of course, no claim is made that there is an algorithm for computing w_i from i. Such an algorithm can only exist if the language L is r.e.

Now, let $\alpha_0, \alpha_1, \alpha_2, \ldots$ be an enumeration of the set of all $\mathscr{A}$-formulas. (By the enumeration principle, such an enumeration must exist.) Let Ω be a given finitely satisfiable set of $\mathscr{A}$-formulas. We define the sequence

$$\Omega_0 = \Omega$$

$$\Omega_{n+1} = \begin{cases} \Omega_n \cup \{\alpha_n\} & \text{if this set is finitely satisfiable} \\ \Omega_n \cup \{\neg\, \alpha_n\} & \text{otherwise.} \end{cases}$$

By Theorem 7.1, we have

Lemma 1. Each Ω_n is finitely satisfiable.

Let $\tilde{\Omega} = \bigcup_{n=0}^{\infty} \Omega_n$. Then, we have

Lemma 2. $\tilde{\Omega}$ is finitely satisfiable.

Proof. Let us be given a finite set $\Delta \subseteq \tilde{\Omega}$. For each $\gamma \in \Delta$, $\gamma \in \Omega_n$ for some n. Hence $\Delta \subseteq \Omega_m$, where m is the maximum of those n. By Lemma 1, Δ is truth-functionally satisfiable. ■

Lemma 3. For each $\mathscr{A}$-formula α either $\alpha \in \tilde{\Omega}$ or $\neg\, \alpha \in \tilde{\Omega}$, but not both.

Proof. Let $\alpha = \alpha_n$. Then $\alpha \in \Omega_{n+1}$ or $\neg\, \alpha \in \Omega_{n+1}$, so that α or $\neg\, \alpha$ belongs to $\tilde{\Omega}$. If $\alpha, \neg\, \alpha \in \tilde{\Omega}$, then by Lemma 2, the finite set $\{\alpha, \neg\, \alpha\}$ would have to be truth-functionally satisfiable. But this is impossible. ■

Now we define an assignment v by letting $v(\lambda) = 1$ if $\lambda \in \tilde{\Omega}$ and $v(\lambda) = 0$ if $\lambda \notin \tilde{\Omega}$ for every atom λ. We have

Lemma 4. For each $\mathscr{A}$-formula α, $\alpha^v = 1$ if and only if $\alpha \in \tilde{\Omega}$.

Proof. As we already know, it suffices to restrict ourselves to formulas using the connectives $\neg$, $\vee$, $\wedge$. And, in fact, the De Morgan relation

$$(\beta_1 \vee \beta_2) = \neg(\neg\, \beta_1 \wedge \neg\, \beta_2)$$

shows that we can restrict ourselves even further, to the connectives $\neg$, $\wedge$. So, we assume that α is an $\mathscr{A}$-formula expressed in terms of the connectives $\neg$, $\wedge$. Our proof will be by induction on the total number of occurrences of these connectives in α.

If this total number is 0, then α is an atom, and the result follows from our definition of v. Otherwise we must have either $\alpha = \neg\beta$ or $\alpha = (\beta \wedge \gamma)$, where by the induction hypothesis we can assume the desired result for β and γ.

Case 1. $\alpha = \neg\beta$
Then, using Lemma 3,

$$\begin{aligned} \alpha^v = 1 \quad &\text{if and only if} \quad \beta^v \neq 1 \\ &\text{if and only if} \quad \beta^v \notin \tilde{\Omega} \\ &\text{if and only if} \quad \alpha \in \tilde{\Omega}. \end{aligned}$$

Case 2. $\alpha = (\beta \wedge \gamma)$
If $\alpha^v = 1$, then $\beta^v = \gamma^v = 1$, so by the induction hypothesis, $\beta, \gamma \in \tilde{\Omega}$. If $\alpha \notin \tilde{\Omega}$, then by Lemma 3, $\neg\alpha \in \tilde{\Omega}$. But the finite set $\{\beta, \gamma, \neg\alpha\}$ is not satisfiable, contradicting Lemma 2. Thus, $\alpha \in \tilde{\Omega}$.

Conversely, if $\alpha \in \tilde{\Omega}$, then neither $\neg\beta$ nor $\neg\gamma$ can belong to $\tilde{\Omega}$, because the finite sets $\{\alpha, \neg\beta\}$, $\{\alpha, \neg\gamma\}$ are not satisfiable. Thus, by Lemma 3, $\beta, \gamma \in \tilde{\Omega}$. By the induction hypothesis $\beta^v = \gamma^v = 1$. Therefore, $\alpha^v = 1$. ■

Now, since $\Omega \subseteq \tilde{\Omega}$, we see that $\alpha^v = 1$ for each $\alpha \in \Omega$. Hence, Ω is truth-functionally satisfiable. Since we began with an arbitrary finitely satisfiable set of $\mathscr{A}$-formulas Ω, we have proved

Theorem 7.2 (Compactness Theorem for Propositional Calculus). Let Ω be a finitely satisfiable set of $\mathscr{A}$-formulas. Then Ω is truth-functionally satisfiable.

Exercises

1. Is the set of clauses

$$\{(p_i \vee \neg p_{i+1}) \mid i = 1, 2, 3, \ldots\}$$

satisfiable? Why?

2. The same for the set

$$\{(p_i \vee \neg p_{i+1}), (\neg p_i \vee p_{i+1}) \mid i = 1, 2, 3, \ldots\}.$$

3.* Let us be given a plane map containing infinitely many countries. Suppose there is no way to color this map with k colors so that adjacent countries are colored with different colors. Prove that there is a finite submap for which the same is true.

4.* Let Γ be a (not necessarily finite) set of $\mathscr{A}$-formulas, and let α be an $\mathscr{A}$-formula. We can generalize the notion of tautological consequence by writing $\Gamma \vDash \alpha$ to mean this: for every assignment v on $\mathscr{A}$ such that $\gamma^v = 1$ for all $\gamma \in \Gamma$, we also have $\alpha^v = 1$.

(a) Show that $\Gamma \vDash \alpha$ if and only if $\gamma_1, \ldots, \gamma_n \vDash \alpha$ for some $\gamma_1, \ldots, \gamma_n \in \Gamma$.

(b) Show that if Γ is an r.e. set, then $\{\alpha \mid \Gamma \vDash \alpha\}$ is also r.e.

(c) Give an r.e. set Γ such that $\{\alpha \mid \Gamma \vDash \alpha\}$ is not recursive.

(d) Let Γ be an r.e. set of $\mathscr{A}$-formulas such that for some $\mathscr{A}$-formula α, both $\Gamma \vDash \alpha$ and $\Gamma \vDash \neg \alpha$. Show that $\{\alpha \mid \Gamma \vDash \alpha\}$ is recursive.

(e) Let Γ be an r.e. set of $\mathscr{A}$-formulas such that for every $\mathscr{A}$-formula α, either $\Gamma \vDash \alpha$ or $\Gamma \vDash \neg \alpha$ but not both. Show that $\{\alpha \mid \Gamma \vDash \alpha\}$ is recursive.

13

Quantification Theory

1. The Language of Predicate Logic

Although a considerable part of logical inference is contained in the propositional calculus, it is only with the introduction of the apparatus of *quantifiers* that one can encompass the full scope of logical deduction as it occurs in mathematics, and in science generally. We begin with an alphabet called a *vocabulary* consisting of two kinds of symbols, *relation symbols* and *function symbols*. Let **W** be a vocabulary. For each symbol $t \in \mathbf{W}$, we assume there is an integer $\delta(t)$ called the *degree of* t. For t a function symbol, $\delta(t) \geq 0$, while for t a relation symbol, $\delta(t) > 0$. A function symbol t whose degree is 0 is also called a *constant symbol*. We assume that **W** contains at least one relation symbol. (What we are calling a *vocabulary* is often called a *language* in the literature of mathematical logic. Obviously this terminology is not suitable for a book on theoretical computer science.) In addition to **W** we shall use the alphabet

$$Q = \{\neg, \wedge, \vee, \supset, \leftrightarrow, \forall, \exists, (,), x, y, z, u, v, w, \mathbf{I}, \mathbf{,}\},$$

where the boldface comma , is one of the symbols that belong to Q. The words that belong to the language

$$\{x\mathbf{I}^{[i]}, y\mathbf{I}^{[i]}, z\mathbf{I}^{[i]}, u\mathbf{I}^{[i]}, v\mathbf{I}^{[i]}, w\mathbf{I}^{[i]} \mid i \in N\}$$

are called *variables*. Again we think of strings of the form $\mathbf{I}^{[i]}$, $i > 0$, as subscripts, e.g., writing x_5 for $x\mathbf{IIIII}$. By a **W**-*term* (or when the vocabulary **W** is understood, simply a *term*) we mean an element of $(Q \cup \mathbf{W})^*$ that either is a constant symbol $c \in \mathbf{W}$ or a variable, or is obtained from constant symbols and variables by repeated application of the operation on $(Q \cup \mathbf{W})^*$ that transforms $\mu_1, \mu_2, \ldots, \mu_n$ into

$$f(\mu_1, \mu_2, \ldots, \mu_n),$$

where f is a function symbol in **W** and $\delta(f) = n > 0$.

An *atomic* **W**-*formula* is an element of $(Q \cup \mathbf{W})^*$ of the form

$$r(\mu_1, \mu_2, \ldots, \mu_n),$$

where $r \in \mathbf{W}$ is a relation symbol, $\delta(r) = n$, and $\mu_1, \mu_2, \ldots, \mu_n$ are terms. Finally, a **W**-*formula* (or simply a *formula*) is either an atomic **W**-formula or is obtained from atomic **W**-formulas by repeated application of the following operations on $(Q \cup \mathbf{W})^*$:

1. transform α into $\neg\, \alpha$;
2. transform α and β into $(\alpha \wedge \beta)$;
3. transform α and β into $(\alpha \vee \beta)$;
4. transform α and β into $(\alpha \supset \beta)$;
5. transform α and β into $(\alpha \leftrightarrow \beta)$;
6. transform α into $(\forall b)\alpha$, where b is a variable;
7. transform α into $(\exists b)\alpha$, where b is a variable.

If b is a variable, the expressions

$$(\forall b) \qquad \text{and} \qquad (\exists b)$$

are called *universal quantifiers* and *existential quantifiers*, respectively.

Let b be a variable, let λ be a formula *or* a term, and suppose that we have the decomposition $\lambda = rbs$, where the leftmost symbol of s is not $\mathbf{I}$. (This means that b is not part of a longer variable. In fact, because λ is a formula or a term, s will have to begin either with , or with).) Then we say that the variable b *occurs in* λ. If more than one such decomposition is possible for a given variable b we speak, in an obvious sense, of the *first occurrence of b in* λ, *the second occurrence of b in* λ, etc., reading from left to right.

Next suppose that α is a formula and that we have the decomposition

$$\alpha = r(\forall b)\beta s \qquad \text{or} \qquad \alpha = r(\exists b)\beta s,$$

where β *is itself a formula*. Then *the occurrence of b in the quantifiers shown, as well as all occurrences of b in* β, *are called* **bound** *occurrences of b in* α. Any occurrence of b in α that is not bound is called a **free** *occurrence* of b in α. A **W**-formula α containing no free occurrences of variables is called a **W**-*sentence*, or simply a *sentence*. Any occurrence of a variable in a term is considered to be a *free occurrence*.

Thus, in the formula

$$(r(x) \supset (\exists y)s(u, y)),$$

x and u each have one occurrence, and it is free; y has two occurrences, and they are both bound. The formula

$$(\forall x)(\exists u)(r(x) \supset (\exists y)s(u, y))$$

is a sentence.

Exercises

1. Let $\mathbf{W} = \{0, s, <\}$, where $0, s$ are function symbols with $\delta(0) = 0$, $\delta(s) = 1$, and $<$ is a relation symbol with $\delta(<) = 2$. Describe the set of **W**-terms and the set of atomic **W**-formulas.
2. (a) Define the *height* of a **W**-term t, denoted $\mathrm{Ht}(t)$, as follows:

$$\mathrm{Ht}(x) = 1 \quad \text{for all variables } x$$
$$\mathrm{Ht}(c) = 1 \quad \text{for all constant symbols } c$$
$$\mathrm{Ht}(f(t_1, \ldots, t_n)) = \max\{\mathrm{Ht}(t_i) \mid 1 \le i \le n\} + 1.$$

Show by induction on height that all **W**-terms have an equal number of left and right parentheses.

(b) Do the same for **W**-formulas.

2. Semantics

In analogy with the propositional calculus, we wish to associate the truth values, 1 and 0, with sentences. To do this for a given sentence α will require an "interpretation" of the function and relation symbols in α.

By an *interpretation I of a vocabulary* **W**, we mean a nonempty set D, called the *domain of I*, together with the following:

1. an element c_I of D, for each constant symbol $c \in \mathbf{W}$;
2. a function f_I from $D^{\delta(f)}$ into[1] D, for each function symbol $f \in \mathbf{W}$ for which $\delta(f) > 0$; and
3. a function r_I from $D^{\delta(r)}$ into $\{0, 1\}$, for each relation symbol $r \in \mathbf{W}$.

[1] Recall from Chapter 1, Section 1, that D^n is the set of n-tuples of elements of D.

Let λ be a *term* or a *formula* and let $b_1, b_2, \ldots, b_n$ be a *list of distinct variables which includes all the variables that have free occurrences in* λ. Then, we write $\lambda = \lambda(b_1, \ldots, b_n)$ as a *declaration* of our intention to regard $b_1, \ldots, b_n$ as acting like parameters taking on values. In such a case, if $t_1, \ldots, t_n$ are terms containing no occurrences of variables that have bound occurrences in λ, we write $\lambda(t_1, \ldots, t_n)$ for the term or formula obtained from λ by *simultaneously* replacing b_1 by t_1, b_2 by $t_2, \ldots, b_n$ by t_n.

Now let t be a **W**-term, $t = t(b_1, b_2, \ldots, b_n)$, and let I be an interpretation of **W**, with domain D. Then we shall define a value $t^I[d_1, d_2, \ldots, d_n] \in D$ for all $d_1, d_2, \ldots, d_n \in D$. For the case $n = 0$, we write simply t^I. We define this notion recursively as follows:

1. If $t = t(b_1, b_2, \ldots, b_n)$ and t is a variable, then t must be b_i for some $i, 1 \le i \le n$, and we define $t^I[d_1, d_2, \ldots, d_n] = d_i$;
2. If $t = t(b_1, b_2, \ldots, b_n)$ and t is a constant symbol c in **W**, then we define $t^I[d_1, d_2, \ldots, d_n] = c_I$;
3. If $t = t(b_1, b_2, \ldots, b_n) = g(t_1, t_2, \ldots, t_m)$, where g is a function symbol in **W**, $\delta(g) = m > 0$, then we first set $t_i = t_i(b_1, b_2, \ldots, b_n)$, $i = 1, 2, \ldots, m$, and we let $s_i = t_i^I[d_1, d_2, \ldots, d_n]$, $i = 1, 2, \ldots, m$. Finally, we define

$$t^I[d_1, d_2, \ldots, d_n] = g_I(s_1, s_2, \ldots, s_m).$$

Continuing, if α is a **W**-formula, $\alpha = \alpha(b_1, b_2, \ldots, b_n)$, and I is an interpretation of **W** with domain D, we shall define a value $\alpha^I[d_1, d_2, \ldots, d_n] \in \{0, 1\}$, for all $d_1, d_2, \ldots, d_n \in D$. Again, in the particular case $n = 0$ (which can happen only if α is a sentence), we simply write α^I. The recursive definition is as follows:

1. If $\alpha = \alpha(b_1, b_2, \ldots, b_n) = r(t_1, t_2, \ldots, t_m)$, where r is a relation symbol in **W**, $\delta(r) = m$, then we first set $t_i = t_i(b_1, b_2, \ldots, b_n)$, $i = 1, 2, \ldots, m$, and then let $s_i = t_i^I[d_1, d_2, \ldots, d_n]$, $i = 1, 2, \ldots, m$. Finally, we define $\alpha^I[d_1, d_2, \ldots, d_n] = r_I(s_1, s_2, \ldots, s_m)$.

In 2–6 which follow, let $\beta = \beta(b_1, \ldots, b_n)$, $\gamma = \gamma(b_1, \ldots, b_n)$, where we assume that $\beta^I[d_1, \ldots, d_n] = k$, $\gamma^I[d_1, \ldots, d_n] = l$ with $k, l \in \{0, 1\}$, are already defined for all $d_1, d_2, \ldots, d_n \in D$:

2. If α is $\neg \beta$, $\alpha = \alpha(b_1, \ldots, b_n)$, then we define

$$\alpha^I[d_1, \ldots, d_n] = \begin{cases} 1 & \text{if } k = 0 \\ 0 & \text{if } k = 1. \end{cases}$$

3. If α is $(\beta \wedge \gamma)$, $\alpha = \alpha(b_1, \ldots, b_n)$, then we define

$$\alpha^I[d_1, \ldots, d_n] = \begin{cases} 1 & \text{if} \quad k = l = 1 \\ 0 & \text{otherwise.} \end{cases}$$

4. If α is $(\beta \vee \gamma)$, $\alpha = \alpha(b_1, \ldots, b_n)$, then we define

$$a^I[d_1, \ldots, d_n] = \begin{cases} 0 & \text{if} \quad k = l = 0 \\ 1 & \text{otherwise.} \end{cases}$$

5. If α is $(\beta \supset \gamma)$, $\alpha = \alpha(b_1, \ldots, b_n)$, then we define

$$\alpha^I[d_1, \ldots, d_n] = \begin{cases} 0 & \text{if} \quad k = 1 \text{ and } l = 0 \\ 1 & \text{otherwise.} \end{cases}$$

6. If α is $(\beta \leftrightarrow \gamma)$, $\alpha = \alpha(b_1, \ldots, b_n)$, then we define

$$\alpha^I[d_1, \ldots, d_n] = \begin{cases} 1 & \text{if} \quad k = l \\ 0 & \text{otherwise.} \end{cases}$$

In 7 and 8 let $\beta = \beta(b_1, \ldots, b_n, b)$, where we assume that $\beta^I[d_1, \ldots, d_n, e]$ is already defined for all $d_1, \ldots, d_n, e \in D$:

7. If α is $(\forall b)\beta$, $\alpha = \alpha(b_1, \ldots, b_n)$, then we define

$$\alpha^I[d_1, \ldots, d_n] = \begin{cases} 1 & \text{if} \quad \beta^I[d_1, \ldots, d_n, e] = 1 \text{ for all } e \in D \\ 0 & \text{otherwise.} \end{cases}$$

8. If α is $(\exists b)\beta$, $\alpha = \alpha(b_1, \ldots, b_n)$, then we define

$$\alpha^I[d_1, \ldots, d_n] = \begin{cases} 1 & \text{if} \quad \beta^I[d_1, \ldots, d_n, e] = 1 \text{ for some } e \in D \\ 0 & \text{otherwise.} \end{cases}$$

It is important to be aware of the entirely nonconstructive nature of 7 and 8 of this definition. When the set D is infinite, the definition provides no algorithm for carrying out the required searches, and, indeed, in many important cases no such algorithm exists.

Let us consider some simple examples.

EXAMPLE 1. $\mathbf{W} = \{c, r, s\}$, where c is a constant symbol, and r and s are relation symbols, $\delta(r) = 3$, $\delta(s) = 2$. Let I have the domain $D = \{0, 1, 2, 3, \ldots,\}$, let $c_I = 0$, and let

$$r_I(x, y, z) = \begin{cases} 1 & \text{if} \quad x + y = z \\ 0 & \text{otherwise,} \end{cases} \qquad s_I(x, y) = \begin{cases} 1 & \text{if} \quad x \le y \\ 0 & \text{otherwise.} \end{cases}$$

If α is the sentence

$$(\forall x)(\forall y)(\forall z)(r(x, y, z) \supset s(x, z)),$$

then it is easy to see that $\alpha^I = 1$. For if $u, v, w \in D$ and $r_I(u, v, w) = 1$, then $u + v = w$, so that $u \leq w$ and therefore $s_I(u, w) = 1$. So if $\gamma = \gamma(x, y, z)$ is the formula $(r(x, y, z) \supset s(x, z))$, then $\gamma^I[u, v, w] = 1$.

On the other hand, if β is the sentence

$$(\forall x)(\exists y) r(x, y, c),$$

then $\beta^I = 0$. This is because $r_I(1, v, 0) = 0$ for all $v \in D$. Therefore

$$r(x, y, c)^I[1, v] = 0$$

for all $v \in D$. Thus, $(\exists y) r(x, y, c)^I[1] = 0$, and therefore, finally, $\beta^I = 0$.

EXAMPLE 2. $\mathbf{W}$, α, β are as in Example 1. I has the domain

$$\{\ldots, -3, -2, -1, 0, 1, 2, 3, \ldots,\},$$

the set of all integers. c_I, r_I, s_I are defined as in Example 1. In this case, it is easy to see that $\alpha^I = 0$ and $\beta^I = 1$.

An interpretation I of the vocabulary $\mathbf{W}$ is called a *model* of a $\mathbf{W}$-sentence α if $\alpha^I = 1$; I is called a *model* of the set Ω of $\mathbf{W}$-sentences if I is a model of each $\alpha \in \Omega$. Ω is said to be *satisfiable* if it has at least one model. An individual $\mathbf{W}$-sentence α is called *satisfiable* if $\{\alpha\}$ is satisfiable, i.e., if α has a model. α is called *valid* if *every* interpretation of $\mathbf{W}$ is a model of α.

If $\alpha = \alpha(b_1, \ldots, b_n)$, $\beta = \beta(b_1, \ldots, b_n)$ are $\mathbf{W}$-formulas, we write $\alpha = \beta$ to mean that α and β are *semantically equivalent*, that is,

$$\alpha^I[d_1, \ldots, d_n] = \beta^I[d_1, \ldots, d_n]$$

for all interpretations I of $\mathbf{W}$ and all $d_1, \ldots, d_n \in D$, the domain of I. Then, as is readily verified, all of the equations from Section 1 of Chapter 12 hold true as well in the present context. We also note the *quantificational De Morgan laws:*

$$\neg (\forall b)\alpha = (\exists b) \neg \alpha; \neg (\exists b)\alpha = (\forall b) \neg \alpha. \tag{2.1}$$

Again, as in the case of the propositional calculus, we may eliminate the connectives $\supset$ and $\leftrightarrow$ by using appropriate equations from Chapter 12, Section 1. Once again, there is a "general principle of duality," but we omit the details.

Now, let $\beta = \beta(b_1, \ldots, b_n, b)$, and let the variable a have no occurrences in β. Then it is quite obvious that

$$\begin{aligned}(\exists b)\beta(b_1, \ldots, b_n, b) &= (\exists a)\beta(b_1, \ldots, b_n, a),\\ (\forall b)\beta(b_1, \ldots, b_n, b) &= (\forall a)\beta(b_1, \ldots, b_n, a).\end{aligned} \tag{2.2}$$

Continuing to assume that a has no occurrences in β, we have

$$\begin{aligned}((\forall a)\alpha \wedge \beta) &= (\forall a)(\alpha \wedge \beta),\\ ((\exists a)\alpha \wedge \beta) &= (\exists a)(\alpha \wedge \beta),\\ ((\forall a)\alpha \vee \beta) &= (\forall a)(\alpha \vee \beta),\\ ((\exists a)\alpha \vee \beta) &= (\exists a)(\alpha \vee \beta).\end{aligned} \tag{2.3}$$

Exercises

1. Let **W** be as in Example 1. For each of the following **W**-sentences give an interpretation that is a model of the sentence as well as one that is not.
   (a) $(\forall x)(\exists y)(\forall z)(s(x, c) \supset r(x, y, z))$.
   (b) $(\exists y)(\forall x)(\forall z)(s(x, c) \supset r(x, y, z))$.
   (c) $(\forall x)(\forall y)(s(x, y) \supset s(y, x))$.
2. Give an interpretation that is a model of (a) in Exercise 1 but not of (b).
3. Let $\mathbf{W} = \{ca, cb, cat, eq\}$, let interpretation I have domain $\{a, b\}^*$, and let $ca_I = a$, $cb_I = b$, $cat_I(u, v) = \widehat{uv}$, and

$$eq_I(u, v) = \begin{cases} 1 & \text{if } u = v \\ 0 & \text{otherwise.} \end{cases}$$

   For each of the following formulas α, calculate α^I.
   (a) $(\forall x)(\exists y)eq(cat(ca, x), y)$.
   (b) $(\exists y)(\forall x)eq(cat(ca, x), y)$.
   (c) $(\forall x)(\exists y)eq(cat(x, y), x)$.
   (d) $(\forall x)(\exists y)(eq(cat(ca, y), x) \vee eq(cat(cb, y), x))$.
   (e) $(\exists x)eq(cat(ca, x), cat(x, cb))$.
4. For each of the following formulas, tell whether it is (i) satisfiable, (ii) valid, (iii) unsatisfiable.
   (a) $((\exists x)p(x) \wedge (\forall y) \neg p(y))$.
   (b) $(\forall x)(\exists y)r(f(a), b)$.

(c) $((\forall x)(\exists y)r(x, y) \supset (\exists y)(\forall x)r(x, y))$.
(d) $((\exists y)(\forall x)r(x, y) \supset (\forall x)(\exists y)r(x, y))$.
(e) $(\exists x)(\forall y){<}(x, y)$.
(f) $((\exists x)p(x) \supset (\exists x)(\forall y)p(x))$.

5. Let $\mathbf{W} = \{0, s, +, eq\}$, let interpretation I have domain N, and let $0_I = 0$, s_I be the successor function, $+_I$ be the addition function, and eq_I be equality (as in Exercise 3). For each of the following sets S, give a formula α such that

$$S = \{(m_1, \dots, m_n) \in N^n \mid \alpha^I[m_1, \dots, m_n] = 1\}.$$

(a) $S = N$.
(b) $S = \{(x, y, z) \in N^3 \mid x + y = z\}$.
(c) $S = \{(x, y) \in N^2 \mid x \leq y\}$.
(d) $S = \{(x, y, z) \in N^3 \mid z \dot{-} y = x\}$.
(e) $S = \{x \in N \mid x \text{ is even}\}$.

6. For a set of sentences Ω, let $\mathrm{Mod}(\Omega)$ be the collection of all models of Ω. Prove that

$$\Omega_1 \subseteq \Omega_2 \quad \textit{implies} \quad \mathrm{Mod}(\Omega_2) \subseteq \mathrm{Mod}(\Omega_1).$$

3. Logical Consequence

We are now ready to use the *semantics* just developed to define the notion of *logical consequence*. Let $\mathbf{W}$ be a vocabulary, let Γ be a set of $\mathbf{W}$-sentences, and let γ be a $\mathbf{W}$-sentence. Then we write

$$\Gamma \models \gamma$$

and call γ a *logical consequence* of the *premises* Γ if every model of Γ is also a model of γ. If $\Gamma = \{\gamma_1, \dots, \gamma_n\}$, then we omit the braces { , }, and write simply

$$\gamma_1, \gamma_2, \dots, \gamma_n \models \gamma.$$

Note that $\gamma_1, \gamma_2, \dots, \gamma_n \models \gamma$ if and only if for every interpretation I of $\mathbf{W}$ for which

$$\gamma_1^I = \gamma_2^I = \cdots = \gamma_n^I = 1,$$

we also have $\gamma^I = 1$. (Intuitively, we may think of the various interpretations as "possible worlds." Then our definition amounts to saying that γ is a logical consequence of some premises if γ is true in every possible world in which the premises are all true.) As in the case of the propositional calculus, logical consequence can be determined by considering a single sentence. The proof of the corresponding theorem is virtually identical to that of Theorem 2.1 in Chapter 12 and is omitted.

Theorem 3.1. The relation $\gamma_1, \gamma_2, \ldots, \gamma_n \vDash \gamma$ is equivalent to each of the following:

1. the sentence $((\gamma_1 \wedge \cdots \wedge \gamma_n) \supset \gamma)$ is valid;
2. the sentence $(\gamma_1 \wedge \cdots \wedge \gamma_n \wedge \neg \gamma)$ is unsatisfiable.

Once again we are led to a problem of satisfiability. We will focus our efforts on computational methods for demonstrating the unsatisfiability of a given sentence. We begin by showing how to obtain a suitable normal form for any given sentence.

As in Chapter 12, Section 3, we begin with the procedures

(I) ELIMINATE $\supset$ and $\leftrightarrow$.
(II) MOVE $\neg$ INWARD.

Procedure (I) is carried out exactly as in Chapter 12. For (II), we also need to use the quantificational De Morgan laws (2.1). Ultimately all $\neg$s will come to immediately precede relation symbols.

(III) RENAME VARIABLES.

Rename bound variables as necessary to ensure that no variables occur in two different quantifiers, using (2.2). Thus, the sentence

$$((\forall x)(\forall y)r(x, y) \vee ((\forall x)s(x) \wedge (\exists y)s(y))$$

might become

$$(\forall x)(\forall y)r(x, y) \vee ((\forall u)s(u) \wedge (\exists v)s(v)).$$

(IV) PULL QUANTIFIERS

Using (2.3), bring all quantifiers to the left of the sentence. *Where possible, do so with existential quantifiers preceding universal quantifiers*. Thus, to continue our example, we would get successively

$$\begin{aligned}&(\forall x)(\forall y)r(x, y) \vee (\exists v)((\forall u)s(u) \wedge s(v))\\&\quad = (\exists v)((\forall x)(\forall y)r(x, y) \vee ((\forall u)s(u) \wedge s(v)))\\&\quad = (\exists v)(\forall x)(\forall y)(\forall u)(r(x, y) \vee (s(u) \wedge s(v))).\end{aligned}$$

After applying (IV) as many times as possible, we obtain a sentence consisting of *a string of quantifiers followed by a formula containing no quantifiers*. Such a sentence is called a *prenex sentence*. A prenex sentence is also said to be in *prenex normal form*.

Let γ be a sentence of the form

$$(\forall b_1) \cdots (\forall b_n)(\exists b)\alpha,$$

where $n \geq 0$, and $\alpha = \alpha(b_1, b_2, \ldots, b_n, b)$. Let g be a function symbol *which is not in* α with $\delta(g) = n$. If necessary, we enlarge the vocabulary **W** to include this new symbol g. Then we write

$$\gamma_s = (\forall b_1) \cdots (\forall b_n)\alpha(b_1, b_2, \ldots, b_n, g(b_1, b_2, \ldots, b_n)).$$

γ_s is called the *Skolemization* of γ. [In the case $n = 0$, g is a constant symbol and the term

$$g(b_1, b_2, \ldots, b_n)$$

is to be simply understood as standing for g.] Skolemization is important because of the following theorem.

Theorem 3.2. Let γ be a **W**-sentence and let γ_s be its Skolemization. Then

1. every model of γ_s is a model of γ;
2. if γ has a model, then so does γ_s;
3. γ is satisfiable if and only if γ_s is satisfiable.

Proof. Condition 3 obviously follows from 1 and 2.

To prove 1, let α, γ be as previously and let

$$\beta = \beta(b_1, \ldots, b_n) = \alpha(b_1, \ldots, b_n, g(b_1, \ldots, b_n)).$$

Let I be a model of γ_s so that $\gamma_s^I = 1$, and let the domain of I be D. Then, if $d_1, \ldots, d_n$ are arbitrary elements of D, we have $\beta^I[d_1, \ldots, d_n] = 1$. Let $e = g_I(d_1, \ldots, d_n)$. Thus $\alpha^I[d_1, \ldots, d_n, e] = 1$, so that

$$(\exists b)\alpha(b_1, \ldots, b_n, b)^I[d_1, \ldots, d_n] = 1.$$

Hence finally, $\gamma^I = 1$.

To prove 2, let $\gamma^I = 1$, where I has the domain D. Again let $d_1, \ldots, d_n$ be any elements of D. Then, writing β for the formula $(\exists b)\alpha$, so that we may write $\beta = \beta(b_1, \ldots, b_n)$, we have $\beta^I[d_1, \ldots, d_n] = 1$. Thus, there is an element $e \in D$ such that $\alpha^I[d_1, \ldots, d_n, e] = 1$. Hence, we have shown that for each $d_1, \ldots, d_n \in D$, there is at least one element $e \in D$ such that $\alpha^I[d_1, \ldots, d_n, e] = 1$. Thus, we may extend the interpretation I to the new function symbol g by defining $g_I(d_1, \ldots, d_n)$ to be such an element[2] e, for each $d_1, \ldots, d_n \in D$. Thus, for all $d_1, d_2, \ldots, d_n \in D$, we have

$$\beta^I[d_1, \ldots, d_n] = \alpha^I[d_1, \ldots, d_n, g_I(d_1, \ldots, d_n)] = 1.$$

Hence, finally, $\gamma_S^I = 1$. ■

Since Theorem 3.2 shows that the leftmost existential quantifier in a prenex formula may be eliminated without affecting satisfiability, we can, by iterated Skolemization, obtain a sentence containing no existential quantifiers. We write this

(V) ELIMINATE EXISTENTIAL QUANTIFIERS.

In the example discussed under (IV), this would yield simply

$$(\forall x)(\forall y)(\forall u)(r(x, y) \vee (s(u) \wedge s(c))), \tag{3.1}$$

where c is a constant symbol.

For another example consider the sentence

$$(\forall x)(\exists u)(\forall y)(\forall z)(\exists v)r(x, y, z, u, v),$$

where r is a relation symbol, $\delta(r) = 5$. Then two Skolemizations yield

$$(\forall x)(\forall y)(\forall z)r(x, y, z, g(x), h(x, y, z)). \tag{3.2}$$

A sentence α is called *universal* if it has the form $(\forall b_1)(\forall b_2) \cdots (\forall b_n)\gamma$, where the formula γ *contains no quantifiers*. We may summarize the procedure (I)–(V) in

Theorem 3.3. There is an algorithm that will transform any given sentence β into a universal sentence α such that β is satisfiable if and only if α is satisfiable. Moreover, any model of α is also a model of β.

In connection with our procedure (I)–(V) consider the example

$$((\forall x)(\exists y)r(x, y) \wedge (\forall u)(\exists v)s(u, v)),$$

where r and s are relation symbols. By varying the order in which the quantifiers are pulled, we can obtain the prenex sentences

1. $(\forall x)(\exists y)(\forall u)(\exists v)(r(x, y) \wedge s(u, v))$,
2. $(\forall u)(\exists v)(\forall x)(\exists y)(r(x, y) \wedge s(u, v))$,

[2] Here we are using a nonconstructive set-theoretic principle known as the *axiom of choice*.

3. $(\forall x)(\forall u)(\exists y)(\exists v)(r(x,y) \wedge s(u,v))$.

Skolemizations will then yield the corresponding universal sentences:

1. $(\forall x)(\forall u)(r(x, g(x)) \wedge s(u, h(x,u)))$,
2. $(\forall u)(\forall x)(r(x, g(u,x)) \wedge s(u, h(u)))$,
3. $(\forall x)(\forall u)(r(x, g(x,u)) \wedge s(u, h(x,u)))$.

But, for this example, one would expect that y should "depend" only on x and v only on u. In other words, we would expect to be able to use a universal sentence such as

4. $(\forall x)(\forall u)(r(x, g(x)) \wedge s(u, h(u)))$.

As we shall see, it is important to be able to justify such simplifications.

Proceeding generally, let γ be a sentence of the form

$$\delta \wedge (\forall b_1) \cdots (\forall b_n)(\exists b)\alpha,$$

where $n \geq 0$ and $\alpha = \alpha(b_1, b_2, \ldots, b_n, b)$. Let g be a function symbol which does not occur in γ with $\delta(g) = n$. Then we write

$$\gamma_s = \delta \wedge (\forall b_1) \cdots (\forall b_n)\alpha(b_1, \ldots, b_n, g(b_1, \ldots, b_n)).$$

γ_s is called a *generalized Skolemization* of γ. Then we have the following generalization of Theorem 3.2.

Theorem 3.4. Let γ be a **W**-sentence and let γ_s be a generalized Skolemization of γ. Then we have 1–3 of Theorem 3.2.

Proof. Again we need verify only 1 and 2. Let α, γ, δ be as above. To prove 1, let I be a model of γ_s with domain D. Let β be defined as in the proof of Theorem 3.2. Then $\delta^I = 1$ and $(\forall b_1) \cdots (\forall b_n)\beta^I = 1$. As in the proof of Theorem 3.2, we conclude that

$$(\forall b_1) \cdots (\forall b_n)(\exists b)\alpha^I = 1,$$

and so $\gamma^I = 1$.

Conversely, let $\gamma^I = 1$, where I has domain D. Then $\delta^I = 1$ and

$$(\forall b_1) \cdots (\forall b_n)(\exists b)\alpha^I = 1.$$

Precisely as in the proof of Theorem 3.2, we can extend the interpretation I to the symbol g in such a way that

$$(\forall b_1) \cdots (\forall b_n)\beta^I = 1.$$

Hence, $\gamma_s^I = 1$. ■

Henceforth we will consider the steps (IV) PULL QUANTIFIERS and (V) ELIMINATE EXISTENTIAL QUANTIFIERS to permit the use of generalized Skolemizations. Moreover, as we have seen, Theorem 3.3 remains correct if the universal sentence is obtained using generalized Skolemizations.

Exercises

1. Consider the inference

$$(\forall x)(p(x) \supset (\forall y)(s(y, x) \supset u(x))),$$
$$(\exists x)(p(x) \wedge (\exists y)(s(y, x) \wedge h(y, x)))$$
$$\models (\exists x)(\exists y)(u(x) \wedge h(y, x) \wedge s(y, x)).$$

   (a) Find a universal sentence whose unsatisfiability is equivalent to the correctness of this inference. Can you do this so that Skolemization introduces only constant symbols?

   (b) Using (a), show that the inference is correct.

2. (a) Using generalized Skolemization find a universal sentence whose unsatisfiability is equivalent to the correctness of the inference

$$(\exists x)(\forall y)r(x, y) \models (\forall y)(\exists x)r(x, y).$$

   (b) Show that the inference is correct.

3. The same as Exercise 2(a) for the inference

$$(\forall x)(\forall y)(\forall z)(\forall u)(\forall V)(\forall w)((P(x, y, u) \wedge P(y, z, v) \wedge P(x, v, w)) \supset P(u, z, w)),$$
$$(\forall x)(\forall y)(\exists z)P(z, x, y),$$
$$(\forall x)(\forall y)(\exists z)P(x, z, y) \models (\exists x)(\forall y)P(y, x, y).$$

4. Prove Theorem 3.1.

5. For each sentence α in Exercise 2.1, perform the following.

   (a) Transform α into a prenex normal form sentence.

   (b) Give the Skolemization γ_s of γ.

   (c) Give a model I of α.

   (d) Extend I to a model of γ_s.

6. Let γ be a **W**-sentence, for some vocabulary **W**, and let γ_s be its Skolemization. Prove or disprove each of the following statements.

   (a) If γ is valid then γ_s is valid.

   (b) If γ_s is valid then γ is valid.

7. Let **W** be a vocabulary, Γ a set of **W**-sentences, and α, β **W**-sentences. Prove each of the following statements.
   (a) (Deduction Theorem) $\Gamma \cup \{\alpha\} \models \beta$ if and only if $\Gamma \models (\alpha \supset \beta)$.
   (b) (Contraposition) $\Gamma \cup \{\alpha\} \models \neg \beta$ if and only if $\Gamma \cup \{\beta\} \models \neg \alpha$.
   (c) (Reductio ad absurdum) $\Gamma \cup \{\alpha\} \models (\beta \wedge \neg \beta)$ if and only if $\Gamma \models \neg \alpha$.

4. Herbrand's Theorem

We have seen that the problem of logical inference is reducible to the problem of satisfiability, which in turn is reducible to the problem of satisfiability of universal sentences. In this section, we will prove Herbrand's theorem, which can be used together with algorithms for truth-functional satisfiability (discussed in Chapter 12) to develop procedures for this purpose.

Let α be a *universal* **W**-sentence for some vocabulary **W**, where we assume that α contains all the symbols in **W**. If α contains at least one constant symbol, we call the set of all constant symbols in α the *constant set* of α. If α contains no constant symbols, we let a be some new constant symbol, which we add to **W**, and we call $\{a\}$ the *constant set* of α. Then the language which consists of all **W**-terms containing no variables is called the *Herbrand universe* of α. The set $\mathscr{A}$ of atomic **W**-formulas containing no variables is called the *atom set* of α. We will work with the set of *propositional formulas over* $\mathscr{A}$, i.e., of $\mathscr{A}$-formulas in the sense of Chapter 12, Section 1. *Each of these $\mathscr{A}$-formulas is also a* **W***-sentence that contains no quantifiers.*

Returning to the universal sentence (3.1), we see that its constant set is $\{c\}$, its Herbrand universe is likewise $\{c\}$, and its atom set is $\{r(c,c), s(c)\}$.

Next, examining the universal sentence (3.2), its constant set is $\{a\}$, but its Herbrand universe is infinite:

$$\mathbf{H} = \{a, g(a), h(a,a,a), g(g(a)), g(h(a,a,a)), h(a,a,g(a)), \ldots\}.$$

Its atom set is likewise infinite:

$$\mathscr{A} = \{r(t_1, t_2, t_3, t_4, t_5) | t_1, t_2, t_3, t_4, t_5 \in \mathbf{H}\}.$$

Theorem 4.1. Let $\zeta = \zeta(b_1, b_2, \ldots, b_n)$ be a **W**-formula containing no quantifiers, so that the sentence

$$\gamma = (\forall b_1)(\forall b_2) \cdots (\forall b_n)\zeta$$

is universal. Let **H** be the Herbrand universe of γ and let $\mathscr{A}$ be its atom set. Then, γ is satisfiable if and only if the set

$$\Omega = \{\zeta(t_1, t_2, \ldots, t_n) | t_1, t_2, \ldots, t_n \in \mathbf{H}\} \tag{4.1}$$

of $\mathscr{A}$-formulas is truth-functionally satisfiable.

Proof. First let γ be satisfiable, say, $\gamma^I = 1$, and let D be the domain of I. We now define an assignment v on $\mathscr{A}$. Let r be a relation symbol of **W**, $\delta(r) = m$, so that $r(t_1, \ldots, t_m) \in \mathscr{A}$ for all $t_1, \ldots, t_m \in \mathbf{H}$. Then we define

$$v(r(t_1, \ldots, t_m)) = r_I(t_1^I, \ldots, t_m^I).$$

We have

Lemma 1. For all $\mathscr{A}$-formulas α, $\alpha^I = \alpha^v$.

Proof. As in Chapter 12, we may assume that α contains only the connectives $\neg$, $\wedge$. Proceeding by induction, we see that if α is an atom, the result is obvious from our definition of v. Thus, we may suppose that $\alpha = \neg \beta$ or $\alpha = (\beta \wedge \gamma)$, where the result is known for β or, for β and γ, respectively.

In the first case, we have

$$\begin{aligned} \alpha^I = 1 \quad & \text{if and only if} \quad \beta^I = 0 \\ & \text{if and only if} \quad \beta^v = 0 \\ & \text{if and only if} \quad \alpha^v = 1. \end{aligned}$$

Similarly, in the second case

$$\begin{aligned} \alpha^I = 1 \quad & \text{if and only if} \quad \beta^I = \gamma^I = 1 \\ & \text{if and only if} \quad \beta^v = \gamma^v = 1 \\ & \text{if and only if} \quad \alpha^v = 1. \end{aligned}$$ ∎

Returning to the proof of the theorem, we wish to show that for all $\alpha \in \Omega$, $\alpha^v = 1$. By Lemma 1, it will suffice to show that $\alpha^I = 1$ for $\alpha \in \Omega$. Now, since $\gamma^I = 1$, we have

$$\zeta^I[d_1, \ldots, d_n] = 1 \qquad \text{for all} \quad d_1, \ldots, d_n \in D.$$

But clearly, for $t_1, \ldots, t_n \in \mathbf{H}$,

$$\zeta(t_1, \ldots, t_n)^I = \zeta^I[t_1^I, \ldots, t_n^I] = 1.$$

We conclude that Ω is truth-functionally satisfiable.

Conversely, let us be given an assignment v on $\mathscr{A}$ such that $\alpha^v = 1$ for all $\alpha \in \Omega$. We shall use v to construct an interpretation I of $\mathbf{W}$. The domain of I is simply the Herbrand universe $\mathbf{H}$. Furthermore,

1. If $c \in \mathbf{W}$ is a constant symbol, then $c_I = c$. (That is, a constant symbol is interpreted as itself.)
2. If $f \in \mathbf{W}$ is a function symbol, $\delta(f) = n > 0$, and $t_1, t_2, \ldots, t_n \in \mathbf{H}$, then
$$f_I(t_1, t_2, \ldots, t_n) = \mathbf{f}(t_1, t_2, \ldots, t_n) \in \mathbf{H}.$$
(Note carefully the use of boldface.)
3. If $r \in \mathbf{W}$ is a relation symbol, $\delta(r) = n$, and $t_1, t_2, \ldots, t_n \in \mathbf{H}$, then
$$r_I(t_1, t_2, \ldots, t_n) = v(r(t_1, t_2, \ldots, t_n)).$$
(Note that the assignment v is only used in 3.) We have

Lemma 2. For every $t \in \mathbf{H}$, $t^I = t$.

Proof. Immediate from 1 and 2. ■

Lemma 3. For every $\mathbf{W}$-formula $\alpha = \alpha(b_1, \ldots, b_n)$ containing no quantifiers, and all $t_1, \ldots, t_n \in \mathbf{H}$, we have
$$\alpha^I[t_1, \ldots, t_n] = v(\alpha(t_1, \ldots, t_n)).$$

Proof. If α is an atom, the result follows at once from 3 and Lemma 2. For the general case it now follows because the same recursive rules are used for the propositional connectives, whether we are evaluating interpretations or assignments. ■

Returning to the proof of the theorem, we wish to show that $\gamma^I = 1$. For this, recalling that $\mathbf{H}$ is the domain of I, it suffices to show that
$$\zeta^I[t_1, \ldots, t_n] = 1 \qquad \text{for all} \quad t_1, \ldots, t_n \in \mathbf{H}.$$
By Lemma 3, this amounts to showing that
$$v(\zeta(t_1, \ldots, t_n)) = 1 \qquad \text{for all} \quad t_1, \ldots, t_n \in \mathbf{H}.$$
But this last is precisely what we have assumed about v. ■

The usefulness of the theorem we have just proved results from combining it with the compactness theorem (Theorem 7.2 in Chapter 12).

Theorem 4.2 (Herbrand's Theorem). Let ζ, γ, $\mathbf{H}$, $\mathscr{A}$, and Ω be as in Theorem 4.1. Then γ is unsatisfiable if and only if there is a truth-functionally unsatisfiable **W**-formula of the form $\bigwedge_{\beta \in \Sigma} \beta$ for some finite subset Σ of Ω.

Proof. If there is a truth-functionally unsatisfiable $\mathscr{A}$-formula $\bigwedge_{\beta \in \Sigma} \beta$, where $\Sigma \subseteq \Omega$, then for every assignment v on $\mathscr{A}$, there is some $\beta \in \Sigma$ such that $\beta^v = 0$. Hence Σ, and therefore also Ω, is not truth-functionally satisfiable; hence by Theorem 4.1, γ is unsatisfiable.

Conversely, if γ is unsatisfiable, then by Theorem 4.1, Ω is not truth-functionally satisfiable. Thus, by the compactness theorem (Theorem 7.2 in Chapter 12), Ω is not finitely satisfiable; i.e., there is a finite set $\Sigma \subseteq \Omega$ such that Σ is not truth-functionally satisfiable. Then, the sentence $\bigwedge_{\beta \in \Sigma} \beta$ is truth-functionally unsatisfiable. ■

This theorem leads at once to a family of procedures for demonstrating the unsatisfiability of a universal sentence γ. Write $\Omega = \bigcup_{n=0}^{\infty} \Sigma_n$, where $\Sigma_0 = \emptyset$, $\Sigma_n \subseteq \Sigma_{n+1}$, the Σ_n are all finite, and where there is an algorithm that transforms each Σ_n into Σ_{n+1}. (This can easily be managed, e.g., by simply writing the elements of Ω as an infinite sequence.) Then we have the procedure

$n \leftarrow 0$

WHILE $\bigwedge_{\beta \in \Sigma_n} \beta$ IS TRUTH-FUNCTIONALLY SATISFIABLE DO

$n \leftarrow n + 1$

END

If γ is unsatisfiable, the procedure will eventually terminate; otherwise it will continue forever. The test for truth-functional satisfiability of $\bigwedge_{\beta \in \Sigma_n} \beta$ can be performed using the methods of Chapter 12, e.g., the Davis–Putnam rules. Using this discussion, we are able to conclude

Theorem 4.3. For every vocabulary **W** the set of unsatisfiable sentences is recursively enumerable. Likewise the set of valid sentences is r.e.

Proof. Given a sentence α, we apply our algorithms to obtain a universal sentence γ that is satisfiable if and only if α is. We then apply the preceding procedure based on Herbrand's theorem. It will ultimately halt if and only if α is unsatisfiable. This procedure shows that the set of unsatisfiable sentences is r.e.

Since a sentence α is valid if and only if $\neg\, \alpha$ is unsatisfiable, the same procedure shows that the set of valid sentences is r.e. ■

One might have hoped that the set of unsatisfiable **W**-sentences would in fact be recursive. But as we shall see later (Theorem 8.1), this is not the case. Thus, as we shall see, we *cannot hope* for an algorithm that, beginning with sentences $\gamma_1, \gamma_2, \ldots, \gamma_n, \gamma$ as input, will return YES if $\gamma_1, \gamma_2, \ldots, \gamma_n \vDash \gamma$, and NO otherwise. The best we can hope for is a general procedure that will halt and return YES whenever the given logical inference is correct, but that may fail to terminate otherwise. And in fact, using Theorem 3.1 and an algorithm of the kind used in the proof of Theorem 4.3, we obtain just such a procedure.

Now let us consider what is involved in testing the truth-functional satisfiability of $\bigwedge_{\beta \in \Sigma} \beta$, where Σ is a finite subset of the set Ω defined in (4.1). If we wish to use the methods developed in Chapter 12, we need to obtain a CNF of $\bigwedge_{\beta \in \Sigma} \beta$. But, if for each $\beta \in \Sigma$, we have a CNF formula β^0 such that $\beta = \beta^0$, then $\bigwedge_{\beta \in \Sigma} \beta^0$ is clearly a CNF of $\bigwedge_{\beta \in \Sigma} \beta$. *This fact makes CNF useful in this context.*

In fact we can go further. We can apply the algorithms of Chapter 12, Section 3, to obtain CNF formulas directly for $\zeta = \zeta(b_1, \ldots, b_n)$. When we do this we are in effect enlarging the set of formulas to which we apply the methods of Chapter 12, by allowing atoms that contain variables. Each formula can then be thought of as representing all of the **W**-formulas obtained by replacing each variable by an element of the Herbrand universe **H**. In this context formulas containing no variables are called *ground formulas*. We also speak of *ground literals*, *ground clauses*, etc.

If the CNF formula obtained in this manner from $\zeta(b_1, \ldots, b_n)$ is given by the set of clauses

$$\{\kappa_i(b_1, \ldots, b_n) \mid i = 1, 2, \ldots, r\}, \tag{4.2}$$

then each $\beta \in \Sigma$ will have a CNF

$$\{\kappa_i(t_1, \ldots, t_n) \mid i = 1, 2, \ldots, r\},$$

where $t_1, \ldots, t_n$ are suitable elements of **H**. Hence, there will be a CNF of $\bigwedge_{\beta \in \Sigma} \beta$ representable in the form

$$\left\{\kappa_i(t_1^j, \ldots, t_n^j) \mid i = 1, \ldots, r, j = 1, \ldots, s\right\}, \tag{4.3}$$

where $t_1^j, \ldots, t_n^j \in \mathbf{H}$, $j = 1, 2, \ldots, s$. Thus, what we are seeking is an unsatisfiable set of clauses of the form (4.3). Of course, such a set can be unsatisfiable without being *minimally* unsatisfiable in the sense of Chapter 12, Section 5. In fact, there is no reason to expect a minimally unsatisfiable set of clauses which contains, say, $\kappa_1(t_1, \ldots, t_n)$ to also contain $\kappa_2(t_1, \ldots, t_n)$. Thus, we are led to treat the clauses in the set (4.2) independently of one another, seeking substitutions of elements of **H** for the variables $b_1, \ldots, b_n$ so as to obtain a truth-functionally inconsistent set

R of clauses. Each of the clauses in (4.2) can give rise by substitution to one or more of the clauses of **R**.

Let us consider some examples.

EXAMPLE 1. Consider this famous inference: *All men are mortal; Socrates is a man;* therefore, *Socrates is mortal.* An appropriate vocabulary would be $\{m, t, s\}$, where m, t are relation symbols of degree 1 (which we think of as standing for the properties of *being a man*, and of *being mortal*, respectively), and s is a constant symbol (which we think of as naming Socrates). The inference becomes

$$(\forall x)(m(x) \supset t(x)), m(s) \vDash t(s).$$

Thus, we wish to prove the unsatisfiability of the sentence

$$((\forall x)(m(x) \supset t(x)) \wedge m(s) \wedge \neg t(s)).$$

Going to prenex form, we see that no Skolemization is needed:

$$(\forall x)((\neg m(x) \vee t(x)) \wedge m(s) \wedge \neg t(s)).$$

The Herbrand universe is just $\{s\}$. In this simple case, Herbrand's theorem tells us that we have to prove the truth-functional unsatisfiability of

$$((\neg m(s) \vee t(s)) \wedge m(s) \wedge \neg t(s));$$

that is, we are led directly to a ground formula in CNF. Using the set representation of Chapter 12, Section 4, we are dealing with the set of clauses

$$\{\{\bar{m}(s), t(s)\}, \{m(s)\}, \{\bar{t}(s)\}\}.$$

Using the Davis–Putnam rules (or, in this case equivalently, resolution), we obtain successively

$$\{\{t(s)\}, \{\bar{t}(s)\}\}, \qquad \text{and} \qquad \{\square\};$$

hence the original inference was valid.

EXAMPLE 2. Another inference: *Every shark eats a tadpole; all large white fish are sharks; some large white fish live in deep water; any tadpole eaten by a deep water fish is miserable;* therefore, *some tadpoles are miserable.*

Our vocabulary is $\{s, b, t, r, m, e\}$, where all of these are relation symbols of degree 1, except e, which is a relation symbol of degree 2. $e(x, y)$ is to represent "x *eats* y." s stands for the property of *being a shark*, b of *being a large white fish*, t of *being a tadpole*, r of *living in deep water*, and m of *being miserable*. The inference translates as

$$(\forall x)(s(x) \supset (\exists y)(t(y) \wedge e(x, y))),$$
$$(\forall x)(b(x) \supset s(x)),$$
$$(\exists x)(b(x) \wedge r(x)),$$
$$(\forall x)(\forall y)((r(x) \wedge t(y) \wedge e(x, y)) \supset m(y)) \models (\exists y)(t(y) \wedge m(y)).$$

Thus, we need to demonstrate the unsatisfiability of the sentence

$$\begin{aligned}((&\forall x)(s(x) \supset (\exists y)(t(y) \wedge e(x, y)))\\ &\wedge (\forall x)(b(x) \supset s(x))\\ &\wedge (\exists x)(b(x) \wedge r(x))\\ &\wedge (\forall x)(\forall y)((r(x) \wedge t(y) \wedge e(x, y)) \supset m(y))\\ &\wedge \neg(\exists y)(t(y) \wedge m(y))).\end{aligned}$$

We proceed as follows.

I. ELIMINATE $\supset$:

$$\begin{aligned}((&\forall x)(\neg s(x) \vee (\exists y)(t(y) \wedge e(x, y)))\\ &\wedge (\forall x)(\neg b(x) \vee s(x))\\ &\wedge (\exists x)(b(x) \wedge r(x))\\ &\wedge (\forall x)(\forall y)(\neg(r(x) \wedge t(y) \wedge e(x, y)) \vee m(y))\\ &\wedge \neg(\exists y)(t(y) \wedge m(y))).\end{aligned}$$

II. MOVE $\neg$ INWARD:

$$\begin{aligned}((&\forall x)(\neg s(x) \vee (\exists y)(t(y) \wedge e(x, y)))\\ &\wedge (\forall x)(\neg b(x) \vee s(x))\\ &\wedge (\exists x)(b(x) \wedge r(x))\\ &\wedge (\forall x)(\forall y)(\neg r(x) \vee \neg t(y) \vee \neg e(x, y) \vee m(y))\\ &\wedge (\forall y)(\neg t(y) \vee \neg m(y))).\end{aligned}$$

III. RENAME VARIABLES;

$$\begin{aligned}((&\forall x)(\neg s(x) \vee (\exists y_1)(t(y_1) \wedge e(x, y_1)))\\ &\wedge (\forall z)(\neg b(z) \vee s(z))\\ &\wedge (\exists u)(b(u) \wedge r(u))\\ &\wedge (\forall v)(\forall w)(\neg r(v) \vee \neg t(w) \vee \neg e(v, w) \vee m(w))\\ &\wedge (\forall y)(\neg t(y) \vee \neg m(y))).\end{aligned}$$

IV. PULL QUANTIFIERS (trying to pull existential quantifiers first):

$$
\begin{aligned}
&(\exists u)(\forall x)(\exists y_1)(\forall z)(\forall v)(\forall w)(\forall y)\\
&\qquad((\neg s(x) \vee (t(y_1) \wedge e(x, y_1)))\\
&\qquad\wedge(\neg b(z) \vee s(z))\\
&\qquad\wedge b(u) \wedge r(u)\\
&\qquad\wedge(\neg r(v) \vee \neg t(w) \vee \neg e(v,w) \vee m(w))\\
&\qquad\wedge(\neg t(y) \vee \neg m(y))).
\end{aligned}
$$

V. ELIMINATE EXISTENTIAL QUANTIFIERS:

$$
\begin{aligned}
&(\forall x)(\forall z)(\forall v)(\forall w)(\forall y)\\
&\qquad((\neg s(x) \vee (t(g(x)) \wedge e(x, g(x))))\\
&\qquad\wedge(\neg b(z) \vee s(z))\\
&\qquad\wedge b(c) \wedge r(c)\\
&\qquad\wedge(\neg r(v) \vee \neg t(w) \vee \neg e(v,w) \vee m(w))\\
&\qquad\wedge(\neg t(y) \vee \neg m(y))).
\end{aligned}
$$

Thus we are led to the clauses

$$
\begin{gathered}
\{\bar{s}(x), t(g(x))\},\\
\{\bar{s}(x), e(x, g(x))\},\\
\{\bar{b}(z), s(z)\},\\
\{b(c)\},\\
\{r(c)\},\\
\{\bar{r}(v), \bar{t}(w), \bar{e}(v,w), m(w)\},\\
\{\bar{t}(y), \overline{m}(y)\}.
\end{gathered}
$$

The Herbrand universe is

$$\mathbf{H} = \{c, g(c), g(g(c)), \ldots\}.$$

To find substitutions for the variables in **H**, we have recourse to Theorem 5.2 (2) in Chapter 12. To search for a minimally unsatisfiable set of ground clauses, we should seek substitutions that will lead to every literal having a mate (in another clause). By inspection, we are led to the substitution

$$x = c, \qquad z = c, \qquad v = c, \qquad w = g(c), \qquad y = g(c).$$

We thus obtain the set of ground clauses

$$\{\bar{s}(c), t(g(c))\},$$
$$\{\bar{s}(c), e(c, g(c))\},$$
$$\{\bar{b}(c), s(c)\},$$
$$\{b(c)\},$$
$$\{r(c)\},$$
$$\{\bar{r}(c), \bar{t}(g(c)), \bar{e}(c, g(c)), m(g(c))\},$$
$$\{\bar{t}(g(c)), \overline{m}(g(c))\}.$$

Although this set of clauses is linked, we must still test for satisfiability. Using the Davis–Putnam rules we obtain, first using the unit rule on $\{b(c)\}$,

$$\{\bar{s}(c), t(g(c))\},$$
$$\{\bar{s}(c), e(c, g(c))\},$$
$$\{s(c)\},$$
$$\{r(c)\},$$
$$\{\bar{r}(c), \bar{t}(g(c)), \bar{e}(c, g(c)), m(g(c))\},$$
$$\{\bar{t}(g(c)), \overline{m}(g(c))\}.$$

Using the unit rule on $\{s(c)\}$ and then on $\{r(c)\}$ gives

$$\{t(g(c))\},$$
$$\{e(c, g(c))\},$$
$$\{\bar{t}(g(c)), \bar{e}(c, g(c)), m(g(c))\},$$
$$\{\bar{t}(g(c)), \overline{m}(g(c))\}.$$

Using the unit rule on $\{t(g(c))\}$ and then on $\{e(c, g(c))\}$ gives

$$\{m(g(c))\},$$
$$\{\overline{m}(g(c))\}.$$

Finally, we obtain the set of clauses consisting of the empty clause:

$$\square.$$

In Examples 1 and 2 each clause of (4.2) gave rise to just one clause in the truth-functionally unsatisfiable set of clauses obtained. That is, we obtain a truth-functionally unsatisfiable set of clauses of the form (4.3) with $s = 1$. Our next example will be a little more complicated.

EXAMPLE 3. We consider the inference

$$(\forall x)(\exists y)(r(x, y) \vee r(y, x)),$$
$$(\forall x)(\forall y)(r(x, y) \supset r(y, y)) \vDash (\exists z)r(z, z).$$

Thus, we wish to demonstrate the unsatisfiability of the sentence

$$(\forall x)(\exists y)(r(x, y) \vee r(y, x)) \wedge (\forall x)(\forall y)(r(x, y) \supset r(y, y)) \wedge \neg(\exists z)r(z, z).$$

We proceed as follows:

I, II, III. ELIMINATE $\supset$; MOVE $\neg$ INWARD; RENAME VARIABLES:

$$(\forall x)(\exists y)(r(x, y) \vee r(y, x)) \wedge (\forall u)(\forall v)(\neg r(u, v) \vee r(v, v)) \wedge (\forall z)\neg r(z, z).$$

IV. PULL QUANTIFIERS:

$$(\forall x)(\exists y)(\forall u)(\forall v)(\forall z)((r(x, y) \vee r(y, x)) \wedge (\neg r(u, v) \vee r(v, v)) \wedge \neg r(z, z)).$$

V. ELIMINATE EXISTENTIAL QUANTIFIERS:

$$(\forall x)(\forall u)(\forall v)(\forall z)((r(x, g(x)) \vee r(g(x), x)) \wedge (\neg r(u, v) \vee r(v, v)) \wedge \neg r(z, z)).$$

We thus obtain the set of clauses

$$\{r(x, g(x)), r(g(x), x)\},$$
$$\{\bar{r}(u, v), r(v, v)\},$$
$$\{\bar{r}(z, z)\}.$$

The Herbrand universe is

$$\mathbf{H} = \{a, g(a), g(g(a)), \ldots\}.$$

How can we find a mate for $r(x, g(x))$? Not by using $\bar{r}(z, z)$—whichever element $t \in \mathbf{H}$ we substitute for x, $r(x, g(x))$ will become $r(t, g(t))$, which cannot be obtained from $r(z, z)$ by replacing z by any element of $\mathbf{H}$.

Thus the only potential mate for $r(x, g(x))$ is $\bar{r}(u, v)$. We tentatively set $u = x$, $v = g(x)$ so that the second clause becomes

$$\{\bar{r}(x, g(x)), r(g(x), g(x))\}.$$

But now, $\bar{r}(u, v)$ is also the only available potential mate for $r(g(x), x)$. Thus, we are led to also substitute $v = x$, $u = g(x)$ in the second clause, obtaining

$$\{\bar{r}(g(x), x), r(x, x)\}.$$

Both $r(g(x), g(x))$ and $r(x, x)$ can be matched with $\bar{r}(z, z)$ to produce mates. We thus arrive at the set of clauses

$$\{r(x, g(x)), r(g(x), x)\},$$

$$\{\bar{r}(x, g(x)), r(g(x), g(x))\},$$

$$\{\bar{r}(g(x), x), r(x, x)\},$$

$$\{\bar{r}(x, x)\},$$

$$\{\bar{r}(g(x), g(x))\}.$$

Now we can replace x by any element of $\mathbf{H}$ to obtain a linked set of ground clauses. For example, we can set $x = a$; but any other substitution for x will do. Actually, it is just as easy to work with the nonground clauses as listed, since the propositional calculus processing is quite independent of which element of $\mathbf{H}$ we substitute for x. In fact after four applications of the unit rule (or of resolution) we obtain $\square$, which shows that the original inference was correct.

Exercises

1. Describe the Herbrand universe and the atom set of the universal sentence obtained in Exercise 3.1.
2. Do the same for Exercise 3.2.
3. Do the same for Exercise 3.3.
4. Let $\mathbf{W} = \{c, f, p\}$, where c is a constant symbol, f is a function symbol with $\delta(f) = 1$, and p is a relation symbol with $\delta(p) = 1$. Show that $\{(\exists x)p(x), \neg p(c), \neg p(f(c)), \neg p(f(f(c))), \ldots\}$ is satisfiable.

5. Unification

We continue our consideration of Example 3 of the previous section. Let us analyze what was involved in attempting to "mate" our literals. Suppose we want to mate $r(x, g(x))$ with $\bar{r}(z, z)$. The first step is to observe that both literals have the same relation symbol r, and that r is negated in one and only one of the two literals. Next we were led to the equations

$$x = z, \qquad g(x) = z.$$

The first equation is easily satisfied by setting $x = z$. But then the second equation becomes $g(z) = z$, and clearly no substitution from the Herbrand universe can satisfy this equation. Thus, we were led to consider instead the pair of literals $r(x, g(x))$, $\bar{r}(u, v)$. The equations we need to solve are then

$$x = u, \qquad g(x) = v.$$

Again we satisfy the first equation by letting $x = u$; the second equation becomes $g(u) = v$, which can be satisfied by letting $v = g(u)$. So the literals become $r(u, g(u))$ and $\bar{r}(u, g(u))$.

This example illustrates the so-called *unification algorithm* for finding substitutions which will transform given literals $r(\lambda_1, \ldots, \lambda_n)$, $\bar{r}(\mu_1, \ldots, \mu_n)$ into mates of one another. The procedure involves comparing two terms μ, λ and distinguishing four cases:

1. One of μ, λ (say, μ) is a variable and λ does not contain this variable. Then replace μ by λ throughout.
2. One of μ, λ (say, μ) is a variable, $\lambda \neq \mu$, but λ contains μ. Then report: NOT UNIFIABLE.
3. μ, λ both begin with function symbols, but not with the same function symbol. Again report: NOT UNIFIABLE.
4. μ, λ begin with the same function symbol, say

   $$\mu = g(\nu_1, \ldots, \nu_k), \qquad \lambda = g(\eta_1, \ldots, \eta_k).$$

   Then use this same procedure recursively on the pairs

   $$\nu_1 = \eta_1, \qquad \nu_2 = \eta_2, \qquad \ldots, \qquad \nu_k = \eta_k.$$

In applying the unification algorithm to

$$r(\lambda_1, \ldots, \lambda_n), \qquad \bar{r}(\mu_1, \ldots, \mu_n),$$

we begin with the pairs of terms

$$\lambda_1 = \mu_1, \qquad \lambda_2 = \mu_2, \qquad \ldots, \qquad \lambda_n = \mu_n$$

and apply the preceding procedure to each. Naturally, substitutions called for by step 1 must be made in all of the terms before proceeding.

To see that the process always terminates, it is necessary to note only that whenever step 1 is applied, the total number of variables present decreases.

EXAMPLE Let us attempt to unify

$$r(g(x), y, g(g(z))) \qquad \text{with} \quad \bar{r}(u, g(u), g(v)).$$

We are led to the equations

$$g(x) = u, \qquad y = g(u), \qquad g(g(z)) = g(v).$$

The first equation leads to letting

$$u = g(x),$$

and the remaining equations then become

$$y = g(g(x)) \qquad \text{and} \quad g(g(z)) = g(v).$$

The second is satisfied by letting

$$y = g(g(x)),$$

which does not affect the third equation. The third equation leads recursively to

$$g(z) = v,$$

which is satisfied by simply setting v equal to the left side of this equation. The final result is

$$r(g(x), g(g(x)), g(g(z))), \qquad \bar{r}(g(x), g(g(x)), g(g(z))).$$

Numerous systematic procedures for showing sentences to be unsatisfiable based on the unification algorithm have been studied. These procedures work directly with clauses containing variables and do not require that substitutions from the Herbrand universe actually be carried out. In particular, there are *linked conjunct procedures* that are based on searches for a linked set of clauses, followed by a test for truth-functional unsatisfiability. However, most computer implemented procedures have been based on *resolution*. In these procedures, when a pair of literals have been mated by an appropriate substitution, they are immediately eliminated by resolution. We illustrate the use of resolution on Examples 2 and 3 of the previous section.

Beginning with the clauses of Example 2, applying the unification algorithm to the pair of literals $s(z), \bar{s}(x)$, and then using resolution, we get

$$\{\bar{b}(x), t(g(x))\},$$
$$\{\bar{b}(x), e(x, g(x))\},$$
$$\{b(c)\},$$
$$\{r(c)\},$$
$$\{\bar{r}(v), \bar{t}(w), \bar{e}(v, w), m(w)\},$$
$$\{\bar{t}(y), \overline{m}(y)\}.$$

Next, unifying

$$e(x, g(x)) \qquad \text{and} \qquad \bar{e}(v, w)$$

and using resolution, we get

$$\{\bar{b}(x), t(g(x))\},$$
$$\{b(c)\},$$
$$\{r(c)\},$$
$$\{\bar{b}(x), \bar{r}(x), \bar{t}(g(x)), m(g(x))\},$$
$$\{\bar{t}(y), \overline{m}(y)\}.$$

Another stage of unification and resolution yields

$$\{t(g(c))\},$$
$$\{r(c)\},$$
$$\{\bar{r}(c), \bar{t}(g(c)), m(g(c))\},$$
$$\{\bar{t}(y), \overline{m}(y)\},$$

and then

$$\{r(c)\},$$
$$\{\bar{r}(c), m(g(c))\},$$
$$\{\overline{m}(g(c))\}.$$

Finally, we get

$$\{r(c)\},$$
$$\{\bar{r}(c)\},$$

and, then, to complete the proof,

$$\square .$$

The combination of unification with resolution can be thought of as a single step constituting a kind of generalized resolution. Thus, resolution in the sense of Chapter 12, that is, resolution involving only ground clauses, will now be called *ground resolution*, while the unmodified word *resolution* will be used to represent this more general operation. In the ground case we used the notation $\text{res}_\lambda(\kappa_1, \kappa_2)$ for the *resolvent* of κ_1, κ_2 with respect to the literal λ, namely,

$$(\kappa_1 - \{\lambda\}) \cup (\kappa_2 - \{\neg \lambda\}).$$

In the general case, let $\lambda \in \kappa_2$, $\neg \mu \in \kappa_2$, where the unification algorithm can be successfully applied to λ and $\neg \mu$. Thus, there are substitutions for the variables which yield new clauses $\tilde{\kappa}_1, \tilde{\kappa}_2$ such that if the substitutions transform λ into $\tilde{\lambda}$, they also transform $\neg \mu$ into $\neg \tilde{\lambda}$. Then we write

$$\text{res}_{\lambda,\mu}(\kappa_1, \kappa_2) = \left(\tilde{\kappa}_1 - \{\tilde{\lambda}\}\right) \cup \left(\tilde{\kappa}_2 - \{\neg \tilde{\lambda}\}\right).$$

Let α be a finite set of clauses. Then a sequence of clauses $\kappa_1, \kappa_2, \ldots, \kappa_n$ is called a *resolution derivation of* $\kappa_n = \kappa$ *from* α if for each $i, 1 \le i \le n$, either $\kappa_i \in \alpha$ or there are $j, k < i$ and literals λ, μ such that $\kappa_i = \text{res}_{\lambda,\mu}(\kappa_j, \kappa_k)$. As in Chapter 12, a resolution derivation of $\square$ from α is called a *resolution refutation of* α. The key theorem is

Theorem 5.1 (J. A. Robinson's General Resolution Theorem). Let $\zeta = \zeta(b_1, \ldots, b_n)$ be a **W**-formula containing no quantifiers, and let ζ be in CNF. Let

$$\gamma = (\forall b_1)(\forall b_2) \cdots (\forall b_n)\zeta.$$

Then, the sentence γ is unsatisfiable if and only if there is a resolution refutation of the clauses of ζ.

We shall not prove this theorem here, but will content ourselves with showing how it applies to Example 3 of the previous section. The clauses were

1. $\{r(x, g(x)), r(g(x), x)\}$
2. $\{\bar{r}(u, v), r(v, v)\}$
3. $\{\bar{r}(z, z)\}$.

A resolution refutation is obtained as follows:

4. $\{r(g(x), x), r(g(x), g(x))\}$ (resolving 1 and 2);
5. $\{r(x, x), r(g(x), g(x))\}$ (resolving 2 and 4);
6. $\{r(g(x), g(x))\}$ (resolving 3 and 5);
7. $\square$ (resolving 3 and 6).

Exercises

1. Indicate which of the following pairs of terms are unifiable.
   (a) $x, g(y)$.
   (b) $x, g(x)$.
   (c) $f(x), g(y)$.
   (d) $f(x, h(a)), f(g(y), h(y))$.
   (e) $f(x, x), f(g(y), a)$.
   (f) $f(x, y, z), f(g(w, w), g(x, x), g(y, y))$.
2. Prove the correctness of the inferences of Exercises 3.1–3.3 by obtaining minimally unsatisfiable sets of clauses.
3. Prove the correctness of the inferences of Exercises 3.1–3.3 by obtaining resolution refutations.
4. (a) Prove that the problem of the validity of the sentence

$$(\exists x)(\exists y)(\forall z)((r(x, y) \supset (r(y, z) \wedge r(z, z))) \\ \wedge ((r(x, y) \wedge s(x, y)) \supset (s(x, z) \wedge s(z, z))))$$

   leads to the list of clauses

$$\{r(x, y)\},$$
$$\{s(x, y), \bar{r}(y, h(x, y)), \bar{r}(h(x, y), h(x, y))\},$$
$$\{\bar{r}(y, h(x, y)), \bar{r}(h(x, y), h(x, y)), \\ \bar{s}(x, h(x, y)), \bar{s}(h(x, y), h(x, y))\}.$$

   [*Hint:* Use Theorem 5.1 in Chapter 12.]

   (b) Prove the validity of the sentence in (a) by giving a resolution refutation.
5.* A conventional notation for describing a substitution is $\{x_1/t_1, \ldots, x_n/t_n\}$, where $x_1, \ldots, x_n$ are distinct variables and $t_1, \ldots, t_n$ are terms. If λ is a term or a formula and θ is a substitution, then $\lambda\theta$ denotes the result of simultaneously replacing each occurrence of x_i in λ by t_i, $1 \leq i \leq n$. A *unifier* of two terms or formulas λ, μ is a substitution θ such that $\lambda\theta$ and $\mu\theta$ are identical. Modify the unification algorithm so that if λ, μ are unifiable, it returns a unifier of λ, μ. Apply the modified algorithm to Exercise 1.
6.* An $\vee$-clause with at most one literal that is not negated is called a *Horn clause*. Horn clauses are the basis of *logic programming languages* such as Prolog. Horn clauses of the form λ or $(\neg\lambda_1 \vee \cdots \vee \neg\lambda_n \vee \lambda)$, where the latter is sometimes written $(\lambda_1 \wedge \cdots \wedge \lambda_n \supset \lambda)$, are

called *program clauses*, and a *Horn program* is a set (or conjunction) of program clauses. The input to a Horn program $\mathscr{P}$ is a clause of the form $(\neg \lambda_1 \vee \cdots \vee \neg \lambda_n)$, called a *goal clause*, and the output is a substitution θ, called an *answer substitution*, such that

$$(\forall x_1) \cdots (\forall x_l)\mathscr{P} \models (\forall y_1) \cdots (\forall y_k)[(\lambda_1 \wedge \cdots \wedge \lambda_n)\theta],$$

where $x_1, \ldots, x_l$ are all of the variables which occur free in $\mathscr{P}$ and $y_1, \ldots, y_k$ are all of the variables which occur free in $(\lambda_1 \wedge \cdots \wedge \lambda_n)\theta$. (If there is no such answer substitution then the program can either stop and return NO or it can run forever.) If $(\lambda_1 \wedge \cdots \wedge \lambda_n)\theta$ has no free variable occurrences, then θ is a *ground answer substitution.*

(a) Let θ be a substitution such that $(\lambda_1 \vee \cdots \vee \lambda_n)\theta$ has no free variable occurrences. Show that θ is a ground answer substitution if and only if

$$(\forall x_1) \cdots (\forall x_l)[\mathscr{P} \cup \{(\neg \lambda_1 \vee \cdots \vee \neg \lambda_n)\theta\}]$$

is unsatisfiable.

(b) Let $\mathscr{P}$ be the Horn program with clauses

$$\{\text{edge}(a, b), \text{edge}(b, c), \text{edge}(x, y) \supset \text{connected}(x, y),$$
$$\text{edge}(x, y) \wedge \text{connected}(y, z) \supset \text{connected}(x, z)\}.$$

For each of the following goal clauses, use resolution and the modified unification algorithm from Exercise 5 to find all possible answer substitutions.

(i) $\neg\text{edge}(a, y)$.

(ii) $\neg\text{edge}(x, a)$.

(iii) $\neg\text{edge}(x, y)$.

(iv) $\neg\text{connected}(b, y)$.

(v) $\neg\text{connected}(a, y)$.

6. Compactness and Countability

In this section we give two applications of the circle of ideas surrounding Herbrand's theorem that are extremely important in mathematical logic. It will be interesting to see if they have a role to play in the application of logic to computer science.

Theorem 6.1 (Compactness Theorem for Predicate Logic). Let Ω be a set of **W**-sentences each finite subset of which is satisfiable. Then Ω is satisfiable.

Proof. If Ω is finite, there is nothing to prove. If Ω is infinite, we can use the enumeration principle from Chapter 12, Section 7, to obtain an enumeration $\beta_0, \beta_1, \beta_2, \ldots$ of the elements of Ω. Let us write

$$\gamma_n = \bigwedge_{i \le n} \beta_i, \qquad n = 0, 1, 2, \ldots .$$

Let steps (I)–(V) of Section 3 be applied to each of $\beta_0, \beta_1, \beta_2, \ldots$ to obtain universal sentences

$$\alpha_i = (\forall b_1^{(i)}) \cdots (\forall b_{m_i}^{(i)}) \zeta_i(b_1^{(i)}, \ldots, b_{m_i}^{(i)}).$$

Then by Theorem 3.3, for each i, α_i is satisfiable if and only if β_i is satisfiable, and moreover any model of α_i is also a model of β_i. Now let us apply the same steps (I)–(V) to the sentence γ_n. We see that *if we use generalized Skolemization* we can do this in such a way that the universal sentence δ_n we obtain, corresponding to γ_n in the sense of Theorem 3.3, consists of universal quantifiers followed by the formula

$$\bigwedge_{i \le n} \zeta_i .$$

Now, by hypothesis, each γ_n is satisfiable. Hence, by Theorem 3.3, so is each δ_n. For each n, let $\mathbf{H}_n$ be the Herbrand universe of δ_n. Thus,

$$\mathbf{H}_0 \subseteq \mathbf{H}_1 \subseteq \mathbf{H}_2 \cdots .$$

Let $\mathbf{H} = \bigcup_{n \in N} \mathbf{H}_n$. By Theorem 4.1, the sets

$$\Sigma_n = \left\{ \bigwedge_{i \le n} \zeta_i\left(t_1^{(i)}, \ldots, t_{m_i}^{(i)}\right) \middle| t_1^{(i)}, \ldots, t_{m_i}^{(i)} \in \mathbf{H}_n, i = 0, 1, \ldots, n \right\}$$

are truth-functionally satisfiable. We wish to show that the set

$$\Gamma = \left\{ \zeta_i(t_1, \ldots, t_{m_i}) \mid t_1, t_2, \ldots \in \mathbf{H} \right\}$$

is itself truth-functionally satisfiable. By the compactness theorem for propositional calculus (Theorem 7.2 in Chapter 12) it suffices to prove this for every finite subset Δ of Γ. But for any finite subset Δ of Γ, there is a largest value of the subscript i which occurs, and all the t_j which occur are in some $\mathbf{H}_k$. Let l be the larger of this subscript k and this largest value of i. Then Δ is itself a subset of

$$\Lambda_l = \left\{ \zeta_i(t_1, \ldots, t_{m_i}) \mid t_1, t_2, \ldots \in \mathbf{H}_l, 0 \le i \le l \right\}.$$

Moreover, since Σ_l is truth-functionally satisfiable, so is Λ_l, and therefore Δ. This shows that Γ is truth-functionally satisfiable.

Now, let $\mathscr{A}$ be the set of all atoms which occur in the formulas that belong to Γ. Let v be an assignment on $\mathscr{A}$ such that $\beta^v = 1$ for all $\beta \in \Gamma$. Then we use v to construct an interpretation I of **W** with domain **H** precisely as in the proof of Theorem 4.1. Then Lemmas 2 and 3 of that proof hold and precisely as in that case we have

$$\zeta_i^I[t_1, \ldots, t_{m_i}] = 1 \qquad \text{for all} \quad t_1, \ldots, t_{m_i} \in \mathbf{H} \text{ and } i \in N.$$

Hence, $\alpha_i^I = 1$ for all $i \in N$. Since any model of α_i is also a model of β_i, we have $\beta_i^I = 1$ for all $i \in N$. Thus, I is a model of Ω. ■

Now let us begin with a set Ω of **W**-sentences which has a model I. Then of course I is a model of every finite subset of Ω. Thus, the *method of proof* of the previous theorem can be applied to Ω. Of course, this would be pointless if our aim were merely to obtain a model of Ω; we already have a model I of Ω. But the method of proof of Theorem 6.1 gives us a model of Ω *whose domain* **H** *is a language on an alphabet*. Thus, we have proved

Theorem 6.2 (Skolem–Löwenheim Theorem). Let Ω be a satisfiable set of **W**-sentences. Then Ω has a model whose domain is a language on some alphabet.

What makes this important and interesting is that any language satisfies the enumeration principle of Chapter 12, Section 7. Infinite sets that possess an enumeration are called *countably infinite*. This brings us to the usual form of the Skolem–Löwenheim theorem.

Corollary 6.3. Let Ω be a satisfiable set of **W**-sentences. Then Ω has a model whose domain is countably infinite.

Many infinite sets that occur in mathematics are *not* countable. In fact, the diagonal method, which was used in obtaining unsolvability results in Part 1 of this book, was originally developed by Cantor to prove that the set of real numbers is not countable. What the Skolem–Löwenheim theorem shows is that no set of sentences can characterize an infinite uncountable set in the sense of excluding countable models.

We close this section with another useful form of the compactness theorem.

Theorem 6.4. If $\Gamma \vDash \gamma$, then there is a finite subset Δ of Γ such that $\Delta \vDash \gamma$.

Proof. Since every model of Γ is a model of γ, the set $\Gamma \cup \{\neg \gamma\}$ has no models; that is, it is not satisfiable. Thus, by Theorem 6.1, there is a finite subset Δ of Γ such that $\Delta \cup \{\neg \gamma\}$ is unsatisfiable. Thus every model of Δ is a model of γ, i.e., $\Delta \models \gamma$. ■

Exercises

2. Let **W** be a vocabulary with relation symbol =, where $\delta(=) = 2$, and let Ω be a set of **W**-sentences containing EQ_W, where EQ_W consists of the sentence $(\forall x)(x=x)$ and all sentences of the form

$$(\forall x_1)\cdots(\forall x_{2i})((x_1=x_{i+1} \wedge \cdots \wedge x_i=x_{2i}) \supset f(x_1,\ldots,x_i)=(x_{i+1},\ldots,x_{2i})),$$
$$(\forall x_1)\cdots(\forall x_{2j})((x_1=x_{j+1} \wedge \cdots \wedge x_j=x_{2j} \wedge p(x_1,\ldots,x_j)) \supset p(x_{j+1},\ldots,x_{2j}))$$

where f is a function symbol in **W** with $\delta(f) = i$, and p is a predicate symbol in **W** with $\delta(p) = j$. A model I of Ω is *normal* if $=^I(x,y) = 1$ if and only if x, y are the same element. Show that Ω has a model if and only if it has a normal model. [*Hint:* Let D be the domain of a model of Ω. Create a normal model using domain elements $[a] = \{x \in D \mid =^I (x, a) = 1\}$, where $a \in D$.]

3. Let **W** and Ω be as in Exercise 2. Show that if Ω has arbitrarily large finite normal models, then it has an infinite normal model. [*Hint:* Show that $\Omega \cup \{(\exists x_1) \cdots (\exists x_n)\bigwedge_{1\le i<j\le n} \neg x_i=x_j \mid n \in N\}$ has a normal model.]

*7. Gödel's Incompleteness Theorem

Let Γ be a recursive set of **W**-sentences for some given vocabulary **W**. We think of Γ as being considered for use as a set of "axioms" for some part of mathematics. The requirement that Γ be recursive is natural, because, by Church's thesis, it simply amounts to requiring that there be some algorithmic method of determining whether or not an alleged "axiom" really is one. Often Γ will be finite. We define $\mathbf{T}_\Gamma = \{\gamma \mid \Gamma \models \gamma\}$ and call $\mathbf{T}_\Gamma$ the *axiomatizable theory* on **W** whose axioms are the sentences belonging to the set Γ. Of course, it is quite possible to have different sets of axioms which define the same theory.

If **T** is an axiomatizable theory, we write

$$\vdash_{\mathbf{T}} \gamma$$

(read: "**T** *proves* γ") to mean that $\gamma \in \mathbf{T}$. We also write $\nvdash_{\mathbf{T}} \gamma$ to mean that $\gamma \notin \mathbf{T}$. The most important fact about axiomatizable theories is given by the following theorem.

Theorem 7.1. An axiomatizable theory is r.e.

Proof. By Theorems 3.1 and 6.4, $\gamma \in \mathbf{T}_\Gamma$ if and only if

$$(\gamma_1 \wedge \gamma_2 \wedge \cdots \wedge \gamma_n \wedge \neg \gamma)$$

is unsatisfiable for some $\gamma_1, \gamma_2, \ldots, \gamma_n \in \Gamma$. Since Γ is recursive, it is certainly r.e. Thus, by Theorem 4.11 in Chapter 4, there is a recursive function g on N whose range is Γ. For a given sentence γ, let

$$\delta(n, \gamma) = (g(0) \wedge g(1) \wedge \cdots \wedge g(n) \wedge \neg \gamma)$$

for all $n \in N$. Clearly, $\delta(n, \gamma)$ is a recursive function of n and γ. Moreover, the sentence γ belongs to $\mathbf{T}_\Gamma$ if and only if there is an $n \in N$ such that $\delta(n, \gamma)$ is unsatisfiable. But by Theorem 4.3, the set of unsatisfiable **W**-sentences is r.e. Hence there is a partially computable function h which is defined for a given input if and only if that input is an unsatisfiable **W**-sentence. Let h be computed by program $\mathscr{P}$ and let $p = \#(\mathscr{P})$. Then the following "dovetailing" program halts if and only if the input γ belongs to $\mathbf{T}_\Gamma$, thereby showing that $\mathbf{T}_\Gamma$ is r.e.:

$$\begin{array}{ll} [A] & Z \leftarrow \delta(l(T), \gamma) \\ & T \leftarrow T + 1 \\ & \text{IF} \sim \text{STP}^{(1)}(Z, p, r(T)) \text{ GOTO } A \end{array}$$

■

We shall see in the next section that there is a Γ such that $\mathbf{T}_\Gamma$ is not recursive.

Now let **W** be some vocabulary intended for use in expressing properties of the natural numbers. By a *numeral system* for **W**, we mean a *recursive function* ν on N such that for each $n \in N$, $\nu(n)$ is a **W**-term containing no variables, and such that for all $n, m \in N$, $n \neq m$ implies $\nu(n) \neq \nu(m)$. When ν can be understood from the context, we write $\bar{n}$ for $\nu(n)$. $\bar{n}$ is called the *numeral* corresponding to n and may be thought of as a notation for n using the vocabulary **W**. A popular choice is

$$\bar{n} = S(S(\cdots S(0)) \cdots),$$

where S is a function symbol of degree 1, 0 is a constant symbol, and the number of occurrences of S is n.

Let $\alpha = \alpha(b)$ be a **W**-formula and let **T** be an axiomatizable theory on **W**. Then, given a numeral system for **W**, we can associate with α the set

$$U = \{n \in N \mid \vdash_{\mathbf{T}} \alpha(\bar{n})\}. \tag{7.1}$$

In this case, we say that the formula α *represents the set U in* **T**. If we begin with a set $U \subseteq N$, we can ask the question: is there a **W**-formula α which represents U in **T**? We have

Theorem 7.2. If there is a formula α which represents the set U in an axiomatizable theory **T**, then U is r.e.

Proof. Let **T** be an axiomatizable theory, and let α represent U in **T**. By Theorem 7.1, we know that there is a program $\mathscr{P}$ that will halt for given input γ if and only if $\vdash_{\mathbf{T}} \gamma$. Given $n \in N$, we need only compute $\alpha(\bar{n})$ [which we can do because $\nu(n) = \bar{n}$ is recursive], and feed it as input to $\mathscr{P}$. The new program thus defined halts for given input $n \in N$ if and only if $\vdash_{\mathbf{T}} \alpha(\bar{n})$. By (7.1), U is r.e. ∎

In fact, there are many axiomatizable theories in which all r.e. sets are representable. To see the negative force of Theorem 7.2, we rewrite it as follows.

Corollary 7.3. Let **T** be an axiomatizable theory. Then if $U \subseteq N$ is not r.e., there is no formula which represents U in **T**.

This corollary is a form of Gödel's incompleteness theorem. To obtain a more striking form of the theorem, let us say that the formula α *quasi-represents the set U in* **T** if

$$\{n \in N \mid \vdash_{\mathbf{T}} \alpha(\bar{n})\} \subseteq U. \tag{7.2}$$

We can think of such a formula α as intended to express the proposition "$n \in U$" using the vocabulary **W**. Comparing (7.1) and (7.2) and considering Corollary 7.3, we have

Corollary 7.4. Let **T** be an axiomatizable theory and let $U \subseteq N$ be a set that is not r.e. Let the formula α quasi-represent U in **T**. Then, there is a number n_0 such that $n_0 \in U$ but $\nvdash_{\mathbf{T}} \alpha(\bar{n}_0)$.

As we can say loosely, the sentence $\alpha(\bar{n}_0)$ is "true" but not provable. Corollary 7.4 is another form of Gödel's incompleteness theorem. We conclude with our final version.

Theorem 7.5. Let **T** be an axiomatizable theory, and let S be an r.e. set that is not recursive. Let $\alpha = \alpha(\boldsymbol{x})$ be a formula such that α represents S in **T**, and $\neg\,\alpha$ quasi-represents $\bar{S}$ in **T**. Then there is a number n_0 such that $\nvdash_{\mathbf{T}} \alpha(\bar{n}_0)$ and $\nvdash_{\mathbf{T}} \neg\,\alpha(\bar{n}_0)$.

Proof. We take $U = \bar{S}$ in Corollary 7.4 to obtain a number n_0 such that $n_0 \in \bar{S}$, but $\nvdash_{\mathbf{T}} \neg\,\alpha(\bar{n}_0)$. Since $n_0 \notin S$ and α represents S in **T**, we must also have $\nvdash_{\mathbf{T}} \alpha(\bar{n}_0)$. ∎

In this last case, it is usual to say that $\alpha(\bar{n}_0)$ is undecidable in **T**.

Exercises

1. Let Γ be an r.e. set of **W**-sentences for some vocabulary **W**. Show that $\{\gamma \mid \Gamma \vDash \gamma\}$ is r.e.

2. Let **T** be an axiomatizable theory on some vocabulary **W**. **T** is *consistent* if there is no **W**-sentence α such that both $\vdash_{\mathbf{T}} \alpha$ and $\vdash_{\mathbf{T}} \neg \alpha$, and **T** is *inconsistent* otherwise.
   (a) Show that if **T** is inconsistent then $\vdash_{\mathbf{T}} \alpha$ for all **W**-sentences α.
   (b) Show that if there is a formula which represents some nonrecursive set in **T**, then **T** is consistent.
   (c) Show that if **T** is consistent and the formula α represents some r.e. set U in **T**, then $\neg \alpha$ quasi-represents $\overline{U}$ in **T**.
3. An axiomatizable theory **T** on vocabulary **W** is *complete* if for all **W**-sentences α, either $\vdash_{\mathbf{T}} \alpha$ or $\vdash_{\mathbf{T}} \neg \alpha$. Show that if **T** is complete then it is recursive. [See also Exercise 2.]
4. An axiomatizable theory **T** on some vocabulary **W** is *ω-consistent* if the following holds for all **W**-formulas $\alpha(b)$: If $\vdash_{\mathbf{T}} \neg \alpha(\bar{n})$ for all $n \in N$, then $\nvdash_{\mathbf{T}} (\exists \boldsymbol{x})\alpha(\boldsymbol{x})$. Show that if **T** is ω-consistent then it is consistent. [See Exercise 2 for the definition of consistency.]
5. Let **W** be a vocabulary with relation symbol =, where $\delta(=) = 2$. A function $f(x_1, \ldots, x_n)$ is *representable* in an axiomatizable theory **T** containing $EQ_{\mathbf{W}}$ [see Exercise 6.2] if there is a formula $\alpha(b_1, \ldots, b_n, b)$ such that if $f(m_1, \ldots, m_n) = k$ then

$$\vdash_{\mathbf{T}} \alpha(\overline{m_1}, \ldots, \overline{m_n}, \bar{k}) \text{ and } \vdash_{\mathbf{T}} (\forall y)(\alpha(\overline{m_1}, \ldots, \overline{m_n}, y) \supset y = \bar{k}).$$

   We say that α *represents* $f(x_1, \ldots, x_n)$ in **T**. Let **T** be a consistent axiomatizable theory [see Exercise 2] such that (i) $EQ_{\mathbf{W}} \subseteq \mathbf{T}$, (ii) $\vdash_{\mathbf{T}} \neg\bar{0}=\bar{1}$, and (iii) every primitive recursive function is representable in **T**.
   (a) Let $\alpha(\boldsymbol{x}, \boldsymbol{y}, \boldsymbol{t}, \boldsymbol{z})$ represent the function $STP^{(1)}(x, y, t)$ in **T**, and for every r.e. set W_m, let $\beta_m(\boldsymbol{x})$ be the formula $(\exists \boldsymbol{t})\alpha(\boldsymbol{x}, \bar{m}, \boldsymbol{t}, \bar{1})$. Show that if $n \in W_m$ then $\vdash_{\mathbf{T}} \beta_m(\bar{n})$.
   (b) Show that if $n \notin W_m$ then $\vdash_{\mathbf{T}} \neg \alpha(\bar{n}, \bar{m}, \bar{t}, \bar{1})$ for all $t \in N$.
   (c) Show that if **T** is ω-consistent then $n \notin W_m$ implies $\nvdash_{\mathbf{T}} \beta_m(\bar{n})$. [See Exercise 4.]
   (d) Conclude that if **T** is an ω-consistent axiomatizable theory in which every primitive recursive function is representable and if $\vdash_{\mathbf{T}} \neg\bar{0}=\bar{1}$, then **T** has an undecidable sentence.

*8. Unsolvability of the Satisfiability Problem in Predicate Logic

In 1928, the great mathematician David Hilbert called the problem of finding an algorithm for testing a given sentence to determine whether it is satisfiable "the main problem of mathematical logic." This was because experience had shown that all of the inferences in mathematics could be expressed within the logic of quantifiers. Thus, an algorithm meeting Hilbert's requirements would have provided, in principle, algorithmic solutions to all the problems in mathematics. So, when unsolvable problems were discovered in the 1930s, it was only to be expected that Hilbert's satisfiability problem would also turn out to be unsolvable.

Theorem 8.1 (Church–Turing). There is a vocabulary **W** such that there is no algorithm for testing a given **W**-sentence to determine whether it is satisfiable.

Proof. Our plan will be to translate the word problem for a Thue process into predicate logic in such a way that a solution to Hilbert's satisfiability problem would also yield a solution to the word problem for the given process.

Thus, using Theorem 3.5 in Chapter 7, let Π be a Thue process on the alphabet $\{a, b\}$ with an unsolvable word problem. Let Π have the productions $g_i \to h_i, i = 1, 2, \ldots, K$, together with their inverses, where we may assume that for each $i, g_i, h_i \neq 0$ (recall Theorem 3.5 in Chapter 7). We introduce the vocabulary $\mathbf{W} = \{\mathbf{a}, \mathbf{b}, \bullet, \cong\}$, where **a**, **b** are constant symbols, $\bullet$ is a function symbol, and $\cong$ is a relation symbol, with $\delta(\bullet) = \delta(\cong) = 2$. We will make use of the interpretation I with domain $\{a, b\}^* - \{0\}$ which is defined as follows:

$$\mathbf{a}_I = a,$$

$$\mathbf{b}_I = b,$$

$$\bullet_I(u, v) = uv,$$

$$\cong_I(u, v) = 1 \qquad \text{if and only if} \quad u \underset{\Pi}{\overset{*}{\Rightarrow}} v.$$

For ease of reading, we shall write $\bullet$ and $\cong$ in "infix" position. Thus, we shall write, for example,

$$((x \bullet \mathbf{a}) \cong y) \qquad \text{instead of} \qquad \cong(\bullet(x, \mathbf{a}), y).$$

For each word $w \in \{a, b\}^* - \{0\}$, we now define a **W**-term $w^\#$ as follows:

$$\begin{aligned} a^\# &= \mathbf{a}, & b^\# &= \mathbf{b}, \\ (ua)^\# &= (u^\# \bullet \mathbf{a}), & (ub)^\# &= (u^\# \bullet \mathbf{b}). \end{aligned} \tag{8.1}$$

We have

Lemma 1. For every word $w \in \{a, b\}^* - \{0\}$, we have $(w^\#)^I = w$.

Proof. The proof is by an easy induction on $|w|$, using (8.1) and the definition of the interpretation I. ■

Let Γ be the set of **W**-sentences obtained by prefixing the appropriate universal quantifiers to each **W**-formula in the following list:

1. $(x \cong x)$,
2. $((x \cong y) \supset (y \cong x))$,
3. $(((x \cong y) \wedge (y \cong z)) \supset (x \cong z))$,
4. $(((x \cong y) \wedge (u \cong v)) \supset ((x \bullet u) \cong (y \bullet v)))$,
5. $(((x \bullet y) \bullet z) \cong (x \bullet (y \bullet z)))$,

5 + i. $(g_i^\# \cong h_i^\#)$, $1 \leq i \leq K$.

We have

Lemma 2. The interpretation I is a model of the set of sentences Γ.

Proof. The sentences of Γ all express in logical notation basic facts about concatenation of strings and about derivations in Thue processes. Detailed verification is left to the reader. ■

Lemma 3. If $\Gamma \vDash (u^\# \cong v^\#)$, then $u \overset{*}{\underset{\Pi}{\Rightarrow}} v$.

Proof. By the definition of logical inference and Lemma 2, we have $(u^\# \cong v^\#)^I = 1$. Hence

$$u = (u^\#)^I \overset{*}{\underset{\Pi}{\Rightarrow}} (v^\#)^I = v.$$ ■

We next wish to establish the converse of Lemma 3. For this it will suffice to show that if $u \overset{*}{\underset{\Pi}{\Rightarrow}} v$, then the sentence

$$\bigwedge_{\alpha \in \Gamma} \alpha \wedge \neg (u^\# \cong v^\#)$$

is unsatisfiable (recall Theorem 3.1). The Herbrand universe is

$$\mathbf{H} = \{\mathbf{a}, \mathbf{b}, \mathbf{a} \bullet \mathbf{a}, \mathbf{a} \bullet \mathbf{b}, \mathbf{b} \bullet \mathbf{a}, \mathbf{b} \bullet \mathbf{b}, \mathbf{a} \bullet (\mathbf{a} \bullet \mathbf{a}), \ldots\}.$$

Let us call a **W**-sentence α a *Herbrand instance* of a **W**-formula β if α can be obtained from β by replacing each of its free variables by an element of **H**. α is said to be *rooted* if it is a tautological consequence of the sentences 5 + i together with Herbrand instances of the formulas listed in 1–5. Obviously, if the sentence β is rooted, then $\Gamma \vDash \beta$.

Lemma 4. If $w = uv$, where $u \neq 0$ and $v \neq 0$, then

$$(w^\# \cong (u^\# \bullet v^\#)) \tag{8.2}$$

is rooted.

Proof. The proof is by induction on $|v|$. If $|v| = 1$, we can assume without loss of generality that $v = a$. But in this case, the sentence (8.2) is a Herbrand instance of formula 1.

Supposing the result known for v, we need to establish it for va and vb. We give the proof for va, that for vb being similar. So let $w = uv$, where we can assume that (8.2) is rooted. We need to show that the sentence

$$((wa)^\# \cong (u^\# \bullet (va)^\#))$$

is likewise rooted. By (8.1) this amounts to showing that

$$((w^\# \bullet \mathbf{a}) \cong (u^\# \bullet (v^\# \bullet \mathbf{a})))$$

is rooted. But this follows from the induction hypothesis, noting that the following sentences are rooted. (For each of these sentences, the number of the corresponding formula of which it is a Herbrand instance is given.)

$$(\mathbf{a} \cong \mathbf{a}) \tag{1}$$

$$(((w^\# \cong (u^\# \bullet v^\#)) \wedge (\mathbf{a} \cong \mathbf{a})) \supset ((w^\# \bullet \mathbf{a}) \cong ((u^\# \bullet v^\#) \bullet \mathbf{a}))) \tag{4}$$

$$(((u^\# \bullet v^\#) \bullet \mathbf{a}) \cong (u^\# \bullet (v^\# \bullet \mathbf{a}))) \tag{5}$$

$$\begin{aligned}((((w^\# \bullet \mathbf{a}) \cong ((u^\# \bullet v^\#) \bullet \mathbf{a})) \wedge (((u^\# \bullet v^\#) \bullet \mathbf{a}) \cong (u^\# \bullet (v^\# \bullet \mathbf{a}))))\\ \supset (((w^\# \bullet \mathbf{a}) \cong (u^\# \bullet (v^\# \bullet \mathbf{a})))).\end{aligned} \tag{3}$$

■

Lemma 5. If $u \underset{\Pi}{\Rightarrow} v$, then $(u^\# \cong v^\#)$ is rooted.

Proof. For some i, $1 \le i \le K$, we have either $u = pg_iq$, $v = ph_iq$, or $u = ph_iq$, $v = pg_iq$, where $p, q \in \{a, b\}^*$. We may assume that in fact $u = pg_iq$, $v = ph_iq$, because in the other case we could use the following Herbrand instance of formula 2:

$$((v^\# \cong u^\#) \supset (u^\# \cong v^\#)).$$

The proof now divides into three cases.

Case I. $p = q = 0$. Then the sentence $(u^\# \cong v^\#)$ is just $5 + i$ and is therefore in Γ.

Case II. $p = 0$, $q \neq 0$. Using $5 + i$ and the following Herbrand instance of formula 4:

$$(((g_i^\# \cong h_i^\#) \wedge (q^\# \cong q^\#)) \supset ((g_i^\# \bullet q^\#) \cong (h_i^\# \bullet q^\#))),$$

we see that the sentence

$$((g_i^\# \bullet q^\#) \cong (h_i^\# \bullet q^\#))$$

is rooted. Using Lemma 4 and Herbrand instances of formulas 2 and 3 we obtain the result.

Case III. $p, q \neq 0$. Using Case II, the sentence $((g_iq)^\# \cong (h_iq)^\#)$ is rooted. Using the Herbrand instance of formula 4:

$$(((p^\# \cong p^\#) \wedge ((g_iq)^\# \cong (h_iq)^\#)) \supset ((p^\# \bullet (g_iq)^\#) \cong (p^\# \bullet (h_iq)^\#))),$$

we see that

$$((p^\# \bullet (g_iq)^\#) \cong (p^\# \bullet (h_iq)^\#))$$

is rooted. The result now follows using Lemma 4 and Herbrand instances of formulas 2 and 3. ■

Lemma 6. If $u \overset{*}{\underset{\Pi}{\Rightarrow}} v$, then $(u^\# \cong v^\#)$ is rooted.

Proof. The proof is by induction on the length of a derivation of v from u. If this length is 1, then $v = u$, and we may use a Herbrand instance of formula 1. To complete the proof, we may assume that $u \overset{*}{\underset{\Pi}{\Rightarrow}} w \underset{\Pi}{\Rightarrow} v$, where it is known that $(u^\# \cong w^\#)$ is rooted. By Lemma 5, $(w^\# \cong v^\#)$ is rooted. We then get the result by using the following Herbrand instance of formula 3:

$$(((u^\# \cong w^\#) \wedge (w^\# \cong v^\#)) \supset (u^\# \cong v^\#)). \quad ■$$

Combining Lemmas 3 and 6, we obtain

Lemma 7. $u \overset{*}{\underset{\Pi}{\Rightarrow}} v$ if and only if $\Gamma \models (u^\# \cong v^\#)$.

Now it is easy to complete the proof of our theorem. If we possessed an algorithm for testing a given **W**-sentence for satisfiability, we could use it to test the sentence

$$\bigwedge_{\alpha \in \Gamma} \alpha \wedge \neg (u^\# \cong v^\#)$$

and therefore, by Theorem 3.1, to test the correctness of the logical inference $\Gamma \vDash (u^\# \cong v^\#)$. This would in turn lead to an algorithm for solving the word problem for Π, which we know is unsolvable. ■

A final remark: We really have been working with the axiomatizable theory $\mathbf{T}_\Gamma$. Thus what Lemma 7 states is just that

$$u \overset{*}{\underset{\Pi}{\Rightarrow}} v \quad \text{if and only if} \quad \vdash_\Gamma (u^\# \cong v^\#). \tag{8.3}$$

Hence we conclude that the theory $\mathbf{T}_\Gamma$ is not recursive. [If it were, we could use (8.3) to solve the word problem for Π.] Thus we have proved

Theorem 8.2. There are axiomatizable theories that are not recursive.

Exercises

1. Prove Lemma 2.
2. Let $\mathbf{W}$ be the vocabulary used in this section. Show that for every deterministic Turing machine $\mathscr{M}$ there is a finite set Γ of $\mathbf{W}$-sentences and a computable function $f(x)$ such that for any string w, $\mathscr{M}$ accepts w if and only if $\Gamma \vDash f(w)$. [*Hint:* See Theorems 3.3 and 3.4 in Chapter 7.]

Part 4

Complexity

14

Abstract Complexity

1. The Blum Axioms

In this chapter we will develop an abstract theory of the amount of resources needed to carry out computations. In practical terms resources can be measured in various ways: storage space used, time, some weighted average of central processor time and peripheral processor time, some combinations of space and time used, or even monetary cost. The theorems proved in this chapter are quite independent of which of these "measures" we use. We shall work with two very simple assumptions known as the *Blum axioms* after Manuel Blum, who introduced them in his doctoral dissertation. These assumptions are satisfied by any of the "measures" mentioned above (if given precise definitions in any natural manner) as well as by many others.

Definition. A 2-ary partial function C on N is called a *complexity measure* if it satisfies the *Blum axioms:*

1. $C(x, i)\downarrow$ if and only if $\Phi_i(x)\downarrow$;
2. The predicate $C(x, i) \le y$ is recursive. (This predicate is of course false if $C(x, i)\uparrow$.)

We write $C_i(x) = C(x, i)$. We think of $C_i(x)$ as the *complexity of the computation* that occurs when the program whose number is i is fed the input x. It is not very difficult to see that various natural ways of measuring complexity of computation do satisfy the Blum axioms. What is remarkable is that some very interesting and quite nontrivial results can be derived from such meager assumptions.

Let us examine some examples of proposed complexity measures:

1. $C_i(x)$ = *the number of steps in a computation by program number i on input x*. The first axiom is clearly satisfied; the second follows from the computability of the step-counter predicate $\mathrm{STP}^{(1)}$.
2. $M_i(x)$ = *the largest value assumed by any variable in program number i when this program is given input x, if* $\Phi_i(x)\downarrow$; $M_i(x)\uparrow$ otherwise. The definition forces the first axiom to be true. The truth of the second axiom is a more subtle matter. The key observation is that, for a given program, there are only finitely many *different* snapshots[1] in which all variables have values less than or equal to a given number y. Hence, given numbers i, x, y we can test the condition $M_i(x) \le y$ by "running" program number i on the input x until one of the following occurs:

   I. *A snapshot is reached in which some variable has a value* $> y$. Then we return the value FALSE.

   II. *The computation halts with all variables having values* $\le y$. Then we return the value TRUE.

   III. *The same snapshot is reached twice*. (By the pigeon-hole principle this must happen eventually if neither I nor II occurs.) Then, recognizing that the computation is in an "infinite" loop and so will never terminate, we return the value FALSE. (The reader should note that this algorithm in no way contradicts the unsolvability of the halting problem. Case I can include both halting and nonhalting computations.)

   We will make important use of this "maximum-space" complexity measure, and we reserve the notation $M_i(x)$ for it.
3. $C_i(x) = \Phi_i(x)$. Although the first Blum axiom is satisfied, the second is *certainly not;* namely, choose i so that

$$\Phi_i(x) = \begin{cases} 0 & \text{for } x \in S \\ \uparrow & \text{otherwise,} \end{cases}$$

   where S is any given r.e. nonrecursive set. Then the condition $\Phi_i(x) \le 0$ is equivalent to $x \in S$ and hence is not recursive.

[1] The definition of *snapshot* is in Chapter 2, Section 3.

If $P(x)$ is any predicate on N, we write

$$P(x) \qquad \text{a.e.},$$

and say that $P(x)$ is true *almost everywhere*, to mean that there exists $m_0 \in N$ such that $P(x)$ is true for all $x > m_0$. Equivalently, $P(x)$ is true for all but a finite set of numbers. We may think of a partial function on N as a total function with values in the set $N \cup \{\infty\}$. That is, we write $g(x) = \infty$ to mean that $g(x)\uparrow$. We extend the meaning of $<$ so that $n < \infty$ for all $n \in N$. $x \leq y$ continues to mean $x < y$ or $x = y$, so that $n \leq \infty$ for $n \in N$ but also $\infty \leq \infty$.

The second Blum axiom can be written in the equivalent forms:

2′. The predicate $C_i(x) = y$ is recursive.
2″. The predicate $C_i(x) < y$ is recursive.

To see that 2, 2′ and 2″ are all equivalent we note that

$$C_i(x) = y \Leftrightarrow (C_i(x) \leq y \ \& \sim (C_i(x) \leq y \dot{-} 1)) \vee (y = 0 \ \& \ C_i(x) \leq y),$$

so that 2 implies 2′. 2′ implies 2″ because

$$C_i(x) < y \Leftrightarrow (\exists z)_{<y}(C_i(x) = z).$$

Finally, 2″ implies 2 because

$$C_i(x) \leq y \Leftrightarrow C_i(x) < y + 1.$$

Let us call a *recursive* function $r(x)$ a *scaling factor* if

1. r is increasing, i.e., $r(x + 1) \geq r(x)$, and
2. $\lim_{x \to \infty} r(x) = \infty$, i.e. r assumes arbitrarily large values.

Condition 1 is obviously equivalent to the statement: $x \leq y$ implies $r(x) \leq r(y)$. Then we have

Theorem 1.1. Let $C_i(x)$ be a complexity measure and let $r(x)$ be a scaling factor. Let $D_i(x) = r(C_i(x))$. Then $D_i(x)$ is a complexity measure.

Proof. It is clear that D satisfies the first Blum axiom. To test $D_i(x) \leq y$, note that if $y < r(0)$ then $D_i(x) = r(C_i(x)) \geq r(0) > y$. Otherwise, find the number t for which

$$r(0) \leq r(1) \leq r(2) \leq \cdots \leq r(t) \leq y < r(t + 1).$$

We claim that $D_i(x) \leq y$ if and only if $C_i(x) \leq t$. It remains only to verify this claim. If $C_i(x) \leq t$, then

$$D_i(x) = r(C_i(x)) \leq r(t) \leq y.$$

Otherwise, if $t + 1 \leq C_i(x)$, then

$$y < r(t + 1) \leq r(C_i(x)) = D_i(x). \qquad \blacksquare$$

This theorem is hardly surprising. Naturally, if $C_i(x)$ is a plausible complexity measure, we would expect $2^{C_i(x)}$ to be one as well. What is surprising is that any pair of complexity measures are related to each other in a manner not so different from C and D in Theorem 1.1.

Theorem 1.2 (Recursive Relatedness Theorem). Let C and D be arbitrary complexity measures. Then there is a recursive function $r(x, y)$ such that $r(x, y) < r(x, y + 1)$, and for all i

$$C_i(x) \leq r(x, D_i(x)) \qquad \text{a.e.}$$

and (1.1)

$$D_i(x) \leq r(x, C_i(x)) \qquad \text{a.e.}$$

[where we let $r(x, \infty) = \infty$ for all x].

Proof. Note that by the first Blum axiom

$$\begin{array}{ll} C_i(x)\downarrow & \textit{if and only if} \\ \Phi_i(x)\downarrow & \textit{if and only if} \quad D_i(x)\downarrow. \end{array}$$

By the second Blum axiom (in the form 2′), the predicate

$$C_i(x) = y \quad \vee \quad D_i(x) = y$$

is recursive. Hence the function h defined as follows is recursive:

$$h(i, x, y) = \begin{cases} \max(C_i(x), D_i(x)) & \text{if} \quad C_i(x) = y \quad \vee \quad D_i(x) = y \\ 0 & \text{otherwise.} \end{cases}$$

Let

$$r(x, y) = y + \max_{j \leq x} \max_{z \leq y} h(j, x, z),$$

so that $r(x, y)$ is recursive. Then

$$\begin{aligned} r(x, y + 1) &= (y + 1) + \max_{j \leq x} \max_{z \leq y+1} h(j, x, z) \\ &> y + \max_{j \leq x} \max_{z \leq y} h(j, x, z) \\ &= r(x, y) \end{aligned}$$

since maximizing over a larger set of numbers cannot result in a smaller outcome. Moreover, using this same principle, and assuming that $x \geq i$,

$$\begin{aligned} r(x, D_i(x)) &\geq \max_{j \leq x} \max_{z \leq D_i(x)} h(j, x, z) \\ &\geq \max_{j \leq x} h(j, x, D_i(x)) \\ &\geq h(i, x, D_i(x)) \qquad (\text{since } x \geq i) \\ &= \max(C_i(x), D_i(x)) \\ &\geq C_i(x). \end{aligned}$$

Thus, the inequality

$$r(x, D_i(x)) \geq C_i(x)$$

holds for all $x \geq i$ and hence *almost everywhere*. Since the definition of h is symmetric in C and D, the same argument shows that

$$r(x, C_i(x)) \geq D_i(x) \qquad \text{a.e.}$$ ■

As we shall see, one use of the recursive relatedness theorem is in enabling us to proceed, in some cases, from the knowledge that a theorem is true for one particular complexity measure to the truth of that theorem for all complexity measures.

Exercises

1. Which of the following are complexity measures?

(**a**) $C_i(x) = 0$ for all i, x. (That is, all computation is "free.")

(**b**) $C_i(x) = \begin{cases} M_i(x) & \text{for } i \notin A \\ 0 & \text{for } i \in A, \end{cases}$

where A is some given finite set such that Φ_i is total for all $i \in A$. (That is, the programs whose numbers belong to A can be run "free.")

(**c**) $C_i(x) = 2^{\Phi_i(x)}$.

(**d**) $C_i(x) = \begin{cases} M_i(x) & \text{if } i \text{ is even} \\ \text{the number of steps in computing } \Phi_i(x) & \text{if } i \text{ is odd.} \end{cases}$

2. Prove that if C is a complexity measure and

$$D_i(x) = \begin{cases} C_i(x) & \text{for } i \notin A \\ 0 & \text{for } i \in A, \end{cases}$$

where A is as in Exercise 1(b), then D is a complexity measure.

3. Let $C_i(x)$ be the number of steps in the computation on input x by $\mathscr{S}$ program $\mathscr{P}$, where $\#(\mathscr{P}) = i$. For some fixed $n > 0$, let $D_i(x)$ be the number of steps in the computation on input x by $\mathscr{S}_n$ program $\mathscr{P}'$, where $\#(\mathscr{P}) = i$ and $\mathscr{P}'$ is constructed from $\mathscr{P}$ as in Section 3 of Chapter 5, by treating each $\mathscr{S}$ instruction as a macro in $\mathscr{S}_n$.
   (a) Show that D is a complexity measure.
   (b) Give a function $r(x, y)$ that satisfies the recursive relatedness theorem for C and D. [See Exercise 3.2 in Chapter 5.]

4. Let C be a complexity measure.
   (a) Show that for every i, $C_i(x)$ is partially computable.
   (b) Show that if $\Phi_i(x)$ is total, then $C_i(x)$ is computable.

5. Let C be a complexity measure. Show that the predicate $P(i)$, defined

$$P(i) \Leftrightarrow (\forall x)(\exists y \in N) C_i(x) \le y,$$

is not computable.

6. Let C be an arbitrary complexity measure. Show that there is a recursive function t such that

$$\Phi_i(x) \le t(x, C_i(x)) \qquad \text{a.e.}$$

[*Hint:* Use the complexity measure $M_i(x)$ and the recursive relatedness theorem.]

7. Can the result of the previous problem be improved so that t is a unary recursive function such that

$$\Phi_i(x) \le t(C_i(x)) \qquad \text{a.e.?}$$

Prove that your answer is correct.

8. **(a)** Let C be the complexity measure in Example 1. Show that for any computable function $f(x)$ there is a program number i such that $\Phi_i(x) = 0$ and $C_i(x) > f(x)$ for all x. Conclude that there are arbitrarily (with respect to computable lower bounds) slow $\mathscr{S}$ programs that compute constant functions.

(b) Let D be an arbitrary complexity measure. Show that for any computable function $f(x)$ there is a program number i such that $\Phi_i(x) = 0$ for all x and $D_i(x) > f(x)$ a.e. [*Hint:* Use (a) and the recursive relatedness theorem.]

9. Let C be a complexity measure. Show that there is no computable function $g(x, y)$ such that for all i, x, if $\Phi_i(x)\downarrow$ then $C_i(x) \leq g(x, \Phi_i(x))$. Compare with Exercise 6. [*Hint:* Use Exercise 8.]

10. Let C be a complexity measure. Show that for any computable function $f(x)$ there is a computable function $g(x)$ such that $g(x) \leq 1$ for all x and such that for any i, if $\Phi_i = g$ then $C_i(x) > f(x)$ for infinitely many x. Conclude that there are arbitrarily (with respect to computable lower bounds) complex "small" computable functions. [*Hint:* Define

$$g(x) = \begin{cases} 1 & \text{if } C_{l(x)}(x) \leq f(x) \text{ and } \Phi_{l(x)}(x) \neq 1 \\ 0 & \text{otherwise.]} \end{cases}$$

11. Let C be a complexity measure. Show that for any computable function $f(x)$ there is a computable function $g(x)$ such that $g(x) \leq x$ for all x and such that for any i, if $\Phi_i = g$ then $C_i(x) > f(x)$ for all $x > i$. Compare with Exercise 10.

2. The Gap Theorem

In this section C *is some given fixed complexity measure*. Suppose that $t(x)$ is a complexity bound. That is, assume that we are restricted to computations for which $C_i(x) \leq t(x)$ whenever $\Phi_i(x)\downarrow$. Then, in response to our complaints, the bound is increased enormously to $g(t(x))$, where g is some recursive, rapidly increasing function, e.g., $g(x) = 2^x$ or

$$g(x) = 2^{2^{\cdot^{\cdot^{\cdot^{2}}}}}\Big\}x \qquad \text{or} \qquad g(x) = 2^{2^{\cdot^{\cdot^{\cdot^{2}}}}}\Big\}2^x$$

Then, we can carry out far more computations. Right? Wrong! If the original function $t(x)$ is sufficiently tricky, it is possible that for every i, there are only finitely many values of x for which

$$C_i(x) \leq g(t(x)), \qquad \textit{but not} \quad C_i(x) \leq t(x).$$

This surprising assertion is a consequence of the gap theorem.

Theorem 2.1 (Gap Theorem). Let $g(x, y)$ be any recursive function such that $g(x, y) > y$. Then, there is a recursive function $t(x)$ such that if $x > i$ and $C_i(x) < g(x, t(x))$, then $C_i(x) \le t(x)$. (See Fig. 2.1.)

Proof. Consider the predicate

$$P(x, y) \leftrightarrow (\forall i)_{<x}(C_i(x) \le y \vee g(x, y) \le C_i(x)).$$

By the second Blum axiom, the predicate $C_i(x) \le y$ is computable. So is the predicate

$$g(x, y) \le C_i(x) \Leftrightarrow \sim (\exists z)_{<g(x,y)}(z = C_i(x)).$$

Hence, $P(x, y)$ is also recursive. We define

$$t(x) = \min_y P(x, y), \tag{2.1}$$

so that t is a partially computable function. We will show that t is total.

Let x be a given number. Consider the set $Q = \{C_i(x) \mid i < x \ \&\ \Phi_i(x)\downarrow\}$. Let $y_0 = 0$ if $Q = \varnothing$ and let y_0 be the largest element of Q otherwise. We

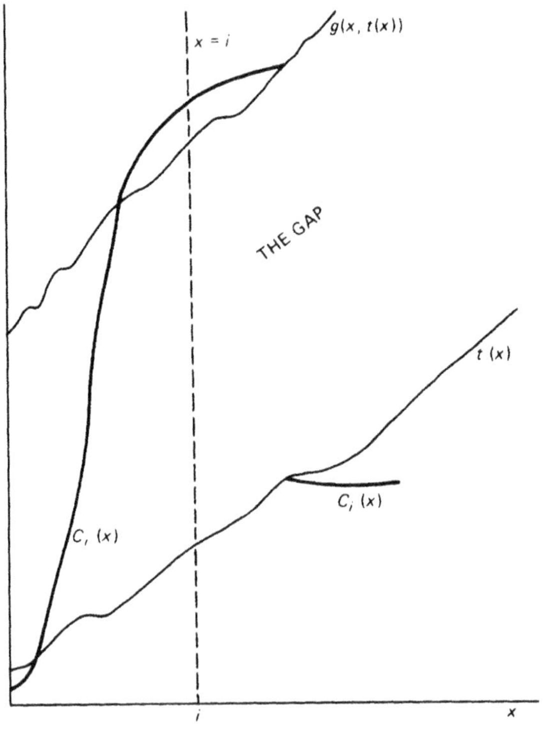

Figure 2.1. For $x > i$, $C_i(x)$ cannot enter the "gap."

claim that $P(x, y_0)$ is true. To see this, choose $i < x$. Then if $\Phi_i(x)\downarrow$, then $C_i(x)\downarrow$ and therefore $C_i(x) \le y_0$. If, on the other hand, $\Phi_i(x)\uparrow$ then $C_i(x)\uparrow$. Since $g(x, y_0)\downarrow$, the predicate $g(x, y_0) \le C_i(x)$ is true. Thus, we have $P(x, y_0)$. We have shown that for every $x \in N$ there is a number y such that $P(x, y)$. Thus, $t(x)$ defined by (2.1) is total and therefore recursive.

Now let $x > i$ and $C_i(x) < g(x, t(x))$. Since $P(x, t(x))$ is true, and $i < x$, we have $C_i(x) \le t(x) \vee g(x, t(x)) \le C_i(x)$. But $C_i(x) < g(x, t(x))$. Hence $C_i(x) \le t(x)$. ■

In their fine book, Machtey and Young (see "Suggestions for Further Reading") give an amusing interpretation of the gap theorem. Let us imagine two computers, one of which is very much faster than the other. We think of each computer equipped with a reasonably efficient interpreter for our programming language $\mathscr{S}$ so that we can speak of running a program of $\mathscr{S}$ on one or another of the computers. Let $C_i(x)$ be the computation time of the slow computer running program number i on input x. Similarly for $D_i(x)$ and the fast computer. Clearly, C and D satisfy the Blum axioms. By the recursive relatedness theorem, there is a recursive function r satisfying (1.1). If we let $g(x, y) = r(x, y) + y + 1$, then we have $g(x, y) > y$, $g(x, y + 1) > g(x, y)$ and

$$C_i(x) \le r(x, D_i(x)) < g(x, D_i(x)) \qquad \text{a.e.}$$

Now let $t(x)$ satisfy the gap theorem for the complexity measure C with respect to this function g. And consider a program $\mathscr{P}$ with number i such that $D_i(x) \le t(x)$ a.e. That is, for sufficiently large inputs x, $\mathscr{P}$ runs on the fast machine in time bounded by $t(x)$. Then on the slow computer, $\mathscr{P}$ will run in time

$$C_i(x) < g(x, D_i(x)) \le g(x, t(x)) \qquad \text{a.e.}$$

But now the gap theorem comes into play to assure us that

$$C_i(x) \le t(x) \qquad \text{a.e.}$$

Conclusion: *Any program that runs in time $t(x)$ on the fast computer also runs in time $t(x)$ (for sufficiently large x) on the slow computer!*

Exercises

1. Let C be a complexity measure. Does the gap theorem imply that there is no program number i such that $|x|^2 \le C_i(x) \le |x|^3$ a.e.? Explain.

2. Let C be a complexity measure. We will say that a total function $f(x)$ is *C-constructible* if there is a program number i such that $C_i(x) = f(x)$ for all x. Prove or disprove that every computable function is C-constructible.

3. Preliminary Form of the Speedup Theorem

Computer scientists often seek programs that will obtain a desired result using minimum resources. The speedup theorem, which is the deepest theorem in this chapter, tells us that it is possible for there to be no best program for this purpose. Roughly speaking, the theorem states that there exists a recursive function that is so badly behaved that for every program to compute it, there is another program that computes the same function but which uses much less resources. The proof of the speedup theorem is quite intricate. In this section we will prove a preliminary version. Then in the next section we will use this preliminary version to obtain the full speedup theorem. The proof of the speedup theorem will use the parameter theorem and the recursion theorem from Chapter 4 (Theorems 5.1 and 8.1).

We define a particular complexity measure $M_i(x)$ as follows. If $\Phi_i(x)\uparrow$, then $M_i(x)\uparrow$. If $\Phi_i(x)\downarrow$, then $M_i(x)$ is the largest value assumed by any variable in program number i when computing with input x. Thus $M_i(x)$ is just the complexity measure in Example 2 of Section 1. We will also work with $M_i^{(2)}(x_1, x_2)$, which is defined exactly like $M_i(x)$ except that program number i is given the pair of inputs x_1, x_2. $M_i(x)$ and $M_i^{(2)}(x_1, x_2)$ are related by

Theorem 3.1. $M_i^{(2)}(x, y) = M_{S_1^1(y,i)}(x)$, where S_1^1 is the function defined in the parameter theorem.

Proof. Let $i = \#(\mathscr{P}_0)$. Then, examining the *proof* of Theorem 5.1 in Chapter 4, we see that $S_1^1(y, i) = \#(\mathscr{P})$, where $\mathscr{P}$ is a program consisting of y copies of the instruction $X_2 \leftarrow X_2 + 1$ followed by the program $\mathscr{P}_0$. The result is now obvious. ■

Our preliminary form of the speedup theorem is as follows.

Theorem 3.2. Let $g(x, y)$ be any given recursive function. Then there is a recursive function $f(x)$ such that $f(x) \leq x$ and, whenever $\Phi_i = f$, there is a j such that

$$\Phi_j(x) = f(x) \qquad \text{a.e.} \tag{3.1}$$

and

$$g(x, M_j(x)) \le M_i(x) \qquad \text{a.e.} \tag{3.2}$$

Discussion. To see the force of the theorem take $g(x, y) = 2^y$. Then, given $\Phi_i = f$, there is a j satisfying (3.1) such that

$$2^{M_j(x)} \le M_i(x) \qquad \text{a.e.,}$$

i.e.,

$$M_j(x) \le \log_2 M_i(x) \qquad \text{a.e.}$$

Thus program number j computes f a.e. and uses far less resources than program number i. In Section 4 we shall improve this preliminary version of the speedup theorem by eliminating the "a.e." condition in (3.1) and by obtaining (3.2) for an arbitrary complexity measure, not merely for M.

The proof of Theorem 3.2 will use a diagonal argument, but one far more complex than we have encountered so far. Let us recall how a simple diagonal argument works. When we write

$$\bar{K} = \{n \in N \mid n \notin W_n\}$$

we know that $\bar{K}$ is not r.e. because it differs from each r.e. set W_i with respect to the number i, namely, $i \in W_i$ if and only if $i \notin \bar{K}$. More generally, a diagonal argument constructs an object that is guaranteed not to belong to a given class by systematically ensuring that the object differs in some way from each member of the class. More intricate diagonal arguments often are carried out in an infinite sequence of stages; at each stage one seeks to ensure that the object being constructed is different from some particular member of the class. The proof of the speedup theorem is of this character.

Proof of Theorem 3.2. We will proceed through "stages" $x = 0, 1, 2, 3, \ldots$. At each stage x and for certain $n, w \in N$, we will define a set $C(n, w, x) \subseteq N$. We think of the members of $C(n, w, x)$ as numbers of programs which are *cancelled* at stage x with respect to n and w. $C(n, w, x)$ is defined recursively by the equation

$$C(n, w, x) = \Big\{ i \in N \mid w \le i < x \;\&\; i \notin \bigcup_{y<x} C(n, w, y) \\ \& \; M_i(x) < g(x, M_n^{(2)}(x, i+1)) \Big\}. \tag{3.3}$$

We think of C as a 3-ary partial function on N. (The fact that the values of C are finite subsets of N instead of numbers is of no importance. Naturally, if we wished, we could use some coding device to represent each finite subset of N by a particular number.) The three conditions in (3.3) connected by & are to be tested in order with the understanding that if the first or second condition is false, the succeeding conditions are simply not tested. Thus we have

$$w \ge x \quad \textit{implies} \quad C(n, w, x) = \varnothing \quad \textit{for all } n. \tag{3.4}$$

Moreover, we have obviously

Lemma 1. If $C(n, w, y)\downarrow$ for all $y < x$ and $M_n^{(2)}(x, i+1)\downarrow$ for all i such that $w \le i < x$, then $C(n, w, x)\downarrow$.

Indeed, when the conditions of Lemma 1 are satisfied, we can explicitly compute $C(n, w, x)$ given knowledge of $C(n, w, y)$ for $y < x$. Now clearly, when the conditions of Lemma 1 are not satisfied, $C(n, w, x)\uparrow$. Thus (3.3) can be used to give an algorithm for computing C and we may conclude that C is a partially computable function.

Lemma 2. If $i \in C(n, w, x)$, then $M_i(x)\downarrow$ and $\Phi_i(x)\downarrow$.

Proof. The truth of the condition

$$M_i(x) < g(x, M_n^{(2)}(x, i+1))$$

implies that $M_i(x)\downarrow$, and by the Blum axioms, this implies $\Phi_i(x)\downarrow$. ■

We shall now define a 3-ary partially computable function k on N such that if $C(n, w, x)\downarrow$, then for each $i \in C(n, w, x)$, we will have $k(x, w, n) \ne \Phi_i(x)$. k is computed by using the following procedure:

Compute $C(n, w, x)$. If this computation terminates, compute $\Phi_i(x)$ for each $i \in C(n, w, x)$. [By Lemma 2, each such $\Phi_i(x)\downarrow$.] Finally, set $k(x, w, n)$ equal to the least number which is not a member of the finite set

$$\{\Phi_i(x) \mid i \in C(n, w, x)\}.$$

It is to this function k that we apply the recursion theorem. Thus, we obtain a number e such that

$$\Phi_e^{(2)}(x, w) = k(x, w, e). \tag{3.5}$$

Lemma 3. If $x \leq w$, then $k(x, w, e) = 0$.

Proof. Let $x \leq w$. By (3.4), $C(e, w, x) = \varnothing$. Hence, by definition, $k(x, w, e)$ is the least number which does not belong to $\varnothing$, namely, 0. ∎

Lemma 4. If $k(x, w, e)\downarrow$, then $k(x, w, e) \leq x$.

Proof. The largest possible value for $k(x, w, e)$ would be obtained if the values $\Phi_i(x)$ for $i \in C(e, w, x)$ were all different and were consecutive numbers beginning with 0. In this "worst" case, there would be as many values of $\Phi_i(x)$ as in the set $C(e, w, x)$. But,

$$\begin{aligned} C(e, w, x) &\subseteq \{i \in N \mid w \leq i < x\} \\ &\subseteq \{0, 1, 2, \ldots, x - 1\}. \end{aligned}$$

Thus, all the values of $\Phi_i(x)$ would be $< x$ and hence $k(x, w, e) \leq x$. ∎

Lemma 5. Let $x > w$. Suppose that

$$\Phi_e^{(2)}(x, w + 1)\downarrow, \Phi_e^{(2)}(x, w + 2)\downarrow, \ldots, \Phi_e^{(2)}(x, x)\downarrow \tag{3.6}$$

and

$$\Phi_e^{(2)}(0, w)\downarrow, \Phi_e^{(2)}(1, w)\downarrow, \ldots, \Phi_e^{(2)}(x - 1, w)\downarrow. \tag{3.7}$$

Then, $\Phi_e^{(2)}(x, w)\downarrow$, i.e., $k(x, w, e)\downarrow$.

The reader is referred to Fig. 3.1 in connection with this lemma. In effect, Lemma 5 states that if $\Phi_e^{(2)}$ is defined along both the horizontal and vertical "pincers" shown pointing at (x, w), then it must also be defined at (x, w).

Proof of Lemma 5. By (3.7), $\Phi_e^{(2)}(y, w)\downarrow$ for all $y < x$. By definition of $k(y, w, e) = \Phi_e^{(2)}(y, w)$, we have that $C(e, w, y)\downarrow$ for all $y < x$. By (3.6), $\Phi_e^{(2)}(x, i + 1)\downarrow$ for all i such that $w \leq i < x$. Hence, likewise, $M_e^{(2)}(x, i + 1)\downarrow$ for these i. By Lemma 1, $C(e, w, x)\downarrow$. But now, by definition of k, $k(x, w, e)\downarrow$. ∎

Lemma 6. $\Phi_e^{(2)}$ is total.

Proof. We shall prove by induction on x the assertion

$$\textit{For all } w, \qquad \Phi_e^{(2)}(x, w)\downarrow. \tag{3.8}$$

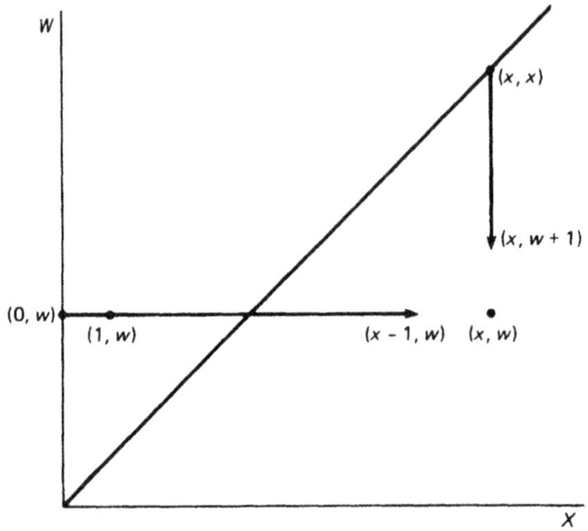

Figure 3.1. Horizontal and vertical "pincers" pointing at (x, w). (See Lemma 5.)

By Lemma 3, we have

$$\Phi_e^{(2)}(0, w) = k(0, w, e) = 0,$$

which gives the result for $x = 0$. Suppose that $x > 0$, and it is known that

$$\Phi_e^{(2)}(y, w)\downarrow$$

for all $y < x$ and all w. We shall show that (3.8) then follows.

By Lemma 3, (3.8) holds for all $w \geq x$. Thus, we need show only that (3.8) holds for $w < x$. That is, it suffices to show that

$$\Phi_e^{(2)}(x, x - 1)\downarrow, \dots, \Phi_e^{(2)}(x, 0)\downarrow.$$

We will prove each of these in succession by using Lemma 5. That is, in Lemma 5, we successively set $w = x - 1, x - 2, \dots, 0$. In each case (3.7) (the horizontal "pincer") is satisfied by the induction hypothesis. For $w = x - 1$, (3.6) requires only that $\Phi_e^{(2)}(x, x)\downarrow$, and this last follows at once from Lemma 3. Thus by Lemma 5, $\Phi_e^{(2)}(x, x - 1)\downarrow$. But this means that (3.6) is now satisfied with $w = x - 2$. Hence once again Lemma 5 shows that $\Phi_e^{(2)}(x, x - 2)\downarrow$. Continuing, we eventually obtain $\Phi_e^{(2)}(x, 0)\downarrow$. ■

For the remainder of the proof of Theorem 3.1, we will use the notation

$$I_w = \{i \in N \mid i < w\} = \{0, 1, \dots, w - 1\}.$$

Lemma 7. $C(n,w,x) = C(n,0,x) - I_w$.

Proof. The proof is by induction on x. $C(n,w,0) = \emptyset$ for all n,w. Hence the result for $x = 0$ is trivially true. Suppose the result known for all $y < x$. We obtain the result for x as follows (noting $\{i \in N \mid w \leq i < x\} \cap I_w = \emptyset$):

$$\begin{aligned}
C(n,w,x) &= \Big\{ i \in N \mid w \leq i < x \ \&\ i \notin \bigcup_{y<x} C(n,w,y) \\
&\qquad \&\ M_i(x) < g(x, M_n^{(2)}(x, i+1)) \Big\} \\
&= \Big\{ i \in N \mid w \leq i < x \ \&\ i \notin \bigcup_{y<x} (C(n,0,y) - I_w) \\
&\qquad \&\ M_i(x) < g(x, M_n^{(2)}(x, i+1)) \Big\} \\
&= \Big\{ i \in N \mid w \leq i < x \ \&\ i \notin \bigcup_{y<x} C(n,0,y) \\
&\qquad \&\ M_i(x) < g(x, M_n^{(2)}(x, i+1)) \Big\} \\
&= \Big\{ i \in N \mid 0 \leq i < x \ \&\ i \notin \bigcup_{y<x} C(n,0,y) \\
&\qquad \&\ M_i(x) < g(x, M_n^{(2)}(x, i+1)) \Big\} - I_w \\
&= C(n,0,x) - I_w .
\end{aligned}$$

■

Lemma 8. For each $w \in N$, there is a number m_w such that for all $x > m_w$, we have

$$\Phi_e^{(2)}(x,w) = \Phi_e^{(2)}(x,0).$$

Proof. By (3.3) [the definition of $C(n,w,x)$], we have $C(e,0,x) \cap C(e,0,y) = \emptyset$ for $x \neq y$. [Numbers in $C(e,0,y)$ for $y < x$ are automatically excluded from $C(e,0,x)$.] Hence each number in I_w belongs to at most one of the sets $C(e,0,x)$. If we let m_w be the largest such value of x, then for $x > m_w$,

$$C(e,0,x) \cap I_w = \emptyset.$$

Hence, using Lemma 7, for $x > m_w$,

$$C(e,w,x) = C(e,0,x) - I_w = C(e,0,x).$$

Hence, by the definition of the function k we have for $x > m_w$,

$$\Phi_e^{(2)}(x, w) = k(x, w, e) = k(x, 0, e) = \Phi_e^{(2)}(x, 0). \qquad \blacksquare$$

Note that there is no claim being made that m_w is a computable function of w, and indeed it is not!

We are now ready to define the function $f(x)$ whose existence is asserted in Theorem 3.2. We set

$$f(x) = \Phi_e^{(2)}(x, 0).$$

Lemma 9. If $\Phi_i = f$ and $x > i$, then

$$g(x, M_e^{(2)}(x, i+1)) \leq M_i(x).$$

Proof. Suppose otherwise. Choose the *least value of* $x > i$ *with*

$$g(x, M_e^{(2)}(x, i+1)) > M_i(x). \tag{3.9}$$

Then we claim that for $y < x$, $i \notin C(e, 0, y)$. This is because

$$C(e, 0, y) = \left\{ j \in N \mid j < y \ \&\ j \notin \bigcup_{z<y} C(e, 0, z) \right.$$
$$\left. \&\ M_j(y) < g(y, M_e^{(2)}(y, j+1)) \right\},$$

so that, if $i \in C(e, 0, y)$, we would have $i < y < x$, and

$$g(y, M_e^{(2)}(y, i+1)) > M_i(y),$$

contradicting the choice of x as the least number $> i$ satisfying (3.9). Thus, we have

$$i \notin \bigcup_{y<x} C(e, 0, y).$$

Hence,

$$i \in C(e, 0, x) = \left\{ j \in N \mid j < x \ \&\ j \notin \bigcup_{y<x} C(e, 0, y) \right.$$
$$\left. \&\ M_j(x) < g(x, M_e^{(2)}(x, j+1)) \right\}.$$

Now $k(x, 0, e)$ was defined to be different from all $\Phi_j(x)$ for which $j \in C(e, 0, x)$. Hence, $k(x, 0, e) \neq \Phi_i(x)$. But

$$k(x, 0, e) = \Phi_e^{(2)}(x, 0) = f(x) = \Phi_i(x),$$

This contradiction completes the proof.

Proof of Theorem 3.2 Concluded. Let $\Phi_i = f$, and set $j = S_1^1(i+1, e)$. Then, by Theorem 3.1 and Lemma 9, we have for $x > i$,

$$g(x, M_j(x)) = g(x, M_e^{(2)}(x, i+1)) \leq M_i(x),$$

which proves (3.2). Finally, using the parameter theorem (Theorem 5.1 in Chapter 4) and Lemma 8 we have for $x > m_{i+1}$,

$$\Phi_j(x) = \Phi_e^{(2)}(x, i+1) = \Phi_e^{(2)}(x, 0) = f(x),$$

which proves (3.1). ∎

4. The Speedup Theorem Concluded

We will begin by showing how to eliminate the a.e. from Eq. (3.1) in Theorem 3.2. The technique we will use is a general one; to change a condition

$$\Phi_j(x) = f(x) \qquad \text{a.e.}$$

into an equation valid everywhere, we need only modify program number j to agree with $f(x)$ at a finite number of values. We can do this by patching in a "table look-up" program. More precisely, we have

Theorem 4.1. There is a recursive function $t(u, w)$ such that

$$\Phi_{t(u,w)}(x) = \begin{cases} \Phi_u(x) & \text{if} \quad x > l(w) \\ (r(w))_{x+1} & \text{if} \quad x \leq l(w), \end{cases}$$

$$M_{t(u,w)}(x) = M_u(x) \qquad \text{if} \quad x > l(w).$$

Here, once again we are using the pairing functions and Gödel numbers as coding devices (Chapter 3, Section 8).

Proof. Let the numbers u, w be given. Let P_u be program number u of the language $\mathscr{S}$, if this program begins with a labeled statement. Otherwise let P_u be program number u modified by having its initial statement labeled by a label not otherwise occurring in the program. In either case let L be the label with which P_u begins.

Let $Q_{u,w}$ be a program of $\mathscr{S}$ which computes the primitive recursive function $(r(w))_{x+1}$, which always terminates using a branch instruction, and which has no labels in common with P_u. Let V be a local variable that occurs neither in P_u nor in $Q_{u,w}$. Let $t(u, w)$ be the number of the program indicated in Fig. 4.1. Note that $V \leftarrow X$ is to be replaced by a

$$
\begin{array}{l}
V \leftarrow X \\
\left.\begin{array}{l} V \leftarrow V - 1 \\ V \leftarrow V - 1 \\ \vdots \\ V \leftarrow V - 1 \end{array}\right\} l(w) \\
\text{IF } V \neq 0 \text{ GOTO } L \\
Q_{u,w} \\
P_u
\end{array}
$$

Figure 4.1

suitable macro expansion as in Chapter 2 and that there are $l(w)$ statements $V \leftarrow V - 1$. Clearly this can all be done with t a recursive (even primitive recursive) function.

Now, let $x > l(w)$. Then after the $l(w)$ decrement instructions $V \leftarrow V - 1$ have been executed, V will have the value $x - l(w) > 0$. Hence, the branch shown will be taken and program P_u will be executed. Hence, $\Phi_{t(u,w)}(x) = \Phi_u(x)$. To compare the value of $M_{t(u,w)}(x)$ and $M_u(x)$ we need to be concerned about the maximum value assumed by variables in the macro expansion of $V \leftarrow X$. Examining this macro expansion as given in (c) in Chapter 2, Section 2, we see that the only possibility for a number $> x$ to arise is in the case $x = 0$. This is because local variables need to be incremented to 1 in this macro expansion in order to force a branch to be taken.[2] However, we are assuming $x > l(w) \geq 0$, so that $x \neq 0$. Hence, $M_{t(u,w)}(x) = M_u(x)$.

Finally, let $x \leq l(w)$. Then after $l(w)$ executions of $V \leftarrow V - 1$, V has the value 0. Thus $Q_{u,w}$ is executed. Hence, $\Phi_{t(u,w)}(x) = (r(w))_x$. ■

Now we can easily prove

Theorem 4.2. Let $g(x, y)$ be any given recursive function. Then there is a recursive $f(x)$ such that $f(x) \leq x$ and, whenever $\Phi_i = f$, there is a j such that

$$\Phi_j(x) = f(x)$$

and

$$g(x, M_j(x)) \leq M_i(x) \qquad \text{a.e.}$$

[2] Actually, if each unconditional branch statement in program (c), Chapter 2, Section 2, is directly expanded, some of the local variables used in this expansion will reach values > 1. The simplest way to get around this is to place the single statement $Z_2 \leftarrow Z_2 + 1$ at the beginning of this program and then to replace each of the four unconditional branch statements GOTO L by the corresponding conditional branch statement IF $Z_2 \neq 0$ GOTO L.

Proof. Let f be as in Theorem 3.2, and suppose $\Phi_i = f$. Then there is $j \in N$ such that (3.1) and (3.2) hold. Let $\Phi_j(x) = f(x)$ for $x > x_0$. Let

$$w = \langle x_0, [f(0), \ldots, f(x_0)] \rangle.$$

Finally, let $\bar{j} = t(j, w)$. Then using Theorem 4.1,

$$\Phi_{\bar{j}}(x) = \Phi_{t(j,w)}(x) = \begin{cases} \Phi_j(x) & \text{if } \; x > x_0 \\ f(x) & \text{if } \; x \le x_0, \end{cases}$$

i.e., $\Phi_{\bar{j}} = f$. Theorem 4.1 also implies that $M_{\bar{j}}(x) = M_j(x)$ a.e. Hence, using (3.2), we have almost everywhere

$$g\big(x, M_{\bar{j}}(x)\big) = g\big(x, M_j(x)\big) \le M_i(x). \qquad \blacksquare$$

Finally, we are ready to give the speedup theorem for arbitrary complexity measures.

Theorem 4.3 (Blum Speedup Theorem). Let $g(x, y)$ be any given recursive function and let C be any complexity measure. Then there is a recursive function $f(x)$ such that $f(x) \le x$ and whenever $\Phi_i = f$, there is a j such that

$$\Phi_j(x) = f(x)$$

and

$$g\big(x, C_j(x)\big) \le C_i(x) \qquad \text{a.e.}$$

Proof. Using the recursive relatedness theorem (Theorem 1.2), there is a recursive function $r(x, y)$ such that

$$r(x, y) < r(x, y + 1),$$

$$C_i(x) \le r(x, M_i(x)) \qquad \text{a.e.}$$

$$M_i(x) \le r(x, C_i(x)) \qquad \text{a.e.}$$

Let

$$h(x, y) = \sum_{z \le y} g(x, z),$$

so that h is recursive,

$$h(x, y) \ge g(x, y),$$

and

$$h(x, y + 1) \ge h(x, y).$$

Finally, let

$$\bar{g}(x, y) = r(x, h(x, r(x, y))).$$

Now, we apply Theorem 4.2 using $\bar{g}$ as the given function g. Let $f(x)$ be the recursive function obtained, so that $f(x) \leq x$. Let $\Phi_i = f$. Then there is a j such that $\Phi_j = f$ and

$$\bar{g}(x, M_j(x)) \leq M_i(x) \qquad \text{a.e.}$$

Hence, we have, almost everywhere,

$$\begin{aligned} r(x, g(x, C_j(x))) &\leq r(x, h(x, C_j(x))) \\ &\leq r(x, h(x, r(x, M_j(x)))) \\ &= \bar{g}(x, M_j(x)) \\ &\leq M_i(x) \leq r(x, C_i(x)). \end{aligned}$$

Now, if $C_i(x) < g(x, C_j(x))$ for any value of x, we would have, for that value of x,

$$r(x, C_i(x)) < r(x, g(x, C_j(x))).$$

Hence, we must have, almost everywhere,

$$g(x, C_j(x)) \leq C_i(x). \qquad \blacksquare$$

Exercises

1. Show that for all $i \in N$ there is a j such that $\Phi_j(x) = M_i(x)$ and $\Phi_j(x) = M_j(x)$ for all x. Conclude that every function $M_i(x)$ has an "optimal" program with respect to complexity measure M.
2. Let L be the set of all strings that are syntactically correct Pascal programs, and let

$$P(x) = \begin{cases} 1 & \text{if } x \in L \\ 0 & \text{otherwise.} \end{cases}$$

Does the speedup theorem imply that there is no fastest $\mathscr{S}$ program that computes $P(x)$? Explain.

15

Polynomial–Time Computability

1. Rates of Growth

In this chapter we will be working with functions f such that $f(n) \in N$ for all sufficiently large $n \in N$, but which may be undefined or have negative values for some finite number of values of n. We refer to such functions briefly, and slightly inaccurately, as functions from N to N. These functions f will typically have the additional property

$$\lim_{n \to \infty} f(n) = \infty. \tag{1.1}$$

Examples of such functions are n^2, 2^n, and $\lfloor \log_2 n \rfloor$. It will be important for us to understand in what sense we can say that 2^n grows faster than n^2 and that n^2 grows faster than $\lfloor \log_2 n \rfloor$. Although in practice, the definitions we are about to give are of interest only for functions that satisfy (1.1), our definitions will not assume that this is the case.

Definition. Let f, g be functions from N to N. Then, we say that $f(n) = O(g(n))$ if there are numbers c and n_0 such that $f(n) \leq cg(n)$ for all $n \geq n_0$. If these conditions do not hold we say that $f(n) \neq O(g(n))$.

If $f(n) = O(g(n))$ and $g(n) = O(f(n))$ we say that f and g have *the same rate of growth*. On the other hand, if $f(n) = O(g(n))$ but $g(n) \neq O(f(n))$, we say that $g(n)$ *grows faster than* $f(n)$.

An example should help clarify these notions. We have

$$n^2 = O(3n^2 - 6n + 5)$$

since

$$\frac{n^2}{3n^2 - 6n + 5} = \frac{1}{3 - 6/n + 5/n^2} \to \frac{1}{3}$$

as $n \to \infty$, and therefore there is a number n_0 such that for all $n \geq n_0$,

$$\frac{n^2}{3n^2 - 6n + 5} \leq 1.$$

Likewise $3n^2 - 6n + 5 = O(n^2)$, so that these two functions have the same rate of growth.

Clearly, it is also true that $3n^2 - 6n + 5 = O(n^3)$; however,

$$n^3 \neq O(3n^2 - 6n + 5)$$

because

$$\frac{n^3}{3n^2 - 6n + 5} = n \cdot \frac{1}{3 - 6/n + 5/n^2} \to \infty$$

as $n \to \infty$. Thus, we can say that n^3 grows faster than $3n^2 - 6n + 5$.

More generally, we can prove

Theorem 1.1. Let f, g be functions from N to N, and let

$$\lim_{n \to \infty} \frac{f(n)}{g(n)} = \beta, \tag{1.2}$$

where β is a positive real number. Then $f(n) = O(g(n))$ and $g(n) = O(f(n))$, so that f and g have the same rate of growth.

If, on the other hand,

$$\lim_{n \to \infty} \frac{f(n)}{g(n)} = \infty, \tag{1.3}$$

then $g(n) = O(f(n))$ but $f(n) \neq O(g(n))$, so that $f(n)$ grows faster than $g(n)$.

Proof. If (1.2) holds, then there is a number n_0 such that for all $n \geq n_0$,

$$\frac{f(n)}{g(n)} \leq \beta + 1.$$

Hence, $f(n) = O(g(n))$. Since (1.2) implies that

$$\lim_{n \to \infty} \frac{g(n)}{f(n)} = \frac{1}{\beta},$$

the same reasoning can be used to show that $g(n) = O(f(n))$.

Next, (1.3) implies that

$$\lim_{n \to \infty} \frac{g(n)}{f(n)} = 0.$$

Therefore, there is a number n_0 such that $n \geq n_0$ implies

$$\frac{g(n)}{f(n)} \leq 1.$$

Hence, $g(n) = O(f(n))$. If we had also $f(n) = O(g(n))$, then for numbers c, n_0 we should have for $n \geq n_0$,

$$\frac{f(n)}{g(n)} \leq c;$$

on the other hand, (1.3) implies that there is a number n_1 such that $n \geq n_1$ implies

$$\frac{f(n)}{g(n)} > c,$$

which is a contradiction. ■

A *polynomial* is a function p from N to N that is defined by a formula of the form

$$p(n) = a_0 + a_1 n + a_2 n^2 + \cdots + a_r n^r, \tag{1.4}$$

where $a_0, a_1, \ldots, a_{r-1}$ are integers, positive, negative, or zero, while a_r is a positive integer. In this case the number r is called the *degree* of the polynomial p. The degree of a polynomial determines its rate of growth in the following precise sense.

Theorem 1.2. Let p be a polynomial of degree r. Then p and n^r have the same rate of growth. Moreover, p grows faster than n^m if $m < r$, and n^m grows faster than p if $m > r$.

Proof. Letting p be as in (1.4), we have

$$\frac{p(n)}{n^r} = \frac{a_0}{n^r} + \frac{a_1}{n^{r-1}} + \cdots + a_r \to a_r$$

as $n \to \infty$. Also,

$$\frac{p(n)}{n^m} = \frac{p(n)}{n^r} \cdot n^{r-m},$$

so that

$$\frac{p(n)}{n^m} \to \infty \quad \text{if} \quad r > m, \qquad \text{and} \qquad \frac{p(n)}{n^m} \to 0 \quad \text{if} \quad r < m.$$

The result then follows from Theorem 1.1. ∎

Next we shall see that exponential functions grow faster than any fixed power.

Theorem 1.3. The function k^n, with $k > 1$, grows faster than any polynomial.

Proof. It clearly suffices to prove that for any $r \in N$,

$$\lim_{n \to \infty} \frac{k^n}{n^r} = \infty.$$

One way to obtain this result is to use L'Hospital's rule from calculus; on differentiating the numerator and denominator of this fraction r times, a fraction is obtained whose numerator approaches infinity and whose denominator is a constant (in fact, $r!$). To obtain the result directly, we first prove the following lemma.

Lemma. Let g be a function from N to N such that

$$\lim_{n \to \infty} \frac{g(n+1)}{g(n)} = \beta > 1.$$

Then $g(n) \to \infty$ as $n \to \infty$.

Proof of Lemma. Let γ be a number strictly between 1 and β, for example, $\gamma = (1 + \beta)/2$. Then there is a number n_0 such that $n \geq n_0$ implies

$$\frac{g(n+1)}{g(n)} \geq \gamma.$$

Thus, for each m,

$$g(n_0 + m) \geq \gamma g(n_0 + m - 1) \geq \cdots \geq \gamma^m g(n_0).$$

Since $\gamma^m \to \infty$ as $m \to \infty$, the result follows. ∎

Proof of Theorem 1.3 *Concluded.* Setting

$$g(n) = k^n/n^r,$$

we have

$$\frac{g(n+1)}{g(n)} = \frac{k}{\left(1 + \frac{1}{n}\right)^r} \to k \qquad \text{as} \quad n \to \infty,$$

which, by the lemma, gives the result. ∎

Exercises

1. Suppose we have a computer that executes 1 million instructions per second.
   - (a) For each of the following functions $f(x)$, give the length of the longest string that can be processed in one hour if $f(|w|)$ instructions are required to process a string w: $f(x) = x$; $f(x) = x^2$; $f(x) = x^4$; $f(x) = 2^x$.
   - (b) For the same functions, approximately how long would it take to process w if $|w| = 100$?
2. What is the least $x \in N$ such that $10000x^2 \leq 2^x$?
3. For each of the following functions $f(x)$, give a function $g(x)$ such that some Turing machine on a two-symbol alphabet can calculate $f(x)$ in $O(g(|x|))$ steps: $f(x) = 2x$; $f(x) = x^2$; $f(x) = 2^x$; $f(x) = 2^{(2^x)}$.
4. (a) Show that if $p(n)$ is defined by (1.4), then $p(n)$ is positive for n sufficiently large, so that p is a function from N to N in the sense defined at the beginning of this chapter.
   - (b) Show that if $p(n)$ is as in (a) with $r > 0$, then $p(n) \to \infty$ as $n \to \infty$.
5. Show that n grows faster than $\lfloor \log_2 n \rfloor$.
6. Show that for any $k \geq 1$ and any polynomials $p(x)$, $q(x)$, there is a polynomial $r(x)$ such that $q(x) \cdot k^{p(x)} = O(2^{r(x)})$.

2. P versus NP

Computability theory has enabled us to distinguish clearly and precisely between problems for which there are algorithms and those for which there are none. However, there is a great deal of difference between solvability "in principle," with which computability theory deals, and solvability "in practice," which is a matter of obtaining an algorithm that can be implemented to run using space and time resources likely to be available. It has become customary to speak of problems that are solvable, not only in principle but also in practice, as *tractable*; problems that may be solvable in principle but are not solvable in practice are then called *intractable*.

The satisfiability problem, discussed in Chapter 12, is an example that is illuminating in this connection and will, in fact, play a central role in this chapter. The satisfiability problem is certainly *solvable*; in Chapter 12, we discussed algorithms for testing a given formula in CNF for satisfiability based on truth tables, on converting to DNF, on resolution, and on the Davis–Putnam rules. However, we cannot claim that the satisfiability problem is tractable on the basis of any of these algorithms or, for that matter, on the basis of any known algorithm. As we have seen, procedures based on truth tables or DNF require a number of steps which is an *exponential* function of the length of the expression representing a given formula in CNF. It is because of the rapid growth of the exponential function that these procedures can quickly exhaust available resources. Procedures based on resolution or on the Davis–Putnam rules can be designed that work well on "typical" formulas. However, no one has succeeded in designing such a procedure for which it can be proved that exponential behavior never arises, and it is widely believed (for reasons that will be indicated later) that every possible procedure for the satisfiability problem behaves exponentially in some cases. Thus the satisfiability problem is regarded as a prime candidate for intractability, although the matter remains far from being settled.

This association of intractability with the exponential function, coupled with the fact (Theorem 1.3) that an exponential function grows faster than any polynomial function, suggests that a problem be regarded as tractable if there is an algorithm that solves it which requires a number of steps bounded by some polynomial in the length of the input.

To make these ideas precise, we have recourse to the Turing machine model of computation as developed in Chapter 6. In particular, we shall use the terms *configuration* and *computation* as in Chapter 6.

Definition. A language L on an alphabet A is said to be *polynomial–time decidable* if there is a Turing machine $\mathscr{M}$ that accepts L, and a polynomial $p(n)$, such that the number of steps in an accepting computation by $\mathscr{M}$ with input x is $\leq p(|x|)$. When the alphabet is understood, we write **P** for the class of polynomial–time decidable languages.

Definition. A total function f on A^*, where A is an alphabet, is said to be *polynomial-time computable* if there is a Turing machine $\mathcal{M}$ that computes f, and a polynomial $p(n)$, such that the number of steps in the computation by $\mathcal{M}$ with input x is $\leq p(|x|)$.

With respect to both of these definitions, we note

1. It suffices that there exist a polynomial $p(n)$ such that the number of steps in the computation by $\mathcal{M}$ with input x is $\leq p(|x|)$ *for all but a finite number of input strings* x. For, in such a case, to include the finite number of omitted cases as well, we let c be the largest number of steps used by $\mathcal{M}$ in these cases, and replace $p(n)$ by the polynomial $p(n) + c$.
2. Using 1 and Theorem 1.2, it suffices that the number of steps be $O(|x|^r)$ for some $r \in N$.

The discussion leading to these definitions suggests that in analogy with Church's thesis, we consider the

Cook–Karp Thesis. The problem of determining membership of strings in a given language L is tractable if and only if $L \in \mathbf{P}$.

The evidence supporting the Cook–Karp thesis is much weaker than that supporting Church's thesis. Nevertheless, it has gained wide acceptance. Later, we shall discuss some of the reasons for this.

The following simple result is quite important.

Theorem 2.1. Let $L \in \mathbf{P}$, let f be a polynomial-time computable function on A^*, and let $Q = \{x \in A^* \mid f(x) \in L\}$. Then $Q \in \mathbf{P}$.

Proof. Let $\mathcal{M}$ accept L using a number of steps which is $O(|x|^r)$, and let $\mathcal{N}$ compute $f(x)$ in a number of steps which is $O(|x|^s)$. A Turing machine $\mathcal{R}$ that accepts Q is easily constructed that, in effect, first runs $\mathcal{N}$ on x to compute $f(x)$ and then runs $\mathcal{M}$ on $f(x)$ to determine whether $f(x) \in L$. Since a Turing machine cannot print more symbols in the course of a computation then there are steps in that computation, we have

$$|f(x)| \leq |x| + p(|x|), \qquad \text{where} \quad p(n) = O(n^s).$$

By Theorem 1.2, it follows that $|f(x)| = O(|x|^s)$. Hence, the number of steps required by $\mathcal{R}$ on input x is $O(|x|^{sr})$. ■

Theorem 2.2. Let f, g be polynomial-time computable functions, and let $h(x) = f(g(x))$. Then h is polynomial-time computable.

Proof. The proof is similar to that of the previous theorem. ■

It has turned out to be extremely difficult to prove that specific languages do not belong to **P**, although there are many likely candidates. An important example is the satisfiability problem discussed in Chapter 12. To make matters definite, we assume a set of atoms $\mathscr{A} = \{\alpha_2, \alpha_2, \ldots\}$, where subscripts are understood as in Section 1 of Chapter 12. We use the symbols

$$\alpha_1, \alpha_2, \ldots, \overline{\alpha}_1, \overline{\alpha}_2, \ldots$$

for the atoms and their negations, simply using concatenation for disjunction. Finally, we use the symbol / to begin a clause. Then, any string on the alphabet $C = \{\alpha, \overline{\alpha}, \mathsf{I}, /\}$ which begins / and in which / is *never* immediately followed by I, stands for a CNF formula (where in the interest of simplicity we are permitting empty and tautologous clauses and repetitions of literals in a clause). Thus the CNF formula

$$(p \vee q \vee \bar{r} \vee s) \wedge (\bar{q} \vee \bar{p} \vee \bar{r} \vee s) \wedge (\bar{q} \vee \bar{p} \vee \bar{r})$$

from Chapter 12 could be written as

$$/\alpha_1 \alpha_2 \overline{\alpha}_3 \alpha_4 / \overline{\alpha}_2 \overline{\alpha}_1 \overline{\alpha}_3 \alpha_4 / \overline{\alpha}_2 \overline{\alpha}_1 \overline{\alpha}_3 .$$

Any string in C^* which ends / or in which / is repeated represents a CNF formula which contains the empty clause, and hence is unsatisfiable.

Now, we write SAT for the language consisting of all elements of C^* that represent satisfiable CNF formulas. In spite of a great deal of attention to the question, it is still not known whether SAT $\in$ **P**. The starting point of the work on computational complexity that we discuss in this chapter is the observation that the situation changes entirely when we shift our attention from deterministic to nondeterministic computation. Nondeterministically one can discover very rapidly that a formula is satisfiable; it is necessary only that the satisfying assignment be "guessed." That is, instead of constructing an entire truth table, it suffices to construct a single row. To make these ideas precise, we have recourse to nondeterministic Turing machines as discussed in Chapter 6, Section 5.

Definition. A language L is said to belong to the class **NP** if there is a nondeterministic Turing machine $\mathscr{M}$ that accepts L, and a polynomial $p(n)$, such that for each $x \in L$, there is an accepting computation $\gamma_1, \gamma_2, \ldots, \gamma_m$ by $\mathscr{M}$ for x with $m \leq p(|x|)$.

We then have readily

Theorem 2.3. **P** $\subseteq$ **NP**. If $L \in$ **NP**, then L is recursive.

Proof. The first inclusion is obvious, since an ordinary Turing machine is a nondeterministic Turing machine.

For the rest, let $L \in \mathbf{NP}$, let $\mathcal{M}$ be a nondeterministic Turing machine which accepts L, with corresponding polynomial $p(n)$. We set γ_1 to be the configuration

$$\begin{array}{c} s_0 x \\ \uparrow \\ q_1 \end{array}$$

Next, by examining the quadruples of $\mathcal{M}$, we find all configurations γ_2 such that $\gamma_1 \vdash \gamma_2$. Continuing in this manner, we determine all possible sequences $\gamma_1, \gamma_2, \ldots, \gamma_m$ with $m \leq p(|x|)$ such that

$$\gamma_1 \vdash \gamma_2 \vdash \cdots \vdash \gamma_m .$$

Then, $x \in L$ if and only if at least one of these sequences is an accepting computation by $\mathcal{M}$ for x. This gives an algorithm for determining whether $x \in L$, and so, invoking Church's thesis, we conclude that L is recursive. (Methods like those used in Chapter 7 could be used to prove that L is recursive without using Church's thesis.) ■

In line with our discussion of the satisfiability problem viewed nondeterministically, we can prove

Theorem 2.4. SAT $\in$ **NP**.

Proof. Without providing all the rather messy details, we indicate how to construct a nondeterministic Turing machine $\mathcal{M}$ that accepts SAT.

$\mathcal{M}$ will begin by checking that a given input string $x \in C^*$ really does represent a CNF formula. Such a check requires only verifying that x begins with the symbol / and that no / is immediately followed by **|**. This can clearly be accomplished by $\mathcal{M}$ in a single pass over x, and therefore it can be done in $O(|x|)$ steps.

The remainder of the computation will involve successive passes over the string x in which truth values are assigned to literals, and clauses thus satisfied are labeled as being such. When a clause has been satisfied, the symbol / that introduces it is replaced by ! (so the fact that a clause still begins / indicates that it has not yet been satisfied). Also, when a literal $\alpha \mathbf{|}^{[i]}$ is assigned the value 1, all occurrences of the literal $\bar{\alpha} \mathbf{|}^{[i]}$ in clauses not yet satisfied will be replaced by $\varphi \mathbf{|}^{[i]}$ (so that they will not be assigned the value 1 in a subsequent pass). Likewise, when the literal $\alpha \mathbf{|}^{[i]}$ is assigned the value 0, all occurrences of that literal in clauses not yet satisfied will be replaced by $\varphi \mathbf{|}^{[i]}$.

We will speak of $\mathscr{M}$ as being in one of two modes: *search* or *update*. After verifying that the input string x does represent a CNF formula, $\mathscr{M}$ enters search mode. In search mode, $\mathscr{M}$ begins by finding the first occurrence of / remaining in x, starting from the left. If no / remains, then the formula has been satisfied and the computation halts. Otherwise, $\mathscr{M}$ has found an / and seeks to satisfy the clause that it heads. $\mathscr{M}$ scans the clause, moving to the right. When the symbol α or $\bar{\alpha}$ is encountered, $\mathscr{M}$ is scanning the first symbol of a literal $\alpha \mathbf{I}^{[i]}$ or $\bar{\alpha} \mathbf{I}^{[i]}$, as the case may be. $\mathscr{M}$ thus has the opportunity to satisfy the clause by making this literal true, assigning $\alpha \mathbf{I}^{[i]}$ the value 1 in the first case and 0 in the second. $\mathscr{M}$ *nondeterministically* decides whether to make this assignment. (This is the only respect in which $\mathscr{M}$ behaves nondeterministically.) If $\mathscr{M}$ does not make the assignment, then it continues its scan. If it reaches the end of the clause without having made an assignment, $\mathscr{M}$ enters an infinite loop. If $\mathscr{M}$ does make such an assignment, it enters update mode.

In update mode, $\mathscr{M}$ begins by marking the newly assigned literal, replacing α by ρ, or $\bar{\alpha}$ by $\bar{\rho}$, respectively. $\mathscr{M}$ then moves left to the / that begins the clause and replaces it by !. Finally, $\mathscr{M}$ moves to the right end of x, and then scans from right to left, checking all literals in subsequent clauses to see whether they match the newly assigned literal. This can be done by checking each block of $\mathbf{I}$s against the block that follows ρ (or $\bar{\rho}$). For literals that have been made true by the new assignment, the clause containing them is marked as satisfied, by replacing the / at its head by !. For literals that have been made false, the α or $\bar{\alpha}$ is replaced by φ. When the update is complete, $\mathscr{M}$ reenters search mode.

This completes the description of how $\mathscr{M}$ operates. It remains to estimate the number of steps that $\mathscr{M}$ requires for a successful computation. The number of steps between $\mathscr{M}$ entering and leaving each of search and update mode is clearly $O(|x|)$. Since this will happen no more than $|x|$ times, we conclude that the time for the entire computation is $O(|x|^2)$. ■

It is natural to ask whether the inclusion $\mathbf{P} \subseteq \mathbf{NP}$ is proper, i.e., whether there is a language L such that $L \in \mathbf{NP} - \mathbf{P}$. As we shall see, using the notion of **NP**-*completeness* to be defined below, it can be shown that if there were such a language, then it would follow that SAT $\in \mathbf{NP} - \mathbf{P}$. Unfortunately, this remains an open question.

Definition.[1] Let L, Q be languages. Then we write

$$Q \leq_p L,$$

[1] For a general discussion of reducibility, see Chapter 8.

and say that Q is *polynomial-time reducible* to L, if there is a polynomial-time computable function f such that

$$x \in Q \quad \Leftrightarrow \quad f(x) \in L.$$

Theorem 2.5. Let $R \leq_p Q$ and $Q \leq_p L$. Then $R \leq_p L$.

Proof. This follows at once from Theorem 2.2. ∎

Definition. A language L is called **NP**-*hard* if for every $Q \in$ **NP**, we have $Q \leq_p L$. L is called **NP**-*complete* if $L \in$ **NP** and L is **NP**-hard.

The significance of **NP**-completeness can be appreciated from the following result.

Theorem 2.6. If there is an **NP**-complete language L such that $L \in$ **P**, then **NP** = **P**.

Proof. We need to show that if $Q \in$ **NP**, then $Q \in$ **P**. Let $Q \subseteq A^*$. Since L is **NP**-hard, there is a polynomial-time computable function f such that

$$Q = \{x \in A^* \mid f(x) \in L\}.$$

The result now follows from Theorem 2.1. ∎

Intuitively, one can thus think of the **NP**-complete languages as the "hardest" languages in **NP**. As we shall see in the next section, SAT is **NP**-complete. Thus, if it should turn out that SAT $\in$ **P**, then every **NP**-complete problem would also be in **P**. It is considerations like these that have led to the tentative conclusion that **NP**-complete problems should be regarded as being intractable. To date, however, although very many problems are known to be **NP**-complete, there is no language known to be in **NP** − **P**, and it thus remains possible that **NP** = **P**.

Exercises

1. Show that Theorem 2.1 still holds when **P** is replaced by **NP**.
2. Show that if $\emptyset \subset L, M \subset A^*$ for some alphabet A, and if $L, M \in$ **P**, then $L \leq_p M$.
3. Show that $L \leq_p M$ does not necessarily imply that $M \leq_p L$.
4. Let $L, M \in$ **P** be languages on some alphabet A. Show that each of the following languages are in **P**: $A^* - L$, $L \cap M$, $L \cup M$.
5. Let $L, M \in$ **NP** be languages on some alphabet A. Show that each of the following languages are in **NP**: $L \cap M$, $L \cup M$.

6. Show that every regular language is polynomial–time decidable. [See Chapter 9.]

7. Show that every context-free language is polynomial–time decidable. [See Chapter 10.]

8. Give a language that is not polynomial–time decidable.

9. Give a function that is not polynomial–time computable.

10. Let A be an alphabet and set

$$\textbf{co-NP} = \{L \subseteq A^* \mid A^* - L \in \textbf{NP}\}.$$

Show that if there is a language L such that L is **NP**-complete and $L \in$ **co-NP**, then **NP** = **co-NP**.

11. Prove Theorem 2.3 without using Church's thesis.

12.* Let f be a total function on N, and let A be an alphabet. A total unary function $g(x)$ on A^* is computed in $\mathbf{DTIME}(f)$ if it is computed by some Turing machine that always runs in $\leq f(|x|)$ steps on input x. A language $L \subseteq A^*$ belongs to $\mathbf{DTIME}(f)$ if L is accepted by some Turing machine that runs in $\leq f(|x|)$ steps for every $x \in L$. L belongs to $\mathbf{NTIME}(f)$ if L is accepted by some nondeterministic Turing machine that has an accepting computation with $\leq f(|x|)$ steps for every $x \in L$. For languages $L, M \subseteq A^*$, we will write $L \leq_f M$ to indicate that there is a function g computable in $\mathbf{DTIME}(f)$ such that $x \in L$ if and only if $g(x) \in M$.

(a) Show that $\mathbf{P} = \bigcup_{n \geq 0} \mathbf{DTIME}(x^n)$.

(b) Show that $\mathbf{NP} = \bigcup_{n \geq 0} \mathbf{NTIME}(x^n)$.

(c) Prove that if $L \in \mathbf{DTIME}(x^2)$ and $M \leq_f L$, where $f(x) = x$, then $M \in \mathbf{DTIME}(4x^2 + x)$.

(d) Prove that if $L \in \mathbf{NTIME}(x^2)$ and $M \leq_f L$, where $f(x) = x$, then $M \in \mathbf{NTIME}(4x^2 + x)$.

(e) Let $f(x) = x^2$. Give a function $g(x)$ such that if $L \in \mathbf{DTIME}(x^2)$ and $M \leq_f L$, then $M \in \mathbf{DTIME}(g)$.

13.* A language L belongs to **EXPTIME** if there is a Turing machine $\mathscr{M}$ that accepts L and a polynomial $p(n)$ such that for every $x \in L$, $\mathscr{M}$ runs for no more than $2^{p(|x|)}$ steps.

(a) Let $\mathscr{N}$ be a nondeterministic Turing machine with k states. For a function $f(x)$, what is the maximum number of distinct computations that $\mathscr{N}$ can carry out in $\leq f(x)$ steps?

(b) Show that $\mathbf{NP} \subseteq \mathbf{EXPTIME}$. [See Exercise 1.6.]

14.* A language L belongs to **PSPACE** if there is a Turing machine $\mathscr{M}$ that accepts L and a polynomial $p(n)$ such that for every $x \in L$, $\mathscr{M}$

scans at most $p(|x|)$ different squares on its tape. L belongs to **NPSPACE** if there is a nondeterministic Turing machine $\mathcal{N}$ that accepts L and a polynomial $q(n)$ such that for every $x \in L$, $\mathcal{N}$ has some accepting computation in which at most $p(|x|)$ different tape squares are scanned.

(a) Show that **PSPACE** = **NPSPACE**.

(b) Show that **NP** $\subseteq$ **PSPACE**.

15.* (a) Let $\mathcal{M}$ be a Turing machine with states $q_1, \ldots, q_k$ and alphabet $\{s_1, \ldots, s_n\}$. How many distinct configurations of $\mathcal{M}$ are there with m tape squares?

(b) Show that **PSPACE** $\subseteq$ **EXPTIME**. [*Hint:* Use the pigeon-hole principle. See the discussion in Section 1 of Chapter 14.]

3. Cook's Theorem

We now prove the main theorem of this chapter.

Theorem 3.1 (Cook's Theorem). SAT is **NP**-complete.

Proof. Since we know, by Theorem 2.4, that SAT $\in$ **NP**, it remains to show that SAT is **NP**-hard. That is, we need to show that if $L \in$ **NP**, then $L \leq_{\mathrm{p}}$ SAT. Thus, let $L \in$ **NP**, and let $\mathcal{M}$ be a nondeterministic Turing machine that accepts L, with $p(n)$ the polynomial that furnishes a bound on the number of steps $\mathcal{M}$ requires to accept an input string. Without loss of generality, we assume that $p(n) \geq n$ for all n. We must show that there is a polynomial–time computable function that translates any input string u for $\mathcal{M}$ into a CNF formula δ_u such that u is accepted by $\mathcal{M}$ if and only if δ_u is satisfiable. For a given input u, let $t = p(|u|)$.

We know that if $\mathcal{M}$ accepts input u, it does so in $\leq t$ steps. Therefore, in order to determine whether $\mathcal{M}$ accepts u, we need only run it on u for at most t steps and check to see whether the final configuration is terminal. Since at each step of the computation, $\mathcal{M}$ can move at most one square to the left or right of the square currently being scanned, it follows that after t steps, the scanned square can be at most t squares to the left or t squares to the right of its original position. Since we have chosen the polynomial $p(n)$ so that $t \geq |u|$, for our present purposes it suffices to consider $2t + 1$ squares of tape. Thus, since we are considering only t steps of the computation, we can completely exhibit all of the information on $\mathcal{M}$'s tape, using a t by $(2t + 1)$ array (see Fig. 3.1).

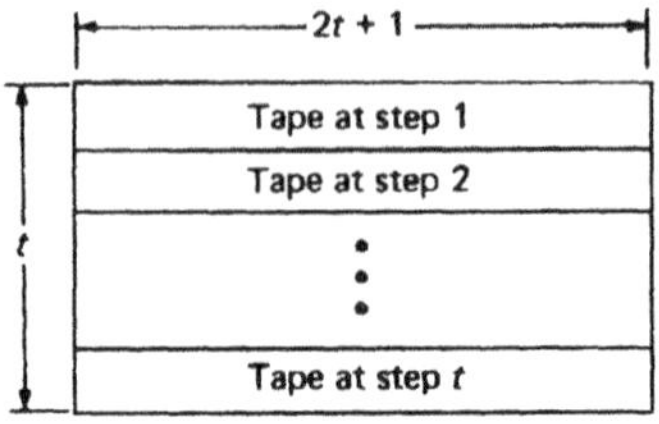

Figure 3.1

The first line of this array, corresponding to the initial tape contents, will then have the form

$$s_0^{[t+1]}us_0^{[t-|u|]}$$

where $\mathscr{M}$ begins in state q_1 scanning the $(t+1)$th symbol in this string, the s_0 immediately preceding u.

We will find it convenient, in this proof, to use the Turing machine model used in Theorem 4.2 in Chapter 6, in which acceptance of an input is by arrival in a unique accepting state q_m. We assume, therefore, that $\mathscr{M}$ is a Turing machine of this type. Let the set of states of $\mathscr{M}$ be $Q = \{q_1, q_2, \ldots, q_m\}$ and let the set of tape symbols be $S = \{s_0, s_1, \ldots, s_r\}$. It will simplify matters if we need to check only configuration number t to determine acceptance. Thus, we alter our definition of accepting computation to permit any number of repetitions of consecutive configurations; hence we may assume that our accepting computation consists of *exactly t* steps.

We will define a CNF formula δ_u that is satisfiable if and only if u is accepted by $\mathscr{M}$. Our set of atoms (each of length $O(t^2)$) will be

$$\mathscr{A} = \left\{ \rho_{h,j,k}, \sigma_{i,j,k} \mid 1 \leq h \leq m, 0 \leq i \leq r, 1 \leq j \leq 2t+1, 1 \leq k \leq t \right\}.$$

We first assume that u is accepted by $\mathscr{M}$, so that we have an accepting computation by $\mathscr{M}$ for u. We assume that the above t by $2t+1$ array has been constructed correspondingly. We will construct the CNF formula δ_u so that $\delta_u^v = 1$, where v is the assignment on $\mathscr{A}$ defined by

$$v(\rho_{h,j,k}) = \begin{cases} 1 & \text{if } \mathscr{M} \text{ is in state } q_h \text{ scanning the } j\text{th position at the} \\ & k\text{th step of the computation} \\ 0 & \text{otherwise,} \end{cases}$$

$$\tag{3.1}$$

$$v(\sigma_{i,j,k}) = \begin{cases} 1 & \text{if tape symbol } s_i \text{ is in the } j\text{th position of the } k\text{th} \\ & \text{row of the array} \\ 0 & \text{otherwise.} \end{cases}$$

In constructing δ_u, we will find the following abbreviation useful:

$$\nabla\{x_e \mid 1 \le e \le l\} = \bigwedge_{1 \le e < f \le l} (\neg x_e \vee \neg x_f) \wedge \bigvee_{1 \le e \le l} x_e,$$

where $\{x_e \mid 1 \le e \le l\}$ is a set of formulas. Thus,

$$\nabla\{x_e \mid 1 \le e \le l\} \tag{3.2}$$

is a formula whose value is TRUE (i.e., 1) under a given assignment if and only if *exactly* one of the formulas $x_1, x_2, \ldots, x_l$ has the value TRUE under that assignment. In the particular case that $x_1, x_2, \ldots, x_l$ are atoms, (3.2) is a CNF formula. We will need to calculate $|\nabla\{x_e \mid 1 \le e \le l\}|$ in this case. Formula (3.2) contains a clause consisting of two literals for each pair (e, f) with $1 \le e < f \le l$, followed by a single clause of l literals. Since there are $l(l-1)/2$ such pairs (e, f), and since in our notation, with / being used to separate clauses, each clause is of length 1 plus the number of its literals, we have

$$|\nabla\{x_e \mid 1 \le e \le l\}| = \left(\frac{l(l-1)}{2} \cdot 3 + (l+1)\right) \cdot O(t^2) = O(l^2 t^2).$$

Let us write

$$u = s_{u_1} s_{u_2} \cdots s_{u_z}, \qquad \text{where} \quad |u| = z.$$

We present a sequence of CNF formulas whose conjunction δ_u (which is then also a CNF formula) may be thought of as simulating the behavior of $\mathcal{M}$ in accepting u. Each of these formulas has the value TRUE under the assignment v. We precede each formula with an English sentence in quotes, which may be thought of as expressing a corresponding property of the accepting computation by $\mathcal{M}$ for u; each such sentence is intended to make it clear that the corresponding formula is indeed true under the assignment v. In some cases the formula as written will not be in CNF; in these cases the formula written is intended to stand for a formula in CNF obtained from it by using the methods of Chapter 12, Section 3.

(1) "The initial configuration has tape contents corresponding to the first row of the array, with $\mathcal{M}$ in state q_1 scanning the symbol s_0 immediately to the left of the first symbol of u."

$$\bigwedge_{0 < j \le t+1} \sigma_{0,j,1} \wedge \bigwedge_{0 < j \le z} \sigma_{u_j, t+j+1, 1} \wedge \bigwedge_{0 < j \le t-z} \sigma_{0, t+z+j+1, 1} \wedge \rho_{1, t+1, 1}.$$

This expression is clearly of length $O(t^3)$.

(2) "At each step of the computation there is a unique state and a unique scanned square."

$$\bigwedge_{1 \le k \le t} \nabla\{\rho_{h,j,k} \mid 1 \le h \le m, 1 \le j \le 2t+1\}.$$

By the preceding remarks, the length of this expression is $O(t^5)$.

(3) "Each entry of the array contains exactly one symbol."

$$\bigwedge_{1 \le k \le t} \; \bigwedge_{1 \le j \le 2t+1} \nabla\{\sigma_{i,j,k} \mid 0 \le i \le r\}.$$

r is a constant, so that this expression is of length $O(t^4)$.

(4) "Each configuration in the computation, after the first, is identical to the preceding configuration, or is obtained from it by applying one of the quadruples of $\mathscr{M}$."

This formula will be the most complicated. Let the quadruples of $\mathscr{M}$ be as follows:

$$\{q_{i_a}\ s_{j_a}\ s_{k_a}\ q_{l_a} \mid a = 1,2,\ldots,\bar{a}\}, \tag{3.3a}$$

$$\{q_{i_b}\ s_{j_b}\ R\, q_{l_b} \mid b = 1,2,\ldots,\bar{b}\}, \tag{3.3b}$$

$$\{q_{i_c}\ s_{j_c}\ L\, q_{l_c} \mid c = 1,2,\ldots,\bar{c}\}. \tag{3.3c}$$

To make the formula easier to understand, we write it in the form

$$\bigwedge_{1 \le k < t} \; \bigwedge_{1 \le j \le 2t+1} (\text{NOTHEAD}(j,k) \vee \text{IDENT}(j,k)$$

$$\vee A(j,k) \vee B(j,k) \vee C(j,k)),$$

where each of these five disjuncts will be explained below. It will turn out that each disjunct has length $O(t^2)$; hence we may conclude that the length of the entire formula will be $O(t^4)$.

We define

$$\text{NOTHEAD}(j,k) = \bigvee_{0 \le i \le r} (\sigma_{i,j,k} \wedge \sigma_{i,j,k+1}) \wedge \bigwedge_{1 \le h \le m} \neg\, \rho_{h,j,k}$$

so that $\text{NOTHEAD}(j,k)^v = 1$ for given j,k if and only if $\mathscr{M}$ is not scanning the jth position at the kth step of the computation.

Next we set

$$\text{IDENT}(j,k) = \bigvee_{1 \le h \le m} \; \bigvee_{0 \le i \le r} (\rho_{h,j,k} \wedge \sigma_{i,j,k} \wedge \rho_{h,j,k+1} \wedge \sigma_{i,j,k+1}),$$

so that IDENT$(j, k)^v = 1$ for given j, k if and only if $\mathcal{M}$ is scanning the jth position at both the kth and the $(k + 1)$th steps of the computation, and both the state and the symbol are the same in both of these configurations.

Next,

$$A(j,k) = \bigvee_{1 \le a \le \bar{a}} (\rho_{i_a, j, k} \wedge \sigma_{j_a, j, k} \wedge \sigma_{k_a, j, k+1} \wedge \rho_{l_a, j, k+1}),$$

where $A(j, k)^v = 1$ if and only if the $(k + 1)$th step results from the kth by one of the quadruples of (3.3a).

Similarly, we will define $B(j, k)$ so that $B(j, k)^v = 1$ if and only if the $(k + 1)$th step results from the kth by one of the quadruples of (3.3b). For $j \neq 2t + 1$, we can define

$$B(j,k) = \bigvee_{1 \le b \le \bar{b}} (\rho_{i_b, j, k} \wedge \sigma_{j_b, j, k} \wedge \sigma_{j_b, j, k+1} \wedge \rho_{l_b, j+1, k+1}).$$

This definition will not work for $j = 2t + 1$ because there are no atoms $\rho_{h, 2t+2, k}$. But since the computation cannot proceed beyond the boundaries of our array, it suffices to take $B(2t + 1, k)$ to be any unsatisfiable formula, e.g., the empty clause.

Finally, we will define $C(j, k)$ so that $C(j, k)^v = 1$ if and only if the $(k + 1)$th step results from the kth by one of the quadruples of (3.3c). For $j \neq 1$, we can define

$$C(j,k) = \bigvee_{1 \le c \le \bar{c}} (\rho_{i_c, j, k} \wedge \sigma_{j_c, j, k} \wedge \sigma_{j_c, j, k+1} \wedge \rho_{l_c, j-1, k+1}).$$

This definition will not work for $j = 1$ because there are no atoms $\rho_{h, 0, k}$. But since the computation cannot proceed beyond the boundaries of our array, it suffices to let $C(1, k)$ be any unsatisfiable formula, e.g., the empty clause.

(5) "The tth configuration is a terminal configuration." Equivalently, "At the tth step, $\mathcal{M}$ is in state q_m."

$$\bigvee_{1 \le j \le 2t+1} \rho_{m, j, t}.$$

This expression is clearly of length $O(t^3)$.

Now, we take δ_u to be simply the conjunction of the CNF formulas (1) through (5) above. It is clear from what has already been said that if $\mathcal{M}$ accepts u, then δ_u is satisfiable; in fact, $\delta_u^v = 1$.

Conversely, let v be an assignment such that $\delta_u^v = 1$. We will show that $\mathcal{M}$ accepts u. By (3), we see that for each $1 \leq j \leq 2t + 1$, $1 \leq k \leq t$, there is a unique i such that $v(\sigma_{i,j,k}) = 1$. Hence we can uniquely reconstruct our t by $2t + 1$ array. By (2), for each row of the array there is a unique state q_h and position j in the row such that $v(\rho_{h,j,k}) = 1$. Thus, each row can be made into a configuration of $\mathcal{M}$ so that (3.1) is satisfied. By (1), the configuration corresponding to the first row of the array is an initial configuration for $\mathcal{M}$ with input u. By (4), for each row of the array after the first, the corresponding configuration is identical to the preceding configuration or results from it using one of the quadruples of $\mathcal{M}$. Finally, by (5) the entire sequence of configurations constitutes an accepting computation by $\mathcal{M}$ for u. Thus, u is accepted by $\mathcal{M}$.

It remains to be shown that there is a polynomial-time computable function that maps each string u onto the corresponding CNF formula δ_u. Now, the CNF formulas of (2)–(5) do not depend on u, and a Turing machine can easily be constructed to write these on a tape in a number of steps proportional to the length of the expression, which, as we have seen, is $O(t^5)$, and hence polynomial in $|u|$. It remains to consider (1), which is a conjunction of atoms. Some of these atoms do not depend directly on u; producing this part of (1) simply involves writing $O(t^3)$ symbols. The remaining atoms of (1) correspond in a one-one manner to the symbols making up u; they can obviously be produced by a Turing machine in a number of steps proportional to $|u|$. This completes the proof. ■

Using Theorem 2.6, we have at once

Corollary 3.2. **P** = **NP** if and only if SAT $\in$ **P**.

Exercises

1. Let $\mathcal{M}$ be the Turing machine with the single tuple $q_1\ B\ a\ q_2$, and let u be the string a. Give δ_u for $t = 1$.
2. For any set $\mathcal{A}$ of atoms, show that the set of all propositional DNF formulas over $\mathcal{A}$ that are not tautologies is **NP**-complete.
3. For any set $\mathcal{A}$ of atoms, show that the set of all satisfiable propositional formulas over $\mathcal{A}$ is **NP**-complete.
4. The HALF-SAT problem is this: given a propositional CNF formula γ, determine if there is an assignment v on the atoms in $\gamma^v = 1$ and such that $\alpha^v = 1$ for exactly half of the atoms α in γ. [*Hint:* Show that SAT $\leq_p$ HALF-SAT. Given a CNF formula γ, create a new atom α' for each atom α in γ and add clauses of the form $\{\alpha, \alpha'\}$, $\{\bar{\alpha}, \overline{\alpha'}\}$.]

4. Other NP-Complete Problems

The principal technique for proving a problem to be **NP**-complete is given by the following result:

Theorem 4.1. Let Q be an **NP**-complete problem, and let $Q \leq_p L$. Then L is **NP**-hard.

Proof. Let R be any language such that $R \in$ **NP**. Since Q is **NP**-complete, we have $R \leq_p Q$. By Theorem 2.5, $R \leq_p L$. Thus, L is **NP**-hard. ∎

Corollary 4.2. Let Q be an **NP**-complete problem, let $L \in$ **NP**, and let $Q \leq_p L$. Then L is **NP**-complete.

Thus, once it has been shown that a problem is **NP**-complete, it can be used to show that other problems are **NP**-complete. In this way many problems have been shown to be **NP**-complete. It is this fact that constitutes the main evidence for regarding **NP**-complete problems as being intractable. Since the existence of a polynomial-time algorithm for even a single one of these problems would imply that there is a polynomial-time algorithm for every one of them, and, since it is argued that it is most unlikely that this could be the case without even one of these algorithms having been discovered, it is concluded that in all likelihood none of these problems have polynomial-time algorithms, and so they should all be regarded as intractable.

We will present a very small sample of this work, showing that a few problems are **NP**-complete. We begin with a restricted form of the satisfiability problem.

The 3-SAT problem is to determine whether a formula in CNF in which no clause contains more than three literals is satisfiable. We show that 3-SAT is **NP**-complete by showing that any CNF formula ζ can be transformed in polynomial time to a CNF formula ζ' containing at most three literals per clause such that ζ is satisfiable if and only if ζ' is satisfiable.

Theorem 4.3. 3-SAT is **NP**-complete.

Proof. Since 3-SAT is a special case of SAT, and SAT is in **NP**, it follows that 3-SAT is in **NP**. Let

$$/\alpha_1\alpha_2 \cdots \alpha_k, \qquad k \geq 4, \tag{4.1}$$

be any one of the clauses of ζ containing more than three literals. Let $\beta_1, \beta_2, \ldots, \beta_{k-3}$ be atoms which do not appear in ζ. We construct ζ' by replacing (4.1) by the conjunction

$$/\alpha_1 \alpha_2 \beta_1 / \alpha_3 \bar{\beta}_1 \beta_2 / \alpha_4 \bar{\beta}_2 \beta_3 / \cdots / \alpha_{k-2} \bar{\beta}_{k-4} \beta_{k-3} / \alpha_{k-1} \alpha_k \bar{\beta}_{k-3} .$$

It is easy to see that ζ is satisfiable if and only if ζ' is satisfiable. Moreover, since the length of ζ' is bounded by a constant times the length of ζ, the transformation can be performed in linear time. ■

It is interesting that there are problems which superficially appear to be unrelated, but between which we can readily find a polynomial-time transformation. Our next example is known as the COMPLETE-SUBGRAPH problem. A *graph* G consists of a finite nonempty set of *vertices* $V = \{v_1, \ldots, v_n\}$ and a finite set of *edges* E. Each edge is a pair of vertices. The *size* of the graph is simply the number of vertices it contains. A *subgraph* of a graph $G = (V, E)$ is a graph $G' = (V', E')$ where $V' \subseteq V$, and $E' \subseteq E$. A graph $G = (V, E)$ is *complete* if there is an edge in E between every pair of distinct vertices in V.

The COMPLETE-SUBGRAPH problem is this: given a graph and a number k, does the graph have a complete subgraph of size k?

Theorem 4.4. COMPLETE-SUBGRAPH is **NP**-complete.

Proof. We show informally that COMPLETE-SUBGRAPH is in **NP**. Let the number k and a list of the vertices and of the edges of the given graph be written on the tape of a Turing machine in any reasonable notation. The procedure begins by nondeterministically selecting a vertex and then decrementing k. By continuing this process until k has been decremented to 0, a list of k vertices is obtained. The procedure then tests [in time $O(k^2)$] whether the graph has a complete subgraph in those vertices. Since

$$k \le n \le \text{length of the string representing } G \text{ on the tape,}$$

where G is the given graph, this shows that COMPLETE-SUBGRAPH $\in$ **NP**.

To show that COMPLETE-SUBGRAPH is **NP**-hard, we show that SAT $\le_{\text{p}}$ COMPLETE-SUBGRAPH. Thus, we must show how to map each CNF formula γ into a pair consisting of a number k and a graph G so that γ is satisfiable if and only if G has a complete subgraph of size k. If $\gamma = /\gamma_1/\gamma_2 \cdots /\gamma_k$ is a CNF formula, where $\gamma_1, \gamma_2, \ldots, \gamma_k$ are clauses, then we take the number k to be simply the number of clauses in γ and construct the graph $G = (V, E)$, where

$$V = \{(\alpha, i) \mid \alpha \text{ is a literal in } \gamma_i\},$$

$$E = \{((\alpha, i), (\beta, j)) \mid \alpha \ne \neg \beta \text{ and } i \ne j\}.$$

Thus we have a vertex for each occurrence of each literal in γ. Edges join pairs of vertices that represent literals in different clauses provided one is not the negation of the other. This means that these literals can both be assigned the value "TRUE" at the same time. If γ is satisfiable, there is some way to assign truth values to the atoms so that γ evaluates to "TRUE." Thus at least one literal of each clause of γ must be assigned the value "TRUE," and in G there will be an edge connecting each pair of "true literals." This means that the nodes of G corresponding to the "true literals" of γ form a complete subgraph of size k. Conversely, if γ contains a complete subgraph of size k, then since edges join pairs of literals in different clauses that can be true at the same time, there is a way to make each clause of γ true at the same time. Thus γ is satisfiable. Furthermore, G can clearly be obtained from γ by a polynomial–time computable function. ■

A *clique* in a given graph is a maximal complete subgraph of that graph; that is, a clique is a complete subgraph of a given graph that is *not* a subgraph of any other complete subgraph of that graph. The MAX-CLIQUE problem is to find the size of the largest clique in a given graph. Of course, in this form, MAX-CLIQUE is not a language but rather a function, and so it does not make sense in terms of our definitions to ask whether it is in **NP**. However, since removing a vertex and all edges containing it from a complete subgraph yields another complete subgraph, we see that any algorithm for the MAX-CLIQUE problem that could actually be implemented using reasonable resources could easily be transformed into an equally usable algorithm for the COMPLETE-SUBGRAPH problem. Hence, to the extent that **NP**-completeness can be regarded as implying intractability, we are entitled to conclude that MAX-CLIQUE is likewise intractable.

We next consider a closely related graph-theoretic problem, known as VERTEX-COVER. A set S is a *vertex cover* for a graph $G = (V, E)$ if $S \subseteq V$ and for every $(x, y) \in E$, either $x \in S$ or $y \in S$. The VERTEX-COVER problem is to determine for a given graph G and integer k whether G has a vertex cover of size k.

Theorem 4.5. Let $G = (V, E)$ be a graph and let

$$E' = \{(x, y) \mid x, y \in V, x \neq y, \text{and } (x, y) \notin E\}.$$

Let us consider the graph $G' = (V, E')$ (sometimes called the *complement graph* of G). Then $S \subseteq V$ is the set of vertices of a complete subgraph of G if and only if $V - S$ is a vertex cover in G'.

Proof. Let S be the set of vertices of a complete subgraph of G. Then, by definition, for any $(x, y) \in E'$, either $x \in V - S$ or $y \in V - S$. Thus, $V - S$ is a vertex cover of G'. Conversely, if $V - S$ is a vertex cover of G', then for any $(x, y) \in E'$, either $x \in V - S$ or $y \in V - S$. Thus no edge of G' connects two vertices in S. Thus for every $u, v \in S$, $u \neq v$, we have $(u, v) \in E$, and so S is the set of vertices of a complete subgraph of G. ■

Corollary 4.6. VERTEX-COVER is **NP**-complete.

The SET-COVER problem is to determine for a family of sets $\Delta = \{S_1, S_2, \ldots, S_n\}$, and number k, whether there exists a subfamily Γ of Δ of size k, $\Gamma = \{S_{m_1}, S_{m_2}, \ldots, S_{m_k}\}$, such that

$$\bigcup_{i \leq n} S_i = \bigcup_{j \leq k} S_{m_j}.$$

Corollary 4.7. SET-COVER is **NP**-complete.

Proof. Let $G = (V, E)$ be a graph with $V = \{v_1, v_2, \ldots, v_n\}$. For $i = 1, 2, \ldots, n$, let

$$S_i = \{(v_i, v_j) \mid (v_i, v_j) \in E\} \cup \{(v_j, v_i) \mid (v_i, v_j) \in E\}.$$

Clearly $\Gamma = \{S_{i_1}, S_{i_2}, \ldots, S_{i_k}\}$ is a set cover for $\Delta = \{S_1, S_2, \ldots, S_n\}$ if and only if $\{v_{i_1}, v_{i_2}, \ldots, v_{i_k}\}$ is a vertex cover for G. ■

Many hundreds of **NP**-complete problems have been identified in quite diverse areas. We conclude this section with a few more examples. For each we indicate in brackets the nature of some known proof of **NP**-hardness.

1. HAMILTONIAN-CIRCUIT (HC): given a graph $G = (V, E)$ with k vertices, determine if there is an ordering $v_1, \ldots, v_k$ of the vertices in V such that $(v_i, v_{i+1}) \in E$, $1 \leq i < k$, and $(v_k, v_1) \in E$. [VERTEX-COVER $\leq_{\text{p}}$ HC.]
2. 3-DIMENSIONAL-MATCHING (3DM): given a set $S \subseteq A \times B \times C$, where A, B, C are disjoint finite sets each with q elements, determine if there is a subset $M \subseteq S$ with q elements such that for any (a, b, c), $(a', b', c') \in M$, $a \neq a'$, $b \neq b'$, and $c \neq c'$. [3SAT $\leq_{\text{p}}$ 3DM.]
3. PARTITION: given a set $A = \{a_1, \ldots, a_n\}$ of positive integers, determine if there is a subset $S \subseteq A$ such that $\Sigma_{a \in S}\, a = \Sigma_{a \in A - S}\, a$. [3DM $\leq_{\text{p}}$ PARTITION.]
4. INTEGER-PROGRAMMING (IP): given a finite set

$$X = \{((x_1^i, \ldots, x_n^i), z_i) \mid 1 \leq i \leq m\},$$

where all x_j^i, z_i are integers, and given a tuple $(c_1, \ldots, c_n)$ of integers and an integer b, determine if there is a tuple $(y_1, \ldots, y_n)$ of integers such that $(x_1^i, \ldots, x_n^i) \cdot (y_1, \ldots, y_n) \le z_i$, $1 \le i \le n$, and $(c_1, \ldots, c_n) \cdot (y_1, \ldots, y_n) \ge b$. (The dot product of any two n-tuples is defined $(x_1, \ldots, x_n) \cdot (y_1, \ldots, y_n) = \sum_{i=1}^{n} x_i \cdot y_i$.) [3SAT $\le_p$ IP.]

5. QUADRATIC-DIOPHANTINE-EQUATIONS (QDE): given positive integers a, b, c, determine if there are positive integers x, y such that $ax^2 + by = c$. [3SAT $\le_p$ QDE.]
6. STRAIGHTLINE-PROGRAM-INEQUIVALENCE (SPI): given a set of variables $\{X_1, \ldots, X_n\}$, two programs $\mathscr{P}$, $\mathscr{Q}$ each being a sequence of assignments of the form

$$V \leftarrow \text{IF } W = X \text{ THEN } Y \text{ ELSE } Z,$$

where $V, W, X, Y, Z \in \{X_1, \ldots, X_n\}$, and given a set of values $\{v_1, \ldots, v_m\}$, determine if there is an initial state

$$\{X_1 = v_{i_1}, \ldots, X_n = v_{i_n}\},$$

where each $v_{i_j} \in \{v_1, \ldots, v_m\}$, such that $\mathscr{P}$, $\mathscr{Q}$ end with a different value for some variable. [3SAT $\le_p$ SPI.]

Exercises

1. The CHROMATIC-NUMBER problem is to determine for a given graph $G = (V, E)$ and integer k whether there is a function f from V to $\{1, 2, \ldots, k\}$ such that if $(x, y) \in E$, then $f(x) \ne f(y)$. (Intuitively, this problem amounts to determining whether or not it is possible to "color" the vertices of G using k colors in such a way that no two adjacent vertices are colored the same.) Show that CHROMATIC-NUMBER is **NP**-complete. [*Hint:* Show 3-SAT $\le_p$ CHROMATIC-NUMBER.] [*Further hint:* Assume $\gamma = /\gamma_1/\gamma_2 \cdots /\gamma_m$ is a CNF formula such that no γ_i contains more than three literals. Assume there are n atoms $\alpha_1, \alpha_2, \ldots, \alpha_n$ that appear either negated or unnegated in γ. Construct a graph G with $3n + m$ vertices such that G is $n + 1$ colorable if and only if γ is satisfiable.]

2. The 2-COLORABILITY problem is to determine whether a given graph can be colored using only two colors. Show that 2-COLORABILITY is in **P**.

3. The 2-SAT problem is to determine whether a CNF formula in which no clause contains more than two literals is satisfiable. It is known that 2-SAT $\in$ **P**. Show why a technique like the one used to show

3-SAT is **NP**-complete does not work for 2-SAT. Show that 2-SAT is in **P**.

4. The EXACT-COVER problem is to determine for a finite family of sets $\Delta = \{S_1, S_2, \ldots, S_n\}$ whether there exists a set cover Γ of Δ such that the elements of Γ are pairwise disjoint. Show that EXACT-COVER is **NP**-complete. [*Hint:* Show that

$$\text{CHROMATIC-NUMBER} \le_{\text{p}} \text{EXACT-COVER.}]$$

5. The SUBGRAPH-ISOMORPHISM (SI) problem is, given graphs $G_1 = (V_1, E_1)$, $G_2 = (V_2, E_2)$, to determine if there is a one–one function f from V_1 to V_2 such that $(v_i, v_j) \in E_1$ if and only if $(f(v_i), f(v_j)) \in E_2$. Show that SUBGRAPH-ISOMORPHISM is **NP**-complete. [*Hint:* Show COMPLETE-SUBGRAPH $\le_{\text{p}}$ SI.]

6. The LONGEST-COMMON-SUBSEQUENCE (LCS) problem is, given an alphabet A, a set $\{w_1, \ldots, w_n\}$ of strings on A, and a positive integer k, to determine if there is a string $y \in A^*$ with $|y| \ge k$ such that, for $1 \le i \le n$, $w_i = x_0 y_1 x_1 y_2 x_2 \cdots y_l x_l$ and $y = y_1, \ldots, y_l$ for some $x_0, \ldots, x_l, y_1, \ldots, y_l \in A^*$. Show that LCS is **NP**-complete. [*Hint:* Show VERTEX-COVER $\le_{\text{p}}$ LCS. Let $G = (V, E)$ be a graph, where $V = \{v_1, \ldots, v_n\}$ and $E = \{(v_{i_1}, v_{j_1}), \ldots, (v_{i_m}, v_{j_m})\}$, where $i_l \le j_l$, $1 \le l \le m$. For each edge (v_{i_l}, v_{j_l}), create the string

$$w_l = v_1 \cdots v_{i_l - 1} v_{i_l + 1} \cdots v_n v_1 \cdots v_{j_l - 1} v_{j_l + 1} \cdots v_n,$$

and also create the string $u = v_1 \cdots v_n$. Show that G has a vertex cover of size k if and only if $\{w_1, \ldots, w_m, u\}$ has a common subsequence of size $n - k$.]

7. The TRAVELING-VENDOR (TV) problem is, given a set $C = \{c_1, \ldots, c_n\}$ of cities, a positive integer distance $d(c_i, c_j)$ for each pair of cities, and a positive integer b, to determine if there is a Hamiltonian circuit $\langle c_{i_1}, \ldots, c_{i_m} \rangle$ such that

$$\sum_{j=1}^{m-1} d(c_{i_j}, c_{i_{j+1}}) + d(c_{i_m}, c_{i_1}) \le b.$$

Show that TV is **NP**-complete. [*Hint:* show HC $\le_{\text{p}}$ TV.]

8. The SUBSET-SUM problem is, given a set $\{a_1, \ldots, a_n\}$ of positive integers and positive integer b, to determine if there is a subset $\{b_1, \ldots, b_m\} \subseteq \{a_1, \ldots, a_n\}$ such that $\sum_{i=1}^{m} b_i = b$. Show that SUBSET-SUM is **NP**-complete. [*Hint:* Show PARTITION $\le_{\text{p}}$ SUBSET-SUM.]

9. The KNAPSACK problem is, given a set $S = \{(s_1, v_1), \ldots, (s_n, v_n)\}$ of pairs of positive integers, where s_i is a size and v_i is a value, $1 \le i \le n$, and given positive integers b, k, to determine if there is a subset $A \subseteq S$ such that

$$\sum_{(s,v)\in A} s \le b \quad \text{and} \quad \sum_{(s,v)\in A} v \ge k.$$

Show that KNAPSACK is **NP**-complete. [*Hint:* Show PARTITION $\le_{\mathrm{p}}$ KNAPSACK.]

10. The MULTIPROCESSOR-SCHEDULING (MS) problem is, given a set $T = \{t_1, \ldots, t_n\}$ of positive integers (task times) and positive integers m (number of processors) and d (deadline), to determine if there is a partition of T into disjoint sets $T_1, \ldots, T_m$ such that for $1 \le i \le m$, $\sum_{t \in T_i} t \le d$. Show that MS is **NP**-complete. [*Hint:* Show PARTITION $\le_{\mathrm{p}}$ MS.]

11. The RECORD-ALLOCATION (RA) problem is, given a set $L = \{l_1, \ldots, l_n\}$ of positive integers (record lengths) and positive integers t (track length) and k (number of tracks), to determine if there is a partition of L into disjoint sets $L_1, \ldots, L_k$ such that for $1 \le i \le k$, $\sum_{l \in L_i} l \le t$. Show that RA is **NP**-complete. [*Hint:* Show PARTITION $\le_{\mathrm{p}}$ RA.]

12. The TASK-SEQUENCING (TS) problem is, given a set

$$\{(t_1, d_1, p_1), \ldots, (t_n, d_n, p_n)\}$$

of triples of positive integers (where for $1 \le i \le n$, t_i is the amount of time necessary to complete task i, d_i is the deadline for task i, and p_i is the penalty for failing to complete task i by its deadline) and given a positive integer b, to determine if there is a sequence $\langle i_1, \ldots, i_n \rangle$ of tasks such that $\sum_{i \in L} p_i \le b$, where $L \subseteq \{1, \ldots, n\}$ is the set of late tasks, i.e., those i_j with $\sum_{r=1}^{j} t_{i_r} > d_{i_j}$. Show that TS is **NP**-complete. [*Hint:* Show PARTITION $\le_{\mathrm{p}}$ TS.]

Part 5

Semantics

16

Approximation Orderings

1. Programming Language Semantics

In Part 1 of this book we studied various classes of functions, principally the class of partially computable functions. In Part 5 we also investigate classes of functions but from a different perspective. One of the key results from Part 1 is that the partially computable functions can be defined by way of any number of substantially different formalisms. Once the equivalence of $\mathscr{S}$ programs, Turing machines, etc., has been demonstrated, it becomes clear that the definition of partially computable functions in terms of $\mathscr{S}$ programs is an artifact of our particular exposition, and in results like Theorem 2.1 of Chapter 4 concerning the HALT predicate, the role of $\mathscr{S}$ programs recedes to the formal background. In direct proportion to the accumulation of equivalent formal systems, the class of partially computable numeric functions takes on an independent, absolute status, and the status of each particular formal system declines. It is fair to say that computability is about a certain class of functions, however they are defined.

On the more practical side of computer science, however, the formal description of functions has blossomed into the elaborate field of programming languages, where we find thousands of formal systems far richer and more complex than any we describe in this book. The differences between the two fields are entirely appropriate. In the *theory* of computation, where we are interested in the abstract mathematical properties of functions and classes of functions, it is appropriate to eliminate all but the most essential components of our formal systems. In the areas of computer science that support the *practice* of solving problems, it is appropriate to elaborate a wide range of programming languages to support the needs of various problem domains, programming styles, and philosophies of language design.

When the various programming languages are important in their own right, the business of associating a function to each program is more than a means to the end of defining a class of functions. It becomes the subject of *programming language semantics.*

The issue came up already in Chapter 2 when we carefully defined a semantics for $\mathscr{S}$ programs, associating a partial function $\psi_{\mathscr{P}}$ with each program $\mathscr{P}$. In the course of defining the semantics of $\mathscr{S}$ programs we defined the notion of a computation, which characterizes a mechanical process of deriving a numerical output value from an input value. In essence we have defined an abstract machine that stores the values of the X, Y, and Z variables and updates those values by performing various operations as specified by a program. A computation describes the sequence of states assumed by the machine in the process of deriving an output (if there is one), and the meaning of a program is characterized by all of the computations it performs, one for each possible input. This style of defining the meaning of programs is called *operational semantics* because it depends on the operation of some kind of machine.

It is clear that programs and computations are very different sorts of objects. Computations are dynamic in nature; that is, they describe a process that evolves over time. Without the benefit of a semantics to give them meaning, programs are simply static sequences of syntactic symbols. Now, the goal of creating a semantics is to associate a function with a program, and at least from the perspective of set theory, a function is a static entity: it is simply a set of ordered pairs. We could argue, then, that it is a diversion to interpose the conceptual complication of computations between a program and its function. It would be more straightforward to define a function, by purely "mathematical" means, directly from the syntactic structure of the program. Of course, the concept of "purely 'mathematical' means" is not precisely defined. In the present context it implies, at the very least, an absence of operational detail. Its connotation will become clearer as we proceed. This alternative to operational semantics is called *denotational semantics*.

The denotational approach might be preferable for its conceptual simplicity, but we do not mean to imply that it is "better" than, or a replacement for, an operational semantics. In a practical setting the two are complementary. A denotational semantics can provide a succinct description of the meaning of a programming language, abstracted from the level of pragmatic details, and an operational semantics approaches more closely an actual implementation. In the theoretical area of computability, the operational style is crucial, preceding the denotational style both historically and conceptually. It is the mechanical nature of an operational semantics that gives sense to the term *computable function*. If computability theory is more about the class of partially computable functions than the particular formal systems for defining them, it is just as much about the concept of mechanical computation which is embodied in the operational semantics of $\mathscr{S}$ programs, Turing machines, Pascal, LISP, etc.

The exposition given here of semantics, both operational and denotational, has two goals.

1. It should broaden and deepen the understanding of computable functions and computation.
2. It is an introduction to some of the ideas found in the theoretical study of programming languages.

There are two ways in which we will extend the theory of computable functions covered in Part 1. One is to expand the class of data objects that are directly covered by the theory. We have accounted for computable functions on the natural numbers and computable functions on strings over arbitrary finite alphabets, but the typical programming language offers a much greater variety of data types like lists, arrays, and in many cases, user-defined data types. Now, natural numbers and strings are both perfectly appropriate data types for a theory of computable functions because of their capacity for encoding more complex structures. We showed in Chapter 3, for example, how finite lists of numbers can be encoded as a single Gödel number. So a theorem like the universality theorem in Chapter 4 implicitly tells us that there is a partially computable universal function for partially computable functions on lists of numbers. However, by explicitly admitting a richer assortment of data types, we can bring the theory closer to the actual practice of computation.

In every model of computation we covered in Part 1, a function computed by some program or machine was considered to be defined for a given input just when there was a finite computation for that input. Indeed, in Chapter 2 we defined a computation as a *finite* sequence of instantaneous descriptions. When a given input leads to an infinite sequence of instantaneous descriptions, we did not consider that sequence a computation, and so we did not consider the program to be doing any useful work in this case. However, there are programs whose sole reason for being is the work they accomplish while they are running, rather than the output they produce at the end of a computation. For example, an operating system, the program that organizes the functioning of a computer, produces little useful output when it is terminated, and, in fact, it is designed to be able to run without ever terminating. The termination of an operating system might just as well indicate a failure rather than the successful completion of a computation.

Our second extension, then, is to try to account for the work done in the course of a computation. Our perspective here is that instantaneous descriptions represent partial results that *approximate* the overall result of a computation. We will admit the possibility of infinite computations as well. In a sense, the result of an infinite computation is the *computation itself*, and each instantaneous description is a piece of the result, approximating the whole. Our data structures, therefore will come equipped with an ordering, an *approximation ordering*, which formalizes the notion of different partial results being more or less complete realizations of the total result.

We will focus here on the equational style of function definition. In particular, we will work with equations like

$$F(X) = H(G(X), X). \tag{1.1}$$

There are two kinds of variables in Eq. (1.1). X is intended to denote individuals, say, natural numbers, and F, G, and H denote functions. There is another important difference in our interpretation of these variables. In the equation $2x = x + x$, with x ranging over the natural numbers, equality holds for all values of x, but the equation $x^2 - 4 = 0$ calls for one or more particular values of x which make equality hold. In (1.1) X has the first interpretation, and F, G, and H have the second. That is, we are looking for functions f, g, and h that make $f(X) = h(g(X), X)$ true for all values of X, in which case the assignment of f, g, and h to variables F, G, and H constitutes a solution to (1.1). In equational programming we define functions by writing sets of equations to be satisfied.

Normally equations are understood to be symmetric in their left and right sides, but the two sides of (1.1) have an important distinction. If F is assigned a function there might be a number of assignments for G and H that solve (1.1), but assigning functions to G and H induces a unique value for F. In this sense we can interpret the right side as a function (sometimes called a *higher order function*) which takes any pair of functions g, h assigned to G and H and yields a unique function for F. We will make essential use of such higher order functions in the denotational semantics of recursion equations.

We need to consider the issue of partial functions. In the simple equation

$$F(X) = G(X), \tag{1.2}$$

if F and G are interpreted as total functions f and g, then there is no ambiguity in the requirement that $f(x) = g(x)$ for all values of x. Suppose, though, that f and g are defined

$$f(x) = g(x) = \begin{cases} 0 & \text{if } x > 0 \\ \text{undefined} & \text{if } x = 0. \end{cases}$$

What should be the meaning of $f(0) = g(0)$? Previously we have interpreted equality to mean

1. either both sides are defined or both sides are undefined, and
2. when both sides are defined they have the same value.

An alternative, which we will now adopt, is to extend the universe of objects with a new element that represents the property of being undefined. For instance, we extend N to $N_\perp = N \cup \{\perp_N\}$, where $\perp_N$ (pronounced "bottom") is different from all natural numbers. Now we can define

$$f_\perp(x) = g_\perp(x) = \begin{cases} 0 & \text{if } x > 0 \\ \perp_N & \text{if } x = 0. \end{cases}$$

f and g are not total functions, but $f_\perp$ and $g_\perp$ are, and $f_\perp(x) = g_\perp(x)$ for all $x \in N$.

There are two distinct kinds of elements in $N_\perp$: numbers and $\perp_N$. We can compare them by saying that numbers are completely defined elements and $\perp_N$ is the unique completely undefined element. This is a simple example of our notion of approximation. In a sense $\perp_N$ is an approximation, a very weak one, to any natural number n. The idea is clearer, perhaps, in a richer set like $N_\perp^3$, where we can say that $(\perp_N, \perp_N, \perp_N)$ and $(3, \perp_N, 5)$ both approximate $(3, 7, 5)$, and that $(3, \perp_N, 5)$ is a better approximation than $(\perp_N, \perp_N, \perp_N)$. Thus, we can think of a sequence like $(\perp_N, \perp_N, \perp_N)$, $(3, \perp_N, 5)$, $(3, 7, 5)$ as a computation, where $(\perp_N, \perp_N, \perp_N)$ and $(3, \perp_N, 5)$ are partial results leading to the final value $(3, 7, 5)$.

In this chapter we investigate the mathematical aspects of approximation orderings and functions defined on them. In the next chapter we apply these ideas to the semantics of recursion equations.

2. Partial Orders

For a set D, a *binary relation on D* is any subset of $D \times D$. If R is a binary relation on some set, we generally write $a\,R\,b$ to mean $(a, b) \in R$. If R is a binary relation on D and $E \subseteq D$, then the binary relation $\{(a,b) \in E \times E \mid a\,R\,b\}$ on E is the *restriction of R to E*.

Definition. Let D be a set and $\sqsubseteq$ a binary relation on D. $\sqsubseteq$ is a *partial ordering of D* if it has the properties of

1. reflexivity: $a \sqsubseteq a$ for all $a \in D$;
2. antisymmetry: $a \sqsubseteq b$ and $b \sqsubseteq a$ implies $a = b$ for all $a, b \in D$;
3. transitivity: $a \sqsubseteq b$ and $b \sqsubseteq c$ implies $a \sqsubseteq c$ for all $a, b, c \in D$.

If $\sqsubseteq$ is a partial ordering of D, then the pair $(D, \sqsubseteq)$ is a *partially ordered set*, or simply a *partial order*. We will sometimes write $a \sqsubset b$ to mean $a \sqsubseteq b$ and $a \neq b$.

It is easy to find examples of partial orders. $(N, \leq)$, where $\leq$ is the usual ordering of N, is a partial order. $(N_\perp, \sqsubseteq_{N_\perp})$, where $\sqsubseteq_{N_\perp}$ is defined

$$m \sqsubseteq_{N_\perp} n \quad \text{if and only if} \quad m = \perp_N \text{ or } m = n,$$

is also a partial order. Note that while $5 \leq 7$, for example, it is not true that $5 \sqsubseteq_{N_\perp} 7$. If D is a set, the *power set* of D, denoted $\mathscr{P}(D)$, is the set of all subsets of D. For any set D, it is easy to see that $(\mathscr{P}(D), \subseteq_{\mathscr{P}(D)})$ is a partial order, where $\subseteq_{\mathscr{P}(D)}$ is the subset relation on the sets in $\mathscr{P}(D)$. Also, $(D, =_D)$ is a partial order for any set D, where $=_D$ is the equality relation on the elements of D. Although $\leq$, $\sqsubseteq_{N_\perp}$, $\subseteq_{\mathscr{P}(N)}$, and $=_N$ are all partial orderings, they are quite different in structure.

Definition. A partial ordering $\sqsubseteq$ of a set D is a *linear ordering of D* if for every $a, b \in D$, either $a \sqsubseteq b$ or $b \sqsubseteq a$. $(D, \sqsubseteq)$ is a *linearly ordered set*, or simply a *linear order*.

So, for example, $(N, \leq)$ is a linear order, but $(N_\perp, \sqsubseteq_{N_\perp})$, $(N, =_N)$ and $(\mathscr{P}(N), \subseteq_{\mathscr{P}(n)})$ are not.

We will often find it useful to create new partial orders from given partial orders.

Definition. Let $(D_1, \sqsubseteq_{D_1}), \ldots, (D_n, \sqsubseteq_{D_n})$ be partial orders. Then $\sqsubseteq_{D_1 \times \cdots \times D_n}$, the *Cartesian product ordering* on $D_1 \times \cdots \times D_n$ determined by $\sqsubseteq_{D_1}, \ldots, \sqsubseteq_{D_n}$, is defined

$$(d_1, \ldots, d_n) \sqsubseteq_{D_1 \times \cdots \times D_n} (e_1, \ldots, e_n) \text{ if and only if } d_i \sqsubseteq_{D_i} e_i \text{ for all } 1 \leq i \leq n.$$

Theorem 2.1. If $(D_1, \sqsubseteq_{D_1}), \ldots, (D_n, \sqsubseteq_{D_n})$ are partial orders, then $(D_1 \times \cdots \times D_n, \sqsubseteq_{D_1 \times \cdots \times D_n})$ is a partial order.

Proof. We will write $\sqsubseteq$ for $\sqsubseteq_{D_1 \times \cdots \times D_n}$ and $\sqsubseteq_i$ for $\sqsubseteq_{D_i}$, $1 \leq i \leq n$. We need to show that $\sqsubseteq$ is reflexive, symmetric, and transitive. For any $(d_1, \ldots, d_n) \in D_1 \times \cdots \times D_n$, we have $d_i \sqsubseteq_i d_i$ by the reflexivity of $\sqsubseteq_i$, $1 \leq i \leq n$, so $(d_1, \ldots, d_n) \sqsubseteq (d_1, \ldots, d_n)$. If $(d_1, \ldots, d_n) \sqsubseteq (e_1, \ldots, e_n)$ and $(e_1, \ldots, e_n) \sqsubseteq (d_1, \ldots, d_n)$, then, for $1 \leq i \leq n$, $d_i \sqsubseteq_i e_i$ and $e_i \sqsubseteq_i d_i$, which implies $d_i = e_i$ by the antisymmetry of $\sqsubseteq_i$, so we have $(d_1, \ldots, d_n) = (e_1, \ldots, e_n)$. Finally, if $(d_1, \ldots, d_n) \sqsubseteq (e_1, \ldots, e_n)$ and $(e_1, \ldots, e_n) \sqsubseteq (f_1, \ldots, f_n)$, then, for $1 \leq i \leq n$, $d_i \sqsubseteq_i e_i$ and $e_i \sqsubseteq_i f_i$, which implies $d_i \sqsubseteq_i f_i$ by the transitivity of $\sqsubseteq_i$, so we have $(d_1, \ldots, d_n) \sqsubseteq (f_1, \ldots, f_n)$. ■

For example, $(N_\perp^n, \sqsubseteq)$ is a partial order, where $\sqsubseteq$ is the Cartesian product ordering on $N_\perp^n$ determined by $\sqsubseteq_{N_\perp}, \ldots, \sqsubseteq_{N_\perp}$.

Definition. Let D and E be sets. A function whose domain is D and whose range is a subset of E is a function *from D into E*. The set of all functions from D into E is denoted $D \to E$. If $\sqsubseteq_E$ is a partial ordering of E, then $\sqsubseteq_{D \to E}$, the *function space ordering* on $D \to E$ determined by $\sqsubseteq_E$, is defined

$$f \sqsubseteq_{D \to E} g \quad \text{if and only if} \quad f(d) \sqsubseteq_E g(d) \text{ for all } d \in D.$$

We sometimes write $f: D \to E$ to indicate that $f \in D \to E$.

Theorem 2.2. If D is a set and $(E, \sqsubseteq_E)$ is a partial order, then $(D \to E, \sqsubseteq_{D \to E})$ is a partial order.

Proof. We will write $\sqsubseteq$ for $\sqsubseteq_{D \to E}$. Let f, g, h be arbitrary functions in $D \to E$. For any $d \in D$, $f(d) \sqsubseteq_E f(d)$ by the reflexivity of $\sqsubseteq_E$, so $f \sqsubseteq f$. If $f \sqsubseteq g$ and $g \sqsubseteq f$, then for any $d \in D$, $f(d) \sqsubseteq_E g(d)$ and $g(d) \sqsubseteq_E f(d)$, which implies by the antisymmetry of $\sqsubseteq_E$ that $f(d) = g(d)$ for all $d \in D$, i.e., $f = g$. If $f \sqsubseteq g$ and $g \sqsubseteq h$, then for any $d \in D$, $f(d) \sqsubseteq_E g(d)$ and $g(d) \sqsubseteq_E h(d)$, which implies by the transitivity of $\sqsubseteq_E$ that $f(d) \sqsubseteq_E h(d)$ for all $d \in D$, i.e., $f \sqsubseteq h$. ■

For example, $(N_\perp^n \to N_\perp, \sqsubseteq)$ is a partial order, where $\sqsubseteq$ is the function space ordering on $N_\perp^n \to N_\perp$ determined by $\sqsubseteq_{N_\perp}$.

Definition. Let A be a set and let $\mathscr{D}$ be a function with domain A such that $\mathscr{D}(a)$ is a partial order for all $a \in A$. We will write $(D_a, \sqsubseteq_a)$ for $\mathscr{D}(a)$. A *$\mathscr{D}$-choice function*[1] is a function f with domain A such that $f(a) \in D_a$ for all $a \in A$. ch($\mathscr{D}$) is the set of all $\mathscr{D}$-choice functions. The *$\mathscr{D}$-choice function ordering* $\sqsubseteq_{\text{ch}(\mathscr{D})}$ is defined

$$f \sqsubseteq_{\text{ch}(\mathscr{D})} g \quad \text{if and only if} \quad f(a) \sqsubseteq_a g(a) \quad \text{for all} \quad a \in A.$$

For example, let $A = \{1,2\}$, let $\mathscr{D}(i) = (N^i \to N_\perp, \sqsubseteq_{N^i \to N_\perp})$ for $i = 1$, 2, and let $f(i) = u_1^i$ for $i = 1, 2$. That is, $f(1)$ is the unary projection function, and $f(2)$ is a binary projection function. (Recall that u_1^1 and u_1^2 are defined in Chapter 3.) Then $f \in \text{ch}(\mathscr{D})$.

Theorem 2.3. Let A be a set and $\mathscr{D}(a)$ a partial order for all $a \in A$. Then $(\text{ch}(\mathscr{D}), \sqsubseteq_{\text{ch}(\mathscr{D})})$ is a partial order.

Proof. The proof is identical to the proof of Theorem 2.2, except that instead of a single ordering $\sqsubseteq_E$, we have a different ordering $\sqsubseteq_a$ for each $a \in A$. ■

Exercises

1. Show that $(N, |)$ is a partial order, where $m \mid n$ is the predicate "m is a divisor of n" defined in Chapter 3. [Note that $0 \mid 0$ is true.] Is it linear?
2. Let $(D, \sqsubseteq_D)$ be a partial order, and let

$$\sqsupseteq_D = \{(x,y) \in D \times D \mid y \sqsubseteq_D x\}.$$

   Show that $(D, \sqsupseteq_D)$ is a partial order.
3. Let $(D, \sqsubseteq_D)$ be a partial order, let $E \subseteq D$, and let $\sqsubseteq_E$ be the restriction of $\sqsubseteq_D$ to E.
   - **(a)** Show that $(E, \sqsubseteq_E)$ is a partial order.
   - **(b)** Show that if $(D, \sqsubseteq_D)$ is a linear order, then $(E, \sqsubseteq_E)$ is a linear order.
4. For which set(s) A is $(A, \varnothing)$ a partial order?

[1] When A is infinite, proving the existence of $\mathscr{D}$-choice functions generally requires an axiom from set theory known as the axiom of choice. However, our treatment of denotational semantics will require only $\mathscr{D}$-choice functions with a finite domain, so we need not be concerned with this issue. The interested reader should consult any introductory text on set theory, e.g., those mentioned in "Suggestions for Further Reading."

5. Let $(D, \sqsubseteq_D)$ be a partial order, and let $d_1, \ldots, d_n \in D$ be such that $d_1 \sqsubseteq_D d_2 \sqsubseteq_D \cdots \sqsubseteq_D d_n \sqsubseteq_D d_1$. Show that $d_1 = d_2 = \cdots = d_n$.
6. **(a)** Show that there are three distinct partial orderings of $\{0, 1\}$.
   (b) Show that there are nineteen distinct partial orderings of $\{0, 1, 2\}$.
7. Let $D = \{a, b, c\}$, $\sqsubseteq_D = \{(a,a), (b,b), (c,c), (a,b), (a,c)\}$, $E = \{d, e\}$, and $\sqsubseteq_E = \{(d,d), (e,e), (d,e)\}$.
   (a) What is $(D \times E, \sqsubseteq_{D \times E})$?
   (b) What is $(D \to E, \sqsubseteq_{D \to E})$?
8. Let $(D, \sqsubseteq_D)$ be a partial order, and let

$$\sqsubset_D = \{(x, y) \in D \times D \mid x \sqsubseteq_D y \text{ and } x \neq y\}.$$

Show that $\sqsubset_D$ is transitive and satisfies the property of *asymmetry*, namely, that $x \sqsubset_D y$ implies $y \not\sqsubset_D x$, for all $x, y \in D$. Is $\sqsubset_D$ reflexive?
9. Let D, E be finite sets with m, n elements, respectively.
   (a) Show by induction on m that $D \times E$ has $m \cdot n$ elements.
   (b) Show by induction on m that $D \to E$ has n^m elements.
10. Give linear orders $(D, \sqsubseteq_D)$ and $(E, \sqsubseteq_E)$ such that $(D \times E, \sqsubseteq_{D \times E})$ and $(D \to E, \sqsubseteq_{D \to E})$ are not linear orders.
11. Give a linear order $(D, \sqsubseteq_D)$ with $D \neq \varnothing$ such that $(N \times D, \sqsubseteq_{N \times D})$ and $(N \to D, \sqsubseteq_{N \to D})$ are linear orders (where $\sqsubseteq_{N \times D}$ is determined by $\leq$, the usual ordering on N, and $\sqsubseteq_D$). Is $(D \times N, \sqsubseteq_{D \times N})$ a linear order? $(D \to N, \sqsubseteq_{D \to N})$?
12. Give distinct functions $f, g \in N_\perp^n \to N_\perp$ such that $f \sqsubseteq_{N_\perp^n \to N_\perp} g$.
13. Let $\mathscr{D}(i) = N_\perp^i \to N_\perp$ for all $i \in N$, $i \neq 0$. Give distinct functions $f, g \in \text{ch}(\mathscr{D})$ such that $f \sqsubseteq_{\text{ch}(\mathscr{D})} g$.

3. Complete Partial Orders

We will be particularly interested in partial orders that are rich in certain kinds of elements.

Definition. Let $(D, \sqsubseteq_D)$ be a partial order, and let $E \subseteq D$. An element $e_0 \in E$ is the *least element* of E with respect to $\sqsubseteq_D$ if $e_0 \sqsubseteq_D e$ for all $e \in E$, and it is the *greatest element* of E with respect to $\sqsubseteq_D$ if $e \sqsubseteq_D e_0$ for all $e \in E$.

If $(D, \sqsubseteq_D)$ is a partial order and $E \subseteq D$, then E can have at most one least element: if e, e' are least elements of E, then $e \sqsubseteq_D e'$ and $e' \sqsubseteq_D e$, so by antisymmetry $e = e'$. Similarly, E can have at most one greatest element. Therefore, we are justified in speaking about *the* least element of E and *the* greatest element of E.

Definition. Let $(D, \sqsubseteq_D)$ be a partial order, and let $E \subseteq D$. An element $d \in D$ is a *lower bound* of E in $(D, \sqsubseteq_D)$ if $d \sqsubseteq_D e$ for all $e \in E$, and it is an *upper bound* of E in $(D, \sqsubseteq_D)$ if $e \sqsubseteq_D d$ for all $e \in E$. Moreover, d is the *least upper bound* of E in $(D, \sqsubseteq_D)$ if it is the least element with respect to $\sqsubseteq_D$ of the set of all upper bounds of E in $(D, \sqsubseteq_D)$, and it is the *greatest lower bound* of E in $(D, \sqsubseteq_D)$ if it is the greatest element with respect to $\sqsubseteq_D$ of the set of all lower bounds of E in $(D, \sqsubseteq_D)$. If the least upper bound of E in $(D, \sqsubseteq_D)$ exists, it is denoted $\sqcup_{(D, \sqsubseteq D)} E$. If the greatest lower bound of E in $(D, \sqsubseteq_D)$ exists, it is denoted $\sqcap_{(D, \sqsubseteq D)} E$.

Suppose $(D, \sqsubseteq_D)$ is a partial order and $E \subseteq D$. Since the set of upper bounds of E in $(D, \sqsubseteq_D)$ can have at most one least element, it follows that E can have at most one least upper bound in $(D, \sqsubseteq_D)$. Similarly, E can have at most one greatest lower bound in $(D, \sqsubseteq_D)$. Note that $\sqcup_{(D, \sqsubseteq D)} E$, if it exists, is not necessarily an element of E, though if it is then it is the greatest element of E. A similar observation holds for $\sqcap_{(D, \sqsubseteq D)} E$.

In our work on semantics we are interested primarily in least upper bounds. We will generally drop the subscript and write $\sqcup E$ when it is apparent to which partial order we are referring. Occasionally we will write $\sqcup_D E$.

Partial orders can differ greatly in the existence of upper and lower bounds of their various subsets. Let $A = \{0, 1, 2\}$ and let $\sqsubseteq_A$ be the usual ordering on $\{0, 1, 2\}$. Then every subset of A has one or more upper bounds and one least upper bound. For example, 1 and 2 are both upper bounds of $\{0, 1\}$, and 1 is the least upper bound. Note that $2 = \sqcap\varnothing$ and $0 = \sqcup\varnothing$. However, consider $(N, \leq)$. Every finite subset of N has a greatest element, and every nonempty subset of N has a finite set of lower bounds, so every nonempty subset of N has a greatest lower bound. Also, every nonempty subset of N has a least element, and every finite subset of N has a nonempty set of upper bounds, so every finite subset of N has a least upper bound. However, an infinite subset of N has no upper bounds. Note that $\sqcap_N\varnothing$ does not exist. (Why?)

A subset of a partial order can fail to have a least upper bound for one of two reasons. Either the set of upper bounds is empty, as in the case of an infinite subset of N, or it is nonempty but has no least element. For example, let $(\mathbf{Q}, \leq_{\mathbf{Q}})$ be the ordered set of the rational numbers. Then

$A = \{q \in \mathbf{Q} \mid q^2 < 2\}$ has plenty of upper bounds, but it has no *rational* least upper bound. On the other hand, A has a least upper bound in $(\mathbf{R}, \leq_{\mathbf{R}})$, the ordered set of the real numbers. In fact, $\sqcup_{\mathbf{R}} A = \sqrt{2}$. For a simpler example, consider $(D, \sqsubseteq_D)$, where $D = \{a, b, c, d\}$ and

$$\sqsubseteq_D = \{(a,a),(b,b),(c,c),(d,d),(a,c),(a,d),(b,c),(b,d)\}.$$

Here, c and d are both upper bounds of $\{a, b\}$, but $\{c, d\}$ has no least element.

If we have a sequence $d_0, d_1, d_2, \ldots$ that represents a finite or infinite computation, where $d_0, d_1, d_2, \ldots$ are elements in some partial order $(D, \sqsubseteq_D)$, then we want $(D, \sqsubseteq_D)$ to contain some element d which represents the result of that computation. The following definition formalizes this idea.

Definition. Let $(D, \sqsubseteq_D)$ be a partial order. A *chain* in $(D, \sqsubseteq_D)$ is a nonempty set $C \subseteq D$ such that $c \sqsubseteq_D c'$ or $c' \sqsubseteq_D c$ for every $c, c' \in C$. $(D, \sqsubseteq_D)$ is a *complete partial order*, or *cpo*, if

1. D has a least element with respect to $\sqsubseteq_D$, and
2. $\sqcup_{(D, \sqsubseteq_D)} C$ exists for every chain C in $(D, \sqsubseteq_D)$.

The least element in a partial order $(D, \sqsubseteq_D)$ is generally written $\perp_D$ or $\perp$, and called the *bottom element* of $(D, \sqsubseteq_D)$, or simply *bottom* of $(D, \sqsubseteq_D)$. Note that if $(D, \sqsubseteq_D)$ is a partial order, C is a chain in $(D, \sqsubseteq_D)$, and $\sqsubseteq_C$ is the restriction of $\sqsubseteq_D$ to C, then $(C, \sqsubseteq_C)$ is a linear order.

Every nonempty subset of N is a chain in $(N, \leq)$, since $(N, \leq)$ is a linear order, and, as we showed previously, no infinite subset of N has a least upper bound in $(N, \leq)$, so $(N, \leq)$ is not a cpo. However, any set can be turned into a cpo, in the same way that we turned N into $(N_\perp, \sqsubseteq_{N_\perp})$. Let D be a set, let $\perp_D$ be some new object not in D, and let $D_\perp = D \cup \{\perp_D\}$. Then $\sqsubseteq_{D_\perp}$, defined

$$d \sqsubseteq_{D_\perp} e \quad \text{if and only if} \quad d = \perp_D \text{ or } d = e,$$

is the *flat partial ordering* of $D_\perp$. Every chain in $(D_\perp, \sqsubseteq_{D_\perp})$ is either $\{\perp_D\}$, or $\{d\}$ for some $d \in D$, or $\{\perp_D, d\}$ for some $d \in D$, so every chain in $(D_\perp, \sqsubseteq_{D_\perp})$ has a least upper bound, and therefore $(D_\perp, \sqsubseteq_{D_\perp})$ is a cpo. We call $(D_\perp, \sqsubseteq_{D_\perp})$ the *flat cpo* on D. For example, $(N_\perp, \sqsubseteq_{N_\perp})$ is the flat cpo on N.

We can generalize this discussion about flat cpos.

Theorem 3.1. Let $(D, \sqsubseteq_D)$ be a partial order, and let C be a finite chain in $(D, \sqsubseteq_D)$. Then $\sqcup C$ exists and $\sqcup C \in C$.

Proof. We argue by induction on the size of finite chains in $(D, \sqsubseteq_D)$. If $C = \{c\}$, then obviously $\sqcup C = c$. If $C = \{c_1, \ldots, c_{n+1}\}$, then $C' = \{c_1, \ldots, c_n\}$ is also a chain, so $\sqcup C' \in C'$ by the induction hypothesis. Now, if $c_{n+1} \sqsubseteq_D \sqcup C'$, then $\sqcup C = \sqcup C' \in C$. Otherwise, $\sqcup C' \sqsubseteq_D c_{n+1}$, since C is a chain, so $\sqcup C = c_{n+1} \in C$. ■

We immediately get

Corollary 3.2. Let $(D, \sqsubseteq_D)$ be a partial order with a bottom element. If every chain in $(D, \sqsubseteq_D)$ is finite, then $(D, \sqsubseteq_D)$ is a cpo.

Corollary 3.3. Every finite partial order with a bottom element is a cpo.

Power sets are another source of cpos. Let D be a set, and let $\mathscr{E} \subseteq \mathscr{P}(D)$. Then the *union* of $\mathscr{E}$, denoted $\cup\mathscr{E}$, is defined

$$\cup\mathscr{E} = \{d \in D \mid d \in E \text{ for some } E \in \mathscr{E}\}.$$

It is a basic mathematical fact that $\cup\mathscr{E}$ exists.

Theorem 3.4. Let D be a set. Then $(\mathscr{P}(D), \subseteq_{\mathscr{P}(D)})$ is a cpo.

Proof. We have already noted that $(\mathscr{P}(D), \subseteq_{\mathscr{P}(D)})$ is a partial order. For any set $E \in \mathscr{P}(D)$, $\varnothing \subseteq E$, so $\varnothing$ is the bottom element of $(\mathscr{P}(D), \subseteq_{\mathscr{P}(D)})$. Let $\mathscr{E} \subseteq \mathscr{P}(D)$. It is clear that $\cup\mathscr{E} \in \mathscr{P}(D)$, and we claim that $\cup\mathscr{E} = \sqcup\mathscr{E}$. For any $E \in \mathscr{E}$, we have $E \subseteq \cup\mathscr{E}$, so $\cup\mathscr{E}$ is an upper bound of $\mathscr{E}$ in $(\mathscr{P}(D), \subseteq_{\mathscr{P}(D)})$. Let $A \in \mathscr{P}(D)$ be any upper bound of $\mathscr{E}$. Then for any $d \in \cup\mathscr{E}$, $d \in E$ for some $E \in \mathscr{E}$, and $E \subseteq A$, so $d \in A$. Therefore, $\cup\mathscr{E} \subseteq A$, which implies $\cup\mathscr{E} = \sqcup\mathscr{E}$. This argument holds, in particular, when $\mathscr{E}$ is a chain, so $(\mathscr{P}(D), \subseteq_{\mathscr{P}(D)})$ is a cpo. ■

Note that in the proof of Theorem 3.4, we actually showed that *every* subset of $\mathscr{P}(D)$ has a least upper bound. (See Exercise 20 for more on this point.)

The constructions of Section 2 can also be used to construct cpos with a richer structure than flat cpos. Let $D_1, \ldots, D_n$ be sets, and let $D = D_1 \times \cdots \times D_n$. We define the *projection functions* $\downarrow 1: D \to D_1, \ldots, \downarrow n: D \to D_n$ as follows. For $1 \le i \le n$,

$$\downarrow i(d_1, \ldots, d_n) = d_i \quad \text{for all} \quad (d_1, \ldots, d_n) \in D.$$

Note that if $D_i = N$, for all $1 \le i \le n$, then $\downarrow j$ is the function u_j^n from Chapter 3, where $1 \le j \le n$. We will write $(d_1, \ldots, d_n)\downarrow i$ instead of $\downarrow i(d_1, \ldots, d_n)$. If $E \subseteq D$, we write $E \downarrow i$ to denote $\{e \downarrow i \mid e \in E\}$.

Theorem 3.5. Let $(D, \sqsubseteq_1), \ldots, (D_n, \sqsubseteq_n)$ be partial orders, and let $E \subseteq D_1 \times \cdots \times D_n$. Then $\sqcup E$ exists if and only if $\sqcup(E \downarrow 1), \ldots, \sqcup(E \downarrow n)$ exist, and if $\sqcup E$ exists, then $\sqcup E = (\sqcup(E \downarrow 1), \ldots, \sqcup(E \downarrow n))$.

Proof. We will write $\sqsubseteq$ for $\sqsubseteq_{D_1 \times \cdots \times D_n}$. Suppose $\sqcup E$ exists, and let $\sqcup E = (e_1, \ldots, e_n)$. Then $e_i \in D_i$, $1 \le i \le n$, and we claim that $e_i = \sqcup(E \downarrow i)$. For $1 \le i \le n$, if $c \in E \downarrow i$, then there is some element

$$(c_1, \ldots, c_{i-1}, c, c_{i+1}, \ldots, c_n) \in E,$$

and $(c_1, \ldots, c_{i-1}, c, c_{i+1}, \ldots, c_n) \sqsubseteq (e_1, \ldots, e_n)$ implies $c \sqsubseteq_i e_i$, so e_i is an upper bound of $E \downarrow i$. Let d be any upper bound of $E \downarrow i$. Then for any $(c_1, \ldots, c_n) \in E$, $c_j \sqsubseteq_j e_j$ for $1 \le j \le n$, $j \ne i$, and $c_i \sqsubseteq_i d$, so $(e_1, \ldots, d, \ldots, e_n)$ is an upper bound of E. But $(e_1, \ldots, e_n)$ is the least upper bound of E, so $(e_1, \ldots, e_n) \sqsubseteq (e_1, \ldots, d, \ldots, e_n)$, and, in particular, $e_i \sqsubseteq_i d$, so $e_i = \sqcup(E \downarrow i)$. Therefore, $\sqcup(E \downarrow i)$ exists, $1 \le i \le n$, and $\sqcup E = (e_1, \ldots, e_n) = (\sqcup(E \downarrow 1), \ldots, \sqcup(E \downarrow n))$.

Now, suppose $\sqcup(E \downarrow 1), \ldots, \sqcup(E \downarrow n)$ exist. Then $(\sqcup(E \downarrow 1), \ldots, \sqcup(E \downarrow n))$ is an element of $D_1 \times \cdots \times D_n$, and we claim that it is $\sqcup E$. If $(c_1, \ldots, c_n) \in E$, then for $1 \le i \le n$, $c_i \in E \downarrow i$, which implies $c_i \sqsubseteq_i \sqcup(E \downarrow i)$, so $(c_1, \ldots, c_n) \sqsubseteq (\sqcup(E \downarrow 1), \ldots, \sqcup(E \downarrow n))$. Therefore, $(\sqcup(E \downarrow 1), \ldots, \sqcup(E \downarrow n))$ is an upper bound of E. Let $(d_1, \ldots, d_n)$ be any upper bound of E. For $1 \le i \le n$, if $c \in E \downarrow i$ then there is some $(c_1, \ldots, c, \ldots, c_n) \in E$, and $(c_1, \ldots, c, \ldots, c_n) \sqsubseteq (d_1, \ldots, d_n)$ implies $c \sqsubseteq_i d_i$, so d_i is an upper bound of $E \downarrow i$. But then $\sqcup(E \downarrow i) \sqsubseteq_i d_i$, $1 \le i \le n$, which implies $(\sqcup(E \downarrow 1), \ldots, \sqcup(E \downarrow n)) \sqsubseteq (d_1, \ldots, d_n)$, so $(\sqcup(E \downarrow 1), \ldots, \sqcup(E \downarrow n)) = \sqcup E$. ∎

Theorem 3.6. If $(D_1, \sqsubseteq_1), \ldots, (D_n, \sqsubseteq_n)$ are cpos, then $(D_1 \times \cdots \times D_n, \sqsubseteq_{D_1 \times \cdots \times D_n})$ is a cpo.

Proof. We will write D for $D_1 \times \cdots \times D_n$ and $\sqsubseteq$ for $\sqsubseteq_{D_1 \times \cdots \times D_n}$. $(D, \sqsubseteq)$ is a partial order by Theorem 2.1. Let $\bot_i$ be the bottom element of $(D_i, \sqsubseteq_i)$, $1 \le i \le n$. Then $(\bot_1, \ldots, \bot_n) \sqsubseteq (d_1, \ldots, d_n)$ for all $(d_1, \ldots, d_n) \in D$, so $(D, \sqsubseteq)$ has a bottom element.

Now, let C be a chain in $(D, \sqsubseteq)$. We must show that $\sqcup C$ exists. For $1 \le i \le n$, if $c_i, c_i' \in C \downarrow i$, then there are $c, c' \in C$ such that $c_i = c \downarrow i$ and $c_i' = c' \downarrow i$. Since C is a chain, either $c \sqsubseteq c'$ or $c' \sqsubseteq c$, which implies that either $c_i \sqsubseteq_i c_i'$ or $c_i' \sqsubseteq_i c_i$. Therefore $C \downarrow i$ is a chain in cpo $(D_i, \sqsubseteq_i)$, so $\sqcup(C \downarrow i)$ exists, $1 \le i \le n$, and by Theorem 3.5, $\sqcup C$ exists. ∎

We can prove a similar result for function space orderings. If $\mathscr{F} \subseteq D \to E$ for some sets D, E, then for any $d \in D$ we write $\mathscr{F}(d)$ to denote the set $\{f(d) \mid f \in \mathscr{F}\}$.

Theorem 3.7. Let D be a set and $(E, \sqsubseteq_E)$ a partial order, and let $\mathcal{F} \subseteq D \to E$. Then $\sqcup\mathcal{F}$ exists if and only if $\sqcup(\mathcal{F}(d))$ exists for all $d \in D$, and if $\sqcup\mathcal{F}$ exists then $(\sqcup\mathcal{F})(d) = \sqcup(\mathcal{F}(d))$ for all $d \in D$.

Proof. We will write $\sqsubseteq$ for $\sqsubseteq_{D\to E}$. Suppose $\sqcup\mathcal{F}$ exists, and let $d \in D$. Then $(\sqcup\mathcal{F})(d)$ is an element of E, and we claim $(\sqcup\mathcal{F})(d) = \sqcup(\mathcal{F}(d))$. For any $f \in \mathcal{F}$, $f \sqsubseteq \sqcup\mathcal{F}$ implies $f(d) \sqsubseteq_E (\sqcup\mathcal{F})(d)$, so $(\sqcup\mathcal{F})(d)$ is an upper bound of $\mathcal{F}(d)$. Let e be any upper bound of $\mathcal{F}(d)$, and let $f_e: D \to E$ be defined

$$f_e(x) = \begin{cases} e & \text{if } x = d \\ (\sqcup\mathcal{F})(x) & \text{otherwise.} \end{cases}$$

Then for any $f \in \mathcal{F}$, $f(x) \sqsubseteq_E (\sqcup\mathcal{F})(x) = f_e(x)$ for $x \in D$ such that $x \neq d$, and $f(d) \sqsubseteq_E e = f_e(d)$, so f_e is an upper bound of $\mathcal{F}$. But then $\sqcup\mathcal{F} \sqsubseteq f_e$, and, in particular, $(\sqcup\mathcal{F})(d) \sqsubseteq_E f_e(d) = e$, so $(\sqcup\mathcal{F})(d) = \sqcup(\mathcal{F}(d))$.

Now, suppose $\sqcup(\mathcal{F}(d))$ exists for all $d \in D$. Then the function

$$g(d) = \sqcup(\mathcal{F}(d)) \quad \text{for all } d \in D$$

belongs to $D \to E$, and we claim that $g = \sqcup\mathcal{F}$. If $f \in \mathcal{F}$, then for any $d \in D$, $f(d) \in \mathcal{F}(d)$, which implies $f(d) \sqsubseteq_E \sqcup(\mathcal{F}(d)) = g(d)$, so $f \sqsubseteq g$. Therefore, g is an upper bound of $\mathcal{F}$. Let h be any upper bound of $\mathcal{F}$. For any $d \in D$, if $e \in \mathcal{F}(d)$, then $e = f(d)$ for some $f \in \mathcal{F}$, and $f \sqsubseteq h$ implies $e = f(d) \sqsubseteq_E h(d)$, so $h(d)$ is an upper bound of $\mathcal{F}(d)$. But then $g(d) = \sqcup(\mathcal{F}(d)) \sqsubseteq_E h(d)$ for all $d \in D$, which implies $g \sqsubseteq h$, so $g = \sqcup\mathcal{F}$. ■

Theorem 3.8. If D is a set and $(E, \sqsubseteq_E)$ a cpo, then $(D \to E, \sqsubseteq_{D\to E})$ is a cpo.

Proof. We will write $\sqsubseteq$ for $\sqsubseteq_{D\to E}$. $(D \to E, \sqsubseteq)$ is a partial order by Theorem 2.2. Define the constant function $\bot_{D\to E}(d) = \bot_E$ for all $d \in D$, where $\bot_E$ is the bottom element of $(E, \sqsubseteq_E)$. Then $\bot_{D\to E} \sqsubseteq f$ for all $f \in D \to E$, so $(D \to E, \sqsubseteq)$ has a bottom element.

Now, let $\mathcal{F}$ be a chain in $(D \to E, \sqsubseteq)$. Then $\mathcal{F}(d)$ is a chain in $(E, \sqsubseteq_E)$ for any $d \in D$, since, for any $f(d)$, $g(d) \in \mathcal{F}(d)$, $f \sqsubseteq g$ implies $f(d) \sqsubseteq_E g(d)$ and $g \sqsubseteq f$ implies $g(d) \sqsubseteq_E f(d)$. Therefore $\sqcup(\mathcal{F}(d))$ exists for all $d \in D$, since $(E, \sqsubseteq_E)$ is a cpo, and by Theorem 3.7, $\sqcup\mathcal{F}$ exists. ■

The proofs of the following two theorems are almost identical to the proofs of Theorem 3.7 and Theorem 3.8.

Theorem 3.9. Let A be a set, let $\mathcal{D}(a)$ be a partial order for each $a \in A$, and let $\mathcal{F} \subseteq \text{ch}(D)$. Then $\sqcup\mathcal{F}$ exists if and only if $\sqcup(\mathcal{F}(a))$ exists for all $a \in A$, and if $\sqcup\mathcal{F}$ exists, then $(\sqcup\mathcal{F})(a) = \sqcup(\mathcal{F}(a))$ for all $a \in A$.

Theorem 3.10. Let A be a set and let $\mathscr{D}(a)$ be a cpo for each $a \in A$. Then $(\text{ch}(\mathscr{D}), \sqsubseteq_{\text{ch}(\mathscr{D})})$ is a cpo.

The iteration of our operations for constructing partial orders quickly gives us partial orders of considerable complexity. Theorems 3.6, 3.8, and 3.10 tell us that if we start with cpos, we end up with cpos. For example,

$$\left((N_\perp^2 \to N_\perp) \to (N_\perp \to N_\perp), \sqsubseteq_{(N_\perp^2 \to N_\perp) \to (N_\perp \to N_\perp)}\right)$$

is the cpo of functions that transform binary functions in $N_\perp^2 \to N_\perp$ into unary functions in $N_\perp \to N_\perp$. Functions that operate on other functions are sometimes called *higher order functions*. For example $\text{Id}_{D \to D}: (D \to D) \to (D \to D)$, defined $\text{Id}_{D \to D}(f) = f$, is an easily described higher order function. One way of defining a higher order function F is to give a definition of the function $F(f)$ for every function f in the domain of F. For example,

$$\text{Id}_{D \to D}(f)(d) = f(d) \quad \text{for all } d \in D.$$

Note that $\text{Id}_{D \to D}(f)(d)$ is to be interpreted as $(\text{Id}_{D \to D}(f))(d)$. Similarly, when we write an expression such as $f(g)(h)(d)$, we mean $((f(g))(h))(d)$. Another example is the composition operator $\circ: (E \to F) \times (D \to E) \to (D \to F)$, for some sets D, E, F, where, for any $f: E \to F$ and $g: D \to E$, $\circ(f, g)$ is defined

$$\circ(f, g)(d) = f(g(d)) \quad \text{for all } d \in D.$$

($\circ(f, g)$ is usually written $f \circ g$.) We will make frequent use of this sort of definition in the next chapter.

One way to show that a partial order $(E, \sqsubseteq_E)$ is a cpo is to build it up explicitly by the constructions we have described. Another is to show that it is contained in another partial order $(D, \sqsubseteq_D)$ known to be a cpo and that the least upper bounds of all chains in $(E, \sqsubseteq_E)$ belong to E.

Theorem 3.11. Let $(D, \sqsubseteq_D)$ be a cpo, let $E \subseteq D$, and let $\sqsubseteq_E$ be the restriction of $\sqsubseteq_D$ to E. If

1. E has a least element with respect to $\sqsubseteq_E$ and
2. $\sqcup_D C \in E$ for all chains C in $(E, \sqsubseteq_E)$,

then $(E, \sqsubseteq_E)$ is a cpo and $\sqcup_E C = \sqcup_D C$ for all chains C in $(E, \sqsubseteq_E)$.

Proof. E has a least element by assumption, so we only need to show that every chain C in $(E, \sqsubseteq_E)$ has a least upper bound in $(E, \sqsubseteq_E)$, i.e., that

$\sqcup_E C$ exists. If C is a chain in $(E, \sqsubseteq_E)$, then it is also a chain in $(D, \sqsubseteq_D)$, and we know that $\sqcup_D C$ exists. We claim that $\sqcup_D C$ is $\sqcup_E C$. For any $c \in C$, we have $c \sqsubseteq_D \sqcup_D C$, which implies $c \sqsubseteq_E \sqcup_D C$, since c and $\sqcup_D C$ are both in E. Therefore, $\sqcup_D C$ is an upper bound of C in $(E, \sqsubseteq_E)$. Let e be any upper bound of C in $(E, \sqsubseteq_E)$. Then for all $c \in C$, $c \sqsubseteq_E e$ implies $c \sqsubseteq_D e$, so e is an upper bound of C in $(D, \sqsubseteq_D)$. Therefore, $\sqcup_D C \sqsubseteq_D e$, which implies $\sqcup_D C \sqsubseteq_E e$ since $\sqcup_D C$ and e are both in E, so $\sqcup_D C = \sqcup_E C$. ∎

We give one application of Theorem 3.11 here and another in the next section. The following construction gives us a way of turning an arbitrary partial order into a cpo.

Definition. Let $(D, \sqsubseteq_D)$ be a partial order. A set $E \subseteq D$ is *downward closed* if for all $e \in E$ and all $d \in D$, if $d \sqsubseteq_D e$, then $d \in E$. E is *directed* if for all $c, d \in E$, there is some $e \in E$ such that $c \sqsubseteq_D e$ and $d \sqsubseteq_D e$. An *ideal* of $(D, \sqsubseteq_D)$ is a nonempty, downward closed, directed subset of D. The set of all ideals of $(D, \sqsubseteq_D)$ is denoted $\mathrm{id}(D, \sqsubseteq_D)$. The ordering $\subseteq_{\mathrm{id}(D, \sqsubseteq_D)}$ is $\subseteq_{\mathscr{P}(D)}$ restricted to $\mathrm{id}(D, \sqsubseteq_D)$.

We will write $\mathrm{id}(D)$ when the ordering $\sqsubseteq_D$ is understood.

Theorem 3.12. Let $(D, \sqsubseteq_D)$ be a partial order with a bottom element. Then $(\mathrm{id}(D), \subseteq_{\mathrm{id}(D)})$ is a cpo.

Proof. It is easy to check that $(\mathrm{id}(D), \subseteq_{\mathrm{id}(D)})$ is a partial order. An ideal is nonempty by definition, so $\perp_D \in I$ for any ideal I, since I is downward closed. Moreover, $\{\perp_D\}$ is an ideal, and $\{\perp_D\} \subseteq I$ for any ideal I, so $\{\perp_D\}$ is the bottom element of $(\mathrm{id}(D), \subseteq_{\mathrm{id}(D)})$. Now, let $\mathscr{I}$ be a chain in $(\mathrm{id}(D), \subseteq_{\mathrm{id}(D)})$. It is obvious that $\mathrm{id}(D) \subseteq \mathscr{P}(D)$, so by Theorem 3.11 we need to show only that $\sqcup_{\mathscr{P}(d)} \mathscr{I} \in \mathrm{id}(D)$, i.e., that $\cup \mathscr{I}$ is an ideal. $\mathscr{I}$ is a nonempty set of nonempty sets, so $\cup \mathscr{I}$ is nonempty. Let $e \in \cup \mathscr{I}$, $d \in D$, and $d \sqsubseteq_D e$. Then $e \in I$ for some $I \in \mathscr{I}$, which implies $d \in I$ since I is an ideal. Therefore, $d \in \cup \mathscr{I}$, so $\cup \mathscr{I}$ is downward closed. Now, if $c, d \in \cup \mathscr{I}$, then $c \in I_1$ and $d \in I_2$ for some $I_1, I_2 \in \mathscr{I}$, which implies $c, d \in I_1 \cup I_2$. But $I_1 \cup I_2 \in \mathscr{I}$, since $I_1 \subseteq I_2$ implies $I_1 \cup I_2 = I_2$ and $I_2 \subseteq I_1$ implies $I_1 \cup I_2 = I_1$. Therefore, there is an $e \in I_1 \cup I_2$ such that $c \sqsubseteq_D e$ and $d \sqsubseteq_D e$, since $I_1 \cup I_2$ is directed, and $I_1 \cup I_2 \subseteq \cup \mathscr{I}$, so $e \in \cup \mathscr{I}$. So $\cup \mathscr{I}$ is directed, and it is an ideal. ∎

Let $(D, \sqsubseteq_D)$ be a partial order. For each $e \in D$, the *principal ideal generated by* e, denoted $\mathrm{pid}(e)$, is the set $\{d \in D \mid d \sqsubseteq_D e\}$. It is easy to see that $\mathrm{pid}(e)$ is an ideal of $(D, \sqsubseteq_D)$. The set of principal ideals of $(D, \sqsubseteq_D)$, denoted pid(D), is $\{\text{pid}(d) \mid d \in D\}$, and $\subseteq_{\text{pid}(D)}$ is the restriction of $\subseteq_{\mathscr{P}(D)}$ to pid(D). We can think of the partial order $(\text{pid}(D), \subseteq_{\text{pid}(D)})$ as a "copy" of $(D, \sqsubseteq_D)$ in $(\text{id}(D), \subseteq_{\text{id}(D)})$, and any chain C in $(D, \sqsubseteq_D)$ has a "copy" $\{\text{pid}(c) \mid c \in C\}$ in $(\text{id}(D), \subseteq_{\text{id}(D)})$. Going from $(D, \sqsubseteq_D)$ to $(\text{id}(D), \subseteq_{\text{id}(D)})$, then, has the effect of guaranteeing the existence of a least upper bound in $(\text{id}(D), \subseteq_{\text{id}(D)})$ for each ("copy" of a) chain in $(D, \sqsubseteq_D)$. For this reason, $(\text{id}(D), \subseteq_{\text{id}(D)})$ is called the *ideal completion* of $(D, \sqsubseteq_D)$. For more on this subject, see Exercise 19.

Exercises

1. Give an example of a partial order that is not a cpo.
2. Give an example of a cpo in which not every chain has a greatest lower bound.
3. Let ω be some object not in N. Give a binary relation $\sqsubseteq$ such that $(N \cup \{\omega\}, \leq \cup \sqsubseteq)$ is a cpo (where $\leq$ is the usual ordering of N).
4. Let $(D, \sqsubseteq_D)$ be a cpo, let C be a chain in $(D, \sqsubseteq_D)$, and let $d \in D$. Show that if $c \sqsubseteq_D d$ for all $c \in C$, then $\sqcup C \sqsubseteq_D d$.
5. Let $(D, \sqsubseteq_D)$ be a cpo, and let $C_1 \cup C_2$ be a chain in $(D, \sqsubseteq_D)$, where $C_1, C_2 \subseteq D$. Show that $\sqcup(C_1 \cup C_2) = \sqcup\{\sqcup C_1, \sqcup C_2\}$.
6. Let $(D, \sqsubseteq_D)$ be a cpo, and let C_1, C_2 be chains in $(D, \sqsubseteq_D)$.
   - (a) Show that if for all $c_1 \in C_1$ there is a $c_2 \in C_2$ such that $c_1 \sqsubseteq_D c_2$, then $\sqcup C_1 \sqsubseteq_D \sqcup C_2$.
   - (b) Show that if for all $c_1 \in C_1$ there is a $c_2 \in C_2$ such that $c_1 \sqsubseteq_D c_2$, and if for all $c_2 \in C_2$ there is a $c_1 \in C_1$ such that $c_2 \sqsubseteq_D c_1$, then $\sqcup C_1 = \sqcup C_2$.
7. Let $(D, \sqsubseteq_D), (E, \sqsubseteq_E)$ be cpos, and let C be a chain in $(D \times E, \sqsubseteq_{D \times E})$. Show that $\sqcup C = \sqcup(C\downarrow 1 \times C\downarrow 2)$.
8. (a) Let $(D, \sqsubseteq_D)$, $(E, \sqsubseteq_E)$ be partial orders such that the largest chain in $(D, \sqsubseteq_D)$ has $m \in N$ elements and the largest chain in $(E, \sqsubseteq_E)$ has $n \in N$ elements. What is the size of the largest chain in $(D \times E, \sqsubseteq_{D \times E})$?
   - (b) Let $(D_1, \sqsubseteq_{D_1}), \ldots, (D_n, \sqsubseteq_{D_n})$ be partial orders such that the largest chain in $(D_i, \sqsubseteq_{D_i})$ has $m_i \in N$ elements, $1 \leq i \leq n$. Prove by induction on n that all chains in

$$(D_1 \times \cdots \times D_n, \sqsubseteq_{D_1 \times \cdots \times D_n})$$

are finite.

9. Let D be a set with $m \in N$ elements, and let $(E, \sqsubseteq_E)$ be a partial order in which the largest chain has $n \in N$ elements. What is the size of the largest chain in $(D \to E, \sqsubseteq_{D \to E})$?

10. Let D be a set and $(E, \sqsubseteq_E)$ a partial order. Show that $(E, \sqsubseteq_E)$ is a cpo if and only if $(D \to E, \sqsubseteq_{D \to E})$ is a cpo.

11. Let $(D, \sqsubseteq_D)$, $(E, \sqsubseteq_E)$ be cpos. Show that

$$\big((D \times E) \to (D \times E), \sqsubseteq_{(D \times E) \to (D \times E)}\big)$$

is a cpo.

12. Let D be a set and $(E, \sqsubseteq_E)$ a cpo. Show that

$$\big((D \to E) \to (D \to E), \sqsubseteq_{(D \to E) \to (D \to E)}\big)$$

is a cpo.

13. Let D be a set, let $E \subseteq D$, let $\mathscr{P}_E(D) = \{A \in \mathscr{P}(D) \mid E \subseteq A\}$, and let $\subseteq_{\mathscr{P}_E(D)}$ be the restriction of $\subseteq_{\mathscr{P}(D)}$ to $\mathscr{P}_E(D)$. Show that $(\mathscr{P}_E(D), \subseteq_{\mathscr{P}_E(D)})$ is a cpo.

14. For sets D, E, let $D \underset{p}{\to} E$ be the set of all partial functions f on D such that the range of f is a subset of E, and let $\subseteq_{D \underset{p}{\to} E}$ be defined as follows: for all $f, g \in D \underset{p}{\to} E$, $f \subseteq_{D \underset{p}{\to} E} g$ if and only if $f \subseteq g$. Show that $(D \underset{p}{\to} E, \subseteq_{D \underset{p}{\to} E})$ is a cpo. [*Hint:* Note that $D \underset{p}{\to} E$ is a subset of $\mathscr{P}(D \times E)$ and $\subseteq_{D \underset{p}{\to} E}$ is the restriction of $\subseteq_{\mathscr{P}(D \times E)}$ to $D \underset{p}{\to} E$.]

15. **(a)** Give a partial order $(D, \sqsubseteq_D)$ and a chain C in $(D, \sqsubseteq_D)$ such that C is not an ideal.

(b) Give a partial order $(D, \sqsubseteq_D)$ and an ideal I of $(D, \sqsubseteq_D)$ such that I is not a chain.

16. Let $(D, \sqsubseteq_D)$ be a partial order.

(a) Show that if $(D, \sqsubseteq_D)$ has a bottom element, and if $\sqcup E$ exists for every directed set $E \subseteq D$, then $(D, \sqsubseteq_D)$ is a cpo.

(b)* Show that if $(D, \sqsubseteq_D)$ is a cpo, then $\sqcup E$ exists for every directed set $E \subseteq D$.

17.* Let $(D, \sqsubseteq_D)$, $(E, \sqsubseteq_E)$ be partial orders. The *lexicographic ordering* on $D \times E$, denoted $\sqsubseteq_{L(D \times E)}$, is defined

$$(d_1, e_1) \sqsubseteq_{L(D \times E)} (d_2, e_2) \quad \text{if and only if}$$

$$d_1 \sqsubseteq_D d_2 \text{ or } (d_1 = d_2 \text{ and } e_1 \sqsubseteq_E e_2).$$

Show that if $(D, \sqsubseteq_D)$, $(E, \sqsubseteq_E)$ are cpos, then $(D \times E, \sqsubseteq_{L(D \times E)})$ is a cpo.

18.* For any sets D, E, a function $f: D \to E$ is *onto* if the range of f is all of E. Let $(D, \sqsubseteq_D)$, $(E, \sqsubseteq_E)$ be partial orders. An *isomorphism* from $(D, \sqsubseteq_D)$ to $(E, \sqsubseteq_E)$ is a one–one, onto function $f: D \to E$ such that $d \sqsubseteq_D d'$ if and only if $f(d) \sqsubseteq_E f(d')$ for all $d, d' \in D$. $(D, \sqsubseteq_D)$, $(E, \sqsubseteq_E)$ are *isomorphic* if there is an isomorphism from $(D, \sqsubseteq_D)$ to $(E, \sqsubseteq_E)$. If $f: D \to E$ is one–one, then the *inverse* of f, denoted f^{-1}, is defined $f^{-1} = \{(e, d) \in E \times D \mid f(d) = e\}$.

(a) Show that if f is an isomorphism from partial order $(D, \sqsubseteq_D)$ to partial order $(E, \sqsubseteq_E)$, then f^{-1} is an isomorphism from $(E, \sqsubseteq_E)$ to $(D, \sqsubseteq_D)$.

(b) Let $(D, \sqsubseteq_D)$, $(E, \sqsubseteq_E)$ be isomorphic partial orders. Show that $(D, \sqsubseteq_D)$ is a cpo if and only if $(E, \sqsubseteq_E)$ is a cpo.

(c) Let D be a set with $n \in N$ elements and let $(E, \sqsubseteq_E)$ be a partial order. Show that $(E^n, \sqsubseteq_{E^n})$ and $(D \to E, \sqsubseteq_{D \to E})$ are isomorphic.

(d) Let $(D_1, \sqsubseteq_{D_1}), \ldots, (D_n, \sqsubseteq_{D_n})$ be partial orders, let $A = \{1, \ldots, n\}$, and let $\mathscr{D}(i) = (D_i, \sqsubseteq_{D_i})$, $1 \le i \le n$. Show that $(D_1 \times \cdots \times D_n, \sqsubseteq_{D_1 \times \cdots \times D_n})$ and $(\text{ch}(\mathscr{D}), \sqsubseteq_{\text{ch}(\mathscr{D})})$ are isomorphic.

19.* Let $(D, \sqsubseteq_D)$ be a partial order.

(a) Let I be a principal ideal generated by some $d \in D$. Show that I is an ideal of $(D, \sqsubseteq_D)$.

(b) Show that $(D, \sqsubseteq_D)$ and $(\text{pid}(D), \subseteq_{\text{pid}(D)})$ are isomorphic. [See Exercise 18 for the definition of *isomorphic partial orders*.]

20.* A partial order $(D, \sqsubseteq_D)$ is a *lattice* if $\sqcup\{d, e\}$ and $\sqcap\{d, e\}$ exist for every $d, e \in D$. It is a *complete lattice* if $\sqcup E$ and $\sqcap E$ exist for every $E \subseteq D$.

(a) Give an example of a cpo that is not a lattice.

(b) Give an example of a lattice that is not a cpo.

(c) Let $(D, \sqsubseteq_D)$ be a lattice. Show that for every nonempty finite set $E \subseteq D$, $\sqcup E$ and $\sqcap E$ exist.

(d) Show that for any set D, $(\mathscr{P}(D), \subseteq_{\mathscr{P}(D)})$ is a complete lattice.

(e) Show that for any set D, $(\mathscr{P}(D), \supseteq_{\mathscr{P}(D)})$ is a complete lattice.

21.* Let $(D, \sqsubseteq_D)$ be a partial order with $D \neq \varnothing$. $(D, \sqsubseteq_D)$ is *bounded-complete* if $\sqcup B$ exists for every $B \subseteq D$ that has an upper bound in $(D, \sqsubseteq_D)$.

(a) Give an example of a cpo that is not bounded-complete.

(b) Give an example of a bounded-complete partial order that is not a cpo.

(c) Show that every bounded-complete partial order has a bottom element. [*Hint:* Consider $\sqcup\varnothing$.]

(d) Let $(D, \sqsubseteq_D)$ be a bounded-complete partial order. Show that for every nonempty $E \subseteq D$, $\sqcap E$ exists. [*Hint:* Consider the least upper bound of $\{d \in D \mid d \sqsubseteq_D e \text{ for all } e \in E\}$.]

(e) $(D, \sqsubseteq_D)$ is *well-founded* if every nonempty subset of D has a least element. Prove that every nonempty well-founded partial order is bounded-complete.

(f) Let $(D, \sqsubseteq_D)$, $(E, \sqsubseteq_E)$ be bounded-complete cpos. Show that $(D \times E, \sqsubseteq_{D \times E})$ is a bounded-complete cpo.

(g) Let D be a set and $(E, \sqsubseteq_E)$ a bounded-complete cpo. Show that $(D \to E, \sqsubseteq_{D \to E})$ is a bounded-complete cpo.

22.* Let $(D, \sqsubseteq_D)$ be a cpo. An element $d \in D$ is *compact* (sometimes called *finite*) if for every chain C in $(D, \sqsubseteq_D)$ such that $d \sqsubseteq_D \sqcup C$, there is a $c \in C$ such that $d \sqsubseteq_D c$. The set of compact elements in $(D, \sqsubseteq_D)$ is denoted $K(D)$. $(D, \sqsubseteq_D)$ is *algebraic* if for every $d \in D$, there is a chain $C \subseteq K(D)$ such that $d = \sqcup C$.

(a) Let $(D, \sqsubseteq_D)$ be a cpo, let $d \in K(D)$, and let C be a chain in $(D, \sqsubseteq_D)$ such that $d = \sqcup C$. Show that $d \in C$.

(b) Let $(D, \sqsubseteq_D)$ be a cpo in which every chain is finite. Show that $(D, \sqsubseteq_D)$ is algebraic.

(c) Let D be a set. Show that $(\mathscr{P}(D), \subseteq_{\mathscr{P}(D)})$ is an algebraic cpo.

(d) Give an example of a cpo $(D, \sqsubseteq_D)$, a compact $d \in K(D)$, and an infinite chain C in $(D, \sqsubseteq_D)$ such that $d = \sqcup C$.

(e) Give an example of a cpo that is not algebraic.

(f) Show that if $(D, \sqsubseteq_D)$, $(E, \sqsubseteq_E)$ are algebraic cpos, then so is $(D \times E, \sqsubseteq_{D \times E})$.

(g) Let $(D, \sqsubseteq_D)$ be a partial order. Show that $(\text{id}(D), \subseteq_{\text{id}(D)})$ is an algebraic cpo and that $K(\text{id}(D)) = \text{pid}(D)$.

(h) Show that if $(D, \sqsubseteq_D)$ is an algebraic cpo, then $(D, \sqsubseteq_D)$ and $(\text{id}(K(D)), \subseteq_{\text{id}(K(D))})$ are isomorphic. [See Exercise 18 for the definition of *isomorphic partial orders*.]

4. Continuous Functions

Consider a computable function f composed with a partially computable function g applied to a number n, where $g(n)\uparrow$. How should we understand the composition $f(g(n))$? One interpretation is that the computation of $g(n)$ never terminates, so f never gets a result from $g(n)$ and $f(g(n))$ must be undefined. In fact, this is the treatment of composition given in Chapter 3.

Definition. Let $(D_1, \sqsubseteq_1), \ldots, (D_n, \sqsubseteq_n), (E, \sqsubseteq_E)$ be cpos with bottom elements $\bot_{D_1}, \ldots, \bot_{D_n}, \bot_E$. A function $f: D_1 \times \cdots \times D_n \to E$ is *strict* if $f(d_1, \ldots, d_n) = \bot_E$ for all $(d_1, \ldots, d_n)$ such that $d_i = \bot_{D_i}$ for one or more $1 \le i \le n$.

Let $D_1, \ldots, D_n, E$ be sets, and let $((D_i)_\bot, \sqsubseteq_{(D_i)_\bot})$, $1 \le i \le n$, and $(E_\bot, \sqsubseteq_{E_\bot})$ be the flat cpos on $D_1, \ldots, D_n, E$, with bottom elements $\bot_{D_i}$, $1 \le i \le n$, and $\bot_E$. For any partial function f on $D_1 \times \cdots \times D_n$ with range contained in E, the *strict extension* $f_\bot : (D_1)_\bot \times \cdots \times (D_n)_\bot \to E_\bot$ of f is defined.

$$f_\bot(x_1, \ldots, x_n)$$
$$= \begin{cases} \bot_E & \text{if } (x_1, \ldots, x_n) \notin D_1 \times \cdots \times D_n \\ \bot_E & \text{if } (x_1, \ldots, x_n) \in D_1 \times \cdots \times D_n \text{ and } f(x_1, \ldots, x_n)\uparrow \\ f(x_1, \ldots, x_n) & \text{otherwise.} \end{cases}$$

Clearly any such $f_\bot$ is strict. We have not defined what it means for a function in $N_\bot^n \to N_\bot$, for example, to be computable, but if f is partially computable, then it certainly would be reasonable to consider $f_\bot$ to be partially computable. In the next chapter we will use strict functions in some situations, but we will not require computable functions to be strict. For example, the function $g(x) = 3$ for all $x \in N_\bot$ is not strict, but it certainly should be considered computable. However, some restrictions on computable functions appear to be reasonable.

Consider the elements $(\bot_N, \bot_N)$, $(\bot_N, 3)$, and $(7, 3)$ of $N_\bot \times N_\bot$. If $(\bot_N, \bot_N)$ approximates $(7,3)$ and $(\bot_N, 3)$ better approximates $(7,3)$, then we should expect a function $f: N_\bot \times N_\bot \to N_\bot$ to behave such that $f(\bot_N, \bot_N)$ approximates $f(7,3)$ and $f(\bot_N, 3)$ better approximates $f(7,3)$. That is, if a function gets more information to compute with, it should be able to give a more informative result. It makes little sense, for our purposes, to consider a function f such that, for example, $f(\bot_N, \bot_N) = 6$, $f(\bot_N, 3) = 8$, and $f(7,3) = \bot_N$. Since 6 and 8 are completely defined, neither approximates the other, and since $\bot_N$ is completely undefined, neither 6 nor 8 approximates $\bot_N$. We formalize this notion as follows.

Definition. Let $(D, \sqsubseteq_D)$ and $(E, \sqsubseteq_E)$ be partial orders. A function $f: D \to E$ is *monotonic* if, for all $d, d' \in D$, $d \sqsubseteq_D d'$ implies $f(d) \sqsubseteq_E f(d')$.

It is easy to see that any strict function $f: N_\bot^n \to N_\bot$ is monotonic, though the reverse is not necessarily true. (See Exercise 1.) Though we will not require computable functions to be strict, we will certainly expect them to be monotonic. For example, let eq: $N_\bot \times N_\bot \to N_\bot$ be the equality predicate on $N_\perp$, that is,

$$\text{eq}(x, y) = \begin{cases} 1 & \text{if } x = y \\ 0 & \text{if } x \neq y, \end{cases}$$

where as before 1 represents TRUE and 0 represents FALSE. Then eq is not monotonic, since $(\perp_N, \perp_N) \sqsubseteq_{N_\perp \times N_\perp} (\perp_N, 0)$ and $\text{eq}(\perp_N, \perp_N) = 1 \not\sqsubseteq_{N_\perp} 0 = \text{eq}(\perp_N, 0)$. Now let $\Phi_\perp(x, y)$ be the strict extension of the universal function $\Phi(x, y)$ defined in Chapter 4, and let $n_\perp$ be the strict extension of the function $n(x) = 0$. Then for all $x, y \in N$,

$$\text{HALT}(x, y) = \text{eq}(0, n_\perp(\Phi_\perp(x, y))).$$

But we showed in Chapter 4 that HALT(x, y) is not computable, so if $n_\perp$ and $\Phi_\perp$ are partially computable, then eq certainly is not.

We will make frequent use of the following simple theorem. First we introduce a new piece of notation. If f is a function with domain D, and $E \subseteq D$, then $f(E)$ denotes the set $\{f(e) \mid e \in E\}$.

Theorem 4.1. Let $(D, \sqsubseteq_D)$, $(E, \sqsubseteq_E)$ be partial orders and $f: D \to E$ monotonic. If C is a chain in $(D, \sqsubseteq_D)$, then $f(C)$ is a chain in $(E, \sqsubseteq_E)$.

Proof. Let $f(d_1), f(d_2) \in f(C)$. Then either $d_1 \sqsubseteq_D d_2$, which implies $f(d_1) \sqsubseteq_E f(d_2)$, or $d_2 \sqsubseteq_D d_1$, which implies $f(d_2) \sqsubseteq_E f(d_1)$. ■

Ordered sets like $(N_\perp^n, \sqsubseteq_{N_\perp^n})$ are fairly simple, since all chains in $(N_\perp^n, \sqsubseteq_{N_\perp^n})$ are finite, but when we go on to consider richer structures, we will require a property that is, in general, stronger than monotonicity. Let $(D, \sqsubseteq_D)$ and $(E, \sqsubseteq_E)$ be cpos, and let C be a chain in $(D, \sqsubseteq_D)$. C might be an infinite set, so that we reach $\sqcup C$ by way of an infinite chain of approximations. For a function $f: D \to E$ we would like to be able to reach $f(\sqcup C)$ by way of the approximations $\{f(c) \mid c \in C\}$.

Definition. Let $(D, \sqsubseteq_D)$ and $(E, \sqsubseteq_E)$ be partial orders. A function $f: D \to E$ is *continuous* if, for any chain C in $(D, \sqsubseteq_D)$ such that $\sqcup C$ exists, $\sqcup f(C)$ exists and $f(\sqcup C) = \sqcup f(C)$. $[D \to E]$ denotes the set of all continuous functions in $D \to E$. The *continuous function space ordering* on $[D \to E]$ determined by $(E, \sqsubseteq_E)$, which is denoted $\sqsubseteq_{[D \to E]}$, is the restriction of $\sqsubseteq_{D \to E}$ to $[D \to E]$.

Note that if $(D, \sqsubseteq_D)$ is a cpo then we can drop the reference to the existence of $\sqcup C$.

Theorem 4.2. Let $(D, \sqsubseteq_D)$, $(E, \sqsubseteq_E)$ be partial orders, and let $f \in D \to E$.

1. If f is continuous then it is monotonic.
2. If f is monotonic and C is a finite chain in $(D, \sqsubseteq_D)$, then $\sqcup C$ and $\sqcup f(C)$ exist and $f(\sqcup C) = \sqcup f(C)$.
3. If all chains in $(D, \sqsubseteq_D)$ are finite, then f is continuous if and only if it is monotonic.

Proof. Let f be continuous, and let $d_1, d_2 \in D$ be such that $d_1 \sqsubseteq_D d_2$. Then $\sqcup\{f(d_1), f(d_2)\} = f(\sqcup\{d_1, d_2\}) = f(d_2)$, so $f(d_1) \sqsubseteq_E f(d_2)$, and therefore f is monotonic.

Now, let f be monotonic, and let C be a finite chain in $(D, \sqsubseteq_D)$. Then $f(C)$ is a finite chain by Theorem 4.1, so $\sqcup C$ and $\sqcup f(C)$ both exist by Theorem 3.1. We have $\sqcup C \in C$ by Theorem 3.1, so $f(\sqcup C) \in f(C)$, which implies $f(\sqcup C) \sqsubseteq_E \sqcup f(C)$. Also, $c \sqsubseteq_D \sqcup C$ for all $c \in C$ implies $f(c) \sqsubseteq_E f(\sqcup C)$ for all $c \in C$ by the monotonicity of f, so that $f(\sqcup C)$ is an upper bound of $f(C)$. Therefore $\sqcup f(C) \sqsubseteq_E f(\sqcup C)$, and we have $f(\sqcup C) = \sqcup f(C)$ by the antisymmetry of $\sqsubseteq_E$.

Finally, part 3 follows immediately from parts 1 and 2. ∎

Suppose $(D, \sqsubseteq_D)$, $(E, \sqsubseteq_E)$ are cpos, $f: D \to E$ is monotonic, and C is a chain in $(D, \sqsubseteq_D)$. Then $\sqcup C$ exists, and $f(C)$ is a chain by Theorem 4.1, so $\sqcup f(C)$ exists. Therefore, in these circumstances we can drop the reference to the existence of both $\sqcup C$ and $\sqcup f(C)$. The following theorem simplifies matters a bit further and suggests a technique for proving continuity that is often more convenient than going to the definition.

Theorem 4.3. Let $(D, \sqsubseteq_D)$, $(E, \sqsubseteq_E)$ be cpos, and let $f \in D \to E$. Then f is continuous if and only if: (1) f is monotonic, and (2) $f(\sqcup C) \sqsubseteq_E \sqcup f(C)$ for all chains C in $(D, \sqsubseteq_D)$.

Proof. Let f be monotonic and let C be a chain in $(D, \sqsubseteq_D)$. Then for all $c \in C$, $c \sqsubseteq_D \sqcup C$ implies $f(c) \sqsubseteq_E f(\sqcup C)$, so $f(\sqcup C)$ is an upper bound of $f(C)$ and $\sqcup f(C) \sqsubseteq_E f(\sqcup C)$. Then by assumption (2) we have $f(\sqcup C) = \sqcup f(C)$, and f is continuous. The other direction follows from Theorem 4.2. ∎

It follows from Theorem 4.2 and Exercise 3.8 that in $N_\perp^n \to N_\perp$, for instance, monotonicity and continuity are equivalent properties. In general, however, they are not. For example, let $T: (N_\perp \to N_\perp) \to N_\perp$ be defined

$$T(f) = \begin{cases} 1 & \text{if } f(n) \neq \perp_N \text{ for all } n \in N \\ \perp_N & \text{otherwise.} \end{cases}$$

That is, $T(f)$ is true just in case f is a total function when its domain and range are restricted to N. We will simply say that f is a total function. T is monotonic since for any $g, h \in N_\perp \to N_\perp$, if $g \sqsubseteq_{N_\perp \to N_\perp} h$ then either g is not total and $T(g) = \perp_N \sqsubseteq_{N_\perp} T(h)$, or g is total, which implies that h must be total, so that $T(g) = 1 = T(h)$. However, T is not continuous. For all $m \in N$, let $n_m: N_\perp \to N_\perp$ be the "step function" defined by

$$n_m(x) = \begin{cases} \perp_N & \text{if } x = \perp_N \\ 0 & \text{if } x \neq \perp_N \text{ and } 0 \leq x \leq m \\ \perp_N & \text{if } x \neq \perp_N \text{ and } x > m. \end{cases}$$

Then $\{n_m \mid m \in N\}$ is a chain, and $T(n_m) = \perp_N$ for all $m \in N$, so $\sqcup\{T(n_m) \mid m \in N\} = \perp_N$. But $(\sqcup\{n_m \mid m \in N\})(x) = 0$ for all $x \in N$, so $T(\sqcup\{n_m \mid m \in N\}) = 1$.

In the next chapter continuity will play a major role in our treatment of computable functions.

With the help of the following lemmas we can prove a version of Theorems 3.7 and 3.8 for continuous functions.

Exchange Lemma. Let $(D, \sqsubseteq_D)$ and $(E, \sqsubseteq_E)$ be partial orders, let $(F, \sqsubseteq_F)$ be a cpo, let $f: D \times E \to F$ be monotonic, and let C_1 and C_2 be chains in $(D, \sqsubseteq_D)$, $(E, \sqsubseteq_E)$, respectively. Then

$$\sqcup\{\sqcup\{f(x,y) \mid y \in C_2\} \mid x \in C_1\}, \sqcup\{\sqcup\{f(x,y) \mid x \in C_1\} \mid y \in C_2\}$$

exist, and they are equal.

Proof. For all $c_1 \in C_1$, $\{f(c_1, y) \mid y \in C_2\}$ is a chain by the monotonicity of the unary function $f(c_1, y)$ and Theorem 4.1, so $\sqcup\{f(c_1, y) \mid y \in C_2\}$ exists. Also, if $c_1, c_1' \in C_1$ and $c_1 \sqsubseteq_D c_1'$, then for all $c_2 \in C_2$,

$$f(c_1, c_2) \sqsubseteq_F f(c_1', c_2) \sqsubseteq_F \sqcup\{f(c_1', y) \mid y \in C_2\},$$

so $\sqcup\{f(c_1, y) \mid y \in C_2\} \sqsubseteq_F \sqcup\{f(c_1', y) \mid y \in C_2\}$. Therefore,

$$\{\sqcup\{f(x,y) \mid y \in C_2\} \mid x \in C_1\}$$

is a chain and $\sqcup\{\sqcup\{f(x,y) \mid y \in C_2\} \mid x \in C_1\}$ exists. A similar argument holds for $\sqcup\{\sqcup\{f(x,y) \mid x \in C_1\} \mid y \in C_2\}$.

Let $c_1 \in C_1$. Then for all $c_2 \in C_2$,

$$f(c_1, c_2) \sqsubseteq_F \sqcup\{f(x, c_2) \mid x \in C_1\} \sqsubseteq_F \sqcup\{\sqcup\{f(x,y) \mid x \in C_1\} \mid y \in C_2\},$$

so

$$\sqcup\{f(c_1, y) \mid y \in C_2\} \sqsubseteq_F \sqcup\{\sqcup\{f(x,y) \mid x \in C_1\} \mid y \in C_2\}. \quad (4.1)$$

But (4.1) is true for all $c_1 \in C_1$, so

$$\sqcup\{\sqcup\{f(x,y) \mid y \in C_2\} \mid x \in C_1\} \sqsubseteq_F \sqcup\{\sqcup\{f(x,y) \mid x \in C_1\} \mid y \in C_2\}.$$

Similarly,

$$\sqcup\{\sqcup\{f(x,y) \mid x \in C_1\} \mid y \in C_2\} \sqsubseteq_F \sqcup\{\sqcup\{f(x,y) \mid y \in C_2\} \mid x \in C_1\},$$

so the lemma follows by the antisymmetry of $\sqsubseteq_F$. ∎

Let $(D, \sqsubseteq_D)$, $(E, \sqsubseteq_E)$ be cpos and define apply: $([D \to E] \times D) \to E$ by

$$\text{apply}(f, d) = f(d).$$

Then apply is monotonic, since if $(f, d_1) \sqsubseteq_{[D \to E] \times D} (g, d_2)$, then

$$\begin{aligned} \text{apply}(f, d_1) &= f(d_1) \\ &\sqsubseteq_E g(d_1) && \text{since } f \sqsubseteq_{[D \to E]} g \\ &\sqsubseteq_E g(d_2) && \text{since } d_1 \sqsubseteq_D d_2 \text{ and } g \text{ is monotonic} \\ &= \text{apply}(g, d_2). \end{aligned}$$

Now we can prove

Theorem 4.4. If $(D, \sqsubseteq_D)$ and $(E, \sqsubseteq_E)$ are cpos, then $([D \to E], \sqsubseteq_{[D \to E]})$ is a cpo. Moreover, if $\mathscr{F}$ is a chain in $([D \to E], \sqsubseteq_{[D \to E]})$, then $(\sqcup \mathscr{F})(d) = \sqcup(\mathscr{F}(d))$ for all $d \in D$.

Proof. The bottom element $\perp_{D \to E}$ defined in the proof of Theorem 3.8 is continuous, so $[D \to E]$ has a bottom element. By Theorem 3.11, we only need to show that for all chains $\mathscr{F}$ in $([D \to E], \sqsubseteq_{[D \to E]})$, $\sqcup_{D \to E} \mathscr{F} \in [D \to E]$, that is, the least upper bound of $\mathscr{F}$ in $(D \to E, \sqsubseteq_{D \to E})$, which we know to exist by Theorem 3.8, is continuous. Let $\mathscr{F}$ be a chain in $([D \to E], \sqsubseteq_{[D \to E]})$ and let C be a chain in $(D, \sqsubseteq_D)$. Then

$$\begin{aligned} &(\sqcup_{D \to E} \mathscr{F})(\sqcup C) \\ &\quad = \sqcup\{f(\sqcup C) \mid f \in \mathscr{F}\} && \text{by Theorem 3.7} \\ &\quad = \sqcup\{\sqcup\{f(c) \mid c \in C\} \mid f \in \mathscr{F}\} && \text{since each } f \in \mathscr{F} \text{ is continuous} \\ &\quad = \sqcup\{\sqcup\{\text{apply}(f, c) \mid c \in C\} \mid f \in \mathscr{F}\} \\ &\quad = \sqcup\{\sqcup\{\text{apply}(f, c) \mid f \in \mathscr{F}\} \mid c \in C\} && \text{by the exchange lemma} \\ &\quad = \sqcup\{\sqcup\{f(c) \mid f \in \mathscr{F}\} \mid c \in C\} \\ &\quad = \sqcup\{(\sqcup_{D \to E} \mathscr{F})(c) \mid c \in C\} && \text{by Theorem 3.7} \\ &\quad = \sqcup((\sqcup_{D \to E} \mathscr{F})(C)), \end{aligned}$$

so $\sqcup_{D \to E} \mathscr{F}$ is continuous, and $([D \to E], \sqsubseteq_{[D \to E]})$ is a cpo. ∎

We conclude this section with a result that will be applied in the next chapter. (Also, see Exercise 5.10.) The proof is very similar to the proof of the exchange lemma, so we leave it to Exercise 13.

Diagonal Lemma. Let $(D, \sqsubseteq_D)$ be a partial order and $(E, \sqsubseteq_E)$ a cpo, let $f: D \times D \to E$ be monotonic, and let C be a chain in $(D, \sqsubseteq_D)$. Then

$$\sqcup\{\sqcup\{f(x,y) \mid y \in C\} \mid x \in C\} = \sqcup\{f(x,x) \mid x \in C\}.$$

Exercises

1. Let $(D_1, \sqsubseteq_{D_1}), \ldots, (D_n, \sqsubseteq_{D_n}), (E, \sqsubseteq_E)$ be flat cpos.
   (a) Show that every strict function in $D_1 \times \cdots \times D_n \to E$ is monotonic.
   (b) Give an example of a monotonic function in $D_1 \times \cdots \times D_n \to E$ that is not strict.
   (c) Give a partial order $(D, \sqsubseteq_D)$ and a strict function in $D \to D$ that is not monotonic.
2. Let $f: N_\perp^n \to N_\perp$ satisfy
   $$f(x_1, \ldots, x_n) = h(g_1(x_1, \ldots, x_n), \ldots, g_m(x_1, \ldots, x_n)),$$
   where $h: N_\perp^m \to N_\perp$ and $g_i: N_\perp^n \to N_\perp$, $1 \le i \le n$, are strict. Show that f is strict.
3. Let $(D, \sqsubseteq_D), (E, \sqsubseteq_E)$ be partial orders, let $e \in E$, and let $f_e: D \to E$ be the constant function $f_e(d) = e$. Show that f_e is continuous.
4. Let $(D, \sqsubseteq_D)$ be a partial order, and let $\mathrm{Id}_D: D \to D$ be the identity function $\mathrm{Id}_D(d) = d$. Show that Id_D is continuous.
5. Let $(D_1, \sqsubseteq_{D_1}), \ldots, (D_n, \sqsubseteq_{D_n})$ be partial orders. Show that for $1 \le i \le n$, the projection function $\downarrow i: D_1 \times \cdots \times D_n \to D_i$ is continuous.
6. Let D, E be sets, let $f \in D \to E$, and define $\hat{f}: \mathscr{P}(D) \to \mathscr{P}(E)$ as $\hat{f}(A) = \{f(a) \mid a \in A\}$. Show that $\hat{f}$ is continuous.
7. Let $(D, \sqsubseteq_D), (E, \sqsubseteq_E), (F, \sqsubseteq_F)$ be cpos, and let
   $$\circ: [E \to F] \times [D \to E] \to (D \to F)$$
   be the composition operator on continuous functions. Show that the composition of continuous functions is a continuous function. That is, for all $f \in [E \to F]$, $g \in [D \to E]$, show that $f \circ g \in [D \to F]$.

8. Let $(D, \sqsubseteq_D)$, $(E_1, \sqsubseteq_{E_1}), \ldots, (E_n, \sqsubseteq_{E_n})$ be cpos, and define the function construct: $[D \to E_1] \times \cdots \times [D \to E_n] \to (D \to (E_1 \times \cdots \times E_n))$ as

$$\text{construct}(f_1, \ldots, f_n)(d) = (f_1(d), \ldots, f_n(d)).$$

Let $f_i \in [D \to E_i]$, $1 \le i \le n$. Show that $\text{construct}(f_1, \ldots, f_n) \in [D \to (E_1 \times \cdots \times E_n)]$.

9. Let $(D, \sqsubseteq_D)$, $(E, \sqsubseteq_E)$ be cpos, and let $\text{apply}_n\colon [D \to E] \to (D^n \to E^n)$ be defined as $\text{apply}_n(f)(d_1, \ldots, d_n) = (f(d_1), \ldots, f(d_n))$.

(a) Let $f \in [D \to E]$. Show that $\text{apply}_n(f) \in [D^n \to E^n]$.

(b) Show that $\text{apply}_n \in [[D \to E] \to [D^n \to E^n]]$, i.e., apply_n is continuous.

10. Let $(D, \sqsubseteq_D)$, $(E, \sqsubseteq_E)$, $(F, \sqsubseteq_F)$ be cpos and define curry: $[D \times E \to F] \to (D \to (E \to F))$ as $\text{curry}(f)(d)(e) = f(d, e)$. Let $f \in [D \times E \to F]$ and $d \in D$.

(a) Show that $\text{curry}(f)(d) \in [E \to F]$.

(b) Show that $\text{curry}(f) \in [D \to [E \to F]]$.

(c) Show that $\text{curry} \in [[D \times E \to F] \to [D \to [E \to F]]]$.

11. Let $(D, \sqsubseteq_D)$, $(E, \sqsubseteq_E)$ be cpos, let $D \underset{s}{\to} E$ be the set of all strict functions in $D \to E$, and let $\sqsubseteq_{D \underset{s}{\to} E}$ be the restriction of $\sqsubseteq_{D \to E}$ to $D \underset{s}{\to} E$. Show that $(D \underset{s}{\to} E, \sqsubseteq_{D \underset{s}{\to} E})$ is a cpo.

12. Let $(D, \sqsubseteq_D)$, $(E, \sqsubseteq_E)$ be cpos, let $D \underset{m}{\to} E$ be the set of all monotonic functions in $D \to E$, and let $\sqsubseteq_{D \underset{m}{\to} E}$ be the restriction of $\sqsubseteq_{D \to E}$ to $D \underset{m}{\to} E$. Show that $(L \underset{m}{\to} E, \sqsubseteq_{D \underset{m}{\to} E})$ is a cpo.

13. Prove the diagonal lemma.

14.* Let $(D, \sqsubseteq_D)$, $(E, \sqsubseteq_E)$ be isomorphic cpos, and let f be an isomorphism from $(D, \sqsubseteq_D)$ to $(E, \sqsubseteq_E)$. Show that f is continuous. [See Exercise 3.18 for the definition of *isomorphic partial orders*.]

15.* Let $(D_1, \sqsubseteq_{D_1}), \ldots, (D_n, \sqsubseteq_{D_n})$, $(E, \sqsubseteq_E)$ be partial orders and let $f \in D_1 \times \cdots \times D_n \to E$. We say that f is *monotonic* (*continuous*) *in the ith position*, $1 \le i \le n$, if for all $d_j \in D_j$, $1 \le j \le n$ and $j \ne i$, the unary function $f(d_1, \ldots, d_{i-1}, x, d_{i+1}, \ldots, d_n)$ is monotonic (respectively, continuous).

(a) Let $(D, \sqsubseteq_D)$, $(E, \sqsubseteq_E)$, $(F, \sqsubseteq_F)$ be cpos, let $f\colon D \times E \to F$ be monotonic, and let C be a chain in $(D \times E, \sqsubseteq_{D \times E})$. Show that $\sqcup\{\sqcup\{f(x, y) \mid x \in C \downarrow 1\} \mid y \in C \downarrow 2\}$ and $\sqcup f(C)$ exist, and that

$$\sqcup\{\sqcup\{f(x, y) \mid x \in C \downarrow 1\} \mid y \in C \downarrow 2\} = \sqcup f(C).$$

(b) Let $(D, \sqsubseteq_D)$, $(E, \sqsubseteq_E)$, $(F, \sqsubseteq_F)$ be cpos, and let $f \in D \times E \to F$. Use part (a) to show that f is continuous if and only if f is continuous in the first and second positions.

(c) Let $(D, \sqsubseteq_D)$ and $(E, \sqsubseteq_E)$ be cpos. Show that the function apply: $([D \to E] \times D) \to E$ is continuous.

(d) Show that $\circ$, defined in Exercise 7, is continuous.

(e) Generalize part (b) so that it applies to n-ary functions, $n \geq 1$.

(f) Show that construct, defined in Exercise 8, is continuous.

16.* A *Scott domain* is a bounded-complete algebraic cpo. [See Exercises 3.21 and 3.22 for the definitions of *bounded-complete* and *algebraic cpos*.]

(a) Show that if $(D, \sqsubseteq_D)$, $(E, \sqsubseteq_E)$ are Scott domains, then so is $(D \times E, \sqsubseteq_{D \times E})$.

(b) Show that $(N_\perp \to N_\perp, \sqsubseteq_{N_\perp \to N_\perp})$ is a Scott domain.

(c) Show that if $(D, \sqsubseteq_D)$, $(E, \sqsubseteq_E)$ are Scott domains, then so is $([D \to E], \sqsubseteq_{[D \to E]})$. [*Hint:* For each $d \in K(D)$, $e \in K(E)$, define $(d \searrow e)$: $D \to E$ as

$$(d \searrow e)(x) = \begin{cases} e & \text{if } d \sqsubseteq_D x \\ \perp_E & \text{otherwise.} \end{cases}$$

Show that every function $(d \searrow e)$ is continuous and compact. Then show that for every $d_1, \ldots, d_n \in K(D)$ and $e_1, \ldots, e_n \in K(E)$, the function $(d_1 \searrow e_1) \sqcup \cdots \sqcup (d_n \searrow e_n)$ is continuous and compact. Use these functions to show that $([D \to E], \sqsubseteq_{[D \to E]})$ is algebraic.]

(d) Give an algebraic cpo $(D, \sqsubseteq_D)$ such that $([D \to D], \sqsubseteq_{[D \to D]})$ is not algebraic.

5. Fixed Points

We will now prove a fundamental theorem that will facilitate our work on denotational semantics.

Definition. Let $(D, \sqsubseteq_D)$ be a partial order, and let $f \in D \to D$. An element $d \in D$ is a *fixed point* of f if $f(d) = d$, and it is the *least fixed point* of f in $(D, \sqsubseteq_D)$ if $d \sqsubseteq_D e$ for every fixed point $e \in D$ of f. The least fixed point of f in $(D, \sqsubseteq_D)$, if it exists, is denoted $\mu_{(D, \sqsubseteq_D)} f$.

We will generally omit the subscript and write μf when the partial order $(D, \sqsubseteq_D)$ is understood.

Fixed points play a fundamental role in denotational semantics, and continuity is important because it allows us to guarantee the existence of least fixed points. Before we can prove the fixed point theorem, we need some notation and a lemma. Let D be a set and let $f \in D \to D$. For each $n \in N$ we define a function $f^n: D \to D$, called the nth *iteration* of f, as follows:

$$f^0(x) = x$$

$$f^{n+1}(x) = f(f^n(x)).$$

Note that these equations can also be understood as defining a single binary function $f^y(x): D \times N \to D$.

Lemma 1. Let $(D, \sqsubseteq_D)$ be a partial order with bottom element $\perp_D$, and let $f: D \to D$ be monotonic. Then $f^n(\perp_D) \sqsubseteq_D f^{n+1}(\perp_D)$ for all $n \in N$, and $\{f^n(\perp_D) \mid n \in N\}$ is a chain in $(D, \sqsubseteq_D)$.

Proof. First we argue by induction on n that $f^n(\perp_D) \sqsubseteq_D f^{n+m}(\perp_D)$ for all $n, m \in N$. If $n = 0$, then $f^0(\perp_D) = \perp_D \sqsubseteq_D f^m(\perp_D)$, so assume $f^n(\perp_D) \sqsubseteq_D f^{n+m}(\perp_D)$ for all $m \in N$. Then for any $m \in N$,

$$\begin{aligned} f^{n+1}(\perp_D) = {} & f(f^n(\perp_D)) \\ \sqsubseteq_D {} & f(f^{n+m}(\perp_D)) && \text{by the induction hypothesis and the monotonicity of } f \\ = {} & f^{n+1+m}(\perp_D). \end{aligned}$$

The lemma now follows immediately. ■

Theorem 5.1 (Fixed Point Theorem for cpos). Let $(D, \sqsubseteq_D)$ be a cpo, and let $f: D \to D$ be continuous. Then the least fixed point μf exists, and $\mu f = \sqcup\{f^n(\perp_D) \mid n \in N\}$.

Proof. By Lemma 1, $\{f^n(\perp_D) \mid n \in N\}$ is a chain, so $\sqcup\{f^n(\perp_D) \mid n \in N\}$ exists. Moreover, $\sqcup\{f^n(\perp_D) \mid n \in N\}$ is a fixed point of f, since

$$\begin{aligned} & f(\sqcup\{f^n(\perp_D) \mid n \in N\}) \\ & = \sqcup f(\{f^n(\perp_D) \mid n \in N\}) && \text{by the continuity of } f \\ & = \sqcup\{f^{n+1}(\perp_D) \mid n \in N\} \\ & = \sqcup\left(\{f^{n+1}(\perp_D) \mid n \in N\} \cup \{f^0(\perp_D)\}\right) && \text{since } f^0(\perp_D) = \perp_D \\ & = \sqcup\{f^n(\perp_D) \mid n \in N\}. \end{aligned}$$

Finally, if e is any fixed point of f, then we argue by induction on n that $f^n(\bot_D) \sqsubseteq_D e$ for all $n \in N$. If $n = 0$ then $f^0(\bot_D) = \bot_D \sqsubseteq_D e$, so assume $f^n(\bot_D) \sqsubseteq_D e$. Then

$$\begin{aligned} f^{n+1}(\bot_D) &= f(f^n(\bot_D)) \\ &\sqsubseteq_D f(e) && \text{by the induction hypothesis and} \\ & && \text{the monotonicity of } f \\ &= e && \text{since } e \text{ is a fixed point of } f, \end{aligned}$$

so e is an upper bound of $\{f^n(\bot_D) \mid n \in N\}$. Therefore,

$$\sqcup\{f^n(\bot_D) \mid n \in N\} \sqsubseteq_D e,$$

and $\sqcup\{f^n(\bot_D) \mid n \in N\}$ is the least fixed point of f. ■

The fixed point theorem has a variety of applications. One, as we will show in the next chapter, is to justify recursive definitions of functions. Another is to justify inductive definitions of sets. Consider, for example, the definition of propositional formulas in Chapter 12. For simplicity we let $\mathscr{A}$, the set of atoms, be $\{\boldsymbol{p}, \boldsymbol{q}\}$, and we consider only formulas with the connectives $\neg$ and $\supset$, so that the alphabet $B = \{\boldsymbol{p}, \boldsymbol{q}, \neg, \supset, (,)\}$. An alternative statement of the definition of propositional formulas over $\mathscr{A}$ is the following: the set of propositional formulas over $\mathscr{A}$ is the smallest (with respect to $\subseteq$) subset of B^* that

1. contains $\mathscr{A}$,
2. is closed under the operation that transforms α to $\neg\,\alpha$, and
3. is closed under the operation that transforms α and β to $(\alpha \supset \beta)$.

In other words, the set of propositional formulas over $\mathscr{A}$ is the smallest set $X \subseteq B^*$ that satisfies

$$\mathscr{A} \cup \{\neg\,\alpha \mid \alpha \in X\} \cup \{(\alpha \supset \beta) \mid \alpha, \beta \in X\} \subseteq X. \tag{5.1}$$

Moreover, since (5.1) would still be satisfied if any element of X not required by 1, 2, or 3 were removed, and since we are looking for the smallest X which satisfies (5.1), we can rewrite (5.1) as the equality

$$X = \mathscr{A} \cup \{\neg\,\alpha \mid \alpha \in X\} \cup \{(\alpha \supset \beta) \mid \alpha, \beta \in X\}. \tag{5.2}$$

One way of looking at this equation is to consider the right side as a function $\Phi: \mathscr{P}(B^*) \to \mathscr{P}(B^*)$ that takes subsets $Z \subseteq B^*$ and transforms them to

$$\Phi(Z) = \mathscr{A} \cup \{\neg\,\alpha \mid \alpha \in Z\} \cup \{(\alpha \supset \beta) \mid \alpha, \beta \in Z\}.$$

A solution to (5.2), then, is some X such that, when Φ is applied to X, the result is still X; that is, X is a fixed point of Φ. For example, let Z be some arbitrary fixed point of Φ. Then

$$Z = \Phi(Z) = \mathscr{A} \cup \{\neg\, \alpha \mid \alpha \in Z\} \cup \{(\alpha \supset \beta) \mid \alpha, \beta \in Z\},$$

so Z is a solution to (5.2).

Now, the definition calls for the smallest such set, i.e., $\mu\Phi$, so for the definition to make sense we need to know that $\mu\Phi$ exists. This is where the fixed point theorem is useful. We have already shown that the partial order $(\mathscr{P}(D), \subseteq_{\mathscr{P}(D)})$ is a cpo for any set D, so $(\mathscr{P}(B^*), \subseteq_{\mathscr{P}(B^*)})$ is a cpo. If Φ is continuous then $\mu\Phi$ exists by the fixed point theorem. Let $\mathscr{C}$ be a chain of subsets of B^*. Then $\sqcup\mathscr{C} = \cup\mathscr{C}$ in $(\mathscr{P}(B^*), \subseteq_{\mathscr{P}(B^*)})$, and

$$\begin{aligned}
\Phi(\cup\mathscr{C}) &= \mathscr{A} \cup \{\neg\, \alpha \mid \alpha \in \cup\mathscr{C}\} \cup \{(\alpha \supset \beta) \mid \alpha, \beta \in \cup\mathscr{C}\} \\
&= \mathscr{A} \cup \big(\cup\big\{\{\neg\, \alpha \mid \alpha \in C\} \mid C \in \mathscr{C}\big\}\big) \\
&\quad \cup\big(\cup\big\{\{(\alpha \supset \beta) \mid \alpha, \beta \in C\} \mid C \in \mathscr{C}\big\}\big) \\
&= \cup\big\{\mathscr{A} \cup \{\neg\, \alpha \mid \alpha \in C\} \cup \{(\alpha \supset \beta) \mid \alpha, \beta \in C\} \mid C \in \mathscr{C}\big\} \\
&= \cup\{\Phi(C) \mid C \in \mathscr{C}\} \\
&= \cup\Phi(\mathscr{C}),
\end{aligned}$$

so Φ is continuous, and $\mu\Phi$ exists.

Note that, although the preceding definition of the set of propositional formulas mentions the operations that transform α to $\neg\, \alpha$ and α, β to $(\alpha \supset \beta)$, no mention is made of a process of building up the set from $\mathscr{A}$ by repeated application of these operations. The set we define is simply a certain solution to a certain equation. On the other hand, the definition given in Chapter 12 does mention repeated applications of these operations. The fixed point theorem, which not only tells us that $\mu\Phi$ exists, but also that $\mu\Phi = \cup\{\Phi^i(\varnothing) \mid i \in N\}$, makes the connection between these two versions of the definition. $\Phi(\varnothing), \Phi^2(\varnothing), \ldots,$ are subsets of $\mu\Phi$ built up by ever more applications of the formula building operations.

Another way of formalizing the notion of "repeated applications" is to give a context-free grammar[2] Γ such that $L(\Gamma)$ is the set of propositional formulas over $\mathscr{A}$. In particular, let Γ consist of the productions

$$\begin{array}{ll} S \to \neg\, S & S \to p \\ S \to (S \supset S) & S \to q, \end{array}$$

[2] The reader who is unfamiliar with Chapter 10 may skip to the definition of admissible predicates.

where S is the start symbol. By definition, $L(\Gamma) = \{u \in B^* \mid S \overset{*}{\Rightarrow} u\}$. Suppose $\alpha, \beta \in L(\Gamma)$. Then $S \overset{*}{\Rightarrow} \alpha$ and $S \overset{*}{\Rightarrow} \beta$, which implies

$$S \Rightarrow (S \supset S) \overset{*}{\Rightarrow} (\alpha \supset S) \overset{*}{\Rightarrow} (\alpha \supset \beta),$$

so $(\alpha \supset \beta) \in L(\Gamma)$. Similarly, if $\alpha \in L(\Gamma)$, then $S \Rightarrow \neg S \overset{*}{\Rightarrow} \neg \alpha$ and $\neg \alpha \in L(\Gamma)$.

It seems, then, that $L(\Gamma)$ is the set of propositional formulas over $\mathscr{A}$, but how can we prove it? By Theorem 1.4 of Chapter 10, $\alpha \in L(\Gamma)$ if and only if there is a derivation tree for α in Γ. We define the *height* of a derivation tree $\mathscr{T}$, denoted $h(\mathscr{T})$, as follows. If $\mathscr{T}$ consists of exactly one vertex, the root, then $h(\mathscr{T}) = 1$. If $\mathscr{T}$ consists of a root with successors $\nu_1, \ldots, \nu_n$, then

$$h(\mathscr{T}) = \max\{h(\mathscr{T}^{\nu_1}), \ldots, h(\mathscr{T}^{\nu_n})\} + 1,$$

where $\mathscr{T}^{\nu_i}$ is the subtree of $\mathscr{T}$ with root ν_i, $1 \leq i \leq n$. For each $n \in N$, we define

$$L_n = \{u \in B^* \mid \text{there is a derivation tree } \mathscr{T} \text{ for } u \text{ in } \Gamma \text{ with } h(\mathscr{T}) \leq n + 1\}.$$

Clearly, $L(\Gamma) = \bigcup_{n \in N} L_n$. If we can show that $\Phi^n(\varnothing) = L_n$ for all $n \in N$, then we will have

$$\mu\Phi = \bigcup_{n \in N} \Phi^n(\varnothing) = \bigcup_{n \in N} L_n = L(\Gamma).$$

If $n = 0$ then $\Phi^0(\varnothing) = \varnothing = L_0$, since the only derivation tree of height 1 with root S does not yield a word in B^*. For $n > 0$ we argue by induction on n. If $n = 1$ then $\Phi^1(\varnothing) = \{\boldsymbol{p}, \boldsymbol{q}\} = L_1$. For $n + 2$ we have

$$\begin{aligned} \Phi^{n+2}(\varnothing) &= \Phi(\Phi^{n+1}(\varnothing)) \\ &= \Phi(L_{n+1}) \qquad \text{by the induction hypothesis} \\ &= \{\boldsymbol{p}, \boldsymbol{q}\} \cup \{\neg \alpha \mid \alpha \in L_{n+1}\} \cup \{(\alpha \supset \beta) \mid \alpha, \beta \in L_{n+1}\}. \end{aligned}$$

Also, it is clear from the definition of L_{n+2} and the nature of Γ that

$$L_{n+2} = L_{n+1} \cup \{\neg \alpha \mid \alpha \in L_{n+1}\} \cup \{(\alpha \supset \beta) \mid \alpha, \beta \in L_{n+1}\}.$$

Since $\{\boldsymbol{p}, \boldsymbol{q}\} \subseteq L_{n+1}$, we have $\Phi^{n+2}(\varnothing) \subseteq L_{n+2}$. On the other hand, by Lemma 1 we have $L_{n+1} = \Phi^{n+1}(\varnothing) \subseteq \Phi^{n+2}(\varnothing)$, which implies $L_{n+2} \subseteq \Phi^{n+2}(\varnothing)$, and so we have $\Phi^{n+2}(\varnothing) = L_{n+2}$, completing the induction and the proof that $\mu\Phi = L(\Gamma)$.

These various treatments of the definition of propositional formulas help to illustrate some of the ideas in the next two chapters. On the one hand, we have an abstract mathematical characterization of the set of propositional formulas as the smallest solution to equation (5.2). On the other hand, we have $L(\Gamma) = \{u \in B^* \mid S \overset{*}{\Rightarrow} u\}$, the set of words generated from S by derivations in Γ. We can give a "deterministic" characterization of $L(\Gamma)$ as the set of words in B^* for which there exists a leftmost derivation from the start symbol. A somewhat more abstract characterization of the same set is $\bigcup_{n \in N} L_n$, which is given in terms of derivation trees, without any reference to the details of the choices made in the construction of derivation sequences. In the terminology of semantics, $\mu\Phi$ is a denotational definition, and $L(\Gamma)$ is an operational definition. The link between $\mu\Phi$ and $L(\Gamma)$ is given by the fixed point theorem and its characterization of $\mu\Phi$ as $\sqcup\{\Phi^i(\emptyset) \mid i \in N\}$.

A useful tool for reasoning about fixed points is embodied in the fixed point induction principle.

Definition. Let $(D, \sqsubseteq_D)$ be a cpo. A predicate $P(x)$ on D is *admissible* if the following holds for all chains C in $(D, \sqsubseteq_D)$:

$$\text{if } P(c) \text{ for all } c \in C, \text{ then } P(\sqcup C).$$

Theorem 5.2 (Fixed Point Induction Principle). Let $(D, \sqsubseteq_D)$ be a cpo, $f: D \to D$ a continuous function, and $P(x)$ an admissible predicate on D. If

1. $P(\perp_D)$, and
2. $P(f^i(\perp_D))$ implies $P(f^{i+1}(\perp_D))$ for all $i \in N$,

then $P(\mu f)$.

Proof. Ordinary induction shows that $P(f^i(\perp_D))$ holds for all $i \in N$. The set $\{f^i(\perp_D) \mid i \in N\}$ is a chain by Lemma 1, so $P(\sqcup\{f^i(\perp_D) \mid i \in N\})$ holds by the admissibility of $P(x)$, and of course $\sqcup\{f^i(\perp_D) \mid i \in N\} = \mu f$. ■

For example, suppose we define Y as the smallest subset of B^* that

1. contains $\{p, q, \neg p, \neg q\}$,
2. is closed under the operation that transforms α to $\neg \alpha$,
3. is closed under the operation that transforms α and β to $(\alpha \supset \beta)$.

Is Y equal to the set of propositional formulas over $\mathscr{A}$? Let

$$\Psi(Z) = \{p, q, \neg p, \neg q\} \cup \{\neg \alpha \mid \alpha \in Z\} \cup \{(\alpha \supset \beta) \mid \alpha, \beta \in Z\}.$$

Ψ is continuous, by an argument almost identical to the argument that Φ is continuous, so $\mu\Psi$ exists. The question, then, is whether $\mu\Psi = \mu\Phi$. We argue by fixed point induction that $\mu\Psi \subseteq \mu\Phi$. Let $P(X)$ be the predicate $X \subseteq \mu\Phi$. $P(X)$ is admissible, since if $\mathscr{C}$ is a chain in $(\mathscr{P}(B^*), \subseteq_{\mathscr{P}(B^*)})$ and $C \subseteq \mu\Phi$ for all $C \in \mathscr{C}$, then clearly $\bigcup\mathscr{C} \subseteq \mu\Phi$. The bottom element of $(\mathscr{P}(B^*), \subseteq_{\mathscr{P}(B^*)})$ is $\varnothing$, and $\varnothing \subseteq \mu\Phi$. Also, if $\Psi^i(\varnothing) \subseteq \mu\Phi$, then

$$\begin{aligned}
\Psi^{i+1}(\varnothing) &= \Psi(\Psi^i(\varnothing)) \\
&\subseteq \Psi(\mu\Phi) \text{ by the induction hypothesis and the monotonicity of } \Psi \\
&= \{\boldsymbol{p}, \boldsymbol{q}, \neg \boldsymbol{p}, \neg \boldsymbol{q}\} \cup \{\neg \alpha \mid \alpha \in \mu\Phi\} \cup \{(\alpha \supset \beta) \mid \alpha, \beta \in \mu\Phi\} \\
&= \{\boldsymbol{p}, \boldsymbol{q}\} \cup \{\neg \alpha \mid \alpha \in \mu\Phi\} \cup \{(\alpha \supset \beta) \mid \alpha, \beta \in \mu\Phi\} \\
&\quad \text{since } \neg \boldsymbol{p}, \neg \boldsymbol{q} \in \{\neg \alpha \mid \alpha \in \mu\Phi\} \\
&= \Phi(\mu\Phi) \\
&= \mu\Phi.
\end{aligned}$$

Therefore, by the fixed point induction principle, we have $\mu\Psi \subseteq \mu\Phi$. A similar induction argument on the admissible predicate $X \subseteq \mu\Psi$ shows that $\mu\Phi \subseteq \mu\Psi$, so we have $\mu\Psi = \mu\Phi$. The point is that both definitions characterize the same set, and the second definition, with its unnecessary reference to $\neg \boldsymbol{p}$ and $\neg \boldsymbol{q}$, can be simplified to the first definition. When fixed points are used to define the meaning of programs, the same technique can be used to show that two programs are equivalent, or to simplify programs.

For sets defined like the set of propositional formulas over $\mathscr{A}$, fixed point induction is closely related to a form of induction known as *structural induction*. Let $P(x)$ be a property of propositional formulas over $\mathscr{A}$ (rather than sets of formulas). If

1. $P(\alpha)$ for every $\alpha \in \mathscr{A}$,
2. $P(\alpha)$ implies $P(\neg \alpha)$ for all propositional formulas α over $\mathscr{A}$,
3. $P(\alpha)$ and $P(\beta)$ implies $P((\alpha \supset \beta))$ for all propositional formulas α, β over $\mathscr{A}$,

then the structural induction principle allows us to conclude $P(\alpha)$ for all propositional formulas α over $\mathscr{A}$. The assumptions $P(\alpha)$ and $P(\beta)$ in 2 and 3 are the *structural induction hypotheses*. To see why the conclusion is valid, let $\hat{P}(X)$ be the property on sets of propositional formulas over $\mathscr{A}$ defined

$$\hat{P}(A) \quad \text{if and only if} \quad P(\alpha) \text{ for all } \alpha \in A.$$

$\hat{P}(X)$ is admissible: if $\mathscr{C}$ is a chain in $(\mathscr{P}(B^*), \subseteq_{\mathscr{P}(B^*)})$ and $\hat{P}(C)$ for all $C \in \mathscr{C}$, then for each $C \in \mathscr{C}$ we have $P(\alpha)$ for all $\alpha \in C$, which implies $P(\alpha)$ for all $\alpha \in \cup\mathscr{C}$, that is, $\hat{P}(\cup\mathscr{C})$. Now, assumptions 1, 2, and 3 enable us to prove

- $\hat{P}(\varnothing)$,
- $\hat{P}(\Phi^i(\varnothing))$ implies $\hat{P}(\Phi^{i+1}(\varnothing))$ for all $i \in N$,

from which we conclude, by fixed point induction, $\hat{P}(\mu\Phi)$, i.e., $P(\alpha)$ for all $\alpha \in \mu\Phi$, the set of propositional formulas over $\mathscr{A}$. We will give several structural induction arguments in the next chapter.

The reader will recall that we proved a theorem in Chapter 4 that was also called a fixed point theorem. The two are closely related. The recursion theorem from Chapter 4 is sometimes called, for historical reasons, the *second recursion theorem*. In fact, the earlier fixed point theorem, which follows from (and just as easily implies) the second recursion theorem, is itself sometimes called the *second recursion theorem*. The fixed point theorem in this chapter is a version of a classical theorem from computability theory that is sometimes called the *first recursion theorem*.

The names of these two recursion theorems come from the fact that they can both be used in proving functions to be partially computable, particularly functions defined by recursion. However, there is a significant distinction between the two theorems. The fixed point theorem in this chapter gives a fixed point for each continuous function on a cpo. In particular, if $F: [N_\perp \to N_\perp] \to [N_\perp \to N_\perp]$ is continuous, then we get a function $f \in [N_\perp \to N_\perp]$ such that $F(f) = f$. On the other hand, the fixed point theorem in Chapter 4 is more directly concerned with programs than functions. A computable function g gets the effect of transforming a function $\Phi_{\#(\mathscr{P})}$ to the function $\Phi_{g(\#(\mathscr{P}))}$ by acting on the (code of the) program that computes $\Phi_{\#(\mathscr{P})}$, and the fixed point theorem in Chapter 4 gives a program $\mathscr{Q}$ such that $\Phi_{\#(\mathscr{Q})} = \Phi_{g(\#(\mathscr{Q}))}$. It would be reasonable, then, to call that earlier theorem a *syntactic* fixed point theorem and to call the current theorem a *semantic* fixed point theorem.

Just as the second recursion theorem gives a partially computable function $\Phi_{\#(\mathscr{Q})}$ that satisfies $\Phi_{\#(\mathscr{Q})} = \Phi_{g(\#(\mathscr{Q}))}$, so too does the first recursion theorem give, for the appropriate kind of F, a partially computable μF. We will say no more about the first recursion theorem in its classical form (other than to direct the reader to Exercise 11), but as we shall see, the main point of the next chapter is to use the fixed point theorem for cpos to define partially computable functions.

Exercises

1. Give functions $f, g, h: N_\perp \to N_\perp$ such that

(a) f has no fixed points;

(b) g has exactly one fixed point;

(c) h has infinitely many fixed points.

2. Give a function $f: N_\perp \to N_\perp$ such that f is not continuous and μf exists.

3. Give a partial order $(D, \sqsubseteq_D)$ and a function $f: D \to D$ such that f is continuous, f has at least one fixed point, and μf does not exist.

4. Give a fixed point characterization of the set of **W**-terms defined at the beginning of Chapter 13, where **W** is some vocabulary.

5. Let $(D, \sqsubseteq_D)$ be a cpo, let $f \in [D \to D]$, let $E = \{e \in D \mid f(e) = e\}$, and let $\sqsubseteq_E$ be the restriction of $\sqsubseteq_D$ to E. Show that $(E, \sqsubseteq_E)$ is a cpo.

6. Let $(D, \sqsubseteq_D)$ be a cpo.

(a) Let $P(X)$ be the predicate on $\mathscr{P}(D)$ defined "X is a finite set." In the context of $(\mathscr{P}(D), \subseteq_{\mathscr{P}(D)})$, is P admissible?

(b) Let $Q(X)$ be the predicate on $\mathscr{P}(D)$ defined "X is an infinite set." In the context of $(\mathscr{P}(D), \subseteq_{\mathscr{P}(D)})$, is Q admissible?

(c) Let $R(f)$ be the predicate on $D \to D$ defined "f is strict." In the context of $(D \to D, \sqsubseteq_{D \to D})$, is R admissible?

7. Let $(D, \sqsubseteq_D)$, $(E, \sqsubseteq_E)$ be cpos, let $P(x)$, $Q(x)$ be admissible predicates on D, and let $R(x, y)$ be an admissible predicate on $D \times E$.

(a) Show that $P(x)$ & $Q(x)$ is admissible.

(b) Show that $P(x) \vee Q(x)$ is admissible.

(c) Show that $(\forall d \in D)R(d, y)$ is an admissible predicate on E.

8.* **(a)** Let Γ be a context-free grammar with variables $\mathscr{V}$ and terminals T. Give a fixed point characterization of $L(\Gamma)$. [*Hint:* Define a function Φ such that $\mu\Phi(V) = \{w \in T^* \mid V \overset{*}{\Rightarrow} w\}$ for all $V \in \mathscr{V}$.]

(b) Let Γ be the grammar with $\mathscr{V} = \{S\}$, $T = \{a\}$, and the single production $S \to aSa$. Show by fixed point induction that $L(\Gamma) = \varnothing$.

9.* Let $(D, \sqsubseteq_D)$ be a complete lattice, and let $f: D \to D$ be monotonic. Show that μf exists and that $\mu f = \sqcap\{d \in D \mid f(d) \sqsubseteq_D d\}$. [See Exercise 3.20 for the definition of *complete lattices*.]

10.* Let $(D, \sqsubseteq_D)$ be a cpo, and define $\mu_D: [D \to D] \to D$ as $\mu_D(f) = \mu f$ for all $f \in [D \to D]$.

(a) Let $f, g \in [D \to D]$, and suppose $f \sqsubseteq_{[D \to D]} g$. Show by induction on n that $f^n \sqsubseteq_{[D \to D]} g^n$ for all $n \in N$.

(b) Let $\mathscr{F}$ be a chain in $([D \to D], \sqsubseteq_{[D \to D]})$. Show by induction on n that $(\sqcup\mathscr{F})^n = \sqcup\{f^n \mid f \in \mathscr{F}\}$. [*Hint:* Use the diagonal lemma.]

(c) Show that μ_D is continuous. [*Hint:* Use parts (a) and (b) and the exchange lemma.]

11.* As in Exercise 3.14, $N \underset{p}{\rightarrow} N$ is the set of all partial functions on N. For each finite function $\theta = \{(x_1, y_1), \ldots, (x_n, y_n)\}$ in $N \underset{p}{\rightarrow} N$, $n \geq 0$, we encode θ as $\tilde{\theta} \in N$, where

$$\tilde{\theta} = \textstyle\prod_{i=1}^{n} p_{x_i}^{y_i+1}.$$

A function $F: (N \underset{p}{\rightarrow} N) \rightarrow (N \underset{p}{\rightarrow} N)$ is a *recursive operator* if there is some partially computable function $h(y, x)$ such that

$$F(g)(x) = z \text{ if and only if } h(\tilde{\theta}, x) = z \text{ for some } \theta \subseteq g.$$

(a) Let $G: (N \underset{p}{\rightarrow} N) \rightarrow (N \underset{p}{\rightarrow} N)$ be defined $G(f)(x) = 2 \cdot f(x)$. Show that G is a recursive operator.

(b) Show that every recursive operator is monotonic and continuous.

(c) Show that, if F is a recursive operator, then there is a computable function f such that $F(\Phi_x) = \Phi_{f(x)}$ for all $x \in N$.

(d) (First Recursion Theorem) Prove that, for every recursive operator F, μF exists and is partially computable.

17

Denotational Semantics of Recursion Equations

1. Syntax

Now that we have developed a theory of approximation orders, we can define recursion equations and give them a denotational semantics. The operational semantics, given in the next chapter, will show that the functions defined by recursion equations are, in a reasonable sense, computable.

As in Chapter 13,[1] where we defined the terms and formulas of quantification theory, we begin with a small alphabet

$$A = \{\mathbf{t}, \mathbf{x}, \mathbf{f}, \mathbf{I}, \times\ , \rightarrow, \#, (,,,), =\}$$

of symbols that are always available. The members of

$$\mathrm{VAR_T} = \{\mathbf{t}\,\mathbf{I}^{[i]} \mid i \in N\}$$

are *type variables*, and a *type* is

- a type variable, or
- $\tau_1 \times \cdots \times \tau_n$, $n \geq 1$, where $\tau_1, \ldots, \tau_n$ are type variables, or
- $\tau_1 \times \cdots \times \tau_n \rightarrow \tau$, $n \geq 1$, where $\tau_1, \ldots, \tau_n, \tau$ are type variables.

[1] Knowledge of Chapter 13 is not assumed, but there is a substantial overlap in the treatment of the syntax and semantics of terms.

These three kinds of types are *individual types*, *product types*, and *function types*, respectively. For a given $\boldsymbol{\tau} \in \mathrm{VAR_T}$, the members of $\mathrm{VAR}_\mathrm{I}^{\boldsymbol{\tau}} = \{\mathbf{x\#\tau\#I}^{[i]} \mid i \in N\}$ are *individual variables of type* $\boldsymbol{\tau}$, and

$$\mathrm{VAR_I} = \bigcup_{\boldsymbol{\tau} \in \mathrm{VAR_T}} \mathrm{VAR}_\mathrm{I}^{\boldsymbol{\tau}}$$

is the set of all individual variables. For a function type $\boldsymbol{\tau}_1 \times \cdots \times \boldsymbol{\tau}_n \rightarrow \boldsymbol{\tau}$, the members of

$$\mathrm{VAR}_\mathrm{F}^{\boldsymbol{\tau}_1 \times \cdots \times \boldsymbol{\tau}_n \rightarrow \boldsymbol{\tau}} = \left\{\mathbf{f\#}\boldsymbol{\tau}_1 \times \cdots \times \boldsymbol{\tau}_n \rightarrow \boldsymbol{\tau}\mathbf{\#I}^{[i]} \mid i \in N\right\}$$

are the *function variables of type* $\boldsymbol{\tau}_1 \times \cdots \times \boldsymbol{\tau}_n \rightarrow \boldsymbol{\tau}$, and

$$\mathrm{VAR_F} = \bigcup_{\boldsymbol{\tau}_1, \ldots, \boldsymbol{\tau}_n, \boldsymbol{\tau} \in \mathrm{VAR_T}} \mathrm{VAR}_\mathrm{F}^{\boldsymbol{\tau}_1 \times \cdots \times \boldsymbol{\tau}_n \rightarrow \boldsymbol{\tau}}$$

is the set of all function variables. Also, $\mathrm{VAR} = \mathrm{VAR_I} \cup \mathrm{VAR_F}$. We will let **X**, **Y**, **Z** (possibly subscripted) stand for individual variables and **F**, **G**, **H** (possibly subscripted) stand for function variables.[2] Occasionally we will write **V** for an arbitrary variable of either kind. We will also use more suggestive names in the examples. If **O** is any of the syntactic objects defined in this section (**W**-terms, **W**-programs, etc.), then $IV(\mathbf{O})$ is the set of all individual variables which occur in **O**, $FV(\mathbf{O})$ is the set of all function variables which occur in **O**, and $V(\mathbf{O}) = IV(\mathbf{O}) \cup FV(\mathbf{O})$.

A *typed vocabulary* is a pair $(\mathbf{W}, \tau)$, where **W** is a finite set of *function symbols* distinct from the symbols in A, and τ is a function on **W** such that for each $\mathbf{f} \in \mathbf{W}$, $\tau(\mathbf{f})$ is either an individual type or a function type. We say that $\tau(\mathbf{f})$ is the *type* of **f**. **f** is a *constant symbol* if $\tau(\mathbf{f})$ is an individual type, and it is a *proper function symbol* otherwise. Given τ, it is easy to determine the *arity* of any $\mathbf{f} \in \mathbf{W}$, denoted ar(**f**). If **f** is a constant symbol, then ar(**f**) = 0, and if $\tau(\mathbf{f}) = \boldsymbol{\tau}_1 \times \cdots \times \boldsymbol{\tau}_n \rightarrow \boldsymbol{\tau}$, then ar(**f**) = n. It will also be useful to supplement τ with the functions δ and ρ, which give the *domain type* and *range type*, respectively, of symbols in **W**. For constant symbols $\mathbf{c} \in \mathbf{W}$, $\delta(\mathbf{c})$ is undefined and $\rho(\mathbf{c}) = \tau(\mathbf{c})$, and for proper function symbols $\mathbf{f} \in \mathbf{W}$ with $\tau(\mathbf{f}) = \boldsymbol{\tau}_1 \times \cdots \times \boldsymbol{\tau}_n \rightarrow \boldsymbol{\tau}$,

$$\delta(\mathbf{f}) = \boldsymbol{\tau}_1 \times \cdots \times \boldsymbol{\tau}_n \quad \text{and} \quad \rho(\mathbf{f}) = \boldsymbol{\tau}.$$

---

[2] Note that the letter **X** is not itself an individual variable. It is what we sometimes call a *metavariable*. That is, it is a variable, which we use in talking about the syntax of recursion equations, whose *values* are individual variables. Similarly, **F** is a metavariable whose values are function variables. We also use metavariables such as $\boldsymbol{\tau}$, **c**, **f**, **t**, and **P**, whose values are type variables, constant symbols, function symbols, terms, and programs, respectively.

We extend τ and ρ to VAR and δ to VAR_F in the obvious way. For example, $\tau(\mathbf{x\#\tau\#I}^{[i]}) = \boldsymbol{\tau}$ and $\delta(\mathbf{f\#\tau}_1 \times \cdots \times \boldsymbol{\tau}_n \rightarrow \boldsymbol{\tau}\mathbf{\#I}^{[i]}) = \boldsymbol{\tau}_1 \times \cdots \times \boldsymbol{\tau}_n$. For a typed vocabulary $(\mathbf{W}, \tau)$, $TV(\mathbf{W}, \tau)$ is the set of all type variables that occur in the types of all of the symbols in $\mathbf{W}$. We will omit τ and write $TV(\mathbf{W})$ for $TV(\mathbf{W}, \tau)$.

Let $(\mathbf{W}, \tau)$ be a typed vocabulary. For any $\boldsymbol{\tau} \in TV(\mathbf{W})$, a **W**-*term of type* $\boldsymbol{\tau}$ is

- an individual variable of type $\boldsymbol{\tau}$, or
- $\mathbf{c} \in \mathbf{W}$, where $\tau(\mathbf{c}) = \boldsymbol{\tau}$, or
- $\mathbf{g}(\mathbf{t}_1, \ldots, \mathbf{t}_n)$, where $\mathbf{g} \in \mathbf{W}$, $\tau(\mathbf{g}) = \boldsymbol{\tau}_1 \times \cdots \times \boldsymbol{\tau}_n \rightarrow \boldsymbol{\tau}$, and $\mathbf{t}_i$ is a **W**-term of type $\boldsymbol{\tau}_i$, $1 \le i \le n$, or
- $\mathbf{F}(\mathbf{t}_1, \ldots, \mathbf{t}_n)$, where $\mathbf{F} \in \text{VAR}_F$, $\tau(\mathbf{F}) = \boldsymbol{\tau}_1 \times \cdots \times \boldsymbol{\tau}_n \rightarrow \boldsymbol{\tau}$, and $\mathbf{t}_i$ is a **W**-term of type $\boldsymbol{\tau}_i$, $1 \le i \le n$.

We extend τ so that $\tau(\mathbf{t}) = \boldsymbol{\tau}$ for any term $\mathbf{t}$ of type $\boldsymbol{\tau}$. For $V_0 \subseteq \text{VAR}$, $\text{TM}_{\mathbf{W}}^{\boldsymbol{\tau}}(V_0)$ is the set of all **W**-terms $\mathbf{t}$ of type $\boldsymbol{\tau}$ such that $V(\mathbf{t}) \subseteq V_0$, and $\text{TM}_{\mathbf{W}}(V_0)$ is the set of all **W**-terms $\mathbf{t}$ such that $V(\mathbf{t}) \subseteq V_0$. Also, we will write $\text{TM}_{\mathbf{W}}^{\boldsymbol{\tau}}$ for $\text{TM}_{\mathbf{W}}^{\boldsymbol{\tau}}(\emptyset)$ and $\text{TM}_{\mathbf{W}}$ for $\text{TM}_{\mathbf{W}}(\emptyset)$. Terms in $\text{TM}_{\mathbf{W}}$, that is, **W**-terms without variables, are sometimes called *ground* **W**-*terms*.

For example, let **N** be a type variable, and let $(\mathbf{W}_1, \tau_1)$ be the typed vocabulary with $\mathbf{W}_1 = \{\mathbf{0}, \mathbf{s}\}$, $\tau_1(\mathbf{0}) = \mathbf{N}$ and $\tau_1(\mathbf{s}) = \mathbf{N} \rightarrow \mathbf{N}$. We have $TV(\mathbf{W}_1) = \{\mathbf{N}\}$ and $\text{TM}_{\mathbf{W}_1} = \{\mathbf{0}, \mathbf{s}(\mathbf{0}), \mathbf{s}(\mathbf{s}(\mathbf{0})), \ldots\}$. This is a vocabulary suitable for naming the natural numbers. We call terms of the form

$$\overbrace{\mathbf{s}(\cdots \mathbf{s}}^{n}(\mathbf{0}) \cdots), \quad n \in N,$$

numerals, which we will generally write as $\mathbf{n}$ or $\mathbf{s}^n(\mathbf{0})$.

Now, let **NL** be a type variable distinct from **N**, and let $(\mathbf{W}_2, \tau_2)$ be the typed vocabulary with $\mathbf{W}_2 = \{\mathbf{0}, \mathbf{s}, \mathbf{nil}, \mathbf{cons}\}$, $\tau_2(\mathbf{0}) = \mathbf{N}$, $\tau_2(\mathbf{s}) = \mathbf{N} \rightarrow \mathbf{N}$, $\tau_2(\mathbf{nil}) = \mathbf{NL}$, and $\tau_2(\mathbf{cons}) = \mathbf{N} \times \mathbf{NL} \rightarrow \mathbf{NL}$. Then $TV(\mathbf{W}_2) = \{\mathbf{N}, \mathbf{NL}\}$ and $\text{TM}_{\mathbf{W}_2}$ is

$$\text{TM}_{\mathbf{W}_1} \cup \{\mathbf{nil}, \mathbf{cons}(\mathbf{0}, \mathbf{nil}), \mathbf{cons}(\mathbf{s}(\mathbf{0}), \mathbf{nil}), \mathbf{cons}(\mathbf{0}, \mathbf{cons}(\mathbf{0}, \mathbf{nil})), \ldots\}.$$

We might use this vocabulary for naming lists of numbers. The idea is that a list is either empty or it is constructed from a first element and a list of all succeeding elements. (The reader familiar with the programming language LISP will recognize **cons** and **nil**.)

A **W**-*recursion equation* is an equation of the form $\mathbf{F}(\mathbf{X}_1, \ldots, \mathbf{X}_n) = \mathbf{t}$, where, for some $\boldsymbol{\tau} \in TV(\mathbf{W})$

1. $\mathbf{X}_1, \ldots, \mathbf{X}_n$ are distinct individual variables, **F** is a function variable, and $\tau(\mathbf{F}) = \tau(\mathbf{X}_1) \times \cdots \times \tau(\mathbf{X}_n) \rightarrow \boldsymbol{\tau}$, and
2. $\mathbf{t} \in \text{TM}_{\mathbf{W}}^{\boldsymbol{\tau}}(\{\mathbf{X}_1, \ldots, \mathbf{X}_n\} \cup \text{VAR}_F)$.

If $\mathbf{E}$ is the $\mathbf{W}$-recursion equation $\mathbf{F}(\mathbf{X}_1, \ldots, \mathbf{X}_n) = \mathbf{t}$, then $\mathbf{F}$ is the *principal function variable* of $\mathbf{E}$, denoted $PF(\mathbf{E})$, and any function variable that occurs in $\mathbf{t}$ is an *auxiliary function variable* of $\mathbf{E}$. Note that a function variable can be both principal and auxiliary in a given equation. $AF(\mathbf{E})$ is the set of auxiliary function variables of $\mathbf{E}$. A *$\mathbf{W}$-recursion program* (or simply *$\mathbf{W}$-program*) is a finite set $\{\mathbf{E}_1, \ldots, \mathbf{E}_n\}$, $n \geq 0$, of $\mathbf{W}$-recursion equations such that

1. $PF(\mathbf{E}_i) \neq PF(\mathbf{E}_j)$ for $1 \leq i < j \leq n$, and
2. $\bigcup_{i=1}^{n} AF(\mathbf{E}_i) \subseteq \{PF(\mathbf{E}_1), \ldots, PF(\mathbf{E}_n)\}$.

If equation $\mathbf{E}$ in $\mathbf{W}$-program $\mathbf{P}$ is $\mathbf{F}(\mathbf{X}_1, \ldots, \mathbf{X}_n) = \mathbf{t}$, then $\mathbf{E}$ is the *defining equation* for $\mathbf{F}$ in $\mathbf{P}$. The first restriction in the definition of $\mathbf{W}$-programs prevents inconsistencies, and the second ensures that every function variable that occurs on the right side of any equation is defined. When some program $\mathbf{P}$ is given and $\mathbf{F}(\mathbf{X}_1, \ldots, \mathbf{X}_n) = \mathbf{t}$ is the defining equation for $\mathbf{F}$ in $\mathbf{P}$, we will sometimes write rhs($\mathbf{F}$) to denote the term $\mathbf{t}$ on the righthand side of the equation.

Note that we require each function to be defined by exactly one equation, while in Chapter 3 we used two equations to define a function by recursion. For example,

$$+(x, 0) = x$$
$$+(x, y + 1) = s(+(x, y))$$

is a (somewhat informal) definition of addition. Another way of describing addition is

$$+(x, y) = \begin{cases} x & \text{if } y = 0 \\ s(+(x, y \dot{-} 1)) & \text{otherwise,} \end{cases} \tag{1.1}$$

which can be construed as a single equation if the if-then-else test is itself a function. That is, given the function

$$\text{if}(b, x, y) = \begin{cases} x & \text{if } b = \text{TRUE} \\ y & \text{otherwise,} \end{cases}$$

we can rewrite (1.1) as

$$+(x, y) = \text{if}(y = 0, x, s(+(x, y \dot{-} 1))). \tag{1.2}$$

Of course, we also need the predicate $y = 0$ and the predecessor function $y \dot{-} 1$ for (1.2) to be meaningful. Therefore, we impose the following conditions on the vocabularies we will use.

Let **Bool** be some particular type variable and let **tt**, **ff** be two new symbols. (It does not matter which type variable we choose for **Bool**, but it will remain fixed throughout.) A *standard constructor vocabulary* is any vocabulary $(\mathbf{W}_c, \tau_c)$ such that $\mathbf{tt}, \mathbf{ff} \in \mathbf{W}_c$, with $\tau_c(\mathbf{tt}) = \tau_c(\mathbf{ff}) = \mathbf{Bool}$, and such that for each $\boldsymbol{\tau} \in TV(\mathbf{W}_c)$ there is at least one constant symbol $\mathbf{c} \in \mathbf{W}_c$ with $\tau_c(\mathbf{c}) = \boldsymbol{\tau}$. The latter requirement is not strictly necessary, but it will turn out to be convenient. We create a set of *built-in function symbols* for a standard constructor vocabulary $(\mathbf{W}_c, \tau_c)$ as follows. Let $\mathbf{W}_c^- = \mathbf{W}_c - \{\mathbf{tt}, \mathbf{ff}\}$ and let ρ_c be the range type function derived from τ_c. For each $\boldsymbol{\tau} \in TV(\mathbf{W}_c)$ we create the new symbol $\mathbf{if}_{\boldsymbol{\tau}}$, for each $\mathbf{f} \in \mathbf{W}_c^-$ we define the set of new symbols

$$B(\mathbf{f}) = \{\mathbf{is\_f}\} \cup \{\mathbf{f}_i^{-1} \mid 1 \le i \le \mathrm{ar}(\mathbf{f})\},$$

and we define

$$B(\mathbf{W}_c) = \{\mathbf{if}_{\boldsymbol{\tau}} \mid \boldsymbol{\tau} \in TV(\mathbf{W}_c)\} \cup \bigcup_{\mathbf{f} \in \mathbf{W}_c^-} B(\mathbf{f}).$$

Note that $\{\mathbf{f}_i^{-1} \mid 1 \le i \le \mathrm{ar}(\mathbf{f})\} = \varnothing$ if $\mathbf{f}$ is a constant symbol. We assign types to these new symbols with $\tau_{B(\mathbf{W}_c)}$:

$$\tau_{B(\mathbf{W}_c)}(\mathbf{if}_{\boldsymbol{\tau}}) = \mathbf{Bool} \times \boldsymbol{\tau} \times \boldsymbol{\tau} \to \boldsymbol{\tau} \qquad \text{for each } \boldsymbol{\tau} \in TV(\mathbf{W}_c)$$

$$\tau_{B(\mathbf{W}_c)}(\mathbf{is\_f}) = \boldsymbol{\tau} \to \mathbf{Bool} \qquad \text{where } \rho_c(\mathbf{f}) = \boldsymbol{\tau}$$

$$\tau_{B(\mathbf{W}_c)}(\mathbf{f}_i^{-1}) = \boldsymbol{\tau} \to \boldsymbol{\tau}_i \qquad \text{where } \tau_c(\mathbf{f}) = \boldsymbol{\tau}_1 \times \cdots \times \boldsymbol{\tau}_n \to \boldsymbol{\tau}.$$

A *standard vocabulary* is any typed vocabulary $(\mathbf{W}, \tau)$ such that

$$(\mathbf{W}, \tau) = \left(\mathbf{W}_c \cup B(\mathbf{W}_c), \tau_c \cup \tau_{B(\mathbf{W}_c)}\right)$$

for some standard constructor vocabulary $(\mathbf{W}_c, \tau_c)$. The symbols in $\mathbf{W}_c$ are *constructor symbols:* they are used to build up data objects. **tt** and **ff**, in particular, will be used to represent TRUE and FALSE. The **is_f** symbols are *discriminator function symbols:* they are used to determine how an object is constructed. The $\mathbf{f}_i^{-1}$ symbols are *selector function symbols:* they are used to decompose compound objects.

For example, we can expand the typed vocabulary $(\mathbf{W}_1, \tau_1)$ given above to the standard constructor vocabulary $(\mathbf{W}_3, \tau_3)$, where $\mathbf{W}_3 = \{\mathbf{tt}, \mathbf{ff}, \mathbf{0}, \mathbf{s}\}$, $\tau_3(\mathbf{tt}) = \tau_3(\mathbf{ff}) = \mathbf{Bool}$, $\tau_3(\mathbf{0}) = \mathbf{N}$, and $\tau_3(\mathbf{s}) = \mathbf{N} \to \mathbf{N}$. Then

$$B(\mathbf{W}_3) = \{\mathbf{if}_{\mathbf{Bool}}, \mathbf{if}_{\mathbf{N}}, \mathbf{is\_0}, \mathbf{is\_s}, \mathbf{s}_1^{-1}\},$$

where

$$\tau_{B(\mathbf{W}_3)}(\mathbf{if}_{\mathbf{Bool}}) = \mathbf{Bool} \times \mathbf{Bool} \times \mathbf{Bool} \to \mathbf{Bool}$$

$$\tau_{B(\mathbf{W}_3)}(\mathbf{if}_{\mathbf{N}}) = \mathbf{Bool} \times \mathbf{N} \times \mathbf{N} \to \mathbf{N}$$

$$\tau_{B(\mathbf{W}_3)}(\mathbf{is\_0}) = \mathbf{N} \to \mathbf{Bool}$$

$$\tau_{B(\mathbf{W}_3)}(\mathbf{is\_s}) = \mathbf{N} \to \mathbf{Bool}$$

$$\tau_{B(\mathbf{W}_3)}(\mathbf{s}_1^{-1}) = \mathbf{N} \to \mathbf{N},$$

and we can rewrite Eq. (1.2) as

$$+(\mathbf{X}, \mathbf{Y}) = \mathbf{if}_{\mathbf{N}}(\mathbf{is\_0}(\mathbf{Y}), \mathbf{X}, \mathbf{s}(+(\mathbf{X}, \mathbf{s}_1^{-1}(\mathbf{Y})))), \tag{1.3}$$

where $+ \in VAR_F$.

Similarly, we can expand $(\mathbf{W}_2, \tau_2)$ to $(\mathbf{W}_4, \tau_4)$, where $\mathbf{W}_4 = \{\mathbf{tt}, \mathbf{ff}, \mathbf{0}, \mathbf{s}, \mathbf{nil}, \mathbf{cons}\}$, $\tau_4(\mathbf{tt}) = \tau_4(\mathbf{ff}) = \mathbf{Bool}$, $\tau_4(\mathbf{0}) = \mathbf{N}$, $\tau_4(\mathbf{s}) = \mathbf{N} \to \mathbf{N}$, $\tau_4(\mathbf{nil}) = \mathbf{NL}$, and $\tau_4(\mathbf{cons}) = \mathbf{N} \times \mathbf{NL} \to \mathbf{NL}$. Then

$$B(\mathbf{W}_4) = B(\mathbf{W}_3) \cup \{\mathbf{if}_{\mathbf{NL}}, \mathbf{is\_nil}, \mathbf{is\_cons}, \mathbf{cons}_1^{-1}, \mathbf{cons}_2^{-1}\}.$$

Henceforth, we let $(\mathbf{W}_{\mathbf{N}}, \tau_{\mathbf{N}})$ be the standard vocabulary based on $(\mathbf{W}_3, \tau_3)$ and we let $(\mathbf{W}_{\mathbf{NL}}, \tau_{\mathbf{NL}})$ be the standard vocabulary based on $(\mathbf{W}_4, \tau_4)$. That is,

$$\mathbf{W}_{\mathbf{N}} = \{\mathbf{tt}, \mathbf{ff}, \mathbf{0}, \mathbf{s}, \mathbf{if}_{\mathbf{Bool}}, \mathbf{if}_{\mathbf{N}}, \mathbf{is\_0}, \mathbf{is\_s}, \mathbf{s}_1^{-1}\}$$

and

$$\mathbf{W}_{\mathbf{NL}} = \mathbf{W}_{\mathbf{N}} \cup \{\mathbf{nil}, \mathbf{cons}, \mathbf{if}_{\mathbf{NL}}, \mathbf{is\_nil}, \mathbf{is\_cons}, \mathbf{cons}_1^{-1}, \mathbf{cons}_2^{-1}\}.$$

Generally we will just write τ for $\tau_{\mathbf{N}}$ or $\tau_{\mathbf{NL}}$.

Note that we intend to interpret $\mathbf{s}_1^{-1}$ as the predecessor function in Eq. (1.3). At this point, of course, (1.3) has no meaning at all. The task of giving a meaning to equations like (1.3) begins in the next section.

Exercises

1. Let $\tau(\mathbf{X}) = \mathbf{N}$ and $\tau(\mathbf{F}) = \mathbf{N} \times \mathbf{N} \to \mathbf{N}$. Describe $TM_{\mathbf{W}_{\mathbf{N}}}(\{\mathbf{X}, \mathbf{F}\})$.
2. Let $\mathbf{W}_c = \{\mathbf{tt}, \mathbf{ff}, \mathbf{0}, \mathbf{s}, \mathbf{leaf}, \mathbf{tree}\}$, and let $\tau_c(\mathbf{tt}) = \tau_c(\mathbf{ff}) = \mathbf{Bool}$, $\tau_c(\mathbf{0}) = \mathbf{N}$, $\tau_c(\mathbf{s}) = \mathbf{N} \to \mathbf{N}$, $\tau_c(\mathbf{leaf}) = \mathbf{T}$, and $\tau_c(\mathbf{tree}) = \mathbf{N} \times \mathbf{T} \times \mathbf{T} \to \mathbf{T}$.
   - **(a)** Describe $TM_{\mathbf{W}_c}^{\boldsymbol{\tau}}$ for each $\boldsymbol{\tau} \in TV(\mathbf{W}_c)$.
   - **(b)** Describe $(B(\mathbf{W}_c), \tau_{B(\mathbf{W}_c)})$.

(c) Let $(\mathbf{W}, \tau) = (\mathbf{W}_c \cup B(\mathbf{W}_c), \tau_c \cup \tau_{B(\mathbf{W}_c)})$. Describe $\mathrm{TM}^{\boldsymbol{\tau}}_{\mathbf{W}}$ for each $\boldsymbol{\tau} \in TV(\mathbf{W})$.

3. Let $\tau(\mathbf{X}) = \mathbf{N}$, $\tau(\mathbf{Y}) = \mathbf{NL}$, $\tau(\mathbf{F}) = \mathbf{N} \times \mathbf{N} \rightarrow \mathbf{N}$, $\tau(\mathbf{G}) = \mathbf{N} \rightarrow \mathbf{N}$, and $\tau(\mathbf{H}) = \mathbf{N} \rightarrow \mathbf{NL}$. Which of the following are $\mathbf{W}_{\mathbf{NL}}$-terms?
   (a) $\mathbf{F(s(X))}$.
   (b) $\mathbf{cons(G(X), cons(X, nil))}$.
   (c) $\mathbf{cons(nil, cons(0, Y))}$.
   (d) $\mathbf{cons}_1^{-1}\mathbf{(H(0))}$.
   (e) $\mathbf{if}_{\mathbf{N}}\mathbf{(ff, cons(0, nil), 0)}$.
   (f) $\mathbf{if}_{\mathbf{NL}}\mathbf{(ff, cons(0, nil), H(if}_{\mathbf{N}}\mathbf{(ff, X, 0)))}$.

4. Assume that each of the following are $\mathbf{W}_{\mathbf{NL}}$-terms. Give the types of the variables in each term.
   (a) $\mathbf{if}_{\mathbf{N}}\mathbf{(F(0), G(0), s(H(0)))}$.
   (b) $\mathbf{s}_1^{-1}\mathbf{(F(cons(X, G(s(Y)))))}$.
   (c) $\mathbf{is\_cons(F(s}_1^{-1}\mathbf{(X), if}_{\mathbf{N}}\mathbf{(G(0), H(G(Y)), 0), X))}$.
   (d) $\mathbf{F(is\_0(F(X)))}$.

5. Describe the values of $\boldsymbol{\tau}$ and the types of **X**, **Y**, **F**, **G**, **H** that make $\mathbf{F(if}_{\boldsymbol{\tau}}\mathbf{(tt, G(F(X), Y), H(s(Y))))}$ a $\mathbf{W}_{\mathbf{NL}}$-term.

6. Let $(\mathbf{W}, \tau)$ be a vocabulary, let $V \subseteq \mathrm{VAR}$, and extend τ to $(A \cup \mathbf{W})^*$ so that $\tau(\mathbf{w}) = \boldsymbol{\tau}$ if and only if $\mathbf{w} = \boldsymbol{\phi}\mathbf{u}$ for some $\boldsymbol{\phi} \in \mathbf{W} \cup V$ with $\rho(\boldsymbol{\phi}) = \boldsymbol{\tau}$.
   (a) Give a fixed point definition of $\mathrm{TM}_{\mathbf{W}}(V)$ in the manner of the definition of the set of propositional formulas given in Section 5 of Chapter 16.
   (b) State and prove a structural induction principle for $\mathrm{TM}_{\mathbf{W}}(V)$.

2. Semantics of Terms

We develop the semantics of **W**-programs in several stages. In this section we work on the semantics of terms, beginning with the semantics of vocabularies. We will work exclusively with standard vocabularies, so throughout the rest of this chapter we take $(\mathbf{W}, \tau)$ to be some arbitrary standard vocabulary based on some standard constructor vocabulary $(\mathbf{W}_c, \tau_c)$. We will generally refer simply to **W** rather than $(\mathbf{W}, \tau)$.

Definition. A *type assignment* for **W** is a function $\mathscr{T}$ with domain $TV(\mathbf{W})$ such that

1. for each $\boldsymbol{\tau} \in TV(\mathbf{W})$, $\mathscr{T}(\boldsymbol{\tau})$ is a partial order $(D_{\mathscr{T}(\boldsymbol{\tau})}, \sqsubseteq_{\mathscr{T}(\boldsymbol{\tau})})$ with bottom element $\bot_{\mathscr{T}(\boldsymbol{\tau})}$, and, in particular,

2. $\mathscr{T}(\mathbf{Bool})$ is the flat cpo on some set with exactly two elements.

$\mathscr{T}$ is a *complete type assignment* for **W** if $\mathscr{T}(\boldsymbol{\tau})$ is a cpo for each $\boldsymbol{\tau} \in TV(\mathbf{W})$.

A type assignment for **W** gives a meaning for each type variable in $TV(\mathbf{W})$. For example, if $\mathscr{T}_{\mathbf{N}}(\mathbf{Bool})$ is the flat cpo on {TRUE, FALSE} and $\mathscr{T}_{\mathbf{N}}(\mathbf{N}) = (N_{\perp}, \sqsubseteq_{N_{\perp}})$, then $\mathscr{T}_{\mathbf{N}}$ is a complete type assignment for $\mathbf{W}_{\mathbf{N}}$.

When $\mathscr{T}$ is understood, we will write (Bool, $\sqsubseteq_{\text{Bool}}$) for $\mathscr{T}(\mathbf{Bool})$ and $\perp_{\text{Bool}}$ for the bottom element of $\mathscr{T}(\mathbf{Bool})$. For an arbitrary $\boldsymbol{\tau} \in TV(\mathbf{W})$, we will often write $(D_{\boldsymbol{\tau}}, \sqsubseteq_{\boldsymbol{\tau}})$ for $(D_{\mathscr{T}(\boldsymbol{\tau})}, \sqsubseteq_{\mathscr{T}(\boldsymbol{\tau})})$ and $\perp_{\boldsymbol{\tau}}$ for $\perp_{\mathscr{T}(\boldsymbol{\tau})}$. Also, we will sometimes write

$$(D_{\boldsymbol{\tau}_1 \times \cdots \times \boldsymbol{\tau}_n}, \sqsubseteq_{\boldsymbol{\tau}_1 \times \cdots \times \boldsymbol{\tau}_n})$$

for

$$\left(D_{\mathscr{T}(\boldsymbol{\tau}_1)} \times \cdots \times D_{\mathscr{T}(\boldsymbol{\tau}_n)}, \sqsubseteq_{D_{\mathscr{T}(\boldsymbol{\tau}_1)} \times \cdots \times D_{\mathscr{T}(\boldsymbol{\tau}_n)}}\right).$$

In particular, $D_{\delta(\mathbf{F})} = D_{\boldsymbol{\tau}_1 \times \cdots \times \boldsymbol{\tau}_n}$ if $\tau(\mathbf{F}) = \boldsymbol{\tau}_1 \times \cdots \times \boldsymbol{\tau}_n \rightarrow \boldsymbol{\tau}$.

It will be useful to define the following notation. For sets D, E and $f: D \rightarrow E$, ran f is the range of f. Also, if $e \in E$, then ran $e = \{e\}$. In effect, we are treating e as a function of 0 arguments.

Definition. Let $\mathscr{T}$ be a type assignment for **W**. A *$\mathscr{T}$-interpretation* for **W** is a function $\mathscr{I}$ with domain **W** that satisfies the following conditions.

1. For all constant symbols $\mathbf{c} \in \mathbf{W}_c$ with $\tau(\mathbf{c}) = \boldsymbol{\tau}$, $\mathscr{I}(\mathbf{c}) \in D_{\boldsymbol{\tau}} - \{\perp_{\boldsymbol{\tau}}\}$. We will write tt for $\mathscr{I}(\mathbf{tt})$ and ff for $\mathscr{I}(\mathbf{ff})$.
2. For all proper function symbols $\mathbf{f} \in \mathbf{W}_c$ with $\tau(\mathbf{f}) = \boldsymbol{\tau}_1 \times \cdots \times \boldsymbol{\tau}_n \rightarrow \boldsymbol{\tau}$,
   a. $\mathscr{I}(\mathbf{f}) \in D_{\boldsymbol{\tau}_1} \times \cdots \times D_{\boldsymbol{\tau}_n} \rightarrow D_{\boldsymbol{\tau}}$;
   b. if $\mathscr{I}(\mathbf{f})(d_1, \ldots, d_n) = \mathscr{I}(\mathbf{f})(e_1, \ldots, e_n) \neq \perp_{\boldsymbol{\tau}}$, then $(d_1, \ldots, d_n) = (e_1, \ldots, e_n)$;
   c. if $d_i \neq \perp_{\boldsymbol{\tau}_i}$, $1 \leq i \leq n$, then $\mathscr{I}(\mathbf{f})(d_1, \ldots, d_n) \neq \perp_{\boldsymbol{\tau}}$.
3. For all $\mathbf{f}, \mathbf{g} \in \mathbf{W}_c$ such that $\rho(\mathbf{f}) = \rho(\mathbf{g})$, ran $\mathscr{I}(\mathbf{f})$ and ran $\mathscr{I}(\mathbf{g})$ can have at most $\perp_{\rho(\mathbf{f})}$ in common; that is,

$$(\text{ran}\,\mathscr{I}(\mathbf{f}) \cap \text{ran}\,\mathscr{I}(\mathbf{g})) - \{\perp_{\rho(\mathbf{f})}\} = \varnothing.$$

4. For all $\boldsymbol{\tau} \in TV(\mathbf{W}), \mathscr{I}(\mathbf{if}_{\boldsymbol{\tau}})$: Bool $\times D_{\boldsymbol{\tau}} \times D_{\boldsymbol{\tau}} \rightarrow D_{\boldsymbol{\tau}}$ is defined

$$\mathscr{I}(\mathbf{if}_{\boldsymbol{\tau}})(b, d, e) = \begin{cases} d & \text{if } b = \text{tt} \\ e & \text{if } b = \text{ff} \\ \perp_{\boldsymbol{\tau}} & \text{if } b = \perp_{\text{Bool}}. \end{cases}$$

5. For all constant symbols $\mathbf{c} \in \mathbf{W}_c^-$ with $\tau(\mathbf{c}) = \boldsymbol{\tau}$, $\mathscr{I}(\mathbf{is\_c})$: $D_{\boldsymbol{\tau}} \to$ Bool is defined

$$\mathscr{I}(\mathbf{is\_c})(d) = \begin{cases} \text{tt} & \text{if } d = \mathscr{I}(\mathbf{c}) \\ \text{ff} & \text{if } d \neq \bot_{\boldsymbol{\tau}} \text{ and } d \neq \mathscr{I}(\mathbf{c}) \\ \bot_{\text{Bool}} & \text{if } d = \bot_{\boldsymbol{\tau}}; \end{cases}$$

for all proper function symbols $\mathbf{f} \in \mathbf{W}_c$ with $\tau(\mathbf{f}) = \boldsymbol{\tau}_1 \times \cdots \times \boldsymbol{\tau}_n \to \boldsymbol{\tau}$, $\mathscr{I}(\mathbf{is\_f})$: $D_{\boldsymbol{\tau}} \to$ Bool is defined

$$\mathscr{I}(\mathbf{is\_f})(d) = \begin{cases} \text{tt} & \text{if } d \neq \bot_{\boldsymbol{\tau}} \text{ and } d \in \text{ran}\,\mathscr{I}(\mathbf{f}) \\ \text{ff} & \text{if } d \neq \bot_{\boldsymbol{\tau}} \text{ and } d \notin \text{ran}\,\mathscr{I}(\mathbf{f}) \\ \bot_{\text{Bool}} & \text{if } d = \bot_{\boldsymbol{\tau}}. \end{cases}$$

6. For all proper function symbols $\mathbf{f} \in \mathbf{W}_c$ with $\tau(\mathbf{f}) = \boldsymbol{\tau}_1 \times \cdots \times \boldsymbol{\tau}_n \to \boldsymbol{\tau}$ and for all $1 \leq i \leq n$, $\mathscr{I}(\mathbf{f}_i^{-1})$: $D_{\boldsymbol{\tau}} \to D_{\boldsymbol{\tau}_i}$ is defined

$$\mathscr{I}(\mathbf{f}_i^{-1})(d) = \begin{cases} d_i & \text{if } d \neq \bot_{\boldsymbol{\tau}} \text{ and } d = \mathscr{I}(\mathbf{f})(d_1, \ldots, d_n) \text{ for} \\ & \text{some } (d_1, \ldots, d_n) \in D_{\boldsymbol{\tau}_1} \times \cdots \times D_{\boldsymbol{\tau}_n} \\ \bot_{\boldsymbol{\tau}_i} & \text{otherwise.} \end{cases}$$

If $\mathscr{I}(\mathbf{f})$ is continuous for all proper function symbols $\mathbf{f} \in \mathbf{W}$, then $\mathscr{I}$ is a *continuous $\mathscr{T}$-interpretation.* A **W**-*structure* is a pair $\Sigma = (\mathscr{T}, \mathscr{I})$, where $\mathscr{T}$ is a type assignment for **W** and $\mathscr{I}$ is a $\mathscr{T}$-interpretation for **W**. Σ is a *complete* **W**-*structure* if $\mathscr{T}$ is a complete type assignment, and it is a *continuous* **W**-*structure* if $\mathscr{I}$ is a continuous $\mathscr{T}$-interpretation.

A $\mathscr{T}$-interpretation for **W** gives a meaning to each symbol in **W**, using the objects made available by $\mathscr{T}$ in the sets $D_{\boldsymbol{\tau}}$. Conditions 4, 5, and 6 require a specific interpretation for the built-in function symbols. Note in particular that $\mathscr{I}(\mathbf{f}_i^{-1})$ is a well-defined function because of condition 2b. Conditions 2c and 3 are imposed to make certain information about the objects of a **W**-structure available at the syntactic level of **W**-terms. Condition 2c implies that the meaning of a ground term is never the bottom element, so that it makes sense, for example, to replace $\mathbf{s}_1^{-1}(\mathbf{s}(\mathbf{0}))$ with **0**, since $\mathscr{I}(\mathbf{s})(\mathscr{I}(\mathbf{0})) \neq \bot_N$. Condition 3 implies that it makes sense to replace a term such as $\mathbf{is\_f}(\mathbf{g}(\mathbf{c}))$ with **ff**. As we will see in the next chapter, the replacement of terms by equivalent terms is the basis of the operational semantics of recursion equations, so these conditions are included to make the operational semantics work correctly.

For an example, let $\mathbf{W}_N$ and $\mathcal{T}_N$ be as before. If

$$\mathcal{I}_N(\mathbf{tt}) = \text{TRUE}$$

$$\mathcal{I}_N(\mathbf{ff}) = \text{FALSE}$$

$$\mathcal{I}_N(\mathbf{0}) = 0$$

$$\mathcal{I}_N(\mathbf{s})(m) = \begin{cases} \bot_N & \text{if } m = \bot_N \\ m+1 & \text{otherwise} \end{cases}$$

$$\mathcal{I}_N(\mathbf{if}_{\mathbf{Bool}})(b,d,e) = \begin{cases} d & \text{if } b = \text{TRUE} \\ e & \text{if } b = \text{FALSE} \\ \bot_{\text{Bool}} & \text{if } b = \bot_{\text{Bool}} \end{cases}$$

$$\mathcal{I}_N(\mathbf{if}_{\mathbf{N}})(b,d,e) = \begin{cases} d & \text{if } b = \text{TRUE} \\ e & \text{if } b = \text{FALSE} \\ \bot_N & \text{if } b = \bot_{\text{Bool}} \end{cases}$$

$$\mathcal{I}_N(\mathbf{is\_0})(d) = \begin{cases} \text{TRUE} & \text{if } d = 0 \\ \text{FALSE} & \text{if } d \neq \bot_N \text{ and } d > 0 \\ \bot_{\text{Bool}} & \text{if } d = \bot_N \end{cases}$$

$$\mathcal{I}_N(\mathbf{is\_s})(d) = \begin{cases} \text{TRUE} & \text{if } d \neq \bot_N \text{ and } d > 0 \\ \text{FALSE} & \text{if } d = 0 \\ \bot_{\text{Bool}} & \text{if } d = \bot_N \end{cases}$$

$$\mathcal{I}_N(\mathbf{s}_1^{-1})(d) = \begin{cases} d-1 & \text{if } d \neq \bot_N \text{ and } d > 0 \\ \bot_N & \text{otherwise,} \end{cases}$$

then $\mathcal{I}_N$ is a $\mathcal{T}_N$-assignment for $\mathbf{W}_N$. We will write Σ_N for $(\mathcal{T}_N, \mathcal{I}_N)$. It is easy to check that Σ_N is complete and continuous.

We now have a way of interpreting the symbols of **W**, but before we can give a meaning to arbitrary terms, we need a way of interpreting variables.

Definition. Let $\mathcal{T}$ be a type assignment for **W** and V a set of variables. A *variable assignment* for V based on $\mathcal{T}$ is a function α with domain V such that

1. $\alpha(\mathbf{X}) \in D_{\boldsymbol{\tau}}$ for each individual variable $\mathbf{X} \in V$ with $\tau(\mathbf{X}) = \boldsymbol{\tau}$, and
2. $\alpha(\mathbf{F}) \in D_{\boldsymbol{\tau}_1} \times \cdots \times D_{\boldsymbol{\tau}_n} \to D_{\boldsymbol{\tau}}$ for each function variable $\mathbf{F} \in V$ with $\tau(\mathbf{F}) = \boldsymbol{\tau}_1 \times \cdots \times \boldsymbol{\tau}_n \rightarrow \boldsymbol{\tau}$.

α is a *continuous variable assignment* for V if $\alpha(\mathbf{F})$ is continuous for each function variable $\mathbf{F} \in V$. $\mathcal{A}_{\mathcal{T}}(V)$ is the set of all variable assignments for V

based on $\mathscr{T}$, and $\mathscr{CA}_{\mathscr{T}}(V)$ is the set of all continuous variable assignments for V based on $\mathscr{T}$.

Let $(\mathscr{T}, \mathscr{I})$ be a **W**-structure and V a set of variables. For any $\alpha \in \mathscr{A}_{\mathscr{T}}(V)$, we extend α to a function $\overline{\alpha}_{\mathscr{I}}$ with domain $\mathrm{TM}_{\mathbf{W}}(V)$ as follows:

$$\overline{\alpha}_{\mathscr{I}}(\mathbf{c}) = \mathscr{I}(\mathbf{c}) \qquad \text{for all constant symbols } \mathbf{c} \in \mathbf{W}$$

$$\overline{\alpha}_{\mathscr{I}}(\mathbf{X}) = \alpha(\mathbf{X}) \qquad \text{for all individual variables } \mathbf{X} \in V$$

$$\overline{\alpha}_{\mathscr{I}}(\mathbf{f}(\mathbf{t}_1, \ldots, \mathbf{t}_n)) = \mathscr{I}(\mathbf{f})(\overline{\alpha}_{\mathscr{I}}(\mathbf{t}_1), \ldots, \overline{\alpha}_{\mathscr{I}}(\mathbf{t}_n)) \qquad \text{where } \mathbf{f} \in \mathbf{W}$$

$$\overline{\alpha}_{\mathscr{I}}(\mathbf{F}(\mathbf{t}_1, \ldots, \mathbf{t}_n)) = \alpha(\mathbf{F})(\overline{\alpha}_{\mathscr{I}}(\mathbf{t}_1), \ldots, \overline{\alpha}_{\mathscr{I}}(\mathbf{t}_n)) \qquad \text{where } \mathbf{F} \in V.$$

$\overline{\alpha}_{\mathscr{I}}$ is a function we can use to assign a meaning to any term in $\mathrm{TM}_{\mathbf{W}}(V)$. Note that $\varnothing$ is the unique assignment in $\mathscr{A}_{\mathscr{T}}(\varnothing)$, and if $\mathbf{t} \in \mathrm{TM}_{\mathbf{W}}$, i.e., $\mathbf{t}$ contains no variables, then $\overline{\varnothing}_{\mathscr{I}}$ is sufficient for interpreting $\mathbf{t}$. When $\mathscr{I}$ is understood, we will often write $\overline{\alpha}$ for $\overline{\alpha}_{\mathscr{I}}$.

For example, let $V = \{\mathbf{X}, \mathbf{Y}, \mathbf{F}\}$, and let

$$\alpha(\mathbf{X}) = 3, \qquad \alpha(\mathbf{Y}) = 5, \qquad \alpha(\mathbf{F}) = +_{\perp},$$

where $+_{\perp}$ is the strict extension of $+$. Then $\alpha \in \mathscr{CA}_{\mathscr{T}_{\mathbf{N}}}(V)$, and

$$\begin{aligned}
\overline{\alpha}_{\mathscr{I}_{\mathbf{N}}}(\mathbf{s}(\mathbf{F}(\mathbf{X}, \mathbf{s}(\mathbf{Y})))) &= \mathscr{I}_{\mathbf{N}}(\mathbf{s})(\overline{\alpha}_{\mathscr{I}_{\mathbf{N}}}(\mathbf{F}(\mathbf{X}, \mathbf{s}(\mathbf{Y})))) \\
&= \mathscr{I}_{\mathbf{N}}(\mathbf{s})(\alpha(\mathbf{F})(\overline{\alpha}_{\mathscr{I}_{\mathbf{N}}}(\mathbf{X}), \overline{\alpha}_{\mathscr{I}_{\mathbf{N}}}(\mathbf{s}(\mathbf{Y})))) \\
&= \mathscr{I}_{\mathbf{N}}(\mathbf{s})(\alpha(\mathbf{F})(\alpha(\mathbf{X}), \mathscr{I}_{\mathbf{N}}(\mathbf{s})(\overline{\alpha}_{\mathscr{I}_{\mathbf{N}}}(\mathbf{Y})))) \\
&= \mathscr{I}_{\mathbf{N}}(\mathbf{s})(\alpha(\mathbf{F})(\alpha(\mathbf{X}), \mathscr{I}_{\mathbf{N}}(\mathbf{s})(\alpha(\mathbf{Y})))) \\
&= \mathscr{I}_{\mathbf{N}}(\mathbf{s})(\alpha(\mathbf{F})(3, \mathscr{I}_{\mathbf{N}}(\mathbf{s})(5))) \\
&= \mathscr{I}_{\mathbf{N}}(\mathbf{s})(\alpha(\mathbf{F})(3, 6)) \\
&= \mathscr{I}_{\mathbf{N}}(\mathbf{s})(9) \\
&= 10.
\end{aligned}$$

The next theorem shows that $\overline{\alpha}_{\mathscr{I}}(\mathbf{t})$ assigns a value to term $\mathbf{t}$ in the appropriate set, namely, $D_{\tau(\mathbf{t})}$. We need it to verify that the definition of $\overline{\alpha}_{\mathscr{I}}$ makes sense. For example, when we define

$$\overline{\alpha}_{\mathscr{I}}(\mathbf{f}(\mathbf{t}_1, \ldots, \mathbf{t}_n)) = \mathscr{I}(\mathbf{f})(\overline{\alpha}_{\mathscr{I}}(\mathbf{t}_1), \ldots, \overline{\alpha}_{\mathscr{I}}(\mathbf{t}_n)),$$

where $\tau(\mathbf{f}) = \boldsymbol{\tau}_1 \times \cdots \times \boldsymbol{\tau}_n \rightarrow \boldsymbol{\tau}$, we have $\mathscr{I}(\mathbf{f}) \in D_{\boldsymbol{\tau}_1} \times \cdots \times D_{\boldsymbol{\tau}_n} \rightarrow D_{\boldsymbol{\tau}}$, so we want to know that $\overline{\alpha}_{\mathscr{I}}(\mathbf{t}_i) \in D_{\boldsymbol{\tau}_i}$, $1 \leq i \leq n$.

Theorem 2.1. Let $(\mathscr{T}, \mathscr{I})$ be a **W**-structure, V a set of variables, $\alpha \in \mathscr{A}_{\mathscr{T}}(V)$, and $\mathbf{t} \in \mathrm{TM}_{\mathbf{W}}^{\boldsymbol{\tau}}(V)$ for some $\boldsymbol{\tau} \in TV(\mathbf{W})$. Then $\bar{\alpha}_{\mathscr{I}}(\mathbf{t}) \in D_{\boldsymbol{\tau}}$.

Proof. We argue by structural induction on $\mathbf{t}$. If $\mathbf{t}$ is a constant symbol $\mathbf{c} \in \mathbf{W}$, then $\bar{\alpha}_{\mathscr{I}}(\mathbf{c}) = \mathscr{I}(\mathbf{c}) \in D_{\boldsymbol{\tau}}$, and if $\mathbf{t}$ is an individual variable $\mathbf{X} \in V$, then $\bar{\alpha}_{\mathscr{I}}(\mathbf{X}) = \alpha(\mathbf{X}) \in D_{\boldsymbol{\tau}}$. If $\mathbf{t}$ is $\mathbf{f}(\mathbf{t}_1, \ldots, \mathbf{t}_n)$, where $\mathbf{f} \in \mathbf{W}$, $\tau(\mathbf{f}) = \boldsymbol{\tau}_1 \times \cdots \times \boldsymbol{\tau}_n \rightarrow \boldsymbol{\tau}$, and $\mathbf{t}_i \in \mathrm{TM}_{\mathbf{W}}^{\boldsymbol{\tau}_i}(V)$, $1 \le i \le n$, then $\bar{\alpha}_{\mathscr{I}}(\mathbf{t}_i) \in D_{\boldsymbol{\tau}_i}$, $1 \le i \le n$, by the induction hypothesis, and $\mathscr{I}(\mathbf{f}) \in D_{\boldsymbol{\tau}_1} \times \cdots \times D_{\boldsymbol{\tau}_n} \rightarrow D_{\boldsymbol{\tau}}$, so

$$\bar{\alpha}_{\mathscr{I}}(\mathbf{f}(\mathbf{t}_1, \ldots, \mathbf{t}_n)) = \mathscr{I}(\mathbf{f})(\bar{\alpha}_{\mathscr{I}}(\mathbf{t}_1), \ldots, \bar{\alpha}_{\mathscr{I}}(\mathbf{t}_n)) \in D_{\boldsymbol{\tau}}.$$

Similarly, if $\mathbf{t}$ is $\mathbf{F}(\mathbf{t}_1, \ldots, \mathbf{t}_n)$, where $\mathbf{F} \in V$, $\tau(\mathbf{F}) = \boldsymbol{\tau}_1 \times \cdots \times \boldsymbol{\tau}_n \rightarrow \boldsymbol{\tau}$, and $\mathbf{t}_i \in \mathrm{TM}_{\mathbf{W}}^{\boldsymbol{\tau}_i}(V)$, $1 \le i \le n$, then $\alpha(\mathbf{F}) \in D_{\boldsymbol{\tau}_1} \times \cdots \times D_{\boldsymbol{\tau}_n} \rightarrow D_{\boldsymbol{\tau}}$ and

$$\bar{\alpha}_{\mathscr{I}}(\mathbf{F}(\mathbf{t}_1, \ldots, \mathbf{t}_n)) = \alpha(\mathbf{F})(\bar{\alpha}_{\mathscr{I}}(\mathbf{t}_1), \ldots, \bar{\alpha}_{\mathscr{I}}(\mathbf{t}_n)) \in D_{\boldsymbol{\tau}}. \qquad \blacksquare$$

Let $\mathscr{T}$ be a complete type assignment for **W** and V a set of variables. Then for each individual variable $\mathbf{X} \in V$ with $\tau(\mathbf{X}) = \boldsymbol{\tau}$, $(D_{\boldsymbol{\tau}}, \sqsubseteq_{\boldsymbol{\tau}})$ is a cpo by assumption, and for each function variable $\mathbf{F} \in V$ with $\tau(\mathbf{F}) = \boldsymbol{\tau}_1 \times \cdots \times \boldsymbol{\tau}_n \rightarrow \boldsymbol{\tau}$,

$$\left([D_{\boldsymbol{\tau}_1} \times \cdots \times D_{\boldsymbol{\tau}_n} \rightarrow D_{\boldsymbol{\tau}}], \sqsubseteq_{[D_{\boldsymbol{\tau}_1} \times \cdots \times D_{\boldsymbol{\tau}_n} \rightarrow D_{\boldsymbol{\tau}}]}\right)$$

is a cpo by Theorems[3] 16.3.6 and 16.4.4. Now, let $\mathscr{D}_V$ be the function with domain V such that

$$\mathscr{D}_V(\mathbf{X}) = (D_{\boldsymbol{\tau}}, \sqsubseteq_{\boldsymbol{\tau}}) \quad \text{for each } \mathbf{X} \in V \text{ with } \tau(\mathbf{X}) = \boldsymbol{\tau}$$

$$\mathscr{D}_V(\mathbf{F}) = \left([D_{\boldsymbol{\tau}_1} \times \cdots \times D_{\boldsymbol{\tau}_n} \rightarrow D_{\boldsymbol{\tau}}], \sqsubseteq_{[D_{\boldsymbol{\tau}_1} \times \cdots \times D_{\boldsymbol{\tau}_n} \rightarrow D_{\boldsymbol{\tau}}]}\right)$$
$$\text{for each } \mathbf{F} \in V \text{ with } \tau(\mathbf{F}) = \boldsymbol{\tau}_1 \times \cdots \times \boldsymbol{\tau}_n \rightarrow \boldsymbol{\tau}.$$

Then a continuous variable assignment for V based on $\mathscr{T}$ is a $\mathscr{D}_V$-choice function and vice versa, so $\mathscr{C}\mathscr{A}_{\mathscr{T}}(V) = \mathrm{ch}(\mathscr{D}_V)$, and $(\mathrm{ch}(\mathscr{D}_V), \sqsubseteq_{\mathrm{ch}(\mathscr{D}_V)})$ is a cpo by Theorem 16.3.10. Writing $\sqsubseteq_{\mathscr{C}\mathscr{A}_{\mathscr{T}}(V)}$ for $\sqsubseteq_{\mathrm{ch}(\mathscr{D}_V)}$, we have proved

Theorem 2.2. Let $\mathscr{T}$ be a complete type assignment for **W** and V a set of variables. Then $(\mathscr{C}\mathscr{A}_{\mathscr{T}}(V), \sqsubseteq_{\mathscr{C}\mathscr{A}_{\mathscr{T}}(V)})$ is a cpo.

Note that the bottom element of $(\mathscr{C}\mathscr{A}_{\mathscr{T}}(V), \sqsubseteq_{\mathscr{C}\mathscr{A}_{\mathscr{T}}(V)})$, which we will

---

[3] We will refer to theorems in Chapter 16 frequently here, so we adopt the convention of writing Theorem 16.3.6, for example, to refer to Theorem 3.6 in Chapter 16.

write $\Omega_{\mathscr{CA}_{\mathscr{T}}(V)}$, or simply Ω when $\mathscr{T}$ and V are understood, satisfies

$$\Omega(\mathbf{X}) = \bot_{\tau(\mathbf{X})} \quad \text{for each individual variable } \mathbf{X} \in V$$

$$\Omega(\mathbf{F})(d_1, \ldots, d_n) = \bot_{\rho(\mathbf{F})} \quad \text{for each function variable } \mathbf{F} \in V \text{ and all } (d_1, \ldots, d_n) \in D_{\delta(\mathbf{F})}.$$

Note that if $V = \varnothing$, then $\Omega_{\mathscr{CA}_{\mathscr{T}}(V)} = \varnothing$. Given a ground term $\mathbf{t} \in \mathrm{TM}_{\mathbf{W}}$, we will generally write $\overline{\Omega}(\mathbf{t})$ to interpret $\mathbf{t}$.

The next theorem says, in effect, that the function that extends assignments α to $\overline{\alpha}_{\mathscr{I}}$ is monotonic and continuous.

Theorem 2.3. Let $(\mathscr{T}, \mathscr{I})$ be a complete, continuous $\mathbf{W}$-structure, V a set of variables, and $\mathbf{t} \in \mathrm{TM}^{\boldsymbol{\tau}}_{\mathbf{W}}(V)$ for some $\boldsymbol{\tau} \in TV(\mathbf{W})$.

1. For $\alpha, \beta \in \mathscr{CA}_{\mathscr{T}}(V)$, $\alpha \sqsubseteq_{\mathscr{CA}_{\mathscr{T}}(V)} \beta$ implies $\overline{\alpha}_{\mathscr{I}}(\mathbf{t}) \sqsubseteq_{\boldsymbol{\tau}} \overline{\beta}_{\mathscr{I}}(\mathbf{t})$.
2. For a chain $\mathscr{A}$ in $(\mathscr{CA}_{\mathscr{T}}(V), \sqsubseteq_{\mathscr{CA}_{\mathscr{T}}(V)})$, $\overline{\sqcup\mathscr{A}}_{\mathscr{I}}(\mathbf{t}) = \sqcup\{\overline{\alpha}_{\mathscr{I}}(\mathbf{t}) \mid \alpha \in \mathscr{A}\}$.

Proof. Both parts can be proven by structural induction on $\mathbf{t}$, and part 1 is straightforward, so we leave it as an exercise and concentrate on part 2. Let $\mathscr{A}$ be a chain in $(\mathscr{CA}_{\mathscr{T}}(V), \sqsubseteq_{\mathscr{CA}_{\mathscr{T}}(V)})$. Then $\sqcup\mathscr{A}$ exists by Theorem 2.2. If $\mathbf{t}$ is a constant symbol $\mathbf{c} \in \mathbf{W}$, then

$$\overline{\sqcup\mathscr{A}}(\mathbf{c}) = \mathscr{I}(\mathbf{c}) = \sqcup\{\mathscr{I}(\mathbf{c}) \mid \alpha \in \mathscr{A}\} = \sqcup\{\overline{\alpha}(\mathbf{c}) \mid \alpha \in \mathscr{A}\},$$

and if $\mathbf{t}$ is an individual variable $\mathbf{X} \in \mathbf{V}$, then

$$\begin{aligned}
\overline{\sqcup\mathscr{A}}(\mathbf{X}) &= (\sqcup\mathscr{A})(\mathbf{X}) \\
&= \sqcup\{\alpha(\mathbf{X}) \mid \alpha \in \mathscr{A}\} \qquad \text{by Theorem 16.3.9} \\
&= \sqcup\{\overline{\alpha}(\mathbf{X}) \mid \alpha \in \mathscr{A}\}.
\end{aligned}$$

If $\mathbf{t}$ is $\mathbf{f}(\mathbf{t}_1, \ldots, \mathbf{t}_n)$, where $\mathbf{f} \in \mathbf{W}$, then

$$\begin{aligned}
&\overline{\sqcup\mathscr{A}}(\mathbf{f}(\mathbf{t}_1, \ldots, \mathbf{t}_n)) \\
&= \mathscr{I}(\mathbf{f})(\overline{\sqcup\mathscr{A}}(\mathbf{t}_1), \ldots, \overline{\sqcup\mathscr{A}}(\mathbf{t}_n)) \\
&= \mathscr{I}(\mathbf{f})(\sqcup\{\overline{\alpha}(\mathbf{t}_1) \mid \alpha \in \mathscr{A}\}, \ldots, \sqcup\{\overline{\alpha}(\mathbf{t}_n) \mid \alpha \in \mathscr{A}\}) \quad \text{by the induction hypothesis} \\
&= \mathscr{I}(\mathbf{f})(\sqcup\{(\overline{\alpha}(\mathbf{t}_1), \ldots, \overline{\alpha}(\mathbf{t}_n)) \mid \alpha \in \mathscr{A}\}) \quad \text{by Theorem 16.3.5} \\
&= \sqcup\{\mathscr{I}(\mathbf{f})(\overline{\alpha}(\mathbf{t}_1), \ldots, \overline{\alpha}(\mathbf{t}_n)) \mid \alpha \in \mathscr{A}\} \quad \text{since } \mathscr{I}(\mathbf{f}) \text{ is continuous and } \{(\overline{\alpha}(\mathbf{t}_1), \ldots, \overline{\alpha}(\mathbf{t}_n)) \mid \alpha \in \mathscr{A}\} \text{ is a chain by part 1} \\
&= \sqcup\{\overline{\alpha}(\mathbf{f}(\mathbf{t}_1, \ldots, \mathbf{t}_n)) \mid \alpha \in \mathscr{A}\}.
\end{aligned}$$

Finally, let $\mathbf{t}$ be $\mathbf{F}(\mathbf{t}_1, \ldots, \mathbf{t}_n)$, where $\mathbf{F}$ is a function variable in V, and let $\Gamma: \mathscr{CA}_{\mathscr{T}}(V) \times \mathscr{CA}_{\mathscr{T}}(V) \to D_{\boldsymbol{\tau}}$ be defined $\Gamma(\alpha, \beta) = \alpha(\mathbf{F})(\overline{\beta}(\mathbf{t}_1), \ldots, \overline{\beta}(\mathbf{t}_n))$. Then Γ is monotonic by part 1 and by the monotonicity of $\alpha(\mathbf{F})$ for all

$\alpha \in \mathscr{A}$, so

$$\begin{aligned}
&\overline{\sqcup\mathscr{A}}(\mathbf{F}(\mathbf{t}_1,\ldots,\mathbf{t}_n))\\
&\quad = (\sqcup\mathscr{A})(\mathbf{F})(\overline{\sqcup\mathscr{A}}(\mathbf{t}_1),\ldots,\overline{\sqcup\mathscr{A}}(\mathbf{t}_n))\\
&\quad = (\sqcup\{\alpha(\mathbf{F}) \mid \alpha \in \mathscr{A}\})(\overline{\sqcup\mathscr{A}}(\mathbf{t}_1),\ldots,\overline{\sqcup\mathscr{A}}(\mathbf{t}_n)) \quad \text{by Theorem 16.3.9}\\
&\quad = \sqcup\{\alpha(\mathbf{F})(\overline{\sqcup\mathscr{A}}(\mathbf{t}_1),\ldots,\overline{\sqcup\mathscr{A}}(\mathbf{t}_n)) \mid \alpha \in \mathscr{A}\} \quad \text{by Theorem 16.3.7}\\
&\quad = \sqcup\{\alpha(\mathbf{F})(\sqcup\{\bar\beta(\mathbf{t}_1) \mid \beta \in \mathscr{A}\},\ldots,\sqcup\{\bar\beta(\mathbf{t}_n) \mid \beta \in \mathscr{A}\}) \mid \alpha \in \mathscr{A}\} \quad \text{by the induction hypothesis}\\
&\quad = \sqcup\{\alpha(\mathbf{F})(\sqcup\{(\bar\beta(\mathbf{t}_1),\ldots,\bar\beta(\mathbf{t}_n)) \mid \beta \in \mathscr{A}\}) \mid \alpha \in \mathscr{A}\} \quad \text{by Theorem 16.3.5}\\
&\quad = \sqcup\{\sqcup\{\alpha(\mathbf{F})(\bar\beta(\mathbf{t}_1),\ldots,\bar\beta(\mathbf{t}_n)) \mid \beta \in \mathscr{A}\} \mid \alpha \in \mathscr{A}\} \quad \text{since } \alpha(\mathbf{F}) \text{ is continuous for each } \alpha \in \mathscr{A} \text{ and } \{(\bar\beta(\mathbf{t}_1),\ldots,\bar\beta(\mathbf{t}_n)) \mid \beta \in \mathscr{A}\} \text{ is a chain by part 1}\\
&\quad = \sqcup\{\sqcup\{\Gamma(\alpha,\beta) \mid \beta \in \mathscr{A}\} \mid \alpha \in \mathscr{A}\}\\
&\quad = \sqcup\{\Gamma(\alpha,\alpha) \mid \alpha \in \mathscr{A}\} \quad \text{by the diagonal lemma}\\
&\quad = \sqcup\{\alpha(\mathbf{F})(\bar\alpha(\mathbf{t}_1),\ldots,\bar\alpha(\mathbf{t}_n)) \mid \alpha \in \mathscr{A}\}\\
&\quad = \sqcup\{\bar\alpha(\mathbf{F}(\mathbf{t}_1,\ldots,\mathbf{t}_n)) \mid \alpha \in \mathscr{A}\}. \qquad \blacksquare
\end{aligned}$$

We also prove one more result about variable assignments that we will use in the next section.

Coincidence Lemma. Let $(\mathscr{T},\mathscr{I})$ be a **W**-structure, let V_1, V_2 be sets of variables, let $\alpha \in \mathscr{A}_{\mathscr{T}}(V_1)$ and $\beta \in \mathscr{A}_{\mathscr{T}}(V_2)$, and let

$$V = \{\mathbf{V} \in V_1 \cap V_2 \mid \alpha(\mathbf{V}) = \beta(\mathbf{V})\}.$$

Then for all $\mathbf{t} \in \mathrm{TM}_{\mathbf{W}}(V)$, $\bar\alpha(\mathbf{t}) = \bar\beta(\mathbf{t})$.

Proof. We argue by structural induction on **t**. If **t** is an individual variable $\mathbf{X} \in V$, then $\bar\alpha(\mathbf{X}) = \alpha(\mathbf{X}) = \beta(\mathbf{X}) = \bar\beta(\mathbf{X})$. If **t** is a constant symbol $\mathbf{c} \in \mathbf{W}$, then $\bar\alpha(\mathbf{c}) = \mathscr{I}(\mathbf{c}) = \bar\beta(\mathbf{c})$. If **t** is $\mathbf{f}(\mathbf{t}_1,\ldots,\mathbf{t}_n)$, where $\mathbf{f} \in \mathbf{W}$, then

$$\begin{aligned}
\bar\alpha(\mathbf{f}(\mathbf{t}_1,\ldots,\mathbf{t}_n)) &= \mathscr{I}(\mathbf{f})(\bar\alpha(\mathbf{t}_1),\ldots,\bar\alpha(\mathbf{t}_n))\\
&= \mathscr{I}(\mathbf{f})(\bar\beta(\mathbf{t}_1),\ldots,\bar\beta(\mathbf{t}_n)) \qquad \text{by the induction hypothesis}\\
&= \bar\beta(\mathbf{f}(\mathbf{t}_1,\ldots,\mathbf{t}_n)),
\end{aligned}$$

and if **t** is $\mathbf{F}(\mathbf{t}_1,\ldots,\mathbf{t}_n)$, where $\mathbf{F} \in V$, then

$$\begin{aligned}
\bar\alpha(\mathbf{F}(\mathbf{t}_1,\ldots,\mathbf{t}_n)) &= \alpha(\mathbf{F})(\bar\alpha(\mathbf{t}_1),\ldots,\bar\alpha(\mathbf{t}_n))\\
&= \alpha(\mathbf{F})(\bar\beta(\mathbf{t}_1),\ldots,\bar\beta(\mathbf{t}_n)) \qquad \text{by the induction hypothesis}\\
&= \beta(\mathbf{F})(\bar\beta(\mathbf{t}_1),\ldots,\bar\beta(\mathbf{t}_n)) \qquad \text{since } \mathbf{F} \in V\\
&= \bar\beta(\mathbf{F}(\mathbf{t}_1,\ldots,\mathbf{t}_n)). \qquad \blacksquare
\end{aligned}$$

Exercises

1. Show that $\Sigma_{\mathbf{N}}$ is a complete, continuous $\mathbf{W_N}$-structure.

2. Let $\mathscr{T}(\mathbf{Bool}) = (\{\perp, 0, 1\}, \{(\perp, \perp), (\perp, 0), (\perp, 1), (0,0), (1,1)\})$, $\mathscr{T}(\mathbf{N}) = (N_\perp, \sqsubseteq_{N_\perp})$, $\mathscr{I}(\mathbf{tt}) = 0$, and $\mathscr{I}(\mathbf{ff}) = 1$. Extend $\mathscr{I}$ to a $\mathscr{T}$-interpretation for $\mathbf{W_N}$.

3. Let $\mathscr{T}$ be a type assignment for $\mathbf{W_{NL}}$ with $\mathscr{T}(\mathbf{N}) = \mathscr{T}(\mathbf{NL}) = (N_\perp, \sqsubseteq_{N_\perp})$.
   (a) Let $\mathscr{I}(\mathbf{nil}) = 0$ and $\mathscr{I}(\mathbf{cons}) = +_\perp$ (the strict extension of $+$). Show that $\mathscr{I}$ cannot be extended to a $\mathscr{T}$-interpretation for $\mathbf{W_{NL}}$.
   (b) Give a continuous $\mathscr{T}$-interpretation $\mathscr{I}'$ for $\mathbf{W_{NL}}$. [*Hint:* Consider the pairing function $\langle x, y\rangle$ from Chapter 3.] What is $\overline{\varnothing}_{\mathscr{I}'}(\mathbf{cons(s(0), nil)})$?

4. Let $\Sigma = (\mathscr{T}, \mathscr{I})$ be a $\mathbf{W}$-structure such that $\mathscr{T}(\boldsymbol{\tau})$ is a flat cpo for all $\boldsymbol{\tau} \in TV(\mathbf{W})$. Show that for every built-in function symbol $\mathbf{f} \in \mathbf{W}$, $\mathscr{I}(\mathbf{f})$ is continuous.

5. Let $\alpha(\mathbf{X}) = 3$, $\alpha(\mathbf{Y}) = 2$, $\alpha(\mathbf{F}) = +_\perp$, and $\alpha(\mathbf{G}) = \cdot_\perp$ (the strict extension of the multiplication function). Calculate $\bar{\alpha}_{\mathscr{I}_{\mathbf{N}}}(\mathbf{t})$, where $\mathbf{t}$ is as follows.
   (a) $\mathbf{F(s(X), G(s(X), F(X, Y)))}$.
   (b) $\mathbf{s_1^{-1}(F(s_1^{-1}(s(X)), Y))}$.
   (c) $\mathbf{if_N(is\_0(G(X, s_1^{-1}(s_1^{-1}(0)))), X, Y)}$.
   (d) $\mathbf{if_{Bool}(is\_s(X), is\_0(X), is\_0(X))}$.

6. Let $\mathscr{I}_e(\mathbf{0}) = 0$ and $\mathscr{I}_e(\mathbf{s}) = e_\perp$, where $e(n) = 2^n$ for all $n \in N$.
   (a) Extend $\mathscr{I}_e$ to a $\mathscr{T}_{\mathbf{N}}$-interpretation for $\mathbf{W_N}$.
   (b) Calculate $\bar{\alpha}_{\mathscr{I}_e}(\mathbf{t})$ for each term $\mathbf{t}$ given in Exercise 5.

7. Let $\mathscr{T}(\mathbf{Bool}) = \mathscr{T}_{\mathbf{N}}(\mathbf{Bool})$, $\mathscr{T}(\mathbf{N}) = \mathscr{T}_{\mathbf{N}}(\mathbf{N})$, and $\mathscr{T}(\mathbf{NL}) = (\mathrm{TUP}_\perp, \sqsubseteq_{\mathrm{TUP}_\perp})$, where TUP is the set of all tuples of natural numbers and $(\mathrm{TUP}_\perp, \sqsubseteq_{\mathrm{TUP}_\perp})$ is the flat cpo on TUP. Give a $\mathscr{T}$-interpretation for $\mathbf{W_{NL}}$.

8. Let $\mathscr{T}(\mathbf{Bool}) = \mathscr{T}_{\mathbf{N}}(\mathbf{Bool})$, $\mathscr{T}(\mathbf{N}) = \mathscr{T}_{\mathbf{N}}(\mathbf{N})$, and $\mathscr{T}(\mathbf{NL}) = (\mathscr{P}_{\mathrm{f}}(N), \subseteq_{\mathscr{P}_{\mathrm{f}}(N)})$, where $\mathscr{P}_{\mathrm{f}}(N)$ consists of all the finite subsets of N, and let $\mathscr{I}(\mathbf{cons})(e, \{d_1, \ldots, d_n\}) = \{e, d_1, \ldots, d_n\}$. Explain why $\mathscr{I}$ cannot be extended to a $\mathscr{T}$-interpretation for $\mathbf{W_{NL}}$.

9. Let $\Sigma = (\mathscr{T}, \mathscr{I})$ be a continuous $\mathbf{W}$-structure, and let $\mathbf{f} \in \mathbf{W}_c$. Show that for all $d, e \in D_{\rho(\mathbf{f})}$ such that $\perp_{\rho(\mathbf{f})} \neq d \sqsubseteq_{\rho(\mathbf{f})} e$, if $d \in \operatorname{ran} \mathscr{I}(\mathbf{f})$ then $e \in \operatorname{ran} \mathscr{I}(\mathbf{f})$. [*Hint:* Use **is_f**.]

10. Let $\Sigma = (\mathscr{T}, \mathscr{I})$ be a continuous $\mathbf{W}$-structure, and let $\mathbf{f} \in \mathbf{W}_c$. Show that for all $d, e \in D_{\delta(\mathbf{f})}$, if $\perp_{\rho(\mathbf{f})} \neq \mathscr{I}(\mathbf{f})(d) \sqsubseteq_{\rho(\mathbf{f})} \mathscr{I}(\mathbf{f})(e)$, then $d \sqsubseteq_{\delta(\mathbf{f})} e$. [*Hint:* Use $\mathbf{f}_i^{-1}$, $1 \le i \le \mathrm{ar}(\mathbf{f})$.]

11. Let $(\mathcal{T},\mathcal{I}) = \Sigma_{\mathbf{N}}$, and let $\mathbf{X} \in \mathrm{VAR}_1^{\mathbf{N}}$. Give a term $\mathbf{t} \in \mathrm{TM}_{\mathbf{W}_{\mathbf{N}}}(\{\mathbf{X}\})$ and $\alpha, \beta \in \mathscr{C}\mathscr{A}_{\mathcal{T}}(\{\mathbf{X}\})$ such that $\overline{\alpha}_{\mathcal{I}}(\mathbf{t}) \sqsubseteq_{N_\perp} \overline{\beta}_{\mathcal{I}}(\mathbf{t})$ and $\beta \sqsubset_{\mathscr{C}\mathscr{A}_{\mathcal{T}}(\{\mathbf{X}\})} \alpha$.

12. Let $(\mathcal{T},\mathcal{I})$ be a **W**-structure, let V be a set of variables, and let $\alpha, \beta \in \mathscr{C}\mathscr{A}_{\mathcal{T}}(V)$. Show that if $\overline{\alpha}_{\mathcal{I}}(\mathbf{t}) \sqsubseteq_{\tau(\mathbf{t})} \overline{\alpha}_{\mathcal{I}}(\mathbf{t})$ for all $\mathbf{t} \in \mathrm{TM}_{\mathbf{W}}(V)$, then $\alpha \sqsubseteq_{\mathscr{C}\mathscr{A}_{\mathcal{T}}(V)} \beta$.

13. Prove part 1 of Theorem 2.3.

14. Let $\Sigma = (\mathcal{T},\mathcal{I})$ be a complete, continuous **W**-structure, and let V be a set of variables.

(**a**) Define a function $\mathscr{D}'_V$ such that $\mathscr{A}_{\mathcal{T}}(V) = \mathrm{ch}(\mathscr{D}'_V)$, and show that $(\mathscr{A}_{\mathcal{T}}(V), \sqsubseteq_{\mathscr{A}_{\mathcal{T}}(V)})$ is a cpo.

(**b**) Show that part 1 of Theorem 2.3 holds for $(\mathscr{A}_{\mathcal{T}}(V), \sqsubseteq_{\mathscr{A}_{\mathcal{T}}(V)})$.

(**c**) Show that part 2 of Theorem 2.3 fails for $(\mathscr{A}_{\mathcal{T}}(V), \sqsubseteq_{\mathscr{A}_{\mathcal{T}}(V)})$.

3. Solutions to W-Programs

Now that we have the tools for giving a meaning to terms, we can take the first step toward defining the denotational semantics of programs. Let

$$\mathbf{P} = \{\mathbf{F}_1(\mathbf{X}_1,\ldots,\mathbf{X}_{n_1}) = \mathbf{t}_1,\ldots,\mathbf{F}_m(\mathbf{X}_1,\ldots,\mathbf{X}_{n_m}) = \mathbf{t}_m\}$$

be a **W**-program, and let $\Sigma = (\mathcal{T},\mathcal{I})$ be a **W**-structure. We want to define the meaning of $\mathbf{F}_i$ in terms of $\mathbf{t}_i$, $1 \le i \le m$. The idea is that we start with a variable assignment $\alpha \in \mathscr{C}\mathscr{A}_{\mathcal{T}}(FV(\mathbf{P}))$ which gives a meaning to each function variable in $\mathbf{t}_i$. Then for any possible input $(d_1,\ldots,d_{n_i}) \in D_{\delta(\mathbf{F}_i)}$ we extend α with the assignment $\beta = \{(\mathbf{X}_1, d_1),\ldots,(\mathbf{X}_{n_i}, d_{n_i})\}$ and use $\alpha \cup \beta$ to interpret $\mathbf{t}_i$, giving us an output value for input $(d_1,\ldots,d_{n_i})$.

It will be convenient to introduce a special notation for the assignment β in the previous paragraph. Given an equation $\mathbf{F}(\mathbf{X}_1,\ldots,\mathbf{X}_n) = \mathbf{t}$ and $d = (d_1,\ldots,d_n) \in D_{\delta(\mathbf{F})}$, the variable assignment $\alpha_{(d_1,\ldots,d_n)}$, also written α_d, is $\{(\mathbf{X}_1, d_1),\ldots,(\mathbf{X}_n, d_n)\}$; that is,

$$\alpha_{(d_1,\ldots,d_n)}(\mathbf{X}_i) = d_i, \qquad 1 \le i \le n.$$

The particular equation that determines the variables in the domain of α_d will always be clear from the context in which α_d is used.

Definition. Let $\Sigma = (\mathcal{T},\mathcal{I})$ be a **W**-structure and **P** a **W**-program. We associate with **P** the higher order function $\Phi_{\mathbf{P}}^{\Sigma}: \mathscr{C}\mathscr{A}_{\mathcal{T}}(FV(\mathbf{P})) \to \mathscr{A}_{\mathcal{T}}(FV(\mathbf{P}))$, defined as follows. For each $\mathbf{F} \in FV(\mathbf{P})$, with defining equation

$\mathbf{F}(\mathbf{X}_1, \ldots, \mathbf{X}_n) = \mathbf{t}$, and for all $(d_1, \ldots, d_n) \in D_{\delta(\mathbf{F})}$,

$$\Phi_{\mathbf{P}}^{\Sigma}(\alpha)(\mathbf{F})(d_1, \ldots, d_n) = \overline{(\alpha \cup \alpha_{(d_1, \ldots, d_n)})_{\mathscr{I}}}(\mathbf{t}).$$

When Σ is understood we will write $\Phi_{\mathbf{P}}$ for $\Phi_{\mathbf{P}}^{\Sigma}$.

It is clear from the definition that $\Phi_{\mathbf{P}}^{\Sigma} \in \mathscr{CA}_{\mathscr{T}}(FV(\mathbf{P})) \to \mathscr{A}_{\mathscr{T}}(FV(\mathbf{P}))$, but if Σ is complete and continuous we can prove something stronger.

Theorem 3.1. Let $\Sigma = (\mathscr{T}, \mathscr{I})$ be a complete, continuous **W**-structure and let **P** be a **W**-program. Then $\Phi_{\mathbf{P}}^{\Sigma} \in \mathscr{CA}_{\mathscr{T}}(FV(\mathbf{P})) \to \mathscr{CA}_{\mathscr{T}}(FV(\mathbf{P}))$.

Proof. We will write $\Phi_{\mathbf{P}}$ for $\Phi_{\mathbf{P}}^{\Sigma}$. Let $\alpha \in \mathscr{CA}_{\mathscr{T}}(FV(\mathbf{P}))$. We need to show that $\Phi_{\mathbf{P}}(\alpha)(\mathbf{F}) \in [D_{\delta(\mathbf{F})} \to D_{\rho(\mathbf{F})}]$ for each $\mathbf{F} \in FV(\mathbf{P})$. It follows from Theorem 2.1 that $\Phi_{\mathbf{P}}(\alpha)(\mathbf{F}) \in D_{\delta(\mathbf{F})} \to D_{\rho(\mathbf{F})}$, so we just need to show that $\Phi_{\mathbf{P}}(\alpha)(\mathbf{F})$ is continuous. Let $\mathbf{F}(\mathbf{X}_1, \ldots, \mathbf{X}_n) = \mathbf{t}$ be an equation in **P** with $\tau(\mathbf{F}) = \boldsymbol{\tau}_1 \times \cdots \times \boldsymbol{\tau}_n \to \boldsymbol{\tau}$, and let C be a chain in $(D_{\boldsymbol{\tau}_1 \times \cdots \times \boldsymbol{\tau}_n}, \sqsubseteq_{\boldsymbol{\tau}_1 \times \cdots \times \boldsymbol{\tau}_n})$. Then $\{\alpha \cup \alpha_c \mid c \in C\}$ is a chain in $(\mathscr{CA}_{\mathscr{T}}(V), \sqsubseteq_{\mathscr{CA}_{\mathscr{T}}(V)})$, where $V = \{\mathbf{X}_1, \ldots, \mathbf{X}_n\} \cup FV(\mathbf{P})$, so $\sqcup\{\alpha \cup \alpha_c | c \in C\}$ exists by Theorem 2.2. Moreover, for and $\mathbf{G} \in FV(\mathbf{P})$.

$$(\alpha \cup \alpha_{\sqcup C})(\mathbf{G}) = \alpha(\mathbf{G}) = \sqcup\{(\alpha \cup \alpha_c)(\mathbf{G}) \mid c \in C\},$$

and for $\mathbf{X}_i$, $1 \le i \le n$,

$$(\alpha \cup \alpha_{\sqcup C})(\mathbf{X}_i) = (\sqcup C)\downarrow i = \sqcup(C \downarrow i) = \sqcup\{(\alpha \cup \alpha_c)(\mathbf{X}_i) \mid c \in C\},$$

so by Theorem 16.3.9, $\alpha \cup \alpha_{\sqcup C} = \sqcup\{\alpha \cup \alpha_c \mid c \in C\}$. Therefore,

$$\begin{aligned}
\Phi_{\mathbf{P}}(\alpha)(\mathbf{F})(\sqcup C) &= \overline{\alpha \cup \alpha_{\sqcup C}}(\mathbf{t}) \\
&= \overline{\sqcup\{\alpha \cup \alpha_c \mid c \in C\}}(\mathbf{t}) \\
&= \sqcup\{\overline{\alpha \cup \alpha_c}(\mathbf{t}) \mid c \in C\} \qquad \text{by Theorem 2.3} \\
&= \sqcup\{\Phi_{\mathbf{P}}(\alpha)(\mathbf{F})(c) \mid c \in C\} \\
&= \sqcup \Phi_{\mathbf{P}}(\alpha)(\mathbf{F})(C),
\end{aligned}$$

and $\Phi_{\mathbf{P}}(\alpha)(\mathbf{F})$ is continuous. ■

Since a program **P** is a set of equations, it makes sense to try to solve these equations to find the meaning of **P**.

Definition. Let $\Sigma = (\mathcal{T}, \mathcal{I})$ be a **W**-structure and **P** a **W**-program. A *solution* to **P** in Σ is any $\alpha \in \mathcal{A}_{\mathcal{T}}(FV(\mathbf{P}))$ such that

$$\alpha(\mathbf{F})(d_1, \ldots, d_n) = \overline{(\alpha \cup \alpha_{(d_1, \ldots, d_n)})_{\mathcal{I}}}(\mathbf{t}) \tag{3.1}$$

for every equation $\mathbf{F}(\mathbf{X}_1, \ldots, \mathbf{X}_n) = \mathbf{t}$ in **P** and every $(d_1, \ldots, d_n) \in D_{\delta(\mathbf{F})}$.

In other words, every function variable in **P** is assigned a function such that every equation in **P** is satisfied for every possible value taken by the individual variables. Note that an equivalent statement of (3.1) is

$$\overline{(\alpha \cup \alpha_{(d_1, \ldots, d_n)})_{\mathcal{I}}}(\mathbf{F}(\mathbf{X}_1, \ldots, \mathbf{X}_n)) = \overline{(\alpha \cup \alpha_{(d_1, \ldots, d_n)})_{\mathcal{I}}}(\mathbf{t})$$

for every equation $\mathbf{F}(\mathbf{X}_1, \ldots, \mathbf{X}_n) = \mathbf{t}$ in **P** and every $(d_1, \ldots, d_n) \in D_{\delta(\mathbf{F})}$.

It is important to understand that for an arbitrary $\alpha \in \mathcal{A}_{\mathcal{T}}(FV(\mathbf{P}))$, $\Phi_{\mathbf{P}}(\alpha)$ is not necessarily a solution to **P**. Consider the $\mathbf{W}_{\mathbf{N}}$-structure $\Sigma_{\mathbf{N}} = (\mathcal{T}_{\mathbf{N}}, \mathcal{I}_{\mathbf{N}})$ and the $\mathbf{W}_{\mathbf{N}}$-program **Q** with equations

$$\mathbf{F}(\mathbf{X}) = \mathbf{G}(\mathbf{X})$$
$$\mathbf{G}(\mathbf{X}) = \mathbf{F}(\mathbf{X}).$$

The problem is that $\Phi_{\mathbf{Q}}(\alpha)(\mathbf{F})$ is defined in terms of $\alpha(\mathbf{G})$, but applying $\Phi_{\mathbf{Q}}$ to α changes the function assigned to **G** from $\alpha(\mathbf{G})$ to $\Phi_{\mathbf{Q}}(\alpha)(\mathbf{G})$. For example, if $\alpha(\mathbf{F})$ is the constant function of $\alpha(\mathbf{F})(x) = 3$ and $\alpha(\mathbf{G})$ is the constant function $\alpha(\mathbf{G})(x) = 7$, then

$$\Phi_{\mathbf{Q}}(\alpha)(\mathbf{F})(0) = \overline{\alpha \cup \alpha_0}(\mathbf{G}(\mathbf{X})) = \alpha(\mathbf{G})(0) = 7,$$

but

$$\begin{aligned}\overline{\Phi_{\mathbf{Q}}(\alpha) \cup \alpha_0}(\mathrm{rhs}(\mathbf{F})) &= \overline{\Phi_{\mathbf{Q}}(\alpha) \cup \alpha_0}(\mathbf{G}(\mathbf{X})) \\ &= \Phi_{\mathbf{Q}}(\alpha)(\mathbf{G})(0) \\ &= \overline{\alpha \cup \alpha_0}(\mathbf{F}(\mathbf{X})) = \alpha(\mathbf{F})(0) = 3,\end{aligned}$$

so $\Phi_{\mathbf{Q}}(\alpha)$ is not a solution to **Q**.

What we need is an α such that $\Phi_{\mathbf{Q}}$ leaves $\alpha(\mathbf{F})$ and $\alpha(\mathbf{G})$ unchanged; that is, we need a fixed point of $\Phi_{\mathbf{Q}}$. If α is some fixed point of $\Phi_{\mathbf{Q}}$, then for any $d \in N_{\perp}$ we have

$$\alpha(\mathbf{F})(d) = \Phi_{\mathbf{Q}}(\alpha)(\mathbf{F})(d) = \overline{\alpha \cup \alpha_d}(\mathrm{rhs}(\mathbf{F}))$$
$$\alpha(\mathbf{G})(d) = \Phi_{\mathbf{Q}}(\alpha)(\mathbf{G})(d) = \overline{\alpha \cup \alpha_d}(\mathrm{rhs}(\mathbf{G})),$$

so α is a solution to **Q**. More generally, we can prove

Theorem 3.2. Let $\Sigma = (\mathcal{T}, \mathcal{I})$ be a **W**-structure, and let **P** be a **W**-program. Then $\alpha \in \mathscr{CA}_{\mathcal{T}}(FV(\mathbf{P}))$ is a solution to **P** in Σ if and only if α is a fixed point of $\Phi_{\mathbf{P}}^{\Sigma}$.

Proof. If $\alpha \in \mathscr{CA}_{\mathcal{T}}(FV(\mathbf{P}))$ is a solution to **P** in Σ, then

$$\alpha(\mathbf{F})(d) = \overline{\alpha \cup \alpha_d}(\text{rhs}(\mathbf{F})) = \Phi_{\mathbf{P}}^{\Sigma}(\alpha)(\mathbf{F})(d)$$

holds for all $\mathbf{F} \in FV(\mathbf{P})$ and all $d \in D_{\delta(\mathbf{F})}$, so $\Phi_{\mathbf{P}}^{\Sigma}(\alpha) = \alpha$. On the other hand, if $\alpha \in \mathscr{CA}_{\mathcal{T}}(FV(\mathbf{P}))$ is a fixed point of $\Phi_{\mathbf{P}}^{\Sigma}$, then

$$\alpha(\mathbf{F})(d) = \Phi_{\mathbf{P}}^{\Sigma}(\alpha)(\mathbf{F})(d) = \overline{\alpha \cup \alpha_d}(\text{rhs}(\mathbf{F}))$$

holds for all $\mathbf{F} \in FV(\mathbf{P})$ and all $d \in D_{\delta(\mathbf{F})}$, so α is a solution to **P** in Σ. ∎

Going back to the example, we still have the problem that $\Phi_{\mathbf{Q}}$ has more than one fixed point. Any α that assigns the same function f to both **F** and **G**, where f could be anything from the everywhere undefined function to (some extension to $N_{\perp}$ of) the total predicate HALT(X, X), is a solution to **Q**. Clearly, there is nothing in program **Q** to indicate that the programmer meant to specify a solution to the halting problem. For that matter, there is no indication that the programmer meant to solve any problem at all in writing program **Q**. The sensible approach is to focus on the *least* fixed point of $\Phi_{\mathbf{Q}}$, which would be $\alpha(\mathbf{F})(d) = \alpha(\mathbf{G})(d) = \perp_N$ for all $d \in N_{\perp}$, i.e., $\alpha = \Omega$.

Of course, we do not know that $\mu\Phi_{\mathbf{P}}^{\Sigma}$ exists in general, for an arbitrary **W**-structure Σ and an arbitrary **W**-program **P**.

Theorem 3.3. Let $\Sigma = (\mathcal{T}, \mathcal{I})$ be a complete, continuous **W**-structure and let **P** be a **W**-program. Then $\mu\Phi_{\mathbf{P}}^{\Sigma}$ exists, and $\mu\Phi_{\mathbf{P}}^{\Sigma} \in \mathscr{CA}_{\mathcal{T}}(FV(\mathbf{P}))$.

Proof. We will write $\Phi_{\mathbf{P}}$ for $\Phi_{\mathbf{P}}^{\Sigma}$. If $\mu\Phi_{\mathbf{P}}$ exists, then Theorem 3.1 implies that $\mu\Phi_{\mathbf{P}} \in \mathscr{CA}_{\mathcal{T}}(FV(\mathbf{P}))$, so we just need to show that it exists. By Theorem 2.2, it is sufficient to show that

$$\Phi_{\mathbf{P}} \in [\mathscr{CA}_{\mathcal{T}}(FV(\mathbf{P})) \to \mathscr{CA}_{\mathcal{T}}(FV(\mathbf{P}))],$$

so by Theorem 3.1 we only need to show that $\Phi_{\mathbf{P}}$ is continuous. Let $\mathscr{A}$ be a chain in $(\mathscr{CA}_{\mathcal{T}}(FV(\mathbf{P})), \sqsubseteq_{\mathscr{CA}_{\mathcal{T}}(FV(\mathbf{P}))})$, $\mathbf{F}(\mathbf{X}_1, \ldots, \mathbf{X}_n) = \mathbf{t}$ an equation in **P**, and $d \in D_{\delta(\mathbf{F})}$. It is easy to see that $\{\alpha \cup \alpha_d \mid \alpha \in \mathscr{A}\}$ is a chain in $(\mathscr{CA}_{\mathcal{T}}(V), \sqsubseteq_{\mathscr{CA}_{\mathcal{T}}(V)})$, where $V = \{\mathbf{X}_1, \ldots, \mathbf{X}_n\} \cup FV(\mathbf{P})$, and that

$$\sqcup\mathscr{A} \cup \alpha_d = \sqcup\{\alpha \cup \alpha_d \mid \alpha \in \mathscr{A}\}, \tag{3.2}$$

so we have

$$\begin{aligned}
&\Phi_{\mathbf{P}}(\sqcup \mathscr{A})(\mathbf{F})(d) \\
&= \overline{\sqcup \mathscr{A} \cup \alpha_d}(\mathbf{t}) \\
&= \overline{\sqcup \{\alpha \cup \alpha_d \mid \alpha \in \mathscr{A}\}}(\mathbf{t}) && \text{by (3.2)} \\
&= \sqcup \{\overline{\alpha \cup \alpha_d}(\mathbf{t}) \mid \alpha \in \mathscr{A}\} && \text{by Theorem 2.2} \\
&= \sqcup \{\Phi_{\mathbf{P}}(\alpha)(\mathbf{F})(d) \mid \alpha \in \mathscr{A}\} \\
&= (\sqcup \{\Phi_{\mathbf{P}}(\alpha)(\mathbf{F}) \mid \alpha \in \mathscr{A}\})(d) && \text{by Theorem 16.3.7} \\
&= (\sqcup \{\Phi_{\mathbf{P}}(\alpha) \mid \alpha \in \mathscr{A}\})(\mathbf{F})(d) && \text{by Theorem 16.3.9} \\
&= (\sqcup \Phi_{\mathbf{P}}(\mathscr{A}))(\mathbf{F})(d).
\end{aligned}$$

(Note that $\mathbf{F}$ and d were arbitrarily chosen, so $\sqcup\{\Phi_{\mathbf{P}}(\alpha)(\mathbf{F})(d) \mid \alpha \in \mathscr{A}\}$ exists for all $d \in D_{\delta(\mathbf{F})}$ and $\sqcup\{\Phi_{\mathbf{P}}(\alpha)(\mathbf{F}) \mid \alpha \in \mathscr{A}\}$ exists for all $\mathbf{F} \in FV(\mathbf{P})$, justifying the use of Theorems 16.3.7 and 16.3.9.) Now, $\Phi_{\mathbf{P}}(\sqcup \mathscr{A})(\mathbf{F})(d) = (\sqcup \Phi_{\mathbf{P}}(\mathscr{A}))(\mathbf{F})(d)$ for all $d \in D_{\delta(\mathbf{F})}$ and all $\mathbf{F} \in FV(\mathbf{P})$, so $\Phi_{\mathbf{P}}(\sqcup \mathscr{A}) = \sqcup \Phi_{\mathbf{P}}(\mathscr{A})$ and $\Phi_{\mathbf{P}}$ is continuous. ■

The fixed point theorem not only tells us that $\mu\Phi_{\mathbf{P}}$ exists, but it also gives us a way of calculating $\mu\Phi_{\mathbf{P}}$, since $\mu\Phi_{\mathbf{P}} = \sqcup\{\Phi_{\mathbf{P}}^i(\Omega) \mid i \in N\}$. For example, let **ADD** be the $\mathbf{W_N}$-program with the equation

$$+(\mathbf{X},\mathbf{Y}) = \mathbf{if_N}(\mathbf{is\_0}(\mathbf{Y}), \mathbf{X}, \mathbf{s}(+(\mathbf{X}, \mathbf{s_1^{-1}}(\mathbf{Y})))).$$

Then, writing $\mathscr{I}$ for $\mathscr{I}_{\mathbf{N}}$, in $\Sigma_{\mathbf{N}}$ we have, for any $d \in N_\perp$ and any $n \in N$,

$$\begin{aligned}
&\Phi_{\mathbf{ADD}}^{n+1}(\Omega)(+)(d, \perp_N) \\
&\quad= \overline{\Phi_{\mathbf{ADD}}^n(\Omega) \cup \alpha_{(d, \perp_N)}}(\mathbf{if_N}(\mathbf{is_0}(\mathbf{Y}), \mathbf{X}, \mathbf{s}(+(\mathbf{X}, \mathbf{s_1^{-1}}(\mathbf{Y}))))) \\
&\quad= \mathscr{I}(\mathbf{if_N})(\mathscr{I}(\mathbf{is_0})(\perp_N), d, \overline{\Phi_{\mathbf{ADD}}^n(\Omega) \cup \alpha_{(d, \perp_N)}}(\mathbf{s}(+(\mathbf{X}, \mathbf{s_1^{-1}}(\mathbf{Y}))))) \\
&\quad= \mathscr{I}(\mathbf{if_N})(\perp_{\mathrm{Bool}}, d, \overline{\Phi_{\mathbf{ADD}}^n(\Omega) \cup \alpha_{(d, \perp_N)}}(\mathbf{s}(+(\mathbf{X}, \mathbf{s_1^{-1}}(\mathbf{Y}))))) \\
&\quad= \perp_N,
\end{aligned}$$

and

$$\begin{aligned}
&\Phi_{\mathbf{ADD}}^{n+1}(\Omega)(+)(d, 0) \\
&\quad= \overline{\Phi_{\mathbf{ADD}}^n(\Omega) \cup \alpha_{(d,0)}}(\mathbf{if_N}(\mathbf{is_0}(\mathbf{Y}), \mathbf{X}, \mathbf{s}(+(\mathbf{X}, \mathbf{s_1^{-1}}(\mathbf{Y}))))) \\
&\quad= \mathscr{I}(\mathbf{if_N})(\mathscr{I}(\mathbf{is_0})(0), d, \overline{\Phi_{\mathbf{ADD}}^n(\Omega) \cup \alpha_{(d,0)}}(\mathbf{s}(+(\mathbf{X}, \mathbf{s_1^{-1}}(\mathbf{Y}))))) \\
&\quad= \mathscr{I}(\mathbf{if_N})(tt, d, \overline{\Phi_{\mathbf{ADD}}^n(\Omega) \cup \alpha_{(d,0)}}(\mathbf{s}(+(\mathbf{X}, \mathbf{s_1^{-1}}(\mathbf{Y}))))) \\
&\quad= d,
\end{aligned}$$

so $\mu\Phi_{\mathbf{ADD}}(\mathbf{+})(d, \perp_N) = \perp_N$ and $\mu\Phi_{\mathbf{ADD}}(\mathbf{+})(d,0) = d$. The situation is more complicated if the second argument is > 0. For example,

$$
\begin{aligned}
&\Phi_{\mathbf{ADD}}(\Omega)(\mathbf{+})(d,1)\\
&\quad = \mathcal{I}(\mathbf{if_N})(\mathcal{I}(\mathbf{is\_0})(1), d, \overline{\Omega \cup \alpha_{(d,1)}}(\mathbf{s}(\mathbf{+}(\mathbf{X}, \mathbf{s_1^{-1}}(\mathbf{Y})))))\\
&\quad = \overline{\Omega \cup \alpha_{(d,1)}}(\mathbf{s}(\mathbf{+}(\mathbf{X}, \mathbf{s_1^{-1}}(\mathbf{Y}))))\\
&\quad = \mathcal{I}(\mathbf{s})(\Omega(\mathbf{+})(d, \mathcal{I}(\mathbf{s_1^{-1}})(1)))\\
&\quad = \mathcal{I}(\mathbf{s})(\perp_N)\\
&\quad = \perp_N,
\end{aligned}
$$

but if we iterate $\Phi_{\mathbf{ADD}}$ $n + 2$ times, $n \geq 0$, we get

$$
\begin{aligned}
\Phi_{\mathbf{ADD}}^{n+2}(\Omega)(\mathbf{+})(d,1) &= \overline{\Phi_{\mathbf{ADD}}^{n+1}(\Omega) \cup \alpha_{(d,1)}}(\mathbf{s}(\mathbf{+}(\mathbf{X}, \mathbf{s_1^{-1}}(\mathbf{Y}))))\\
&= \mathcal{I}(\mathbf{s})(\Phi_{\mathbf{ADD}}^{n+1}(\Omega)(\mathbf{+})(d,0))\\
&= \mathcal{I}(\mathbf{s})(d)\\
&= \begin{cases} d+1 & \text{if } d \in N\\ \perp_N & \text{otherwise.}\end{cases}
\end{aligned}
$$

Similarly,

$$
\begin{aligned}
\Phi_{\mathbf{ADD}}^{2}(\Omega)(\mathbf{+})(d,2) &= \overline{\Phi_{\mathbf{ADD}}(\Omega) \cup \alpha_{(d,2)}}(\mathbf{s}(\mathbf{+}(\mathbf{X}, \mathbf{s_1^{-1}}(\mathbf{Y}))))\\
&= \mathcal{I}(\mathbf{s})(\Phi_{\mathbf{ADD}}(\Omega)(\mathbf{+})(d,1))\\
&= \perp_N,
\end{aligned}
$$

but if we iterate $\Phi_{\mathbf{ADD}}$ $n + 3$ times, $n \geq 0$, we get

$$
\begin{aligned}
\Phi_{\mathbf{ADD}}^{n+3}(\Omega)(\mathbf{+})(d,2) &= \overline{\Phi_{\mathbf{ADD}}^{n+2}(\Omega) \cup \alpha_{(d,2)}}(\mathbf{s}(\mathbf{+}(\mathbf{X}, \mathbf{s_1^{-1}}(\mathbf{Y}))))\\
&= \mathcal{I}(\mathbf{s})(\Phi_{\mathbf{ADD}}^{n+2}(\Omega)(\mathbf{+})(d,1))\\
&= \begin{cases} d+2 & \text{if } d \in N\\ \perp_N & \text{otherwise.}\end{cases}
\end{aligned}
$$

In general, it can be shown by induction on n that, for any $n \in N$,

$$
\Phi_{\mathbf{ADD}}^{n}(\Omega)(\mathbf{+})(d,e) = \begin{cases} d+e & \text{if } d, e \in N \text{ and } 0 \leq e < n\\ \perp_N & \text{otherwise,}\end{cases} \tag{3.3}
$$

so, for all $d, e \in N_\perp$,

$$\mu\Phi_{\mathbf{ADD}}(\mathbf{+})(d,e) = \bigsqcup\{\Phi^n_{\mathbf{ADD}}(\Omega)(\mathbf{+})(d,e) \mid n \in N\}$$

$$= \begin{cases} d + e & \text{if } d, e \in N \\ \perp_N & \text{otherwise.} \end{cases}$$

That is, $\mu\Phi_{\mathbf{ADD}}(\mathbf{+})$ is the strict extension of $+$.

We can also use the fact that $\mu\Phi_{\mathbf{ADD}}$ is a fixed point of $\Phi_{\mathbf{ADD}}$ to verify that $\mu\Phi_{\mathbf{ADD}}(\mathbf{+})(d,e) = d +_\perp e$ for any given $d, e \in N_\perp$. For example,

$$\mu\Phi_{\mathbf{ADD}}(\mathbf{+})(3,2)$$

$$= \Phi_{\mathbf{ADD}}(\mu\Phi_{\mathbf{ADD}})(\mathbf{+})(3,2)$$

$$= \overline{\mu\Phi_{\mathbf{ADD}} \cup \alpha_{(3,2)}}(\mathbf{if_N(is\_0(Y), X, s(+(X, s_I^{-1}(Y))))})$$

$$= \mathscr{I}(\mathbf{s})(\mu\Phi_{\mathbf{ADD}}(\mathbf{+})(3,1))$$

$$= \mathscr{I}(\mathbf{s})(\Phi_{\mathbf{ADD}}(\mu\Phi_{\mathbf{ADD}})(\mathbf{+})(3,1))$$

$$= \mathscr{I}(\mathbf{s})(\overline{\mu\Phi_{\mathbf{ADD}} \cup \alpha_{(3,1)}}(\mathbf{if_N(is\_0(Y), X, s(+(X, s_I^{-1}(Y))))}))$$

$$= \mathscr{I}(\mathbf{s})(\mathscr{I}(\mathbf{s})(\mu\Phi_{\mathbf{ADD}}(\mathbf{+})(3,0)))$$

$$= \mathscr{I}(\mathbf{s})(\mathscr{I}(\mathbf{s})(\Phi_{\mathbf{ADD}}(\mu\Phi_{\mathbf{ADD}})(\mathbf{+})(3,0)))$$

$$= \mathscr{I}(\mathbf{s})(\mathscr{I}(\mathbf{s})(\overline{\mu\Phi_{\mathbf{ADD}} \cup \alpha_{(3,0)}}(\mathbf{if_N(is\_0(Y), X, s(+(X, s_I^{-1}(Y))))})))$$

$$= \mathscr{I}(\mathbf{s})(\mathscr{I}(\mathbf{s})(3))$$

$$= 5.$$

Before we go on we prove the following useful lemma.

Extension Lemma. Let $(\mathscr{T}, \mathscr{I})$ be a complete, continuous **W**-structure, and let **P**, **Q** be **W**-programs such that $\mathbf{P} \subseteq \mathbf{Q}$.

1. For all $\mathbf{F} \in FV(\mathbf{P})$, $\mu\Phi_{\mathbf{P}}(\mathbf{F}) = \mu\Phi_{\mathbf{Q}}(\mathbf{F})$.
2. For all $\mathbf{t} \in \mathrm{TM}_{\mathbf{W}}(FV(\mathbf{P}))$, $\overline{\mu\Phi_{\mathbf{P}}}(\mathbf{t}) = \overline{\mu\Phi_{\mathbf{Q}}}(\mathbf{t})$.

Proof. Let $\mathbf{F}(\mathbf{X}_1, \ldots, \mathbf{X}_n) = \mathbf{t}$ be the defining equation for **F** in **P**. First we prove by induction on i that

$$\Phi^i_{\mathbf{P}}(\Omega)(\mathbf{F}) = \Phi^i_{\mathbf{Q}}(\Omega)(\mathbf{F}) \qquad \text{for all } i \in N. \tag{3.4}$$

If $i = 0$ then $\Phi^0_{\mathbf{P}}(\Omega)(\mathbf{F}) = \Omega(\mathbf{F}) = \Phi^0_{\mathbf{Q}}(\Omega)(\mathbf{F})$, so assume $\Phi^i_{\mathbf{P}}(\Omega)(\mathbf{F}) = \Phi^i_{\mathbf{Q}}(\Omega)(\mathbf{F})$ and let $d \in D_{\delta(\mathbf{F})}$. Then

$$\begin{aligned}
\Phi^{i+1}_{\mathbf{P}}(\Omega)(\mathbf{F})(d) &= \overline{\Phi^i_{\mathbf{P}}(\Omega) \cup \alpha_d}(\mathbf{t}) \\
&= \overline{\Phi^i_{\mathbf{Q}}(\Omega) \cup \alpha_d}(\mathbf{t}) && \text{by the induction hypothesis and the coincidence lemma} \\
&= \Phi^{i+1}_{\mathbf{Q}}(\Omega)(\mathbf{F})(\mathrm{d}),
\end{aligned}$$

and d is an arbitrary element of $D_{\delta(\mathbf{F})}$, so we have $\Phi^{i+1}_{\mathbf{P}}(\Omega)(\mathbf{F}) = \Phi^{i+1}_{\mathbf{Q}}(\Omega)(\mathbf{F})$, concluding the induction. Now,

$$\begin{aligned}
\mu\Phi_{\mathbf{P}}(\mathbf{F}) &= \sqcup\{\Phi^i_{\mathbf{P}}(\Omega) \mid i \in N\}(\mathbf{F}) && \text{by the fixed point theorem} \\
&= \sqcup\{\Phi^i_{\mathbf{P}}(\Omega)(\mathbf{F}) \mid i \in N\} && \text{by Theorem 16.3.9} \\
&= \sqcup\{\Phi^i_{\mathbf{Q}}(\Omega)(\mathbf{F}) \mid i \in N\} && \text{by (3.4)} \\
&= \sqcup\{\Phi^i_{\mathbf{Q}}(\Omega) \mid i \in N\}(\mathbf{F}) && \text{by Theorem 16.3.9} \\
&= \mu\Phi_{\mathbf{Q}}(\mathbf{F}),
\end{aligned}$$

which completes the proof of part 1. Part 2 follows immediately from part 1 by the coincidence lemma. ■

We have one more step to take before we define the denotational semantics of **W**-programs. In the next section we will select from any complete, continuous **W**-structure Σ certain objects, the data objects, to get a data structure system Δ. We will then give the denotational semantics of **W**-programs in terms of Δ. However, it will turn out that Σ_{N} is already a data structure system, which we will also call Δ_{N}, so we can anticipate the next section and give the denotational semantics in Δ_{N} for $\mathbf{W}_{\mathrm{N}}$-programs.[4] The idea is to give a single function that assigns a meaning to all $\mathbf{W}_{\mathrm{N}}$-programs.

Definition. The *denotational meaning function* for Δ_{N}, denoted $\mathscr{D}_{\Delta_{\mathrm{N}}}$, is defined

$$\mathscr{D}_{\Delta_{\mathrm{N}}}(\mathbf{P}) = \mu\Phi^{\Sigma}_{\mathbf{P}}$$

for all $\mathbf{W}_{\mathrm{N}}$-programs $\mathbf{P}$.

[4] The reader who wishes to go on at this point to the chapter on operational semantics will be able to read the first two sections of that chapter as they apply to the particular structure Σ_{N}. We simply need to remark that Σ_{N} is a simple $\mathbf{W}_{\mathrm{N}}$-structure (as defined in Section 5 of the current chapter) and that $\Delta_{\mathrm{N}} = \mathrm{rep}(\Sigma_{\mathrm{N}}) = \Sigma_{\mathrm{N}}$ (as defined in Section 4 of the current chapter) is a simple data structure system for $\mathbf{W}_{\mathrm{N}}$.

Exercises

1. Let $\Sigma = (\mathcal{T}, \mathcal{I})$ be a complete, continuous **W**-structure, let **P** be a **W**-program, and let $\Phi_{\mathbf{P}} = \Phi_{\mathbf{P}}^{\Sigma}$. Prove the following statements.
   (a) $\Phi_{\mathbf{P}}^{i}(\Omega) \in \mathcal{CA}_{\mathcal{T}}(FV(\mathbf{P}))$ for all $i \in N$.
   (b) $\{\Phi_{\mathbf{P}}^{i}(\Omega) \mid i \in N\}$ is a chain in $(\mathcal{CA}_{\mathcal{T}}(FV(\mathbf{P})), \sqsubseteq_{\mathcal{CA}_{\mathcal{T}}(FV(\mathbf{P}))})$. [*Hint:* See Lemma 1 in Section 5 of Chapter 16.]
2. Give a $\mathbf{W_N}$-program **P** such that $\mu\Phi_{\mathbf{P}}^{\Sigma_N} = \Omega$.
3. Give a $\mathbf{W_N}$-program **P** such that $\Phi_{\mathbf{P}}^{\Sigma_N}$ has infinitely many fixed points.
4. Show that $\Phi_{\mathbf{ADD}}^{\Sigma_N}$ has exactly one fixed point.
5. Give a $\mathbf{W_N}$-program **P** with $FV(\mathbf{P}) = \{\mathbf{F}\}$ such that $\mu\Phi_{\mathbf{P}}^{\Sigma_N}(\mathbf{F})$ is not strict.
6. Prove (3.3).
7. Let $\Sigma = \Sigma_N$ and let **P** be the $\mathbf{W_N}$-program with the equation

$$\mathbf{F(X) = if_N(is\_0(X), 2, s(F(s_1^{-1}(X)))).}$$

   (a) Show by induction on n that, for any $n \in N$,

$$\Phi_{\mathbf{P}}^{n}(\Omega)(\mathbf{F})(x) = \begin{cases} x + 2 & \text{if } x \in N \text{ and } 0 \le x < n \\ \bot_N & \text{if } x = \bot_N \text{ or } x \ge n \end{cases}$$

   for all $x \in N_\bot$.
   (b) Show that $\mu\Phi_{\mathbf{P}}(\mathbf{F}) = f_\bot$, where $f(x) = x + 2$ for all $x \in N$.
8. Let **P** be the $\mathbf{W_N}$-program with equations

$$\mathbf{F(X,Y) = if_N(is\_0(Y), X, s(F(X, s_1^{-1}(Y))))}$$
$$\mathbf{G(X) = F(X,X)}$$
$$\mathbf{H(X) = if_N(is\_0(X), s(0), G(H(s_1^{-1}(X))))}$$

   and let $\Phi_{\mathbf{P}} = \Phi_{\mathbf{P}}^{\Sigma_N}$.
   (a) Let $\alpha(\mathbf{F}) = \cdot_\bot$ (the strict extension of the multiplication function), $\alpha(\mathbf{G})(x) = x +_\bot 2$ for all $x \in N_\bot$, and $\alpha(\mathbf{H})(x) = 3$ for all $x \in N_\bot$. What is $\Phi_{\mathbf{P}}(\alpha)(\mathbf{F})(3, 5)$? $\Phi_{\mathbf{P}}(\alpha)(\mathbf{G})(7)$? $\Phi_{\mathbf{P}}(\alpha)(\mathbf{H})(13)$?
   (b) What is $\Phi_{\mathbf{P}}^{2}(\alpha)(\mathbf{F})(3,5)$? $\Phi_{\mathbf{P}}^{2}(\alpha)(\mathbf{G})(7)$? $\Phi_{\mathbf{P}}^{2}(\alpha)(\mathbf{H})(13)$?
   (c) Describe $\Phi_{\mathbf{P}}^{i}(\Omega)(\mathbf{F})$, $\Phi_{\mathbf{P}}^{i}(\Omega)(\mathbf{G})$, and $\Phi_{\mathbf{P}}^{i}(\Omega)(\mathbf{H})$ for all $i \in N$.
   (d) Describe $\mu\Phi_{\mathbf{P}}(\mathbf{F})$, $\mu\Phi_{\mathbf{P}}(\mathbf{G})$, and $\mu\Phi_{\mathbf{P}}(\mathbf{H})$.

9. Let **P** be the $\mathbf{W}_{\mathbf{N}}$-program with equations

$$\mathbf{F(X) = if_N(is\_0(X), s(0), G(X,X))}$$

$$\mathbf{G(X,Y) = if_N(is\_0(Y), F(X), s(G(X, s_1^{-1}(X))))}$$

and let $\Phi_{\mathbf{P}} = \Phi_{\mathbf{P}}^{\Sigma_{\mathrm{N}}}$.

(a) Let $\alpha(\mathbf{F})(x) = 3$ for all $x \in N_{\perp}$, and let $\alpha(\mathbf{G}) = +_{\perp}$. What is $\Phi_{\mathbf{P}}(\alpha)(\mathbf{F})(3)$? $\Phi_{\mathbf{P}}(\alpha)(\mathbf{G})(3,2)$?

(b) What is $\Phi_{\mathbf{P}}^2(\alpha)(\mathbf{F})(3)$? $\Phi_{\mathbf{P}}^2(\alpha)(\mathbf{G})(3,2)$?

(c) Describe $\Phi_{\mathbf{P}}^i(\Omega)$ for each $i \in N$.

(d) Describe $\mu\Phi_{\mathbf{P}}$.

10. Give a $\mathbf{W}_{\mathbf{N}}$-program **P** such that $\mu\Phi_{\mathbf{P}}^{\Sigma_{\mathrm{N}}}(\mathbf{F}) = \cdot_{\perp}$ (the strict extension of the multiplication function) for some $\mathbf{F} \in FV(\mathbf{P})$.

11. Give a $\mathbf{W}_{\mathbf{N}}$-program **P** such that $\mu\Phi_{\mathbf{P}}^{\Sigma_{\mathrm{N}}}(\mathbf{F}) = F_{\perp}$, where $F(n)$ is the nth Fibonacci number, for some $\mathbf{F} \in FV(\mathbf{P})$. [See Exercise 8.3 in Chapter 3 for the definition of *Fibonacci numbers*.]

12. Let &, ∨, ~ be the usual operations on truth values. Give a **W**-program **P** such that, in any **W**-structure, $\mu\Phi_{\mathbf{P}}(\&) = \&_{\perp}$, $\mu\Phi_{\mathbf{P}}(\vee) = \vee_{\perp}$, and $\mu\Phi_{\mathbf{P}}(\sim) = \sim_{\perp}$, where

$$\tau(\&) = \tau(\vee) = \mathbf{Bool} \times \mathbf{Bool} \to \mathbf{Bool}$$

and $\tau(\sim) = \mathbf{Bool} \to \mathbf{Bool}$.

13. Let $\Sigma = (\mathscr{T}, \mathscr{I})$ be a **W**-structure. Suppose we extend the standard vocabulary **W** to **W′** by adding the symbols **is_tt**, **is_ff**, and suppose we give **is_tt**, **is_ff** their natural interpretations $\mathscr{I}(\mathbf{is\_tt})$, $\mathscr{I}(\mathbf{is\_ff})$ as in condition 5 on $\mathscr{T}$-interpretations. Give a **W**-program **P** with function variables **Is_tt**, **Is_ff** such that $\overline{\mu\Phi_{\mathbf{P}}}(\mathbf{Is\_tt}) = \mathscr{I}(\mathbf{is\_tt})$ and $\overline{\mu\Phi_{\mathbf{P}}}(\mathbf{Is\_ff}) = \mathscr{I}(\mathbf{is\_ff})$.

14. Let $\Sigma = (\mathscr{T}_{\mathbf{N}}, \mathscr{I}_e)$, where $\mathscr{I}_e$ is given in Exercise 2.6. What is $\mu\Phi_{\mathbf{ADD}}^{\Sigma}$?

15.* Let $\Sigma = (\mathscr{T}, \mathscr{I})$ be a complete, continuous **W**-structure, and let **P** be a **W**-program. Define $\Psi_{\mathbf{P}}^{\Sigma}: \mathscr{A}_{\mathscr{T}}(FV(\mathbf{P})) \to \mathscr{A}_{\mathscr{T}}(FV(\mathbf{P}))$ exactly like $\Phi_{\mathbf{P}}^{\Sigma}$ except that its domain is $\mathscr{A}_{\mathscr{T}}(FV(\mathbf{P}))$. [This exercise requires the results of Exercise 2.14.]

(a) Show that $\alpha \in \mathscr{A}_{\mathscr{T}}(FV(\mathbf{P}))$ is a solution to **P** in Σ if and only if α is a fixed point of $\Psi_{\mathbf{P}}^{\Sigma}$.

(b) Show that $\Psi_{\mathbf{P}}^{\Sigma}$ is monotonic.

(c) Give a **W**-program **Q** such that $\Psi_{\mathbf{Q}}^{\Sigma}$ is not continuous. [*Hint:* Let f be a function and C a chain such that $f(\sqcup C) \neq \sqcup f(C)$. For all $c \in C$, let $\alpha_c(\mathbf{F}) = f$ and let $\alpha_c(\mathbf{G})$ be the constant function $\alpha_c(\mathbf{G})(x) = c$. Put the equation $\mathbf{H}(\mathbf{X}) = \mathbf{F}(\mathbf{G}(\mathbf{X}))$ in $\mathbf{Q}$.]

(d) Show that $(\Psi_{\mathbf{P}}^{\Sigma})^i(\Omega) \in \mathscr{CA}_{\mathscr{T}}(FV(\mathbf{P}))$ for all $i \in N$. [*Hint:* Use Theorem 2.3 in the induction step.]

(e) Show that $\sqcup\{(\Psi_{\mathbf{P}}^{\Sigma})^i(\Omega) \mid i \in N\}$ exists and is $\mu\Psi_{\mathbf{P}}^{\Sigma}$.

4. Denotational Semantics of W-Programs

Next we turn to the treatment of data structures. There are two properties that they should satisfy:

1. Since the semantics of program **P** is to be based on the function $\Phi_{\mathbf{P}}$, we want data structures to be rich enough to guarantee the existence of $\mu\Phi_{\mathbf{P}}$ for every program **P**.
2. Since we need to be able to specify the inputs to a program, we want every element in a data structure to be the meaning of some term.

These two properties may seem to be contradictory. Property 1 requires data structures to have enough elements to give meanings to programs, and property 2 requires that data structures not have too many elements. We deal with these requirements in two steps. Theorem 3.3 guarantees that $\mu\Phi_{\mathbf{P}}^{\Sigma}$ exists when Σ is a complete, continuous **W**-structure, so we begin with such structures and pare them down so that property 2 is satisfied.

Definition. Let $\Sigma = (\mathscr{T}, \mathscr{I})$ be a complete, continuous **W**-structure.

1. An element $d \in D_{\mathscr{T}(\boldsymbol{\tau})}$, for some $\boldsymbol{\tau} \in TV(\mathbf{W})$, is *representable* in **W** if there is some **W**-program **P** and some term $\mathbf{t} \in \mathrm{TM}_{\mathbf{W}}^{\boldsymbol{\tau}}(FV(\mathbf{P}))$ such that $d = \overline{(\mu\Phi_{\mathbf{P}}^{\Sigma})_{\mathscr{I}}}(\mathbf{t})$. $\mathrm{rep}(D_{\mathscr{T}(\boldsymbol{\tau})})$ is the set
$$\left\{d \in D_{\mathscr{T}(\boldsymbol{\tau})} \mid d \text{ is representable in } \mathbf{W}\right\},$$
$\sqsubseteq_{\mathrm{rep}(\mathscr{T}(\boldsymbol{\tau}))}$ is the restriction of $\sqsubseteq_{\mathscr{T}(\boldsymbol{\tau})}$ to $\mathrm{rep}(D_{\mathscr{T}(\boldsymbol{\tau})})$, and $\mathrm{rep}(\mathscr{T})$ is defined
$$\mathrm{rep}(\mathscr{T})(\boldsymbol{\tau}) = \left(\mathrm{rep}(D_{\mathscr{T}(\boldsymbol{\tau})}), \sqsubseteq_{\mathrm{rep}(\mathscr{T}(\boldsymbol{\tau}))}\right)$$
for all $\boldsymbol{\tau} \in TV(\mathbf{W})$.
2. For any function
$$f \in D_{\mathscr{T}(\boldsymbol{\tau}_1)} \times \cdots \times D_{\mathscr{T}(\boldsymbol{\tau}_n)} \to D_{\mathscr{T}(\boldsymbol{\tau})},$$
where $\boldsymbol{\tau}_1, \ldots, \boldsymbol{\tau}_n, \boldsymbol{\tau} \in TV(\mathbf{W})$, let
$$\mathrm{rep}(f) \in \mathrm{rep}(D_{\mathscr{T}(\boldsymbol{\tau}_1)}) \times \cdots \times \mathrm{rep}(D_{\mathscr{T}(\boldsymbol{\tau}_n)}) \to D_{\mathscr{T}(\boldsymbol{\tau})}$$

be defined

$$\text{rep}(f)(d_1, \ldots, d_n) = f(d_1, \ldots, d_n)$$

for all $(d_1, \ldots, d_n) \in \text{rep}(D_{\mathscr{T}(\tau_1)}) \times \cdots \times \text{rep}(D_{\mathscr{T}(\tau_n)})$. Then $\text{rep}(\mathscr{I})$ is defined

$\text{rep}(\mathscr{I})(\mathbf{c}) = \mathscr{I}(\mathbf{c})$	for all constant symbols $\mathbf{c} \in \mathbf{W}$
$\text{rep}(\mathscr{I})(\mathbf{f}) = \text{rep}(\mathscr{I}(\mathbf{f}))$	for all proper function symbols $\mathbf{f} \in \mathbf{W}$.

3. Finally, $\text{rep}(\Sigma) = (\text{rep}(\mathscr{T}), \text{rep}(\mathscr{I}))$ is the *data structure system for* **W** *based on* Σ.

The point is that an arbitrary **W**-structure Σ might contain objects that we can never use as data since there is no way to refer to them. Therefore, in defining functions that we wish to consider computable, we will restrict our attention to the representable objects in $\text{rep}(\Sigma)$. We might call these the *data objects* of Σ.

It is important to understand that even if $d_1, \ldots, d_n$ are representable, $\text{rep}(f)(d_1, \ldots, d_n)$ may not be representable if f is some arbitrary function. However, we will show that $\text{rep}(f)(d_1, \ldots, d_n)$ is representable when f is the interpretation $\mathscr{I}(\mathbf{f})$ of some $\mathbf{f} \in \mathbf{W}$.

When $\mathscr{T}$ is understood, we will generally write $(D_{r(\tau)}, \sqsubseteq_{r(\tau)})$ for $\text{rep}(\mathscr{T})(\tau)$,

$D_{r(\tau_1) \times \cdots \times r(\tau_n)}$	for $\text{rep}(D_{\mathscr{T}(\tau_1)}) \times \cdots \times \text{rep}(D_{\mathscr{T}(\tau_n)})$,
$D_{r(\tau_1) \times \cdots \times r(\tau_n) \to r(\tau)}$	for $\text{rep}(D_{\mathscr{T}(\tau_1)}) \times \cdots \times \text{rep}(D_{\mathscr{T}(\tau_n)}) \to \text{rep}(D_{\mathscr{T}(\tau)})$,

and, when $\delta(\mathbf{F}) = \tau_1 \times \cdots \times \tau_n \to \tau$, $D_{r(\delta(\mathbf{F}))}$ for $D_{r(\tau_1) \times \cdots \times r(\tau_n)}$.

Let $\Sigma = (\mathscr{T}, \mathscr{I})$ be a **W**-structure. Then for every constant symbol $\mathbf{c} \in \mathbf{W}$, $\mathscr{I}(\mathbf{c})$ is representable since $\overline{\mu\Phi_{\mathbf{P}}^{\Sigma}(\mathbf{c})} = \mathscr{I}(\mathbf{c})$ for any **W**-program **P**. Moreover, for every $\tau \in TV(\mathbf{W})$, $\bot_\tau$ is representable: let $\mathbf{c} \in \mathbf{W}_c$ be a constant symbol with $\tau(\mathbf{c}) = \tau$, and let **P** be the **W**-program with equation $\mathbf{B}(\mathbf{X}) = \mathbf{B}(\mathbf{X})$, where $\tau(\mathbf{X}) = \tau$ and $\tau(\mathbf{B}) = \tau \to \tau$. Then $\overline{\mu\Phi_{\mathbf{P}}^{\Sigma}(\mathbf{B}(\mathbf{c}))} = \overline{\Omega}(\mathbf{B}(\mathbf{c})) = \bot_\tau$. (This explains, by the way, our requirement that $\mathbf{W}_c$ contain a constant symbol of type τ for every $\tau \in TV(\mathbf{W}_c)$.)

In $\Sigma_{\mathbf{N}}$, then, it is clear that every element in Bool is representable in $\mathbf{W}_{\mathbf{N}}$. Moreover, every element in $N_\bot$ is representable in $\mathbf{W}_{\mathbf{N}}$: for all $n \in N$, $\overline{\mu\Phi_{\mathbf{P}}(\mathbf{n})} = n$, where **P** is any $\mathbf{W}_{\mathbf{N}}$-program, e.g., the empty program. (In a case like this we can simply say $\overline{\Omega}(\mathbf{n}) = n$.) Therefore, $\text{rep}(\mathscr{T}_{\mathbf{N}}) = \mathscr{T}_{\mathbf{N}}$, $\text{rep}(\mathscr{I}_{\mathbf{N}}) = \mathscr{I}_{\mathbf{N}}$, and $\text{rep}(\Sigma_{\mathbf{N}}) = \Sigma_{\mathbf{N}}$. We will write $\Delta_{\mathbf{N}}$ for $\Sigma_{\mathbf{N}}$ when we want to emphasize that $\Sigma_{\mathbf{N}}$ is a data structure system for $\mathbf{W}_{\mathbf{N}}$.

Now, let V_f be a set of function variables, and let $d = \overline{\mu\Phi_{\mathbf{P}}^{\Sigma}(\mathbf{t})}$. It is useful to note that simply by changing the function variables in **P** and **t**, we can always find a **W**-program **Q** and a term $\mathbf{u} \in \text{TM}_{\mathbf{W}}(FV(\mathbf{Q}))$ such that

$FV(\mathbf{Q}) \cap V_f = \varnothing$ and $\overline{\mu\Phi_{\mathbf{Q}}^{\Sigma}(\mathbf{u})} = \overline{\mu\Phi_{\mathbf{P}}^{\Sigma}(\mathbf{t})}$. Therefore, given representable elements $d_1, \ldots, d_n$, we can always find **W**-programs $\mathbf{P}_1, \ldots, \mathbf{P}_n$ such that $FV(\mathbf{P}_i) \cap FV(\mathbf{P}_j) = \varnothing$, $1 \le i < j \le n$, and terms $\mathbf{t}_1, \ldots, \mathbf{t}_n$ such that $\mathbf{t}_i \in \mathrm{TM}_{\mathbf{W}}(FV(\mathbf{P}_i))$ and $d_i = \overline{\mu\Phi_{\mathbf{P}_i}^{\Sigma}(\mathbf{t}_i)}$, $1 \le i \le n$. We will say that $\mathbf{P}_1, \ldots, \mathbf{P}_n$ are *consistent* if $FV(\mathbf{P}_i) \cap FV(\mathbf{P}_j) = \varnothing$, $1 \le i < j \le n$.

The first thing we need to do is show that data structure systems are **W**-structures. We begin with a lemma that shows that, for all proper function symbols $\mathbf{f} \in \mathbf{W}$ with $\tau(\mathbf{f}) = \boldsymbol{\tau}_1 \times \cdots \times \boldsymbol{\tau}_n \to \boldsymbol{\tau}$, $\mathrm{rep}(\mathscr{I})(\mathbf{f}) \in D_{r(\boldsymbol{\tau}_1)} \times \cdots \times D_{r(\boldsymbol{\tau}_n)} \to D_{r(\boldsymbol{\tau})}$. In other words, data structure systems are closed under the interpretations of the function symbols.

Lemma 1. Let $\Sigma = (\mathscr{T}, \mathscr{I})$ be a complete, continuous **W**-structure, and let $\mathbf{f} \in \mathbf{W}$ with $\tau(\mathbf{f}) = \boldsymbol{\tau}_1 \times \cdots \times \boldsymbol{\tau}_n \to \boldsymbol{\tau}$. If $(d_1, \ldots, d_n) \in D_{r(\boldsymbol{\tau}_1) \times \cdots \times r(\boldsymbol{\tau}_n)}$, then $\mathrm{rep}(\mathscr{I})(\mathbf{f})(d_1, \ldots, d_n) \in D_{r(\boldsymbol{\tau})}$.

Proof. Let $d_i = \overline{(\mu\Phi_{\mathbf{P}_i})_{\mathscr{I}}}(\mathbf{t}_i)$, $1 \le i \le n$, where $\mathbf{P}_1, \ldots, \mathbf{P}_n$ are consistent. Then $\mathbf{P} = \bigcup_{i=1}^{n} \mathbf{P}_i$ is a **W**-program, and

$$\begin{aligned}
&\overline{(\mu\Phi_{\mathbf{P}})_{\mathscr{I}}}(\mathbf{f}(\mathbf{t}_1, \ldots, \mathbf{t}_n)) \\
&\quad = \mathscr{I}(\mathbf{f})\big(\overline{(\mu\Phi_{\mathbf{P}})_{\mathscr{I}}}(\mathbf{t}_1), \ldots, \overline{(\mu\Phi_{\mathbf{P}})_{\mathscr{I}}}(\mathbf{t}_n)\big) \\
&\quad = \mathscr{I}(\mathbf{f})\big(\overline{(\mu\Phi_{\mathbf{P}_1})_{\mathscr{I}}}(\mathbf{t}_1), \ldots, \overline{(\mu\Phi_{\mathbf{P}_n})_{\mathscr{I}}}(\mathbf{t}_n)\big) \qquad \text{by the extension lemma} \\
&\quad = \mathscr{I}(\mathbf{f})(d_1, \ldots, d_n) \\
&\quad = \mathrm{rep}(\mathscr{I})(\mathbf{f})(d_1, \ldots, d_n) \qquad \text{since } (d_1, \ldots, d_n) \in D_{r(\boldsymbol{\tau}_1) \times \cdots \times r(\boldsymbol{\tau}_n)}.
\end{aligned}$$

■

We will use the next lemma when considering the interpretations of the built-in function symbols **is_f** and $\mathbf{f}_i^{-1}$.

Lemma 2. Let $\Sigma = (\mathscr{T}, \mathscr{I})$ be a complete, continuous **W**-structure, and let $\mathbf{f} \in \mathbf{W}_c$ with $\tau(\mathbf{f}) = \boldsymbol{\tau}_1 \times \cdots \times \boldsymbol{\tau}_n \to \boldsymbol{\tau}$. If $(d_1, \ldots, d_n) \in D_{\boldsymbol{\tau}_1 \times \cdots \times \boldsymbol{\tau}_n}$ and $\mathscr{I}(\mathbf{f})(d_1, \ldots, d_n) \in D_{r(\boldsymbol{\tau})} - \{\bot_{\boldsymbol{\tau}}\}$, then $(d_1, \ldots, d_n) \in D_{r(\boldsymbol{\tau}_1) \times \cdots \times r(\boldsymbol{\tau}_n)}$.

Proof. Let $\mathscr{I}(\mathbf{f})(d_1, \ldots, d_n) = \overline{(\mu\Phi_{\mathbf{P}})_{\mathscr{I}}}(\mathbf{t})$. Then for $1 \le i \le n$,

$$\begin{aligned}
\overline{(\mu\Phi_{\mathbf{P}})_{\mathscr{I}}}(\mathbf{f}_i^{-1}(\mathbf{t})) &= \mathscr{I}(\mathbf{f}_i^{-1})\big(\overline{(\mu\Phi_{\mathbf{P}})_{\mathscr{I}}}(\mathbf{t})\big) \\
&= \mathscr{I}(\mathbf{f}_i^{-1})(\mathscr{I}(\mathbf{f})(d_1, \ldots, d_n)) \\
&= d_i.
\end{aligned}$$

■

Theorem 4.1. Let $\Sigma = (\mathscr{T}, \mathscr{I})$ be a complete, continuous **W**-structure. Then $\mathrm{rep}(\Sigma)$ is a **W**-structure.

Proof. For any $\tau \in TV(\mathbf{W})$, $(D_{r(\tau)}, \sqsubseteq_{r(\tau)})$ is clearly a partial order with bottom element $\perp_\tau$, and $(D_{r(\mathbf{Bool})}, \sqsubseteq_{r(\mathbf{Bool})}) = \mathscr{T}(\mathbf{Bool})$, so rep$(\mathscr{T})$ is a type assignment for **W**.

Now we need to show that rep$(\mathscr{I})$ is a rep$(\mathscr{T})$-interpretation for **W**. Note that in the context of the type assignment rep$(\mathscr{T})$, each reference to a set D_τ in the definition of $\mathscr{T}$-interpretations should be understood as referring to rep$(D_{\mathscr{T}(\tau)})$, i.e., $D_{r(\tau)}$. Also, each reference there to $\mathscr{I}$ should be understood as a reference to rep$(\mathscr{I})$. It is clear that rep$(\mathscr{I})(\mathbf{c}) = \mathscr{I}(\mathbf{c})$ for all constant symbols $\mathbf{c} \in \mathbf{W}_c$, so condition 1 in the definition of $\mathscr{T}$-interpretations is satisfied. If $\mathbf{f} \in \mathbf{W}_c$ is a proper function symbol with $\tau(\mathbf{f}) = \tau_1 \times \cdots \times \tau_n \to \tau$, and $d_i \in D_{r(\tau_i)}$, $1 \le i \le n$, then rep$(\mathscr{I})(\mathbf{f})(d_1, \ldots, d_n) \in D_{r(\tau)}$ by Lemma 1, so condition 2a is satisfied. Conditions 2b, 2c, and 3 follow immediately from the definition of rep$(\mathscr{I})$. Condition 4 follows immediately from Lemma 1 and the definition of rep$(\mathscr{I})$, as does condition 5 for all constant symbols in $\mathbf{W}_c^-$, so let $\mathbf{f} \in \mathbf{W}_c$ with $\tau(\mathbf{f}) = \tau_1 \times \cdots \times \tau_n \to \tau$, and let $d \in D_{r(\tau)}$. If $d = \perp_\tau$ then rep$(\mathscr{I})(\mathbf{is_f})(d) = \mathscr{I}(\mathbf{is_f})(d) = \perp_{\text{Bool}}$, so assume $d \neq \perp_\tau$. If $d \in$ ran(rep$(\mathscr{I})(\mathbf{f})$), then obviously $d \in \operatorname{ran} \mathscr{I}(\mathbf{f})$, so that rep$(\mathscr{I})(\mathbf{is_f})(d) = \mathscr{I}(\mathbf{is_f})(d) =$ tt. Now, if $d \in \operatorname{ran} \mathscr{I}(\mathbf{f})$, then $d = \mathscr{I}(\mathbf{f})(d_1, \ldots, d_n)$ for some $(d_1, \ldots, d_n) \in D_{r(\delta(\mathbf{f}))}$ by Lemma 2, so $d \in$ ran(rep$(\mathscr{I})(\mathbf{f})$). Therefore, if $d \notin$ ran(rep $(\mathscr{I})(\mathbf{f})$), then $d \notin \operatorname{ran} \mathscr{I}(\mathbf{f})$, and rep $(\mathscr{I})(\mathbf{is\_f})(d) = \mathscr{I}(\mathbf{is\_f})(d) =$ ff, so condition 5 is satisfied. Finally, condition 6 is satisfied by a similar argument, so rep$(\mathscr{I})$ is a rep$(\mathscr{T})$-interpretation, and rep(Σ) is a **W**-structure. ∎

We can now define the denotational semantics of recursion programs. For a complete, continuous **W**-structure $(\mathscr{T}, \mathscr{I})$ and a variable assignment $\alpha \in \mathscr{A}_{\mathscr{T}}(V)$, where V is a set of function variables, let rep(α) be the function on V defined by

$$\text{rep}(\alpha)(\mathbf{F}) = \text{rep}(\alpha(\mathbf{F})) \qquad \text{for all } \mathbf{F} \in V.$$

Note that the domain of $\alpha(\mathbf{F})$ is $D_{r(\delta(\mathbf{F}))}$ for all $\mathbf{F} \in V$, but rep(α) is a variable assignment in $\mathscr{A}_{\text{rep}(\mathscr{T})}(V)$ if and only if

$$\text{rep}(\alpha)(\mathbf{F})(d) \in D_{r(\rho(\mathbf{F}))} \qquad \text{for all } \mathbf{F} \in V \text{ and all } d \in D_{r(\delta(\mathbf{F}))}.$$

Definition. Let Σ be a complete, continuous **W**-structure, and let $\Delta =$ rep(Σ). The *denotational meaning function* for Δ, denoted $\mathscr{D}_\Delta$, is defined

$$\mathscr{D}_\Delta(\mathbf{P}) = \text{rep}(\mu\Phi_{\mathbf{P}}^{\Sigma})$$

for all **W**-programs **P**.

For a **W**-program **P** and $\mathbf{F} \in FV(\mathbf{P})$, we have

$$\mathscr{D}_\Delta(\mathbf{P})(\mathbf{F}) = \text{rep}(\mu\Phi_{\mathbf{P}}^{\Sigma})(\mathbf{F}) = \text{rep}(\mu\Phi_{\mathbf{P}}^{\Sigma}(\mathbf{F})).$$

The point is that, rather than taking $\mu\Phi_{\mathbf{P}}^{\Sigma}(\mathbf{F})$ as the function assigned to $\mathbf{F}$, we assign to $\mathbf{F}$ a function whose domain consists only of representable objects.

We showed that $\mu\Phi_{\mathbf{P}}^{\Sigma}$ is a solution to $\mathbf{P}$ in any complete, continuous $\mathbf{W}$-structure Σ. We now show that $\mathscr{D}_\Delta(\mathbf{P})$ is a solution to $\mathbf{P}$ in $\Delta = \text{rep}(\Sigma)$. That is, we still have a solution when we restrict our attention to the data structure system based on Σ. We begin with three lemmas that let us ignore nonrepresentable objects when applying $\mu\Phi_{\mathbf{P}}^{\Sigma}$ to terms. In particular, we want to show that if $\beta \in \mathscr{A}_{\text{rep}(\mathscr{T})}(V)$, where V is a set of individual variables, then for any term $\mathbf{t} \in \text{TM}_{\mathbf{W}}(FV(\mathbf{P}) \cup V)$,

$$\overline{(\mu\Phi_{\mathbf{P}}^{\Sigma} \cup \beta)_{\mathscr{I}}}(\mathbf{t}) = \overline{(\text{rep}(\mu\Phi_{\mathbf{P}}^{\Sigma}) \cup \beta)_{\text{rep}(\mathscr{I})}}(\mathbf{t}).$$

It will follow easily, then, that $\mathscr{D}_\Delta(\mathbf{P}) = \text{rep}(\mu\Phi_{\mathbf{P}}^{\Sigma})$ is a solution to $\mathbf{P}$ in Δ.

Lemma 3. Let $\Sigma = (\mathscr{T}, \mathscr{I})$ be a complete, continuous $\mathbf{W}$-structure, let $\mathbf{P}$ be a $\mathbf{W}$-program, let $V \subseteq \text{VAR}_{\text{I}}$, and let $\alpha \in \mathscr{A}_{\text{rep}(\mathscr{T})}(V)$. Then for any term $\mathbf{t} \in \text{TM}_{\mathbf{W}}(V \cup FV(\mathbf{P}))$, $\overline{(\mu\Phi_{\mathbf{P}}^{\Sigma} \cup \alpha)_{\mathscr{I}}}(\mathbf{t}) \in D_{r(\tau(\mathbf{t}))}$.

Proof. We have $\mu\Phi_{\mathbf{P}}^{\Sigma} \cup \alpha \in \mathscr{A}_{\mathscr{T}}(V \cup FV(\mathbf{P}))$, which implies $\overline{(\mu\Phi_{\mathbf{P}}^{\Sigma} \cup \alpha)_{\mathscr{I}}}(\mathbf{t}) \in D_{\tau(\mathbf{t})}$ by Theorem 2.1, so we need to show only that $\overline{(\mu\Phi_{\mathbf{P}}^{\Sigma} \cup \alpha)_{\mathscr{I}}}(\mathbf{t})$ is representable. We argue by structural induction on $\mathbf{t}$. If $\mathbf{t}$ is a constant symbol $\mathbf{c} \in \mathbf{W}$, then $\overline{(\mu\Phi_{\mathbf{P}}^{\Sigma} \cup \alpha)_{\mathscr{I}}}(\mathbf{c})$ is clearly representable, and if $\mathbf{t}$ is $\mathbf{X} \in V$, then $\overline{(\mu\Phi_{\mathbf{P}}^{\Sigma} \cup \alpha)_{\mathscr{I}}}(\mathbf{X}) = \alpha(\mathbf{X})$ is representable by assumption. If $\mathbf{t}$ is $\mathbf{f}(\mathbf{t}_1, \ldots, \mathbf{t}_n)$, where $\mathbf{f} \in \mathbf{W}$, then

$$\overline{(\mu\Phi_{\mathbf{P}}^{\Sigma} \cup \alpha)_{\mathscr{I}}}(\mathbf{f}(\mathbf{t}_1, \ldots, \mathbf{t}_n))$$

$$= \mathscr{I}(\mathbf{f})\Big(\overline{(\mu\Phi_{\mathbf{P}}^{\Sigma} \cup \alpha)_{\mathscr{I}}}(\mathbf{t}_1), \ldots, \overline{(\mu\Phi_{\mathbf{P}}^{\Sigma} \cup \alpha)_{\mathscr{I}}}(\mathbf{t}_n)\Big)$$

$$= \mathscr{I}(\mathbf{f})\Big(\overline{(\mu\Phi_{\mathbf{P}_1}^{\Sigma})_{\mathscr{I}}}(\mathbf{u}_1), \ldots, \overline{(\mu\Phi_{\mathbf{P}_n}^{\Sigma})_{\mathscr{I}}}(\mathbf{u}_n)\Big)$$

for some $\mathbf{P}_i, \mathbf{u}_i$, $1 \le i \le n$, by the induction hypothesis, where $\mathbf{P}_1, \ldots, \mathbf{P}_n$ are consistent

$$= \mathscr{I}(\mathbf{f})\Big(\overline{(\mu\Phi_{\mathbf{P}_0}^{\Sigma})_{\mathscr{I}}}(\mathbf{u}_1), \ldots, \overline{(\mu\Phi_{\mathbf{P}_0}^{\Sigma})_{\mathscr{I}}}(\mathbf{u}_n)\Big)$$

by the extension lemma, where $\mathbf{P}_0 = \bigcup_{i=1}^{n} \mathbf{P}_i$

$$= \overline{(\mu\Phi_{\mathbf{P}_0}^{\Sigma})_{\mathscr{I}}}(\mathbf{f}(\mathbf{u}_1, \ldots, \mathbf{u}_n))$$

$$\in D_{\tau(\mathbf{t})} \quad \text{by Theorem 2.1,}$$

and $\overline{(\mu\Phi_{\mathbf{P}_0}^{\Sigma})_{\mathscr{I}}}(\mathbf{f}(\mathbf{u}_1, \ldots, \mathbf{u}_n))$ is representable, so it is in $D_{r(\tau(\mathbf{t}))}$. The argument is similar if $\mathbf{t}$ is $\mathbf{F}(\mathbf{t}_1, \ldots, \mathbf{t}_n)$ with $\mathbf{F} \in FV(\mathbf{P})$. ■

Lemma 4. Let $\Sigma = (\mathcal{T}, \mathcal{I})$ be a complete, continuous **W**-structure, let $\Delta = \text{rep}(\Sigma)$, and let **P** be a **W**-program. Then $\text{rep}(\mu\Phi_{\mathbf{P}}^{\Sigma}) \in \mathscr{A}_{\text{rep}(\mathcal{T})}(FV(\mathbf{P}))$.

Proof. Let $\mathbf{F}(\mathbf{X}_1, \ldots, \mathbf{X}_n) = \mathbf{t}$ be an equation in **P**, and let $d \in D_{r(\delta(\mathbf{F}))}$. Then

$$\begin{aligned}
\text{rep}(\mu\Phi_{\mathbf{P}}^{\Sigma})(\mathbf{F})(d) &= \text{rep}(\mu\Phi_{\mathbf{P}}^{\Sigma}(\mathbf{F}))(d) \\
&= \mu\Phi_{\mathbf{P}}^{\Sigma}(\mathbf{F})(d) && \text{since } d \in D_{r(\delta(\mathbf{F}))} \\
&= \overline{(\mu\Phi_{\mathbf{P}}^{\Sigma} \cup \alpha_d)_{\mathcal{I}}}(\mathbf{t}) && \text{since } \mu\Phi_{\mathbf{P}}^{\Sigma} \text{ is a solution to } \mathbf{P} \\
&\in D_{r(\tau(\mathbf{t}))} && \text{by Lemma 3.}
\end{aligned}$$

■

Lemma 5. Let $\Sigma = (\mathcal{T}, \mathcal{I})$ be a complete, continuous **W**-structure, let $V_i \subseteq \text{VAR}_\text{I}$ and $V_f \subseteq \text{VAR}_\text{F}$, let $\alpha \in \mathscr{A}_{\mathcal{T}}(V_f)$ be such that $\text{rep}(\alpha) \in \mathscr{A}_{\text{rep}(\mathcal{T})}(V_f)$, and let $\beta \in \mathscr{A}_{\text{rep}(\mathcal{T})}(V_i)$. Then for any term $\mathbf{t} \in \text{TM}_{\mathbf{W}}(V_f \cup V_i)$,

$$\overline{(\alpha \cup \beta)_{\mathcal{I}}}(\mathbf{t}) = \overline{(\text{rep}(\alpha) \cup \beta)_{\text{rep}(\mathcal{I})}}(\mathbf{t}).$$

Proof. Note that $\text{rep}(\alpha) \cup \beta \in \mathscr{A}_{\text{rep}(\mathcal{T})}(V_f \cup V_i)$, so by Theorems 2.1 and 4.1,

$$\overline{(\text{rep}(\alpha) \cup \beta)_{\text{rep}(\mathcal{I})}}(\mathbf{t}) \in D_{r(\tau(\mathbf{t}))} \qquad \text{for any } \mathbf{t} \in \text{TM}_{\mathbf{W}}(V_f \cup V_i). \tag{4.1}$$

We argue by structural induction on **t**. If **t** is a constant symbol $\mathbf{c} \in \mathbf{W}$, then

$$\overline{(\alpha \cup \beta)_{\mathcal{I}}}(\mathbf{c}) = \mathcal{I}(\mathbf{c}) = \text{rep}(\mathcal{I})(\mathbf{c}) = \overline{(\text{rep}(\alpha) \cup \beta)_{\text{rep}(\mathcal{I})}}(\mathbf{c}),$$

and if **t** is $\mathbf{X} \in V_i$, then

$$\overline{(\alpha \cup \beta)_{\mathcal{I}}}(\mathbf{X}) = \beta(\mathbf{X}) = \overline{(\text{rep}(\alpha) \cup \beta)_{\text{rep}(\mathcal{I})}}(\mathbf{X}).$$

If **t** is $\mathbf{f}(\mathbf{t}_1, \ldots, \mathbf{t}_n)$, where $\mathbf{f} \in \mathbf{W}$, then

$$\begin{aligned}
&\overline{(\alpha \cup \beta)_{\mathcal{I}}}(\mathbf{f}(\mathbf{t}_1, \ldots, \mathbf{t}_n)) \\
&\quad = \mathcal{I}(\mathbf{f})\big(\overline{(\alpha \cup \beta)_{\mathcal{I}}}(\mathbf{t}_1), \ldots, \overline{(\alpha \cup \beta)_{\mathcal{I}}}(\mathbf{t}_n)\big) \\
&\quad = \mathcal{I}(\mathbf{f})\big(\overline{(\text{rep}(\alpha) \cup \beta)_{\text{rep}(\mathcal{I})}}(\mathbf{t}_1), \ldots, \overline{(\text{rep}(\alpha) \cup \beta)_{\text{rep}(\mathcal{I})}}(\mathbf{t}_n)\big) \\
&\qquad\qquad \text{by the induction hypothesis} \\
&\quad = \text{rep}(\mathcal{I})(\mathbf{f})\big(\overline{(\text{rep}(\alpha) \cup \beta)_{\text{rep}(\mathcal{I})}}(\mathbf{t}_1), \ldots, \overline{(\text{rep}(\alpha) \cup \beta)_{\text{rep}(\mathcal{I})}}(\mathbf{t}_n)\big) \\
&\qquad\qquad \text{by (4.1)} \\
&\quad = \overline{(\text{rep}(\alpha) \cup \beta)_{\text{rep}(\mathcal{I})}}(\mathbf{f}(\mathbf{t}_1, \ldots, \mathbf{t}_n)).
\end{aligned}$$

The argument is similar if **t** is $\mathbf{F}(\mathbf{t}_1, \ldots, \mathbf{t}_n)$, where $\mathbf{F} \in V_f$. ■

Theorem 4.2. Let $\Sigma = (\mathcal{T}, \mathcal{I})$ be a complete, continuous **W**-structure, and let $\Delta = \text{rep}(\Sigma)$. Then for any **W**-program **P**, $\mathcal{D}_\Delta(\mathbf{P})$ is a solution to **P** in Δ.

Proof. Let **P** be a **W**-program, let $\mathbf{F}(\mathbf{X}_1, \ldots, \mathbf{X}_n) = \mathbf{t}$ be an equation in **P**, and let $d \in D_{r(\delta(\mathbf{F}))}$. Then

$$\begin{aligned}
\mathcal{D}_\Delta(\mathbf{P})(\mathbf{F})(d) &= \text{rep}(\mu\Phi_{\mathbf{P}}^{\Sigma})(\mathbf{F})(d) \\
&= \overline{(\mu\Phi_{\mathbf{P}}^{\Sigma} \cup \alpha_d)_{\mathcal{I}}}(\mathbf{t}) && \text{as in the proof of Lemma 4} \\
&= \overline{(\text{rep}(\mu\Phi_{\mathbf{P}}^{\Sigma}) \cup \alpha_d)_{\text{rep}(\mathcal{I})}}(\mathbf{t}) && \text{by Lemmas 4 and 5} \\
&= \overline{(\mathcal{D}_\Delta(\mathbf{P}) \cup \alpha_d)_{\text{rep}(\mathcal{I})}}(\mathbf{t}). && \blacksquare
\end{aligned}$$

Let Δ be a data structure system for **W**. Now that we have a meaning in Δ for every **W**-program, it makes sense to ask if a given **W**-program **P** defines the functions we want it to define. That is, we can ask if **P** is correct. Determining that a program is correct is known as *program verification.*

Definition. Let Δ be a data structure system for **W**, let **P** be a **W**-program, and let $f \in D_{r(\boldsymbol{\tau}_1) \times \cdots \times r(\boldsymbol{\tau}_n) \to r(\boldsymbol{\tau})}$ for some $\boldsymbol{\tau}_1, \ldots, \boldsymbol{\tau}_n, \boldsymbol{\tau} \in TV(\mathbf{W})$. We say that **P** is *partially correct* with respect to f if

$$\mathcal{D}_\Delta(\mathbf{P})(\mathbf{F}) \sqsubseteq_{r(\boldsymbol{\tau}_1) \times \cdots \times r(\boldsymbol{\tau}_n) \to r(\boldsymbol{\tau})} f$$

for some $\mathbf{F} \in \mathbf{P}$, and we say that **P** is *totally correct* with respect to f if $\mathcal{D}_\Delta(\mathbf{P})(\mathbf{F}) = f$ for some $\mathbf{F} \in \mathbf{P}$.

For example, we indicated in the previous section that $\mu\Phi_{\mathbf{ADD}}^{\Sigma_{\mathbf{N}}}(+) = +_\perp$, and $\mathcal{D}_{\Delta_{\mathbf{N}}}(\mathbf{ADD}) = \mu\Phi_{\mathbf{ADD}}^{\Sigma_{\mathbf{N}}}$ since $\Delta_{\mathbf{N}} = \Sigma_{\mathbf{N}}$, so **ADD** is totally correct with respect to $+_\perp$. Recall that the correctness argument for **ADD** was based on ordinary induction. We will now give an application of fixed point induction in establishing a partial correctness result.

Let eq: $N_\perp \times N_\perp \to \text{Bool}$ (where $\text{Bool} = \{\perp_{\text{Bool}}, \text{TRUE}, \text{FALSE}\}$ here) be the strict function defined by

$$\text{eq}(x, y) = \begin{cases} \text{TRUE} & \text{if } x, y \in N \text{ and } x = y \\ \text{FALSE} & \text{if } x, y \in N \text{ and } x \neq y \\ \perp_{\text{Bool}} & \text{otherwise,} \end{cases}$$

and let **EQ** be the $\mathbf{W_N}$-program with the equation

$$\mathbf{E(X,Y) = if_{Bool}(is\_0(X), if_{Bool}(is\_0(Y), tt, ff), E(s_1^{-1}(X), s_1^{-1}(Y)))}.$$

Writing $\mathscr{CA}$ for $\mathscr{CA}_{\mathscr{I}_N}(\{\mathbf{E}\})$, let $\alpha_{eq} \in \mathscr{CA}$ be the assignment $\alpha_{eq}(\mathbf{E}) = \text{eq}$, and let $P(x)$ be the predicate on $\mathscr{CA}$ defined by

$$P(\alpha) = \begin{cases} \text{TRUE} & \text{if } \alpha \sqsubseteq_{\mathscr{CA}} \alpha_{eq} \\ \text{FALSE} & \text{otherwise.} \end{cases}$$

If $\mathscr{A}$ is a chain in $(\mathscr{CA}, \sqsubseteq_{\mathscr{CA}})$ such that $P(\alpha)$ holds for all $\alpha \in \mathscr{A}$, then clearly $P(\sqcup \mathscr{A})$ holds, so $P(x)$ is admissible. We want to show by fixed point induction that $P(\mu\Phi_{\mathbf{EQ}})$ holds. Let $\Omega = \Omega_{\mathscr{CA}}$. It is obvious that $P(\Omega)$ holds, so we assume $P(\Phi_{\mathbf{EQ}}^i(\Omega))$ and show that $P(\Phi_{\mathbf{EQ}}^{i+1}(\Omega))$ holds. We have $\Phi_{\mathbf{EQ}}^i(\Omega) \sqsubseteq_{\mathscr{CA}} \alpha_{eq}$ by the induction hypothesis, which implies $\Phi_{\mathbf{EQ}}^{i+1}(\Omega) \sqsubseteq_{\mathscr{CA}} \Phi_{\mathbf{EQ}}(\alpha_{eq})$ by the monotonicity of $\Phi_{\mathbf{EQ}}$, so if we can show that $\Phi_{\mathbf{EQ}}(\alpha_{eq}) \sqsubseteq_{\mathscr{CA}} \alpha_{eq}$, then we will have $\Phi_{\mathbf{EQ}}^{i+1}(\Omega) \sqsubseteq_{\mathscr{CA}} \alpha_{eq}$, i.e., $P(\Phi_{\mathbf{EQ}}^{i+1}(\Omega))$. If $x = \bot_N$ then it is easy to see that

$$\Phi_{\mathbf{EQ}}(\alpha_{eq})(\mathbf{E})(x, y) = \bot_{\text{Bool}} = \alpha_{eq}(\mathbf{E})(x, y),$$

so assume that $x \in N$. If $x = 0$, then

$$\begin{aligned} \Phi_{\mathbf{EQ}}(\alpha_{eq})(\mathbf{E})(x, y) &= \begin{cases} \text{TRUE} & \text{if } y = x \\ \text{FALSE} & \text{if } y \in N \text{ and } y \neq x \\ \bot_{\text{Bool}} & \text{if } y = \bot_N \end{cases} \\ &= \alpha_{eq}(\mathbf{E})(x, y), \end{aligned}$$

and if $x > 0$, then

$$\begin{aligned} \Phi_{\mathbf{EQ}}(\alpha_{eq})(\mathbf{E})(x, y) &= \alpha_{eq}(\mathbf{E})\big(\mathscr{I}(\mathbf{s}_1^{-1})(x), \mathscr{I}(\mathbf{s}_1^{-1})(y)\big) \\ &= \begin{cases} \text{TRUE} & \text{if } x = y \\ \text{FALSE} & \text{if } y \in N - \{0\} \text{ and } x \neq y \\ \bot_{\text{Bool}} & \text{otherwise} \end{cases} \\ &\sqsubseteq_{\text{Bool}} \alpha_{eq}(\mathbf{E})(x, y). \end{aligned}$$

This concludes the proof of $P(\Phi_{\mathbf{EQ}}^{i+1}(\Omega))$, so $P(\mu\Phi_{\mathbf{EQ}})$ holds by fixed point induction. Therefore,

$$\mathscr{D}_{\Delta_N}(\mathbf{EQ})(\mathbf{E}) = \mu\Phi_{\mathbf{EQ}}(\mathbf{E}) \sqsubseteq_{N_\bot \times N_\bot \to \text{Bool}} \text{eq},$$

and **EQ** is partially correct with respect to eq.

Exercises

1. Let $\mathbf{P}_1 = \{\mathbf{F}(\mathbf{X}) = \mathbf{F}(\mathbf{s}(\mathbf{X}))\}$, $\mathbf{P}_2 = \{\mathbf{F}(\mathbf{X}) = \mathbf{s}(\mathbf{F}(\mathbf{X}))\}$ be $\mathbf{W}_\mathbf{N}$-programs. Give consistent $\mathbf{W}_\mathbf{N}$-programs $\mathbf{Q}_1, \mathbf{Q}_2$ such that, in any $\mathbf{W}_\mathbf{N}$-structure Σ, $\mu\Phi^{\Sigma}_{\mathbf{Q}_1 \cup \mathbf{Q}_2}(\mathbf{G}_1) = \mu\Phi^{\Sigma}_{\mathbf{P}_1}(\mathbf{F})$ and $\mu\Phi^{\Sigma}_{\mathbf{Q}_1 \cup \mathbf{Q}_2}(\mathbf{G}_2) = \mu\Phi^{\Sigma}_{\mathbf{P}_2}(\mathbf{F})$.
2. Give a standard vocabulary $\mathbf{W} = \mathbf{W}_c \cup B(\mathbf{W}_c)$ and a $\mathbf{W}$-structure $\Sigma = (\mathscr{T}, \mathscr{I})$ such that, for some $\boldsymbol{\tau} \in TV(\mathbf{W})$, there is an element $d \in D_\tau - \{\perp_\tau\}$ that is not in the range of $\mathscr{I}(\mathbf{f})$ for any $\mathbf{f} \in \mathbf{W}_c$.
3. Give a $\mathbf{W}_\mathbf{N}$-structure Σ such that $\text{rep}(\Sigma) \neq \Sigma$.
4. Show that Lemma 2 is not necessarily true for all $\mathbf{f} \in \mathbf{W}$.
5. (a) Give a standard vocabulary $\mathbf{W}$, a $\mathbf{W}$-structure $\Sigma = (\mathscr{T}, \mathscr{I})$, and a function $f \in D_{\tau_1} \to D_{\tau_2}$, for some $\boldsymbol{\tau}_1, \boldsymbol{\tau}_2 \in TV(\mathbf{W})$, such that $\text{rep}(f) \notin D_{r(\tau_1)} \to D_{r(\tau_2)}$.
   (b) Give an assignment $\alpha \in \mathscr{A}_{\mathscr{T}}(\{\mathbf{F}\})$, where $\mathbf{F}$ is a function variable, such that $\text{rep}(\alpha) \notin \mathscr{A}_{\text{rep}(\mathscr{T})}(\{\mathbf{F}\})$.
6. Let $\Sigma = (\mathscr{T}, \mathscr{I})$ be a complete, continuous $\mathbf{W}$-structure, let V be a set of function variables, and let $\alpha, \beta \in \mathscr{C}\mathscr{A}_{\mathscr{T}}(V)$. Show that if $\alpha \sqsubseteq_{\mathscr{C}\mathscr{A}_{\mathscr{T}}(V)} \beta$, then $\text{rep}(\alpha)(\mathbf{F}) \sqsubseteq_{r(\delta(\mathbf{F})) \to \rho(\mathbf{F})} \text{rep}(\beta)(\mathbf{F})$ for all $\mathbf{F} \in V$.
7. Show that **EQ** is not totally correct with respect to eq.
8. Use fixed point induction to show that **ADD** is partially correct with respect to $+_\perp$.
9. Let $\Sigma = (\mathscr{T}, \mathscr{I})$ be a complete, continuous $\mathbf{W}$-structure such that $\text{rep}(\Sigma) = \Sigma$, let $\mathbf{P}$ be a $\mathbf{W}$-program, and let $\alpha \in \mathscr{C}\mathscr{A}_{\mathscr{T}}(FV(\mathbf{P}))$ be a solution to $\mathbf{P}$ in Σ.
   (a) Show that $\mu\Phi^{\Sigma}_{\mathbf{P}} \sqsubseteq_{\mathscr{C}\mathscr{A}_{\mathscr{T}}(FV(\mathbf{P}))} \alpha$.
   (b) Show that for all $\mathbf{F} \in FV(\mathbf{P})$, $\mathbf{P}$ is partially correct with respect to $\alpha(\mathbf{F})$.
10.* Let $\Sigma = (\mathscr{T}, \mathscr{I})$ be a complete, continuous $\mathbf{W}$-structure, and let $\Omega_1 = \Omega_{\mathscr{C}\mathscr{A}_{\mathscr{T}}(\text{VAR}_1)}$. For any $\boldsymbol{\tau} \in TV(\mathbf{W})$, we will say that an element $d \in D_\tau$ is *constructed* if $d = \overline{\Omega_I}(\mathbf{t})$ for some $\mathbf{t} \in \text{TM}_{\mathbf{W}_c}(\text{VAR}_1)$.
   (a) Let $\mathbf{P}$ be a $\mathbf{W}$-program. Show by induction on i that for all $V \subseteq \text{VAR}_1$, all $\beta \in \mathscr{A}_{\mathscr{T}}(V)$ such that $\beta(\mathbf{X})$ is constructed for all $\mathbf{X} \in V$, and all $\mathbf{t} \in \text{TM}_{\mathbf{W}}(FV(\mathbf{P}) \cup V)$, $\overline{\Phi^i_{\mathbf{P}}(\Omega) \cup \beta}(\mathbf{t})$ is constructed. [*Hint:* For cases $i = 0$ and $i = k + 1$, argue by structural induction on $\mathbf{t}$.]
   (b) Let $\boldsymbol{\tau} \in TV(\mathbf{W})$, and let $d \in D_{r(\tau)}$. Show that $d = \sqcup C$ for some chain C of constructed elements. [*Hint:* Use part (a) with $V = \emptyset$.]
   (c) Let $\boldsymbol{\tau} \in TV(\mathbf{W})$. Show that for every $d \in D_{r(\tau)} - \{\perp_\tau\}$, $d \in \text{ran}\,\mathscr{I}(\mathbf{f})$ for some $\mathbf{f} \in \mathbf{W}_c$. [*Hint:* See Exercise 2.9.] Compare with Exercise 2.

5. Simple Data Structure Systems

So far Δ_N is the only data structure system we have seen, so in this section we give some more examples. In particular, we look at a rather simple form of data structure system.

Definition. Let $\Sigma = (\mathcal{T}, \mathcal{I})$ be a **W**-structure. Σ is a *simple* **W**-structure if

1. $\mathcal{T}(\tau)$ is a flat cpo for all $\tau \in TV(\mathbf{W})$, and
2. $\mathcal{I}(\mathbf{f})$ is strict for every proper constructor function symbol $\mathbf{f} \in \mathbf{W}_c$.

$\Delta = \text{rep}(\Sigma)$ is a *simple data structure system* for **W** if Σ is a simple **W**-structure.

It is easy to see that any simple **W**-structure is complete and continuous. Clearly, Δ_N is simple. For another example, we extend Δ_N with tuples of natural numbers to create a simple $\mathbf{W_{NL}}$-structure, where $\mathbf{W_{NL}}$ is the vocabulary for lists described in Section 1. Let $\text{TUP}(N)$ be the set of all tuples of natural numbers, including the "empty" tuple ( ), and let $(\text{TUP}(N)_\perp, \sqsubseteq_{\text{TUP}(N)_\perp})$ be the flat cpo on $\text{TUP}(N)$. We define $\Sigma_{\mathbf{NL}} = (\mathcal{T}_{\mathbf{NL}}, \mathcal{I}_{\mathbf{NL}})$ as follows:

$$\mathcal{T}_{\mathbf{NL}}(\mathbf{Bool}) = \mathcal{T}_{\mathbf{N}}(\mathbf{Bool})$$

$$\mathcal{T}_{\mathbf{NL}}(\mathbf{N}) = \mathcal{T}_{\mathbf{N}}(\mathbf{N})$$

$$\mathcal{T}_{\mathbf{NL}}(\mathbf{NL}) = \left(\text{TUP}(N)_\perp, \sqsubseteq_{\text{TUP}(N)_\perp}\right)$$

$$\mathcal{I}_{\mathbf{NL}}(\mathbf{0}) = 0$$

$$\mathcal{I}_{\mathbf{NL}}(\mathbf{s}) = s_\perp$$

$$\mathcal{I}_{\mathbf{NL}}(\mathbf{nil}) = (\ )$$

$$\mathcal{I}_{\mathbf{NL}}(\mathbf{cons}) = \text{cons}_\perp \qquad \text{(the strict extension of cons)}$$

where cons: $N \times \text{TUP}(N) \to \text{TUP}(N)$ is defined

$$\text{cons}(m, (m_1, \ldots, m_n)) = (m, m_1, \ldots, m_n).$$

For built-in function symbols $\mathbf{f}$, $\mathcal{I}_{\mathbf{NL}}(\mathbf{f})$ is defined according to conditions 4–6 on $\mathcal{T}$-interpretations. It is easy to check that $\Sigma_{\mathbf{NL}}$ is a simple **W**-structure. Moreover, we have $\overline{\Omega}(\mathbf{nil}) = (\ )$, and for any tuple $(m_1, \ldots, m_n)$ we have

$$\overline{\Omega}(\mathbf{cons}(\mathbf{m_1}, \mathbf{cons}(\mathbf{m_2}, \cdots \mathbf{cons}(\mathbf{m_n}, \mathbf{nil}) \cdots))) = (m_1, \ldots, m_n).$$

Therefore, every element in $\text{TUP}(N)_\perp$ is representable, so we have $\Sigma_{\mathbf{NL}} = \text{rep}(\Sigma_{\mathbf{NL}})$, and $\Sigma_{\mathbf{NL}}$ is a simple data structure system for $\mathbf{W_{NL}}$. We will write Δ_{NL} for Σ_{NL} when we are interested in Σ_{NL} as a data structure system.

Now, let **LIST** be the $\mathbf{W}_{NL}$-program with equations

$$\begin{aligned}
\mathbf{Length(X)} &= \mathbf{if_N(is_nil(X), 0, s(Length(cons_2^{-1}(X))))} \\
\mathbf{Nth(X,Y)} &= \mathbf{if_N(is_0(X), cons_1^{-1}(Y), Nth(s_1^{-1}(X), cons_2^{-1}(Y)))} \\
\mathbf{Cat(X,Y)} &= \mathbf{if_{NL}(is_nil(X), Y, cons(cons_1^{-1}(X), Cat(cons_2^{-1}(X), Y)))} \\
\mathbf{Rev(X)} &= \mathbf{if_{NL}(is_nil(X),} \\
&\qquad \mathbf{X, Cat(Rev(cons_2^{-1}(X)),\ cons(cons_1^{-1}(X), nil)))}.
\end{aligned}$$

Then $\mathscr{D}_{\Delta_{NL}}$(**LIST**)(**Length**) evaluates the length of a list. More precisely, if $\text{len}((m_1, \ldots, m_n)) = n$ for any list $(m_1, \ldots, m_n) \in \text{TUP}(N)$, then $\mathscr{D}_{\Delta_{NL}}$(**LIST**)(**Length**) $= \text{len}_\perp$. It is clear that $\mathscr{D}_{\Delta_{NL}}$(**LIST**)(**Length**) is strict, so we argue by induction on the length of lists:

$$\begin{aligned}
&\mathscr{D}_{\Delta_{NL}}(\mathbf{LIST})(\mathbf{Length})((\,)) \\
&\quad = \mu\Phi_{\mathbf{LIST}}(\mathbf{Length})((\,)) \\
&\quad = \Phi_{\mathbf{LIST}}(\mu\Phi_{\mathbf{LIST}})(\mathbf{Length})((\,)) \\
&\quad = \overline{\mu\Phi_{\mathbf{LIST}} \cup \alpha_{(\,)}}(\mathbf{if_N(is_nil(X), 0, s(Length(cons_2^{-1}(X))))}) \\
&\quad = \mathscr{I}_{\mathbf{NL}}(\mathbf{if_N})(\mathscr{I}_{\mathbf{NL}}(\mathbf{is_nil})((\,)), 0, \overline{\mu\Phi_{\mathbf{LIST}} \cup \alpha_{(\,)}}(\mathbf{s(Length(cons_2^{-1}(X)))})) \\
&\quad = 0,
\end{aligned}$$

and

$$\begin{aligned}
&\mathscr{D}_{\Delta_{NL}}(\mathbf{LIST})(\mathbf{Length})((m_1, \ldots, m_{n+1})) \\
&\quad = \overline{\mu\Phi_{\mathbf{LIST}} \cup \alpha_{(m_1, \ldots, m_{n+1})}}(\mathbf{if_N(is_nil(X), 0, s(Length(cons_2^{-1}(X))))}) \\
&\quad = \mathscr{I}_{\mathbf{NL}}(\mathbf{s})(\mu\Phi_{\mathbf{LIST}}(\mathbf{Length})(\mathscr{I}_{\mathbf{NL}}(\mathbf{cons_2^{-1}})((m_1, \ldots, m_{n+1})))) \\
&\quad = \mathscr{I}_{\mathbf{NL}}(\mathbf{s})(\mu\Phi_{\mathbf{LIST}}(\mathbf{Length})((m_2, \ldots, m_{n+1}))) \\
&\quad = \mathscr{I}_{\mathbf{NL}}(\mathbf{s})(n) \qquad \text{by the induction hypothesis} \\
&\quad = n + 1.
\end{aligned}$$

We leave it to the reader to verify that $\mathscr{D}_{\Delta_{NL}}$(**LIST**)(**Nth**)$(n, l)$ returns the nth element, starting from 0, of list l (if it exists); $\mathscr{D}_{\Delta_{NL}}$(**LIST**)(**Cat**) concatenates two lists; and $\mathscr{D}_{\Delta_{NL}}$(**LIST**)(**Rev**) reverses a list.

In the previous paragraph we referred to lists rather than tuples. Why? The real question is, what is a list? The vocabulary $\mathbf{W}_{\mathbf{NL}}$ was created to let us name lists of numbers, based on our intuitive understanding of the nature of lists. Our point of view is that a list of numbers is just an element of $\mathscr{T}(\mathbf{NL})$, where $(\mathscr{T}, \mathscr{I})$ is *any* $\mathbf{W}_{\mathbf{NL}}$-structure. This is essentially an *axiomatic* approach, where we express the properties we expect from our data objects, without specifying just what those objects are. So $\mathrm{TUP}(N)_{\perp}$ is one set of objects that can serve as lists of numbers, but there are others. An alternative $\mathbf{W}_{\mathbf{NL}}$-structure will be given by the construction preceding Theorem 5.1.

For a third example, we show that $\mathscr{S}$ programs, defined in Chapter 2, can be incorporated into a simple data structure system. We start with type variables **Bool**, **N**, **V**, **L**, **S**, **I**, and **P**, to be assigned truth values, numbers, variables (of $\mathscr{S}$), labels, statements, instructions, and $\mathscr{S}$ programs, respectively. For a constructor vocabulary we take

$$\mathbf{W}_c = \{\mathbf{tt}, \mathbf{ff}, \mathbf{0}, \mathbf{s}, \mathbf{var}, \mathbf{lab}, \mathbf{skip}, \mathbf{incr}, \mathbf{decr}, \mathbf{goto}, \mathbf{unlab\_instr}, \mathbf{lab\_instr}, \mathbf{empty}, \mathbf{cons}\},$$

where

$$\begin{aligned} \tau(\mathbf{0}) &= \mathbf{N} & \tau(\mathbf{s}) &= \mathbf{N} \to \mathbf{N} \\ \tau(\mathbf{var}) &= \mathbf{N} \to \mathbf{V} & \tau(\mathbf{goto}) &= \mathbf{V} \times \mathbf{L} \to \mathbf{S} \\ \tau(\mathbf{lab}) &= \mathbf{N} \to \mathbf{L} & \tau(\mathbf{unlab\_instr}) &= \mathbf{S} \to \mathbf{I} \\ \tau(\mathbf{skip}) &= \mathbf{V} \to \mathbf{S} & \tau(\mathbf{lab\_instr}) &= \mathbf{L} \times \mathbf{S} \to \mathbf{I} \\ \tau(\mathbf{incr}) &= \mathbf{V} \to \mathbf{S} & \tau(\mathbf{empty}) &= \mathbf{P} \\ \tau(\mathbf{decr}) &= \mathbf{V} \to \mathbf{S} & \tau(\mathbf{cons}) &= \mathbf{I} \times \mathbf{P} \to \mathbf{P}, \end{aligned}$$

and we set $\mathbf{W}_{\mathscr{S}} = \mathbf{W}_c \cup B(\mathbf{W}_c)$. Now we set $\mathscr{T}_{\mathscr{S}}(\boldsymbol{\tau}) = \mathscr{T}_{\mathbf{N}}(\boldsymbol{\tau})$ for $\boldsymbol{\tau} = \mathbf{Bool}, \mathbf{N}$, and we set $\mathscr{T}_{\mathscr{S}}(\boldsymbol{\tau})$ to be the flat cpo on $\mathscr{S}$ variables, labels, statements, instructions, and $\mathscr{S}$ programs for $\boldsymbol{\tau} = \mathbf{V}, \mathbf{L}, \mathbf{S}, \mathbf{I}, \mathbf{P}$, respectively. Let $V_0, V_1, \ldots$ and $L_0, L_1, \ldots$ enumerate the $\mathscr{S}$ variables and labels, as in the beginning of Chapter 4 (except that we begin counting at 0 rather than 1). Then for the constructor symbols we define $\mathscr{I}_{\mathscr{S}}$ as follows:

$$\begin{aligned} \mathscr{I}_{\mathscr{S}}(\mathbf{0}) &= 0 & \mathscr{I}_{\mathscr{S}}(\mathbf{s}) &= s_{\perp} \\ \mathscr{I}_{\mathscr{S}}(\mathbf{var}) &= \mathrm{var}_{\perp} & \mathscr{I}_{\mathscr{S}}(\mathbf{goto}) &= \mathrm{goto}_{\perp} \\ \mathscr{I}_{\mathscr{S}}(\mathbf{lab}) &= \mathrm{lab}_{\perp} & \mathscr{I}_{\mathscr{S}}(\mathbf{unlab\_instr}) &= \mathrm{unlab\_instr}_{\perp} \\ \mathscr{I}_{\mathscr{S}}(\mathbf{skip}) &= \mathrm{skip}_{\perp} & \mathscr{I}_{\mathscr{S}}(\mathbf{lab\_instr}) &= \mathrm{lab\_instr}_{\perp} \\ \mathscr{I}_{\mathscr{S}}(\mathbf{incr}) &= \mathrm{incr}_{\perp} & \mathscr{I}_{\mathscr{S}}(\mathbf{empty}) &= \text{empty program} \\ \mathscr{I}_{\mathscr{S}}(\mathbf{decr}) &= \mathrm{decr}_{\perp} & \mathscr{I}_{\mathscr{S}}(\mathbf{cons}) &= \mathrm{cons}_{\perp}, \end{aligned}$$

where

$$\begin{aligned} \text{var}(n) &= V_n & \text{goto}(V, L) &= \text{IF } V \neq 0 \text{ GOTO } L \\ \text{lab}(n) &= L_n & \text{unlab\_instr}(S) &= S \\ \text{skip}(n) &= V \leftarrow V & \text{lab\_instr}(L, S) &= [L]\ S \\ \text{incr}(V) &= V \leftarrow V + 1 & & \\ \text{decr}(V) &= V \leftarrow V - 1 & \text{cons}(I, (I_1, \ldots, I_n)) &= (I, I_1, \ldots, I_n). \end{aligned}$$

For example,

$$\begin{aligned} &\overline{\Omega}(\mathbf{cons}(\mathbf{lab\_instr}(\mathbf{lab}(0), \mathbf{decr}(\mathbf{var}(1))), \\ &\qquad \mathbf{cons}(\mathbf{unlab\_instr}(\mathbf{incr}(\mathbf{var}(0))), \\ &\qquad\qquad \mathbf{cons}(\mathbf{unlab\_instr}(\mathbf{goto}(\mathbf{var}(1), \mathbf{lab}(0))) \\ &\qquad\qquad\quad \mathbf{empty})))) \\ &\qquad [A] \quad X \leftarrow X - 1 \\ &= \qquad\quad Y \leftarrow Y + 1 \\ &\qquad\qquad\quad \text{IF } X \neq 0 \text{ GOTO } A \end{aligned}$$

$\Delta_{\mathscr{S}} = (\text{rep}(\mathscr{T}_{\mathscr{S}}), \text{rep}(\mathscr{I}_{\mathscr{S}})) = (\mathscr{T}_{\mathscr{S}}, \mathscr{I}_{\mathscr{S}})$ is the simple data structure system of $\mathscr{S}$ programs.

One of the standard problems in programming language theory is to show the existence of structures that satisfy a given set of conditions. We can ask, for example, is there a strict data structure system for every standard vocabulary? We will show that there is. We define $\Sigma_{\mathscr{H}(\mathbf{W})} = (\mathscr{T}_{\mathscr{H}(\mathbf{W})}, \mathscr{I}_{\mathscr{H}(\mathbf{W})})$, the *simple Herbrand* **W**-*structure*, as follows.[5] For each $\boldsymbol{\tau} \in TV(\mathbf{W})$, let $\mathscr{T}_{\mathscr{H}(\mathbf{W})}(\boldsymbol{\tau})$ be the flat cpo on $\text{TM}^{\boldsymbol{\tau}}_{\mathbf{W}_c}$. We will write $\bot_{\boldsymbol{\tau}}$ for the bottom element of $\mathscr{T}_{\mathscr{H}(\mathbf{W})}(\boldsymbol{\tau})$. Next we define $\mathscr{I}_{\mathscr{H}(\mathbf{W})}$:

$$\mathscr{I}_{\mathscr{H}(\mathbf{W})}(\mathbf{c}) = \mathbf{c} \quad \text{for each constant symbol } \mathbf{c} \in \mathbf{W}_c$$

$$\mathscr{I}_{\mathscr{H}(\mathbf{W})}(\mathbf{f})(\mathbf{t}_1, \ldots, \mathbf{t}_n) = \begin{cases} \mathbf{f}(\mathbf{t}_1, \ldots, \mathbf{t}_n) & \text{if } \mathbf{t}_i \neq \bot_{\boldsymbol{\tau}_i}, 1 \leq i \leq n \\ \bot_{\boldsymbol{\tau}} & \text{otherwise,} \end{cases}$$

for each $\mathbf{f} \in \mathbf{W}_c$ with $\tau(\mathbf{f}) = \boldsymbol{\tau}_1 \times \cdots \times \boldsymbol{\tau}_n \to \boldsymbol{\tau}$. For each $\mathbf{f} \in B(\mathbf{W}_c)$, $\mathscr{I}_{\mathscr{H}(\mathbf{W})}(\mathbf{f})$ is defined according to conditions 4–6 on $\mathscr{T}$-interpretations. It is clear that $\Sigma_{\mathscr{H}(\mathbf{W})}$ is a simple **W**-structure, and an easy structural induction shows that

[5] The idea of creating structures based on terms comes from the field of mathematical logic. For example, Herbrand *universes* are defined and play a significant role in Chapter 13.

$\overline{\Omega}(\mathbf{t}) = \mathbf{t}$ for all $\mathbf{t} \in \mathrm{TM}_{\mathbf{W}_c}$, so every element in $\Sigma_{\mathscr{H}(\mathbf{W})}$ is representable. Therefore, $\Delta_{\mathscr{H}(\mathbf{W})} = \mathrm{rep}(\Sigma_{\mathscr{H}(\mathbf{W})}) = \Sigma_{\mathscr{H}(\mathbf{W})}$ is a simple data structure system for **W**, and we have proved

Theorem 5.1. There is a simple data structure system for every standard vocabulary.

Exercises

1. Show that any simple **W**-structure is complete and continuous.
2. Let $\Sigma = (\mathscr{T}, \mathscr{I})$ be a simple **W**-structure, and let $\mathbf{f} \in \mathbf{W}_c$ with $\tau(\mathbf{f}) = \boldsymbol{\tau} \rightarrow \boldsymbol{\tau}$. What is $\mu\mathscr{I}(\mathbf{f})$?
3. Show that Σ_{NL} is a simple $\mathbf{W}_{\mathrm{NL}}$-structure.
4. Show by induction on n that

$$\mathscr{D}_{\Delta_{\mathrm{NL}}}(\mathbf{LIST})(\mathbf{Nth})(i, (m_0, \ldots, m_n)) = \begin{cases} m_i & \text{if } 0 \le i \le n \\ \bot_N & \text{if } i > n. \end{cases}$$

5. Show by induction on n that

$$\mathscr{D}_{\Delta_{\mathrm{NL}}}(\mathbf{LIST})(\mathbf{Cat})((l_1, \ldots, l_n), (m_1, \ldots, m_r)) = (l_1 \ldots l_n, m_1, \ldots, m_r).$$

6. Using Exercise 5, show by induction on n that

$$\mathscr{D}_{\Delta_{\mathrm{NL}}}(\mathbf{LIST})(\mathbf{Rev})((m_1, \ldots, m_n)) = (m_n, \ldots, m_1).$$

7. Give a $\mathbf{W}_{\mathrm{NL}}$-program **P** such that

$$\mathscr{D}_{\Delta_{\mathrm{NL}}}(\mathbf{P})(\mathbf{F})((m_1, \ldots, m_n)) = \sum_{i=1}^{n} m_i .$$

8. Give the $\mathbf{W}_{\mathscr{S}}$-term **t** such that $\overline{\Omega}(\mathbf{t})$ is

$$\begin{array}{ll} [C_1] & Z_2 \leftarrow Z_3 - 1 \\ & \text{IF } Z_3 \neq 0 \text{ GOTO } C_1 \end{array}$$

9. Describe $\Sigma_{\mathscr{H}(\mathbf{W}_{\mathrm{NL}})}$.
10. Complete the proof of Theorem 5.1 by showing that $\mathrm{rep}(\Sigma_{\mathscr{H}(\mathbf{W})}) = \Sigma_{\mathscr{H}(\mathbf{W})}$ for any standard vocabulary **W**.
11.* Let $\Sigma = (\mathscr{T}, \mathscr{I})$ be a **W**-structure. For any $\boldsymbol{\tau} \in TV(\mathbf{W})$, we will say that an element $d \in D_{\boldsymbol{\tau}}$ is *ground* if $d = \bot_{\boldsymbol{\tau}}$ or $d = \overline{\Omega}(\mathbf{t})$ for some

$\mathbf{t} \in TM_{\mathbf{W}_c}$, and we will say that Σ is *term-generated* if for all $\tau \in TV(\mathbf{W})$, every element of D_τ is ground.

(a) Show that there is a term-generated **W**-structure for every standard vocabulary **W**.

(b) Give a standard vocabulary **W** and a simple **W**-structure that is not term-generated.

(c) Let $\Sigma = (\mathscr{T}, \mathscr{I})$ be a simple **W**-structure, and let **P** be a **W**-program. Show by induction on i that for all $V \subseteq \text{VAR}_1$, all $\beta \in \mathscr{A}_{\mathscr{T}}(V)$ such that $\beta(\mathbf{X})$ is ground for all $\mathbf{X} \in V$, and all $\mathbf{t} \in TM_{\mathbf{W}}(FV(\mathbf{P}) \cup V)$, $\overline{\Phi_{\mathbf{P}}^{i}(\Omega) \cup \beta}(\mathbf{t})$ is ground. [*Hint:* For cases $i = 0$ and $i = k + 1$, argue by structural induction on **t**.]

(d) Let $\Sigma = (\mathscr{T}, \mathscr{I})$ be a simple **W**-structure. Show that rep(Σ) is term-generated. [*Hint:* Use part (c) with $V = \varnothing$.]

12.* Let $\Sigma = (\mathscr{T}, \mathscr{I})$, $\Sigma' = (\mathscr{T}', \mathscr{I}')$ be **W**-structures. We say that Σ, Σ' are *isomorphic* if there is a set of functions $\{f_\tau \mid \tau \in TV(\mathbf{W})\}$ such that

- for all $\tau \in TV(\mathbf{W})$, f_τ is an isomorphism from $\mathscr{T}(\tau)$ to $\mathscr{T}'(\tau)$,
- for all constant symbols $\mathbf{c} \in \mathbf{W}_c$ with $\tau(\mathbf{c}) = \tau$, $f_\tau(\mathscr{I}(\mathbf{c})) = \mathscr{T}'(\mathbf{c})$, and
- for all proper function symbols $\mathbf{f} \in \mathbf{W}_c$ with $\tau(\mathbf{f}) = \tau_1 \times \cdots \times \tau_n \to \tau$,

$$f_\tau(\mathscr{I}(\mathbf{f})(d_1, \ldots, d_n)) = \mathscr{I}'(\mathbf{f})\big(f_{\tau_1}(d_1), \ldots, f_{\tau_n}(d_n)\big)$$

for all $(d_1, \ldots, d_n) \in D_{\delta(\mathbf{f})}$.

[See Exercise 3.18 in Chapter 16 for the definition of *isomorphic partial orders*.]

(a) Show that $\Sigma_{\mathbf{N}}$, $\Sigma_{\mathscr{H}(\mathbf{W_N})}$ are isomorphic.

(b) Show that $\Sigma_{\mathbf{NL}}$, $\Sigma_{\mathscr{H}(\mathbf{W_{NL}})}$ are isomorphic.

(c) Let Σ be a simple **W**-structure. Use Exercise 11 to show that rep(Σ), $\Sigma_{\mathscr{H}(\mathbf{W})}$ are isomorphic.

6. Infinitary Data Structure Systems

Simple data structure systems are too elementary to demonstrate the power of the framework we developed in Chapter 16, so in this section we look at more complex systems.

Definition. Let $\Sigma = (\mathscr{T}, \mathscr{I})$ be a complete, continuous **W**-structure. Σ is an *infinitary* **W**-structure if for every proper constructor function symbol

$\mathbf{f} \in \mathbf{W}_c$, $\perp_{\rho(\mathbf{f})} \notin \operatorname{ran} \mathscr{I}(\mathbf{f})$. $\Delta = \operatorname{rep}(\Sigma)$ is an *infinitary data structure system* for $\mathbf{W}$ if Σ is an infinitary $\mathbf{W}$-structure.

Suppose we try to define, as an interpretation $\mathscr{I}(\mathbf{s})$ for $\mathbf{s} \in \mathbf{W}_{\mathbf{N}}$, a continuous successor function s^∞ for $N_\perp$ such that $\perp_N \notin \operatorname{ran} s^\infty$. If s^∞ and s, the ordinary successor function, agree on N then we still get the positive natural numbers $s^\infty(0), s^\infty(s^\infty(0)), \ldots$ by repeated application of s^∞ to 0, but we get other objects as well. Since $\perp_N \notin \operatorname{ran} s^\infty$, we get $\perp_N \sqsubset s^\infty(\perp_N)$, which implies $s^\infty(\perp_N) \sqsubset s^\infty(s^\infty(\perp_N))$ by the monotonicity of s^∞ and by condition 2b on $\mathscr{T}$-interpretations. The idea is that we know that $s^\infty(\perp_N)$ is the successor of *something*, but that is all we know. Now, $s^\infty(s^\infty(\perp_N))$ is also the successor of something, but we also know that it is the successor of something that is the successor of something, so $s^\infty(s^\infty(\perp_N))$ is more defined than $s^\infty(\perp_N)$. Moreover, an object like $s^\infty(\perp_N)$ must be different than every natural number: $s^\infty(\perp_N) = 0$ would violate condition 3 on $\mathscr{T}$-interpretations, and $s^\infty(\perp_N) = n + 1 = s^\infty(n)$ would violate condition 2b. What we get, therefore, is an infinite chain of distinct new objects:

$$\perp_N \sqsubset s^\infty(\perp_N) \sqsubset s^\infty(s^\infty(\perp_N)) \sqsubset s^\infty(s^\infty(s^\infty(\perp_N))) \sqsubset \cdots.$$

But then we need yet another new object $\sqcup\{(s^\infty)^i(\perp_N) \mid i \in N\}$ if we are to have a cpo.

It is not at all obvious, then, that infinitary $\mathbf{W}$-structures exist. In fact, we will show that they do. We begin by defining a variation on the Herbrand structures of Section 5. For each $\boldsymbol{\tau} \in TV(\mathbf{W})$, we create a new constant symbol[6] $\boldsymbol{\perp}_{\boldsymbol{\tau}}$ with $\tau(\boldsymbol{\perp}_{\boldsymbol{\tau}}) = \boldsymbol{\tau}$, and we set

$$\mathbf{W}^+ = \mathbf{W}_c \cup \{\boldsymbol{\perp}_{\boldsymbol{\tau}} \mid \boldsymbol{\tau} \in TV(\mathbf{W})\}.$$

For each $\boldsymbol{\tau} \in TV(\mathbf{W})$ we define the ordering $\sqsubseteq_{\boldsymbol{\tau}^+}$ on $\mathrm{TM}_{\mathbf{W}^+}^{\boldsymbol{\tau}}$ as follows:

$$\mathbf{t} \sqsubseteq_{\boldsymbol{\tau}^+} \mathbf{u} \quad \text{if and only if} \quad \begin{cases} \mathbf{t} = \boldsymbol{\perp}_{\boldsymbol{\tau}} \text{ or } \mathbf{t} = \mathbf{u} \text{ or} \\ [\mathbf{t} = \mathbf{f}(\mathbf{t}_1, \ldots, \mathbf{t}_n) \text{ and } \mathbf{u} = \mathbf{f}(\mathbf{u}_1, \ldots, \mathbf{u}_n), \text{ for} \\ \text{some } \mathbf{f} \in \mathbf{W}_c \text{ with } \tau(\mathbf{f}) = \boldsymbol{\tau}_1 \times \cdots \times \boldsymbol{\tau}_n \to \boldsymbol{\tau}, \text{ and} \\ \mathbf{t}_i \sqsubseteq_{\boldsymbol{\tau}_i^+} \mathbf{u}_i, 1 \le i \le n]. \end{cases}$$

For $\mathbf{W}_{\mathbf{N}}$ we have $\mathbf{W}_{\mathbf{N}}^+ = \{\mathbf{tt}, \mathbf{ff}, \mathbf{0}, \mathbf{s}, \boldsymbol{\perp}_{\mathbf{Bool}}, \boldsymbol{\perp}_{\mathbf{N}}\}$, $\mathrm{TM}_{\mathbf{W}_{\mathbf{N}}^+}^{\mathbf{Bool}} = \{\boldsymbol{\perp}_{\mathbf{Bool}}, \mathbf{tt}, \mathbf{ff}\}$, and

[6] Note that the symbol $\boldsymbol{\perp}_{\boldsymbol{\tau}}$ is introduced into the *semantics* of $\mathbf{W}$-programs. The vocabulary $\mathbf{W}$ remains the same. That is, $\boldsymbol{\perp}_{\boldsymbol{\tau}}$ cannot appear in a $\mathbf{W}$-program.

$TM^{N}_{W_N^+} = \{s^i(0) \mid i \in N\} \cup \{s^i(\perp_N) \mid i \in N\}$. The definition of $\sqsubseteq_{N^+}$ implies, for example,

$$\perp_N \sqsubseteq_{N^+} s^i(0) \quad \text{for all } i \in N,$$

$$\perp_N \sqsubseteq_{N^+} s(\perp_N) \sqsubseteq_{N^+} s(s(\perp_N)) \sqsubseteq_{N^+} s(s(0)), \text{ and}$$

$$\perp_N \sqsubseteq_{N^+} s(\perp_N) \sqsubseteq_{N^+} s(s(\perp_N)) \sqsubseteq_{N^+} s(s(s(\perp_N))) \sqsubseteq_{N^+} \cdots.$$

Now, for each $\tau \in TV(W)$, $(TM^{\tau}_{W^+}, \sqsubseteq_{\tau^+})$ is a partial order with bottom element $\perp_\tau$, but we still have the problem that $(TM^{\tau}_{W^+}, \sqsubseteq_{\tau^+})$ is not, in general, a cpo. For example, $\{s^i(\perp_N) \mid i \in N\}$ is a chain in $(TM^{N}_{W_N^+}, \sqsubseteq_{N^+})$ without a least upper bound. Here is where we apply the ideal construction.

Definition. The *Herbrand ideal type assignment* for **W**, denoted $\mathcal{T}_{\mathcal{H}^\infty(W)}$, is defined, for all $\tau \in TV(W)$,

$$\mathcal{T}_{\mathcal{H}^\infty(W)}(\tau) = \left(\text{id}(TM^{\tau}_{W^+}), \subseteq_{\text{id}(TM^{\tau}_{W^+})}\right).$$

(As usual, we are writing $\text{id}(TM^{\tau}_{W^+})$ for $\text{id}(TM^{\tau}_{W^+}, \sqsubseteq_{\tau^+})$.) The *Herbrand ideal $\mathcal{T}_{\mathcal{H}^\infty(W)}$-interpretation*, denoted $\mathcal{I}_{\mathcal{H}^\infty(W)}$, is defined

- for all constant symbols $\mathbf{c} \in W_c$, with $\tau(\mathbf{c}) = \tau$,

$$\mathcal{I}_{\mathcal{H}^\infty(W)}(\mathbf{c}) = \{\perp_\tau, \mathbf{c}\};$$

- for all proper function symbols $\mathbf{f} \in W_c$, with $\tau(\mathbf{f}) = \tau_1 \times \cdots \times \tau_n \to \tau$,

$$\mathcal{I}_{\mathcal{H}^\infty(W)}(\mathbf{f})(I_1, \ldots, I_n) = \{\mathbf{f}(\mathbf{t}_1, \ldots, \mathbf{t}_n) \mid \mathbf{t}_i \in I_i, 1 \le i \le n\} \cup \{\perp_\tau\}$$

  for all $(I_1, \ldots, I_n) \in \text{id}(TM^{\tau_1}_{W^+}) \times \cdots \times \text{id}(TM^{\tau_n}_{W^+})$;
- for all $\mathbf{f} \in B(W_c)$, $\mathcal{I}_{\mathcal{H}^\infty(W)}(\mathbf{f})$ is defined according to conditions 4–6 on $\mathcal{T}$-interpretations.

The *Herbrand ideal* **W**-*structure*, denoted $\Sigma_{\mathcal{H}^\infty(W)}$, is $(\mathcal{T}_{\mathcal{H}^\infty(W)}, \mathcal{I}_{\mathcal{H}^\infty(W)})$.

When **W** is understood we will write $(\mathcal{T}^\infty, \mathcal{I}^\infty)$ for $(\mathcal{T}_{\mathcal{H}^\infty(W)}, \mathcal{I}_{\mathcal{H}^\infty(W)})$, $(D_{\tau^\infty}, \sqsubseteq_{\tau^\infty})$ for $\mathcal{T}_{\mathcal{H}^\infty(W)}(\tau)$, and $\perp_{\tau^\infty}$ for the bottom element of $(D_{\tau^\infty}, \sqsubseteq_{\tau^\infty})$. As usual, we will write tt for $\mathcal{I}^\infty(\mathbf{tt})$ and ff for $\mathcal{I}^\infty(\mathbf{ff})$.

For a first example we consider $\Sigma_{\mathcal{H}^\infty(W_N)} = (\mathcal{T}_{\mathcal{H}^\infty(W_N)}, \mathcal{I}_{\mathcal{H}^\infty(W_N)})$, which we will write as Σ^∞_N. There are three kinds of elements in D_{N^∞}, including two kinds of principal ideals. For each numeral **n** there is $\text{pid}(\mathbf{n}) = \{s^i(\perp_N) \mid i \le n\} \cup \{\mathbf{n}\}$, and for each term of the form $s^n(\perp_N)$ there is $\text{pid}(s^n(\perp_N)) = \{s^i(\perp_N) \mid i \le n\}$. For all numerals **n** we will write n for $\text{pid}(\mathbf{n})$, e.g., $3 = \text{pid}(\mathbf{3})$, where $\text{pid}(\mathbf{n})$ is distinguished from the natural number n by context. Somewhat ambiguously, perhaps, we will call these objects *numbers*. [Indeed, by our discussion about lists in Section 5, we could make a case that pid(**3**) *is* the natural number 3.] Also, we will write $n_\perp$ for $\mathrm{pid}(\mathbf{s}^n(\perp_\mathbf{N}))$. These two kinds of objects look very similar, but the significant difference is that no object n can occur in an infinite chain. This is because **n** is the greatest element of n, so n cannot be a proper subset of any larger ideal. Therefore, there is no object d distinct from n such that $n \sqsubseteq_{\mathbf{N}^\infty} d$. In other words, n is not an approximation of any other element, so we can say that n is *completely defined.* On the other hand, $\{n_\perp \mid n \in N\}$ is an infinite chain, and

$$\begin{aligned} \sqcup\{n_\perp \mid n \in N\} &= \cup\{n_\perp \mid n \in N\} \\ &= \cup\{\mathrm{pid}(\mathbf{s}^n(\perp_\mathbf{N})) \mid n \in N\} \\ &= \{\mathbf{s}^n(\perp_\mathbf{N}) \mid n \in N\}. \end{aligned}$$

In fact, $\{\mathbf{s}^n(\perp_\mathbf{N}) \mid n \in N\}$ is the unique infinite ideal of $\mathcal{T}^\infty(\mathbf{N})$, and we will write it as ω. Clearly, ω is not a principal ideal.

We will now show that, for any standard vocabulary **W**, $\Sigma_{\mathcal{I}^\infty(\mathbf{W})}$ is an infinitary **W**-structure.

Lemma 1. Let $I \in D_{\tau^\infty}$ for some $\tau \in TV(\mathbf{W})$.

1. If $\mathbf{c} \in I$, where $\mathbf{c} \in \mathbf{W}_c$ is some constant symbol, then $I = \mathcal{I}^\infty(\mathbf{c})$.
2. If $\mathbf{f}(\mathbf{t}_1, \ldots, \mathbf{t}_n) \in I$, where $\mathbf{f} \in \mathbf{W}_c$ is some proper function symbol, then $I = \mathcal{I}^\infty(\mathbf{f})(I_1, \ldots, I_n)$, where, for $1 \le i \le n$,

$$I_i = \{\mathbf{u} \in TM_{\mathbf{W}^+}^{\tau_i} \mid \mathbf{f}(\mathbf{t}_1, \ldots, \mathbf{t}_{i-1}, \mathbf{u}, \mathbf{t}_{i+1}, \ldots, \mathbf{t}_n) \in I\}.$$

Proof. Let $\mathbf{c} \in I$ be some constant symbol. I is directed, so no term $\mathbf{u} \in TM_{\mathbf{W}^+}^{\tau}$ of the form **g** or $\mathbf{g}(\mathbf{u}_1, \ldots, \mathbf{u}_m)$, where $\mathbf{g} \in \mathbf{W}_c$ is distinct from **c**, can be in I since, by definition of $\sqsubseteq_{\tau^+}$, there is no term $\mathbf{v} \in TM_{\mathbf{W}^+}^{\tau}$ such that $\mathbf{c}, \mathbf{u} \sqsubseteq_{\tau^+} \mathbf{v}$. Also, $\perp_\tau \in I$ since I is downward closed, so $I = \{\perp_\tau, \mathbf{c}\} = \mathcal{I}^\infty(\mathbf{c})$.

Now let $\mathbf{f}(\mathbf{t}_1, \ldots, \mathbf{t}_n) \in I$, for some $\mathbf{f} \in \mathbf{W}_c$ with $\tau(\mathbf{f}) = \tau_1 \times \cdots \times \tau_n \rightarrow \tau$, and for $1 \le i \le n$ let $I_1, \ldots, I_n$ be as defined in the statement of the lemma. First we show that $I_1, \ldots, I_n$ are ideals. For $1 \le i \le n$, $\mathbf{t}_i \in I_i$, so I_i is nonempty. If $\mathbf{u} \in I_i$, $\mathbf{v} \in TM_{\mathbf{W}^+}^{\tau_i}$, and $\mathbf{v} \sqsubseteq_{\tau_i^+} \mathbf{u}$, then

$$\mathbf{f}(\mathbf{t}_1, \ldots, \mathbf{t}_{i-1}, \mathbf{u}, \mathbf{t}_{i+1}, \ldots, \mathbf{t}_n) \in I$$

and

$$\mathbf{f}(\mathbf{t}_1, \ldots, \mathbf{t}_{i-1}, \mathbf{v}, \mathbf{t}_{i+1}, \ldots, \mathbf{t}_n) \sqsubseteq_{\tau^+} \mathbf{f}(\mathbf{t}_1, \ldots, \mathbf{t}_{i-1}, \mathbf{u}, \mathbf{t}_{i+1}, \ldots, \mathbf{t}_n),$$

which implies $\mathbf{f}(\mathbf{t}_1,\ldots,\mathbf{t}_{i-1},\mathbf{v},\mathbf{t}_{i+1},\ldots,\mathbf{t}_n) \in I$ since I is downward closed. Therefore, $\mathbf{v} \in I_i$ and I_i is downward closed. If $\mathbf{u},\mathbf{v} \in I_i$, there is a term $\mathbf{f}(\mathbf{t}_1,\ldots,\mathbf{t}_{i-1},\mathbf{w},\mathbf{t}_{i+1},\ldots,\mathbf{t}_n) \in I$ such that

$$\mathbf{f}(\mathbf{t}_1,\ldots,\mathbf{t}_{i-1},\mathbf{u},\mathbf{t}_{i+1},\ldots,\mathbf{t}_n) \sqsubseteq_{\tau^+} \mathbf{f}(\mathbf{t}_1,\ldots,\mathbf{t}_{i-1},\mathbf{w},\mathbf{t}_{i+1},\ldots,\mathbf{t}_n),$$

$$\mathbf{f}(\mathbf{t}_1,\ldots,\mathbf{t}_{i-1},\mathbf{v},\mathbf{t}_{i+1},\ldots,\mathbf{t}_n) \sqsubseteq_{\tau^+} \mathbf{f}(\mathbf{t}_1,\ldots,\mathbf{t}_{i-1},\mathbf{w},\mathbf{t}_{i+1},\ldots,\mathbf{t}_n),$$

since I is directed. Then $\mathbf{w} \in I_i$ and $\mathbf{u},\mathbf{v} \sqsubseteq_{\tau_i^+} \mathbf{w}$, so I_i is directed. Therefore, I_i is an ideal, $1 \le i \le n$, i.e., $I_i \in D_{\tau_i^\infty}$.

We claim that $I = \mathscr{I}^\infty(\mathbf{f})(I_1,\ldots,I_n)$. Let $\mathbf{u} \in I$. If $\mathbf{u} = \perp_\tau$, then $\mathbf{u} \in \mathscr{I}^\infty(\mathbf{f})(I_1,\ldots,I_n)$ by definition of $\mathscr{I}^\infty(\mathbf{f})$. Otherwise, $\mathbf{u}$ must be of the form $\mathbf{f}(\mathbf{u}_1,\ldots,\mathbf{u}_n)$ since I is directed. Then there is a term $\mathbf{f}(\mathbf{v}_1,\ldots,\mathbf{v}_n) \in I$ such that

$$\mathbf{f}(\mathbf{t}_1,\ldots,\mathbf{t}_n), \mathbf{f}(\mathbf{u}_1,\ldots,\mathbf{u}_n) \sqsubseteq_{\tau^+} \mathbf{f}(\mathbf{v}_1,\ldots,\mathbf{v}_n),$$

which implies, for $1 \le i \le n$,

$$\mathbf{f}(\mathbf{t}_1,\ldots,\mathbf{t}_{i-1},\mathbf{u}_i,\mathbf{t}_{i+1},\ldots,\mathbf{t}_n) \sqsubseteq_{\tau^+} \mathbf{f}(\mathbf{v}_1,\ldots,\mathbf{v}_n),$$

so that

$$\mathbf{f}(\mathbf{t}_1,\ldots,\mathbf{t}_{i-1},\mathbf{u}_i,\mathbf{t}_{i+1},\ldots,\mathbf{t}_n) \in I.$$

Therefore, we have $\mathbf{u}_i \in I_i$, $1 \le i \le n$, which implies $\mathbf{f}(\mathbf{u}_1,\ldots,\mathbf{u}_n) \in \mathscr{I}^\infty(\mathbf{f})(I_1,\ldots,I_n)$, and so $I \subseteq \mathscr{I}^\infty(\mathbf{f})(I_1,\ldots,I_n)$. Now, let $\mathbf{u} \in \mathscr{I}^\infty(\mathbf{f})(I_1,\ldots,I_n)$. If $\mathbf{u} = \perp_\tau$ then $\mathbf{u} \in I$. Otherwise, $\mathbf{u}$ is of the form $\mathbf{f}(\mathbf{u}_1,\ldots,\mathbf{u}_n)$, where $\mathbf{u}_i \in I_i$, $1 \le i \le n$. Therefore, by definition of I_i,

$$\mathbf{f}(\mathbf{t}_1,\ldots,\mathbf{t}_{i-1},\mathbf{u}_i,\mathbf{t}_{i+1},\ldots,\mathbf{t}_n) \in I, \quad 1 \le i \le n,$$

and since I is directed, a simple induction on n shows that there is a term $\mathbf{f}(\mathbf{v}_1,\ldots,\mathbf{v}_n) \in I$ such that

$$\mathbf{f}(\mathbf{t}_1,\ldots,\mathbf{t}_{i-1},\mathbf{u}_i,\mathbf{t}_{i+1},\ldots,\mathbf{t}_n) \sqsubseteq_{\tau^+} \mathbf{f}(\mathbf{v}_1,\ldots,\mathbf{v}_n), \quad 1 \le i \le n.$$

Then $\mathbf{u}_i \sqsubseteq_{\tau_i^+} \mathbf{v}_i$, $1 \le i \le n$, which implies $\mathbf{f}(\mathbf{u}_1,\ldots,\mathbf{u}_n) \sqsubseteq_{\tau^+} \mathbf{f}(\mathbf{v}_1,\ldots,\mathbf{v}_n)$, so $\mathbf{f}(\mathbf{u}_1,\ldots,\mathbf{u}_n) \in I$, since I is downward closed. Therefore, $\mathscr{I}^\infty(\mathbf{f})(I_1,\ldots,I_n) \subseteq I$, and so $I = \mathscr{I}^\infty(\mathbf{f})(I_1,\ldots,I_n)$. ■

Theorem 6.1. $\Sigma_{\mathscr{H}^\infty(\mathbf{W})}$ is an infinitary **W**-structure.

Proof. It is clear that $\mathscr{T}^{+\infty}$ is a type assignment for **W**, and it is complete by Theorem 16.3.12. Therefore, we begin by showing that $\mathscr{I}^\infty$ is a $\mathscr{T}^\infty$-interpretation for **W**. Note that $\perp_\tau$ occurs in every ideal of $(TM^\tau_{\mathbf{W}^+}, \sqsubseteq_{\tau^+})$, and $\{\perp_\tau\}$ is an ideal, so $\{\perp_\tau\}$ is the bottom element $\perp_{\tau^\infty}$

of $(D_{\tau^\infty}, \sqsubseteq_{\tau^\infty})$. Now, for any constant symbol $\mathbf{c} \in \mathbf{W}_c$ with $\tau(\mathbf{c}) = \boldsymbol{\tau}$, $\mathcal{I}^\infty(\mathbf{c}) = \{\boldsymbol{\perp}_{\boldsymbol{\tau}}, \mathbf{c}\} \neq \perp_{\tau^\infty}$ is an ideal of $(\mathrm{TM}^{\boldsymbol{\tau}}_{\mathbf{W}^+}, \sqsubseteq_{\tau^+})$, so condition 1 is satisfied. Let $\mathbf{f} \in \mathbf{W}_c$ be a proper function symbol with $\tau(\mathbf{f}) = \boldsymbol{\tau}_1 \times \cdots \times \boldsymbol{\tau}_n \rightarrow \boldsymbol{\tau}$, let $(I_1, \ldots, I_n) \in D_{\tau_1^\infty} \times \cdots \times D_{\tau_n^\infty}$, and let $I = \mathcal{I}^\infty(\mathbf{f})(I_1, \ldots, I_n)$. I is nonempty since $\boldsymbol{\perp}_{\boldsymbol{\tau}} \in I$. Suppose that $\mathbf{f}(\mathbf{t}_1, \ldots, \mathbf{t}_n) \in I$, $\mathbf{u} \in \mathrm{TM}^{\boldsymbol{\tau}}_{\mathbf{W}^+}$, and $\mathbf{u} \sqsubseteq_{\tau^+} \mathbf{f}(\mathbf{t}_1, \ldots, \mathbf{t}_n)$. If $\mathbf{u} = \boldsymbol{\perp}_{\boldsymbol{\tau}}$ or $\mathbf{u} = \mathbf{f}(\mathbf{t}_1, \ldots, \mathbf{t}_n)$ then $\mathbf{u} \in I$. Otherwise, by the definition of $\sqsubseteq_{\tau^+}$, $\mathbf{u}$ must be a term of the form $\mathbf{f}(\mathbf{u}_1, \ldots, \mathbf{u}_n)$, where $\mathbf{u}_i \sqsubseteq_{\tau_i^+} \mathbf{t}_i$, $1 \leq i \leq n$. Then $\mathbf{u}_i \in I_i$, $1 \leq i \leq n$, since I_i is downward closed, so $\mathbf{f}(\mathbf{u}_1, \ldots, \mathbf{u}_n) \in I$ and I is downward closed. Suppose $\mathbf{t}, \mathbf{u} \in I$. If $\mathbf{t} \sqsubseteq_{\tau^+} \mathbf{u}$ or $\mathbf{u} \sqsubseteq_{\tau^+} \mathbf{t}$ then either $\mathbf{t}, \mathbf{u} \sqsubseteq_{\tau^+} \mathbf{t} \in I$ or $\mathbf{t}, \mathbf{u} \sqsubseteq_{\tau^+} \mathbf{u} \in I$, so suppose otherwise. Then $\mathbf{t}, \mathbf{u}$ must be terms of the form $\mathbf{f}(\mathbf{t}_1, \ldots, \mathbf{t}_n), \mathbf{f}(\mathbf{u}_1, \ldots, \mathbf{u}_n)$, respectively, and $\mathbf{t}_i, \mathbf{u}_i \in I_i$ implies there is a $\mathbf{w}_i \in I_i$ such that $\mathbf{t}_i, \mathbf{u}_i \sqsubseteq_{\tau_i^+} \mathbf{w}_i$, $1 \leq i \leq n$, so $\mathbf{f}(\mathbf{t}_1, \ldots, \mathbf{t}_n), \mathbf{f}(\mathbf{u}_1, \ldots, \mathbf{u}_n) \sqsubseteq_{\tau^+} \mathbf{f}(\mathbf{w}_1, \ldots, \mathbf{w}_n) \in I$. Therefore, I is directed and it is an ideal, so $I \in D_{\tau^\infty}$ and condition 2a is satisfied. If $(I_1, \ldots, I_n), (J_1, \ldots, J_n) \in D_{\tau_1^\infty} \times \cdots \times D_{\tau_n^\infty}$ are distinct, then clearly $\mathcal{I}^\infty(\mathbf{f})(I_1, \ldots, I_n) \neq \mathcal{I}^\infty(\mathbf{f})(J_1, \ldots, J_n)$, so $\mathcal{I}^\infty(\mathbf{f})$ is one–one and condition 2b is satisfied. Conditions 2c, 3, 4, 5, and 6 follow immediately from the definition of $\mathcal{I}^\infty(\mathbf{f})$, so $(\mathcal{T}^\infty, \mathcal{I}^\infty)$ is a complete **W**-structure.

It is immediate from the definition that $\perp_{\rho(\mathbf{f})} \notin \operatorname{ran} \mathcal{I}^\infty(\mathbf{f})$ for every proper constructor function symbol **f**, so it remains for us only to show that $\mathcal{I}^\infty(\mathbf{f})$ is continuous for every proper $\mathbf{f} \in \mathbf{W}$. Let $\mathbf{f} \in \mathbf{W}_c$ be a proper function symbol with $\tau(\mathbf{f}) = \boldsymbol{\tau}_1 \times \cdots \times \boldsymbol{\tau}_n \rightarrow \boldsymbol{\tau}$, and let $\mathscr{C}$ be a chain in the cpo $(D_{\tau_1^\infty \times \cdots \times \tau_n^\infty}, \sqsubseteq_{\tau_1^\infty \times \cdots \times \tau_n^\infty})$. It is easy to see $\mathcal{I}^\infty(\mathbf{f})$ is monotonic, so by Theorem 16.4.3 we just need to show that $\mathcal{I}^\infty(\mathbf{f})(\sqcup \mathscr{C}) \sqsubseteq_{\tau^\infty} \sqcup \mathcal{I}^\infty(\mathbf{f})(\mathscr{C})$; that is,

$$\mathcal{I}^\infty(\mathbf{f})(\sqcup \mathscr{C}) \subseteq \sqcup \mathcal{I}^\infty(\mathbf{f})(\mathscr{C}).$$

Let $\mathbf{t} \in \mathcal{I}^\infty(\mathbf{f})(\sqcup \mathscr{C}) = \mathcal{I}^\infty(\mathbf{f})(\cup(\mathscr{C} \downarrow 1), \ldots, \cup(\mathscr{C} \downarrow n))$. If $\mathbf{t} = \boldsymbol{\perp}_{\boldsymbol{\tau}}$ then $\mathbf{t} \in \mathcal{I}^\infty(\mathbf{f})(I)$ for all $I \in \mathscr{C}$, so that $\mathbf{t} \in \sqcup \mathcal{I}^\infty(\mathbf{f})(\mathscr{C})$. Otherwise, $\mathbf{t}$ is some term $\mathbf{f}(\mathbf{t}_1, \ldots, \mathbf{t}_n)$, where $\mathbf{t}_i \in \cup(\mathscr{C} \downarrow i)$, $1 \leq i \leq n$. Now, for $1 \leq i \leq n$, if $\mathbf{t}_i \in \cup(\mathscr{C} \downarrow i)$, then $\mathbf{t}_i \in (I_1^i, \ldots, I_n^i) \downarrow i$ for some $(I_1^i, \ldots, I_n^i) \in C$, and since $\{(I_1^i, \ldots, I_n^i) \mid 1 \leq i \leq n\}$ is a finite subset of the chain $\mathscr{C}$, there is some $(I_1, \ldots, I_n) \in \mathscr{C}$ such that

$$(I_1^i, \ldots, I_n^i) \sqsubseteq_{\tau_1^\infty \times \cdots \times \tau_n^\infty} (I_1, \ldots, I_n), \; 1 \leq i \leq n.$$

Then $\mathbf{t}_i \in I_i$, $1 \leq i \leq n$, which implies $\mathbf{f}(\mathbf{t}_1, \ldots, \mathbf{t}_n) \in \mathcal{I}^\infty(\mathbf{f})(I_1, \ldots, I_n)$, so that $\mathbf{f}(\mathbf{t}_1, \ldots, \mathbf{t}_n) \in \cup \mathcal{I}^\infty(\mathbf{f})(\mathscr{C}) = \sqcup \mathcal{I}^\infty(\mathbf{f})(\mathscr{C})$. Therefore, $\mathcal{I}^\infty(\mathbf{f})(\sqcup \mathscr{C}) \subseteq \sqcup \mathcal{I}^\infty(\mathbf{f})(\mathscr{C})$.

We turn now to the built-in function symbols. Again, it is easy to see that each $\mathcal{I}^\infty(\mathbf{if}_{\boldsymbol{\tau}})$, $\mathcal{I}^\infty(\mathbf{is\_f})$, and $\mathcal{I}^\infty(\mathbf{f}_i^{-1})$ is monotonic, using Lemma 1 in the latter two cases. Let $\boldsymbol{\tau} \in TV(\mathbf{W})$, and let $\mathscr{C}$ be a chain in

$(D_{\mathbf{Bool}^\infty \times \tau^\infty \times \tau^\infty}, \sqsubseteq_{\mathbf{Bool}^\infty \times \tau^\infty \times \tau^\infty})$. If $(\sqcup\mathscr{C})\downarrow 1 = \bot_{\mathbf{Bool}^\infty}$ then $I\downarrow 1 = \bot_{\mathbf{Bool}^\infty}$ for all $I \in \mathscr{C}$ and we have

$$\mathscr{I}^\infty(\mathbf{if}_\tau)(\sqcup\mathscr{C}) = \bot_{\tau^\infty} = \sqcup\mathscr{I}^\infty(\mathbf{if}_\tau)(\mathscr{C}).$$

If $(\sqcup\mathscr{C})\downarrow 1 = \text{tt}$ then $\mathscr{I}^\infty(\mathbf{if}_\tau)(\sqcup\mathscr{C}) = (\sqcup\mathscr{C})\downarrow 2 = \cup(\mathscr{C}\downarrow 2)$. Now, if $\mathbf{t} \in \cup(\mathscr{C}\downarrow 2)$, then $\mathbf{t} \in I$ for some $(b, I, J) \in \mathscr{C}$, and there is some $(\text{tt}, I', J') \in \mathscr{C}$ such that $I \sqsubseteq_{\tau^\infty} I'$, so

$$\mathbf{t} \in I' = \mathscr{I}^\infty(\mathbf{if}_\tau)(\text{tt}, I', J') \subseteq \sqcup\mathscr{I}^\infty(\mathbf{if}_\tau)(\mathscr{C}).$$

Therefore, $\mathscr{I}^\infty(\mathbf{if}_\tau)(\sqcup\mathscr{C}) \subseteq \sqcup\mathscr{I}^\infty(\mathbf{if}_\tau)(\mathscr{C})$. Similarly, if $(\sqcup\mathscr{C})\downarrow 1 = \text{ff}$ then $\mathscr{I}^\infty(\mathbf{if}_\tau)(\sqcup\mathscr{C}) = (\sqcup\mathscr{C})\downarrow 3 \subseteq \sqcup\mathscr{I}^\infty(\mathbf{if}_\tau)(\mathscr{C})$, so $\mathscr{I}^\infty(\mathbf{if}_\tau)$ is continuous.

Next, let $\mathbf{f} \in \mathbf{W}_c$ with $\rho(\mathbf{f}) = \tau$, and let $\mathscr{C}$ be a chain in $(D_{\tau^\infty}, \sqsubseteq_{\tau^\infty})$. If $\mathscr{C} = \{\bot_{\tau^\infty}\}$ then $\mathscr{I}^\infty(\mathbf{is\_f})(\sqcup\mathscr{C}) = \bot_{\mathbf{Bool}^\infty} = \sqcup\mathscr{I}^\infty(\mathbf{is\_f})(\mathscr{C})$, so assume $\mathscr{C}$ contains some $I \neq \bot_{\tau^\infty}$. Then there is some term in $I \subseteq \sqcup\mathscr{C}$ of the form $\mathbf{g}$ or $\mathbf{g}(\mathbf{u}_1, \ldots, \mathbf{u}_m)$, where $\mathbf{g} \in \mathbf{W}_c$, so $I, \sqcup\mathscr{C} \in \operatorname{ran}\mathscr{I}^\infty(\mathbf{g})$ by Lemma 1. Moreover, for all $J \in \mathscr{C}$, if $J \neq \bot_{\tau^\infty}$ then $J \subseteq \sqcup\mathscr{C}$ implies that J also contains a term of the form $\mathbf{g}$ or $\mathbf{g}(\mathbf{v}_1, \ldots, \mathbf{v}_m)$, so $J \in \operatorname{ran}\mathscr{I}^\infty(\mathbf{g})$. Therefore, if $\mathbf{f}$, $\mathbf{g}$ are the same then $\mathscr{I}^\infty(\mathbf{is\_f})(\sqcup\mathscr{C}) = \text{tt} = \sqcup\mathscr{I}^\infty(\mathbf{is\_f})(\mathscr{C})$, and if they are distinct then, by condition 3 on $\mathscr{T}$-interpretations, $\mathscr{I}^\infty(\mathbf{is\_f})(\sqcup\mathscr{C}) = \text{ff} = \sqcup\mathscr{I}^\infty(\mathbf{is\_f})(\mathscr{C})$.

We conclude with the functions $\mathscr{I}^\infty(\mathbf{f}_i^{-1})$. Let $\mathbf{f} \in \mathbf{W}_c$ with $\tau(\mathbf{f}) = \tau_1 \times \cdots \times \tau_n \to \tau$, and let $\mathscr{C}$ be a chain in $(D_{\tau^\infty}, \sqsubseteq_{\tau^\infty})$. If $\sqcup\mathscr{C} \notin \operatorname{ran}\mathscr{I}^\infty(\mathbf{f})$, then $\mathscr{I}^\infty(\mathbf{f}_i^{-1})(\sqcup\mathscr{C}) = \bot_{\tau_i^\infty}$, and, for any $I \in \mathscr{C}$, if there were a term $\mathbf{f}(\mathbf{t}_1, \ldots, \mathbf{t}_n) \in I$, then $\mathbf{f}(\mathbf{t}_1, \ldots, \mathbf{t}_n) \in \sqcup\mathscr{C}$ would imply by Lemma 1 that $\sqcup\mathscr{C} \in \operatorname{ran}\mathscr{I}^\infty(\mathbf{f})$, so $\mathscr{I}^\infty(\mathbf{f}_i^{-1})(I) = \bot_{\tau_i^\infty}$ and $\sqcup\mathscr{I}^\infty(\mathbf{f}_i^{-1})(\mathscr{C}) = \bot_{\tau_i^\infty}$. Suppose, then, that $\sqcup\mathscr{C} = \mathscr{I}^\infty(\mathbf{f})(I_1, \ldots, I_n)$ for some $(I_1, \ldots, I_n) \in D_{\tau_1^\infty \times \cdots \times \tau_n^\infty}$, so that $\mathscr{I}^\infty(\mathbf{f}_i^{-1})(\sqcup\mathscr{C}) = I_i$, and let $\mathbf{t} \in I_i$. Then there is a term $\mathbf{u}$ of the form $\mathbf{f}(\mathbf{u}_1, \ldots, \mathbf{u}_{i-1}, \mathbf{t}, \mathbf{u}_{i+1}, \ldots, \mathbf{u}_n) \in \sqcup\mathscr{C}$, which implies $\mathbf{u} \in I$ for some $I \in \mathscr{C}$. Therefore, by Lemma 1, $\mathbf{t} \in \mathscr{I}^\infty(\mathbf{f}_i^{-1})(I) \subseteq \sqcup\mathscr{I}^\infty(\mathbf{f}_i^{-1})(\mathscr{C})$, and we have $\mathscr{I}^\infty(\mathbf{f}_i^{-1})(\sqcup\mathscr{C}) \subseteq \sqcup\mathscr{I}^\infty(\mathbf{f}_i^{-1})(\mathscr{C})$, so $\mathscr{I}^\infty(\mathbf{f}_i^{-1})$ is continuous. ■

We immediately get

Corollary 6.2. There is an infinitary data structure system for every standard vocabulary.

For example, $\Delta_{\mathbf{N}}^\infty = \operatorname{rep}(\Sigma_{\mathbf{N}}^\infty)$ is an infinitary data structure system for $\mathbf{W}_{\mathbf{N}}$. It is clear that each element in $D_{\mathbf{Bool}^\infty}$ is representable. It turns out, moreover, that every element in $D_{\mathbf{N}^\infty}$ is representable. Certainly each

$n \in D_{N^\infty}$ is representable, since $\overline{\Omega}(\mathbf{n}) = n$. For example,

$$\begin{aligned}\overline{\Omega}(\mathbf{2}) = \mathcal{I}^\infty(\mathbf{s})(\mathcal{I}^\infty(\mathbf{s})(\mathcal{I}^\infty(\mathbf{0}))) &= \mathcal{I}^\infty(\mathbf{s})(\mathcal{I}^\infty(\mathbf{s})(\{\bot_N, \mathbf{0}\}))\\ &= \mathcal{I}^\infty(\mathbf{s})(\{\bot_N, \mathbf{s}(\bot_N), \mathbf{s}(\mathbf{0})\})\\ &= \{\bot_N, \mathbf{s}(\bot_N), \mathbf{s}(\mathbf{s}(\bot_N)), \mathbf{s}(\mathbf{s}(\mathbf{0}))\}\\ &= 2.\end{aligned}$$

Now, let $\mathbf{P}$ be the $\mathbf{W}_N$-program with equations

$$\mathbf{B}(\mathbf{X}) = \mathbf{B}(\mathbf{X})$$
$$\mathbf{G}(\mathbf{X}) = \mathbf{s}(\mathbf{G}(\mathbf{X})).$$

Then for any $n_\bot \in D_{N^\infty}$, $n_\bot = \overline{\mu\Phi_{\mathbf{P}}}(\mathbf{s}^n(\mathbf{B}(\mathbf{0})))$. For example,

$$\begin{aligned}\overline{\mu\Phi_{\mathbf{P}}}(\mathbf{s}(\mathbf{s}(\mathbf{B}(\mathbf{0})))) &= \mathcal{I}^\infty(\mathbf{s})(\mathcal{I}^\infty(\mathbf{s})(\Omega(\mathbf{B})(\mathcal{I}^\infty(\mathbf{0}))))\\ &= \mathcal{I}^\infty(\mathbf{s})(\mathcal{I}^\infty(\mathbf{s})(\{\bot_N\}))\\ &= \mathcal{I}^\infty(\mathbf{s})(\{\bot_N, \mathbf{s}(\bot_N)\})\\ &= \{\bot_N, \mathbf{s}(\bot_N), \mathbf{s}(\mathbf{s}(\bot_N))\}\\ &= 2_\bot .\end{aligned}$$

Also, an easy induction on $n \in N$ shows that $\Phi^n_{\mathbf{P}}(\Omega)(\mathbf{G})(0) = n_\bot$ for all $n \in N$ (where $0 = \{\bot_N, \mathbf{0}\}$), so

$$\begin{aligned}\overline{\mu\Phi_{\mathbf{P}}}(\mathbf{G}(\mathbf{0})) &= \sqcup\{\overline{\Phi^n_{\mathbf{P}}(\Omega)}(\mathbf{G}(\mathbf{0})) \mid n \in N\}\\ &= \sqcup\{\Phi^n_{\mathbf{P}}(\Omega)(\mathbf{G})(0) \mid n \in N\}\\ &= \sqcup\{n_\bot \mid n \in N\}\\ &= \omega.\end{aligned}$$

Therefore, every element of D_{N^∞} is representable, and $\Delta^\infty_N = \Sigma^\infty_N$.

The situation is more interesting when we consider

$$\Sigma^\infty_{NL} = (\mathcal{T}_{\mathcal{H}^\infty(\mathbf{W}_{NL})}, \mathcal{I}_{\mathcal{H}^\infty(\mathbf{W}_{NL})})$$

and $\Delta^\infty_{NL} = \text{rep}(\Sigma^\infty_{NL})$. As in the previous example, $\mathcal{T}^\infty(\mathbf{N}) = \text{rep}(\mathcal{T}^\infty)(\mathbf{N})$, but now $\mathcal{T}^\infty(\mathbf{NL}) \neq \text{rep}(\mathcal{T}^\infty)(\mathbf{NL})$. Again there are three kinds of elements in D_{NL^∞}. For each term of the form $\mathbf{cons}(\mathbf{t}_1, \cdots \mathbf{cons}(\mathbf{t}_n, \mathbf{nil}) \cdots)$, $n \geq 0$, where $\mathbf{t}_i \in \mathrm{TM}^{\mathrm{N}}_{\mathrm{W}^+_{\mathrm{NL}}}$, $1 \le i \le n$, there is a principal ideal. For example,

$$\begin{aligned}&\mathrm{pid}(\mathbf{cons}(\mathbf{1}, \mathbf{cons}(\bot_{\mathrm{N}}, \mathbf{nil})))\\ &\quad = \{\bot_{\mathrm{NL}}, \mathbf{cons}(\bot_{\mathrm{N}}, \bot_{\mathrm{NL}}), \mathbf{cons}(\mathbf{s}(\bot_{\mathrm{N}}), \bot_{\mathrm{NL}}), \mathbf{cons}\ (\mathbf{1}, \bot_{\mathrm{NL}}),\\ &\qquad \mathbf{cons}(\bot_{\mathrm{N}}, \mathbf{cons}(\bot_{\mathrm{N}}, \bot_{\mathrm{NL}})), \mathbf{cons}(\mathbf{s}(\bot_{\mathrm{N}}), \mathbf{cons}(\bot_{\mathrm{N}}, \bot_{\mathrm{NL}})),\\ &\qquad \mathbf{cons}(\mathbf{1}, \mathbf{cons}(\bot_{\mathrm{N}}, \bot_{\mathrm{NL}})), \mathbf{cons}(\bot_{\mathrm{N}}, \mathbf{cons}(\bot_{\mathrm{N}}, \mathbf{nil})),\\ &\qquad \mathbf{cons}(\mathbf{s}(\bot_{\mathrm{N}}), \mathbf{cons}(\bot_{\mathrm{N}}, \mathbf{nil})), \mathbf{cons}(\mathbf{1}, \mathbf{cons}(\bot_{\mathrm{N}}, \mathbf{nil}))\}.\end{aligned}$$

Note that $\mathrm{pid}(\bot_{\mathrm{N}}) = \{\bot_{\mathrm{N}}\}$, $\mathrm{pid}(\mathbf{nil}) = \{\bot_{\mathrm{NL}}, \mathbf{nil}\} = \mathscr{I}^{\infty}(\mathbf{nil})$,

$$\begin{aligned}\mathrm{pid}(\mathbf{cons}(\bot_{\mathrm{N}}, \mathbf{nil})) &= \{\bot_{\mathrm{NL}}, \mathbf{cons}(\bot_{\mathrm{N}}, \bot_{\mathrm{NL}}), \mathbf{cons}(\bot_{\mathrm{N}}, \mathbf{nil})\}\\ &= \mathscr{I}^{\infty}(\mathbf{cons})(\mathrm{pid}(\bot_{\mathrm{N}}), \mathrm{pid}(\mathbf{nil})),\end{aligned}$$

and, as the reader can verify,

$$\mathrm{pid}(\mathbf{cons}(\mathbf{1}, \mathbf{cons}(\bot_{\mathrm{N}}, \mathbf{nil}))) = \mathscr{I}^{\infty}(\mathbf{cons})(\mathrm{pid}(\mathbf{1}), \mathrm{pid}(\mathbf{cons}(\bot_{\mathrm{N}}, \mathbf{nil}))).$$

That is, $\mathrm{pid}(\mathbf{cons}(\mathbf{1}, \mathbf{cons}(\bot_{\mathrm{N}}, \mathbf{nil})))$ is built up from $\mathrm{pid}(\mathbf{nil})$, the list with no elements, by applying $\mathscr{I}^{\infty}(\mathbf{cons})$ to $(\mathrm{pid}(\bot_{\mathrm{N}}), \mathrm{pid}(\mathbf{nil}))$ to get $\mathrm{pid}(\mathbf{cons}(\bot_{\mathrm{N}}, \mathbf{nil}))$, a list with one element (namely, $\mathrm{pid}(\bot_{\mathrm{N}})$), and then applying $\mathscr{I}^{\infty}(\mathbf{cons})$ again to $(\mathrm{pid}(\mathbf{1}), \mathrm{pid}(\mathbf{cons}(\bot_{\mathrm{N}}, \mathbf{nil})))$ to get a list with two elements (namely, $\mathrm{pid}(\mathbf{1})$ and $\mathrm{pid}(\bot_{\mathrm{N}})$). Therefore, we write $\mathrm{pid}(\mathbf{cons}(\mathbf{1}, \mathbf{cons}(\bot_{\mathrm{N}}, \mathbf{nil})))$ as $\langle \mathrm{pid}(\mathbf{1}), \mathrm{pid}(\bot_{\mathrm{N}})\rangle$, or simply $\langle 1, \bot_{\mathrm{N}^{\infty}}\rangle$. In general, we write $\mathrm{pid}(\mathbf{nil})$ as the *empty list* $\langle\ \rangle$ and $\mathrm{pid}(\mathbf{cons}(\mathbf{t}_1, \cdots \mathbf{cons}(\mathbf{t}_n, \mathbf{nil}) \cdots))$ as the *finite list* $\langle d_1, \ldots, d_n\rangle$, where $d_i = \mathrm{pid}(\mathbf{t}_i)$, $1 \le i \le n$.

For each term of the form $\mathbf{cons}(\mathbf{t}_1, \cdots \mathbf{cons}(\mathbf{t}_n, \bot_{\mathrm{NL}}) \cdots)$, $n \ge 0$, where $\mathbf{t}_i \in \mathrm{TM}^{\mathrm{N}}_{\mathrm{W}^+_{\mathrm{NL}}}$, $1 \le i \le n$, there is also a principal ideal. For example, $\mathrm{pid}(\mathbf{cons}(\mathbf{1}, \mathbf{cons}(\bot_{\mathrm{N}}, \bot_{\mathrm{NL}})))$ is

$$\begin{aligned}&\{\bot_{\mathrm{NL}}, \mathbf{cons}(\bot_{\mathrm{N}}, \bot_{\mathrm{NL}}), \mathbf{cons}(\mathbf{s}(\bot_{\mathrm{N}}), \bot_{\mathrm{NL}}),\\ &\qquad \mathbf{cons}(\mathbf{1}, \bot_{\mathrm{NL}}), \mathbf{cons}(\bot_{\mathrm{N}}, \mathbf{cons}(\bot_{\mathrm{N}}, \bot_{\mathrm{NL}})),\\ &\qquad \mathbf{cons}(\mathbf{s}(\bot_{\mathrm{N}}), \mathbf{cons}(\bot_{\mathrm{N}}, \bot_{\mathrm{NL}})), \mathbf{cons}(\mathbf{1}, \mathbf{cons}(\bot_{\mathrm{N}}, \bot_{\mathrm{NL}}))\}.\end{aligned}$$

Here we have

$$\mathrm{pid}(\mathbf{cons}(\mathbf{1}, \mathbf{cons}(\bot_{\mathrm{N}}, \bot_{\mathrm{NL}}))) = \mathscr{I}^{\infty}(\mathbf{cons})(1, \mathscr{I}^{\infty}(\mathbf{cons})(\bot_{\mathrm{N}^{\infty}}, \bot_{\mathrm{NL}^{\infty}})),$$

which we write $\langle 1, \bot_{\mathrm{N}^{\infty}}\rangle_{\bot}$. In general we write elements of the form $\mathrm{pid}(\mathbf{cons}(\mathbf{t}_1, \cdots \mathbf{cons}(\mathbf{t}_n, \bot_{\mathrm{NL}}) \cdots))$ as $\langle d_1, \ldots, d_n\rangle_{\bot}$, where $d_i = \mathrm{pid}(\mathbf{t}_i)$,

$1 \le i \le n$. We call these objects *prefix lists*, since

$$\langle d_1, \ldots, d_n \rangle_\perp \sqsubseteq_{\mathbf{NL}^\infty} \langle d_1, \ldots, d_n, e_1, \ldots, e_m \rangle$$

for any $e_1, \ldots, e_m \in D_{\mathbf{N}^\infty}$.

We also have elements in $D_{\mathbf{NL}^\infty}$ which are nonprincipal ideals. The difference between $\langle d_1, \ldots, d_n \rangle$ and $\langle d_1, \ldots, d_n \rangle_\perp$ is that the former is built up from the completely defined empty list $\mathscr{I}^\infty(\mathbf{nil})$ and the latter is built up from the completely undefined list $\perp_{\mathbf{NL}^\infty}$. Now, we can have infinite chains of finite lists, e.g., $\{\langle n_\perp \rangle \mid n \in N\}$, where we write $\sqcup\{\langle n_\perp \rangle \mid n \in N\}$ as $\langle \omega \rangle$, but the least upper bound is always a finite list (though not a principal ideal). On the other hand, chains of prefix lists can lead to infinite lists. For example, $\{\langle 0, \ldots, n \rangle_\perp \mid n \in N\}$ is a chain, since

$$\langle 0 \rangle_\perp \sqsubseteq_{\mathbf{NL}^\infty} \langle 0, 1 \rangle_\perp \sqsubseteq_{\mathbf{NL}^\infty} \langle 0, 1, 2 \rangle_\perp \sqsubseteq_{\mathbf{NL}^\infty} \cdots,$$

and the least upper bound of $\{\langle 0, \ldots, n \rangle_\perp \mid n \in N\}$ is the infinite ideal which we write as the *infinite list* $\langle 0, 1, 2, 3, \ldots \rangle$. There are other interesting kinds of nonprincipal ideals, such as $\sqcup\{\langle n_\perp \rangle_\perp \mid n \in N\}$ and

$$\sqcup\left\{\langle \overbrace{n_\perp, \ldots, n_\perp}^{n} \rangle_\perp \mid n \in N\right\},$$

which we leave to the reader to explore.

It is interesting to observe that even these infinite lists, or the representable ones, more precisely, can be quite useful. In fact, a family of programming languages known as *lazy functional languages* has been developed based on the use of such objects. The typical use of infinite lists in these languages is to define an infinite list of desired objects and then to select some particular object from the list. The word *lazy* refers to the fact that, in practice, it is not necessary to generate an entire infinite list before performing the selection: it is necessary to generate only *enough* of the list so that the desired object appears. For a simple example, let **P** be the program with equations

$$\mathbf{F(X) = cons(X, F(s(X)))}$$

$$\mathbf{Nth(X, Y) = if_{NL}(is\_0(X), cons_1^{-1}(Y), Nth(s_1^{-1}(X), cons_2^{-1}(Y)))}.$$

Then

$$\overline{\mu\Phi_{\mathbf{P}}}(\mathbf{F(0)}) = \langle 0, 1, 2, \ldots \rangle \text{ and } \overline{\mu\Phi_{\mathbf{P}}}(\mathbf{Nth(n, F(0))}) = n \text{ for all } n \in N.$$

In the next chapter we will show that the infinite list $\langle p_1, p_2, \ldots \rangle$ of all primes is also representable.

Unlike $\Delta_{\mathbf{N}}^{\infty}$, we now have objects that are not representable. For example, the infinite list $L = \langle \text{HALT}(0,0), \text{HALT}(1,1), \ldots \rangle$ is not representable, where HALT(x, x) is the predicate defined in Chapter 4. If it were representable, say $L = \overline{\mu\Phi_{\mathbf{Q}}}(\mathbf{t})$, then we would have

$$\overline{\mu\Phi_{\mathbf{P} \cup \mathbf{Q}}}(\mathbf{Nth}(\mathbf{n}, \mathbf{t})) = \text{HALT}(n, n) \quad \text{for all } n \in N,$$

which, informally at least, would imply that HALT(x, x) is computable.

For our final example, we take a vocabulary, $\mathbf{W_R}$, suitable for representing the decimal expansions of real numbers x in the interval $0 \leq x < 1$. (This interval is usually written $[0, 1)$.) This time we begin with constant symbols $\boldsymbol{d}_0, \boldsymbol{d}_1, \ldots, \boldsymbol{d}_9$, with $\tau(\boldsymbol{d}_0) = \cdots = \tau(\boldsymbol{d}_9) = \mathbf{D}$, to represent decimal digits. We will build up lists of decimal digits with **dnil** and **dcons**, where $\tau(\mathbf{dnil}) = \mathbf{DL}$ and $\tau(\mathbf{dcons}) = \mathbf{D} \times \mathbf{DL} \rightarrow \mathbf{DL}$. We also include in $\mathbf{W_R}$ the symbols of $\mathbf{W_{NL}}$. We call the elements of $D_{\mathbf{D}^\infty}$ *decimal digits*, which we write as $\perp_{\mathbf{D}^\infty}, 0, 1, \ldots, 9$. Again we have three kinds of elements in $D_{\mathbf{DL}^\infty}$: (1) finite lists of decimal digits, which we write $.d_1 d_2 \ldots d_n$, $n \geq 0$; (2) prefix lists of decimal digits, which we write $.d_1 d_2 \ldots d_{n\perp}$, $n \geq 0$; and (3) infinite lists of decimal digits, which we write $.d_1 d_2 \ldots$. It is clear that there is an object in $D_{\mathbf{DL}^\infty}$ for every real number in $[0, 1)$. (Actually, there is more than one object for some real numbers, since, for example, $.29999\ldots$ and $.3$ are distinct in $D_{\mathbf{DL}^\infty}$, but that problem will not concern us here.) We call the elements of $D_{\mathbf{DL}^\infty}$ *computable real numbers*[7] in $[0, 1)$. It is a basic mathematical fact that there are more real numbers than there are $\mathbf{W_R}$-programs and $\mathbf{W_R}$-terms, so there are certainly objects in $D_{\mathbf{DL}^\infty}$ that are not computable real numbers. It is clear that every nonrepeating rational number in $[0, 1)$ is computable, e.g., $.33 = \overline{\Omega}(\mathbf{dcons}(\boldsymbol{d}_3, \mathbf{dcons}(\boldsymbol{d}_3, \mathbf{dnil})))$. It is also clear, intuitively, that we could write a $\mathbf{W_R}$-program to define long division, so every repeating rational number in $[0, 1)$ is computable. In the next chapter we will show, moreover, that some irrational numbers are computable as well.

Exercises

1. Let $\mathbf{W}_c = \{\mathbf{tt}, \mathbf{ff}, \mathbf{c}, \mathbf{f}\}$ be a standard constructor vocabulary with $\tau_c(\mathbf{c}) = \boldsymbol{\tau}$ and $\tau_c(\mathbf{f}) = \boldsymbol{\tau} \times \boldsymbol{\tau} \rightarrow \boldsymbol{\tau}$, and let $\mathbf{W} = \mathbf{W}_c \cup B(\mathbf{W}_c)$.
   (a) Give an infinite chain in $(\text{TM}_{\mathbf{W}^+}^{\boldsymbol{\tau}}, \sqsubseteq_{\boldsymbol{\tau}^+})$.
   (b) Let $\mathscr{I}^\infty = \mathscr{I}_{\mathscr{H}^\infty(\mathbf{W})}$. What is $\overline{\Omega}_{\mathscr{I}^\infty}(\mathbf{f}(\mathbf{X}, \mathbf{f}(\mathbf{c}, \mathbf{c})))$?

[7] Really we should call these objects *representable real numbers* at this point, but the operational semantics in the next chapter will justify the more traditional name.

2. What is pid(3) in $\Delta_{\mathbf{N}}^{\infty}$? What is pid($\mathbf{s}^3(\perp_{\mathbf{N}})$)?
3. Show that $n_{\perp} \sqsubseteq_{\mathbf{N}^{\infty}} n$ for all $n \in N$. [Note that n has two different meanings here.]
4. Let $\mathscr{I}^{\infty} = \mathscr{I}_{\mathscr{H}^{\infty}(\mathbf{W}_{\mathbf{N}})}$.
   (a) Show that $\mathscr{I}^{\infty}(\mathbf{s})(2) = 3$.
   (b) Show that $\mathscr{I}^{\infty}(\mathbf{s})(2_{\perp}) = 3_{\perp}$.
   (c) Show that $\mathscr{I}^{\infty}(\mathbf{s}_1^{-1})(3_{\perp}) = 2_{\perp}$.
5. (a) What is $\mathscr{D}_{\Delta_{\mathbf{N}}^{\infty}}(\mathbf{ADD})(+)(3, 2)$?
   (b) What is $\mathscr{D}_{\Delta_{\mathbf{N}}^{\infty}}(\mathbf{ADD})(+)(3_{\perp}, 2)$?
   (c) What is $\mathscr{D}_{\Delta_{\mathbf{N}}^{\infty}}(\mathbf{ADD})(+)(3, 2_{\perp})$?
6. What is $\langle \perp_{\mathbf{N}^{\infty}}, 1\rangle$ in $\Delta_{\mathbf{NL}}^{\infty}$? What is $\langle \perp_{\mathbf{N}^{\infty}}, 1\rangle_{\perp}$?
7. Show that $\langle n\rangle_{\perp} \sqsubseteq_{\mathbf{NL}^{\infty}} \langle n\rangle$ for all $n \in N$.
8. Let $\mathscr{I}^{\infty} = \mathscr{I}_{\mathscr{H}^{\infty}(\mathbf{W}_{\mathbf{NL}})}$.
   (a) Show that $\mathscr{I}^{\infty}(\mathbf{cons})(0, \langle 1\rangle) = \langle 0, 1\rangle$.
   (b) Show that $\mathscr{I}^{\infty}(\mathbf{cons})(0, \langle 1\rangle_{\perp}) = \langle 0, 1\rangle_{\perp}$.
   (c) Show that $\mathscr{I}^{\infty}(\mathbf{cons}_2^{-1})(\langle 0, 1\rangle_{\perp}) = \langle 1\rangle_{\perp}$.
9. (a) What is $\mathscr{D}_{\Delta_{\mathbf{NL}}^{\infty}}(\mathbf{LIST})(\mathbf{Length})(\langle 2, 2\rangle)$?
   (b) What is $\mathscr{D}_{\Delta_{\mathbf{NL}}^{\infty}}(\mathbf{LIST})(\mathbf{Length})(\langle 2, 2, 2, \ldots\rangle)$?
10. (a) What is $\mathscr{D}_{\Delta_{\mathbf{NL}}^{\infty}}(\mathbf{LIST})(\mathbf{Cat})(\langle 2, 3\rangle, \langle 4, 5\rangle)$?
    (b) What is $\mathscr{D}_{\Delta_{\mathbf{NL}}^{\infty}}(\mathbf{LIST})(\mathbf{Cat})(\langle 2, 3\rangle, \langle 4, 5, 6, \ldots\rangle)$?
11. Write a $\mathbf{W}_{\mathbf{NL}}$-program $\mathbf{P}$ with $\mathbf{F} \in FV(\mathbf{P})$ such that

$$\mathscr{D}_{\Delta_{\mathbf{NL}}^{\infty}}(\mathbf{P})(\mathbf{F})(\langle n_1, n_2, \ldots\rangle) = \langle n_1 + 1, n_2 + 1, \ldots\rangle$$

    for all representable lists of numbers $\langle n_1, n_2, \ldots\rangle \in D_{\mathbf{NL}^{\infty}}$.
12. Write a $\mathbf{W}_{\mathbf{NL}}$-program $\mathbf{P}$ with $\mathbf{F} \in FV(\mathbf{P})$ such that

$$\mathscr{D}_{\Delta_{\mathbf{NL}}^{\infty}}(\mathbf{P})(\mathbf{F})(\langle m_1, m_2, \ldots\rangle, \langle n_1, n_2, \ldots\rangle) = \langle m_1 + n_1, m_2 + n_2, \ldots\rangle$$

    for all representable lists of numbers $\langle m_1, m_2, \ldots\rangle, \langle n_1, n_2, \ldots\rangle \in D_{\mathbf{NL}^{\infty}}$.
13. Show that $\langle 1, 1, 1 \ldots\rangle$ is representable.
14. Show that $\langle 1_{\perp}, 1_{\perp}, 1_{\perp} \ \ldots\rangle$ is representable.
15. Show that for any $n \in N$, $\langle n, n + 1, n + 2, \ldots\rangle$ is representable.
16. Let $\langle \omega\rangle_{\perp} = \sqcup\{\langle n_{\perp}\rangle_{\perp} \mid n \in N\}$. Describe $\langle \omega\rangle_{\perp}$, and show that it is representable.

17. Let

$$\langle \omega, \omega, \ldots \rangle = \sqcup \left\{ \langle \overbrace{n_\perp, \ldots, n_\perp}^{n} \rangle_\perp \mid n \in N \right\}.$$

Describe $\langle \omega, \omega, \ldots \rangle$, and show that it is representable.

18. Prove the assertion made in the proof of Theorem 6.1: for every $\mathbf{f} \in \mathbf{W}$, $\mathscr{I}^{\infty}(\mathbf{f})$ is monotonic.

18

Operational Semantics of Recursion Equations

1. Operational Semantics for Simple Data Structure Systems

The definition of $\mathscr{D}_\Delta$ accomplishes our goal of directly assigning a meaning to programs without the intermediary notion of a computation. On the other hand, at this point we have no *a priori* reason for believing that the functions defined by recursion programs are (partially) computable. It turns out, though, that they are computable in a very reasonable sense. In this chapter we go back to basics and define a notion of computation, one appropriate for recursion programs, which has much in common with computations of $\mathscr{S}$ programs. This new kind of computation will be the basis for the operational semantics of **W**-programs. The idea is that the operational semantics will give us a way to compute the functions defined by the denotational semantics. We say that an operational semantics is *correct* with respect to the denotational semantics $\mathscr{D}_\Delta$ for a data structure system Δ if it gives every **W**-program **P** the same meaning given by $\mathscr{D}_\Delta$, that is, $\mathscr{D}_\Delta(\mathbf{P})$. The precise details depend on the nature of the particular data structure systems in which we wish to compute, so in this section we concentrate on an operational semantics appropriate for simple data structure systems.

An $\mathscr{S}$ program computation is a sequence of snapshots, and the relation between a snapshot and its successor is easily described. Computations of recursion programs are similar in nature. The idea, as before, is that we treat an equation like $\mathbf{F}(\mathbf{X}) = \mathbf{G}(\mathbf{H}(\mathbf{X}))$ as a definition of **F** in terms of $\mathbf{G}(\mathbf{H}(\mathbf{X}))$. Given a term $\mathbf{F}(3)$, for example, we attempt to determine its value by replacing it with $\mathbf{G}(\mathbf{H}(3))$. Of course, **G** and **H** should also be defined, so we replace them by their definitions as well, and continue replacing until we get a numeral. We formalize this idea as follows.

Again we let **W** be some arbitrary standard vocabulary throughout the chapter. A **W**-*substitution* is a finite function $\{(\mathbf{X}_1, \mathbf{t}_1), \ldots, (\mathbf{X}_n, \mathbf{t}_n)\}$ such that $\mathbf{X}_1, \ldots, \mathbf{X}_n$ are distinct individual variables and $\mathbf{t}_1, \ldots, \mathbf{t}_n$ are **W**-terms such that $\tau(\mathbf{t}_i) = \tau(\mathbf{X}_i)$, $1 \le i \le n$. The application of a **W**-substitution θ to a **W**-term **t** is written $\mathbf{t}\theta$, and the result of applying θ to **t** is the **W**-term obtained from **t** by simultaneously replacing each occurrence of $\mathbf{X}_i$ by $\mathbf{t}_i$, $1 \le i \le n$. Note that each variable in the domain of θ is replaced by a term of the same type, so if **t** is a **W**-term, then $\mathbf{t}\theta$ is also a **W**-term. We can give a more formal definition as follows. Let θ be a **W**-substitution. Then

$$
\begin{array}{lll}
\mathbf{c}\theta & = \mathbf{c} & \text{for constant symbols } \mathbf{c} \in \mathbf{W} \\
\mathbf{X}\theta & = \mathbf{t} & \text{for } \mathbf{X} \in \mathrm{VAR}_\mathrm{I} \text{ such that } (\mathbf{X}, \mathbf{t}) \in \theta \\
\mathbf{X}\theta & = \mathbf{X} & \text{for } \mathbf{X} \in \mathrm{VAR}_\mathrm{I} \text{ such that } \mathbf{X} \notin \text{the domain of } \theta \\
\mathbf{f}(\mathbf{t}_1, \ldots, \mathbf{t}_n)\theta & = \mathbf{f}(\mathbf{t}_1\theta, \ldots, \mathbf{t}_n\theta) & \text{where } \mathbf{f} \in \mathbf{W} \\
\mathbf{F}(\mathbf{t}_1, \ldots, \mathbf{t}_n)\theta & = \mathbf{F}(\mathbf{t}_1\theta, \ldots, \mathbf{t}_n\theta) & \text{where } \mathbf{F} \in \mathrm{VAR}_\mathrm{F}.
\end{array}
$$

The following useful lemma shows that we can sometimes trade in part of a variable assignment for a substitution.

Substitution Lemma. Let $\Sigma = (\mathcal{T}, \mathcal{I})$ be a **W**-structure, let $V_i \subseteq \mathrm{VAR}_\mathrm{I}$ and $V_f \subseteq \mathrm{VAR}_\mathrm{F}$, let $\alpha \in \mathcal{A}_\mathcal{T}(V_f)$, and let $\beta \in \mathcal{A}_\mathcal{T}(V_i)$ be a variable assignment and θ a substitution such that, for all $\mathbf{X} \in V_i$, $V(\mathbf{X}\theta) \subseteq V_f$ and $\beta(\mathbf{X}) = \overline{\alpha}(\mathbf{X}\theta)$. Then for all $\mathbf{t} \in \mathrm{TM}_\mathbf{W}(V_i \cup V_f)$, $\overline{\alpha \cup \beta}(\mathbf{t}) = \overline{\alpha}(\mathbf{t}\theta)$.

Proof. We argue by structural induction on **t**. If **t** is a constant symbol $\mathbf{c} \in \mathbf{W}$, then

$$\overline{\alpha \cup \beta}(\mathbf{c}) = \mathcal{I}(\mathbf{c}) = \mathcal{I}(\mathbf{c}\theta) = \overline{\alpha}(\mathbf{c}\theta),$$

and if **t** is $\mathbf{X} \in V_i$, then $\overline{\alpha \cup \beta}(\mathbf{X}) = \beta(\mathbf{X}) = \overline{\alpha}(\mathbf{X}\theta)$ by assumption. If **t** is $\mathbf{f}(\mathbf{t}_1, \ldots, \mathbf{t}_n)$, where $\mathbf{f} \in \mathbf{W}$, then

$$
\begin{aligned}
&\overline{\alpha \cup \beta}(\mathbf{f}(\mathbf{t}_1, \ldots, \mathbf{t}_n)) \\
&\quad = \mathcal{I}(\mathbf{f})(\overline{\alpha \cup \beta}(\mathbf{t}_1), \ldots, \overline{\alpha \cup \beta}(\mathbf{t}_n)) \\
&\quad = \mathcal{I}(\mathbf{f})(\overline{\alpha}(\mathbf{t}_1\theta), \ldots, \overline{\alpha}(\mathbf{t}_n\theta)) \qquad \text{by the induction hypothesis} \\
&\quad = \overline{\alpha}(\mathbf{f}(\mathbf{t}_1\theta, \ldots, \mathbf{t}_n\theta)) \\
&\quad = \overline{\alpha}(\mathbf{f}(\mathbf{t}_1, \ldots, \mathbf{t}_n)\theta).
\end{aligned}
$$

If $\mathbf{t}$ is $\mathbf{F}(\mathbf{t}_1,\ldots,\mathbf{t}_n)$, where $\mathbf{F} \in V_f$, then

$$\begin{aligned}
&\overline{\alpha \cup \beta}(\mathbf{F}(\mathbf{t}_1,\ldots,\mathbf{t}_n)) \\
&\quad = \alpha(\mathbf{F})(\overline{\alpha \cup \beta}(\mathbf{t}_1),\ldots,\overline{\alpha \cup \beta}(\mathbf{t}_n)) \\
&\quad = \alpha(\mathbf{F})(\bar{\alpha}(\mathbf{t}_1\theta),\ldots,\bar{\alpha}(\mathbf{t}_n\theta)) \qquad \text{by the induction hypothesis} \\
&\quad = \bar{\alpha}(\mathbf{F}(\mathbf{t}_1\theta,\ldots,\mathbf{t}_n\theta)) \\
&\quad = \bar{\alpha}(\mathbf{F}(\mathbf{t}_1,\ldots,\mathbf{t}_n)\theta).
\end{aligned}$$ ■

A **W**-*term rewrite rule* is a pair of **W**-terms, written $\mathbf{u} \to \mathbf{v}$, such that $IV(\mathbf{v}) \subseteq IV(\mathbf{u})$ and such that no individual variable occurs more than once in **u**. A **W**-*term rewriting system* is a set of **W**-term rewrite rules.[1] We say that a **W**-term **t** *matches* a rewrite rule $\mathbf{u} \to \mathbf{v}$ with substitution θ if $\mathbf{t} = \mathbf{u}\theta$. A **W**-term rewriting system **T** is *deterministic* if no **W**-term matches more than one rewrite rule in **T**.

In order to use a **W**-term rewriting system **T**, we associate with **T** a *rewriting strategy* σ which selects, for every **W**-term **t**, a (possibly empty) set of occurrences[2] of subterms of **t**. These are the $(\mathbf{T}, \sigma)$*-redexes* of **t**. A **W**-term **w** is a **T**-*rewrite* of **W**-term **t** if **t** matches some rewrite rule $\mathbf{u} \to \mathbf{v}$ in **T** with substitution θ, and $\mathbf{w} = \mathbf{v}\theta$. Given a strategy σ, a **W**-term **w** is a $(\mathbf{T}, \sigma)$*-rewrite* of a **W**-term **t**, denoted

$$\mathbf{t} \underset{\mathbf{T},\sigma}{\Longrightarrow} \mathbf{w},$$

if **w** is the result of replacing every $(\mathbf{T}, \sigma)$-redex $\mathbf{t}'$ of **t** by a **T**-rewrite of $\mathbf{t}'$. A **W**-term **t** is $(\mathbf{T}, \sigma)$*-normal* if the set of $(\mathbf{T}, \sigma)$-redexes is empty. A $(\mathbf{T}, \sigma)$*-computation* for **W**-term **t** is a (possibly infinite) sequence of **W**-terms $\mathbf{t}_0, \mathbf{t}_1, \ldots$ such that

1. $\mathbf{t}_0$ is **t**,
2. $\mathbf{t}_i \underset{\mathbf{T},\sigma}{\Longrightarrow} \mathbf{t}_{i+1}$ for all $\mathbf{t}_i$ occurring in the sequence, and
3. for all $\mathbf{t}_i$ in the sequence, $\mathbf{t}_i$ is $(\mathbf{T}, \sigma)$-normal if and only if $\mathbf{t}_i$ is the last term in the sequence.

For our definition of a computation to be reasonable, it is crucial that the process of finding the term $\mathbf{t}_{i+1}$ that follows term $\mathbf{t}_i$ should itself be "mechanical," that is, computable in some sense. Moreover, the test that a

---

[1] Although the definition permits rules of the form $\mathbf{c} \to \mathbf{d}$, where $\mathbf{c}, \mathbf{d}$ are constant symbols, they play no role in our treatment of operational semantics, so we assume that such rules do not occur in any rewriting system referred to in this chapter.

[2] Note that we distinguish between subterms and *occurrences* of subterms. For example, $\mathbf{F}(\mathbf{G}(\mathbf{0}), \mathbf{G}(\mathbf{0}))$ has two occurrences of the subterm $\mathbf{G}(\mathbf{0})$, and a strategy σ might select the occurrence on the left without selecting the occurrence on the right.

term is $(\mathbf{T}, \sigma)$-normal must also be computable, so that we know when a computation terminates. Now, three kinds of steps are involved in finding $\mathbf{t}_{i+1}$:

- finding the set of $(\mathbf{T}, \sigma)$-redexes of $\mathbf{t}_i$;
- for each $(\mathbf{T}, \sigma)$-redex of $\mathbf{t}_i$, finding a rule $\mathbf{u} \to \mathbf{v}$ in $\mathbf{T}$ and a substitution θ such that $\mathbf{t}_i$ matches $\mathbf{u} \to \mathbf{v}$ with θ; and
- applying θ to $\mathbf{v}$.

It is clear that finding and applying substitutions are fairly simple operations. (For a more detailed treatment see the unification algorithm in Chapter 13.) Therefore, we must look more closely at rewriting strategies and sets of rewrite rules. We begin with four commonly defined rewriting strategies.

Definition. Let **T** be a **W**-term rewriting system. A **W**-term is **T**-*rewritable* if it matches some rule in **T**. Let **t** be a **W**-term. An *innermost* occurrence of a **T**-rewritable subterm of **t** is one which has no **T**-rewritable subterms.[3] An *outermost* occurrence of a **T**-rewritable subterm of **t** is one which is not a subterm of any **T**-rewritable subterm of **t**. The *leftmost innermost* strategy, denoted σ_{li}, selects the leftmost of the innermost occurrences of **T**-rewritable subterms of **t**. If there is no such subterm then σ_{li} selects the empty set. The *parallel innermost* strategy, denoted σ_{pi}, selects the (possibly empty) set of all innermost occurrences of **T**-rewritable subterms of **t**. The *leftmost outermost* strategy, denoted σ_{lo}, selects the leftmost of the outermost occurrences of **T**-rewritable subterms of **t**, or the empty set if there is no such subterm. The *parallel outermost* strategy, denoted σ_{po}, selects the (possibly empty) set of all outermost occurrences of **T**-rewritable subterms of **t**.

Note that in each of the four strategies, the choice of $(\mathbf{T}, \sigma)$-redexes depends on the particular set **T**. Therefore, the computability of applying σ depends on the nature of **T**.

To illustrate, let **T** consist of the rewrite rules

$$\begin{aligned}
&\mathbf{F(X)} \to \mathbf{s(F(s(X)))}\\
&\mathbf{F(X)} \to \mathbf{X}\\
&\mathbf{G(0,Y)} \to \mathbf{Y}\\
&\mathbf{G(s(X),s(Y))} \to \mathbf{s(G(X,Y))}.
\end{aligned}$$

Then the underlined subterm of $\mathbf{G(s(\underline{F(0)}),s(F(0)))}$ is the $(\mathbf{T}, \sigma_{li})$-redex, and the underlined subterm of $\mathbf{G(\underline{F(G(0,0))},F(0))}$ is the $(\mathbf{T}, \sigma_{lo})$-redex.

[3] Note that we are not considering a term to be a subterm of itself.

Each of

$$\underline{\mathbf{F(2)}} \underset{\mathbf{T},\,\sigma_{li}}{\Longrightarrow} \mathbf{2}$$

$$\underline{\mathbf{F(2)}} \underset{\mathbf{T},\,\sigma_{li}}{\Longrightarrow} \mathbf{s(\underline{F(3)})} \underset{\mathbf{T},\,\sigma_{li}}{\Longrightarrow} \mathbf{s(s(\underline{F(4)}))} \underset{\mathbf{T},\,\sigma_{li}}{\Longrightarrow} \mathbf{s(s(4)) = 6}$$

$$\underline{\mathbf{F(2)}} \underset{\mathbf{T},\,\sigma_{li}}{\Longrightarrow} \mathbf{s(\underline{F(3)})} \underset{\mathbf{T},\,\sigma_{li}}{\Longrightarrow} \mathbf{s(s(\underline{F(4)}))} \underset{\mathbf{T},\,\sigma_{li}}{\Longrightarrow} \mathbf{s(s(s(\underline{F(5)})))} \underset{\mathbf{T},\,\sigma_{li}}{\Longrightarrow} \cdots$$

is a $(\mathbf{T}, \sigma_{li})$-computation for $\mathbf{F(2)}$, where the underlined terms are the $(\mathbf{T}, \sigma_{li})$-redexes, and

$$\begin{aligned}
\mathbf{G(\underline{F(F(0))},F(2))} &\underset{\mathbf{T},\,\sigma_{po}}{\Longrightarrow} \underline{\mathbf{G(s(F(s(F(0)))),s(F(3)))}} \\
&\underset{\mathbf{T},\,\sigma_{po}}{\Longrightarrow} \mathbf{s(G(\underline{F(s(F(0)))},\underline{F(3)}))} \\
&\underset{\mathbf{T},\,\sigma_{po}}{\Longrightarrow} \mathbf{s(\underline{G(s(F(0)),s(F(4)))})} \\
&\underset{\mathbf{T},\,\sigma_{po}}{\Longrightarrow} \mathbf{s(s(G(\underline{F(0)},\underline{F(4)})))} \\
&\underset{\mathbf{T},\,\sigma_{po}}{\Longrightarrow} \mathbf{s(s(\underline{G(0,s(F(5)))}))} \\
&\underset{\mathbf{T},\,\sigma_{po}}{\Longrightarrow} \mathbf{s(s(s(\underline{F(5)})))} \\
&\underset{\mathbf{T},\,\sigma_{po}}{\Longrightarrow} \mathbf{s(s(s(5))) = 8}
\end{aligned}$$

is a $(\mathbf{T}, \sigma_{po})$-computation for $\mathbf{G(F(F(0)),F(2))}$, where the underlined terms
are the $(\mathbf{T}, \sigma_{po})$-redexes.

It is clear from the example that computations for a given term **t** are not necessarily unique, which would make them unsuitable for the definition of functions. However, we can associate with each **W**-program a deterministic **W**-term rewriting system, which does give unique computations.

Definition. Let **P** be a **W**-program. The **W**-*term rewriting system associated with* **P** *for simple data structure systems*, denoted $\mathbf{T}_s(\mathbf{P})$, consists of

$$\mathbf{F(X_1,\ldots,X_n) \to u} \quad \text{for each equation } \mathbf{F(X_1,\ldots,X_n) = u} \text{ in } \mathbf{P},$$

together with

- for each $\tau \in TV(\mathbf{W})$,

$$\mathbf{if_\tau(tt,X,Y) \to X}$$
$$\mathbf{if_\tau(ff,X,Y) \to Y}$$

(The choice of particular individual variables **X, Y** is unimportant, as long as they are distinct and of the appropriate type.)

- for each constant symbol $\mathbf{c} \in \mathbf{W}_c^-$, with $\tau(\mathbf{c}) = \boldsymbol{\tau}$,

$$\mathbf{is\_c(c)} \to \mathbf{tt}$$
$$\mathbf{is\_c(t)} \to \mathbf{ff} \qquad \text{for each } \mathbf{t} \in \mathrm{TM}^{\boldsymbol{\tau}}_{\mathbf{W}_c} - \{\mathbf{c}\}$$

- for each proper function symbol $\mathbf{f} \in \mathbf{W}_c$, with $\mathrm{ar}(\mathbf{f}) = n$ and $\rho(\mathbf{f}) = \boldsymbol{\tau}$,

$$\mathbf{is\_f(f(t_1, \ldots, t_\mathit{n}))} \to \mathbf{tt}$$
$$\text{for each } \mathbf{f(t_1, \ldots, t_\mathit{n})} \in \mathrm{TM}^{\boldsymbol{\tau}}_{\mathbf{W}_c}$$
$$\mathbf{is\_f(u)} \to \mathbf{ff}$$
$$\text{for each } \mathbf{u} \in \mathrm{TM}^{\boldsymbol{\tau}}_{\mathbf{W}_c} \text{ not of the form } \mathbf{f(t_1, \ldots, t_\mathit{n})}$$
$$\mathbf{f}_i^{-1}\mathbf{(f(t_1, \ldots, t_\mathit{n}))} \to \mathbf{t}_i$$
$$\text{for each } \mathbf{f(t_1, \ldots, t_\mathit{n})} \in \mathrm{TM}^{\boldsymbol{\tau}}_{\mathbf{W}_c}$$

It is clear that for any **W**-program **P**, $\mathbf{T}_s(\mathbf{P})$ is deterministic, so $\mathbf{T}_s(\mathbf{P})$ gives a unique $(\mathbf{T}_s(\mathbf{P}), \sigma)$-computation for any **W**-term **t**, where σ is any rewriting strategy. It is also easy to see that for any **W**-term **t**, the process of finding a matching rewrite rule in $\mathbf{T}_s(\mathbf{P})$, if one exists, is straightforward. It follows that finding the set of $(\mathbf{T}_s(\mathbf{P}), \sigma)$-redexes for **t**, where σ is any of the strategies just defined, is also straightforward. Moreover, the test that a term is $(\mathbf{T}_s(\mathbf{P}), \sigma)$-normal is easy, so we have four reasonable notions of a computation.

To illustrate, $\mathbf{T}_s(\mathbf{ADD})$ has the rules

$$\mathbf{+(X, Y)} \to \mathbf{if_N(is\_0(Y), X, s(+(X, s_1^{-1}(Y))))}$$
$$\mathbf{if_{Bool}(tt, X, Y)} \to \mathbf{X}$$
$$\mathbf{if_{Bool}(ff, X, Y)} \to \mathbf{Y}$$
$$\mathbf{if_N(tt, X, Y)} \to \mathbf{X}$$
$$\mathbf{if_N(ff, X, Y)} \to \mathbf{Y}$$
$$\mathbf{is\_0(0)} \to \mathbf{tt}$$
$$\mathbf{is\_0(s(n))} \to \mathbf{ff} \qquad \text{for all } n \in N$$
$$\mathbf{is\_s(s(n))} \to \mathbf{tt} \qquad \text{for all } n \in N$$
$$\mathbf{is\_s(0)} \to \mathbf{ff}$$
$$\mathbf{s_1^{-1}(s(n))} \to \mathbf{n} \qquad \text{for all } n \in N,$$

and

$$
\begin{aligned}
&\underline{+(3,2)}\\
&\xrightarrow[\mathrm{T}_s(\mathrm{ADD}),\,\sigma_{lo}]{} \mathrm{if}_{\mathrm{N}}(\underline{\mathrm{is\_0}(2)}, 3, \mathrm{s}(+(3, \mathrm{s}_1^{-1}(2))))\\
&\xrightarrow[\mathrm{T}_s(\mathrm{ADD}),\,\sigma_{lo}]{} \underline{\mathrm{if}_{\mathrm{N}}(\mathrm{ff}, 3, \mathrm{s}(+(3, \mathrm{s}_1^{-1}(2))))}\\
&\xrightarrow[\mathrm{T}_s(\mathrm{ADD}),\,\sigma_{lo}]{} \mathrm{s}(\underline{+(3, \mathrm{s}_1^{-1}(2))})\\
&\xrightarrow[\mathrm{T}_s(\mathrm{ADD}),\,\sigma_{lo}]{} \mathrm{s}(\mathrm{if}_{\mathrm{N}}(\mathrm{is\_0}(\underline{\mathrm{s}_1^{-1}(2)}), 3, \mathrm{s}(+(3, \mathrm{s}_1^{-1}(\mathrm{s}_1^{-1}(2))))))\\
&\xrightarrow[\mathrm{T}_s(\mathrm{ADD}),\,\sigma_{lo}]{} \mathrm{s}(\mathrm{if}_{\mathrm{N}}(\underline{\mathrm{is\_0}(1)}, 3, \mathrm{s}(+(3, \mathrm{s}_1^{-1}(\mathrm{s}_1^{-1}(2))))))\\
&\xrightarrow[\mathrm{T}_s(\mathrm{ADD}),\,\sigma_{lo}]{} \mathrm{s}(\underline{\mathrm{if}_{\mathrm{N}}(\mathrm{ff}, 3, \mathrm{s}(+(3, \mathrm{s}_1^{-1}(\mathrm{s}_1^{-1}(2)))))})\\
&\xrightarrow[\mathrm{T}_s(\mathrm{ADD}),\,\sigma_{lo}]{} \mathrm{s}(\mathrm{s}(\underline{+(3, \mathrm{s}_1^{-1}(\mathrm{s}_1^{-1}(2)))}))\\
&\xrightarrow[\mathrm{T}_s(\mathrm{ADD}),\,\sigma_{lo}]{} \mathrm{s}(\mathrm{s}(\mathrm{if}_{\mathrm{N}}(\mathrm{is\_0}(\mathrm{s}_1^{-1}(\underline{\mathrm{s}_1^{-1}(2)})), 3, \mathrm{s}(+(3, \mathrm{s}_1^{-1}(\mathrm{s}_1^{-1}(\mathrm{s}_1^{-1}(2)))))))))\\
&\xrightarrow[\mathrm{T}_s(\mathrm{ADD}),\,\sigma_{lo}]{} \mathrm{s}(\mathrm{s}(\mathrm{if}_{\mathrm{N}}(\mathrm{is\_0}(\underline{\mathrm{s}_1^{-1}(1)}), 3, \mathrm{s}(+(3, \mathrm{s}_1^{-1}(\mathrm{s}_1^{-1}(\mathrm{s}_1^{-1}(2)))))))))\\
&\xrightarrow[\mathrm{T}_s(\mathrm{ADD}),\,\sigma_{lo}]{} \mathrm{s}(\mathrm{s}(\mathrm{if}_{\mathrm{N}}(\underline{\mathrm{is\_0}(0)}, 3, \mathrm{s}(+(3, \mathrm{s}_1^{-1}(\mathrm{s}_1^{-1}(\mathrm{s}_1^{-1}(2)))))))))\\
&\xrightarrow[\mathrm{T}_s(\mathrm{ADD}),\,\sigma_{lo}]{} \mathrm{s}(\mathrm{s}(\underline{\mathrm{if}_{\mathrm{N}}(\mathrm{tt}, 3, \mathrm{s}(+(3, \mathrm{s}_1^{-1}(\mathrm{s}_1^{-1}(\mathrm{s}_1^{-1}(2)))))))}))\\
&\xrightarrow[\mathrm{T}_s(\mathrm{ADD}),\,\sigma_{lo}]{} \mathrm{s}(\mathrm{s}(3)) = 5
\end{aligned}
$$

is the $(\mathbf{T}_s(\mathbf{ADD}), \sigma_{lo})$-computation for $+(3,2)$.

Now, in this example, the leftmost outermost strategy gives us exactly what we want with respect to Δ_{N}, computing $\mathbf{5}$ from $+(3,2)$. The leftmost innermost strategy, on the other hand, is a different story. Consider the simple program

$$\mathbf{P} = \{\mathbf{G}(\mathbf{X}) = \mathbf{G}(\mathbf{X}), \mathbf{H}(\mathbf{X}) = \mathbf{3}\}.$$

For $\mathbf{G}(\mathbf{0})$ we get the infinite $(\mathbf{T}_s(\mathbf{P}), \sigma_{li})$-computation

$$\underline{\mathbf{G}(\mathbf{0})} \xrightarrow[\mathrm{T}_s(\mathrm{P}),\,\sigma_{li}]{} \underline{\mathbf{G}(\mathbf{0})} \xrightarrow[\mathrm{T}_s(\mathrm{P}),\,\sigma_{li}]{} \cdots,$$

which is entirely appropriate since $\mathscr{D}_{\Delta_{\mathrm{N}}}(\mathbf{P})(\mathbf{G})$ is the everywhere undefined function and $\mathscr{D}_{\Delta_{\mathrm{N}}}(\mathbf{P})(\mathbf{G})(0) = \bot_N$. However, σ_{li} also gives the infinite computation

$$\mathbf{H}(\underline{\mathbf{G}(\mathbf{0})}) \xrightarrow[\mathrm{T}_s(\mathrm{P}),\,\sigma_{li}]{} \mathbf{H}(\underline{\mathbf{G}(\mathbf{0})}) \xrightarrow[\mathrm{T}_s(\mathrm{P}),\,\sigma_{li}]{} \mathbf{H}(\underline{\mathbf{G}(\mathbf{0})}) \xrightarrow[\mathrm{T}_s(\mathrm{P}),\,\sigma_{li}]{} \cdots,$$

which is not what we want, with respect to $\Delta_{\mathbf{N}}$, since

$$\begin{aligned} D_{\Delta_{\mathbf{N}}}(\mathbf{P})(\mathbf{H})(\perp_N) &= \mu\Phi_{\mathbf{P}}(\mathbf{H})(\perp_N) \\ &= \sqcup\left(\{\Phi_{\mathbf{P}}^{i+1}(\Omega)(\mathbf{H})(\perp_N) \mid i \in N\} \cup \{\Phi_{\mathbf{P}}^{0}(\Omega)(\mathbf{H})(\perp_N)\}\right) \\ &= \sqcup\left(\left\{\overline{\Phi_{\mathbf{P}}^{i}(\Omega) \cup \alpha_{\perp_N}}(3) \mid i \in N\right\} \cup \{\perp_N\}\right) \\ &= 3. \end{aligned}$$

The problem is that the nonstrict function assigned to **H** by $\mathscr{D}_{\Delta_{\mathbf{N}}}(\mathbf{P})$ can completely *ignore* its input and produce an output value, but the leftmost (or parallel) innermost strategy requires the computation to try forever to compute **G(0)**. On the other hand, the leftmost (or parallel) outermost strategy gives the finite computation

$$\underline{\mathbf{H}(\mathbf{G}(0))} \xrightarrow[\mathbf{T}_s(\mathbf{P}),\ \sigma_{lo}]{} 3,$$

which is exactly what we want.

The point is that innermost strategies may be fine in a context where all functions are strict, but they are not successful in general. For our purposes they are not appropriate since, even if we interpret all constructor function symbols with strict functions, the interpretations of the $\mathbf{if}_\tau$ symbols are necessarily not strict. This is quite sensible, since we do not want $\mathscr{I}(\mathbf{if}_\tau)(b, d, e)$ to depend on both d and e, but only on (at most) one of d or e, according to the value of b. We choose a strategy which is neither purely innermost nor outermost, but which is closer in spirit to an outermost strategy, since it does not depend on completing the computation of innermost subterms. It will be convenient to define it only for deterministic **W**-term rewriting systems.

Definition. Let **T** be a deterministic **W**-term rewriting system, and let **t** be a **W**-term. In the *full rewriting strategy*, denoted σ_f, the $(\mathbf{T}, \sigma_f)$*-rewrite*[4] of **t**

[4] Technically, the definition of $(\mathbf{T}, \sigma_f)$-rewrites varies somewhat from the general definition of $(\mathbf{T}, \sigma)$-rewrites given earlier, since we replace subterms $\boldsymbol{\phi}(\mathbf{t}_1, \ldots, \mathbf{t}_n)$ by the **T**-rewrite of $\boldsymbol{\phi}(\mathrm{rr}_\mathbf{T}(\mathbf{t}_1), \ldots, \mathrm{rr}_\mathbf{T}(\mathbf{t}_n))$, i.e., $\mathrm{r}_\mathbf{T}(\boldsymbol{\phi}(\mathrm{rr}_\mathbf{T}(\mathbf{t}_1), \ldots, \mathrm{rr}_\mathbf{T}(\mathbf{t}_n)))$, rather than by the **T**-rewrite of $\boldsymbol{\phi}(\mathbf{t}_1, \ldots, \mathbf{t}_n)$. The difference is of no concern, however, and the definition of $(\mathbf{T}, \sigma_f)$-computations, which depends only on $(\mathbf{T}, \sigma_f)$-rewrites and $(\mathbf{T}, \sigma_f)$-normality, conforms to the general definition of $(\mathbf{T}, \sigma)$-computations.

is $\mathbf{rr_T(t)}$, where $\mathbf{rr_T(t)}$ is defined in two stages:

$$\mathbf{r_T}(\boldsymbol{\phi}(\mathbf{t}_1,\ldots,\mathbf{t}_n)) = \begin{cases} \mathbf{u}\theta & \text{if } \boldsymbol{\phi}(\mathbf{t}_1,\ldots,\mathbf{t}_n) \text{ matches a rule} \\ & \boldsymbol{\phi}(\mathbf{u}_1,\ldots,\mathbf{u}_n) \to \mathbf{u} \text{ in } \mathbf{T} \text{ with} \\ & \text{substitution } \theta \\ \boldsymbol{\phi}(\mathbf{t}_1,\ldots,\mathbf{t}_n) & \text{otherwise} \end{cases}$$

$$\mathbf{rr_T(c)} = \mathbf{c} \quad \text{for all constant symbols } \mathbf{c} \in \mathbf{W}$$

$$\mathbf{rr_T}(\boldsymbol{\phi}(\mathbf{t}_1,\ldots,\mathbf{t}_n)) = \mathbf{r_T}(\boldsymbol{\phi}(\mathbf{rr_T}(\mathbf{t}_1),\ldots,\mathbf{rr_T}(\mathbf{t}_n))).$$

The **W**-term **t** is $(\mathbf{T}, \sigma_f)$-*normal* if there is no subterm $\boldsymbol{\phi}(\mathbf{t}_1,\ldots,\mathbf{t}_n)$ of **t** such that $\boldsymbol{\phi}(\mathbf{rr_T}(\mathbf{t}_1),\ldots,\mathbf{rr_T}(\mathbf{t}_n))$ matches a rewrite rule in **T**.

Throughout this section we will write $\mathbf{r_P}$ and $\mathbf{rr_P}$ for $\mathbf{r}_{\mathbf{T}_s(\mathbf{P})}$ and $\mathbf{rr}_{\mathbf{T}_s(\mathbf{P})}$, respectively. Just as with σ_{li}, σ_{pi}, σ_{lo}, and σ_{po}, the computability of applying σ_f depends on the rewriting system **T**, and $\mathbf{T}_s(\mathbf{P})$ and σ_f give us another reasonable notion of a computation. Note that if $\mathbf{t}_0, \mathbf{t}_1, \mathbf{t}_2, \ldots$ is an infinite $(\mathbf{T}_s(\mathbf{P}), \sigma_f)$-computation, then $\mathbf{rr}^i_\mathbf{P}(\mathbf{t}_0) = \mathbf{t}_i$ for all $i \in N$. In other words, $\mathbf{rr}^0_\mathbf{P}(\mathbf{t}_0), \mathbf{rr}^1_\mathbf{P}(\mathbf{t}_0), \mathbf{rr}^2_\mathbf{P}(\mathbf{t}_0), \ldots$ is by definition the $(\mathbf{T}_s(\mathbf{P}), \sigma_f)$-computation for $\mathbf{t}_0$ when the computation is infinite. If $\mathbf{t}_0, \mathbf{t}_1, \ldots, \mathbf{t}_n$ is a finite $(\mathbf{T}_s(\mathbf{P}), \sigma_f)$-computation, then $\mathbf{rr}^i_\mathbf{P}(\mathbf{t}_0) = \mathbf{t}_i$ for $1 \le i \le n$, and $\mathbf{rr}^i_\mathbf{P}(\mathbf{t}_0) = \mathbf{t}_n$ for $i > n$. That is, $\mathbf{rr_P(t)} = \mathbf{t}$ if **t** is $(\mathbf{T}_s(\mathbf{P}), \sigma_f)$-normal.

It is clear that for any term $\mathbf{f}(\mathbf{t}_1,\ldots,\mathbf{t}_n)$, where **f** is a constructor symbol,

$$\mathbf{rr_P}(\mathbf{f}(\mathbf{t}_1,\ldots,\mathbf{t}_n)) = \mathbf{r_P}(\mathbf{f}(\mathbf{rr_P}(\mathbf{t}_1),\ldots,\mathbf{rr_P}(\mathbf{t}_n))) = \mathbf{f}(\mathbf{rr_P}(\mathbf{t}_1),\ldots,\mathbf{rr_P}(\mathbf{t}_n)),$$

since there are no rewrite rules in $\mathbf{T}_s(\mathbf{P})$ for **f**. In $\mathbf{W_N}$, for example, we have $\mathbf{rr_P(s(t))} = \mathbf{s(rr_P(t))}$, and, in particular, $\mathbf{rr_P(n)} = \mathbf{n}$ for any numeral **n**. To illustrate σ_f we give the $(\mathbf{T}_s(\mathbf{ADD}), \sigma_f)$-computation for $\mathbf{+(3,2)}$:

$$\begin{aligned}
\mathbf{+(3,2)} \underset{\mathbf{T}_s(\mathbf{ADD}),\,\sigma_f}{\Longrightarrow} & \mathbf{rr_P(+(3,2))} \\
&= \mathbf{r_P(+(rr_P(3), rr_P(2)))} \\
&= \mathbf{r_P(+(3,2))} \\
&= \mathbf{if_N(is_0(2), 3, s(+(3, s_1^{-1}(2))))} \\
\underset{\mathbf{T}_s(\mathbf{ADD}),\,\sigma_f}{\Longrightarrow} & \mathbf{rr_P(if_N(is_0(2), 3, s(+(3, s_1^{-1}(2)))))} \\
&= \mathbf{r_P(if_N(rr_P(is_0(2)), rr_P(3), rr_P(s(+(3, s_1^{-1}(2))))))} \\
&= \mathbf{r_P(if_N(r_P(is_0(rr_P(2))), 3, r_P(s(rr_P(+(3, s_1^{-1}(2)))))))} \\
&= \mathbf{r_P(if_N(r_P(is_0(2)), 3, r_P(s(r_P(+(rr_P(3), rr_P(s_1^{-1}(2))))))))} \\
&= \mathbf{r_P(if_N(ff, 3, r_P(s(r_P(+(3, r_P(s_1^{-1}(rr_P(2))))))))}
\end{aligned}$$

$$= \mathbf{r_P}(\mathbf{if_N}(\mathbf{ff}, 3, \mathbf{r_P}(\mathbf{s}(\mathbf{r_P}(+(3, \mathbf{r_P}(\mathbf{s_1^{-1}}(2))))))))$$
$$= \mathbf{r_P}(\mathbf{if_N}(\mathbf{ff}, 3, \mathbf{r_P}(\mathbf{s}(\mathbf{r_P}(+(3, 1))))))$$
$$= \mathbf{r_P}(\mathbf{if_N}(\mathbf{ff}, 3, \mathbf{r_P}(\mathbf{s}(\mathbf{if_N}(\mathbf{is\_0}(1), 3, \mathbf{s}(+(3, \mathbf{s_1^{-1}}(1))))))))$$
$$= \mathbf{r_P}(\mathbf{if_N}(\mathbf{ff}, 3, \mathbf{s}(\mathbf{if_N}(\mathbf{is\_0}(1), 3, \mathbf{s}(+(3, \mathbf{s_1^{-1}}(1)))))))$$
$$= \mathbf{s}(\mathbf{if_N}(\mathbf{is\_0}(1), 3, \mathbf{s}(+(3, \mathbf{s_1^{-1}}(1)))))$$

$$\xRightarrow[\mathbf{T}_s(\mathbf{ADD}),\ \sigma_f]{} \mathbf{rr_P}(\mathbf{s}(\mathbf{if_N}(\mathbf{is\_0}(1), 3, \mathbf{s}(+(3, \mathbf{s_1^{-1}}(1))))))$$
$$= \cdots = \mathbf{s}(\mathbf{s}(\mathbf{if_N}(\mathbf{is\_0}(0), 3, \mathbf{s}(+(3, \mathbf{s_1^{-1}}(0))))))$$

$$\xRightarrow[\mathbf{T}_s(\mathbf{ADD}),\ \sigma_f]{} \mathbf{rr_P}(\mathbf{s}(\mathbf{s}(\mathbf{if_N}(\mathbf{is\_0}(0), 3, \mathbf{s}(+(3, \mathbf{s_1^{-1}}(0)))))))$$
$$= \cdots = \mathbf{5}.$$

We use $(\mathbf{T}_s(\mathbf{P}), \sigma_f)$-computations to define the operational semantics of **W**-programs with respect to simple data structure systems. In particular, we use the final terms in finite computations to determine the functions defined by **P**. Moreover, the value of these terms should be independent of the denotational semantics of **P**, so we use the least informative variable assignment, Ω, to interpret them.

Definition. Let $\Delta = \mathrm{rep}(\Sigma)$ be a simple data structure system for **W**. The *operational meaning function* for Δ, denoted $\mathscr{O}_\Delta$, is defined as follows. For all **W**-programs $\mathbf{P}_0$, all $\mathbf{F} \in FV(\mathbf{P}_0)$, and all $(d_1, \ldots, d_n) \in D_{r(\delta(\mathbf{F}))}$,

$$\mathscr{O}_\Delta(\mathbf{P}_0)(\mathbf{F})(d_1, \ldots, d_n) = \begin{cases} \overline{\Omega}(\mathbf{t}) & \text{if the } (\mathbf{T}_s(\mathbf{P}), \sigma_f)\text{-computation for} \\ & \mathbf{F}(\mathbf{t}_1, \ldots, \mathbf{t}_n) \text{ is finite and ends with } \mathbf{t} \\ \bot_{\rho(\mathbf{F})} & \text{otherwise,} \end{cases}$$

where

- $d_i = \overline{\mu\Phi^{\Sigma}_{\mathbf{P}_i}}(\mathbf{t}_i)$, $1 \le i \le n$,
- $\mathbf{P}_0, \mathbf{P}_1, \ldots, \mathbf{P}_n$ are consistent, and
- $\mathbf{P} = \bigcup_{i=0}^{n} \mathbf{P}_i$.

The idea is that we can compute $\mathscr{O}_\Delta(\mathbf{P}_0)(\mathbf{F})(d_1, \ldots, d_n)$ by extending program $\mathbf{P}_0$ with programs $\mathbf{P}_1, \ldots, \mathbf{P}_n$ and then carrying out the $(\mathbf{T}_s(\bigcup_{i=0}^{n} \mathbf{P}_i), \sigma_f)$-computation for $\mathbf{F}(\mathbf{t}_1, \ldots, \mathbf{t}_n)$. In $\Delta_\mathbf{N}$, of course, $n = \overline{\Omega}(\mathbf{n})$ for all $n \in N$, so to compute a function on $(m_1, \ldots, m_n) \in N^n$, we can simply let $\mathbf{P}_i = \varnothing$, $1 \le i \le n$. Moreover, to represent $\bot_N$ we can just include the equation $\mathbf{B}(\mathbf{X}) = \mathbf{B}(\mathbf{X})$.

Note, however, that in general there are many different choices of programs $\mathbf{P}_1, \ldots, \mathbf{P}_n$ and terms $\mathbf{t}_1, \ldots, \mathbf{t}_n$ which characterize the same tuple

$(d_1, \ldots, d_n)$, and these might give different computations. It is not obvious, then, that the definition of $\mathscr{O}_\Delta$ makes sense. We need to know that we get the same *result* in all cases, even if the computations differ. Theorem 1.1 later in this section, which shows that the operational and denotational semantics are equivalent for simple data structure systems, implies that we do.

We should note that the purist might object that the operational semantics of **W**-programs is not really independent of the denotational semantics since the initial term $\mathbf{F}(\mathbf{t}_1, \ldots, \mathbf{t}_n)$ and the program **P** depend on the condition $d_i = \overline{\mu\Phi^{\Sigma}_{\mathbf{P}_i}(\mathbf{t}_i)}$, $1 \le i \le n$. Indeed, we could give an alternative operational semantics in which the input to a program **P** is simply a sequence such as $((\mathbf{P}_1, \mathbf{t}_1), \ldots, (\mathbf{P}_n, \mathbf{t}_n))$ and the output is the final term in the $(\mathbf{T}_s(\bigcup_{i=0}^{n} \mathbf{P}_i), \sigma_f)$-computation for $\mathbf{F}(\mathbf{t}_1, \ldots, \mathbf{t}_n)$. Of course, Theorem 1.1 would need to be reformulated in a suitable way. For our purposes, however, the important thing is that there is *some* term $\mathbf{F}(\mathbf{t}_1, \ldots, \mathbf{t}_n)$ from which the correct value of $\mathscr{D}_\Delta(\mathbf{P}_0)(\mathbf{F})(d_1, \ldots, d_n)$ can be computed. That fact is sufficient to justify calling the function $\mathscr{D}_\Delta(\mathbf{P}_0)(\mathbf{F})$ *computable.*

We now turn to the proof of Theorem 1.1, beginning with four lemmas. In Lemma 1 we finally apply condition 2c on $\mathscr{T}$-interpretations. Lemma 3, which is proved by an induction based on Lemma 2, is the heart of the argument. It shows that the terms of a computation (interpreted by Ω) correctly approximate the value of the function being computed. Lemma 4 guarantees that, if a function has a non-bottom value for some given input, then the computation for that input will eventually terminate.

Lemma 1. Let $\Sigma = (\mathscr{T}, \mathscr{I})$ be a **W**-structure and let α be any variable assignment based on $\mathscr{T}$. Then for any term $\mathbf{t} \in \mathrm{TM}_{\mathbf{W}_c}$, $\bar{\alpha}(\mathbf{t}) \ne \bot_{\tau(\mathbf{t})}$.

Proof. We argue by structural induction on **t**. If **t** is a constant symbol $\mathbf{c} \in \mathbf{W}_c$, then $\bar{\alpha}(\mathbf{c}) = \mathscr{I}(\mathbf{c}) \in D_{\tau(\mathbf{c})} - \{\bot_{\tau(\mathbf{c})}\}$ by condition 1 on $\mathscr{T}$-interpretations. Otherwise, **t** is of the form $\mathbf{f}(\mathbf{t}_1, \ldots, \mathbf{t}_n)$, where $\mathbf{f} \in \mathbf{W}_c$. Then $\bar{\alpha}(\mathbf{f}(\mathbf{t}_1, \ldots, \mathbf{t}_n)) = \mathscr{I}(\mathbf{f})(\bar{\alpha}(\mathbf{t}_1), \ldots, \bar{\alpha}(\mathbf{t}_n)) \ne \bot_{\tau(\mathbf{t})}$ by the induction hypothesis and condition 2c on $\mathscr{T}$-interpretations. ■

Lemma 2. Let $\Sigma = (\mathscr{T}, \mathscr{I})$ be a complete, continuous **W**-structure, and let **P** be a **W**-program. Then for all $\mathbf{t} \in \mathrm{TM}_{\mathbf{W}}(FV(\mathbf{P}))$ and for all $i \in N$, $\overline{\Phi^{i+1}_{\mathbf{P}}(\Omega)}(\mathbf{t}) = \overline{\Phi^{i}_{\mathbf{P}}(\Omega)}(\mathrm{rr}_{\mathbf{P}}(\mathbf{t}))$.

Proof. Let $i \in N$. We argue by structural induction on **t**. If **t** is a constant symbol $\mathbf{c} \in \mathbf{W}$, then

$$\overline{\Phi^{i+1}_{\mathbf{P}}(\Omega)}(\mathbf{c}) = \mathscr{I}(\mathbf{c}) = \mathscr{I}(\mathrm{rr}_{\mathbf{P}}(\mathbf{c})) = \overline{\Phi^{i}_{\mathbf{P}}(\Omega)}(\mathrm{rr}_{\mathbf{P}}(\mathbf{c})).$$

If $\mathbf{t}$ is of the form $\mathbf{f}(\mathbf{t}_1,\ldots,\mathbf{t}_n)$, where $\mathbf{f}\in\mathbf{W}$, then $\mathrm{rr}_{\mathbf{P}}(\mathbf{f}(\mathbf{t}_1,\ldots,\mathbf{t}_n)) = \mathrm{r}_{\mathbf{P}}(\mathbf{f}(\mathrm{rr}_{\mathbf{P}}(\mathbf{t}_1),\ldots,\mathrm{rr}_{\mathbf{P}}(\mathbf{t}_n)))$. If $\mathbf{f}(\mathrm{rr}_{\mathbf{P}}(\mathbf{t}_1),\ldots,\mathrm{rr}_{\mathbf{P}}(\mathbf{t}_n))$ does not match any rewrite rule in $\mathbf{T}_s(\mathbf{P})$, then $\mathrm{rr}_{\mathbf{P}}(\mathbf{f}(\mathbf{t}_1,\ldots,\mathbf{t}_n)) = \mathbf{f}(\mathrm{rr}_{\mathbf{P}}(\mathbf{t}_1),\ldots,\mathrm{rr}_{\mathbf{P}}(\mathbf{t}_n))$, and

$$\overline{\Phi_{\mathbf{P}}^{i+1}(\Omega)}(\mathbf{f}(\mathbf{t}_1,\ldots,\mathbf{t}_n))$$

$$=\mathscr{I}(\mathbf{f})(\overline{\Phi_{\mathbf{P}}^{i+1}(\Omega)}(\mathbf{t}_1),\ldots,\overline{\Phi_{\mathbf{P}}^{i+1}(\Omega)}(\mathbf{t}_n))$$

$$=\mathscr{I}(\mathbf{f})(\overline{\Phi_{\mathbf{P}}^{i}(\Omega)}(\mathrm{rr}_{\mathbf{P}}(\mathbf{t}_1)),\ldots,\overline{\Phi_{\mathbf{P}}^{i}(\Omega)}(\mathrm{rr}_{\mathbf{P}}(\mathbf{t}_n))) \qquad \text{by the induction hypothesis}$$

$$=\overline{\Phi_{\mathbf{P}}^{i}(\Omega)}(\mathbf{f}(\mathrm{rr}_{\mathbf{P}}(\mathbf{t}_1),\ldots,\mathrm{rr}_{\mathbf{P}}(\mathbf{t}_n)))$$

$$=\overline{\Phi_{\mathbf{P}}^{i}(\Omega)}(\mathrm{rr}_{\mathbf{P}}(\mathbf{f}(\mathbf{t}_1,\ldots,\mathbf{t}_n))).$$

Otherwise, $\mathbf{f}(\mathrm{rr}_{\mathbf{P}}(\mathbf{t}_1),\ldots,\mathrm{rr}_{\mathbf{P}}(\mathbf{t}_n))$ does match some rewrite rule in $\mathbf{T}_s(\mathbf{P})$. Suppose $\mathbf{t}$ is $\mathbf{if}_\tau(\mathbf{u},\mathbf{v},\mathbf{w})$. If $\mathrm{rr}_{\mathbf{P}}(\mathbf{u}) = \mathbf{tt}$ then

$$\mathrm{rr}_{\mathbf{P}}(\mathbf{if}_\tau(\mathbf{u},\mathbf{v},\mathbf{w})) = \mathrm{r}_{\mathbf{P}}(\mathbf{if}_\tau(\mathrm{rr}_{\mathbf{P}}(\mathbf{u}),\mathrm{rr}_{\mathbf{P}}(\mathbf{v}),\mathrm{rr}_{\mathbf{P}}(\mathbf{w}))) = \mathrm{rr}_{\mathbf{P}}(\mathbf{v}),$$

and

$$\overline{\Phi_{\mathbf{P}}^{i+1}(\Omega)}(\mathbf{if}_\tau(\mathbf{u},\mathbf{v},\mathbf{w}))$$

$$=\mathscr{I}(\mathbf{if}_\tau)(\overline{\Phi_{\mathbf{P}}^{i+1}(\Omega)}(\mathbf{u}),\overline{\Phi_{\mathbf{P}}^{i+1}(\Omega)}(\mathbf{v}),\overline{\Phi_{\mathbf{P}}^{i+1}(\Omega)}(\mathbf{w}))$$

$$=\mathscr{I}(\mathbf{if}_\tau)(\overline{\Phi_{\mathbf{P}}^{i}(\Omega)}(\mathrm{rr}_{\mathbf{P}}(\mathbf{u})),\overline{\Phi_{\mathbf{P}}^{i}(\Omega)}(\mathrm{rr}_{\mathbf{P}}(\mathbf{v})),\overline{\Phi_{\mathbf{P}}^{i}(\Omega)}(\mathrm{rr}_{\mathbf{P}}(\mathbf{w})))$$

by the induction hypothesis

$$=\mathscr{I}(\mathbf{if}_\tau)(\mathrm{tt},\overline{\Phi_{\mathbf{P}}^{i}(\Omega)}(\mathrm{rr}_{\mathbf{P}}(\mathbf{v})),\overline{\Phi_{\mathbf{P}}^{i}(\Omega)}(\mathrm{rr}_{\mathbf{P}}(\mathbf{w})))$$

$$=\overline{\Phi_{\mathbf{P}}^{i}(\Omega)}(\mathrm{rr}_{\mathbf{P}}(\mathbf{v}))$$

$$=\overline{\Phi_{\mathbf{P}}^{i}(\Omega)}(\mathrm{rr}_{\mathbf{P}}(\mathbf{if}_\tau(\mathbf{u},\mathbf{v},\mathbf{w}))).$$

Similarly, if $\mathrm{rr}_{\mathbf{P}}(\mathbf{u}) = \mathbf{ff}$ then

$$\overline{\Phi_{\mathbf{P}}^{i+1}(\Omega)}(\mathbf{if}_\tau(\mathbf{u},\mathbf{v},\mathbf{w})) = \overline{\Phi_{\mathbf{P}}^{i}(\Omega)}(\mathrm{rr}_{\mathbf{P}}(\mathbf{w})) = \overline{\Phi_{\mathbf{P}}^{i}(\Omega)}(\mathrm{rr}_{\mathbf{P}}(\mathbf{if}_\tau(\mathbf{u},\mathbf{v},\mathbf{w}))).$$

Next, suppose, $\mathbf{t}$ is $\mathbf{is\_c}(\mathbf{u})$ for some constant symbol $\mathbf{c}\in\mathbf{W}_c^-$. If $\mathrm{rr}_{\mathbf{P}}(\mathbf{u}) = \mathbf{c}$ then

$$\mathrm{rr}_{\mathbf{P}}(\mathbf{is\_c}(\mathbf{u})) = \mathrm{r}_{\mathbf{P}}(\mathbf{is\_c}(\mathrm{rr}_{\mathbf{P}}(\mathbf{u}))) = \mathrm{r}_{\mathbf{P}}(\mathbf{is\_c}(\mathbf{c})) = \mathbf{tt},$$

and

$$\overline{\Phi_{\mathbf{P}}^{i+1}(\Omega)}(\mathbf{is\_c}(\mathbf{u}))$$

$$= \mathscr{I}(\mathbf{is\_c})(\overline{\Phi_{\mathbf{P}}^{i+1}(\Omega)}(\mathbf{u}))$$

$$= \mathscr{I}(\mathbf{is\_c})(\overline{\Phi_{\mathbf{P}}^{i}(\Omega)}(\mathrm{rr}_{\mathbf{P}}(\mathbf{u}))) \qquad \text{by the induction hypothesis}$$

$$= \mathscr{I}(\mathbf{is\_c})(\mathscr{I}(\mathbf{c}))$$

$$= \mathrm{tt} \qquad \text{by conditions 1 and 5 on } \mathscr{T}\text{-interpretations}$$

$$= \overline{\Phi_{\mathbf{P}}^{i}(\Omega)}(\mathrm{rr}_{\mathbf{P}}(\mathbf{is\_c}(\mathbf{u}))).$$

Otherwise, $\mathrm{rr}_{\mathbf{P}}(\mathbf{u}) = \mathbf{g}$ or $\mathbf{g}(\mathbf{u}_1, \ldots, \mathbf{u}_m) \in \mathrm{TM}_{\mathbf{W}_c}$ for some constant symbol or proper function symbol $\mathbf{g}$ distinct from $\mathbf{c}$, so that

$$\mathrm{rr}_{\mathbf{P}}(\mathbf{is\_c}(\mathbf{u})) = \mathrm{r}_{\mathbf{P}}(\mathbf{is\_c}(\mathrm{rr}_{\mathbf{P}}(\mathbf{u}))) = \mathbf{ff}.$$

Suppose $\mathrm{rr}_{\mathbf{P}}(\mathbf{c}) = \mathbf{g}$. Then

$$\overline{\Phi_{\mathbf{P}}^{i+1}(\Omega)}(\mathbf{is\_c}(\mathbf{u}))$$

$$= \mathscr{I}(\mathbf{is\_c})(\overline{\Phi_{\mathbf{P}}^{i+1}(\Omega)}(\mathbf{u}))$$

$$= \mathscr{I}(\mathbf{is\_c})(\overline{\Phi_{\mathbf{P}}^{i}(\Omega)}(\mathrm{rr}_{\mathbf{P}}(\mathbf{u}))) \qquad \text{by the induction hypothesis}$$

$$= \mathscr{I}(\mathbf{is\_c})(\mathscr{I}(\mathbf{g}))$$

$$= \mathrm{ff} \qquad \text{by conditions 1, 3, and 5 on } \mathscr{T}\text{-interpretations}$$

$$= \overline{\Phi_{\mathbf{P}}^{i}(\Omega)}(\mathrm{rr}_{\mathbf{P}}(\mathbf{is\_c}(\mathbf{u}))).$$

Similarly, if $\mathrm{rr}_{\mathbf{P}}(\mathbf{u}) = \mathbf{g}(\mathbf{u}_1, \ldots, \mathbf{u}_m)$ then

$$\overline{\Phi_{\mathbf{P}}^{i+1}(\Omega)}(\mathbf{is\_c}(\mathbf{u}))$$

$$= \mathscr{I}(\mathbf{is\_c})(\overline{\Phi_{\mathbf{P}}^{i+1}(\Omega)}(\mathbf{u}))$$

$$= \mathscr{I}(\mathbf{is\_c})(\overline{\Phi_{\mathbf{P}}^{i}(\Omega)}(\mathrm{rr}_{\mathbf{P}}(\mathbf{u}))) \qquad \text{by the induction hypothesis}$$

$$= \mathscr{I}(\mathbf{is\_c})(\mathscr{I}(\mathbf{g})(\overline{\Phi_{\mathbf{P}}^{i}(\Omega)}(\mathbf{u}_1), \ldots, \overline{\Phi_{\mathbf{P}}^{i}(\Omega)}(\mathbf{u}_m)))$$

$$= \mathrm{ff} \qquad \text{by Lemma 1 and conditions 2c, 3, and 5 on } \mathscr{T}\text{-interpretations}$$

$$= \overline{\Phi_{\mathbf{P}}^{i}(\Omega)}(\mathrm{rr}_{\mathbf{P}}(\mathbf{is\_c}(\mathbf{u}))).$$

The argument is nearly the same if $\mathbf{t}$ is $\mathbf{is\_f(u)}$, where $\mathbf{f}$ is a proper function symbol in $\mathbf{W}_c$.

Now suppose $\mathbf{t}$ is $\mathbf{f}_j^{-1}(\mathbf{u})$ and $\mathrm{rr}_{\mathbf{P}}(\mathbf{u}) = \mathbf{f}(\mathbf{t}_1, \ldots, \mathbf{t}_n) \in \mathrm{TM}_{\mathbf{W}_c}$. Then

$$\mathrm{rr}_{\mathbf{P}}\big(\mathbf{f}_j^{-1}(\mathbf{u})\big) = \mathrm{r}_{\mathbf{P}}\big(\mathbf{f}_j^{-1}(\mathrm{rr}_{\mathbf{P}}(\mathbf{u}))\big) = \mathbf{t}_j,$$

and

$$\overline{\Phi_{\mathbf{P}}^{i+1}(\Omega)}(\mathbf{f}_j^{-1}(\mathbf{u}))$$

$$= \mathscr{I}(\mathbf{f}_j^{-1})(\overline{\Phi_{\mathbf{P}}^{i+1}(\Omega)}(\mathbf{u}))$$

$$= \mathscr{I}(\mathbf{f}_j^{-1})(\overline{\Phi_{\mathbf{P}}^{i}(\Omega)}(\mathrm{rr}_{\mathbf{P}}(\mathbf{u}))) \qquad \text{by the induction hypothesis}$$

$$= \mathscr{I}(\mathbf{f}_j^{-1})(\mathscr{I}(\mathbf{f})(\overline{\Phi_{\mathbf{P}}^{i}(\Omega)}(\mathbf{t}_1), \ldots, \overline{\Phi_{\mathbf{P}}^{i}(\Omega)}(\mathbf{t}_n)))$$

$$= \overline{\Phi_{\mathbf{P}}^{i}(\Omega)}(\mathbf{t}_j) \qquad \text{by Lemma 1 and conditions 2c and 6 on } \mathscr{T}\text{-interpretations}$$

$$= \overline{\Phi_{\mathbf{P}}^{i}(\Omega)}(\mathrm{rr}_{\mathbf{P}}(\mathbf{f}_j^{-1}(\mathbf{u}))).$$

Finally, suppose $\mathbf{t}$ is $\mathbf{F}(\mathbf{t}_1, \ldots, \mathbf{t}_n)$, where $\mathbf{F} \in FV(\mathbf{P})$, and let $\mathbf{F}(\mathbf{X}_1, \ldots, \mathbf{X}_n) = \mathbf{u}$ be the defining equation for $\mathbf{F}$ in $\mathbf{P}$. Then

$$\mathrm{rr}_{\mathbf{P}}(\mathbf{F}(\mathbf{t}_1, \ldots, \mathbf{t}_n)) = \mathrm{r}_{\mathbf{P}}(\mathbf{F}(\mathrm{rr}_{\mathbf{P}}(\mathbf{t}_1), \ldots, \mathrm{rr}_{\mathbf{P}}(\mathbf{t}_n))) = \mathbf{u}\theta,$$

where $\mathbf{X}_j\theta = \mathrm{rr}_{\mathbf{P}}(\mathbf{t}_j)$, $1 \le j \le n$, and

$$\overline{\Phi_{\mathbf{P}}^{i+1}(\Omega)}(\mathbf{F}(\mathbf{t}_1, \ldots, \mathbf{t}_n))$$

$$= \Phi_{\mathbf{P}}^{i+1}(\Omega)(\mathbf{F})(\overline{\Phi_{\mathbf{P}}^{i+1}(\Omega)}(\mathbf{t}_1), \ldots, \overline{\Phi_{\mathbf{P}}^{i+1}(\Omega)}(\mathbf{t}_n))$$

$$= \Phi_{\mathbf{P}}^{i+1}(\Omega)(\mathbf{F})(\overline{\Phi_{\mathbf{P}}^{i}(\Omega)}(\mathrm{rr}_{\mathbf{P}}(\mathbf{t}_1)), \ldots, \overline{\Phi_{\mathbf{P}}^{i}(\Omega)}(\mathrm{rr}_{\mathbf{P}}(\mathbf{t}_n)))$$

by the induction hypothesis

$$= \overline{\Phi_{\mathbf{P}}^{i}(\Omega) \cup \alpha}(\mathbf{u}) \qquad \text{where} \quad \alpha(\mathbf{X}_j) = \overline{\Phi_{\mathbf{P}}^{i}(\Omega)}\big(\mathrm{rr}_{\mathbf{P}}(\mathbf{t}_j)\big),\ 1 \le j \le n$$

$$= \overline{\Phi_{\mathbf{P}}^{i}(\Omega)}(\mathbf{u}\theta) \qquad \text{by the substitution lemma}$$

$$= \overline{\Phi_{\mathbf{P}}^{i}(\Omega)}(\mathrm{rr}_{\mathbf{P}}(\mathbf{F}(\mathbf{t}_1, \ldots, \mathbf{t}_n))).$$

■

Lemma 3. Let $\Sigma = (\mathcal{T}, \mathcal{I})$ be a complete, continuous **W**-structure, and let **P** be a **W**-program. Then for all $\mathbf{t} \in TM_{\mathbf{W}}(FV(\mathbf{P}))$ and for $i \in N$, $\overline{\Phi^i_{\mathbf{P}}(\Omega)}(\mathbf{t}) = \overline{\Omega}(\mathrm{rr}^i_{\mathbf{P}}(\mathbf{t}))$.

Proof. We argue by induction on i. If $i = 0$ then for all $\mathbf{t} \in TM_{\mathbf{W}}(FV(\mathbf{P}))$,

$$\overline{\Phi^0_{\mathbf{P}}(\Omega)}(\mathbf{t}) = \overline{\Omega}(\mathbf{t}) = \overline{\Omega}(\mathrm{rr}^0_{\mathbf{P}}(\mathbf{t})),$$

so assume the lemma is true for $i = k$. Then for all $\mathbf{t} \in TM_{\mathbf{W}}(FV(\mathbf{P}))$,

$$\begin{aligned}
\overline{\Phi^{k+1}_{\mathbf{P}}(\Omega)}(\mathbf{t}) &= \overline{\Phi^k_{\mathbf{P}}(\Omega)}(\mathrm{rr}_{\mathbf{P}}(\mathbf{t})) && \text{by Lemma 2} \\
&= \overline{\Omega}(\mathrm{rr}^k_{\mathbf{P}}(\mathrm{rr}_{\mathbf{P}}(\mathbf{t}))) && \text{by the induction hypothesis} \\
&= \overline{\Omega}(\mathrm{rr}^{k+1}_{\mathbf{P}}(\mathbf{t})). && \text{(See Exercise 12.)}
\end{aligned}$$

■

Lemma 4. Let $\Sigma = (\mathcal{T}, \mathcal{I})$ be a simple **W**-structure, let **P** be a **W**-program, and let $\mathbf{t} \in TM_{\mathbf{W}}(FV(\mathbf{P}))$. If $\overline{\Omega}(\mathbf{t}) \neq \perp_{\tau(\mathbf{t})}$, then $\overline{\Omega}(\mathrm{rr}_{\mathbf{P}}(\mathbf{t})) = \overline{\Omega}(\mathbf{t})$ and $\mathrm{rr}_{\mathbf{P}}(\mathbf{t}) \in TM_{\mathbf{W}_c}$.

Proof. Let $\tau(\mathbf{t}) = \boldsymbol{\tau}$. We argue by structural induction on **t**, assuming throughout that $\overline{\Omega}(\mathbf{t}) \neq \perp_{\boldsymbol{\tau}}$. If **t** is a constant symbol $\mathbf{c} \in \mathbf{W}$, then $\mathrm{rr}_{\mathbf{P}}(\mathbf{c}) = \mathbf{c} \in TM_{\mathbf{W}_c}$. If **t** is $\mathbf{f}(\mathbf{t}_1, \ldots, \mathbf{t}_n)$, where $\mathbf{f} \in \mathbf{W}_c$ and $\tau(\mathbf{f}) = \boldsymbol{\tau}_1 \times \cdots \times \boldsymbol{\tau}_n \rightarrow \boldsymbol{\tau}$, then $\overline{\Omega}(\mathbf{t}_i) \neq \perp_{\boldsymbol{\tau}_i}$, $1 \leq i \leq n$, by the strictness of $\mathcal{I}(\mathbf{f})$, so $\overline{\Omega}(\mathrm{rr}_{\mathbf{P}}(\mathbf{t}_i)) = \overline{\Omega}(\mathbf{t}_i)$ and $\mathrm{rr}_{\mathbf{P}}(\mathbf{t}_i) \in TM_{\mathbf{W}_c}$ by the induction hypothesis, $1 \leq i \leq n$, and we have

$$\begin{aligned}
\overline{\Omega}(\mathbf{f}(\mathbf{t}_1, \ldots, \mathbf{t}_n)) &= \mathcal{I}(\mathbf{f})(\overline{\Omega}(\mathbf{t}_1), \ldots, \overline{\Omega}(\mathbf{t}_n)) \\
&= \mathcal{I}(\mathbf{f})(\overline{\Omega}(\mathrm{rr}_{\mathbf{P}}(\mathbf{t}_1)), \ldots, \overline{\Omega}(\mathrm{rr}_{\mathbf{P}}(\mathbf{t}_n))) \\
&= \overline{\Omega}(\mathbf{f}(\mathrm{rr}_{\mathbf{P}}(\mathbf{t}_1), \ldots, \mathrm{rr}_{\mathbf{P}}(\mathbf{t}_n))) \\
&= \overline{\Omega}(\mathrm{rr}_{\mathbf{P}}(\mathbf{f}(\mathbf{t}_1, \ldots, \mathbf{t}_n))),
\end{aligned}$$

where $\mathbf{f}(\mathbf{t}_1, \ldots, \mathbf{t}_n) \in TM_{\mathbf{W}_c}$.

Suppose **t** is $\mathbf{if}_{\boldsymbol{\tau}}(\mathbf{u}, \mathbf{v}, \mathbf{w})$. Then $\overline{\Omega}(\mathbf{u}) \neq \perp_{\mathrm{Bool}}$, so $\overline{\Omega}(\mathrm{rr}_{\mathbf{P}}(\mathbf{u})) = \overline{\Omega}(\mathbf{u})$ and $\mathrm{rr}_{\mathbf{P}}(\mathbf{u}) \in TM_{\mathbf{W}}$ by the induction hypothesis. If $\overline{\Omega}(\mathbf{u}) = tt$ then $\mathrm{rr}_{\mathbf{P}}(\mathbf{u})$ must be **tt** by condition 3 on $\mathcal{T}$-interpretations, and $\mathrm{rr}_{\mathbf{P}}(\mathbf{if}_{\boldsymbol{\tau}}(\mathbf{u}, \mathbf{v}, \mathbf{w})) = \mathrm{rr}_{\mathbf{P}}(\mathbf{v})$.

Therefore,

$$\begin{aligned}
&\overline{\Omega}(\mathbf{if}_\tau(\mathbf{u},\mathbf{v},\mathbf{w}))\\
&\quad = \mathscr{I}(\mathbf{if}_\tau)(\overline{\Omega}(\mathbf{u}), \overline{\Omega}(\mathbf{v}), \overline{\Omega}(\mathbf{w}))\\
&\quad = \overline{\Omega}(\mathbf{v})\\
&\quad = \overline{\Omega}(\mathrm{rr}_\mathbf{P}(\mathbf{v})) \qquad \text{by the induction hypothesis, since}\\
&\qquad\qquad\qquad\qquad \overline{\Omega}(\mathbf{v}) = \overline{\Omega}(\mathbf{if}_\tau(\mathbf{u},\mathbf{v},\mathbf{w})) \neq \bot_\tau\\
&\quad = \overline{\Omega}(\mathrm{rr}_\mathbf{P}(\mathbf{if}_\tau(\mathbf{u},\mathbf{v},\mathbf{w}))),
\end{aligned}$$

and by the induction hypothesis, $\mathrm{rr}_\mathbf{P}(\mathbf{if}_\tau(\mathbf{u},\mathbf{v},\mathbf{w})) = \mathrm{rr}_\mathbf{P}(\mathbf{v}) \in \mathrm{TM}_{\mathbf{W}_c}$. If $\overline{\Omega}(\mathbf{u}) = \mathrm{ff}$ then we get $\overline{\Omega}(\mathbf{if}_\tau(\mathbf{u},\mathbf{v},\mathbf{w})) = \overline{\Omega}(\mathrm{rr}_\mathbf{P}(\mathbf{if}_\tau(\mathbf{u},\mathbf{v},\mathbf{w})))$ and $\mathrm{rr}_\mathbf{P}(\mathbf{if}_\tau(\mathbf{u},\mathbf{v},\mathbf{w})) = \mathrm{rr}_\mathbf{P}(\mathbf{w}) \in \mathrm{TM}_{\mathbf{W}_c}$ by a similar argument.

Next, suppose $\mathbf{t}$ is $\mathbf{is\_f}(\mathbf{u})$ for some $\mathbf{f} \in \mathbf{W}_c$ with $\tau(\mathbf{f}) = \tau_1 \times \cdots \times \tau_n \to \tau$. If $\overline{\Omega}(\mathbf{is\_f}(\mathbf{u})) = \mathrm{tt}$, then $\bot_\tau \neq \overline{\Omega}(\mathbf{u}) \in \operatorname{ran} \mathscr{I}(\mathbf{f})$, so $\overline{\Omega}(\mathrm{rr}_\mathbf{P}(\mathbf{u})) = \overline{\Omega}(\mathbf{u})$ and $\mathrm{rr}_\mathbf{P}(\mathbf{u}) \in \mathrm{TM}_{\mathbf{W}_c}$ by the induction hypothesis. Moreover, $\overline{\Omega}(\mathrm{rr}_\mathbf{P}(\mathbf{u})) \in \operatorname{ran} \mathscr{I}(\mathbf{f})$ implies $\mathrm{rr}_\mathbf{P}(\mathbf{u})$ must be of the form $\mathbf{f}(\mathbf{t}_1,\ldots,\mathbf{t}_n)$ by condition 3 on $\mathscr{T}$-interpretations, so $\mathrm{rr}_\mathbf{P}(\mathbf{is\_f}(\mathbf{u})) = \mathbf{tt} \in \mathrm{TM}_{\mathbf{W}_c}$ and $\overline{\Omega}(\mathrm{rr}_\mathbf{P}(\mathbf{is\_f}(\mathbf{u}))) = \mathrm{tt} = \overline{\Omega}(\mathrm{is\_}\mathbf{f}(\mathbf{u}))$. If $\overline{\Omega}(\mathbf{is\_f}(\mathbf{u})) = \mathrm{ff}$, then $\bot_\tau \neq \overline{\Omega}(\mathbf{u}) \notin \operatorname{ran} \mathscr{I}(\mathbf{f})$, so $\overline{\Omega}(\mathrm{rr}_\mathbf{P}(\mathbf{u})) = \overline{\Omega}(\mathbf{u})$ and $\mathrm{rr}_\mathbf{P}(\mathbf{u}) \in \mathrm{TM}_{\mathbf{W}_c}$ by the induction hypothesis. Moreover, $\overline{\Omega}(\mathrm{rr}_\mathbf{P}(\mathbf{u})) \notin \operatorname{ran} \mathscr{I}(\mathbf{f})$ implies that $\mathrm{rr}_\mathbf{P}(\mathbf{u})$ cannot be of the form $\mathbf{f}(\mathbf{t}_1,\ldots,\mathbf{t}_n)$, so $\mathrm{rr}_\mathbf{P}(\mathbf{is\_f}(\mathbf{u})) = \mathbf{ff} \in \mathrm{TM}_{\mathbf{W}_c}$ and $\overline{\Omega}(\mathrm{rr}_\mathbf{P}(\mathbf{is\_f}(\mathbf{u}))) = \mathrm{ff} = \overline{\Omega}(\mathbf{is\_f}(\mathbf{u}))$. The argument is similar if $\mathbf{t}$ is $\mathbf{is\_c}(\mathbf{u})$ for some constant symbol $\mathbf{c} \in \mathbf{W}_c^-$.

Now suppose $\mathbf{t}$ is $\mathbf{f}_i^{-1}(\mathbf{u})$, where $\tau(\mathbf{f}) = \tau_1 \times \cdots \times \tau_n \to \tau$. Then $\bot_\tau \neq \overline{\Omega}(\mathbf{u}) \in \operatorname{ran} \mathscr{I}(\mathbf{f})$, so $\overline{\Omega}(\mathrm{rr}_\mathbf{P}(\mathbf{u})) = \overline{\Omega}(\mathbf{u})$ and $\mathrm{rr}_\mathbf{P}(\mathbf{u}) \in \mathrm{TM}_{\mathbf{W}_c}$ by the induction hypothesis. Again, $\mathrm{rr}_\mathbf{P}(\mathbf{u})$ must be of the form $\mathbf{f}(\mathbf{t}_1,\ldots,\mathbf{t}_n)$, so $\mathrm{rr}_\mathbf{P}(\mathbf{f}_i^{-1}(\mathbf{u})) = \mathbf{t}_i \in \mathrm{TM}_{\mathbf{W}_c}$, and

$$\begin{aligned}
\overline{\Omega}(\mathbf{f}_i^{-1}(\mathbf{u})) &= \mathscr{I}(\mathbf{f}_i^{-1})(\overline{\Omega}(\mathbf{u}))\\
&= \mathscr{I}(\mathbf{f}_i^{-1})(\overline{\Omega}(\mathrm{rr}_\mathbf{P}(\mathbf{u})))\\
&= \mathscr{I}(\mathbf{f}_i^{-1})(\overline{\Omega}(\mathbf{f}(\mathbf{t}_1,\ldots,\mathbf{t}_n)))\\
&= \mathscr{I}(\mathbf{f}_i^{-1})(\mathscr{I}(\mathbf{f})(\overline{\Omega}(\mathbf{t}_1),\ldots,\overline{\Omega}(\mathbf{t}_n)))\\
&= \overline{\Omega}(\mathbf{t}_i) \quad \text{since } \mathscr{I}(\mathbf{f})(\overline{\Omega}(\mathbf{t}_1),\ldots,\overline{\Omega}(\mathbf{t}_n)) \neq \bot_\tau\\
&= \overline{\Omega}(\mathrm{rr}_\mathbf{P}(\mathbf{f}_i^{-1}(\mathbf{u}))).
\end{aligned}$$

Finally, if $\mathbf{t}$ is of the form $\mathbf{F}(\mathbf{t}_1,\ldots,\mathbf{t}_n)$, where $\mathbf{F} \in FV(\mathbf{P})$, then $\overline{\Omega}(\mathbf{F}(\mathbf{t}_1,\ldots,\mathbf{t}_n)) = \bot_\tau$ and there is nothing to prove in this case. ■

Theorem 1.1. Let $\Delta = \text{rep}(\Sigma)$ be a simple data structure system for **W**. Then $\mathscr{O}_\Delta = \mathscr{D}_\Delta$.

Proof. Let $\mathbf{P}_0$ be a **W**-program, let $\mathbf{F} \in FV(\mathbf{P}_0)$, and let $(d_1, \ldots, d_n) \in D_{r(\delta(\mathbf{F}))}$. For $1 \le i \le n$ let $d_i = \overline{\mu\Phi_{\mathbf{P}_i}}(\mathbf{t}_i)$ for some **W**-program $\mathbf{P}_i$ and some $\mathbf{t}_i \in \text{TM}_\mathbf{W}(FV(\mathbf{P}_i))$ such that $\mathbf{P}_0, \mathbf{P}_1, \ldots, \mathbf{P}_n$ are consistent, and let $\mathbf{P} = \bigcup_{i=0}^n \mathbf{P}_i$. Then

$$\begin{aligned}
&\mathscr{D}_\Delta(\mathbf{P}_0)(\mathbf{F})(d_1, \ldots, d_n)\\
&\quad = \mathscr{D}_\Delta(\mathbf{P}_0)(\mathbf{F})(\overline{\mu\Phi_{\mathbf{P}_1}}(\mathbf{t}_1), \ldots, \overline{\mu\Phi_{\mathbf{P}_n}}(\mathbf{t}_n))\\
&\quad = \mu\Phi_{\mathbf{P}_0}(\mathbf{F})(\overline{\mu\Phi_{\mathbf{P}_1}}(\mathbf{t}_1), \ldots, \overline{\mu\Phi_{\mathbf{P}_n}}(\mathbf{t}_n))\\
&\quad = \mu\Phi_{\mathbf{P}}(\mathbf{F})(\overline{\mu\Phi_{\mathbf{P}}}(\mathbf{t}_1), \ldots, \overline{\mu\Phi_{\mathbf{P}}}(\mathbf{t}_n)) \qquad \text{by the extension lemma}\\
&\quad = \overline{\mu\Phi_{\mathbf{P}}}(\mathbf{F}(\mathbf{t}_1, \ldots, \mathbf{t}_n))\\
&\quad = \overline{\sqcup\{\Phi_{\mathbf{P}}^i(\Omega) \mid i \in N\}}(\mathbf{F}(\mathbf{t}_1, \ldots, \mathbf{t}_n))\\
&\quad = \sqcup\left\{\overline{\Phi_{\mathbf{P}}^i(\Omega)}(\mathbf{F}(\mathbf{t}_1, \ldots, \mathbf{t}_n)) \mid i \in N\right\} \qquad \text{by Theorem 17.2.3.}
\end{aligned}$$

By Theorem 17.2.3 the set $\{\overline{\Phi_{\mathbf{P}}^i(\Omega)}(\mathbf{F}(\mathbf{t}_1, \ldots, \mathbf{t}_n)) \mid i \in N\}$ is a chain in the flat cpo $(D_{\rho(\mathbf{F})}, \sqsubseteq_{\rho(\mathbf{F})})$, so if $\sqcup\{\overline{\Phi_{\mathbf{P}}^i(\Omega)}(\mathbf{F}(\mathbf{t}_1, \ldots, \mathbf{t}_n)) \mid i \in N\} \neq \bot_{\rho(\mathbf{F})}$, then $\sqcup\{\overline{\Phi_{\mathbf{P}}^i(\Omega)}(\mathbf{F}(\mathbf{t}_1, \ldots, \mathbf{t}_n)) \mid i \in N\} = \overline{\Phi_{\mathbf{P}}^{i_0}(\Omega)}(\mathbf{F}(\mathbf{t}_1, \ldots, \mathbf{t}_n))$ for some smallest $i_0 \in N$. Now,

$$\begin{aligned}
\overline{\Phi_{\mathbf{P}}^{i_0}(\Omega)}(\mathbf{F}(\mathbf{t}_1, \ldots, \mathbf{t}_n)) &= \overline{\Omega}(\text{rr}_{\mathbf{P}}^{i_0}(\mathbf{F}(\mathbf{t}_1, \ldots, \mathbf{t}_n))) \qquad \text{by Lemma 3}\\
&= \overline{\Omega}(\text{rr}_{\mathbf{P}}^{i_0+1}(\mathbf{F}(\mathbf{t}_1, \ldots, \mathbf{t}_n))) \qquad \text{by Lemma 4,}
\end{aligned}$$

where $\text{rr}_{\mathbf{P}}^{i_0+1}(\mathbf{F}(\mathbf{t}_1, \ldots, \mathbf{t}_n)) \in \text{TM}_{\mathbf{W}_c}$, so that $\text{rr}_{\mathbf{P}}^{i_0+1}(\mathbf{F}(\mathbf{t}_1, \ldots, \mathbf{t}_n))$ is a $(\mathbf{T}_s(\mathbf{P}), \sigma_f)$-normal term, and

$$\begin{aligned}
\mathscr{D}_\Delta(\mathbf{P}_0)(\mathbf{F})(d_1, \ldots, d_n) &= \overline{\Omega}(\text{rr}_{\mathbf{P}}^{i_0+1}(\mathbf{F}(\mathbf{t}_1, \ldots, \mathbf{t}_n)))\\
&= \mathscr{O}_\Delta(\mathbf{P}_0)(\mathbf{F})(d_1, \ldots, d_n).
\end{aligned}$$

Otherwise, $\sqcup\{\overline{\Phi_{\mathbf{P}}^i(\Omega)}(\mathbf{F}(\mathbf{t}_1, \ldots, \mathbf{t}_n)) \mid i \in N\} = \bot_{\rho(\mathbf{F})}$, and

$$\sqcup\left\{\overline{\Phi_{\mathbf{P}}^i(\Omega)}(\mathbf{F}(\mathbf{t}_1, \ldots, \mathbf{t}_n)) \mid i \in N\right\} = \sqcup\left\{\overline{\Omega}(\text{rr}_{\mathbf{P}}^i(\mathbf{F}(\mathbf{t}_1, \ldots, \mathbf{t}_n))) \mid i \in N\right\}$$

by Lemma 3, so either the $(\mathbf{T}_s(\mathbf{P}), \sigma_f)$-computation for $\mathbf{F}(\mathbf{t}_1, \ldots, \mathbf{t}_n)$ is infinite or it ends with a $(\mathbf{T}_s(\mathbf{P}), \sigma_f)$-normal term $\text{rr}_{\mathbf{P}}^{i_0}(\mathbf{F}(\mathbf{t}_1, \ldots, \mathbf{t}_n))$ such that $\overline{\Omega}(\mathrm{rr}_{\mathbf{P}}^{i_0}(\mathbf{F}(\mathbf{t}_1, \ldots, \mathbf{t}_n))) = \bot_{\rho(\mathbf{F})}$, and in either case

$$\mathscr{D}_\Delta(\mathbf{P}_0)(\mathbf{F})(d_1, \ldots, d_n) = \bot_{\rho(\mathbf{F})} = \mathscr{O}_\Delta(\mathbf{P}_0)(\mathbf{F})(d_1, \ldots, d_n).$$

The choice of $\mathbf{P}_0$, $\mathbf{F}$, and $(d_1, \ldots, d_n)$ was arbitrary, so $\mathscr{D}_\Delta = \mathscr{O}_\Delta$. ■

Thus $\mathscr{O}_\Delta$ is correct with respect to $\mathscr{D}_\Delta$ for every simple data structure system Δ. Moreover, there is no ambiguity in the definition of $\mathscr{D}_\Delta$, so Theorem 1.1 implies that the definition of $\mathscr{O}_\Delta$ is independent of the choice of programs and terms used to denote values in Δ. Theorem 1.1 also justifies the following

Definition. Let Δ be a simple data structure system for **W**. A function f is Δ-*computable* if there is a **W**-program **P** and $\mathbf{F} \in FV(\mathbf{P})$ such that $f = \mathscr{D}_\Delta(\mathbf{P})(\mathbf{F})$.

It follows, then, from our work in the previous chapter that $+_\bot$ is a Δ_{N}-computable function. We have also seen some examples of Δ_{NL}-computable functions. In the next section we will examine the Δ_{N}-computable functions more closely.

Exercises

1. Let $\theta = \{(\mathbf{X}_1, \mathbf{F}(\mathbf{0})), (\mathbf{X}_3, \mathbf{s}(\mathbf{X}_2))\}$. What is $\mathbf{G}(\mathbf{s}(\mathbf{X}_2), \mathbf{X}_3)\theta$?
2. Let α be a variable assignment such that $\alpha(\mathbf{X}) = 1$, $\alpha(\mathbf{Y}) = 2$. Give a substitution θ such that $\overline{\mu\Phi_{\mathbf{ADD}} \cup \alpha}(+(\mathbf{s}(\mathbf{X}), \mathbf{Y})) = \overline{\mu\Phi_{\mathbf{ADD}}}(+(\mathbf{s}(\mathbf{X}), \mathbf{Y})\theta)$ in Σ_{N}.
3. We have left open the possibility of a **W**-term rewriting system **T**, a strategy σ, and a **W**-term **t** such that **t** is not $(\mathbf{T}, \sigma)$-normal but there is no $(\mathbf{T}, \sigma)$-rewrite of **t**. Verify that this situation does not occur for $\mathbf{T}_s(\mathbf{P})$ and σ, where **P** is any **W**-program and σ is any of the five strategies we have defined.
4. Let **T** be the $\mathbf{W}_{\mathrm{N}}$-term rewriting system with rewrite rules
$$\mathbf{F}(\mathbf{X}, \mathbf{Y}) \to \mathbf{s}(\mathbf{Y})$$
$$\mathbf{F}(\mathbf{X}, \mathbf{Y}) \to \mathbf{F}(\mathbf{Y}, \mathbf{Y}),$$
and let $\mathbf{t} = \mathbf{F}(\mathbf{F}(\mathbf{2}, \mathbf{3}), \mathbf{F}(\mathbf{4}, \mathbf{5}))$. Give two distinct $(\mathbf{T}, \sigma)$-computations for **t**, where σ is (a) σ_{li}; (b) σ_{lo}; (c) σ_{pi}; (d) σ_{po}.
5. Give the $(\mathbf{T}_s(\mathbf{ADD}), \sigma)$-computation for $+(\mathbf{3}, \mathbf{2})$, where σ is (a) σ_{li}; (b) σ_{pi}; (c) σ_{po}.
6. Let **P** be a **W**-program and let $\mathbf{t} \in \mathrm{TM}_{\mathbf{W}}(FV(\mathbf{P}))$. Show that for all $i \in N$, $\mathrm{rr}_{\mathbf{P}}^i(\mathbf{t}) \in \mathrm{TM}_{\mathbf{W}}(FV(\mathbf{P}))$.

7. Let **P** be the $\mathbf{W_N}$-program with the equation

$$\mathbf{F(X) = if_N(is\_0(X), 1, s(F(s_1^{-1}(X))))}.$$

(a) Describe $\mathbf{T}_s(\mathbf{P})$.
(b) Give the $(\mathbf{T}_s(\mathbf{P}), \sigma_f)$-computation for $\mathbf{F(2)}$.

8. Let $\mathbf{P} = \mathbf{ADD} \cup \{\mathbf{B(X) = B(X)}\}$, where $\tau(\mathbf{B}) = \mathbf{N \to N}$.
(a) Give the $(\mathbf{T}_s(\mathbf{P}), \sigma_f)$-computation for $\mathbf{+(3, s(s(B(0))))}$.
(b) Give the $(\mathbf{T}_s(\mathbf{P}), \sigma_f)$-computation for $\mathbf{+(s(s(s(B(0)))), 2)}$.

9. Without using Theorem 1.1, give the value of each of the following.
(a) $\mathscr{O}_{\Delta_N}(\mathbf{ADD})(\mathbf{+})(3, 2)$.
(b) $\mathscr{O}_{\Delta_N}(\mathbf{ADD})(\mathbf{+})(\perp_N, 2)$.
(c) $\mathscr{O}_{\Delta_N}(\mathbf{ADD})(\mathbf{+})(3, \perp_N)$.

10. Let **t** be the $\mathbf{W_{NL}}$-term $\mathbf{cons(1, cons(2, nil))}$. Describe $\mathbf{T}_s(\mathbf{LIST})$, and give the $(\mathbf{T}_s(\mathbf{LIST}), \sigma_f)$-computation for each of the following $\mathbf{W_{NL}}$-terms. [**LIST** is defined in Section 5 of Chapter 17.]
(a) $\mathbf{Length(t)}$.
(b) $\mathbf{Nth(1, t)}$.
(c) $\mathbf{Cat(t, t)}$.
(d) $\mathbf{Rev(t)}$.

11. Without using Theorem 1.1, give the value of each of the following.
(a) $\mathscr{O}_{\Delta_{NL}}(\mathbf{LIST})(\mathbf{Length})(\langle 2, 3\rangle)$.
(b) $\mathscr{O}_{\Delta_{NL}}(\mathbf{LIST})(\mathbf{Length})(\perp_{NL})$.
(c) $\mathscr{O}_{\Delta_{NL}}(\mathbf{LIST})(\mathbf{Nth})(2, \langle 2, 3, 4\rangle)$.
(d) $\mathscr{O}_{\Delta_{NL}}(\mathbf{LIST})(\mathbf{Nth})(\perp_{NL}, \langle 2, 3, 4\rangle)$.
(e) $\mathscr{O}_{\Delta_{NL}}(\mathbf{LIST})(\mathbf{Cat})(\langle 2, 3\rangle, \langle 4, 5\rangle)$.
(f) $\mathscr{O}_{\Delta_{NL}}(\mathbf{LIST})(\mathbf{Cat})(\perp_{NL}, \langle 4, 5\rangle)$.
(g) $\mathscr{O}_{\Delta_{NL}}(\mathbf{LIST})(\mathbf{Cat})(\langle 2, 3\rangle, \perp_{NL})$.
(h) $\mathscr{O}_{\Delta_{NL}}(\mathbf{LIST})(\mathbf{Rev})(\langle 2, 3\rangle)$.

12. Let D be a set, let $f \in D \to D$, and let $d \in D$. Show that for all $n \in N$, $f^{n+1}(d) = f^n(f(d))$.

2. Computable Functions

Now that we have a new class of computable numeric functions, it is reasonable to compare it to the partially computable functions defined in Part 1 of the book. We will show that they are essentially the same, just as we showed that a function is partially computable if and only if it is computable by Turing machines or Post–Turing programs. The difference here is that, technically, the two classes contain different kinds of functions, since in this chapter we have defined computable functions on $N_{\perp}^{n}$, $n \geq 1$, rather than partially computable functions on N^n. However, this distinction is easily overcome.

We begin with the primitive recursive functions.

Lemma 1. If f is primitive recursive, then $f_{\perp}$ is $\Delta_{\mathbf{N}}$-computable.

Proof. We argue by induction on the number of compositions and recursions by which f is obtained from the initial functions. The $\Delta_{\mathbf{N}}$-computability of the initial functions is given by the following programs:

$$\mathbf{P}_s = \{\mathbf{S}(\mathbf{X}) = \mathbf{s}(\mathbf{X})\}$$

$$\mathbf{P}_n = \{\mathbf{N}(\mathbf{X}) = \mathbf{if}_{\mathbf{N}}(\mathbf{is\_0}(\mathbf{X}), \mathbf{0}, \mathbf{0})\}$$

$$\mathbf{P}_{u_i^n} = \{\mathbf{U}_i^n(\mathbf{X}_1, \ldots, \mathbf{X}_n) = \mathbf{if}_{\mathbf{N}}(\mathbf{Def}^n(\mathbf{X}_1, \ldots, \mathbf{X}_n), \mathbf{X}_i, \mathbf{0})\} \cup \mathbf{P}_{\mathrm{Def}^n},$$

where $\mathbf{P}_{\mathrm{Def}^1}$ is

$$\{\mathbf{Def}^1(\mathbf{X}) = \mathbf{if}_{\mathbf{Bool}}(\mathbf{is\_0}(\mathbf{X}), \mathbf{tt}, \mathbf{tt})\}$$

and, for $n \geq 1$, $\mathbf{P}_{\mathrm{Def}^{n+1}}$ is $\mathbf{P}_{\mathrm{Def}^n}$ together with the equation

$$\mathbf{Def}^{n+1}(\mathbf{X}_1, \ldots, \mathbf{X}_{n+1}) = \mathbf{if}_{\mathbf{Bool}}(\mathbf{Def}^n(\mathbf{X}_1, \ldots, \mathbf{X}_n), \mathbf{Def}^1(X_{n+1}), \mathbf{tt}).$$

Note that we include $\mathbf{Def}^i$, $1 \leq i \leq n$, to enforce the strictness of $\mu\Phi_{\mathbf{P}_{u_i^n}}(\mathbf{U}_i^n)$.

Now let

$$h(x_1, \ldots, x_n) = f(g^1(x_1, \ldots, x_n), \ldots, g^k(x_1, \ldots, x_n)),$$

where $f, g^1, \ldots, g^k$ are primitive recursive. By the induction hypothesis there are programs $\mathbf{P}_0, \mathbf{P}_1, \ldots, \mathbf{P}_k$ with function variables $\mathbf{F}, \mathbf{G}_1, \ldots, \mathbf{G}_k$ such that $f_{\perp} = \mu\Phi_{\mathbf{P}_0}(\mathbf{F})$ and $g_{\perp}^i = \mu\Phi_{\mathbf{P}_i}(\mathbf{G}_i)$, $1 \leq i \leq n$. We assume that $\mathbf{P}_0, \mathbf{P}_1, \ldots, \mathbf{P}_k$ are consistent and do not contain the function variable $\mathbf{H}$, and we set $\mathbf{P}$ to

$$\bigcup_{i=0}^{n} \mathbf{P}_i \cup \{\mathbf{H}(\mathbf{X}_1, \ldots, \mathbf{X}_n) = \mathbf{F}(\mathbf{G}_1(\mathbf{X}_1, \ldots, \mathbf{X}_n), \ldots, \mathbf{G}_k(\mathbf{X}_1, \ldots, \mathbf{X}_n))\}.$$

Then for any $(x_1, \ldots, x_n) \in N_\perp^n$ we have

$$\begin{aligned}
&\mu\Phi_{\mathbf{P}}(\mathbf{H})(x_1, \ldots, x_n) \\
&\quad = \Phi_{\mathbf{P}}(\mu\Phi_{\mathbf{P}})(\mathbf{H})(x_1, \ldots, x_n) \\
&\quad = \mu\Phi_{\mathbf{P}}(\mathbf{F})(\mu\Phi_{\mathbf{P}}(\mathbf{G}_1)(x_1, \ldots, x_n), \ldots, \mu\Phi_{\mathbf{P}}(\mathbf{G}_k)(x_1, \ldots, x_n)) \\
&\quad = \mu\Phi_{\mathbf{P}_0}(\mathbf{F})(\mu\Phi_{\mathbf{P}_1}(\mathbf{G}_1)(x_1, \ldots, x_n), \ldots, \mu\Phi_{\mathbf{P}_k}(\mathbf{G}_k)(x_1, \ldots, x_n)) \\
&\qquad\qquad \text{by the extension lemma} \\
&\quad = f_\perp(g_\perp^1(x_1, \ldots, x_n), \ldots, g_\perp^k(x_1, \ldots, x_n)).
\end{aligned}$$

If $(x_1, \ldots, x_n) \in N^n$ then

$$\begin{aligned}
&f_\perp(g_\perp^1(x_1, \ldots, x_n), \ldots, g_\perp^k(x_1, \ldots, x_n)) \\
&\quad = f(g^1(x_1, \ldots, x_n), \ldots, g^k(x_1, \ldots, x_n)) \\
&\quad = h(x_1, \ldots, x_n),
\end{aligned}$$

and if not then

$$f_\perp(g_\perp^1(x_1, \ldots, x_n), \ldots, g_\perp^k(x_1, \ldots, x_n)) = \perp_N,$$

so $\mu\Phi_{\mathbf{P}}(\mathbf{H}) = h_\perp$.

Finally, let

$$\begin{aligned}
h(x_1, \ldots, x_n, 0) &= f(x_1, \ldots, x_n) \\
h(x_1, \ldots, x_n, y + 1) &= g(y, h(x_1, \ldots, x_n, y), x_1, \ldots, x_n),
\end{aligned}$$

where g, h are primitive recursive. By the induction hypothesis there are programs $\mathbf{P}_f, \mathbf{P}_g$ with function variables $\mathbf{F}, \mathbf{G}$ such that $\mu\Phi_{\mathbf{P}_f}(\mathbf{F}) = f_\perp$ and $\mu\Phi_{\mathbf{P}_g}(\mathbf{G}) = g_\perp$. We assume that $\mathbf{P}_f, \mathbf{P}_g$ are consistent and do not contain the function variable $\mathbf{H}$, and we set $\mathbf{P}$ to $\mathbf{P}_f \cup \mathbf{P}_g \cup \{\mathbf{H}(\mathbf{X}_1, \ldots, \mathbf{X}_n, \mathbf{Y}) = \mathbf{t}\}$, where $\mathbf{t}$ is

$$\begin{aligned}
&\mathbf{if_N}(\mathbf{is_0}(\mathbf{Y}), \\
&\quad \mathbf{F}(\mathbf{X}_1, \ldots, \mathbf{X}_n), \\
&\quad \mathbf{G}(\mathbf{s}_1^{-1}(\mathbf{Y}), \mathbf{H}(\mathbf{X}_1, \ldots, \mathbf{X}_n, \mathbf{s}_1^{-1}(\mathbf{Y})), \mathbf{X}_1, \ldots, \mathbf{X}_n)).
\end{aligned}$$

Let $(x_1, \ldots, x_n) \in N_\perp^n$. It is clear that $\mu\Phi_{\mathbf{P}}(\mathbf{H})(x_1, \ldots, x_n, \perp_N) = \perp_N$, so to conclude the proof we argue by induction on y that

$\mu\Phi_{\mathbf{P}}(\mathbf{H})(x_1,\ldots,x_n,y) = h_\perp(x_1,\ldots,x_n,y)$ for all $y \in N$. If $y = 0$ then

$$\begin{aligned}\mu\Phi_{\mathbf{P}}(\mathbf{H})(x_1,\ldots,x_n,0) &= \Phi_{\mathbf{P}}(\mu\Phi_{\mathbf{P}})(\mathbf{H})(x_1,\ldots,x_n,0)\\ &= \overline{\mu\Phi_{\mathbf{P}} \cup \alpha_{(x_1,\ldots,x_n,0)}}(\mathbf{t})\\ &= \mu\Phi_{\mathbf{P}}(\mathbf{F})(x_1,\ldots,x_n)\\ &= \mu\Phi_{\mathbf{P}_f}(\mathbf{F})(x_1,\ldots,x_n)\\ &= f_\perp(x_1,\ldots,x_n)\\ &= h_\perp(x_1,\ldots,x_n,0).\end{aligned}$$

Assume, now, that $\mu\Phi_{\mathbf{P}}(\mathbf{H})(x_1,\ldots,x_n,y) = h_\perp(x_1,\ldots,x_n,y)$. Then

$$\begin{aligned}&\mu\Phi_{\mathbf{P}}(\mathbf{H})(x_1,\ldots,x_n,y+1)\\ &\quad= \overline{\mu\Phi_{\mathbf{P}} \cup \alpha_{(x_1,\ldots,x_n,y+1)}}(\mathbf{t})\\ &\quad= \mu\Phi_{\mathbf{P}}(\mathbf{G})(y, \mu\Phi_{\mathbf{P}}(\mathbf{H})(x_1,\ldots,x_n,y), x_1,\ldots,x_n)\\ &\quad= g_\perp(y, h_\perp(x_1,\ldots,x_n,y), x_1,\ldots,x_n)\\ &\quad= h_\perp(x_1,\ldots,x_n,y+1).\end{aligned}$$

■

Theorem 2.1. If f is partially computable, then $f_\perp$ is Δ_{N}-computable.

Proof. Let f be a partially computable n-ary function. By Theorem 3.3 in Chapter 4, there is a primitive recursive predicate $R(x_1,\ldots,x_n,y)$ such that

$$f(x_1,\ldots,x_n) = l(\min_z R(x_1,\ldots,x_n,z)),$$

and by Lemma 1 there are $\mathbf{W}_{\mathrm{N}}$-programs $\mathbf{P}_R, \mathbf{P}_l$ with function variables $\mathbf{R}$ and $\mathbf{L}$ such that $\mu\Phi_{\mathbf{P}_R}(\mathbf{R}) = R_\perp$ and $\mu\Phi_{\mathbf{P}_l}(\mathbf{L}) = l_\perp$. We assume that $\mathbf{P}_R, \mathbf{P}_l$ are consistent and do not include function variables $\mathbf{F}, \mathbf{G}$, and we set $\mathbf{P}$ to $\mathbf{P}_R \cup \mathbf{P}_l$ together with the equations

$$\mathbf{F}(\mathbf{X}_1,\ldots,\mathbf{X}_n) = \mathbf{L}(\mathbf{G}(\mathbf{X}_1,\ldots,\mathbf{X}_n,\mathbf{0}))$$

$$\mathbf{G}(\mathbf{X}_1,\ldots,\mathbf{X}_n,\mathbf{Y}) = \mathbf{if}_{\mathrm{N}}(\mathbf{is\_s}(R(\mathbf{X}_1,\ldots,\mathbf{X}_n,\mathbf{Y})), \mathbf{Y}, \mathbf{G}(\mathbf{X}_1,\ldots,\mathbf{X}_n, \mathrm{s}(\mathbf{Y}))).$$

Let $(x_1, \ldots, x_n) \in N_\perp^n$. It is clear that, for all $y \in N$,

$$\mu\Phi_{\mathbf{P}}(\mathbf{G})(x_1, \ldots, x_n, y)$$

$$= \begin{cases} \perp_N & \text{if } (x_1, \ldots, x_n, y) \notin N^{n+1} \\ y & \text{if } R(x_1, \ldots, x_n, y) \\ \mu\Phi_{\mathbf{P}}(\mathbf{G})(x_1, \ldots, x_n, y+1) & \text{otherwise,} \end{cases}$$

and so

$$\mu\Phi_{\mathbf{P}}(\mathbf{G})(x_1, \ldots, x_n, 0)$$

$$= \begin{cases} \perp_N & \text{if } (x_1, \ldots, x_n, 0) \notin N^{n+1} \\ \perp_N & \text{if } f(x_1, \ldots, x_n)\uparrow \\ \min_z R(x_1, \ldots, x_n, z) & \text{otherwise.} \end{cases}$$

Therefore, if $(x_1, \ldots, x_n) \in N^n$ and $f(x_1, \ldots, x_n)\downarrow$, then

$$\begin{aligned} \mu\Phi_{\mathbf{P}}(\mathbf{F})(x_1, \ldots, x_n) &= l_\perp(\min_z R(x_1, \ldots, x_n, z)) \\ &= l(\min_z R(x_1, \ldots, x_n, z)) \\ &= f(x_1, \ldots, x_n), \end{aligned}$$

and

$$\mu\Phi_{\mathbf{P}}(\mathbf{F})(x_1, \ldots, x_n) = l_\perp(\perp_N) = \perp_N,$$

otherwise, so that

$$\mu\Phi_{\mathbf{P}}(\mathbf{F})(x_1, \ldots, x_n) = f_\perp(x_1, \ldots, x_n)$$

for all $(x_1, \ldots, x_n) \in N_\perp^n$, i.e., $\mu\Phi_{\mathbf{P}}(\mathbf{F}) = \mathrm{f}_\perp$. ■

To prove a result in the other direction, we need a way of going from functions on $N_\perp^n$ to partial functions on N^n.

Definition. For any function f: $N_\perp^n \to N_\perp$, let f_π be the partial n-ary function on N^n defined

$$f_\pi(x_1, \ldots, x_n) = \begin{cases} f(x_1, \ldots, x_n) & \text{if } f(x_1, \ldots, x_n) \neq \perp_N \\ \uparrow & \text{otherwise.} \end{cases}$$

Theorem 2.2. If f is $\Delta_{\mathbf{N}}$-computable, then f_π is partially computable.

Proof. The proof is similar to the proof in Chapter 4 that the $\mathrm{STP}^{(n)}$ predicates are primitive recursive. We encode $\mathbf{W_N}$-terms, $\mathbf{W_N}$-programs, and $\mathbf{W_N}$-substitutions as numbers and give a numeric version of the $\mathrm{rr}_{\mathbf{P}}$

functions. All $\mathbf{W_N}$-terms are words in the 20 symbol alphabet $A \cup \mathbf{W_N}$, so for any word $\mathbf{w} \in (A \cup \mathbf{W_N})^*$, we let $\#_{20}(\mathbf{w})$ be the numeric value of $\mathbf{w}$ in base 20 notation, as defined in Chapter 5. We will use $\#_{20}$ to encode variables and function symbols. We encode each $\mathbf{W_N}$-term $\mathbf{t}$ as $\#(\mathbf{t})$, where

$$\#(\boldsymbol{\phi}) = \langle \#_{20}(\boldsymbol{\phi}), 0\rangle \qquad \text{if } \boldsymbol{\phi} \text{ is a constant symbol or individual variable}$$

$$\#(\boldsymbol{\phi}(\mathbf{t}_1, \ldots, \mathbf{t}_n)) = \langle \#_{20}(\boldsymbol{\phi}), [\#(\mathbf{t}_1), \ldots, \#(\mathbf{t}_n)]\rangle \qquad \text{if } \boldsymbol{\phi} \text{ is a proper function symbol or function variable.}$$

If $\mathbf{P}$ is a $\mathbf{W_N}$-program

$$\left\{\mathbf{F}_1(\mathbf{X}_1^1, \ldots, \mathbf{X}_{n_1}^1) = \mathbf{t}_1, \ldots, \mathbf{F}_m(\mathbf{X}_1^m, \ldots, \mathbf{X}_{n_m}^m) = \mathbf{t}_m\right\},$$

then we associate with $\mathbf{P}$ the finite numeric function

$$\phi_{\mathbf{P}} = \left\{(\#_{20}(\mathbf{F}_i), \langle[\#(\mathbf{X}_1^i), \ldots, \#(\mathbf{X}_{n_i}^i)], \#(\mathbf{t}_i)\rangle) \mid 1 \le i \le m\right\},$$

and if θ is a $\mathbf{W_N}$-substitution $\{(\mathbf{X}_1, \mathbf{t}_1), \ldots, (\mathbf{X}_n, \mathbf{t}_n)\}$, then we associate with θ the finite numeric function

$$\phi_\theta = \{(\#(\mathbf{X}_1), \#(\mathbf{t}_1)), \ldots, (\#(\mathbf{X}_n), \#(\mathbf{t}_n))\}.$$

We encode any finite numeric function $\phi = \{(x_1, y_1), \ldots, (x_n, y_n)\}$ as

$$\tilde{\phi} = \prod_{i=1}^{n} p_{x_i}^{y_i + 1},$$

and we set $\#(\mathbf{P}) = \tilde{\phi}_{\mathbf{P}}$ and $\#(\theta) = \tilde{\phi}_\theta$ for any $\mathbf{W_N}$-program $\mathbf{P}$ and $\mathbf{W_N}$-substitution θ.

Now we define some numeric functions for handling $\mathbf{W_N}$-terms, their values, and their encoding numbers:

$$\begin{aligned}
\text{NUM}(0) &= \#(\mathbf{0})\\
\text{NUM}(x+1) &= \langle \#_{20}(\mathbf{s}), [\text{NUM}(x)]\rangle\\
\text{TERM}(z, x_1, \ldots, x_n) &= \langle z, [\text{NUM}(x_1), \ldots, \text{NUM}(x_n)]\rangle
\end{aligned}$$

$$\text{EVAL}(x) = \begin{cases} 0 & \text{if } x = \#(\mathbf{0})\\ \text{EVAL}((r(x))_1) + 1 & \text{if } l(x) = \#_{20}(\mathbf{s})\\ \uparrow & \text{otherwise}\end{cases}$$

$$\text{IS\_NUM}(x) = \begin{cases} 1 & \text{if } x = \#(\mathbf{0})\\ \text{IS\_NUM}((r(x))_1) & \text{if } l(x) = \#_{20}(\mathbf{s})\\ 0 & \text{otherwise.}\end{cases}$$

It is clear that $\mathrm{NUM}(n) = \#(\mathbf{n})$ and $\mathrm{EVAL}(\#(\mathbf{n})) = n$ for any $n \in N$; $\mathrm{TERM}(\#_{20}(\mathbf{F}), m_1, \ldots, m_n) = \#(\mathbf{F}(\mathbf{m}_1, \ldots, \mathbf{m}_n))$ for any function variable $\mathbf{F}$; and $\mathrm{IS\_NUM}(x)$ is the predicate that tests whether $x = \#(\mathbf{n})$ for some numeral $\mathbf{n}$.

Next we define two functions for handling $\mathbf{W_N}$-substitutions:

$$\mathrm{MAKE\_SUB}(y, t) = \prod_{i=1}^{\mathrm{Lt}(r(t))} p_{(l((y)_{l(t)} \dot{-} 1))_i}^{(r(t))_i + 1}$$

$$\mathrm{APPLY\_SUB}(s, t) = \begin{cases} (s)_t \dot{-} 1 & \text{if } r(t) = 0 \text{ and } (s)_t \neq 0 \\ t & \text{if } r(t) = 0 \text{ and } (s)_t = 0 \\ \langle l(t), \prod_{i=1}^{\mathrm{Lt}(r(t))} p_i^{\mathrm{APPLY\_SUB}(s, (r(t))_i)} \rangle & \\ & \text{otherwise.} \end{cases}$$

If $\mathbf{P}$ is a $\mathbf{W_N}$-program and $\mathbf{F}(\mathbf{X}_1, \ldots, \mathbf{X}_n) = \mathbf{u}$ is an equation in $\mathbf{P}$, then

$$\mathrm{MAKE\_SUB}(\#(\mathbf{P}), \#(\mathbf{F}(\mathbf{t}_1, \ldots, \mathbf{t}_n))) = \#(\{(\mathbf{X}_1, \mathbf{t}_1), \ldots, (\mathbf{X}_n, \mathbf{t}_n)\}).$$

Also, if θ is a $\mathbf{W_N}$-substitution and $\mathbf{t}$ is a $\mathbf{W_N}$-term, then

$$\mathrm{APPLY\_SUB}(\#(\theta), \#(\mathbf{t})) = \#(\mathbf{t}\theta).$$

Finally, we define some functions for handling $(\mathbf{T}_s(\mathbf{P}), \sigma_f)$-rewriting:

$$\mathrm{RP}(y, t) = \begin{cases} \mathrm{APPLY\_SUB}(\mathrm{MAKE\_SUB}(y, t), r((y)_{l(t)} \dot{-} 1)) & \\ & \text{if } (y)_{l(t)} \neq 0 \\ (r(t))_2 & \text{if } l(t) = \#_{20}(\mathbf{if_{BOOL}}) \text{ and } (r(t))_1 = \#(\mathbf{tt}) \\ (r(t))_3 & \text{if } l(t) = \#_{20}(\mathbf{if_{BOOL}}) \text{ and } (r(t))_1 = \#(\mathbf{ff}) \\ (r(t))_2 & \text{if } l(t) = \#_{20}(\mathbf{if_N}) \text{ and } (r(t))_1 = \#(\mathbf{tt}) \\ (r(t))_3 & \text{if } l(t) = \#_{20}(\mathbf{if_N}) \text{ and } (r(t))_1 = \#(\mathbf{ff}) \\ \#(\mathbf{tt}) & \text{if } l(t) = \#_{20}(\mathbf{is\_0}) \text{ and } (r(t))_1 = \#(\mathbf{0}) \\ \#(\mathbf{ff}) & \text{if } l(t) = \#_{20}(\mathbf{is\_0}) \text{ and } l((r(t))_1) = \#_{20}(\mathbf{s}) \\ & \qquad \text{and } \mathrm{IS\_NUM}(t) \\ \#(\mathbf{tt}) & \text{if } l(t) = \#_{20}(\mathbf{is\_s}) \text{ and } l((r(t))_1) = \#_{20}(\mathbf{s}) \\ & \qquad \text{and } \mathrm{IS\_NUM}(t) \\ \#(\mathbf{ff}) & \text{if } l(t) = \#_{20}(\mathbf{is\_s}) \text{ and } (r(t))_1 = \#(\mathbf{0}) \\ (r((r(t))_1))_1 & \text{if } l(t) = \#_{20}(\mathbf{s_1^{-1}}) \text{ and } l((r(t))_1) = \#_{20}(\mathbf{s}) \\ & \qquad \text{and } \mathrm{IS\_NUM}(t) \\ t & \text{otherwise} \end{cases}$$

$$\begin{aligned}
\mathrm{RRP}(y,t) &= \mathrm{RP}\left(y, \langle l(t), \prod_{i=1}^{\mathrm{Lt}(r(t))} p_i^{\mathrm{RRP}(y,(r(t))_i)}\rangle\right)\\
\mathrm{RRP}^*(y,t,0) &= t\\
\mathrm{RRP}^*(y,t,r+1) &= \mathrm{RRP}(y,\mathrm{RRP}^*(y,t,r))\\
\mathrm{RRPT}^*(y,z,x_1,\ldots,x_n,r) &= \mathrm{RRP}^*(y,\mathrm{TERM}(z,x_1,\ldots,x_n),r)\\
\mathrm{END}(y,z,x_1,\ldots,x_n) &= \min_r [\mathrm{RRPT}^*(y,z,x_1,\ldots,x_n,r)\\
&\qquad = \mathrm{RRPT}^*(y,z,x_1,\ldots,x_n,r+1)].
\end{aligned}$$

If **P** is a $\mathbf{W}_\mathrm{N}$-program and **t** is a $\mathbf{W}_\mathrm{N}$-term, then RP(#(**P**), #(**t**)) = #($\mathbf{r}_\mathbf{P}$(**t**)), RRP(#(**P**), #(**t**)) = #($\mathbf{rr}_\mathbf{P}$(**t**)), and RRP*(#(**P**), #(**t**), i) = #($\mathbf{rr}_\mathbf{P}^i$(**t**)). Also, if **F** is a function variable and $(m_1, \ldots, m_n) \in N^n$, then

$$\mathrm{RRPT}^*(\#(\mathbf{P}), \#_{20}(\mathbf{F}), m_1, \ldots, m_n, i) = \#(\mathbf{rr}_\mathbf{P}^i(\mathbf{F}(\mathbf{m}_1, \ldots, \mathbf{m}_n))),$$

and END(#(**P**), $\#_{20}$(**F**), $m_1, \ldots, m_n$) is the smallest i such that $\mathbf{rr}_\mathbf{P}^i(\mathbf{F}(\mathbf{m}_1, \ldots, \mathbf{m}_n))$ is $(\mathbf{T}_s(\mathbf{P}), \sigma_f)$-normal, if such an i exists, and is undefined otherwise.

Now, let f: $N_\perp^n \to N_\perp$ be a Δ_N-computable function, let $f = \mu\Phi_\mathbf{P}(\mathbf{F})$, let $a = \#_{20}(\mathbf{P})$, and let $b = \#_{20}(\mathbf{F})$. Then it is clear that

$$f_\pi(x_1, \ldots, x_n) = \mathrm{EVAL}(\mathrm{RRPT}^*(a, b, x_1, \ldots, x_n, \mathrm{END}(a, b, x_1, \ldots, x_n)))$$

for all $(x_1, \ldots, x_n) \in N^n$. EVAL, RRPT*, and END are partially computable, so f_π is partially computable. ■

Exercises

1. Show that for all $n > 0$ and all $(x_1, \ldots, x_n) \in N_\perp^n$,

$$\mathscr{D}_{\Delta_\mathrm{N}}(\mathbf{P}_{\mathrm{Def}^n})(\mathbf{Def}^n)(x_1, \ldots, x_n) = \begin{cases} \mathrm{TRUE} & \text{if } (x_1, \ldots, x_n) \in N^n \\ \perp_{\mathrm{Bool}} & \text{otherwise.} \end{cases}$$

2. For each of the following functions f from Chapter 3, give a $\mathbf{W}_\mathrm{N}$-program **P** with $\mathbf{F} \in FV(\mathbf{P})$ such that $\mathscr{D}_{\Delta_\mathrm{N}}(\mathbf{P})(\mathbf{F}) = f_\perp$.
   (a) $f(x, y) = x \cdot y$.
   (b) $f(x) = x!$.
   (c) $f(x, y) = x^y$.
   (d) $f(x) = p(x)$.
   (e) $f(x, y) = x \dot{-} y$.
   (f) $f(x, y) = |x - y|$.
   (g) $f(x) = \alpha(x)$.

3. Let $P(x), Q(x)$ be primitive recursive predicates, and let $\mathbf{P}, \mathbf{Q}$ be $\mathbf{W}_\mathrm{N}$-programs such that $\mathscr{D}_{\Delta_\mathrm{N}}(\mathbf{P})(\mathbf{F_P}) = P_\perp$ and $\mathscr{D}_{\Delta_\mathrm{N}}(\mathbf{P})(\mathbf{F_Q}) = Q_\perp$. For each of the following predicates $R(x)$, give a $\mathbf{W}_\mathrm{N}$-program $\mathbf{R}$ with $\mathbf{F_R} \in FV(\mathbf{R})$ such that $\mathscr{D}_{\Delta_\mathrm{N}}(\mathbf{R})(\mathbf{F_R}) = R_\perp$. [Also see Exercise 3.12 in Chapter 17.]
   (a) $R(x) \Leftrightarrow \sim P(x)$.
   (b) $R(x) \Leftrightarrow P(x) \ \& \ Q(x)$.
   (c) $R(x) \Leftrightarrow P(x) \vee Q(x)$.
   (d) $R(x) \Leftrightarrow (\exists z)_{\leq x} P(z)$.
   (e) $R(x) \Leftrightarrow (\forall z)_{\leq x} P(z)$.

4. Let $P(x, y)$ be a primitive recursive predicate, and let $\mathbf{P}$ be a $\mathbf{W}_\mathrm{N}$-program such that $\mathscr{D}_{\Delta_\mathrm{N}}(\mathbf{P})(\mathbf{F_P}) = P_\perp$. Give a $\mathbf{W}_\mathrm{N}$-program $\mathbf{R}$ with $\mathbf{Min_R} \in FV(\mathbf{R})$ such that $\mathscr{D}_{\Delta_\mathrm{N}}(\mathbf{R})(\mathbf{Min_R})$ is the strict extension of $\min_{z \leq x} P(z, y)$.

5. For each of the following predicates P from Chapter 3, give a $\mathbf{W}_\mathrm{N}$-program $\mathbf{P}$ with $\mathbf{F_P} \in FV(\mathbf{P})$ such that $\mathscr{D}_{\Delta_\mathrm{N}}(\mathbf{P})(\mathbf{F_P}) = P_\perp$.
   (a) $P(x, y) \Leftrightarrow x = y$.
   (b) $P(x, y) \Leftrightarrow x \leq y$.
   (c) $P(x, y) \Leftrightarrow x < y$.
   (d) $P(x, y) \Leftrightarrow x \mid y$.
   (e) $P(x) \Leftrightarrow \mathrm{Prime}(x)$.

6. For each of the following functions f from Chapter 3, give a $\mathbf{W}_\mathrm{N}$-program $\mathbf{P}$ with $\mathbf{F} \in FV(\mathbf{P})$ such that $\mathscr{D}_{\Delta_\mathrm{N}}(\mathbf{P})(\mathbf{F}) = f_\perp$.
   (a) $f(x, y) = \lfloor x/y \rfloor$.
   (b) $f(x, y) = R(x, y)$.
   (c) $f(x) = p_x$.
   (d) $f(x, y) = \langle x, y \rangle$.
   (e) $f(x) = l(x)$.
   (f) $f(x) = r(x)$.
   (g) $f(x_1, \ldots, x_n) = [x_1, \ldots, x_n]$.
   (h) $f(x, y) = (x)_y$.
   (i) $f(x) = \mathrm{Lt}(x)$.

7. Show that RRPT* is primitive recursive.

8. Let $\#$: $\{\mathscr{S}$ programs$\} \to N$ be the coding function for $\mathscr{S}$ programs given in Chapter 4. Give a $\mathbf{W}_{\mathscr{S}}$-program $\mathbf{P}$ with $\mathbf{C} \in FV(\mathbf{P})$ such that $\mathscr{D}_{\Delta_{\mathscr{S}}}(\mathbf{P})(\mathbf{C}) = \#_\perp$. [See Section 5 in Chapter 17 for the definitions of $\mathbf{W}_{\mathscr{S}}$ and $\Delta_{\mathscr{S}}$.]

9.* **(a)** Give a $\mathbf{W_N}$-program **SMN** with $\mathbf{S} \in FV(\mathbf{SMN})$ such that $\mathscr{D}_{\Delta_N}(\mathbf{SMN})(\mathbf{S})$ is the strict extension of S_1^1. [See the parameter theorem in Chapter 4 for the definition of S_1^1.]

(b) Let s_1^1: $N \times \{\mathscr{S}$ programs$\} \to \{\mathscr{S}$ programs$\}$ be defined: For all $\mathscr{S}$ programs $\mathscr{P}$ and all $u \in N$,

$$\Phi^{(2)}(x, u, \#(\mathscr{P})) = \Phi\big(x, \#(s_1^1(u, \mathscr{P}))\big).$$

Give a $\mathbf{W}_{\mathscr{S}}$-program **SMN** with $\mathbf{S} \in FV(\mathbf{SMN})$ such that $\mathscr{D}_{\Delta_{\mathscr{S}}}(\mathbf{SMN})(\mathbf{S})$ is the strict extension of s_1^1.

3. Operational Semantics for Infinitary Data Structure Systems

We turn now to the operational semantics for infinitary data structure systems. It differs in two respects from the operational semantics we gave for simple data structure systems. First, for a term such as $\mathbf{is\_f(f(t))}$, we cannot be sure that $\overline{\mu\Phi_{\mathbf{P}}}(\mathbf{is\_f(f(t))}) = \text{tt}$ when $\mathscr{I}(\mathbf{f})$ is strict because $\overline{\mu\Phi_{\mathbf{P}}}(\mathbf{t})$ might be $\perp_{\tau(\mathbf{t})}$. Therefore, we defined $\mathbf{T}_s(\mathbf{P})$ so that we rewrite $\mathbf{is\_f(f(t))}$ to $\mathbf{tt}$ only when $\mathbf{t}$ is free of variables, which guarantees that $\overline{\mu\Phi_{\mathbf{P}}}(\mathbf{t}) \neq \perp_{\tau(\mathbf{t})}$. Moreover, if $\overline{\mu\Phi_{\mathbf{P}}}(\mathbf{t}) \neq \perp_{\tau(\mathbf{t})}$, then $\overline{\Phi_{\mathbf{P}}^{i_0}(\Omega)}(\mathbf{t}) \neq \perp_{\tau(\mathbf{t})}$ for some smallest i_0, and Lemmas 3 and 4 guarantee that $\mathbf{t}$ will eventually be rewritten to some term which is free of variables. In an infinitary data structure system, however, this problem does not arise because we always have $\overline{\mu\Phi_{\mathbf{P}}}(\mathbf{f(t)}) \neq \perp_{\rho(\mathbf{f})}$. Therefore, we can replace $\mathbf{T}_s(\mathbf{P})$ with a simpler, indeed finite, term rewriting system.

Definition. Let **P** be a **W**-program. The *infinitary* **W**-*term rewriting system associated with* **P**, denoted $\mathbf{T}_i(\mathbf{P})$, consists of

$$\mathbf{F}(\mathbf{X}_1, \ldots, \mathbf{X}_n) \to \mathbf{u} \quad \text{for each equation } \mathbf{F}(\mathbf{X}_1, \ldots, \mathbf{X}_n) = \mathbf{u} \text{ in } \mathbf{P},$$

together with

- for each $\tau \in TV(\mathbf{W})$,

$$\mathbf{if}_\tau(\mathbf{tt}, \mathbf{X}, \mathbf{Y}) \to \mathbf{X}$$
$$\mathbf{if}_\tau(\mathbf{ff}, \mathbf{X}, \mathbf{Y}) \to \mathbf{Y}$$

- for each constant symbol $\mathbf{c} \in \mathbf{W}_c^-$, with $\tau(\mathbf{c}) = \boldsymbol{\tau}$,

$$\begin{array}{ll} \mathbf{is\_c(c)} \to \mathbf{tt} & \\ \mathbf{is\_c(d)} \to \mathbf{ff} & \text{for each } \mathbf{d} \in \mathbf{W}_c^- - \{\mathbf{c}\} \text{ with } \tau(\mathbf{d}) = \boldsymbol{\tau} \\ \mathbf{is\_c(f(X_1, \ldots, X_n))} \to \mathbf{ff} & \text{for each } \mathbf{f} \in \mathbf{W}_c \text{ with} \\ & \tau(\mathbf{f}) = \boldsymbol{\tau}_1 \times \cdots \times \boldsymbol{\tau}_n \to \boldsymbol{\tau} \end{array}$$

- for each proper function symbol $\mathbf{f} \in \mathbf{W}_c$, with $\mathrm{ar}(\mathbf{f}) = n$ and $\rho(\mathbf{f}) = \boldsymbol{\tau}$,

$$\begin{array}{ll} \mathbf{is\_f(f(X_1, \ldots, X_n))} \to \mathbf{tt} & \\ \mathbf{is\_f(c)} \to \mathbf{ff} & \text{for each } \mathbf{c} \in \mathbf{W}_c \text{ with } \tau(\mathbf{c}) = \boldsymbol{\tau} \\ \mathbf{is\_f(g(X_1, \ldots, X_m))} \to \mathbf{ff} & \text{for each } \mathbf{g} \in \mathbf{W}_c - \{\mathbf{f}\} \text{ with} \\ & \mathrm{ar}(\mathbf{g}) = m \text{ and } \rho(\mathbf{g}) = \boldsymbol{\tau} \\ \mathbf{f}_i^{-1}\mathbf{(f(X_1, \ldots, X_n))} \to \mathbf{X}_i & \text{for } 1 \le i \le n. \end{array}$$

Again, the choice of particular individual variables is unimportant, as long as they are of the appropriate type and in each rewrite rule the variables are distinct.

It is easy to see that all term rewriting systems $\mathbf{T}_i(\mathbf{P})$ are deterministic and that $(\mathbf{T}_i(\mathbf{P}), \sigma_{\mathrm{f}})$-computations are a reasonable sort of computation. In this section we will write $\mathrm{r}_{\mathbf{P}}$ and $\mathrm{rr}_{\mathbf{P}}$ for $\mathrm{r}_{\mathbf{T}_i(\mathbf{P})}$ and $\mathrm{rr}_{\mathbf{T}_i(\mathbf{P})}$, respectively.

The other difference we need to address is that there are infinite chains in infinitary data structure systems. In particular, there are terms $\mathbf{t}$ such that $\overline{\Omega}(\mathrm{rr}_{\mathbf{P}}^i(\mathbf{t})) = \overline{\Phi_{\mathbf{P}}^i(\Omega)(\mathbf{t})} \sqsubset_{\tau(\mathbf{t})} \overline{\mu\Phi_{\mathbf{P}}(\mathbf{t})}$ for all $i \in N$, so that we get an infinite computation

$$\mathbf{t} \underset{\mathbf{T}_i(\mathbf{P}),\, \sigma_{\mathrm{f}}}{\Longrightarrow} \mathrm{rr}_{\mathbf{P}}(\mathbf{t}) \underset{\mathbf{T}_i(\mathbf{P}),\, \sigma_{\mathrm{f}}}{\Longrightarrow} \mathrm{rr}_{\mathbf{P}}^2(\mathbf{t}) \underset{\mathbf{T}_i(\mathbf{P}),\, \sigma_{\mathrm{f}}}{\Longrightarrow} \cdots$$

which never reaches the desired value $\overline{\mu\Phi_{\mathbf{P}}}(\mathbf{t})$. Therefore, we cannot expect to base the operational semantics on the final terms of finite computations. Instead, we take the point of view that an infinite computation produces ever better approximations to the actual value and that the entire computation gives the meaning of the function being computed.

Definition. Let $\Delta = \mathrm{rep}(\Sigma)$ be an infinitary data structure system for $\mathbf{W}$. The *operational meaning function* for Δ, denoted $\mathscr{O}_\Delta$, is defined as follows. For all $\mathbf{W}$-programs $\mathbf{P}_0$, all $\mathbf{F} \in FV(\mathbf{P}_0)$, and all $(d_1, \ldots, d_n) \in D_{r(\delta(\mathbf{F}))}$,

$$\mathscr{O}_\Delta(\mathbf{P}_0)(\mathbf{F})(d_1, \ldots, d_n) = \sqcup\left\{\overline{\Omega}(\mathrm{rr}_{\mathbf{P}}^i(\mathbf{F}(\mathbf{t}_1, \ldots, \mathbf{t}_n))) \mid i \in N\right\},$$

where

- $d_i = \overline{\mu\Phi^{\Sigma}_{\mathbf{P}_i}(\mathbf{t}_i)}$, $1 \le i \le n$,
- $\mathbf{P}_0, \mathbf{P}_1, \ldots, \mathbf{P}_n$ are consistent, and
- $\mathbf{P} = \bigcup_{i=0}^{n} \mathbf{P}_i$.

The proof that $\mathscr{O}_\Delta$ is correct with respect to $\mathscr{D}_\Delta$ for all infinitary data structure systems is very much like the proof of Theorem 1.1.

Lemma 1. Let $\Sigma = (\mathscr{T}, \mathscr{I})$ be an infinitary **W**-structure, and let **P** be a **W**-program. Then for all $\mathbf{t} \in \mathrm{TM}_{\mathbf{W}}(FV(\mathbf{P}))$ and for all $i \in N$, $\overline{\Phi^{i+1}_{\mathbf{P}}(\Omega)(\mathbf{t})} = \overline{\Phi^{i}_{\mathbf{P}}(\Omega)}(\mathrm{rr}_{\mathbf{P}}(\mathbf{t}))$.

Proof. The proof is almost identical to the proof of Lemma 2 in Section 1. The only differences occur in the cases where **t** is of the form $\mathbf{is\_f}(\mathbf{u})$ or $\mathbf{f}_i^{-1}(\mathbf{u})$. In particular, if $\mathbf{is\_f}(\mathrm{rr}_{\mathbf{P}}(\mathbf{u}))$ or $\mathbf{f}_i^{-1}(\mathrm{rr}_{\mathbf{P}}(\mathbf{u}))$ match a rewrite rule in $\mathbf{T}_i(\mathbf{P})$, then we do not necessarily have $\mathrm{rr}_{\mathbf{P}}(\mathbf{u}) \in \mathrm{TM}_{\mathbf{W}_c}$, so we cannot appeal to Lemma 1 in Section 1 to show that $\overline{\Phi^{i}_{\mathbf{P}}(\Omega)}(\mathrm{rr}_{\mathbf{P}}(\mathbf{u})) \ne \bot_{\tau(\mathbf{u})}$. However, $\mathrm{rr}_{\mathbf{P}}(\mathbf{u})$ must be of the form **g** or $\mathbf{g}(\mathbf{u}_1, \ldots, \mathbf{u}_m)$ for some $\mathbf{g} \in \mathbf{W}_c$, so that $\overline{\Phi^{i}_{\mathbf{P}}(\Omega)}(\mathrm{rr}_{\mathbf{P}}(\mathbf{u})) \ne \bot_{\tau(\mathbf{u})}$ is certainly true in an infinitary **W**-structure, and the argument goes through unchanged but for the reference to Lemma 1 in Section 1. ■

Repeating the proof of Lemma 3 in Section 1 gives us

Lemma 2. Let $\Sigma = (\mathscr{T}, \mathscr{I})$ be an infinitary **W**-structure, and let **P** be a **W**-program. Then for all $\mathbf{t} \in \mathrm{TM}_{\mathbf{W}}(FV(\mathbf{P}))$ and for all $i \in N$, $\overline{\Phi^{i}_{\mathbf{P}}(\Omega)(\mathbf{t})} = \overline{\Omega}(\mathrm{rr}^{i}_{\mathbf{P}}(\mathbf{t}))$.

Now the proof of Theorem 3.1 is even simpler than the proof of Theorem 1.1.

Theorem 3.1. Let $\Delta = \mathrm{rep}(\Sigma)$ be an infinitary data structure system for **W**. Then $\mathscr{O}_\Delta = \mathscr{D}_\Delta$.

Proof. Let $\mathbf{P}_0$ be a **W**-program, let $\mathbf{F} \in FV(\mathbf{P}_0)$, and let $(d_1, \ldots, d_n) \in D_{r(\delta(\mathbf{F}))}$. For $1 \le i \le n$ let $d_i = \overline{\mu\Phi_{\mathbf{P}_i}(\mathbf{t}_i)}$ for some **W**-program $\mathbf{P}_i$ and some $\mathbf{t}_i \in \mathrm{TM}_{\mathbf{W}}(FV(\mathbf{P}_i))$ such that $\mathbf{P}_0, \mathbf{P}_1, \ldots, \mathbf{P}_n$ are consistent, and let $\mathbf{P} = \bigcup_{i=0}^{n} \mathbf{P}_i$. Then

$$
\begin{aligned}
&\mathscr{D}_\Delta(\mathbf{P}_0)(\mathbf{F})(d_1, \ldots, d_n) \\
&\quad = \sqcup\left\{\overline{\Phi^{i}_{\mathbf{P}}(\Omega)}(\mathbf{F}(\mathbf{t}_1, \ldots, \mathbf{t}_n)) \mid i \in N\right\} && \text{as in the proof of Theorem 1.1} \\
&\quad = \sqcup\left\{\overline{\Omega}(\mathrm{rr}^{i}_{\mathbf{P}}(\mathbf{F}(\mathbf{t}_1, \ldots, \mathbf{t}_n))) \mid i \in N\right\} && \text{by Lemma 2} \\
&\quad = \mathscr{O}_\Delta(\mathbf{P}_0)(\mathbf{F})(d_1, \ldots, d_n).
\end{aligned}
$$

■

As we did for simple data structure systems, we can now define computable functions in infinitary data structure systems.

Definition. Let Δ be an infinitary data structure system for **W**. A function f is *Δ-computable* if there is a **W**-program **P** and $\mathbf{F} \in FV(\mathbf{P})$ such that $f = \mathscr{D}_\Delta(\mathbf{P})(\mathbf{F})$.

We can also use Lemma 2 to justify the name *computable real numbers* used in the previous chapter.

Definition. Let Σ be an infinitary **W**-structure, and let $\boldsymbol{\tau} \in TV(\mathbf{W})$. An element $d \in D_\tau$ is *computable* if there is some **W**-program **P** and some term $\mathbf{t} \in TM_{\mathbf{W}}(FV(\mathbf{P}))$ such that $d = \sqcup\{\overline{\Omega}(\mathrm{rr}^i_{\mathbf{P}}(\mathbf{t})) \mid i \in N\}$.

Now we can easily prove

Theorem 3.2. Let Σ be an infinitary **W**-structure, and let $\boldsymbol{\tau} \in TV(\mathbf{W})$. Then an element $d \in D_\tau$ is representable if and only if it is computable.

Proof. Let $d \in D_\tau$. If d is representable then there is a **W**-program **P** and a term $\mathbf{t} \in \mathrm{TM}_{\mathbf{W}}(FV(\mathbf{P}))$ such that

$$\begin{aligned} d &= \overline{\mu\Phi_{\mathbf{P}}}(\mathbf{t}) \\ &= \overline{\sqcup\{\Phi^i_{\mathbf{P}}(\Omega) \mid i \in N\}}(\mathbf{t}) \\ &= \sqcup\left\{\overline{\Phi^i_{\mathbf{P}}(\Omega)}(\mathbf{t}) \mid i \in N\right\} && \text{by Theorem 17.2.3} \\ &= \sqcup\left\{\overline{\Omega}(\mathrm{rr}^i_{\mathbf{P}}(\mathbf{t})) \mid i \in N\right\} && \text{by Lemma 2,} \end{aligned}$$

so d is computable. Similarly, if d is computable then there is a **W**-program **P** and a term $\mathbf{t} \in \mathrm{TM}_{\mathbf{W}}(FV(\mathbf{P}))$ such that

$$d = \sqcup\left\{\overline{\Omega}(\mathrm{rr}^i_{\mathbf{P}}(\mathbf{t})) \mid i \in N\right\} = \overline{\mu\Phi_{\mathbf{P}}}(\mathbf{t}),$$

and so d is representable. ■

Now that we have available all of the strict extensions of the partially computable functions, we conclude with two promised examples of computing with infinite objects. We will be working in infinitary data structures, but it is not hard to verify that if $f(x_1, \ldots, x_n)$ is partially computable and **P** is a $\mathbf{W}_{\mathrm{N}}$-program obtained in the proof of Theorem 2.1 such that $\mathscr{D}_{\Delta_{\mathrm{N}}}(\mathbf{P})(\mathbf{F}) = f_\perp$, then

$$\mathscr{D}_{\Delta^\infty_{\mathrm{NL}}}(\mathbf{P})(\mathbf{F})(x_1, \ldots, x_n) = f_\perp(x_1, \ldots, x_n) \tag{3.1}$$

$$\mathbf{Primes(X)} = \mathbf{Sieve(Seq(X))}$$

$$\mathbf{Seq(X)} = \mathbf{cons(X, Seq(s(X)))}$$

$$\mathbf{Sieve(L)} = \mathbf{cons(cons_1^{-1}(L), Sieve(Elim(cons_1^{-1}(L), cons_2^{-1}(L))))}$$

$$\begin{aligned}\mathbf{Elim(X,L)} = \mathbf{if}_{\mathrm{NL}}(&\mathbf{X} \mid \mathbf{cons_1^{-1}(L)},\\ &\mathbf{Elim(X, cons_2^{-1}(L))},\\ &\mathbf{cons(cons_1^{-1}(L), Elim(X, cons_2^{-1}(L))))}\end{aligned}$$

Figure 3.1. The main part of program **PR**.

for all $(x_1, \ldots, x_n) \in N^n$. We will now freely write $f(x_1, \ldots, x_n)$ as a macro in $\mathbf{W}_{\mathrm{NL}}$-programs and $\mathbf{W}_{\mathbf{R}}$-programs when $f(x_1, \ldots, x_n)$ is partially computable. If $P(x_1, \ldots, x_n)$ is a computable predicate, then when we write $P(x_1, \ldots, x_n)$ in a program its range should be understood as $\{\perp_{\mathrm{Bool}}, \mathrm{tt}, \mathrm{ff}\}$.

The first example is a $\mathbf{W}_{\mathrm{NL}}$-program for generating the list $\langle p_1, p_2, \ldots \rangle$ of all prime numbers. It is based on the method known as Eratosthenes' sieve, where we start with the list $\langle 2, 3, 4, \ldots \rangle$, eliminate all numbers divisible by 2, then eliminate all numbers divisible by 3, all numbers divisible by 5, etc. Let **PR** be the $\mathbf{W}_{\mathrm{NL}}$-program with the equations in Fig. 3.1 along with the definition of the predicate $x \mid y$, i.e., "x divides y." Then in $\Sigma_{\mathrm{NL}}^{\infty}$ we have

$$\begin{aligned}
&\overline{\mu\Phi_{\mathbf{PR}}}(\mathbf{Primes(2)})\\
&\quad= \mu\Phi_{\mathbf{PR}}(\mathbf{Sieve})(\mu\Phi_{\mathbf{PR}}(\mathbf{Seq})(2))\\
&\quad= \mu\Phi_{\mathbf{PR}}(\mathbf{Sieve})(\langle 2,3,4,\ldots\rangle)\\
&\quad= \langle 2, \mu\Phi_{\mathbf{PR}}(\mathbf{Sieve})(\mu\Phi_{\mathbf{PR}}(\mathbf{Elim})(2, \langle 3,4,5,\ldots\rangle))\rangle\\
&\quad= \langle 2, \mu\Phi_{\mathbf{PR}}(\mathbf{Sieve})(\langle 3,5,7,\ldots\rangle)\rangle\\
&\quad= \langle 2,3, \mu\Phi_{\mathbf{PR}}(\mathbf{Sieve})(\mu\Phi_{\mathbf{PR}}(\mathbf{Elim})(3, \langle 5,7,9,\ldots\rangle))\rangle\\
&\quad= \langle 2,3, \mu\Phi_{\mathbf{PR}}(\mathbf{Sieve})(\langle 5,7,11,\ldots\rangle)\rangle\\
&\qquad\quad\vdots \qquad\qquad\qquad \vdots\\
&\quad= \langle 2,3,5,\ldots\rangle
\end{aligned}$$

where the notation $\langle 2, \mu\Phi_{\mathbf{PR}}(\mathbf{Sieve})(\langle 3,5,7,\ldots\rangle)\rangle$ means the list with 2 followed by the elements of the list $\mu\Phi_{\mathbf{PR}}(\mathbf{Sieve})(\langle 3,5,7,\ldots\rangle)$. Therefore, the list of primes is representable. Moreover, Theorem 3.2 shows that it can be generated by a $(\mathbf{T}_i(\mathbf{PR}), \sigma_{\mathrm{f}})$-computation.

Finally, we show that there are computable irrational numbers. It is a mathematical fact that the well-known irrational number $e = 2.7182\ldots$ can be expressed as $2 + \frac{1}{2!} + \frac{1}{3!} + \cdots$, and we can use this fact to show that $e/10 = .27182\ldots$ is computable.[5] The idea is to consider $\frac{n_1}{2!} + \frac{n_2}{3!} + \cdots$ as a sort of base notation system analogous to the decimal system, where .354, for example, represents $\frac{3}{10} + \frac{5}{10^2} + \frac{4}{10^3}$. All we have to do, then, is to change $2 + \frac{1}{2!} + \frac{1}{3!} + \cdots$ (divided by 10) into its decimal representation. The procedure consists of taking the integer part as the first decimal digit, multiplying the fractional part by 10, normalizing the result (i.e., reducing the numerators by carrying), and then repeating these steps with the normalized result. The correct method for carrying is given by the equations

$$\frac{m}{n+1!} = \frac{(n+1)\lfloor m/(n+1)\rfloor + R(m, n+1)}{n+1!}$$
$$= \frac{\lfloor m/(n+1)\rfloor}{n!} + \frac{R(m, n+1)}{n+1!}.$$

For example, starting with $2 + \frac{1}{2!} + \frac{1}{3!} + \frac{1}{4!}$ we get 2 as the first decimal digit, and then we get

$$10 \cdot (\tfrac{1}{2!} + \tfrac{1}{3!} + \tfrac{1}{4!}) = \tfrac{10}{2!} + \tfrac{10}{3!} + \tfrac{10}{4!}$$
$$= \tfrac{10}{2!} + \tfrac{12}{3!} + \tfrac{2}{4!}$$
$$= \tfrac{14}{2!} + \tfrac{0}{3!} + \tfrac{2}{4!} = 7 + \tfrac{0}{2!} + \tfrac{0}{3!} + \tfrac{2}{4!},$$

so 7 is the second decimal digit. Next we get

$$10 \cdot (\tfrac{0}{2!} + \tfrac{0}{3!} + \tfrac{2}{4!}) = \tfrac{0}{2!} + \tfrac{0}{3!} + \tfrac{20}{4!}$$
$$= \tfrac{0}{2!} + \tfrac{5}{3!} + \tfrac{0}{4!} = 0 + \tfrac{1}{2!} + \tfrac{2}{3!} + \tfrac{0}{4!},$$

so 0 is the third decimal digit. Notice, however, that if we start with $2 + \frac{1}{2!} + \frac{1}{3!} + \frac{1}{4!} + \frac{1}{5!}$, we get .271 instead of .270. That is, we get more precision by starting with more terms. However, if we want to get the decimal expansion of the *infinite* sum $2 + \sum_{n=2}^{\infty} \frac{1}{n!}$, then at any given iteration we certainly cannot perform the entire multiplication by 10 before beginning the carry step. Fortunately, we need to perform only enough of it so that multiplying and normalizing any additional terms would not change the decimal digit produced by the current iteration.

[5] This example is due to D. A. Turner, who credits E. W. Dijkstra with the idea.

$$
\begin{aligned}
\mathbf{K(X)} &= \mathbf{cons(X, K(X))} \\
\mathbf{Convert(L)} &= \mathbf{dcons(Digit(First(L)),} \\
&\qquad \mathbf{Convert(Norm(2, cons(0, Mult10(Rest(L))))))} \\
\mathbf{Norm(C, L)} &= \mathbf{if_{NL}(Second(L) + 9 < C,} \\
&\qquad \mathbf{cons(First(L), Norm(s(C), Rest(L))),} \\
&\qquad \mathbf{Carry(C, cons(First(L), Norm(s(C), Rest(L)))))} \\
\mathbf{Carry(C, L)} &= \mathbf{cons(First(L) + \lfloor Second(L)/C \rfloor,} \\
&\qquad \mathbf{cons(}R\mathbf{(Second(L), C), Rest2(L)))} \\
\mathbf{Mult10(L)} &= \mathbf{cons(10 \cdot First(L), Mult10(Rest(L)))} \\
\mathbf{First(L)} &= \mathbf{cons_1^{-1}(L)} \\
\mathbf{Second(L)} &= \mathbf{cons_1^{-1}(cons_2^{-1}(L))} \\
\mathbf{Rest(L)} &= \mathbf{cons_2^{-1}(L)} \\
\mathbf{Rest2(L)} &= \mathbf{cons_2^{-1}(cons_2^{-1}(L))}
\end{aligned}
$$

Figure 3.2. The main part of program **E**.

Now, the carry procedure leaves fractions of the form $R(m,n)/n!$, where $R(m,n) < n$, so when we multiply by 10 we get $10 \cdot R(m,n)/n!$, and then we carry $\lfloor 10 \cdot R(m,n)/n \rfloor < 10n/n = 10$. So the maximum possible carry is always 9, and if we have a term $m/n!$ such that $m + 9 < n$, there will never be a carry out of $m/n!$, regardless of what is carried into $m/n!$. Therefore, when we reach such a term, we can be sure that we have enough information to produce a correct decimal digit. Let **E** be the program with the equations in Fig. 3.2 together with the appropriate definitions of addition, multiplication, integer division, and remainder. Also, **Digit**, with $\tau(\mathbf{Digit}) = \mathbf{N} \to \mathbf{D}$, must be defined so that $\mu\Phi_{\mathbf{E}}(\mathbf{Digit})$ turns natural numbers $0, \ldots, 9$ into decimal digits $0, \ldots, 9$. Now, $\overline{\mu\Phi_{\mathbf{E}}}(\mathbf{K(1)})$ is the infinite list $\langle 1, 1, \ldots \rangle \in D_{\mathrm{NL}^\infty}$, which we use to represent $\Sigma_{n=2}^{\infty} 1/n!$. We leave it as an exercise to check that $\overline{\mu\Phi_{\mathbf{E}}}(\mathbf{Convert(cons(2, K(1)))}) = e/10 \in D_{\mathrm{DL}^\infty}$. So $e/10$ is representable, and Theorem 3.2 justifies calling $e/10$ a computable real number.

Exercises

1. Let $\mathbf{P} = \mathbf{ADD} \cup \{\mathbf{B(X) = B(X)}\}$, where $\tau(\mathbf{B}) = \mathbf{N} \to \mathbf{N}$. Give $\mathbf{T}_i(\mathbf{P})$, and give the $(\mathbf{T}_i(\mathbf{P}), \sigma_{\mathrm{f}})$-computation for each of the following.
   (a) $\mathbf{+(3, 2)}$.
   (b) $\mathbf{+(3, s(s(B(0))))}$.
   (c) $\mathbf{+(s(s(s(B(0)))), 2)}$.

2. Without using Theorem 3.1, give the value of each of the following.
   (a) $\mathscr{O}_{\Delta_N^\infty}(\mathbf{ADD})(+)(3, 2)$.
   (b) $\mathscr{O}_{\Delta_N^\infty}(\mathbf{ADD})(+)(3, 2_\perp)$.
   (c) $\mathscr{O}_{\Delta_N^\infty}(\mathbf{ADD})(+)(3_\perp, 2)$.
3. Let $\mathbf{P} = \mathbf{LIST} \cup \{\mathbf{B(X) = B(X), BL(X) = BL(X)}\}$, where $\tau(\mathbf{B}) = \mathbf{N} \to \mathbf{N}$ and $\tau(\mathbf{BL}) = \mathbf{NL} \to \mathbf{NL}$, let **t** be **cons(B(0), cons(1, nil))**, and let **u** be **cons(0, cons(1, BL(nil)))**. [**LIST** is defined in Section 5 of Chapter 17.] Give $\mathbf{T}_i(\mathbf{P})$, and give the $(\mathbf{T}_i(\mathbf{P}), \sigma_f)$-computation for each of the following.
   (a) **Length(t)**.
   (b) **Length(u)**.
   (c) **Nth(0, t)**.
   (d) **Nth(1, u)**.
   (e) **Cat(t, t)**.
   (f) **Cat(t, u)**.
   (g) **Cat(u, t)**.
   (h) **Reverse(t)**.
   (i) **Reverse(u)**.
   (j) **Length(Cat(Rev(t), t))**.
4. Let $l_1 = \langle \perp_{\mathbf{N}^\infty}, 1\rangle, l_2 = \langle 0, 1\rangle_\perp$. Without using Theorem 3.1, give the value of each of the following.
   (a) $\mathscr{O}_{\Delta_{NL}^\infty}(\mathbf{LIST})(\mathbf{Length})(l_1)$.
   (b) $\mathscr{O}_{\Delta_{NL}^\infty}(\mathbf{LIST})(\mathbf{Length})(l_2)$.
   (c) $\mathscr{O}_{\Delta_{NL}^\infty}(\mathbf{LIST})(\mathbf{Nth})(0, l_1)$.
   (d) $\mathscr{O}_{\Delta_{NL}^\infty}(\mathbf{LIST})(\mathbf{Nth})(1, l_2)$.
   (e) $\mathscr{O}_{\Delta_{NL}^\infty}(\mathbf{LIST})(\mathbf{Cat})(l_1, l_1)$.
   (f) $\mathscr{O}_{\Delta_{NL}^\infty}(\mathbf{LIST})(\mathbf{Cat})(l_1, l_2)$.
   (g) $\mathscr{O}_{\Delta_{NL}^\infty}(\mathbf{LIST})(\mathbf{Cat})(l_2, l_1)$.
   (h) $\mathscr{O}_{\Delta_{NL}^\infty}(\mathbf{LIST})(\mathbf{Rev})(l_1)$.
   (i) $\mathscr{O}_{\Delta_{NL}^\infty}(\mathbf{LIST})(\mathbf{Rev})(l_2)$.
5. Give a $\mathbf{W_N}$-program with $\mathbf{F} \in FV(\mathbf{P})$ such that $\mathscr{D}_{\Delta_N}(\mathbf{P})(\mathbf{F})(0) = \perp_N$ and $\mathscr{D}_{\Delta_N^\infty}(\mathbf{P})(\mathbf{F})(0) = 0$. Verify that $\mathscr{O}_{\Delta_N}(\mathbf{P})(\mathbf{F})(0) = \perp_N$ and $\mathscr{O}_{\Delta_N^\infty}(\mathbf{P})(\mathbf{F})(0) = 0$.
6. Verify the sentence containing (3.1).
7. Give the $(\mathbf{T}_i(\mathbf{PR} \cup \mathbf{LIST}), \sigma_f)$-computation for **Nth(1, Primes(2))**.
8. Suppose we change the defining equation for **Norm** in **E** to

   Norm(C, L) = Carry(C, cons(First(L), Norm(s(C), Rest(L)))).

   Now what is $\overline{\mu\Phi_E}$(**Convert(cons(2, K(1)))**)?

9. Show that if A is a nonempty r.e. set, then there is a computable list $l = \langle i_0, i_1, \ldots \rangle$ in $D_{\mathbf{NL}^\infty}$ such that $A = \{i \in N \mid i \text{ occurs in } l\}$.

10.* Show that if $l = \langle i_0, i_1, \ldots \rangle$ is a computable list of numbers in $D_{\mathbf{NL}^\infty}$, then $\{i \in N \mid i \text{ occurs in } l\}$ is r.e. [*Hint:* Adapt the proof of Theorem 2.2.]

Suggestions for Further Reading

C. L. Chang and R. C. T. Lee, *Symbolic Logic and Mechanical Theorem Proving*. Academic Press, New York, 1973.

A very readable treatise on resolution-based algorithms for satisfiability in quantification theory.

Martin Davis, *Computability and Unsolvability*. Dover, New York, 1983.

Originally published in 1958. The 1983 reprint includes an appendix on unsolvable problems in number theory.

Martin Davis (editor), *The Undecidable*. Raven, New York, 1965.

A collection of basic papers in computability theory. Included are the original papers in which Church announced his "thesis," in which Turing defined his machines and produced a universal computer, in which Post stated his "problem," and in which Turing introduced "oracles."

Herbert P. Enderton, *A Mathematical Introduction to Logic*. Academic Press, New York, 1972.

An introductory textbook on mathematical logic for mathematically mature readers.

Michael R. Garey and David S. Johnson, *Computers and Intractability: A Guide to the Theory of NP-Completeness*. Freeman, New York, 1979.

This treatise includes a comprehensive list of **NP**-complete problems.

Carl A. Gunter, *Semantics of Programming Languages*. MIT Press, Cambridge, Massachusetts, 1992.

A treatment of denotational semantics for the sophisticated reader.

Paul R. Halmos, *Naive Set Theory*. Van Nostrand, Princeton, New Jersey, 1964.

A short, classic introduction to set theory.

Michael Harrison, *Introduction to Formal Language Theory*. Addison-Wesley, Reading, Massachusetts, 1978.
A comprehensive, up-to-date, readable treatise on formal languages.

Karel Hrbacek and Thomas Jech, *Introduction to Set Theory*, second edition. Marcel Dekker, New York, 1984.
Another introduction to set theory, somewhat more detailed than the book by Halmos.

Harry R. Lewis and Christos H. Papadimitriou, *Elements of the Theory of Computation*. Prentice-Hall, Englewood Cliffs, New Jersey, 1981.
Another introduction to theoretical computer science.

Jacques Loeckx and Kurt Sieber, *The Foundations of Program Verification*, second edition. John Wiley and Sons, New York, 1987.
A well-written treatment of programming language semantics with an emphasis on program verification.

Donald W. Loveland, *Automated Theorem Proving: A Logical Basis*. North-Holland Publ., Amsterdam, 1978.
A well-organized account of resolution theory.

Michael Machtey and Paul Young, *An Introduction to the General Theory of Algorithms*. North-Holland Publ., Amsterdam, 1978.
A well-written account of computability and complexity theory.

Hartley Rogers, *Theory of Recursive Functions and Effective Computability*. McGraw-Hill, New York, 1967.
The classic comprehensive treatise on computability and noncomputability.

David Schmidt, *Denotational Semantics: A Methodology for Language Development*. Wm. C. Brown Publishers, Dubuque, Iowa, 1988.
A good general introduction to the denotational semantics of programming languages.

Joseph R. Shoenfield, *Degrees of Unsolvability*. North-Holland Publ., Amsterdam, 1971.
A short and clearly written monograph on the subject going well beyond the material covered in this book.

Robert I. Soare, *Recursively Enumerable Sets and Degrees: The Study of Computable Functions and Computably Generated Sets*. Springer-Verlag, Berlin and New York, 1987.
A modern treatment of advanced recursive function theory.

Joseph E. Stoy, *Denotational Semantics: The Scott–Strachey Approach to Programming Language Theory*. MIT Press, Cambridge, Massachusetts, 1977.
An early treatment of denotational semantics.

Notation Index

Index

P

Printed and bound by CPI Group (UK) Ltd, Croydon, CR0 4YY
03/10/2024
01040413-0017